Deep Learning for Robot Perception and Cognition

Deep Learning for Robot Perception and Cognition

Edited by

Alexandros Iosifidis

Anastasios Tefas

Academic Press is an imprint of Elsevier
125 London Wall, London EC2Y 5AS, United Kingdom
525 B Street, Suite 1650, San Diego, CA 92101, United States
50 Hampshire Street, 5th Floor, Cambridge, MA 02139, United States
The Boulevard, Langford Lane, Kidlington, Oxford OX5 1GB, United Kingdom

Library of Congress Cataloging-in-Publication Data
A catalog record for this book is available from the Library of Congress

British Library Cataloguing-in-Publication Data
A catalogue record for this book is available from the British Library

ISBN: 978-0-323-85787-1

For information on all Academic Press publications
visit our website at https://www.elsevier.com/books-and-journals

Publisher: Mara Conner
Acquisitions Editor: Tim Pitts
Editorial Project Manager: Isabella C. Silva
Production Project Manager: Debasish Ghosh
Designer: Miles Hitchen

Typeset by VTeX

Contents

List of contributors

Mete Ahishali
Faculty of Information Technology and Communication Sciences, Tampere University, Tampere, Finland

Saad Ahmad
Cognitive Robotics group, Automation Technology and Mechanical Engineering, Tampere University, Tampere, Finland

Alexandre Angleraud
Cognitive Robotics group, Automation Technology and Mechanical Engineering, Tampere University, Tampere, Finland

Deniz Bardakci
Artificial Intelligence in Robotics Laboratory (Air Lab), Department of Electrical and Computer Engineering, Aarhus University, Aarhus, Denmark
Department of Electronics, Information and Bioengineering (DEIB), Polytechnic University of Milan, Milan, Italy

Muhammad E.H. Chowdhury
Department of Electrical Engineering, Qatar University, Doha, Qatar

Kateryna Chumachenko
Faculty of Information Technology and Communication Sciences, Tampere University, Tampere, Finland

Aysen Degerli
Faculty of Information Technology and Communication Sciences, Tampere University, Tampere, Finland

Lukas Esterle
Department of Electrical and Computer Engineering, Aarhus University, Aarhus, Denmark

Moncef Gabbouj
Faculty of Information Technology and Communication Sciences, Tampere University, Tampere, Finland

Lukas Hedegaard
Department of Electrical and Computer Engineering, Aarhus University, Aarhus, Denmark

Negar Heidari
Department of Electrical and Computer Engineering, Aarhus University, Aarhus, Denmark

Juana Valeria Hurtado
Department of Computer Science, University of Freiburg, Freiburg, Germany

Turker Ince
Department of Electrical and Electronics Engineering, Izmir University of Economics, Izmir, Turkey

Alexandros Iosifidis
Department of Electrical and Computer Engineering, Aarhus University, Aarhus, Denmark

Erdal Kayacan
Artificial Intelligence in Robotics Laboratory (Air Lab), Department of Electrical and Computer Engineering, Aarhus University, Aarhus, Denmark

Serkan Kiranyaz
Department of Electrical Engineering, Qatar University, Doha, Qatar

Manos Kirtas
Aristotle University of Thessaloniki, Thessaloniki, Greece

Jonas Le Fevre
Artificial Intelligence in Robotics Laboratory (Air Lab), Department of Electrical and Computer Engineering, Aarhus University, Aarhus, Denmark

Amir Mehman Sefat
Cognitive Robotics group, Automation Technology and Mechanical Engineering, Tampere University, Tampere, Finland

Nikos Nikolaidis
Department of Informatics, Aristotle University of Thessaloniki, Thessaloniki, Greece

Paraskevi Nousi
Department of Informatics, Aristotle University of Thessaloniki, Thessaloniki, Greece

Illia Oleksiienko
Department of Electrical and Computer Engineering, Aarhus University, Aarhus, Denmark

Nikolaos Passalis
Department of Informatics, Aristotle University of Thessaloniki, Thessaloniki, Greece
Aristotle University of Thessaloniki, Thessaloniki, Greece

Huy Xuan Pham
Artificial Intelligence in Robotics Laboratory (Air Lab), Department of Electrical and Computer Engineering, Aarhus University, Aarhus, Denmark

Roel Pieters
Cognitive Robotics group, Automation Technology and Mechanical Engineering, Tampere University, Tampere, Finland

Esa Rahtu
Computer Vision group, Computing Sciences, Tampere University, Tampere, Finland

Jenni Raitoharju
Finnish Environment Institute, Programme for Environmental Information, Jyväskylä, Finland
Faculty of Information Technology and Communication Sciences, Tampere University, Tampere, Finland

Charalampos Symeonidis
Department of Informatics, Aristotle University of Thessaloniki, Thessaloniki, Greece

Anastasios Tefas
Department of Informatics, Aristotle University of Thessaloniki, Thessaloniki, Greece
Aristotle University of Thessaloniki, Thessaloniki, Greece

Pavlos Tosidis
Aristotle University of Thessaloniki, Thessaloniki, Greece

Dat Thanh Tran
Faculty of Information Technology and Communication Sciences, Tampere University, Tampere, Finland

Konstantinos Tsampazis
Aristotle University of Thessaloniki, Thessaloniki, Greece

Avraam Tsantekidis
Department of Informatics, Aristotle University of Thessaloniki, Thessaloniki, Greece

Maria Tzelepi

Department of Informatics, Aristotle University of Thessaloniki, Thessaloniki, Greece

Halil Ibrahim Ugurlu

Artificial Intelligence in Robotics Laboratory (Air Lab), Department of Electrical and Computer Engineering, Aarhus University, Aarhus, Denmark

Abhinav Valada

Department of Computer Science, University of Freiburg, Freiburg, Germany

Mehmet Yamac

Faculty of Information Technology and Communication Sciences, Tampere University, Tampere, Finland

Adamantios Zaras

Department of Informatics, Aristotle University of Thessaloniki, Thessaloniki, Greece

Preface

Deep learning refers to methodologies allowing computers to exploit raw sensor data and learn from experience to achieve high-level goals such as perception, cognition, and action. This is a fundamental ability needed in robotic systems, especially for perception and cognition tasks where the robot needs to understand the environment, manipulate objects, and interact with people, animals, and other robots in a dynamic manner. While traditional robotic systems rely on extensive feature engineering for sensor data and environment state representation followed by data analysis steps that are usually disconnected from the data representation, deep learning-based methodologies combine all these steps in an end-to-end learning-based process optimized through experience. This allows for identifying and exploiting complex patterns in sensor data, achieving higher performance levels, and obtaining robust solutions in real-time.

This edited book introduces a broad range of topics in deep learning for robot perception and cognition, along with end-to-end methodologies. It covers the following topics: 2D/3D object detection and tracking, semantic scene segmentation, human activity recognition, autonomous navigation and planning in drone racing, robotic grasping in agile production, multiagent systems, simulation environments, biosignal, and medical image analysis. To provide a complete description of these topics, the first chapters of the book cover basic concepts of deep learning models and machine learning paradigms, which can be used as a reference in understanding how deep learning models can be effectively trained to achieve high performance. These topics are neural networks and backpropagation, convolutional neural networks, graph convolutional networks, recurrent neural networks, deep reinforcement learning, lightweight deep learning, knowledge distillation, progressive and compressive deep learning, and representation learning and retrieval. Finally, concrete examples of robotics problems solved with the Open Deep Learning Toolkit for Robotics (OpenDR) are provided. The reader can use these examples as a starting point to effectively program robotics solutions with deep learning models.

The book offers the conceptual and mathematical background needed for approaching a large number of robot perception and cognition tasks from an end-to-end machine learning point of view. It can be used as a textbook in undergraduate or graduate courses in deep learning, machine learning, robotics and its subdisciplines such as robotic vision and navigation. As deep learning solutions continue providing high-performing and reliable solutions, the number of relevant courses including the topics covered in the book is increasing, and new courses are included in the curricula of many universities. Furthermore, scientists, engineers, and practitioners in robotic vision, intelligent control, mechatronics, and deep learning focusing on robotic perception and cognition tasks can obtain a complete understanding of deep learning and its application in their studied fields. With this book the reader will be able to:

- understand deep learning principles and methodologies
- understand the principles of applying end-to-end learning in robotics applications
- design and train deep learning models
- apply deep learning in several robot vision tasks such as object recognition, image classification, video analysis, and others
- understand how to use robotic simulation environments for training deep learning models
- understand how to use deep learning for different tasks, ranging from planning and navigation to biosignal analysis

Acknowledgements

This book has been edited based on the collaboration that most of the chapters' authors have in the framework of the European Union's Horizon 2020 research and innovation programme under grant agreement no. 871449 OpenDR (Open Deep Learning toolkit for Robotics). More specifically, all chapters, except Chapter 17, are co-authored by OpenDR researchers. The OpenDR toolkit will be available as an open-source toolkit from January 2022 and many of the methodologies presented in this book can be tested using the OpenDR toolkit.

The editors would like to thank all chapter authors for their contribution in covering a wide range of topics in robotics and deep learning, and for their valuable feedback in serving as reviewers for chapters falling within their field of expertise. Inspiration for editing the book in topics of deep learning for robotics originated by the successful edition of a special issue on deep learning for visual content analysis in the Signal Processing: Image Communication journal. The editors would like to thank Professor Frederic Dufaux, the journal's editor-in-chief in 2018, for his suggestion to edit such a book.

Prof. Alexandros Iosifidis
Department of Electrical and Computer Engineering,
Aarhus University,
Aarhus, Denmark

Prof. Anastasios Tefas
Department of Informatics,
Aristotle University of Thessaloniki,
Thessaloniki, Greece

Editors biographies

Alexandros Iosifidis received the Diploma and M.Sc. degree in electrical and computer engineering in 2008 and 2010, respectively, from Democritus University of Thrace, and the Ph.D. degree in informatics in 2014 from the Aristotle University of Thessaloniki, Greece. In 2017, he joined the Department of Electrical and Computer Engineering at Aarhus University, Denmark, where he is currently a Professor of Machine Learning and Computational Intelligence. He has contributed to more than 20 research projects financed by EU, Finnish, and Danish funding agencies and companies. He has (co)authored 90+ articles in international journals, 120+ papers in international conferences/workshops, and contributed to 4 chapters to edited books in his area of expertise. Professor Iosifidis is a Senior Member of the Institute of Electrical and Electronics Engineers (IEEE) and he served as an Officer of the Finnish IEEE Signal Processing-Circuits and Systems Chapter from 2016 to 2018. As of 2020, he is a member of the Technical Area Committee on Visual Information Processing of the European Association of Signal Processing (EURASIP), and as of 2022 he is a member of the Technical Committee on Machine Learning for Signal Processing of IEEE. He is currently serving as the Associate Editor in Chief for the Neurocomputing journal covering the research area of Neural Networks. He is an Area Editor in the Signal Processing: Image Communication journal, and an Associate Editor in the BMC Bioinformatics journal. His current research interests include statistical machine learning, neural networks, and deep learning, with applications in computer/robot vision, ecology, cyber-physical systems and finance.

Anastasios Tefas received the B.Sc. in informatics in 1997 and the Ph.D. degree in informatics in 2002, both from the Aristotle University of Thessaloniki, Greece. Since 2017, he has been an Associate Professor at the Department of Informatics, Aristotle University of Thessaloniki. From 2008 to 2017, he was a Lecturer, Assistant Professor at the same university. From 2006 to 2008, he was an Assistant Professor at the Department of Information Management, Technological Institute of Kavala. From 2003 to 2004, he was a temporary lecturer in the Department of Informatics, University of Thessaloniki. From 1997 to 2002, he was a researcher and teaching assistant in the Department of Informatics, University of Thessaloniki. Dr. Tefas participated in 20 research projects financed by national and European funds. He is the Coordinator of the H2020 project OpenDR, "Open Deep Learning Toolkit for Robotics." He is the Area Editor in the Signal Processing: Image Communications journal. He has coauthored 136 journal papers, 255 papers in international conferences and contributed 17 chapters to edited books in his area of expertise. Over 7500 citations have been recorded to his publications and his H-index is 44 according to Google scholar. His current research interests include computational intelligence, deep learning, pattern recognition, machine learning, digital signal and image analysis and retrieval, computer vision, and robotics.

CHAPTER

Introduction

1

Alexandros Iosifidis[a] **and Anastasios Tefas**[b]
[a]*Department of Electrical and Computer Engineering, Aarhus University, Aarhus, Denmark*
[b]*Department of Informatics, Aristotle University of Thessaloniki, Thessaloniki, Greece*

1.1 Artificial intelligence and machine learning

Almost everything we hear about artificial intelligence (AI) today is thanks to machine learning (ML) and especially the ML algorithms that use neural networks as baseline inference models. This scientific field is called deep learning (DL). Deep learning algorithms have been proved to be immensely powerful in mimicking human skills such as our ability to see and hear. To a very narrow extent, it can even emulate our ability to reason. These capabilities power Google's search and translation services, Facebook's news feed, Tesla's autopilot features, and Netflix's recommendation engine and are transforming industries like healthcare and education.

In order to better understand the meaning of AI as it is used in this book, we should first explain what are the real world problems that AI can help us solve. In everyday life, we are dealing with many different problems (e.g., brush our teeth, cook, walk, solve a linear system, etc.) that have different levels of complexity and difficulty. Most of the problems that human beings are trying to solve are related to some kind of decision and/or action that has to be taken based on the input from the real world. For example, based on the ingredients that are available in the refrigerator and the available time for cooking decide what to cook for lunch and prepare it. In each problem we are dealing with, there is an input (ingredients, available time, etc.) and one or more outputs (what to prepare for lunch, what actions to take in order to prepare it). There are also cases where humans are performing actions that are not related to solving a specific problem but are more related to trying to build an internal representation of the world that will eventually help in solving problems in the future. For example, reading literature, watching a theater play, etc. can be considered to help in building an internal representation that helps to understand better the environment and eventually might help in solving real problems in the future.

One approach to solve real world problems with machines is by trying to mimic the way humans are solving these problems. To this end, there are research efforts that try to represent the world in a symbolic way that is understandable by the machine and develop algorithms that take decisions using these symbolic representations of the world based on reasoning (e.g., decision trees). This research direction includes all the symbolic AI techniques [1], and it is not in the focus of this book. The major

Deep Learning for Robot Perception and Cognition. https://doi.org/10.1016/B978-0-32-385787-1.00006-3

difficulty of these methods is to solve the world representation problem. That is, how to detect and represent the entities (e.g., persons, objects, emotions, etc.) that appear in the world and how to build an appropriate ontology that will allow for performing complex tasks using reasoning.

The second approach is to perceive the environment using available sensors (e.g., cameras, microphones, etc.) and use this raw data in order to represent the world and make decisions and take actions. This means that someone has to build models that will be able to produce decisions or actions from raw data. The probabilistic approach to the problem of AI considers that the world can be modeled in a probabilistic way (e.g., random variables, vectors, distributions, etc.) and then apply methods from probability and statistics in order to build the models that make decisions and take actions. This approach is mostly used in statistical machine learning and pattern recognition [2].

The deep learning approach is based on building large computational models that use as building block the artificial neuron or perceptron and its variants and are able to adapt to the data by changing their parameters in order to solve real world problems [3]. That is, the core of deep learning is to build end-to-end trainable models that are able to use raw sensor information, build an internal representation of the environment, and perform inference based on this representation. Although this end-to-end training approach has been successfully followed for many different tasks ranging from speech recognition to computer vision and machine translation in the last decade, the big challenge for the next years is to successfully apply the same end-to-end training and deployment approach for robotics, which means to build models that are able to sense and act using a unified deep learning architecture [4].

1.2 Real world problems representation

In this section, we will discuss the way the real world can be represented in order to be able to apply ML for solving specific problems. An actor in the real world is an entity that can make a decision and take an action in order to solve a problem. Of course, the dominant actors are humans but animals are also actors as well as machines (e.g., robots) that are able to perform actions. An actor is able to perform actions usually based on specific input that can be as simple as "someone turned on the machine" or more complex (e.g., video, lidar, etc.) that leads to actions such as "stop the car because a pedestrian was detected to cross the road."

The actor should be able to sense the world and also to represent what sensed in a manner that will make the decisions to follow accurate. The actor uses several sensors (e.g., eyes, nose, etc.) and acquires raw data that are processed, analyzed, and used for training or/and inference. The actor is considered to be able to learn to solve a specific task (e.g., face detection in images) if it can improve its performance, as measured by a specific metric (e.g., detection accuracy), through experience [5]. In the context of machine learning the experience is comprised of the data (e.g., images with faces) acquired by the sensors of the actor along with possible annotations (e.g., the exact location of the face in the image).

The environment from which data are acquired and in which all the actions take place provides the context of the task and in many cases should be also represented for solving complex tasks (e.g., a 3D map can be used for robot navigation). The actor should be able to sense the environment (i.e., acquire data) and represent it in an appropriate manner. For example, a chess player should be able to represent the chess board along with the positions of the pieces and possibly the move history of the game. In another case, an autonomous car should be able to represent the real 3D world along with the entities therein (i.e., roads, pedestrians, signs, cars, buildings, etc.). Finally, a chat bot should be able to represent the language (i.e., language model) and the chat history.

The learning tasks of an actor are related with the real world problems that the actor will have to solve. The first category of learning tasks is the supervised learning where the actor has available data (e.g., images with human faces) and also annotations that usually represent the desirable output (e.g., the location of the face). The actor then can be trained to detect faces based on this data set. Using the same input (facial images) and different annotations (e.g., person identity) the actor can learn to recognize persons. Using both annotations, the actor can learn to both detect and recognize persons in images. Finally, using the same input (facial images) and gender and age annotations, the actor (e.g., a welcome robot in a store) can learn to perform more complex tasks, such as to recognize the gender and age of people visiting the store and give shopping related recommendations.

The second learning paradigm is the so-called unsupervised learning where the actor is only given data that are not annotated and the actor tries to solve previously defined problems that will help in building a better representation of the data. Such tasks are, for example, the clustering task where the actor tries to organize the data in groups. In recent years the task that is mostly used for learning from data in an unsupervised manner (also called self-supervised) is to try to predict part of the data based on the rest of the data [6]. For example, an actor can be trained to predict missing words from sentences using huge text data sets or can be trained to complete the missing part of images that are intentionally masked for providing a training data set [7].

Finally, the third paradigm in learning is called reinforcement learning and represents that learning procedure where the actor is able to acquire data from the environment, to make decisions and take actions and then it receives a feedback on whether the decision/action was helpful for solving a prespecified task [8]. For example, an actor can perceive the daily prices of a specific stock and be able to buy or sell stocks. The feedback can be the profit or loss the actor makes after each trading action. In all of the above cases, the data, the environment, the decisions, the actions, etc. are represented as numbers, vectors, matrices, etc., [9].

In the next section, all of these real world tasks and learning paradigms will be defined in more detail, to better understand how we transform the real world problems to the corresponding machine learning problems.

1.3 Machine learning tasks

Approaching a task through machine learning entails the creation of a machine learning model that is specialized to the task through the use of data. A machine learning model can be seen as a function $f(\cdot)$ that receives as input a data item in the form of a vector $\mathbf{x}$ and defines a mapping from $\mathbf{x}$ to a variable y encoding an answer to the task. We refer to this mapping as $f(\mathbf{x}) = y$, where the sign "=" denotes assignment of the value that $f(\cdot)$ takes when receiving as input $\mathbf{x}$ to the variable y.

For example, the task of image-based scene classification where the goal is to classifying an image I to the set of predefined image classes "indoors" and "outdoors" takes as input a vector $\mathbf{x}$ encoding properties of the image I. The vector $\mathbf{x}$ is commonly referred to as the representation of the image I and can be obtained by various ways. A straightforward way to represent I is to "vectorize" it, that is, to assign each element of the vector $\mathbf{x}$ to have one of the pixel values of I when I is a grayscale image or one of the color values of a pixel when I is a color image. This means that for an image I of size $W \times H$ pixels, $\mathbf{x}$ will be a $(W \cdot H \cdot C)$-dimensional vector, where $C = 1$ or $C = 3$ when I is a grayscale or RGB-color image, respectively. While this approach of representing images has been shown to provide good results for low-resolution images and relatively simple tasks, it leads to poor performance in general. To achieve better performance, multiple image representations have been proposed, notably those based on the Scale Invariant Feature Transform (SIFT) [10], the Local Binary Patterns (LBP) [11], and their extensions combined with the Bag-of-Features encoding scheme [12] and its extensions. Because these types of representations need to be designed by experts, they belong to the category of the so-called handcrafted features.

After obtaining the representation $\mathbf{x}$ of image I, introducing it to a machine learning model expressed by a function $f(\cdot)$ leads to a value y. One needs to ensure that the obtained value of y corresponds to an answer to the specific task, in the previous example of image classification to one of the two classes "indoors" and "outdoors," and not to an image classification problem involving other classes, for example, "day" and "night." To do so, the function $f(\cdot)$ used to perform the mapping from $\mathbf{x}$ to y commonly takes the form of a parametric function equipped with a set of parameters Θ. Such a parametric function can be referred to as $f(\cdot; \Theta)$ or $f_\Theta(\cdot)$ in order to make the existence of the parameters Θ explicit. Machine learning refers to the process of estimating the values of parameters Θ defining an optimal mapping from the input data $\mathbf{x}$ to the output y for solving a specific task. This is achieved through a process called training.

In our previous example, in order to estimate the optimal values of parameters Θ of the function $f(\cdot; \Theta)$ for classifying the vector $\mathbf{x}$ representing image I to one of the two classes "indoors" and "outdoors," one commonly uses a set of N images denoted by I_i, $i = 1, \ldots, N$ known to belong to these two classes, which form the so-called training set. Here, the subscript i is used to denote the ith image in the training set. Each image I_i is followed by a corresponding label $y_i \in \{-1, 1\}$, where the labels are associated with specific classes, for example, label $y_i = -1$ can indicate that image I_i

belongs to class "indoors" and label $y_i = 1$ that image I_i belongs to class "outdoors." Then the values of the parameters Θ can be estimated by optimizing a so-called loss function calculated over the entire training set

$$\mathcal{L} = \sum_{i=1}^{N} l(f(\mathbf{x}_i; \Theta), y_i), \tag{1.1}$$

where $l(\cdot, \cdot)$ is a loss function quantifying the error corresponding to a mismatch between the output of the parametric function $f(\cdot; \Theta)$ when receiving as input $\mathbf{x}_i$ and its (known) class y_i. Several loss functions can be used for solving the above-described problem, including the cross-entropy loss and the hinge loss. The choice of a loss function defines the form of optimality for the parameters Θ, as different loss functions enforce different properties in the optimization process. After the estimation of the parameters Θ, a new image I represented by a vector $\mathbf{x}$ can be introduced to the parametric function $f(\mathbf{x}; \Theta)$. I is classified to class "indoors" if $f(\mathbf{x}; \Theta) < 0$, and to class "outdoors" if $f(\mathbf{x}; \Theta) \geq 0$. This binary form of the decision process leads to the name of binary classification models.

Another way to approach the above image classification problem is to formulate it as a regression problem. In this case, each image I_i in the training set is again represented by a vector $\mathbf{x}_i$ and its label y_i is used to create a two-dimensional class indicator vector $\mathbf{t}_i$. The kth element of $\mathbf{t}_i$ takes the value of $\mathbf{t}_{ik} = 1$ if $y_i = k$, otherwise $\mathbf{t}_{ik} = -1$ (or $\mathbf{t}_{ik} = 0$). Then the parameters of the function $f(\cdot; \Theta)$ are optimized to express the mapping from the input vector $\mathbf{x}_i$ to the (target) indicator vector $\mathbf{t}_i$. One advantage of following this approach for solving the classification problem is that it can be easily extended to tackle problems formed by more than two classes by introducing a new dimension to the class indicator vectors for each new class. Moreover, in several cases the use of target values $\mathbf{t}_{ik} \in \{0, 1\}$ allows for a probabilistic interpretation of the model's outputs. Here, we need to note that regression models need not be solely associated to classification problems, and they can be used to define mappings from an input vector space defined by the training vectors $\mathbf{x}_i$ to a target vector space defined by the target vectors $\mathbf{t}_i$ in general. An example of such a regression problem could be the estimation of the price of a house given a set of qualitative and quantitative indicators and measurements, like its size, the year it was built, its location, and access to transportation, to name a few.

Both the above approaches belong to the supervised machine learning category, where the parameters of the model are estimated based on human supervision, that is, each sample in the training set is followed by an expert-defined target (label or vector). In the case where such human supervision is not available, one can try to identify patterns in the available data. One example of unsupervised machine learning problems is that of data clustering. In this case the goal is to identify groups of similar data items by making use of a similarity measure. A classic data clustering method is that of K-means, where the model parameters correspond to the cluster prototypes $\boldsymbol{\mu}_k, k = 1, \ldots, K$ and the loss function used to estimate them is the within-cluster dispersion. K-means can be considered as a special case of the Gaussian mixture model

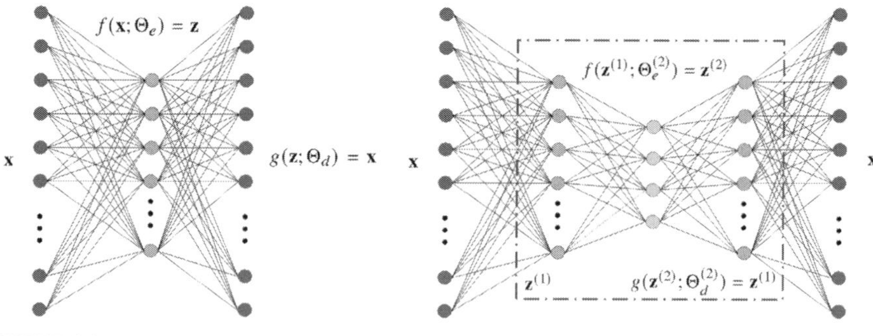

FIGURE 1.1

Representation learning: Autoencoder defines the identity function through a two-step regression process, leading to a learned representation $\mathbf{z}$ of the input data representation $\mathbf{x}$ (left). Nested Autoencoder defines a second-level learned representation $\mathbf{z}^{(2)}$ of the input data representation $\mathbf{x}$ (right).

in which each of the K groups is modeled by a (multidimensional) Gaussian distribution associated with parameters $\{\boldsymbol{\mu}_k, \boldsymbol{\Sigma}_k, \pi_k\}$, where $\boldsymbol{\Sigma}_k$ is the covariance matrix of the Gaussian and π_k is a mixing coefficient defining how small or big the Gaussian is. The parameters of this model are estimated by fitting the data to the model using maximum likelihood by applying the expectation maximization process.

Another example of unsupervised learning models is that of the Autoencoder. An Autoencoder defines the identity function through a two-step regression process (Fig. 1.1 (left)). In the first step, the input vector $\mathbf{x}$ is mapped to an intermediate representation $f(\mathbf{x}; \Theta_e) = \mathbf{z}$, which is then regressed to the target vector which is the same as the input vector, that is, $g(\mathbf{z}; \Theta_d) = g(f(\mathbf{x}; \Theta_e); \Theta_d) = \mathbf{x}$. Θ_e and Θ_d denote the parameters of the encoding and decoding functions, respectively, and are jointly optimized to minimize the so-called reconstruction error. We can see the first processing step as an encoding process, mapping the input vector to another vector (which usually has much lower number of dimensions). Then a decoding process maps the low-dimensional representation of the input vector back to its initial form.

The above-described process is the quintessential example of representation learning. Let us consider again the image classification problem described in the beginning of this section and assume that the image representation vector $\mathbf{x}$ was obtained by vectorizing the input image I. For a relatively low-resolution color image of 250×250 pixels, this leads to a 187,500-dimensional vector. Let us now assume that by using the images I_i, $i = 1, \ldots, N$ in the training set we can train an Autoencoder in which the intermediate representation vector z_i is formed by 1000 dimensions and it can achieve zero reconstruction error. This means that we were able to learn a low-dimensional representation $\mathbf{z}_i$ that effectively encodes all information available in the original image representation $\mathbf{x}_i$ formed by an enormous number of dimensions. Moreover, one can now treat $\mathbf{z}_i$ as the learned image representations of level one, that is, $\mathbf{z}_i^{(1)}$, and proceed to train a second Autoencoder defining the mapping

$g(f(\mathbf{z}_i^{(1)}; \Theta_e^{(2)}); \Theta_d^{(2)}) = \mathbf{z}_i^{(1)}$ the parameters of which are estimated by using the image representations $\mathbf{z}_i^{(1)}$, $i = 1, \ldots, N$ obtained by applying the encoding process once and defining an intermediate representation $\mathbf{z}_i^{(2)}$ for each image I_i formed by 500 dimensions (Fig. 1.1 (right)). This process can be applied multiple times, leading to a cascade of encoding steps followed by the corresponding decoding steps:

$$\mathbf{z}_i^{(L)} = f(\ldots f(f(\mathbf{x}_i; \Theta_e^{(1)}); \Theta_e^{(2)}) \ldots; \Theta_e^{(L)}), \tag{1.2}$$

$$g(\ldots g(g(\mathbf{z}_i^{(L)}; \Theta_d^{(L)}); \Theta_d^{(L-1)}) \ldots; \Theta_d^{(1)}) = \mathbf{x}_i. \tag{1.3}$$

In classifying I, the use of $\mathbf{z}^{(L)}$ instead of the high-dimensional $\mathbf{x}$ has several advantages, as the number of parameters to be estimated for the classification model reduces considerably, leading to easier optimization. Moreover, the representation $\mathbf{z}^{(L)}$ is expected to encode relationships of the input data dimensions obtained by learning the representation through multiple levels of abstraction.

A learning approach that is closely connected to unsupervised learning and has recently gained a lot of attention is that of self-supervised learning. The main idea in self-supervised learning is that an algorithm can use the input data to devise an auxiliary learning task in which supervision is provided by the data itself. Example self-supervised tasks include the prediction of the relative position of an image patch in relation to another (reference) one [13], prediction of the pixels' color from their grayscale intensity values [14], image patch classification to surrogate classes created by performing image transformations to the original image patches [15], prediction of the correct image rotation [16]. Self-supervised based training can be used to exploit large amounts of unlabeled data for optimizing the parameters of the model to gain knowledge about the properties of the data, followed by supervised training using annotated data to specialize on the targeted task.

Another machine learning paradigm that has found application in a wide range of problems is that of reinforcement learning. Contrary to the supervised learning paradigm in which the parameters of a machine learning model are optimized using a training set of data followed by expert-given labels or targets, in reinforcement learning the model can be seen as an agent which is able to interact with its environment, take actions, and receive feedback. When an action taken contributes toward achieving a predefined goal, the positive feedback received is used to update the parameters of the model encouraging it to take similar actions in the future under similar conditions, while when an action taken impedes the achievement of the goal, the negative feedback received is used to update the parameters of the model to avoid taking similar actions in the future. Through trial-and-error, the agent is exploring its environment and exploits the provided feedback to improve its performance. The strategy followed to balance exploration and exploitation plays a crucial role in the final performance of the model, as high exploration leads to very long training in which the model does not effectively exploit the feedback corresponding to relevant to the task locations of its environment, while high exploitation can lead to suboptimal optimization focusing only on some specific locations of the environment without being able to find other locations with more effective feedback. Reinforcement learning and different training strategies are further studied in Chapter 6.

1.4 Shallow and deep learning

Let us now consider an image classification problem defined by the D-dimensional training vectors $\mathbf{x}_i$, $i = 1, \ldots, N$ and the (binary) labels $y_i \in \{-1, 1\}$, and choose to use a linear parametric function. The output of the model when receiving as input the vector $\mathbf{x}_i$ is

$$f(\mathbf{x}_i; \boldsymbol{\theta}) = \theta_0 + \theta_1 \mathbf{x}_{i1} + \cdots + \theta_D \mathbf{x}_{iD}, \tag{1.4}$$

where $\boldsymbol{\theta}$ is the $D + 1$-dimensional parameter vector of the model.

The above function describes the computation performed by the basic computation unit of a neural network, called perceptron neuron. One way to estimate its parameters is to apply the perceptron algorithm. This algorithm randomly initializes the values of the parameters $\boldsymbol{\theta}$ and updates them by applying an iterative optimization process. We refer to the initial parameters by $\boldsymbol{\theta}_0$, and we use the index to denote the iteration of the optimization process. At each iteration t, all the training vectors are introduced to the model, and its outputs $f(\mathbf{x}_i; \boldsymbol{\theta}_t)$, $i = 1, \ldots, N$ are used to calculate its error for updating its parameters. In the context of neural networks, such an iteration is called an epoch of the training process. The perceptron algorithm defines a loss function quantifying the error of the misclassified vectors. To do so, the outputs of the model for all input vectors are compared with a threshold value equal to zero in order to classify them to one of the two classes, and the misclassified samples form the set $\mathcal{X}_t$. Then the loss function is defined by

$$\mathcal{L} = \sum_{\mathbf{x}_i \in \mathcal{X}_t} -y_i f(\mathbf{x}_i; \boldsymbol{\theta}_t). \tag{1.5}$$

Achieving a value of $\mathcal{L} = 0$ leads to correct classification of all training vectors. Thus the gradient descent update rule is followed, and the new parameter values are calculated by

$$\boldsymbol{\theta}_{t+1} = \boldsymbol{\theta}_t - \eta(t) \nabla_{\boldsymbol{\theta}_t} \mathcal{L} = \boldsymbol{\theta}_t + \eta(t) \sum_{\mathbf{x}_i \in \mathcal{X}_t} y_i \hat{\mathbf{x}}_i, \tag{1.6}$$

where we used an augmented version of the input vectors $\hat{\mathbf{x}}_i = [1, \mathbf{x}_i^T]^T$. While the use of the perceptron algorithm can lead to effective training when the two classes are linearly separable, it is not able to converge to a solution when applied to nonlinear classification problems.

An alternative way to optimize the parameters $\boldsymbol{\theta}$ of the model in Eq. (1.4) is to use the mean squared error loss function

$$\mathcal{L} = \sum_{i=1}^{N} \left(\boldsymbol{\theta}^T \hat{\mathbf{x}}_i - y_i \right)^2, \tag{1.7}$$

which gives the solution $\boldsymbol{\theta} = \hat{\mathbf{X}}^\dagger \mathbf{y}$, where $\hat{\mathbf{X}}^\dagger$ is the pseudoinverse of the data matrix $\hat{\mathbf{X}} = [\hat{\mathbf{x}}_1, \ldots, \hat{\mathbf{x}}_N]$ and $\mathbf{y} = [y_1, \ldots, y_N]^T$. The advantages of using the loss function in Eq. (1.7) include the existence of a unique solution for both linear and nonlinear classification problems and its easy extension to multiclass classification problems by the use of class-indicator vectors $\mathbf{t}_i$, $i = 1, \ldots, N$. For a multiclass classification problem formed by C classes $\mathcal{C}_k$, $k = 1, \ldots, C$, this leads to the solution of C binary classification problems (in the one-versus-rest manner) of the form of Eq. (1.7), which can be jointly optimized as $\boldsymbol{\Theta} = \hat{\mathbf{X}}^\dagger \mathbf{T}^T$, where $\boldsymbol{\Theta} = [\boldsymbol{\theta}_1, \ldots, \boldsymbol{\theta}_C]$ and $\mathbf{T} = [\mathbf{t}_1, \ldots, \mathbf{t}_C]$. This case corresponds to the use of multiple perceptron neurons, each dedicated to solve an one-versus-rest binary classification problem by receiving as input the input vectors $\mathbf{x}_i$ and providing an output corresponding to the binary classification problem assigned to it.

By attaching a nonlinear function to a perceptron neuron, a nonlinear mapping is obtained. For a binary classification problem solved by using one neuron, the use of the logistic sigmoid function $\sigma(\alpha) = \frac{1}{1+exp(-\alpha)}$ transforms the output of the model to $f(\mathbf{x}_i; \boldsymbol{\theta}) = \sigma(\boldsymbol{\theta}^T \hat{\mathbf{x}}_i)$, which is always a number in the interval [0, 1]. In statistics, this model is called logistic regression. Logistic regression can be regarded as a probabilistic model, where one employs the class-conditional densities $p(\mathbf{x}|\mathcal{C}_k)$ and class prior probabilities $p(\mathcal{C}_k)$ to compute the class posterior probabilities $p(\mathcal{C}_k|\mathbf{x})$ through Bayes' theorem. For an input vector $\mathbf{x}_i$, the posterior probability of class $\mathcal{C}_1$ is given by

$$p(\mathcal{C}_1|\hat{\mathbf{x}}_i) = \frac{p(\hat{\mathbf{x}}_i|\mathcal{C}_1)p(\mathcal{C}_1)}{\sum_{j=1}^{2} p(\hat{\mathbf{x}}_i|\mathcal{C}_j)p(\mathcal{C}_j)} = \frac{1}{1 + \exp(-\alpha_i)} = \sigma(\alpha_i), \tag{1.8}$$

where $\alpha_i = \ln \frac{p(\hat{\mathbf{x}}_i|\mathcal{C}_1)p(\mathcal{C}_1)}{p(\hat{\mathbf{x}}_i|\mathcal{C}_2)p(\mathcal{C}_2)} = \ln \frac{p(\mathcal{C}_1|\hat{\mathbf{x}}_i)}{1-p(\mathcal{C}_1|\hat{\mathbf{x}}_i)}$ represents the logarithm of the ratio of the posterior probabilities and is known as log odds. Assuming that the class-conditional densities follow Gaussian distributions with a shared covariance matrix, the posterior probability of class $\mathcal{C}_1$ takes the form $p(\mathcal{C}_1|\hat{\mathbf{x}}_i) = \sigma(\boldsymbol{\theta}^T \hat{\mathbf{x}}_i)$. Optimization of the parameters $\boldsymbol{\theta}$ is obtained by assuming that the target values t_i follow a binomial distribution. The negative log-likelihood of the targets given the parameters leads to

$$\mathcal{L} = -\sum_{i=1}^{N} \Big(t_i \ln \hat{y}_i + (1 - t_i) \ln(1 - \hat{y}_i) \Big), \tag{1.9}$$

where $\hat{y}_i = \sigma(\boldsymbol{\theta}^T \hat{\mathbf{x}}_i)$. The loss function in Eq. (1.9) is known as the cross-entropy loss function.

The extension of logistic regression to multiple classes $\mathcal{C}_k,\ k = 1,\dots,C$ is obtained by calculating the posterior probabilities

$$p(\mathcal{C}_k|\hat{\mathbf{x}}_i) = \frac{p(\hat{\mathbf{x}}_i|\mathcal{C}_k)p(\mathcal{C}_k)}{\sum_{j=1}^{C} p(\hat{\mathbf{x}}_i|\mathcal{C}_j)p(\mathcal{C}_j)} = \frac{\exp(\alpha_{ki})}{\sum_{j=1}^{C} \exp(\alpha_{ji})}, \tag{1.10}$$

where $\alpha_{ki} = lnp(\hat{\mathbf{x}}_i|\mathcal{C}_k)p(\mathcal{C}_k)$ and by making similar assumptions to those in the binary case we get $\alpha_{ki} = \boldsymbol{\theta}_k^T \hat{\mathbf{x}}_i$. The normalized exponential function in Eq. (1.10) is also known as the softmax function. One of the properties of the softmax function making it suitable for classification problems is that it compares all its input values and provides probability-like responses, highlighting the maximum of its inputs and suppressing the remaining ones. By using class indicator vectors with values $\mathbf{t}_{ik} \in \{0, 1\}$, the negative log-likelihood of the targets given the parameters leads to

$$\mathcal{L} = -\sum_{i=1}^{N}\sum_{c=1}^{C} \mathbf{t}_{ic} \ln \hat{y}_{ic}, \tag{1.11}$$

where $\hat{y}_{ic} = \sigma(\boldsymbol{\theta}_c^T \hat{\mathbf{x}}_i)$. The loss function in Eq. (1.11) is the cross-entropy loss function for multiclass classification problems.

Optimization of the loss functions in Eqs. (1.9) and (1.11) for updating the parameters in logistic regression is more complicated compared to the linear regression case (Eq. (1.7)), as the nonlinearity of the logistic sigmoid function does not allow to obtain a closed-form solution, and is conducted by applying an iterative reweighted least squares method.

All models described so far correspond to linear classification models. In order to effectively solve problems in which the classes are nonlinearly separable, nonlinear classification models need to be used. One way to devise nonlinear classification models by using the linear models described above is to perform a nonlinear mapping of the input vectors $\mathbf{x}_i$ using a nonlinear function $\phi(\cdot)$ and apply the linear model on the new data representations $\phi(\mathbf{x}_i) = \boldsymbol{\phi}_i,\ i = 1,\dots,N$. In this case, the model in Eq. (1.4) takes the form $f(\boldsymbol{\phi}_i; \boldsymbol{\theta}) = \boldsymbol{\theta}^T \hat{\boldsymbol{\phi}}_i$, where we use again $\hat{\boldsymbol{\phi}}_i = [1, \boldsymbol{\phi}_i^T]^T$. Multiple types of nonlinear mappings can be used for this purpose, notably Radial Basis Functions (RBF) using prototype vectors determined by clustering the training vectors leading to the so-called RBF networks [17] and random mappings used to transform the input vectors to a new feature space before a nonlinear function is applied elementwise to obtain the data representations used as input to a linear regression model [18,19]. Such a processing scheme can be seen as a neural network formed by two layers of neurons; the nonlinear mapping of the input vectors to the new data representations corresponds to one layer of the neural network in which each neuron is equipped with a nonlinear function, while the linear model applied to the new data representations corresponds to a second layer receiving as input the outputs of the first layer. In the context of neural networks, the nonlinear function of each neuron is called activation function. Considering the model from the user's perspective who is introducing the input vectors $\mathbf{x}_i$ to the model and receives its responses in the output, this neural network can be described as a single-hidden layer neural network formed by an input layer corresponding to the input vectors, an output layer providing the responses of the network, and a hidden layer performing the nonlinear transformation of the input vectors.

For such a single-hidden layer network, one needs to determine the number of dimensions of the new data representations, that is, the number of neurons forming the hidden layer, a choice that can affect the performance of the model. An interesting case arises when allowing the number of hidden layer neurons go to infinity and setting a Gaussian prior to randomly sampled parameters of these neurons [20,21]. Then, by adopting an RBF or a sigmoid activation function for the neurons of the hidden layer, the parameters of the output layer can be calculated by solving a regression problem using the so-called Gram matrix expressing dot-products of the training vectors in a different feature space. This leads to a connection of the single-hidden layer neural networks with another paradigm in machine learning, that of kernel methods [22]. Another notable connection between the two paradigms is that of the support vector network [23] and its extensions which determine the parameters of the network's output layer as a linear combination of the some of the columns of the Gram matrix, those corresponding to the input vectors identified as the so-called support vectors. The connection between kernel methods and infinite neural networks can also be observed by considering the similarities in some approximate kernel models [24,25] and randomized single-hidden layer neural networks with a finite number of neurons [26].

Although single-hidden layer networks have been shown to be universal approximators, that is, under mild assumptions they can approximate any continuous function indicated by the targets used in their training, the number of hidden layer neurons required to achieve this tends to be comparable to the number of training vectors. Thus for problems where large data sets are used to train the neural network achieving such an approximation capability becomes impractical. Moreover, as the number of parameters to be estimated in such cases is enormous, single-hidden layer networks tend to memorize the training samples instead of encoding patterns in data, and thus they cannot generalize well on unseen data. The importance of using multiple hidden layers in neural networks, referred to as deep learning models, was studied in [27]. It was recently shown in [28,29] that there exist mappings from D-dimensional feature space to an one-dimensional feature space represented by adequately deep networks with constant width size (i.e., number of neurons per hidden layer), which cannot be approximated by any neural network whose number of layers is smaller. Similar to the universal approximation theorem for neural networks with a single hidden layer, it was shown that width-bounded feed-forward networks (with minimum width size of $D+1$) with additive/affine neurons and Rectified Linear Unit (ReLU) activation function can approximate arbitrarily well any continuous function on the unit cube $[0, 1]^D$ to a given error ϵ [30]. Even though such theoretical results concern neural networks of arbitrary number of layers and they cannot guarantee excellent performance of individual deep neural network implementations, they support the empirical evidence indicating that deep neural networks usually outperform shallow ones formed by one hidden layer.

The architecture of a deep neural network is commonly designed by experts and several deep neural networks targeting specific problems achieving high performance while being efficient in terms of computations have been recently proposed. Lightweight deep neural network architectures are studied in Chapter 7. Metaalgorithms that automatically determine an optimized neural network architecture have also been proposed. Metaalgorithms based on progressive learning are studied in Chapter 9.

The parameters of deep neural networks formed by multiple layers are jointly optimized to minimize a loss, such as the cross-entropy loss function in Eq. (1.11), through gradient-based optimization methods. The data representations obtained in the intermediate layers of a network trained by such an end-to-end optimization process give rise to a different aspect of representation learning. Contrary to the properties of the representations learned using Autoencoders where the objective is to preserve as much information of the input data as possible, representations learned by applying end-to-end tuning of all the parameters of the network to achieve a goal, for example, classification of its inputs to a set of predefined classes, highlight patterns in their inputs suitable for discriminating between samples belonging to different classes while suppressing patterns that may be important for reconstructing the input but reduce classification performance. The optimization process followed to train deep neural networks in an end-to-end manner is described in Chapter 2, while representation learning is further studied in Chapter 10. Moreover, the use of various types of neural layers designed to process different types of data, like convolutional layers suitable for processing images (Fig. 1.2) studied in Chapter 3, graph convolutional layers suitable for processing graph structures studied in Chapter 4, and recurrent neural layers suitable for analyzing time-varying inputs studied in Chapter 5, allows for introducing the raw input data to the neural network and jointly optimize all the intermediate data representations needed to perform the task at hand. Thus, the need of handcrafted features is diminished. It is believed that this is one of the reasons why deep learning models outperform traditional machine learning models exploiting handcrafted data representations by a large margin. Training such deep neural networks on large data sets leads to the estimation of parameter values, which are considered to be detectors of generic patterns, like edges, lines, and curves in the case of convolutional layers placed early in the network's architecture. This property allows using them as feature extractors for solving other tasks the data of which share similar properties with the data the network was trained on, giving rise to the nowadays widely adopted paradigm of transfer learning. Moreover, one can use a high-performing deep neural network to guide the training process of another neural network by means of generating targets at different layers, leading to a process known as knowledge distillation. Knowledge distillation is further studied in Chapter 8.

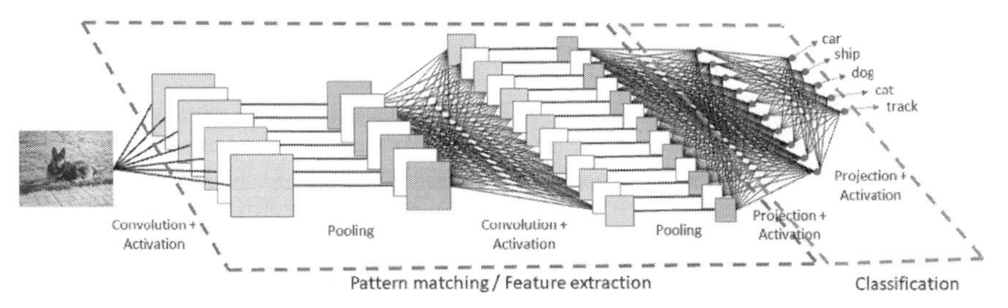

FIGURE 1.2

A convolutional neural network formed by convolutional, pooling, and fully-connected layers. The network can receive as input an image and perform a series of transformations leading to the final output of the network expressing the predicted class label. Jointly optimizing all the parameters of the network corresponding to the feature extraction and the classification layers of the network in an end-to-end manner leads to enhanced performance compared to the use of handcrafted image representations combined with shallow classification models. Convolutional neural networks are further studied in Chapter 3.

1.5 Robotics and deep learning

Deep learning is one of the main research directions we should target in order to achieve autonomy in robotics, that is, to build robots that are able to act without human guidance and control. The application of deep learning in robotics is the major challenge for the years to come as defined by numerous researchers that leads to very specific learning, reasoning and embodiment problems and research questions that are typically not addressed by the computer vision and machine learning communities [4].

Despite the recent successes in robotics, artificial intelligence and computer vision, a complete artificial agent necessarily must include active perception. The reason follows directly from the definition of an agent as an active perceiver if it knows why it wishes to sense, and then chooses what to perceive, and determines how, when, and where to achieve that perception. The computational generation of intelligent behavior has been the goal of all AI, vision and robotics research since its earliest days and agents that know why they behave as they do and choose their behaviors depending on their context clearly would be embodiments of this goal. To be able to build agents with active perception toward improved AI and cognition we should consider how deep learning can be smoothly integrated in the robotics methodologies either for building subsystems (e.g., active object detection) that try to solve a more complex task (e.g., grasping) or for replacing the entire robotic system pipeline leading to end-to-end trainable agents that are able to successfully solve a robotics task (e.g., end-to-end deep learning for navigation). However, integrating deep learning in robotics is not trivial and thus it is still in its infancy compared to the penetration of deep learning to other research areas (e.g., computer vision, search engines, etc.). Some of the obstacles for integrating deep learning in robotics are explained below.

The available deep learning open frameworks (e.g., Tensorflow, PyTorch, etc.) are not easily employed in robotics since they have a long learning curve and radically different methodology (end-to-end data-driven learning, from sensing to acting, etc.) than conventional robotics. This is rapidly changing as DL is used in robotics. There is a great interest from the roboticists to apply deep learning for the tasks they have to solve, but this is not at all easy mainly due to the radically different approach they have to follow in order to design, train, evaluate, and deploy data-driven deep learning based robotic models. In many cases, robotics researchers prefer to use well-known algorithmic implementations (e.g., OpenCV feature based face detection [31]) that are significantly inferior, in terms of performance, to the deep learning alternatives, due to their speed, and easy integration.

The already available deep learning software modules are implemented in order to be deployed on large and expensive GPUs and they rarely perform in real-time even for low resolution input. Current solutions in autonomous mobile systems (e.g., autonomous cars) use multiple GPUs for deploying numerous deep models for the different tasks they have to solve. Most of the state-of-the-art deep learning models for solving difficult perception (e.g., object detection and tracking, semantic scene segmentation, etc.) and manipulation tasks (e.g., grasping) are usually inappropriate for deployment on embedded systems since their analysis capability is a few fps (for vision) and they incorporate large latency in the system. Due to the reduced speed of the deep learning models, researchers are obliged to drop significantly the used input resolution of their sensors. Video resolution of 320×240 pixels and even smaller is in many cases the standard resolution used for autonomous mobile robots and for many computer vision models that are incorporated in robotics.

Another obstacle for applying deep learning in robotics is the importance of simulation in deep robotic learning and the lack of available open-source robotics simulation environments that allow for deep learning training. A robot is an inherently active agent that acts in and interacts with the physical real world. It perceives the world with its different sensors, builds a coherent model of the world and updates this model over time, but ultimately a robot has to make decisions, plan actions, and execute these actions to fulfill a useful task. This is where robotic vision differs from computer vision. For robotic vision, perception is only one part of a more complex, embodied, active, and goal-driven system. Robotic vision therefore has to take into account that its immediate outputs (object detection, segmentation, depth estimates, 3D reconstruction, a description of the scene, and so on), will ultimately result in actions in the real world. In a simplified view, while computer vision takes images and translates them into information, robotic vision translates images into actions. To be able to train and evaluate such agents faster than real-time in order to speed-up the training and convergence, an appropriate robotics simulation environment is needed, since training on real data is rather impossible.

The major tasks of a robotic system can be categorized as follows. In the first category belong the tasks that are related to robot perception. That is, the robot should be able to interact with people and environment, and thus should be able to perceive people and environment and acquire the data the will help in representing them with numbers, vectors, graphs, etc. The most important tasks that are related to person and environment perception are person/object detection and tracking, which will be presented in detail in Chapter 11. Another important task is the semantic scene segmentation, which will be discussed in Chapter 12. The localization and tracking of objects in the 3D space will be presented in Chapter 13. Person activity recognition methods will be presented in Chapter 14.

In the second category, we can find tasks that are related to ability of the robot for action, planning, navigation, manipulation, and cognition. These tasks are in general far more difficult to solve since they use the perception and build upon a useful representation of the environment that will allow for such complex tasks. Methods for autonomous navigation and planning in the context of drone racing will be presented in Chapter 15. Methods for robot grasping in the context of agile production are presented in Chapter 16. Multiactor systems are presented in Chapter 17. The corresponding simulation environments that are needed for training and evaluation of the robotics solutions are presented in Chapter 18. Deep learning for healthcare applications of robotics are presented in Chapter 19 and Chapter 20.

Finally, the Chapter 21 presents several robotics examples that use deep learning and are included in the OpenDR (Open Deep Learning toolkit for Robotics). These tools will help the reader to better understand several methods discussed in this book using the OpenDR toolkit it is easy for anyone to build its own robotic solutions.

References

[1] S. Russell, P. Norvig, Artificial Intelligence: A Modern Approach, Prentice Hall, 2010.

[2] R.O. Duda, P.E. Hart, D.G. Stork, Pattern Classification, Wiley, 2001.

[3] I.J. Goodfellow, Y. Bengio, A. Courville, Deep Learning, MIT Press, Cambridge, MA, USA, 2016.

[4] N. Sünderhauf, O. Brock, W. Scheirer, R. Hadsell, D. Fox, J. Leitner, B. Upcroft, P. Abbeel, W. Burgard, M. Milford, P. Corke, The limits and potentials of deep learning for robotics, The International Journal of Robotics Research 37 (2018) 405–420.

[5] T.M. Mitchell, Machine Learning, McGraw–Hill, 1997.

[6] R. Collobert, J. Weston, L. Bottou, M. Karlen, K. Kavukcuoglu, P. Kuksa, Natural language processing (almost) from scratch, Journal of Machine Learning Research 12 (76) (2011) 2493–2537.

[7] P. Goyal, M. Caron, B. Lefaudeux, M. Xu, P. Wang, V. Pai, M. Singh, V. Liptchinsky, I. Misra, A. Joulin, P. Bojanowski, Self-supervised pretraining of visual features in the wild, arXiv:2103.01988, 2021.

[8] R.S. Sutton, A.G. Barto, Reinforcement Learning: An Introduction, The MIT Press, 2018.

[9] A. Tsantekidis, N. Passalis, A.-S. Toufa, K. Saitas-Zarkias, S. Chairistanidis, A. Tefas, Price trailing for financial trading using deep reinforcement learning, IEEE Transactions on Neural Networks and Learning Systems 32 (7) (2021) 2837–2846.

[10] D.G. Lowe, Object recognition from local scale-invariant features, in: International Conference on Computer Vision, 1999.

[11] T. Ojala, M. Pietikäinen, D. Harwood, Performance evaluation of texture measures with classification based on Kullback discrimination of distributions, in: International Conference on Pattern Recognition, 1994.
[12] G. Csurka, C. Dance, L.X. Fan, J. Willamowski, C. Bray, Visual categorization with bags of keypoints, in: ECCV Workshop on Statistical Learning in Computer Vision, 2004.
[13] C. Doersch, A. Gupta, A.A. Efros, Unsupervised visual representation learning by context prediction, in: International Conference on Computer Vision, 2015.
[14] R. Zhang, P. Isola, A.A. Efros, Colorful image colorization, in: European Conference on Computer Vision, 2016.
[15] A. Dosovitskiy, J.T. Springenberg, M. Riedmiller, T. Brox, Discriminative unsupervised feature learning with convolutional neural networks, in: Advances in Neural Information Processing Systems, 2014.
[16] S. Gidaris, P. Singh, N. Komodakis, Unsupervised representation learning by predicting image rotations, in: International Conference on Learning Representations, 2018.
[17] D.S. Broomhead, D. Lowe, Multivariable functional interpolation and adaptive networks, Complex Systems 2 (1988) 321–355.
[18] Y.-H. Pao, G.-H. Park, D.J. Sobajic, Learning and generalization characteristics of random vector functional-link net, Neurocomputing 6 (1994) 163–180.
[19] G.-B. Huang, Q.-Y. Zhu, C.K. Siew, Extreme learning machine: theory and applications, Neurocomputing 70 (1–3) (2006) 489–501.
[20] R. Neal, Bayesian Learning for Neural Networks, Lecture Notes in Statistics, Springer, 1996.
[21] C. Williams, Computation with infinite neural networks, Neural Computation 10 (5) (1998) 1203–1216.
[22] B. Scholkopf, A. Smola, Learning with Kernels, MIT Press, Cambridge, MA, USA, 2001.
[23] C. Cortes, V. Vapnik, Support-vector networks, Machine Learning 20 (1995) 273–297.
[24] A. Rahimi, B. Recht, Random features for large-scale kernel machines, Advances in Neural Information Processing Systems (2007).
[25] A. Rahimi, B. Recht, Weighted sums of random kitchen sinks: replacing minimization with randomization in learning, Advances in Neural Information Processing Systems (2008).
[26] A. Iosifidis, A. Tefas, I. Pitas, On the kernel extreme learning machine classifier, Pattern Recognition Letters 54 (2015) 11–17.
[27] K. Hornik, Approximation capabilities of multilayer feedforward networks, Neural Networks 4 (2) (1991) 251–257.
[28] R. Eldan, O. Shamir, The power of depth for feedforward neural networks, in: Conference on Learning Theory, 2016.
[29] M. Telgarsky, Benefits of depth in neural networks, in: Conference on Learning Theory, 2016.
[30] B. Hanin, Universal function approximation by deep neural nets with bounded width and ReLU activations, Mathematics 7 (10) (2019) 992.
[31] G. Bradski, The OpenCV library, Dr Dobb's Journal of Software Tools (2000).

CHAPTER

Neural networks and backpropagation

2

Adamantios Zaras, Nikolaos Passalis, and Anastasios Tefas
Department of Informatics, Aristotle University of Thessaloniki, Thessaloniki, Greece

2.1 Introduction

Neural Networks (NNs) are **Machine Learning** (ML) models whose theory has been available for years, but only recently began their widespread use thanks to the evolution of technology and the advent of powerful **Graphics Processing Units** (GPUs) and **Tensor Processing Units** (TPUs). Neural networks consist of a number of *neurons*, also known as *nodes* or *units*, at the input and output of the model, and connections between them. Optionally, there may be additional neurons in between. Each set of neurons at the same level of depth is typically called a *layer*. If there are intermediate layers, they are called *hidden* layers. A model with more than one hidden layer is typically considered as a **Deep Neural Network** (DNN).

Fig. 2.1 demonstrates a typical NN architecture. Every single neuron receives inputs from other neurons via connections, except those in the first layer, which directly accept data, e.g., pixel values. The output neurons compose the final result, known as *decision* or *prediction*. The number of input neurons is the same as the number of data features and receive only a single value, while the number of output neurons is the same as the number of categories to be predicted, known as *classes*. An exception to this rule is the case of binary classification, in which one output neuron can be used, instead of two. It is worth noting that NNs can also be used for regression, i.e., using one neuron for each regressed output, and many other tasks, ranging from clustering and forecasting [1–4] to object detection and panoptic segmentation [5,6]. Each connection between neurons carries a *weight*, while each neuron is equipped with an additional *bias* term. At first, the weights are randomly initialized and they update through an operation called *training*. Training ends up finding the appropriate weights and biases, utilizing the *backpropagation* algorithm, which calculates the derivative of each layer's function after every pass of the data through the network, in order to determine the changes that need to be made to the network's weights. A single pass of a sample (or a batch of samples) is called an *iteration*, while a full pass of all the training data is called an *epoch*. The number of epochs can affect the quality of the predictions, since if it is small the network can underfit the data, while if it is too large, it can lead to *overfitting*. It is also worth mentioning that depending on the architecture, a neuron does not have to be fully connected to all those of the

Deep Learning for Robot Perception and Cognition. https://doi.org/10.1016/B978-0-32-385787-1.00007-5

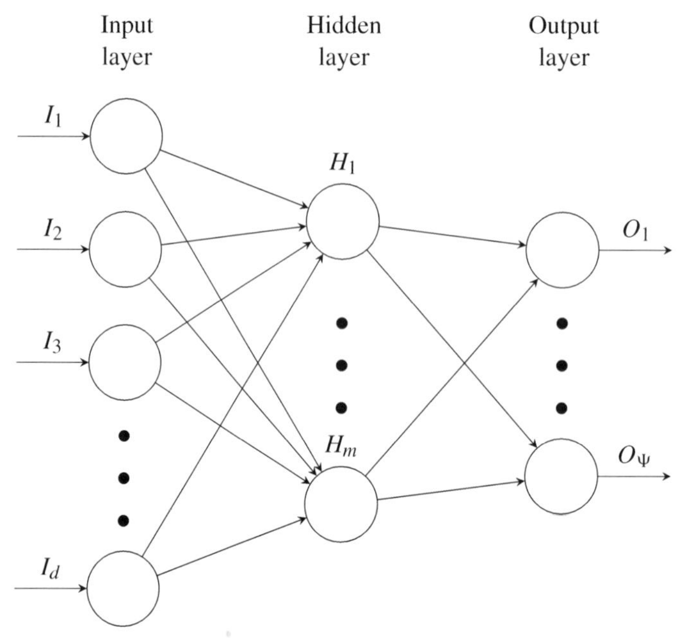

FIGURE 2.1

The architecture of a Neural Network.

next layer, but may be partially connected, e.g., as in the case of **Convolutional Neural Networks** (CNNs) [7–9]. In certain applications, such as in **Recurrent Neural Networks** (RNNs), the output of a layer can be fed back to the input of a previous (or the same) layer in order to capture more complex temporal dynamics of the data [10–13].

The output of a neuron i in a layer ℓ is calculated as follows:

$$o_i^\ell = \alpha_i^\ell(u_i^\ell), \tag{2.1}$$

where α_i^ℓ is an activation function, which calculates the final output of the neuron i in layer ℓ at a given time. Also, u_i^ℓ is called *propagation function* and calculates the total input value at a given time, known as neuron's *state*, by adding all the m individual inputs $\boldsymbol{o}^{\ell-1}$ it receives after first multiplying them with their corresponding weights $\mathbf{W}_i^\ell$ and adding the corresponding bias b_i^ℓ, i.e.,

$$u_i^\ell = \sum_{j=1}^{m} W_{ij}^\ell o_j^{\ell-1} + b_i^\ell. \tag{2.2}$$

In Section 2.2 the activation functions used in NNs are discussed. They are divided in categories, and their advantages and disadvantages are presented, as well as the reason why non-linear functions are the most prevalent in modern use cases. In Section 2.3 the cost functions are presented and in Section 2.4 the backpropagation algorithm is analyzed, which is necessary in order to understand the training procedure of NNs. Training is carried out with the help of optimizers, explained in Section 2.5. Finally, the problem of overfitting is presented, along with solutions to mitigate its effects in Section 2.6.

2.2 Activation functions

The activation functions play a determinant role in the training process and consequently the network's effectiveness, by adjusting the neurons' outputs. A function is attached to each neuron in the network and decides if it should be activated or not, based on whether its input is relevant for the model's prediction. Some activation functions also help normalize the outputs of the neurons. Early NNs employed binary activation functions that were not differentiable, such as the *step* and *sign* functions. It was soon established that the use of differentiable activation functions enables us to use simple, yet effective, training algorithms. For simplicity, in this section, the activation function and state of a single neuron in a single layer are denoted as $\alpha(\cdot)$ and u respectively.

DNNs typically use non-linear activation functions that allow to create complex non-linear mappings between the given inputs and the produced outputs, which are essential for learning and modeling complex data, such as images, video, audio and in general datasets in which the relations between the inputs and the desired outputs are non-linear. Even though linear activation functions, in the form of:

$$\alpha(u) = u, \tag{2.3}$$

can be used, they are not preferred, since they are unable to capture non-linear relations between neural layers and, as a result, do not allow for building deep networks that are composed of multiple layers. A brief description of the most commonly used activation functions in DL follows:

Sigmoid Also known as *logistic function*, it has the advantage of leading to clear divisions, as it tends to produce results near the limits of its range (0, 1). It is also worth mentioning that its output can be directly interpreted as a probability. The sigmoid function is given by:

$$\alpha(u) = \frac{1}{1+e^{-u}}. \tag{2.4}$$

An NN using it can easily be trapped in low gradient regions, leading the process of learning on a plateau, because only small changes are made in each training update.

TanH The hyperbolic tangent function is given by:

$$\alpha(u) = \frac{e^u - e^{-u}}{e^u + e^{-u}}. \tag{2.5}$$

The advantages mentioned for the sigmoid function also apply to this one, as it is a scaled and shifted version of the former:

$$\tanh(u) = 2\sigma(2u) - 1, \tag{2.6}$$

where $\sigma(\cdot)$ is the sigmoid activation function.

Softmax It takes a vector of m neurons' states of a layer ℓ as an input and normalizes it into a probability distribution consisting of m probabilities proportional to the exponentials of the input numbers. The softmax of a single neuron i in a layer ℓ can be calculated as:

$$\alpha(u_i^\ell) = \frac{e^{u_i^\ell}}{\sum_{j=1}^m e^{u_j^\ell}}. \tag{2.7}$$

It has the useful property of generating probabilities for its output because the sum of all the generated results of a layer is always 1:

$$\begin{gathered}
\text{softmax}(u_1^\ell) + \text{softmax}(u_2^\ell) + \ldots + \text{softmax}(u_m^\ell) = \\
\frac{e^{u_1^\ell}}{\sum_{j=1}^m e^{u_j^\ell}} + \frac{e^{u_2^\ell}}{\sum_{j=1}^m e^{u_j^\ell}} + \ldots + \frac{e^{u_m^\ell}}{\sum_{j=1}^m e^{u_j^\ell}} \iff \\
\sum_{i=1}^m \text{softmax}(u_i^\ell) = \frac{\sum_{i=1}^m e^{u_i^\ell}}{\sum_{j=1}^m e^{u_j^\ell}} \iff \\
\sum_{i=1}^m \text{softmax}(u_i^\ell) = 1.
\end{gathered} \tag{2.8}$$

It is able to handle multiple classes, giving the probability of the input value being in a specific class. Thus, it is typically used only in the last layer of an NN.

ReLU It is widely used in NNs, since it allows for efficiently building deep neural networks [14]. If the state is negative, **Rectified Linear Unit** (ReLU) produces 0 as an output, otherwise, the state's value itself is returned:

$$\alpha(u) = \begin{cases} 0 & u < 0 \\ u & u \geq 0 \end{cases}. \tag{2.9}$$

An issue of ReLU is that all the negative values become zero immediately, which decreases the ability of the model to train from the data properly, since all the neurons

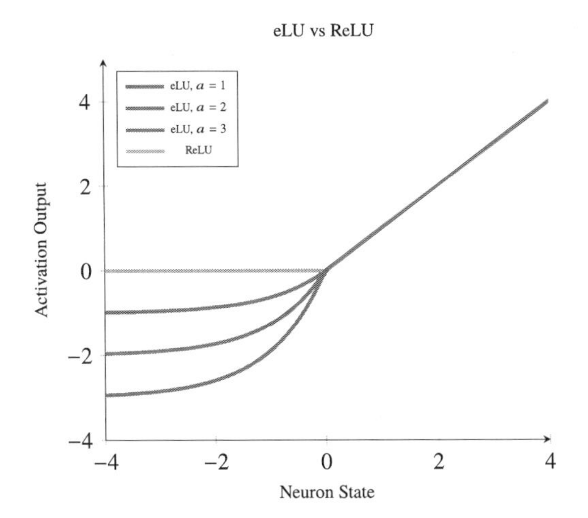

FIGURE 2.2

eLU vs ReLU activation functions' outputs. (For interpretation of the colors in the figure, the reader is referred to the web version of this chapter.)

that have negative states, immediately "turn off". This problem becomes more clear in section 2.4, where the training procedure of NNs and the backpropagation algorithm are explained. In an attempt to overcome the aforementioned limitation, many variations were created, like Leaky ReLU [15], Parametric ReLU [16] and **Exponential Linear Unit** (eLU) [17].

eLU In contrast to ReLU, eLU has negative values which push the mean of the activations closer to zero. As proved in [17], mean activations that are closer to zero enable faster learning as they bring the gradient closer to the natural gradient. Different to other activation functions, it has an extra constant, denoted with a, which should be positive. It is very similar to ReLU except for negative inputs. eLU becomes smooth slowly until its output equals to $-a$, whereas ReLU sharply smooths. eLU is defined as:

$$\alpha(u) = \begin{cases} \mathrm{a}(e^{u} - 1) & u \leq 0 \\ u & u > 0 \end{cases}. \tag{2.10}$$

The differences between ReLU and eLU are shown in Fig. 2.2.

2.3 Cost functions

Deep Learning (DL) algorithms have the objective of finding the most suitable values for the optimization parameters, which are the weights $\mathbf{W}$ and the biases $\boldsymbol{b}$ of each layer of the network. In order to achieve this, they aim to minimize a cost function (also known as *loss* function) $\mathcal{J}$, which receives the predictions of an NN and its expected outputs as parameters and produces a value as an out-turn. This value works as a penalty for the algorithm. It is a measurement which determines to what extent an NN is able to predict a result correctly. More specifically, the goal is to find those parameters that minimize the cost function's output.

Each cost function has a slope, its derivative, which reveals the direction that should be followed during the training process, in order to minimize the cost/error. Given this inclination, a local or even a global minimum can be found in the *optimization space*. In order for a cost function to be suitable for use in an NN, it has to be differentiable and meet the certain conditions [18].

There is a large variety of cost functions. The most basic ones are analyzed below:

Mean Squared Error (MSE) MSE it is defined by the sum of the square of individual errors in the output layer as follows:

$$\mathcal{J}_{MSE}(\boldsymbol{y}_i, \hat{y}_i) = \frac{1}{2}||\boldsymbol{y}_i - \hat{y}_i||_2^2, \tag{2.11}$$

where $||\cdot||_2$ denotes the l_2 norm of a vector, $\hat{y}_i$ are the outputs of the last layer for a sample i, thus the network's predictions, and $\boldsymbol{y}_i$ are the true values that it is trying to predict. For classification problems, $\boldsymbol{y}_i$ are *one-hot* encoded vectors, which means that their components are represented with 0 or 1, indicating true or false. For example, $y_{i,3} = 1$ would mean that the sample i belongs to the 3rd class. The MSE function tends to "punish" the network more for large errors, due to the square. **Mean Absolute Error** (MAE) is an alternative, which employs the l_1 norm instead. It has the feature that it is not sensitive to outliers.

Huber It smooths out the big changes caused by large errors that result from using MSE. It uses both MSE and MAE as follows:

$$\mathcal{J}_{Huber}(y_i, \hat{y}_i) = \begin{cases} \frac{1}{2}(y_i - \hat{y}_i)^2 & \forall |y_i - \hat{y}_i| \leq \delta \\ \delta|y_i - \hat{y}_i| - \frac{1}{2}\delta^2 & \forall |y_i - \hat{y}_i| > \delta \end{cases}, \tag{2.12}$$

where δ defines the proportion by which the error approaches the output of MSE and MAE.

Cross Entropy (CE) It is called *binary* when used for two-class problems and *categorical* when used for multi-class problems in which case it is usually combined with the softmax activation function in the network's output layer. For multi-class problems (categorical cross entropy) is defined as:

$$\mathcal{J}_{CE}(\boldsymbol{y}_i, \hat{\boldsymbol{y}}_i) = -\sum_{j=1}^{\Psi} y_{i,j} \log(\hat{y}_{i,j}), \tag{2.13}$$

with Ψ indicating the number of classes. For binary classification, it is defined as:

$$\mathcal{J}_{CE}(y_i, \hat{y}_i) = -\big(y_i \log(\hat{y}_i) + (1 - y_i)\log(1 - \hat{y}_i)\big). \tag{2.14}$$

Kullback–Leibler (KL) KL is a measure of how one probability distribution is different from another [19]. It calculates the loss of information that results in the attempt of the network to approach the expected values. Evidently, it also needs probabilities as inputs in order to function properly:

$$\mathcal{J}_{KL}(\boldsymbol{y}_i, \hat{\boldsymbol{y}}_i) = \sum_{j=1}^{\Psi} y_{i,j} \log\left(\frac{y_{i,j}}{\hat{y}_{i,j}}\right) \tag{2.15}$$

2.4 Backpropagation

Using the cost functions described in the previous section, the appropriate optimization parameters of an NN can be found by *minimizing* the cost function. To achieve this, the *backpropagation* algorithm is used, which calculates the gradient of the cost function with respect to the bias and the weights of a network. It iterates through the layers backwards – right after the data have been forward passed – calculating the gradients for each neuron of each layer. By applying the chain rule, it is possible to make these calculations for every layer and not only the last one. Thanks to the backpropagation algorithm DNNs can be used, exploiting their ability to represent complicated functions.

Let $\boldsymbol{\epsilon}^L$ be the gradients of the cost function $\mathcal{J}$ with respect to the output vector $\boldsymbol{o}^L$ in the last layer L. This is called *error* of L, because if it converges to 0 it means that the network's predictions are the same as the expected values, and it is given by:

$$\boldsymbol{\epsilon}^L = \nabla_{\boldsymbol{o}^L}\mathcal{J} \odot \alpha'(\boldsymbol{o}^{L-1}), \tag{BP1}$$

where $\alpha'(\cdot)$ denotes the derivative of the activation function which is applied element-wise on the input vector ($\odot$ refers to element-wise multiplication). In a similar way, the error in any of the previous layers can be calculated as:

$$\boldsymbol{\epsilon}^\ell = ((\mathbf{W}^{\ell+1})^T \boldsymbol{\epsilon}^{\ell+1}) \odot \alpha'(\boldsymbol{o}^{\ell-1}) \tag{BP2}$$

where $(\mathbf{W}^{\ell+1})^T$ is the transpose of the weight matrix $\mathbf{W}^{\ell+1}$ for the $\ell+1$ layer. Intuitively, it can be thought as moving the error backwards through the network. Weights indicate to what extent each neuron affects the outcome, so it can be seen as "distributing responsibility", attributing most of it to the neurons that have the largest influence.

Using the above, the gradients with respect to the bias can be easily calculated in any layer, since it is the error itself:

$$\nabla_{\boldsymbol{b}^\ell}\mathcal{J} = \boldsymbol{\epsilon}^\ell \tag{BP3}$$

Finally, the gradients with respect to the weights can be easily calculated as:

$$\nabla_{\mathbf{W}^\ell} \mathcal{J} = \boldsymbol{\epsilon}^\ell (\boldsymbol{o}^{\ell-1})^T \tag{BP4}$$

Therefore, Eqs. (BP1) and (BP2) are auxiliary and they are used in order to achieve the real purpose, i.e., the calculation of the gradients with respect to the network's optimization parameters.

2.5 Optimizers and training

Using the cost functions and the backpropagation algorithm, DNNs can be *trained* with an *optimizer*. The term *training*, also known as *fitting*, refers to finding the appropriate bias and weights for the whole network M, thus its optimization parameters, denoted as θ^M. In order to do this, they need to be updated by exploiting the gradients that the backpropagation algorithm calculates. An initial thought would be to subtract the errors from the parameters. However, this is a mistake that would lead to a model unable to train, oscillating between different values during the training process, since every sample would affect its outputs to a large extent, making it "forget" the previous ones. Thus, changes should be made gradually.

Updates can be applied in many different ways, which has led to the creation of multiple optimizers. In this section, for simplicity, the notation $\boldsymbol{\epsilon}$ represents a vector containing all the gradients of a network, which have been calculated using the backpropagation algorithm. The most basic optimizers are described below:

Gradient Descent It is the simplest form of optimizer. It calculates the new parameters $(\theta^M)'$ by taking steps proportional to the negative of the provided gradients after every iteration [20]:

$$(\theta^M)' = \theta^M - \eta \boldsymbol{\epsilon}, \tag{2.16}$$

where η is called *learning rate* or *step size* and controls how quickly or slowly an NN changes its learning parameters. A large η allows the model to learn faster, usually at the cost of arriving on a sub-optimal final set of θ^M. A smaller η may allow the model to learn a more optimal or even globally optimal set of θ^M, but may take significantly longer to train. At extremes, too large η will result in oscillations of the NN performance over the training epochs. On the other side, an NN using too small η may never converge or may get stuck on a sub-optimal solution. Therewith, η is very important for the training procedure and must be carefully chosen, usually by trial and error.

A full pass of the whole dataset is considered as a single iteration of the optimizer, i.e., the error is computed over the full dataset. The new $(\theta^M)'$ are computed by taking the average of the gradients of all the training examples and then using that mean gradient to update the parameters.

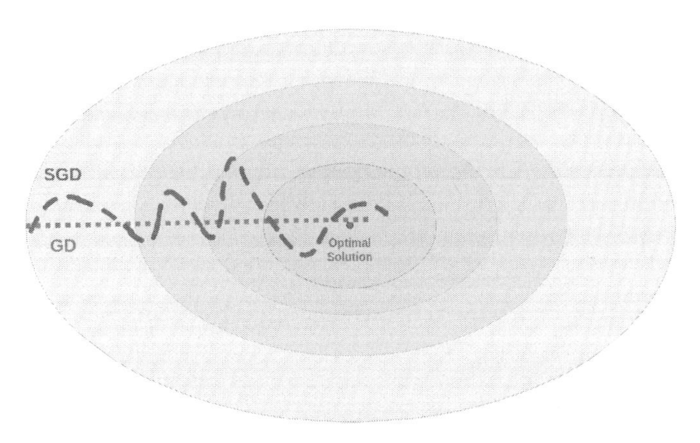

FIGURE 2.3

Gradient Descent vs Stochastic Gradient Descent steps in the optimization space.

Stochastic Gradient Descent (SGD) Time consumption is a well-known problem of the Gradient Descent algorithm. This is because new weights may be calculated only after completing a full epoch, thus converging really slowly, despite the fact that it accurately follows the correct path in the optimization space. For that reason, the SGD algorithm is preferred [21,22]. It overcomes the aforementioned problem, by shortening the duration of a single iteration, updating the θ^M after every training example. It can also be used to learn online. SGD's frequent, noisy updates cause the objective function to fluctuate, which makes the training process less stable, but enables it to reach new and potentially better local minimums.
Usually, a combination of SGD with *Mini-batch Gradient Descent* is used, which means that a subset of the samples is considered in order to estimate the mean gradient on every step. At first, the available data are split in *mini-batches*. Then, for every mini-batch, the errors are calculated and the weights are updated by taking their average. This method is used because a single sample is not always ideal to calculate the new weights, as it results in large fluctuations in the training procedure. Employing mini-batches also reduces the variance of the parameter updates, which can lead to more stable convergence. Lastly, both Mini-batch Gradient Descent and SGD practically help with hardware limitations, since they consume less memory during the training process with the former also making use of highly optimized matrix operations of ML libraries.
Therefore, SGD "assumes" the direction in which it should move in order to optimize the parameters. More often than not, it helps a network to converge faster than the Gradient Descent optimizer, even though it may need more steps. The difference between the two optimizers becomes clearer in Fig. 2.3. A widely used extension of SGD, which differentiates it from Mini-batch Gradient Descent, imitates *momentum*, by considering errors of the past steps [23]:

$$(\theta^M)' = \theta^M - \overline{\epsilon}^{(t)}, \tag{2.17}$$

where

$$\overline{\boldsymbol{\epsilon}}^{(t)} = \eta \boldsymbol{\epsilon}^{(t)} + \beta \overline{\boldsymbol{\epsilon}}^{(t-1)}, \tag{2.18}$$

with (t) indicating the timestep of the current update of the parameters and $(t-1)$ indicating the previous. β is the momentum of the optimizer, which determines the importance of the previous errors in relation to the current ones. An alternative of this version uses the Nesterov momentum instead [24].

Adaptive Gradient (AdaGrad) It is an optimizer with parameter-specific learning rates, which are adapted relative to how frequently a parameter gets updated during training [25]. The more updates a parameter receives, the smaller the learning rate. It achieves this by scaling the latter by the square root of the cumulative sum of squared gradients, using the same update rule with SGD (Eq. (2.17)) where every element i of $\overline{\boldsymbol{\epsilon}}$ is given by:

$$\overline{\epsilon}_i = \eta \frac{\epsilon_i}{\sqrt{\gamma_i^{(t)}} + \varepsilon} \tag{2.19}$$

and

$$\gamma_i^{(t)} = \gamma_i^{(t-1)} + \epsilon_i^2 \tag{2.20}$$

where ε is a tiny value, added in order to ensure that the denominator will not be too close to 0.

Root Mean Square Propagation (RMSProp) Instead of cumulative sum, it uses **Exponential Moving Average** (EMA) [26], replacing Eq. (2.20) with:

$$\gamma_i^{(t)} = (1-\rho)\gamma_i^{(t-1)} + \rho \epsilon_i^2 \tag{2.21}$$

where ρ determines the importance of the current errors in relation to the previous ones.

Adaptive Moment Estimation (Adam) It essentially combines RMSProp with momentum [27]. Let $\boldsymbol{m}^{(t)}$ be the results of Eq. (2.18), called *first moment estimate* and $\boldsymbol{u}^{(t)}$ the results of Eq. (2.21), called *second raw moment estimate* with $\hat{\boldsymbol{m}}^{(t)}$ and $\hat{\boldsymbol{u}}^{(t)}$ denoting their normalized vectors, which are *bias-corrected*. Adam uses the update rule of Eq. (2.17), where every element i of $\overline{\boldsymbol{\epsilon}}$ is given by:

$$\overline{\epsilon}_i = \eta \frac{\hat{m}_i^{(t)}}{\sqrt{\hat{u}_i^{(t)}} + \varepsilon} \tag{2.22}$$

Adam is widely used in DNNs training. The reason for its widespread use is the fact that it tends to converge much faster than its competitors in most problems, while at the same time finding more optimal solutions in the parameters space. It is summarized in Algorithm 1 [27].

Algorithm 1: Adam Optimizer: an algorithm for stochastic optimization. g_t^2 indicates the elementwise square $g_t \odot g_t$. Good default settings for the tested ML problems are $\eta = 0.001$, $\beta_1 = 0.9$, $\beta_2 = 0.999$ and $\varepsilon = 10^{-8}$. All operations on vectors are element-wise. β_1^t and β_2^t denote β_1 and β_2 to the power t.

Data: η: Stepsize
Data: $\beta_1, \beta_2 \in [0, 1)$: Exponential decay rates for the moment estimates
Data: $f(\theta)$: Stochastic objective function with parameters θ
Data: θ^0: Initial parameter vector
Result: θ^t: Resulting parameters

$m_0 \leftarrow 0$; ▷ Initialize 1st moment vector
$u_0 \leftarrow 0$; ▷ Initialize 2nd moment vector
$t \leftarrow 0$; ▷ Initialize timestep
while θ not converged **do**
  $t \leftarrow t + 1$;
  $g_t \leftarrow \nabla_\theta f_t(\theta^{t-1})$; ▷ Get gradients w.r.t. stochastic ▷ objective at timestep t
  $m_t \leftarrow \beta_1 \cdot m_{t-1} + (1 - \beta_1) \cdot g_t$; ▷ Update biased first ▷ moment estimate
  $u_t \leftarrow \beta_2 \cdot u_{t-1} + (1 - \beta_2) \cdot g_t^2$; ▷ Update biased second ▷ raw moment estimate
  $\hat{m}_t \leftarrow \frac{m_t}{1 - \beta_1^t}$; ▷ Compute bias-corrected ▷ first moment estimate
  $\hat{u}_t \leftarrow \frac{u_t}{1 - \beta_2^t}$; ▷ Compute bias-corrected second ▷ raw moment estimate
  $\theta^t \leftarrow \theta^{t-1} - \frac{\eta \cdot \hat{m}_t}{\sqrt{\hat{u}_t} + \varepsilon}$; ▷ Update parameters
end

Any parameter whose value can be chosen by the user and affects the learning process, as for example the η, ρ, etc., as well as the number of layers and neurons, is called *hyperparameter* and must be carefully chosen. A *grid search* can be performed in order to search for the best configuration of the hyperparameters, using a validation set and not the test set to select the best values. It is a very time consuming process and recently, alternatives have been discovered in which DNNs themselves are searching for ideal architectures and parameters [28].

The overall process of a DNN training is composed of the concepts explained up to this section and it is outlined by Algorithm 2. In essence, a DNN model M is trying to find a function $\mathcal{F}^M(\theta^M; X)$, which given data samples X, learns the

Algorithm 2: DNN Training Algorithm.

Data: Untrained DNN model M with $\ell = 0, 1, \ldots, L$ layers, dataset $\mathcal{D}$, optimizer $\mathcal{O}$ with hyperparameters $\mathcal{H}$, cost function $\mathcal{J}$, number of epochs E and mini-batch size $\mathcal{B}$.
Result: Trained DNN model M

```
// Initialize weights and bias.
for ℓ ← 0 to L do
    W^ℓ ← random initialization;
    b^ℓ ← 0;
end
for e ← 0 to E do
    for i ← 0 to size(D)/B do
        // Initialize error.
        ε^L ← 0;
        X, Y ← get_mini_batch(D);
        // Feedforward.
        for ℓ ← 0 to L do
            // Calculate the output of the current layer.
            o^ℓ ← α^ℓ((W^ℓ)^T X_i^ℓ + b^ℓ);
        end
        // Calculate the gradients of the last layer.
        ε^L = ε^L + ∇_{o^L} J(o^L, Y_i) ⊙ α'(o^{L-1});
        // Backpropagate.
        for ℓ ← L − 1 to 0 do
            /* Calculate the gradients of the current layer using
               the chain rule.                                    */
            ε^ℓ = ((W^{ℓ+1})^T ε^{ℓ+1}) ⊙ α'(o^{ℓ-1});
            ∇_{b^ℓ} J = ε^ℓ;
            ∇_{W^ℓ} J = ε^ℓ (o^{ℓ-1})^T;
        end
        /* Update weights and bias using an optimizer, the
           calculated gradients and the optimizer's
           hyperparameters.                                       */
        W, b ← O(∇_b J, ∇_W J, H);
    end
end
```

weights θ^M in order to produce the correct prediction vector $\hat{y}$. It is worth noting that it has been proven that NNs can act as universal approximators, since, as discussed in [29], "...multilayer feedforward networks are, under very general conditions on the hidden unit activation function, universal approximators provided that sufficiently many hidden units are available."

2.6 Overfitting

A well-known problem with NNs is their tendency to *overfit*, that is, to memorize training data, while failing to generalize their knowledge to unseen data. This problem becomes larger when there are not enough samples available. A model with a large number of parameters, can describe an abundance of events. Though, if the model is able to correctly predict the output for the data given as input, this does not mean that it will be able to do the same for unknown samples. The true ability of a model lies in its potential to correctly predict the labels of unseen samples, which is called *generalization* in ML.

One way to find out whether a model is overfitted is to observe its error rate on the training set during the epochs and compare it to that of a second, unknown set of data. In such a case, while the error rate is constantly reducing for both the training and the unknown data, at some point, it begins to grow again, only for the unknown. From that moment on, it is safe to assume that the model begins to overfit.

In addition to carefully selecting hyperparameters, there are many solutions to overfitting, some of which are presented in this section.

2.6.1 Early stopping

Early Stopping is a technique widely used in NNs. First, the data are randomly divided in two subsets, one for training and one for validation. Observing the loss of a network after each epoch for both sets, the training process is interrupted as soon as it stops improving or gets worse, as demonstrated in Fig. 2.4. It is a safe technique that guarantees to tackle overfitting, however it may not provide the best results, since NNs often stop improving for a number of epochs, but then the validation loss starts decreasing again. However, in such a case, the improvement may be negligible. Using the described method, the hyperparameters that lead to the best "point" in the observed diagram can be chosen. The highlighted area to the right of it presents the typical behavior of an overfitted network.

2.6.2 Regularization

Generalization ability can be evaluated using the *bias* and *variance* measurements. Good generalization is achieved when there is the right bias–variance trade-off. Models with a lower bias in parameter estimation tend to have a higher variance of the parameter estimates across samples, and vice versa. Bias measures the expected de-

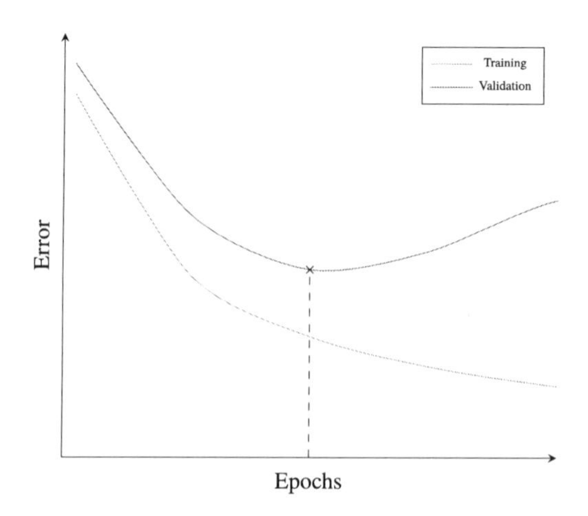

FIGURE 2.4

Early Stopping. (For interpretation of the colors in the figure, the reader is referred to the web version of this chapter.)

viation from the true value. Variance measures the deviation of the model's expected values that any particular input is likely to induce. Low bias and high variance could cause overfitting. In contrast, high bias and low variance could cause *underfitting*, meaning that the model is not able to obtain a sufficiently low error in the training data. Observing the optimal point in Fig. 2.4, if the model were to be trained for more epochs, then it would have a low bias and high variance. Contrariwise, if it were to be trained for fewer epochs, then it would have a high bias and low variance. Whether a model is more likely to overfit or underfit is controlled by altering its complexity.

One way to achieve this is by using regularization techniques. The main idea is to add an extra regularization term, which includes the network's parameters, in the cost function. Hence, the cost function depends on the "power" of the weights and biases, leading in a network that tries to keep its parameter values low, i.e., remaining simple. At the same time, a parameter λ is used, which defines the importance of the regularization term. One option is to consider the sum of the squares of all the weights on each layer [30,31] as:

$$\mathcal{J}' = \mathcal{J} + \lambda \sum_{\ell} \sum_{i,j} {W_{ij}^{\ell}}^2. \tag{2.23}$$

Alternatively, l_1 regularization can be used:

$$\mathcal{J}' = \mathcal{J} + \lambda \sum_{\ell} \sum_{i,j} |W_{ij}^{\ell}|. \tag{2.24}$$

The latter option was proven to provide more sparse results [32]. This can be intuitively explained in the following way. Imagine the l_1 2-sphere, which is a rhombus.

In its vertices, one of the weights is 0 and the other is 1. Also, generalizing this idea in a hypersphere, the probability of the cost function intersecting with a vertex is larger than with another part. Thus, a large number of the weights is going to have values close to 0, leading to sparse results in contrast with the l_2 norm. Regularization should be used in moderation, as over-regularizing an NN can lead to underfitting.

2.6.3 Dropout

Dropout is a regularization technique, where each neuron of the network is "turned off" on every training step by setting its output to 0, with a probability p usually set to 0.5 [33]. This prevents hidden units from co-adapting to others, making them more generally useful and it is achieved with a very small change in the NN's algorithm. Dropout uses a randomly initialized, binary mask vector m^ℓ on each layer's output, with a probability p for its components being 0. This changes Eq. (2.1) to:

$$\boldsymbol{o}^\ell = \alpha^\ell(\boldsymbol{u}^\ell) \odot m^\ell \tag{2.25}$$

At test time, the masks in Eq. (2.25) are replaced by their expectation, which is simply the constant vector $1 - p$. Using dropout leads to longer training time, but also leads to better generalization by preventing overfitting. In essence, dropout is similar with creating an ensemble of weaker models and averaging their predictions, but much less computationally expensive than training them separately.

2.6.4 Batch normalization

Normalizing the inputs is proven to speed up and stabilize training. Batch Normalization, based on this observation, applies normalization at the level of the hidden layers [34]. Each unit's input is subtracted with the mean and divided by the standard deviation of all the unit's inputs $\boldsymbol{o}^{\ell-1}$ for a mini-batch of size $\mathcal{B}$. Backpropagation takes it into account learning two extra parameters γ and ζ, for each unit, which scale and shift the resulting normalized values:

$$(\boldsymbol{o}^{\ell-1})' = BN_{\gamma,\zeta}(\boldsymbol{o}^{\ell-1}) \tag{2.26}$$

where

$$BN_{\gamma,\zeta}(\boldsymbol{o}^{\ell-1}) \equiv \boldsymbol{\gamma} \odot \hat{\boldsymbol{o}}^{\ell-1} + \zeta \tag{2.27}$$

with every normalized unit's input given by:

$$\hat{o}_i^{\ell-1} = \frac{o_i^{\ell-1} - \mu_{\mathcal{B}_i}}{\sqrt{\sigma_{\mathcal{B}_i}^2 + \varepsilon}} \tag{2.28}$$

where

$$\mu_{\mathcal{B}_i} = \frac{1}{\mathcal{B}} \sum_{j=1}^{\mathcal{B}} o_{i,j}^{\ell-1}, \quad \sigma_{\mathcal{B}_i}^2 = \frac{1}{\mathcal{B}} \sum_{j=1}^{\mathcal{B}} (o_{i,j}^{\ell-1} - \mu_{\mathcal{B}_i})^2 \tag{2.29}$$

There are more normalization techniques, such as Layer Normalization [35], Instance Normalization [36], Group Normalization [37] and Deep Adaptive Input Normalization [38].

2.7 Concluding remarks

NNs are highly efficient models, inspired by the way the human brain functions. They work by processing the input using their weights, which can be trained to approach the optimal solution for the task at hand. They create nonlinear mappings between the input and the output with the help of activation functions. Using cost functions in order to calculate the error of their estimations, they utilize backpropagation, which applies the chain rule, in order to determine how the weights should change. Finally, using an optimizer, they update their weights in order to 'calibrate' and correct their predictions. Choosing the network's topology, the activation and cost functions, the optimizer and the hyperparameters to be used is a time-consuming process that requires special care. Nowadays, it is not necessary to manually calculate the results in each step of this complex process, because DL frameworks undertake to do so. These frameworks have the ability to calculate derivatives extremely efficiently. In addition, they implement the necessary functions to speed up the computational process by employing powerful GPUs and/or TPUs. Using DL frameworks is straightforward, since their users just need to know how an NN operates, as well as the advantages and disadvantages of the different components that make it up, in order to make the appropriate choices and create practical networks.

References

[1] B. Kamgar-Parsi, J.A. Gualtieri, J.E. Devaney, B. Kamgar-Parsi, Clustering with neural networks, Biological Cybernetics 63 (3) (1990) 201–208, https://doi.org/10.1007/BF00195859.

[2] R.J. Frank, N. Davey, S.P. Hunt, Time series prediction and neural networks, Journal of Intelligent and Robotic Systems 31 (1) (2001) 91–103, https://doi.org/10.1023/A:1012074215150.

[3] S. Naseer, D.Y. Saleem, S. Khalid, M. Khawar, J. Han, M. Iqbal, K. Han, Enhanced network anomaly detection based on deep neural networks, IEEE Access 6 (2018) 1.

[4] K.S. Zarkias, N. Passalis, A. Tsantekidis, A. Tefas, Deep reinforcement learning for financial trading using price trailing, in: Proceedings of the IEEE International Conference on Acoustics, Speech and Signal Processing, 2019, pp. 3067–3071.

[5] J. Redmon, A. Farhadi, Yolo9000: better, faster, stronger, in: Proceedings of the IEEE Conference on Computer Vision and Pattern Recognition, 2017, pp. 7263–7271.

[6] R. Mohan, A. Valada, Efficientps: efficient panoptic segmentation, arXiv preprint, arXiv:2004.02307, 2020.

[7] I. Goodfellow, Y. Bengio, A. Courville, Y. Bengio, Deep Learning, Vol. 1, MIT Press, Cambridge, 2016.

[8] N. Passalis, A. Tefas, Learning bag-of-features pooling for deep convolutional neural networks, in: Proceedings of the IEEE International Conference on Computer Vision, 2017, pp. 5755–5763.
[9] M. Sun, Z. Song, X. Jiang, J. Pan, Y. Pang, Learning pooling for convolutional neural network, Neurocomputing 224 (2017) 96–104.
[10] D.E. Rumelhart, J.L. McClelland, Learning Internal Representations by Error Propagation, MIT Press, 1987, pp. 318–362.
[11] M. Schuster, K. Paliwal, Bidirectional recurrent neural networks, IEEE Transactions on Signal Processing 45 (11) (1997) 2673–2681, https://doi.org/10.1109/78.650093.
[12] S. Hochreiter, J. Schmidhuber, Long short-term memory, Neural Computation 9 (1997) 1735–1780, https://doi.org/10.1162/neco.1997.9.8.1735.
[13] K. Cho, B. van Merriënboer, C. Gulcehre, D. Bahdanau, F. Bougares, H. Schwenk, Y. Bengio, Learning phrase representations using RNN encoder–decoder for statistical machine translation, in: Proceedings of the Conference on Empirical Methods in Natural Language Processing, 2014, pp. 1724–1734.
[14] V. Nair, G.E. Hinton, Rectified linear units improve restricted Boltzmann machines, in: Proceedings of the International Conference on International Conference on Machine Learning, Omnipress, Madison, WI, USA, 2010, pp. 807–814.
[15] A.L. Maas, A.Y. Hannun, A.Y. Ng, Rectifier nonlinearities improve neural network acoustic models, in: Proc. icml, vol. 30, 2013, p. 3.
[16] K. He, X. Zhang, S. Ren, J. Sun, Delving deep into rectifiers: surpassing human-level performance on imagenet classification, in: Proceedings of the IEEE International Conference on Computer Vision 1502, 02 2015.
[17] D.-A. Clevert, T. Unterthiner, S. Hochreiter, Fast and accurate deep network learning by exponential linear units (elus), CoRR, arXiv:1511.07289, 2016.
[18] M.A. Nielsen, Neural Networks and Deep Learning, Determination Press, 2015.
[19] S. Kullback, R.A. Leibler, On information and sufficiency, The Annals of Mathematical Statistics 22 (1) (1951) 79–86, https://doi.org/10.1214/aoms/1177729694.
[20] A. Cauchy, Louis, Cauchy and the Gradient Method, Méthode générale pour la résolution des systèmes d'équations simultanées (1847) 536–538.
[21] H. Robbins, S. Monro, A stochastic approximation method, The Annals of Mathematical Statistics 22 (3) (1951) 400–407.
[22] J. Kiefer, J. Wolfowitz, Stochastic estimation of the maximum of a regression function, The Annals of Mathematical Statistics 23 (3) (1952) 462–466.
[23] N. Qian, On the momentum term in gradient descent learning algorithms, Neural Networks 12 (1) (1999) 145–151.
[24] Y. Nesterov, Introductory Lectures on Convex Optimization: A Basic Course, Vol. 87, Springer US, 2004.
[25] J. Duchi, E. Hazan, Y. Singer, Adaptive subgradient methods for online learning and stochastic optimization, Journal of Machine Learning Research 12 (2011) 2121–2159.
[26] G. Hinton, N. Srivastava, K. Swersky, rmsprop: divide the gradient by a running average of its recent magnitude, Toronto lecture slides, URL http://www.cs.toronto.edu/~tijmen/csc321/slides/lecture_slides_lec6.pdf.
[27] D. Kingma, J. Ba Adam, A method for stochastic optimization, in: International Conference on Learning Representations, 12 2014.
[28] I. Guyon, L. Sun-Hosoya, M. Boullé, H.J. Escalante, S. Escalera, Z. Liu, D. Jajetic, B. Ray, M. Saeed, M. Sebag, A. Statnikov, W. Tu, E. Viegas, Analysis of the autoML challenge series 2015-2018, in: AutoML, in: Springer Series on Challenges in Machine

Learning, 2019, URL https://www.automl.org/wp-content/uploads/2018/09/chapter10-challenge.pdf, 2019.

[29] K. Hornik, Approximation capabilities of multilayer feedforward networks, Neural Networks 4 (2) (1991) 251–257, https://doi.org/10.1016/0893-6080(91)90009-T.

[30] A.N. Tikhonov, V.Y. Arsenin, Solutions of Ill-Posed Problems, Scripta Series in Mathematics, V. H. Winston & Sons / John Wiley & Sons, Washington, D.C. / New York, 1977, translated from the Russian, Preface by translation editor Fritz John.

[31] A.E. Hoerl, R.W. Kennard, Ridge regression: biased estimation for nonorthogonal problems, Technometrics 42 (2000) 80–86.

[32] R. Tibshirani, Regression shrinkage and selection via the lasso, Journal of the Royal Statistical Society, Series B (Methodological) 58 (1) (1996) 267–288, http://www.jstor.org/stable/2346178.

[33] G. Hinton, N. Srivastava, A. Krizhevsky, I. Sutskever, R. Salakhutdinov, Improving neural networks by preventing co-adaptation of feature detectors, arXiv preprint, arXiv:1207.0580, 07 2012.

[34] S. Ioffe, C. Szegedy, Batch normalization: accelerating deep network training by reducing internal covariate shift, arXiv preprint, arXiv:1502.03167, 2015.

[35] J. Ba, J.R. Kiros, G.E. Hinton, Layer normalization, arXiv preprint, arXiv:1607.06450.

[36] D. Ulyanov, A. Vedaldi, V.S. Lempitsky, Instance normalization: the missing ingredient for fast stylization, arXiv preprint, arXiv:1607.08022, 2016.

[37] Y. Wu, K. He, Group normalization, in: ECCV, 2018.

[38] N. Passalis, A. Tefas, J. Kanniainen, M. Gabbouj, A. Iosifidis, Deep adaptive input normalization for time series forecasting, IEEE Transactions on Neural Networks and Learning Systems (2019).

CHAPTER

Convolutional neural networks

3

Jenni Raitoharju[a,b]

[a]*Finnish Environment Institute, Programme for Environmental Information, Jyväskylä, Finland*

[b]*Faculty of Information Technology and Communication Sciences, Tampere University, Tampere, Finland*

3.1 Introduction

In recent years, deep learning in general and convolutional neural networks (CNNs) in particular have gained huge popularity in numerous fields of science and society varying form agriculture [32] to radiology [68], from autonomous driving [61] to remote sensing [47], and they continue to conquer other fields, such as entomology [29] or historical analysis [8], where the adaptation has been slower. In robotics, CNNs have been applied to various tasks including robot arm guidance [10], defect detection [72], path planning [42], and human-robot interaction [66].

The success of CNNs is based on their ability to process raw data, for example, images [36], audio [25], or surface electromyography (sEMG) signals [10]. For decades, machine learning applications were based on features separately extracted from the data. Designing efficient and representative features for the application at hand required considerable domain expertise and a lot of engineering work. CNNs can learn to extract suitable features directly from the raw data, which alleviates the use of machine learning and has also led to great improvements in the results.

While the basic structure of CNNs [16] and the currently used gradient-based training approach [37] were proposed already in 1980s, their usage long remained scarce and focused on few specific applications such as document reading [38]. In late 1990s and early 2000s, machine learning experts commonly thought that learning to automatically extract features for complex tasks over generic input data was infeasible. In particular, a common believe was that for sufficiently large networks the gradient-based training would inevitably get trapped to a poor local optimum [39]. This assumption was later shown to be incorrect both by theoretical analysis [11,7] and the practical success of CNNs.

The breakthrough of CNNs to public awareness happened in 2012, when Krizhevsky et al. won the well-known ImageNet Large Scale Visual Recognition Challenge [52] with a CNN architecture called AlexNet [36] illustrated in Fig. 3.2. The challenge requires classifying about a million images from 1000 classes, and AlexNet almost halved the existing top 5 error rate to 15.3% by employing 62 million trainable parameters. Some of the key techniques that enabled the unprecedented

Deep Learning for Robot Perception and Cognition. https://doi.org/10.1016/B978-0-32-385787-1.00008-7

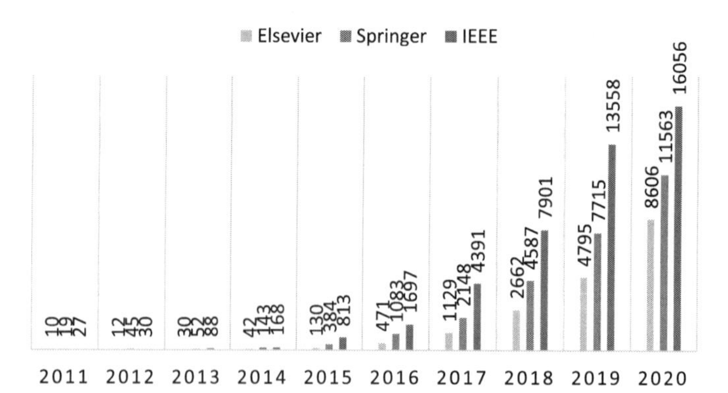

FIGURE 3.1

Yearly amounts of publications on CNNs for the three major publishers Elsevier (publications retrieved from ScienceDirect searching for the term "convolutional neural network" in 05/2021), Springer (publications retrieved from SpringerLink searching *with the exact phrase* "convolutional neural network" in 05/2021), and IEEE (publications retrieved from IEEE Xplore searching for the term "convolutional neural network" in *full text only* in 05/2021).

training effort were the efficient use of graphics processing units (GPUs), *Rectified Linear Unit (ReLU) activation function*, a novel regularization technique called *dropout*, and *data augmentation* [39]. The success of AlexNet raised a wide interest toward CNNs and started a global competition to further improve the related techniques. Three years later, CNNs were able to surpass the human performance level in the same challenge [22] and the number of scientific publications on CNNs started growing rapidly as illustrated in Fig. 3.1 for the three major publishers Elsevier, Springer, and IEEE.

The tasks currently performed by CNNs are numerous. In the simplest *regression* tasks, the CNN should predict a single number, for example, given a picture of a house, predict its price. In *multiclass classification* tasks, the goal is to predict the correct class for a given input, for example, given a picture of a bird, predict its species. In object detection, the CNN should find objects in the image along with their positions typically given as bounding boxes around the objects. In semantic segmentation, the task is to label each image pixel to its class (e.g., sky, building). The prediction may be also a new image, for example, given a facial image, predict how this person will look 30 years later. The task at hand affects many design choices related to CNNs. To focus on the most basic elements, this chapter assumes either regression or classification task, while some of the other tasks are discussed in later chapters of this book.

This chapter describes the most common building blocks and training techniques of modern CNNs. Even though the performance of the ground-breaking AlexNet was soon surpassed by more advanced CNN architectures, AlexNet and the above-mentioned supplemental techniques continue to be illustrative examples of typical design choices for CNNs. They will be discussed in the following sections along with

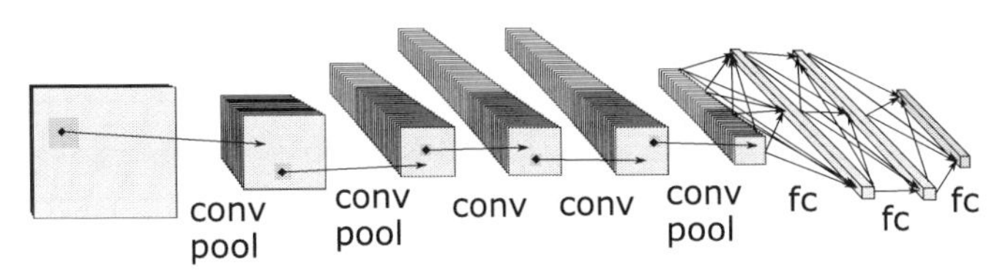

FIGURE 3.2

Illustration of AlexNet architecture.

some later-proposed techniques that have become widely accepted as core design or training choices for CNNs. Section 3.2 concentrates on the structure and building blocks of CNNs, while Section 3.3 focuses on training and techniques to further improve the training efficiency and accuracy. Note that in this chapter CNNs refer specifically to 2D CNNs, which are commonly seen as the default CNN type and applied on 2D inputs, usually images. Nevertheless, 1D and 3D CNNs are actively used for 1D signals and 3D data such as videos, and the basic elements of both 1D and 3D CNNs are very similar to the ones described in the following sections.

3.2 Structure of convolutional neural networks

CNNs are *feedforward* networks, which means that information flows in the inference stage only from lower (closer to input) to higher (closer to output) layers without backward loops from higher to lower layers. Note that during training, as discussed in Section 3.3, information flows also backwards from output toward input, but the term feedforward is used to distinguish from *recurrent* networks, which have loops in the network structure. The famous AlexNet architecture [36] is shown in Fig. 3.2 as an example of a typical CNN that consists of multiple *convolutional layers* alternating with *pooling layers* followed by few *fully connected layers* and an *output layer*. Its more detailed design choices are given in Table 3.1. The meaning of the information shown in Table 3.1 will be clarified in the following subsections.

Convolutional layers (Section 3.2.2), as the name suggests, are the key element in CNNs. They learn to extract features directly from the data using convolution operations, and thus supersede the feature engineering previously needed for applying machine learning algorithms. Earlier commonly used neural networks consisted of only fully connected layers, and for such networks, handling raw data is computationally infeasible. Some of the core factors behind the success of CNNs are *weight sharing* and *local connectivity* of convolutional layers that lead to considerable reduction in the required number of trainable parameters compared to fully connected layers.

Similar to fully connected layers, the final output of convolutional layers is obtained by applying a nonlinear activation function (Section 3.2.3) on the output of the convolution operation. This added nonlinearity is important, because otherwise

Table 3.1 AlexNet layers and corresponding design choices.

Layer	Type	Input size	Filter/ neuron #	Kernel size	Stride	Activation
1	Convolutional	227×227×3	96	11×11	4	ReLU
	Max pooling	55×55×96	96	3×3	2	
2	Convolutional	27×27×96	256	5×5	1	ReLU
	Max pooling	27×27×256	256	3×3	2	
3	Convolutional	13×13×256	384	3×3	1	ReLU
4	Convolutional	13×13×384	384	3×3	1	ReLU
5	Convolutional	13×13×384	256	3×3	1	ReLU
	Max pooling	13×13×256	256	3×3	2	
6	Fully connected	6×6×256 = 9216	4096			ReLU
7	Fully connected	4096	4096			ReLU
8	Output	4096	# of classes			ReLU+softmax

the CNN would be able to learn only linear mappings. While the activation function is an old concept, novel activation function types played an important role in the breakthrough and further development of CNNs [18,22].

Pooling layers (Section 3.2.4) resize feature maps produced by the convolutional layers. This allows to extract features on multiple scales and to reduce the size of the extracted feature maps so that it is feasible to handle them with fully connected layers at the end. The fully connected layers can be seen as a classifier applied on top of the features extracted by the convolutional layers, and the output layer is a fully connected layer that gives the final network prediction in a desired form. Fully connected and output layers are discussed in Section 3.2.5. Finally, Section 3.2.6 discusses the overall CNN structure and introduces some famous CNN architectures.

3.2.1 Notation

All of the notation are introduced when they first appear, but because many notation appear throughout the chapter, the repeating notation are also provided in Table 3.2 for easier access.

3.2.2 Convolutional layers

Convolutional layers are based on a typical image processing operation: 2D convolution between an input image X and a kernel K, which is a small matrix of weights. The output value for an input element $\mathrm{X}[m, n]$ is computed by superimposing the kernel over the image with the center of the kernel on top of the element, taking elementwise products between the kernel values and the corresponding input elements, and summing the values. This is illustrated in Fig. 3.3, where the computation of the upper left corner output element value $\mathrm{Y}[1, 1] = y_{11}$ is shown for a convolution between a 5×5 input image X and 3×3 kernel K. The other elements of the output

Table 3.2 Notation appearing in multiple sections.

Notation	Explanation	Note
a, A	index, scalar	
$\mathbf{a}$	vector	
A	matrix	
$\mathsf{A}[m,n] = a_{mn}$	matrix element	Form a_{mn} used only in figures
$\mathsf{A}[m,n,c] = a^c_{m,n}$	element $[m,n]$ of channel c	Form a^c_{mn} used only in figures
A^l	l is layer index	
A_i	i is iteration or filter index	
b, B	bias	
C	number of channels	
E	error	
$f(.)$	activation function	
K	kernel	
K	kernel size	
L	loss	
$\mathbf{t}_s$	target output for sample s	
x, X	input	
y, Y	layer output before activation	
$\hat{y}$, $\hat{\mathsf{Y}}$	layer output after activation	
$\hat{\mathbf{y}}_s$	predicted output for sample s	
δ, Δ	local error	
η	learning rate	
θ	parameter	
$\nabla_\theta E$	error gradient wrt. θ	

image Y are obtained in the same way by sliding the kernel over the whole input image and repeating the operation. The output elements are also referred to as *features* and the output image as *feature map* for kernel K. A more general definition for 2D convolution[1] can be given as

$$\mathsf{Y}[m,n] = \sum_{i=1}^{K}\sum_{j=1}^{K} \mathsf{K}[i,j]\mathsf{X}[m-1+i, n-1+j], \tag{3.1}$$

where K is the width or height of the kernel K. Here, the width and height are assumed to be the same, that is, all the kernels have a square shape, as is the case in

[1] In signal processing, Eq. (3.1) in fact corresponds to 2D *cross-correlation*, while 2D convolution would be $\mathsf{Y}[m,n] = \sum_{i=1}^{K}\sum_{j=1}^{K} \mathsf{K}[i,j]\mathsf{X}[m+1-i, n+1-j]$, that is, the kernel is flipped around. However, adding the flipping operation would just make the CNN learn the flipped version of the kernels. Therefore the flipping can be omitted, and in the machine learning context, it has become a common practice to talk about Eq. (3.1) convolution.

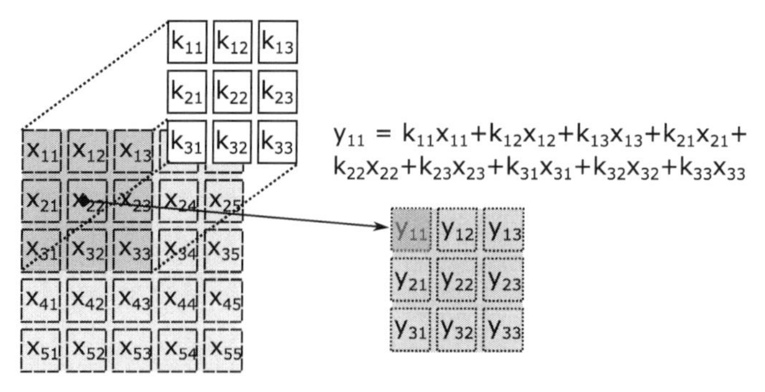

FIGURE 3.3

Computation of a single feature value in 2D convolution.

most CNN implementations. However, this is not strictly necessary. Note that indexing in Eq. (3.1) is somewhat different from the one used in Fig. 3.3, but the outcome of the overall convolution operation is the same. The convolution operation over the whole input image is denoted as $Y = K \star X$.

The area of the input image affecting the value of a particular feature is the *receptive field* of the operation. The receptive field cannot exceed the input image. Therefore, feature values for the edge input elements cannot be computed and the feature map will be smaller than the input image as shown in Fig. 3.3. This can be avoided by *padding* the input image with zeros as shown Fig. 3.4(a). The convolution without any padding is called *valid*, while the convolution padded with enough zeros to keep the feature map size equal to the input size is called *same*. Note that same convolution requires more padding for larger kernel sizes. An even wider zero padding, where in the corners the receptive fields contain only a single element of the original input, gives the third commonly used type, *full convolution*. To obtain full convolution in Fig. 3.4(a), an additional layer of zeros should be added on each side of the input image.

Fig. 3.4(b) illustrates the concept of *strided convolution*. In Figs. 3.3 and 3.4(a), a feature value is computed for every possible location of the kernel over the input image. In this way, the same input elements effect the value of many features due to the overlapping receptive fields. In strided convolution, some of the kernel locations are skipped. Fig. 3.4(b) shows a convolution with stride of 2, that is, each kernel center location is two pixels apart from the neighboring ones. The convolutions in Figs. 3.3 and 3.4(a) are said to have stride of 1 or referred to as *nonstrided convolutions*. As seen in Fig. 3.4(b), striding further decreases the output feature map size with respect to the input image.

Dilated or atrous convolution [69] is a way to increase the receptive field size of the convolution without increasing the computational cost. The input elements affecting a single output element are no longer adjacent, but they are spread around by a dilation factor. This can be thought as inserting zeros between the kernel elements as shown Fig. 3.4(c), which illustrates a dilated convolution with a 3×3 kernel and a

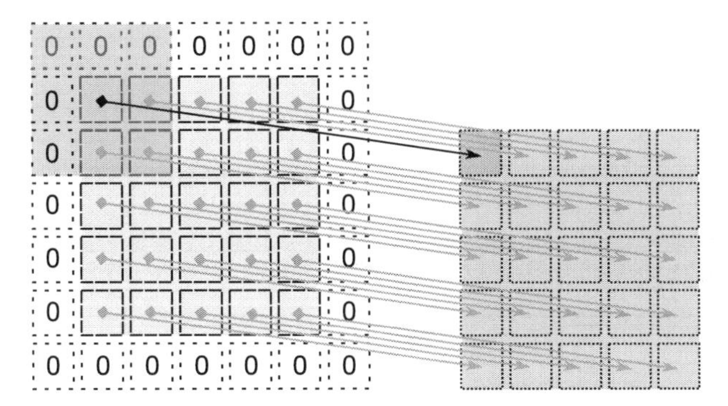

(a) Zero padding in same convolution with 3x3 kernel

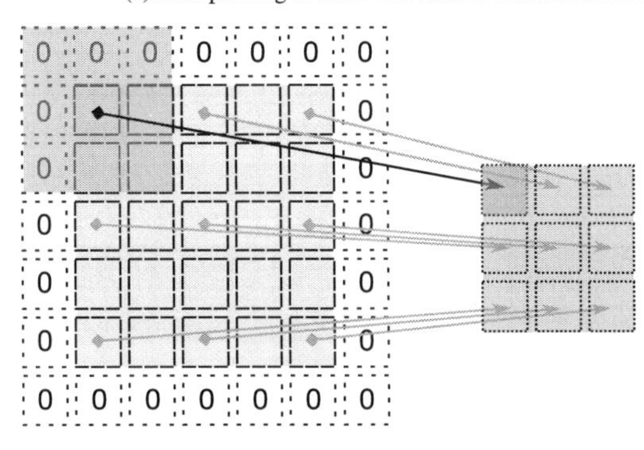

(b) Strided convolution with 3x3 kernel and stride of 2

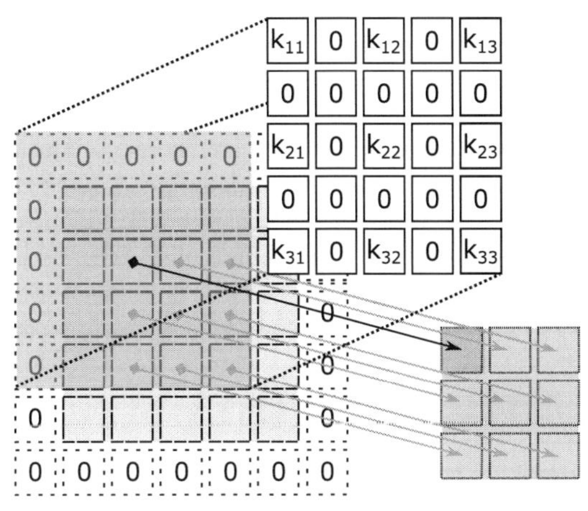

(c) Dilated convolution with 3x3 kernel and dilation factor of 2

FIGURE 3.4

Concepts related to 2D convolution. The shadowed areas show receptive fields and resulting output features for the first operation. Similar operations are repeated for every arrow.

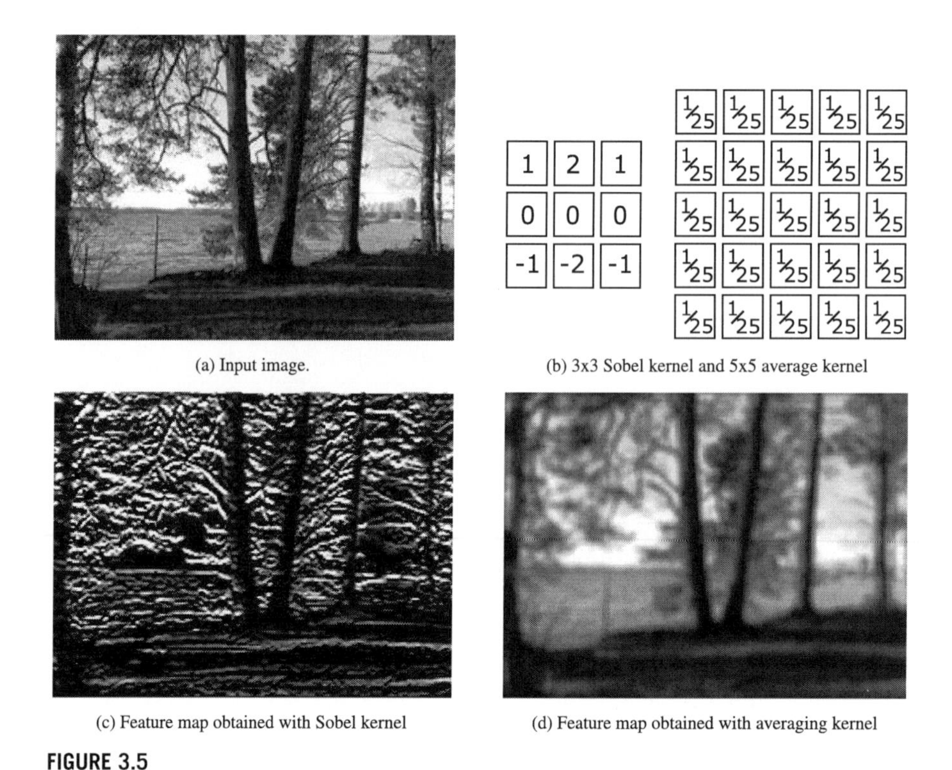

(a) Input image.

(b) 3x3 Sobel kernel and 5x5 average kernel

(c) Feature map obtained with Sobel kernel

(d) Feature map obtained with averaging kernel

FIGURE 3.5

2D convolution of an image with two well-known kernels.

dilation factor of 2. Due to the dilation, the input elements contributing to an output element are now two pixels apart. The kernel still has only 3×3 weights, but the receptive field size for the convolution is 5×5. Dilated convolutions are commonly used in semantic segmentation tasks [5].

2D convolutions with different filters can extract versatile features from the images. Fig. 3.5 illustrates the feature maps (Figs. 3.5(c) and 3.5(d)) obtained from 2D convolution of an input image (Fig. 3.5(a)) with two widely used kernels (Fig. 3.5(b)): 3×3 horizontal Sobel kernel commonly used to extract horizontal edges in the image and 5×5 averaging filter that smooths the image by replacing each pixel with the average value in its 5×5 neighborhood. When machine learning methods on images still required first extracting suitable features, 2D convolutions with different manually predefined kernels were commonly used in the extraction process. Also, CNNs extract features from images using multiple parallel as well as consecutive 2D convolutions. Parallel convolutions of a single convolutional layer extract different feature maps from the same input image, while the following layers apply convolutions on the feature maps extracted in the previous layers. This allows to extract features with different scales. Typically, the feature map size in the higher layers gets smaller, and thus kernels with similar size have larger receptive field sizes with respect to the orig-

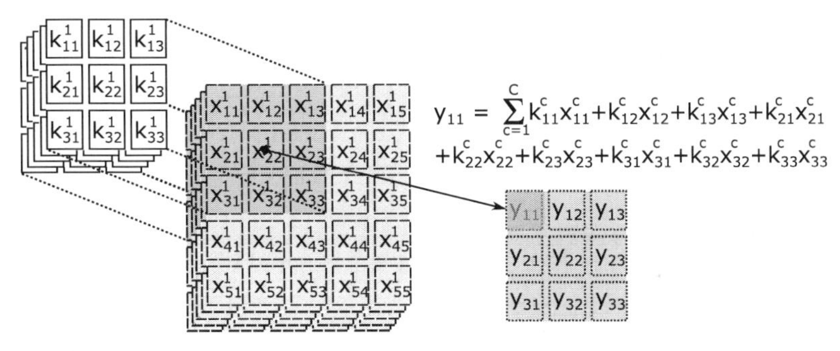

FIGURE 3.6

Multichannel convolution.

inal input size. While manually designing suitable kernels for such a complex system of convolutions would be a very challenging engineering task for any human expert, the main idea of CNNs is that they can learn suitable kernel weights from the training data.

The input for a convolutional layer is typically not a single image, but has multiple *channels*. A typical network input, a color image in RGB format, has three channels and each convolutional layer produces multiple feature maps, which in turn are considered as separate input channels for the following layer. Therefore the *filters* of the convolutional layers consist of multiple kernels, one for each input channel c, and the output feature is formed by summing over all the channels as shown in Fig. 3.6. This can be expressed as

$$\mathrm{Y}[m,n] = \sum_{i=1}^{K}\sum_{j=1}^{K}\sum_{c=1}^{C} \mathrm{K}[i,j,c]\mathrm{X}[m-1+i, n-1+j, c], \tag{3.2}$$

where $\mathrm{K}[i,j,c] = k^c_{ij}$ is a kernel weight for the cth channel and C is the number of channels. Note that Eq. (3.2) shows a nondilated convolution and the input X can be either the original input (valid convolution) or the zero padded version (same or full convolution).

After the convolution operation, the final output of a *convolutional neuron* is obtained by adding a bias value to each element of the feature map and applying an activation function elementwise to the feature map elements:

$$\hat{\mathrm{Y}} = f(\mathrm{K} \star \mathrm{X} + \mathrm{B}) = f(\mathrm{Y} + \mathrm{B}), \tag{3.3}$$

where $\hat{\mathrm{Y}}$ is the final output feature map of the neuron, B matrix adds the same bias value b to every feature map element, and the activation function f is applied element-wise. In deep CNNs, the practical impact of the bias term is small, but the nonlinear activation functions play an important role by making the overall CNN a nonlinear mapping, and thus allowing it to learn a much wider range of different mappings. Different activation functions are introduced in Section 3.2.3.

It should be noted that the concept of a neuron is somewhat arbitrary with CNNs. The definition of Eq. (3.3) associates a neuron with a filter and each neuron, thus produces a full feature map. However, it is possible to explain CNNs also in terms of standard neurons (see Chapter 2). Using this definition, a neuron is associated with single position of the kernel and, thus, produces a single feature map element:

$$\hat{\mathrm{Y}}[m,n] = f\left(\sum_{i=1}^{K}\sum_{j=1}^{K}\sum_{c=1}^{C}\mathrm{K}[i,j,c]\mathrm{X}[m-1+i,n-1+j,c]+b\right), \tag{3.4}$$

where $\hat{\mathrm{Y}}[m,n]$ is a single feature in the final output feature map. This neuron model shows the strong similarity of CNNs with fully connected neural networks, but also highlights some of the major differences. Instead of being connected to every input element as in fully connected neural networks, the neurons in CNNs have a very *local connectivity* to only those input elements, which are within its receptive field. Furthermore, *weight and bias sharing* occurs for the neurons corresponding to the same filter. This shows how convolutional layers significantly reduce the number of trainable parameters compared to fully connected layers that connect every input element with every output element and do not have weight sharing. These factors make it computationally feasible to process raw input data.

We can now again look at Table 3.1 listing the AlexNet design choices. For the first convolutional layer, the input has three channels. The kernel size is 11×11, which means that every filter consists of three 11×11 kernels and there are 96 filters. Thus the total number of weights in the first convolutional layer is $96 \times 3 \times 11 \times 11 = 34848$. Note that the input resolution and stride do not directly affect the number of parameters in a layer. Nevertheless, they determine the output feature map size, and thus directly affect the number of computations needed in the CNN. If the first convolutional layer in AlexNet was replaced by a fully connected layer connecting $227 \times 227 \times 3 = 154587$ input elements with $55 \times 55 \times 96 = 290400$ output elements, this would require $154587 \times 290400 \approx 4.5 * 10^{10}$ weights, which would be an impossible training task for the current computers. Therefore one of the common tasks of the convolutional layers is to reduce the number of elements in the feature maps before they are given to the fully connected layers.

In addition to the different variants of the basic convolution operation discussed above, there are other commonly used variants such as *grouped convolution*, *transposed convolution*, and *separable convolutions*. Grouped convolutions were applied in the original AlexNet [36] to parallelize the convolution operations across two separate GPUs. The main idea is to split filters so that they only operate on a subset of input channels. Later, it was proposed [71] to shuffle the channels between layers of grouped convolutions. Transposed convolutions [15] are used when it is necessary to upsample the feature maps. This is common in tasks such as semantic segmentation or creation of superresolution images, but not in regression or classification. Therefore we do not further discuss transposed convolutions in this chapter. Separable convolutions [6] split the standard convolutional filters into smaller filters so that the resulting operation approximates the original operation, but is computationally more efficient. Separable convolutions are briefly discussed in Section 3.2.6, where some famous CNN architectures are described.

3.2.3 Activation functions

Biological neurons fire or *activate* when they receive enough certain kind of stimulus from the environment. Activation functions in neural networks can be thought to evaluate whether a neuron should activate or not when a specific input is given to the neuron. The simplest activation function would be just an on/off switch that activates when certain conditions are met. However, this kind of activation is not optimal for more complex cases, such as multiclass classification, where it is beneficial to compare the level of activations in different neurons. Therefore activation functions producing more varied output values are used.

While convolutional operations have proven to be efficient and practical for features extracting purposes, they only consist of multiplications and additions, that is, linear operations, as shown in Eq. (3.2). Multiple consecutive linear operations are still only a linear operation. Activation functions are used to add nonlinearity to neural network, which allows them to approximate a much wider range of transformations.

Some activation functions are listed in Table 3.3 and they are illustrated in Fig. 3.7 along with their derivatives. Binary step and linear activation are not used in convolutional layers. *Binary step activation* represents the on/off switch, which is intuitive, but does not allow comparing activation levels. Furthermore, its derivative is zero everywhere, which prevents training the network with the common gradient-based approach. *Linear activation* does not provide the crucial nonlinearity and, therefore, it is not used at the hidden neurons. However, it is commonly used at the output layer for regression tasks.

Table 3.3 Well-known activation functions.

Binary step $\hat{y} = \begin{cases} 0 & \text{if } y \leq 0 \\ 1 & \text{if } y > 0, \end{cases}$	Linear $\hat{y} = ay$
Sigmoid $\hat{y} = \sigma(y) = \dfrac{1}{1+e^{-y}}$	Tangent hyperbolic $\hat{y} = \tanh(y) = \dfrac{e^{y} - e^{-y}}{e^{y} + e^{-y}}$
ReLU $\hat{y} = \begin{cases} 0 & \text{if } y \leq 0 \\ y & \text{if } y > 0 \end{cases} = \max(0, y)$	Leaky ReLU $\hat{y} = \begin{cases} 0.01y & \text{if } y \leq 0 \\ y & \text{if } y > 0 \end{cases} = \max(0.01y, y)$
Parametric ReLU $\hat{y} = \begin{cases} ay & \text{if } y \leq 0 \\ y & \text{if } y > 0 \end{cases} = \max(ay, y)$	Swish $\hat{y} = \dfrac{y}{1+e^{-y}}$

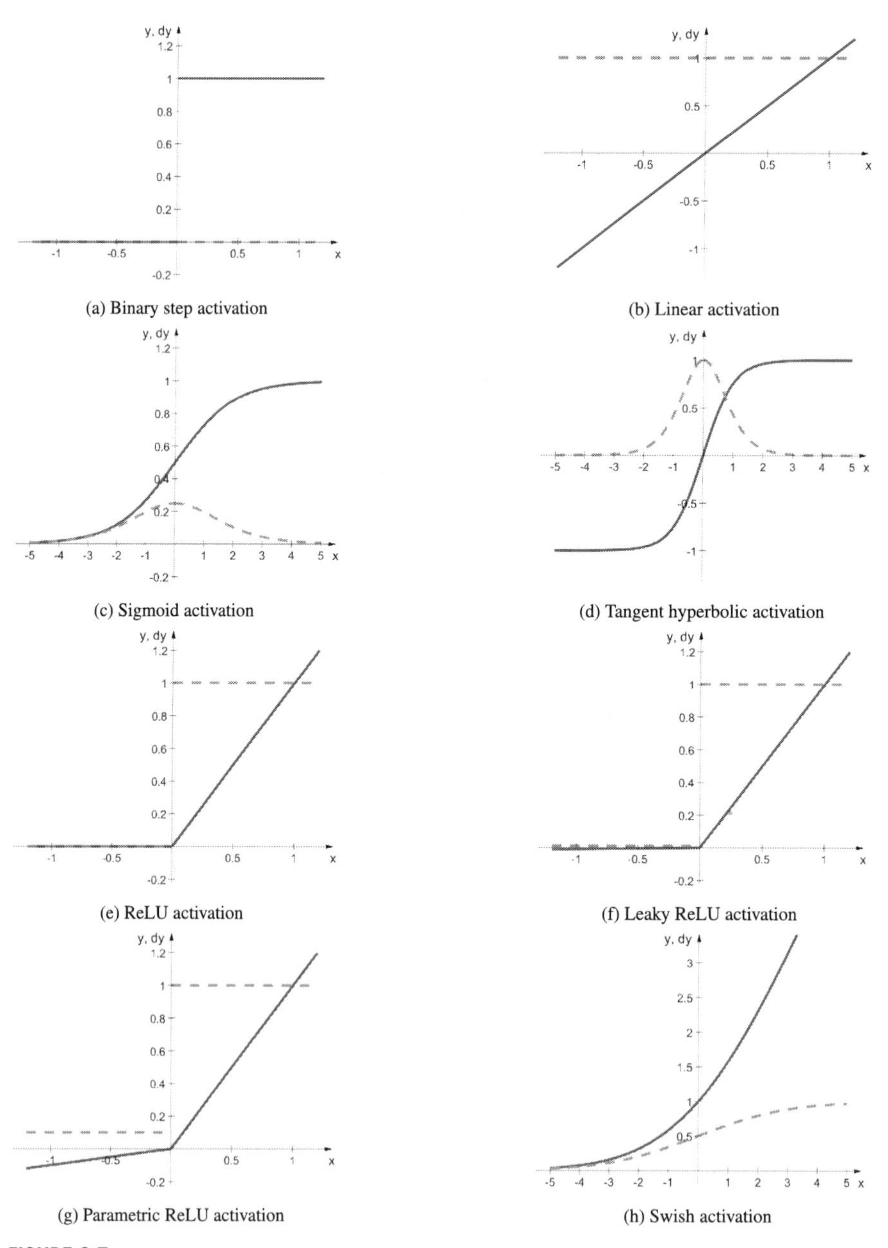

(a) Binary step activation

(b) Linear activation

(c) Sigmoid activation

(d) Tangent hyperbolic activation

(e) ReLU activation

(f) Leaky ReLU activation

(g) Parametric ReLU activation

(h) Swish activation

FIGURE 3.7

Activation functions (blue solid line) and their derivatives (red dashed line).

Sigmoid (logistic) and *tangent hyperbolic* (tanh) are similar activation functions that resemble the binary step function, but are smooth, nonlinear, and their derivatives are nonzero. Both functions bound the output values, which prevents activations from *exploding*. The main difference is that for sigmoid the output values are between 0 and 1 and for tangent hyperbolic between -1 and 1. For input values around zero, function values change rapidly, which makes the predictions clearer. These are positive properties and, indeed, sigmoid and tangent hyperbolic were the two default activation functions before the breakthrough of deep learning. However, with deep learning they lead to the *vanishing gradient* problem (Section 3.3.4.3). Especially toward both ends of the curve, values of the derivative are very small, which makes the gradients tiny as the derivative values are repeatedly multiplied in the backpropagation. This, in turn, leads to very slow and inefficient training.

It was first demonstrated in 2011 [18] that training a very deep network is much faster using *Rectified Linear Unit (ReLU)* activation. Despite its simplicity, ReLU activation was one of the main innovations that enabled record breaking performance of AlexNet and, thereafter, it has been the most commonly used activation for deep networks. The main benefit is that its derivative is one for all the positive values, which helps avoid the vanishing gradient problem. For all the negative inputs, the output is zero, which makes fewer neurons active at the time. This saves some computational costs and makes the networks lighter. A disadvantage is the *dying ReLU problem*, that is, for negative input values the derivative is zero and the weights of the neuron will not get adjusted.

Leaky ReLU [48] and *Parametric ReLU* [22] were proposed to solve the dying ReLU problem. They have a small slope also for negative input values. The main difference between the two activations is that for Parametric ReLU the slope a is also a parameter that can be learned by the training algorithm. However, the performance with Leaky and Parametric ReLU has turned out to be somewhat inconsistent. *Swish* [51] is a newer activation function that combines linear and sigmoid activations. It gives top results more consistently and performs well especially for very deep CNNs.

3.2.4 Pooling layers

Pooling layers reduce the spatial size of the feature maps extracted by convolutional layers. This saves computation costs and allows the following convolutional layer to extract features at a different scale. The two main approaches to pooling are *max pooling* and *average pooling* illustrated in Fig. 3.8. Max pooling selects the maximum value in the receptive field of the pooling kernel, while average pooling takes the average of all the values in the area. Note that average pooling can be presented as 2D convolution between the input and an averaging kernel having all the weights equal to $1/K^2$ as illustrated in Fig. 3.5.

Fig. 3.8 shows nonoverlapping pooling with 2×2 kernels and a stride of 2. Typically in CNNs, the stride is set to a smaller value than the kernel size, which results in overlapping pooling. For example, the pooling layers in AlexNet use max pooling with 3×3 kernel and stride of 2. This reduces the spatial dimensions of the feature maps by half as can be seen in Table 3.1.

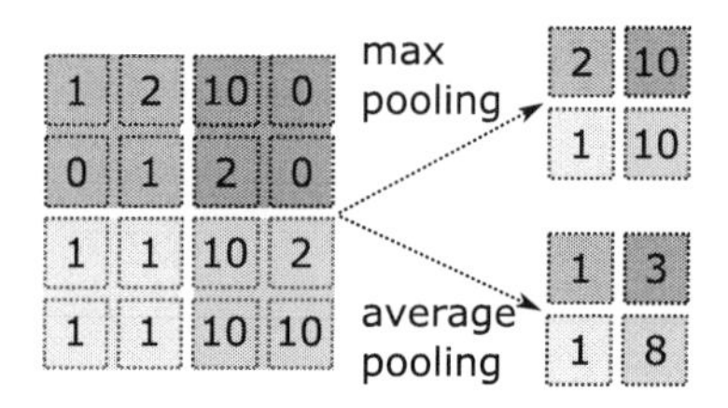

FIGURE 3.8

2×2 max pooling and average pooling with stride of 2.

If the input contains features that lead to a strong activation with a specific convolutional kernel, max pooling keeps the value high instead of blurring it like average pooling. The exact location of the high activation does not matter, which leads to local translation invariance and, through the network, to more global invariance. These are some of the reasons max pooling layers are typically inserted between convolutional layers in most deep CNN architectures.

Pooling can be also global meaning that the kernel size is equal to the input size. Especially *global average pooling* is often used at the end of a CNN to take averages of full feature maps. This significantly reduces the input size and can be used avoid some [21] or even all [44] fully connected layers. Global average pooling makes networks more robust to spatial translations and their decisions easier to interpret [73]. They can also help avoid overfitting (Section 3.3.4.1) as a form of structural regularization.

3.2.5 Fully connected and output layers

Fully connected layers consist of standard neurons (see Chapter 2) and they connect each input element to each output element with a separate weight. Many CNN architectures have few fully connected layers at the end of the network and they are commonly seen as the classifier using the features extracted by the convolutional layers. However, as discussed above, the fully connected layers are computationally costly. For example, in the AlexNet architecture given in Table 3.1, almost 59 million out of the 62 million trainable parameters are in the fully connected layers. Therefore many widely used architectures, for example, ResNet [21], do not have other fully connected layers besides the output layer.

An output layer is a fully connected layer, whose output is the network's prediction for the given input. The output format depends on the task at hand. In simple regression tasks, the output is just a single number. In such cases, the output layer has a single neuron that forms the weighted sum of all the features and a linear activation function is typically used.

While networks trained to a classification task could also output just a single number corresponding to the class index (e.g., 1 = lake, 2 = mountain, 3 = forest, 4 = city, 5 = amusement park), this has practical problems. In training, it is important to be able to consider the size of the error. But if an image represents a lake, is it any

Table 3.4 One-hot encoded target outputs (columns) for 3 images in 5-class classification task.

Lake	0	0	1
Mountain	0	0	0
Forest	1	0	0
City	0	0	0
Amusement park	0	1	0

better to classify it as a mountain than a forest? What should be the output for an image that the CNN sees as borderline case between a lake and a forest? To avoid such problems, the output in classification tasks is commonly a *one-hot encoded* vector, whose length is the total number of classes and each element corresponds to a single class as illustrated in Table 3.4. While the target outputs contain only 0s and 1s, the networks rarely obtain full certainty and a prediction for classification could be something like [0 0 0 0.2 0.8]. Such a prediction shows that the CNN is quite sure that the image depicts an amusement park, but it could be also a city image. One-hot encoding can be also used in *multi-label classification*, where a single image can be classified into multiple classes (e.g., an image can depict both lake and mountain). In this case, the target vector has multiple 1s.

In multilabel classification, a reasonable output layer activation is sigmoid, which confines all the output elements between zero and one. In single-label multiclass classification, most CNN implementations use a special activation function, *softmax*, defined as

$$\hat{\mathbf{y}}[i] = \frac{e^{\mathbf{y}[i]}}{\sum_{c=1}^{C} e^{\mathbf{y}[c]}}, \tag{3.5}$$

where $\mathbf{y}[i]$ denotes the ith element in the vector formed by the outputs of all the neurons and C is the number of output classes. Softmax activation is used only for the output layer and it confines the output element values between zero and one and it also makes the sum of the output elements one, which allows to treat them as probabilities for each class. The exponential function boosts the probability of the most probable class making the predictions clearer. Even though the AlexNet configuration in Table 3.1 has both ReLU and softmax activations for the output layer, it is not necessary to apply ReLU before softmax.

3.2.6 Overall CNN structure

Countless CNN architectures have been applied on different applications, but they typically follow the same main idea: they have consecutive convolutional layers with max pooling layers inserted after every few convolutional layers, followed by one or few fully connected layers before the output. Kernel sizes are often limited to 3×3 to avoid the computational costs of larger kernels and many architectures use global average pooling to reduce the feature map size and computational costs after the convolutional layers.

In general, a better performance can be achieved if the network size is scaled up either by increasing its depth, by adding more channels, by increasing the input resolution, or most efficiently by combining all these, but the gain in performance slows at some point [60] and the designers need to balance the trade-off between performance and computational costs. On the other hand, on shallower data sets (i.e., data sets with more classes and less samples per class) shallower networks may outperform the deeper ones [2].

3.2.6.1 Famous CNN architectures

Some architectures have achieved an iconic position and they have been repeatedly used in multiple applications. After AlexNet shown in Fig. 3.2, the most famous CNN architectures have been VGG16 [53], Inception [58], and ResNet [21]. Simonyan and Zisserman proposed VGG16 [53] that follows the same main ideas as AlexNet in 2014. It has convolutional layers with max pooling layers in between and three fully connected layers at the end. The main novelty was using only 3×3 filters, which allowed to increase the depth up to 16 convolutional layers, which was considered very deep at the time.

Another famous architecture, Inception, was proposed in 2014 by Szegedy et al. [58] and it won the ImageNet Challenge of the year. Inception contains special Inception modules containing parallel 1×1, 3×3, and 5×5 filters, whose outputs are concatenated. The module uses also 1×1 filters to reduce the number of channels before convolutions with larger kernels. This takes a step toward *depthwise separable convolutions* [6], where the channel dimension is handled separately from the spatial dimensions. Later Szegedy et al. proposed an improved version of Inception, coined as Inception v3 [59]. Here, further computational gains were achieved using regularization and *spatial separable convolutions*, where 5×5 filters were replaced by two consecutive 3×3 filters and 3×3 filters with a 1×3 filter followed by a 3×1 filter. The depthwise separable convolutions have become popular [6,28], but spatial separable convolutions are not so actively used due to the limitations to the expression power of the convolution.

The 2015 ImageNet Challenge winner ResNet [21] by He et al. successfully tackled *vanishing gradient problem* (Section 3.3.4.3) using skip connections (Section 3.3.5.7). These concepts are discussed later in the chapter because they require some understanding of the training process. *Residual blocks* formed by convolutional layers and skip connections allowed to increase the number of convolutional layers from tens to hundreds. ResNet had only 3.6% top 5 classification error on ImageNet data set compared to AlexNet's groundbreaking 15.3% only 3 years earlier.

3.3 Training convolutional neural networks

Despite the earlier beliefs that backpropagation would inevitably get stuck to poor local optima with deeper networks, both theoretical and practical evidence show that the local optima are not a significant problem [11,33] and practically all CNNs are trained with backpropagation. However, there are other challenges in the training process. This section first explains how backpropagation is applied on CNNs (Section 3.3.1), in particular on convolutional (Section 3.3.1.1) and pooling layers (Section 3.3.1.2), what the commonly used loss functions are (Section 3.3.2), what batch training is and what commonly used optimizers are (Section 3.3.3). Section 3.3.4 discusses common training challenges and Section 3.3.5 presents some of the most widespread solutions to these challenges.

3.3.1 Backpropagation formulas on CNNs

The backpropagation algorithm trains the network by iteratively alternating between *forward* and *backward passes*. Forward passes compute the network's output for a given input with the current weights. The output is then compared to the target output and a *loss function* is used to compute the network's error E. During backward passes, each trainable network parameter θ is updated using the *gradient descent* approach:

$$\theta_{t+1} = \theta_t - \eta \frac{\partial E_t}{\partial \theta_t} = \theta_t - \eta \nabla_\theta E_t, \tag{3.6}$$

where η is a small value called *learning rate* and $\frac{\partial E_t}{\partial \theta_t} = \nabla_\theta E_t$ is the partial derivative of the error with respect to the parameter θ at iteration t. Despite not being an exact mathematical term, $\nabla_\theta E_t$ is typically called error gradient with respect to θ. The rest of this section focuses on finding the gradients $\nabla_\theta E_t$ at a particular iteration, and we drop the iteration index t for simpler notation.

For the output layer, the gradients are straightforward to compute, but for hidden layers the error must be propagated backwards toward the input layer using the *chain rule* for derivatives. The *local error* (or error/delta error) at layer l, defined as

$$\delta^l = \frac{\partial E}{\partial y^l}, \tag{3.7}$$

where y^l is the output of the layer l before applying the activation function, can be used to find the gradients in layer l as

$$\nabla_{\theta^l} E = \frac{\partial E}{\partial \theta^l} = \frac{\partial E}{\partial y^l} \frac{\partial y^l}{\partial \theta^l} = \delta^l \frac{\partial y^l}{\partial \theta^l}. \tag{3.8}$$

In a similar manner, the local error at layer $l - 1$ can be solved as

$$\delta^{l-1} = \frac{\partial E}{\partial y^{l-1}} = \frac{\partial E}{\partial y^l} \frac{\partial y^l}{\partial y^{l-1}} = \delta^l \frac{\partial y^l}{\partial x^l} f'\left(y^{l-1}\right) = \frac{\partial E}{\partial x^l} f'\left(y^{l-1}\right), \tag{3.9}$$

where f' is the derivative of the activation function f and $x^l = \hat{y}^{l-1} = f\left(y^{l-1}\right)$.

To summarize, the backward pass of the backpropagation algorithm requires the following main steps for every layer l starting from the output layer and proceeding toward input:

- Backpropagate the error to the previous layer using Eq. (3.9) to solve the local error δ^{l-1}. This should be done before updating the parameters of the layer l.
- If there are trainable parameters in layer l, solve the error gradients with respect to them using Eq. (3.8) and update the parameters using Eq. (3.6).

The following Sections 3.3.1.1 and 3.3.1.2 explain how these steps can be applied on convolutional and pooling layers.

3.3.1.1 Backpropagation on convolutional layers

Local error at layer l has the same form as the layer output. As a convolutional layer outputs multiple feature maps, one for each filter, the local error Δ^l similarly has multiple local error maps. Because each convolutional filter affects only a single output feature map, the error is backpropagated only via the corresponding local error map in Δ^l. Here, $\Delta^l_p[m,n] = \frac{\partial E}{\partial \mathrm{Y}^l_p[m,n]} \forall m \in 1,\ldots,M$ and $n \in 1,\ldots,N$, where $M \times N$ is the output feature map size and p refers to the pth convolutional filter. The main things needed to apply backpropagation on convolutional layers are solving the previous layer's local error Δ^{l-1} and error gradients with respect to kernel weights, $\nabla_{K_p} E$ when Δ^l_j is known. Because updating the bias is straightforward and similar as in fully connected networks, it is not discussed here. For simplicity of notation, the layer and filter indices are omitted in the following discussion as we focus only on a single filter of a single layer.

From the backpropagation point of view, the main difference between convolutional layers and fully connected layers is that each kernel weight $\mathrm{K}[i,j,c]$ contributes to multiple values of the feature map Y. When the weight is updated, the error should be backpropagated via all the affected features:

$$\begin{aligned}\nabla_{K[i,j,c]} E = \frac{\partial E}{\partial \mathrm{K}[i,j,c]} &= \sum_{m=1}^{M} \sum_{n=1}^{N} \frac{\partial E}{\partial \mathrm{Y}[m,n]} \frac{\partial \mathrm{Y}[m,n]}{\partial \mathrm{K}[i,j,c]} \\ &= \sum_{m=1}^{M} \sum_{n=1}^{N} \Delta[m,n] \frac{\partial \mathrm{Y}[m,n]}{\partial \mathrm{K}[i,j,c]}. \end{aligned} \tag{3.10}$$

Now looking at Fig. 3.9(a) depicting multichannel 2D convolution, it can be easily seen that

$$\frac{\partial y_{11}}{\partial k^c_{ij}} = x^c_{ij}, \tag{3.11}$$

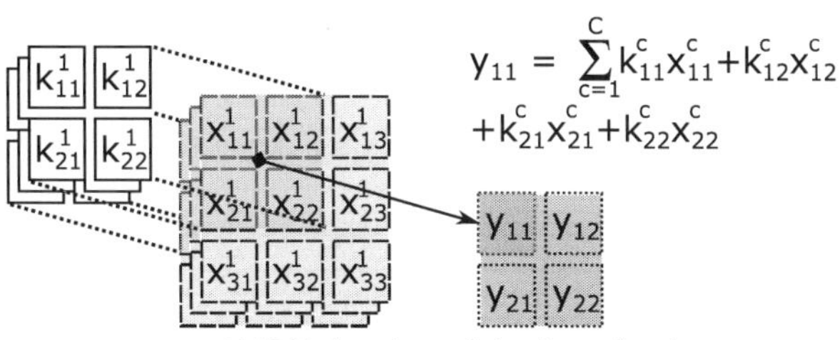

(a) Multi-channel convolution (forward pass)

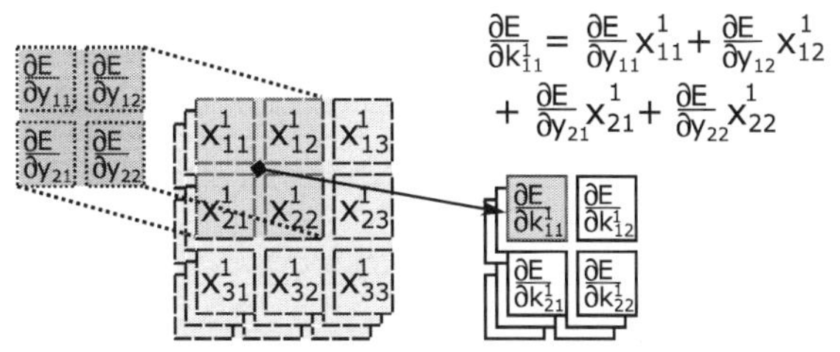

(b) Error gradients with respect to the filter weights (backward pass)

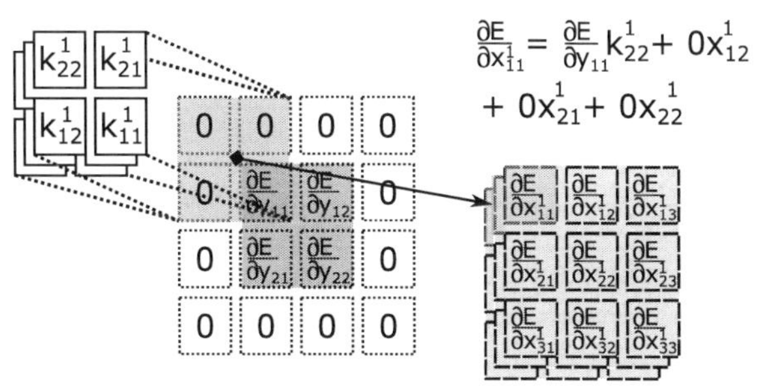

(c) Partial derivatives with respect to the input (backward pass)

FIGURE 3.9

Backpropagation for 2D convolution.

and a more general form for the partial derivatives of the convolution output with respect to kernel weights can be solved from Eq. (3.2):

$$\frac{\partial \mathrm{Y}[m,n]}{\partial \mathrm{K}[i,j,c]} = \mathrm{X}[m-1+i, n-1+j, c] = \mathrm{X}[i-1+m, i-1+m, c]. \tag{3.12}$$

Inserting this to Eq. (3.10), gives

$$\nabla_{K[i,j,c]} E = \sum_{m=1}^{U} \sum_{n=1}^{V} \Delta[m,n] \mathrm{X}[i-1+m, j-1+n, c], \tag{3.13}$$

where $U = M + 1 - K$ and $V = N + 1 - K$. Comparing Eq. (3.1) with Eq. (3.13) shows that the formula for error gradient is 2D convolution between the matrix of local errors and the input. This is illustrated in Fig. 3.9(b) and can be used to update the kernel weights using Eq. (3.6). Note that Δ^{l-1} should be solved before updating the parameters.

To solve Δ^{l-1}, it can be seen from Eq. (3.2) that

$$\frac{\partial \mathrm{Y}[m,n]}{\partial \mathrm{X}[m-1+i,n-1+j,c]} = \mathrm{K}[i,j,c]. \tag{3.14}$$

Changing the indices as $u = m - 1 + i$ $(m = u + 1 - i)$ and $v = n + 1 - j$ $(n = v + 1 - j)$, Eq. (3.14) can be written as

$$\frac{\partial \mathrm{Y}[u+1-i,v+1-j]}{\partial \mathrm{X}[u,v,c]} = \mathrm{K}[i,j,c]. \tag{3.15}$$

This, in turn, can be used to express $\frac{\partial E}{\partial x[u,v,c]}$:

$$\begin{aligned}\frac{\partial E}{\partial \mathrm{X}[u,v,c]} &= \sum_{i=1}^{K}\sum_{j=1}^{K} \frac{\partial E}{\partial \mathrm{Y}[u+1-i,v+1-j]} \frac{\partial \mathrm{Y}[u+1-i,v+1-j]}{\partial \mathrm{X}[u,v,c]} \\ &= \sum_{i=1}^{K}\sum_{j=1}^{K} \Delta[u+1-i,v+1-j]\mathrm{K}[i,j,c]. \end{aligned} \tag{3.16}$$

Eq. (3.16) represents *full* 2D convolution between the matrix of errors and the flipped kernel K (with signal processing terminology this is the 2D convolution, see footnote 1). This is illustrated in Fig. 3.9(c). Finally, the local errors at layer $l-1$ can be obtained by inserting Eq. (3.16) into Eq. (3.9):

$$\begin{aligned}\Delta_c^{l-1}[u,v] &= f'\left(\mathrm{Y}_c^{l-1}[u,v]\right)\frac{\partial E}{\partial \mathrm{X}[u,v,c]} \\ &= f'\left(\mathrm{Y}_c^{l-1}[u,v]\right)\sum_{i=1}^{K}\sum_{j=1}^{K} \Delta[u+1-i,v+1-j]\mathrm{K}[i,j,c], \end{aligned} \tag{3.17}$$

where Δ_c^{l-1} and Y_c^{l-1} refer to the error and output of the cth filter in layer $l-1$. The filter index is shown for layer $l-1$, but not for layer l, because each convolutional filter takes as an input and affects the local errors at layer $l-1$ via all the channels, but produces only a single feature map/channel for the following layer. Eq. (3.17) shows how error is backpropagated to the previous layer via a single filter at layer l. To complete the process, the contributions of all filters to the local error map Δ^{l-1} are summed.

All the above examples assume nonstrided and nondilated convolutions. For convolutions with stride > 1, the receptive fields of neighboring output features are less overlapping. This can be taken into account in backpropagation by inserting zeros to dilate the local error map Δ^l as illustrated in Fig. 3.10 for a local error map corresponding to a feature map resulting from a strided convolution with stride $= 2$ (see

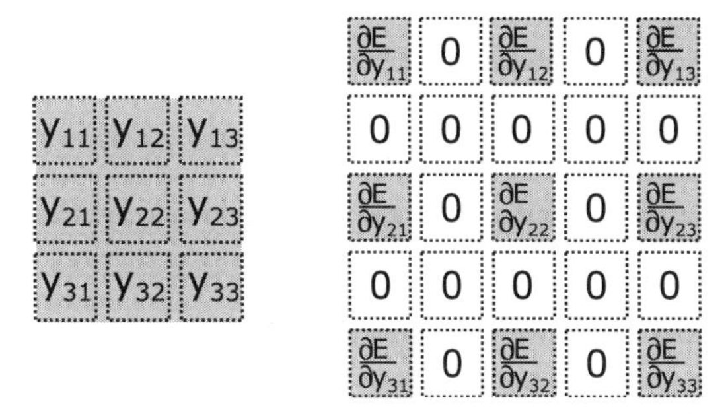

FIGURE 3.10

Left: Feature map produced by convolution with stride of 2, Right: Corresponding local error map padded with zeros.

Fig. 3.4(b)). With this change, the computations of weight updates and local errors can proceed as in Figs. 3.9(b) and 3.9(c): the error gradients $\nabla_{\mathrm{K}[i,j,c]}E$ for updating the kernel weights can be obtained by valid convolution of the zero-dilated local error map and the input and the partial derivatives $\frac{\partial E}{\partial \mathrm{X}[u,v,c]}$ for solving Δ^{l-1} by full convolution of the zero-dilated local error map and the flipped kernel.

For dilated convolutions (see Fig. 3.4(c)), each feature in the output feature map is affected by nonadjacent input elements but for the adjacent output features the same kernel weight is multiplied with adjacent input elements (assuming stride = 1). In the backward pass, these can be taken into account as follows: The error gradients $\nabla_{\mathrm{K}[i,j,c]}E$ can be solved by valid convolution of the input and the local error map with stride equal to the dilating factor. The partial derivatives $\frac{\partial E}{\partial \mathrm{X}[u,v,c]}$ for solving Δ^{l-1} can be solved by full convolution of the flipped filter dilated with zeros and the local error map.

3.3.1.2 Backpropagation on pooling layers

Pooling layers have no parameters to update, but error must be backpropagated to the previous layers. Max pooling passes forward the maximum value within the receptive field of the pooling kernel. Therefore, during training, the error is backpropagated only via the maximum element, but it is not changed.

As average pooling can be presented as a 2D convolution with an averaging kernel having all the weights equal to $1/K^2$, the error update can be obtained directly from Eq. (3.17):

$$\Delta_c^{l-1}[u,v] = f'\left(\mathrm{Y}_c^{l-1}[u,v]\right)\sum_{i=1}^{K}\sum_{j=1}^{K}\Delta[u+1-i,v+1-j]\frac{1}{K^2}. \tag{3.18}$$

Note that for nonoverlapping pooling (stride $\geq K$), this reduces to dividing the local error by K^2 and assigning the same error for each element in the receptive field.

3.3.2 Loss functions

Loss functions are essential in CNN training. They are used to compute the output layer error E_t needed for backpropagation as discussed in following Section 3.3.3. The loss function should match with the true target of the training, for example, correctly classifying as many samples as possible, and at the same time be differentiable and provide a meaningful error landscape for backpropagation. For example, using classification error directly as the loss function is not possible, because it is not differentiable and also because it does not provide any information on whether a sample is classified correctly/wrong with a clear margin or if it is just on the decision border.

As CNNs can be used for multiple tasks, the loss function should be always matched with the task. This chapter focuses only on regression and classification tasks. The loss function commonly used in regression tasks is *mean squared error*:

$$L(\hat{\mathbf{y}}_s, \mathbf{t}_s) = \frac{\sum_{i=1}^{N} (\hat{\mathbf{y}}_s[i] - \mathbf{t}_s[i])^2}{N}, \tag{3.19}$$

where $\hat{\mathbf{y}}_s$ is the network's prediction, $\mathbf{t}_s$ is the known nonbinary groundtruth output vector with length N for the training sample s, and $\hat{\mathbf{y}}_s[i]$, $\mathbf{t}_s[i]$ are the ith elements of the vectors.

Binary cross entropy is used in binary and multilabel classification tasks:

$$L(\hat{\mathbf{y}}_s, \mathbf{t}_s) = -\sum_{i=1}^{N} \left(\mathbf{t}_s[i] \log(\hat{\mathbf{y}}_s[i]) + (1 - \mathbf{t}_s[i]) \log(1 - \hat{\mathbf{y}}_s[i])\right), \tag{3.20}$$

where $\mathbf{t}_s$ is now a binary one-hot encoded (see Section 3.2.5) vector, and thus N is equal to the number of classes. This error function gives an equal weight for every element of the output vector. This is reasonable for binary classification, where it is equally important to decrease the probability of wrong classifications and increase the probability of correct classifications. In multilabel classification, any target vector element can be positive or negative, and thus be seen as a separate binary classification task, where the sample is labeled either as relevant or irrelevant for the particular class.

Categorical cross entropy is used with multiclass single-label tasks, where each sample can belong to only one class:

$$L(\hat{\mathbf{y}}_s, \mathbf{t}_s) = -\sum_{i=1}^{N} \mathbf{t}_s[i] \log(\hat{\mathbf{y}}_s[i]), \tag{3.21}$$

where $\mathbf{t}_s$ is a binary one-hot encoded vector with only a single "1." Categorical cross entropy considers only the predicted output element corresponding to the true class, which is reasonable as this makes the network focus on how to recognize that a sample belongs to a particular class instead of focusing on why it does not belong to all the other classes.

Table 3.5 Suitable output layer activations and loss functions for different tasks.

Task	Activation	Loss function
Regression	Linear	Mean Squared Error
Binary classification	Sigmoid	Binary Cross Entropy
Multiclass, single-label classification	Softmax	Categorical Cross Entropy
Multiclass, multilabel classification	Sigmoid	Binary Cross Entropy

Table 3.5 provides a summary of suitable activations and loss functions for different regression and classification tasks. While these are the default choices, there are also other possibilities for these tasks [45,64]. For example, when a classification task is imbalanced, that is, some classes are much larger than others, focal loss [45] that gives less weight for the samples that have been already learned well can be a good choice.

3.3.3 Batch training and optimizers

3.3.3.1 Batch training

In Section 3.3.1, the backpropagation algorithm was described in terms of the standard gradient descent algorithm, where the parameters are updated as

$$\theta_{t+1} = \theta_t - \eta \nabla_\theta E_t, \tag{3.22}$$

and the error is computed over the whole training set as

$$E_t = \sum_{s=1}^{D} L(\hat{\mathbf{y}}_s, \mathbf{t}_s), \tag{3.23}$$

where D is the size of training data. However, for large data sets of even millions of training samples this approach is not reasonable. It requires presenting all the training samples to the CNN for every backward pass, which will make the training very slow. It also causes memory issues, when the whole data set cannot fit to computer memory.

The other extreme is to do the updates separately for each training sample with

$$E_t = L(\hat{\mathbf{y}}_s, \mathbf{t}_s). \tag{3.24}$$

This approach is known as *stochastic gradient descent (SGD)* and it leads to much faster convergence than the standard gradient descent. However, it may cause some fluctuations in the objective function value due to separate updates for highly varying training samples and possible outliers.

Therefore, a compromise between the two extremes has become the default solution in training CNNs. In *(mini-)batch SGD*, the updates are carried out for each

Table 3.6 Common optimizers.

General form: $\theta_{t+1} = \theta_t - \eta \frac{M_t}{S_t}$		
SGD	$M_t = \nabla_\theta E_t$	$S_t = 1$
SGD with momentum	$m_t = \frac{\beta}{\eta} m_{t-1} + \nabla_\theta E_t$ $M_t = m_t$	$S_t = 1$
Adagrad	$M_t = \nabla_\theta E_t$	$s_t = \sum_{i=1}^{t} \nabla_\theta^2 E_i$ $S_t = \sqrt{s_t} + \epsilon$
RMSprop	$M_t = \nabla_\theta E_t$	$s_t = \beta s_{t-1} + (1-\beta)\nabla_\theta^2 E_t$ $S_t = \sqrt{s_t} + \epsilon$
Adam	$m_t = \beta_1 m_{t-1} + (1-\beta_1)\nabla_\theta E_t$ $\hat{m}_t = \frac{m_t}{1+\beta_1^t}$ $M_t = \hat{m}_t$	$s_t = \beta_2 s_{t-1} + (1-\beta_2)\nabla_\theta^2 E_t$ $\hat{s}_t = \frac{s_t}{1+\beta_2^t}$ $S_t = \sqrt{\hat{s}_t} + \epsilon$

batch, typically 32–256 training samples:

$$E_t = \sum_{s=1}^{B} L(\hat{\mathbf{y}}_s, \mathbf{t}_s), \tag{3.25}$$

where B is the batch size. With the popularity of this approach, it has become common to use the term SGD to refer to it rather than to Eq. (3.24). Note that there is some inconsistency in the terminology used in the literature. In some cases, batch training refers to using the full training set as in Eq. (3.23), while mini-batch SGD refers to using smaller subsets of training data as in Eq. (3.25). In this chapter, batch refers to a subset of data and SGD is assumed to use batches as in Eq. (3.25). Every presentation of a batch to a CNN during training is called an *iteration*, while the whole training set is shown in an *epoch*. The full training requires multiple, even thousands of epochs.

3.3.3.2 Optimizers

The standard update rule of Eq. (3.22) is typically also replaced with a modified version. The different update rules are called *optimizers* and the commonly used optimizers can be described in the following general form:

$$\theta_{t+1} = \theta_t - \eta \frac{M_t}{S_t}, \tag{3.26}$$

where both M_t and S_t are some functions of the error gradients at all time steps so far $\nabla_\theta E_1, ..., \nabla_\theta E_t$. Note that also the basic SGD can be described with this general form by setting $M_t = \nabla_\theta E_t$ and $S_t = 1$. The formulas of some well-known optimizers are provided in Table 3.6, where all β terms are additional hyperparameters and ϵ is a small value typically set to the order of 10^{-8} to avoid any divisions by zero.

The optimizers with a modified M_t term typically use a momentum term. As the physical momentum, the momentum term accelerates gradient descent toward the dominant directions and dampens oscillations in other directions. *SGD with momentum* [50] continues to be one of the best optimizers for CNNs, when combined with a proper adjustment of the learning rate during training.

The term S_t can be seen as a scaling factor for the learning rate. It allows to have different learning rates for different parameters as well as to automatically adjust the learning rates as the training proceeds. An early adaptive learning rate optimizer Adagrad [14] collects the sum of the squares of all the gradients up to the current iteration and uses its square root to divide the original learning rate. As all the values are positive, the sum keeps growing, which eventually makes the learning rate very small. This effectively prevents any further learning. To avoid this phenomenon, in RMSProp optimizer [26], the sum was replaced by an exponential moving average, which gives more weight to the more recent values.

Possibly the most commonly used optimizer for CNNs is Adam [35], which combines the idea of the momentum and the adaptive learning rate of RMSProp. However, while Adam somewhat adapts the learning rate automatically and outperforms SGD with momentum, when a fixed learning rate is applied, it is typical to manually adjust the learning rate during training (see Section 3.3.5.1)—also with Adam. The varying results when combined with different learning rate adaptation techniques have led to a controversy [65,46] on whether the adaptive learning rate really helps or only brings unnecessary computational burden and even harms the performance.

3.3.4 Typical challenges in CNN training

This section describes common challenges in CNN training. Some of the widely spread solutions to these problems are presented in the following Section 3.3.5.

3.3.4.1 Overfitting

Large networks with few training samples tend to overfit the training data. This means that network learns to recognize the training samples very well, but when it is tested with unseen samples the performance is not so good, because it expects something exactly like the training examples, that is, its *generalization* ability is bad. Overfitting can be reduced, for example, by replacing fully connected layer with global average pooling (Section 3.2.4), data augmentation (Section 3.3.5.2), transfer learning (Section 3.3.5.3), weight regularization (Section 3.3.5.4), or dropout (Section 3.3.5.5).

3.3.4.2 Long training time

One of the main challenges in training CNNs with millions of parameters is that training requires a lot of time. The development of computer hardware in general and parallel GPU-based computational resources in particular [12] have made it possible to train significantly larger CNNs and efficient deep learning frameworks, such as TensorFlow and Torch, enable easily implementing CNNs with efficient GPU support [34]. At the same time, the training speed has been also greatly increased by innovations related to the training process.

Carefully optimizing the learning rate is important for the training speed (Section 3.3.5.1). Transfer learning (Section 3.3.5.3) allows to reuse large trained models, and thus saves a lot of time. The challenges described below set constraints to the training speed on their part and, therefore, solutions to these problems have also significantly increased the training speed. One of the possible future directions for speeding up CNN training is to use a greedy approach [3] to solve the *update locking* problem, which means that layers need to wait while updates are carried out in other layers.

3.3.4.3 Vanishing and exploding gradients

As discussed in Section 3.3.1, the gradients are multiplied during training. If the gradients are large, the values will grow exponentially, which makes the network very unstable and it eventually explodes. Small gradients, on the hand make, make the gradients decrease exponentially and finally vanish so that the network will not be able to learn anymore. Vanishing and exploding gradients earlier prevented increasing the number of layers in CNNs and, therefore, solutions preventing this problem have been important for the breakthrough of deep learning. The problem can be eased, for example, by careful initialization of weights [22,17], gradient clipping [49,70], using ReLU instead of sigmoid activation (Section 3.2.3), using an adaptive optimizer (Section 3.3.3.2), normalization (Section 3.3.5.6), or skip connections (Section 3.3.5.7).

3.3.4.4 Internal covariate shift

Covariate shift, that is, a change in the input distribution, harms the performance of learning systems and is typically handled by *domain adaptation*. In a CNN, the updates in the lower layers' weights during training will change the input distribution of the subsequent layers. This phenomenon is called *internal covariate shift* [31]. As CNN layers need to adapt to these continuous shifts, it makes the training much slower. The problem can be eased by different normalization techniques (Section 3.3.5.6).

3.3.5 Solutions to CNN training challenges

There are countless works that propose different solutions to the aforementioned problems or improve the training process in other ways. This chapter covers only some of the most widely spread solutions that have become considered more or less default parts of any CNN implementation. Some additional techniques can be found, for example, in [24].

3.3.5.1 Learning rate scheduling

Learning rate, that is, parameter η in Eq. (3.22), is considered to be the most important hyperparameter in training CNNs [4]. It determines the step size taken into the gradient direction in backpropagation. Too small learning rate will lead to very slow learning or even inability to learn at all, while too large learning rate can lead to exploding or oscillating performance over the training epochs and to a lower final performance. It is typical to start training with a higher learning rate and then decrease it either after a certain number of iterations or gradually at every step. This allows the network to first take faster steps toward the optimal solution, and when converging, focus on finer details without oscillations. The work of Smith [54] showed that it may be beneficial to let the learning rate oscillate up and down during training. This cyclical learning rate also helps avoid manual tuning of the learning rate schedule and it has no additional computational costs.

It should be noted that there is an interdependence between the learning rate and other design choices such as batch size, optimizers (Section 3.3.3.2), or normalization (Section 3.3.5.6). Learning rate for larger batch sizes should be generally larger and Smith et al. [56] proposed increasing batch size during training as an alternative to decreasing learning rate. With this approach, they obtained nearly the same results but with fewer parameter updates. As the name suggests, adaptive learning rate optimizers generally work well with a fixed learning rate η, but still they can further benefit from decaying the learning rate [13]. Normalization, on the other hand, can allow obtaining good results even with exotic learning rates such as an exponentially increasing one [43]. An approach for finding suitable set of training hyperparameters for SGD with momentum was proposed in [55].

3.3.5.2 Data augmentation

One of the commonly used techniques to avoid overfitting is data augmentation, where the amount of training data and its variability is increased by applying different transformations, such as rotation, scaling, cropping, and adding noise, to the original images. Such augmentation is often performed manually, but also techniques for learning suitable augmentation policies automatically have been proposed [9].

3.3.5.3 Transfer learning

In transfer learning [1], some of the CNN layers are pretrained using a larger similar dataset, which can help the network learn better features that are suitable also for the new task and then the network is only finetuned by training shortly with the data set at hand. This can help the network avoid overfitting and improve the performance. It is common to use as a starting point a publicly available network model that has been pretrained with one of the large data sets, such as ImageNet [52]. This makes the training much faster and more economic as the most time-consuming part can be reused in multiple applications.

3.3.5.4 Weight regularization

Networks with larger weights are more likely to overfit and, therefore, CNNs are typically encouraged to learn smaller weights by weight regularization. A popular regularization technique of *weight decay* [20] changes the backpropagation update in Eq. (3.22) to

$$\theta_{t+1} = (1 - \lambda)\theta_t - \eta \nabla_\theta E_t, \tag{3.27}$$

where λ is a small value. *L1/L2 regularization* adds to the loss function a term that depends on L1/L2 norm of the weights. For SGD optimizer, weight decay is equal to L2 regularization, but this is not the case for adaptive optimizers for which L2 regularization is not recommended [46]. Recent studies show that regularization is most helpful at the beginning of the training process and not efficient close to convergence [19].

3.3.5.5 Dropout

Dropout [27,57] was proposed as a simple cure for overfitting. In dropout, some network units and their connections are randomly ignored during training every time when a batch of samples is presented to the network. Dropout can be applied on any network layer except the output layer and the probability of dropping the units is typically set to 0.5 for hidden layers and to a larger value, such as 0.8, for the input layer. Dropout increases the number of required training epochs and it makes the network weights larger, so they are typically scaled down at the end of training by multiplying with the dropout rate. Dropout can prevent network units from co-adapting too much by making the presence of other units unreliable, which makes the network more robust. DropConnect [63] is a generalization of dropout, where network connections are dropped out independently. The results are typically similar.

3.3.5.6 Normalization

Ioffe and Szegedy suggested *batch normalization* as a cure for internal covariate shift in [31]. As the problem is caused by parameter updates in lower layers changing the input distribution of the subsequent layers, they proposed to cure it by normalizing the input elements. During training, it is not feasible to compute the averages and variances needed for normalization over the whole training set and, therefore, they approximate them using the estimates from each batch separately for each channel:

$$x' = \frac{x - \mu_B}{\sqrt{\sigma_B^2 + \epsilon}}, \tag{3.28}$$

where x' is the normalized input, μ_B and σ_B^2 are the average and variance computed over the full feature maps $(H \times W)$ of a batch containing B samples. To compensate for the possible loss of representation ability, batch normalization includes also a linear transformation

$$x'' = \gamma x' + \beta, \tag{3.29}$$

where γ and β are learnable parameters. During inference, μ_B and σ_B^2 are replaced by the average and variance computed over the whole training set on the final trained network. In the original work [31], batch normalization was applied before activation for $\mathrm{Y}[m, n, c]$, but it has also become a common practice to apply it after activation for $\hat{\mathrm{Y}}[m, n, c]$ (i.e., the inputs of the following layer). There is no clear consensus on which of these approaches is better.

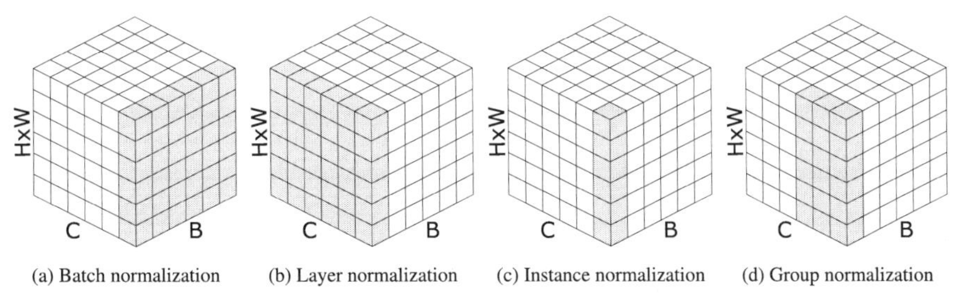

FIGURE 3.11

Illustration of different normalization strategies. $H \times W$ is the spatial dimension, C channels, and B instances.

Batch normalization turned out to improve the efficiency of CNN training significantly. It helps avoid vanishing or exploding gradient by keeping the activations more stable during training, which in turn allows to use larger learning rates. It also makes the training less dependent on the initialization and works as a regularizer that in some cases eliminates the need for dropout.

A problem with batch normalization is that it becomes unstable with small batch sizes, which are needed when the inputs are large, which is typically the case, for example, in segmentation and detection tasks. Therefore, different normalization approaches that mainly differ in how the average and mean estimates for Eq. (3.28) are computed have been proposed: *Instance normalization* [62] handles each input sample separately, *layer normalization* [40] also handles input samples separately but computes the average over all channels, and *group normalization* [67] groups some of the channels together for computing the statistics. Instance and layer normalization can be seen as extreme cases of group normalization, where the group sizes are 1 or all C channels. Fig. 3.11 illustrates the differences between these normalization strategies. Layer and instance normalization have become popular for sequential models (e.g., recurrent neural networks (RNNs)) or for generative models (e.g., generative adversarial networks (GANs)), while group normalization works well in visual recognition tasks that require using small batch sizes.

3.3.5.7 Skip connections

Skip connections are an efficient cure to the vanishing gradient problem. They were the factor that allowed He et al. to significantly increase the number of layers in their winning ResNet [21] architecture and have been later used in numerous other architectures. The main idea is to add to the CNN additional forward shortcuts that send the feature maps from an earlier layer unchanged to a later point in the CNN. When the error is backpropagated, the gradient of this shortcut is one, which helps prevent the local error from shrinking. There have been efforts to apply simple multiplicative operations, such as scaling, gating, 1×1 convolutions, or dropout, on shortcuts, but experimental evidence supports simple identity shortcuts [23]. Another advantage of skip connections is that they help preserve any useful information discovered by the

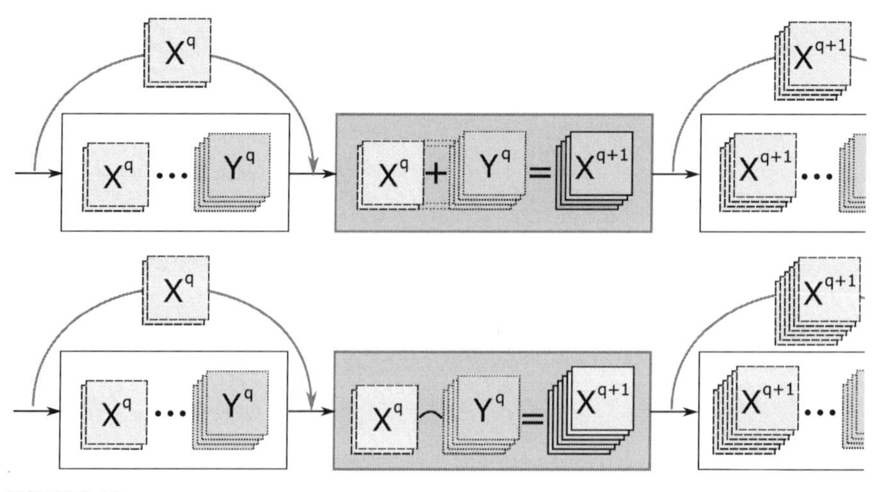

FIGURE 3.12

Skip connections via addition (top) and via concatenation (bottom). The white squares are convolutional blocks and purple color (mid gray in print version) is used to illustrate the skip connection strategies.

early network layers, while the original path via multiple convolutions and activations can easily obscure it. Li et al. demonstrated that skip connections make the error surface smoother [41].

The skip connections are typically implemented via addition as in ResNet [21] or via concatenation as in DenseNet [30]. Skip connections via addition can be expressed as

$$\mathrm{X}^{q+1} = g^q(\mathrm{X}^q) + \mathrm{X}^q = \mathrm{Y}^q + \mathrm{X}^q, \tag{3.30}$$

where the superscripts now refer to indices of convolutional blocks and g^q denotes the operations performed by qth convolutional block. The convolutional blocks can contain multiple convolutional layers along with their activations and even pooling layers, but the feature maps in Y^q and X^q should have the same spatial dimensions (height and width). The number of channels, on the other hand, can vary. In this case, only the corresponding feature maps are summed together. Similarly, skip connections via concatenation can be expressed as

$$\mathrm{X}^{q+1} = g^q(\mathrm{X}^q)^\frown \mathrm{X}^q = \mathrm{Y}^{q\frown} \mathrm{X}^q, \tag{3.31}$$

where $\frown$ denotes the concatenation of all the feature maps to a single tensor with more channels. These two ways to implement skip connections are illustrated in Fig. 3.12.

Note that the equations and the figure both ignore the placement of activation functions. The original ResNet had activations after addition of Y^q and X^q, but same authors later observed that it is better apply activations only in the convolutional blocks and keep the shortcuts clear from activations [23]. The figure also shows that additive shortcuts change the feature maps, while skip connections via concatenation keep them unchanged, but increase the number of channels. This can be advantageous for passing useful information across the CNN. As skip connections are typically used densely throughout the network, this approach also quickly increases the number of channels even with much lower numbers of convolutional filters (and trainable weights) as in DenseNet [30].

3.4 Conclusions

CNNs have revolutionized the field of machine learning during the last decade. This chapter introduced the basic concepts including the network structure and the training procedure along with some of its main challenges and widely spread solutions to these problems. While many great results have been achieved using the presented concepts, they are still just a starting point for many other great innovations that borrow and further develop these concepts for different application domains. Many of these innovations are discussed in the subsequent chapters of this book.

References

[1] Muhammad Jamal Afridi, Arun Ross, Erik M. Shapiro, On automated source selection for transfer learning in convolutional neural networks, Pattern Recognition (ISSN 0031-3203) 73 (2018) 65–75.

[2] S.H. Shabbeer Basha, Shiv Ram Dubey, Viswanath Pulabaigari, Snehasis Mukherjee, Impact of fully connected layers on performance of convolutional neural networks for image classification, Neurocomputing (ISSN 0925-2312) 378 (2020) 112–119.

[3] Eugene Belilovsky, Michael Eickenberg, Edouard Oyallon, Decoupled greedy learning of CNNs, in: Hal Daumé III, Aarti Singh (Eds.), International Conference on Machine Learning (ICML), vol. 119, 2020, pp. 736–745.

[4] Yoshua Bengio, Practical Recommendations for Gradient-Based Training of Deep Architectures, Springer Berlin Heidelberg, Berlin, Heidelberg, ISBN 978-3-642-35289-8, 2012, pp. 437–478.

[5] L. Chen, G. Papandreou, I. Kokkinos, K. Murphy, A.L. Yuille DeepLab, Semantic image segmentation with deep convolutional nets, atrous convolution, and fully connected CRFs, IEEE Transactions on Pattern Analysis and Machine Intelligence 40 (4) (2018) 834–848.

[6] Francois Chollet, Xception: deep learning with depthwise separable convolutions, in: IEEE Conference on Computer Vision and Pattern Recognition (CVPR), 2017.

[7] Anna Choromanska, Mikael Henaff, Michael Mathieu, Gérard Ben Arous, Yann LeCun, The loss surfaces of multilayer networks, in: International Conference on Artificial Intelligence and Statistics (AISTATS), 2015.

[8] K. Chumachenko, A. Männistö, A. Iosifidis, J. Raitoharju, Machine learning based analysis of Finnish World War II photographers, IEEE Access 8 (2020) 144184–144196.

[9] Ekin D. Cubuk, Barret Zoph, Dandelion Mane, Vijay Vasudevan, Quoc V. Le, Autoaugment: learning augmentation strategies from data, in: IEEE/CVF Conference on Computer Vision and Pattern Recognition (CVPR), June 2019.

[10] U. Côté Allard, F. Nougarou, C.L. Fall, P. Giguère, C. Gosselin, F. Laviolette, B. Gosselin, A convolutional neural network for robotic arm guidance using semg based frequency-features, in: IEEE/RSJ International Conference on Intelligent Robots and Systems (IROS), 2016, pp. 2464–2470.

[11] Yann N. Dauphin, Razvan Pascanu, Caglar Gulcehre, Kyunghyun Cho, Surya Ganguli, Yoshua Bengio, Identifying and attacking the saddle point problem in high-dimensional non-convex optimization, in: Advances in Neural Information Processing Systems (NIPS), vol. 27, 2014, pp. 2933–2941.

[12] J. Dean, 1.1 The deep learning revolution and its implications for computer architecture and chip design, in: IEEE International Solid-State Circuits Conference (ISSCC), 2020, pp. 8–14.

[13] Michael Denkowski, Graham Neubig, Stronger baselines for trustable results in neural machine translation, in: Workshop on Neural Machine Translation, 2017, pp. 18–27.

[14] John Duchi, Elad Hazan, Yoram Singer, Adaptive subgradient methods for online learning and stochastic optimization, Journal of Machine Learning Research 12 (61) (2011) 2121–2159.

[15] Vincent Dumoulin, Francesco Visin, A guide to convolution arithmetic for deep learning, arXiv preprint, arXiv:1603.07285, 2016.

[16] K. Fukushima, Neocognitron: A self-organizing neural network model for a mechanism of pattern recognition unaffected by shift in position, Biological Cybernetics 36 (1980) 193–202.

[17] Xavier Glorot, Yoshua Bengio, Understanding the difficulty of training deep feedforward neural networks, in: International Conference on Artificial Intelligence and Statistics, 2010, pp. 249–256.

[18] Xavier Glorot, Antoine Bordes, Yoshua Bengio, Deep sparse rectifier neural networks, in: International Conference on Artificial Intelligence and Statistics, in: Proceedings of Machine Learning Research, vol. 15, 2011, pp. 315–323.

[19] Aditya Sharad Golatkar, Alessandro Achille, Stefano Soatto, Time matters in regularizing deep networks: weight decay and data augmentation affect early learning dynamics, matter little near convergence, in: Advances in Neural Information Processing Systems (NIPS), 2019, pp. 10678–10688.

[20] Stephen José Hanson, Lorien Y. Pratt, Comparing biases for minimal network construction with back-propagation, in: Advances in Neural Information Processing Systems (NIPS), 1989, pp. 177–185.

[21] K. He, X. Zhang, S. Ren, J. Sun, Deep residual learning for image recognition, in: IEEE Conference on Computer Vision and Pattern Recognition (CVPR), 2016, pp. 770–778.

[22] Kaiming He, Xiangyu Zhang, Shaoqing Ren, Jian Sun, Delving deep into rectifiers: surpassing human-level performance on imagenet classification, in: IEEE International Conference on Computer Vision (ICCV), 2015, pp. 1026–1034.

[23] Kaiming He, Xiangyu Zhang, Shaoqing Ren, Jian Sun, Identity mappings in deep residual networks, in: European Conference on Computer Vision (ECCV), Springer International Publishing, 2016, pp. 630–645.

[24] Tong He, Zhi Zhang, Hang Zhang, Zhongyue Zhang, Junyuan Xie, Mu Li, Bag of tricks for image classification with convolutional neural networks, in: IEEE/CVF Conference on Computer Vision and Pattern Recognition (CVPR), June 2019, pp. 558–567.

[25] S. Hershey, S. Chaudhuri, D.P.W. Ellis, J.F. Gemmeke, A. Jansen, R.C. Moore, M. Plakal, D. Platt, R.A. Saurous, B. Seybold, M. Slaney, R.J. Weiss, K. Wilson, Cnn architectures for large-scale audio classification, in: IEEE International Conference on Acoustics, Speech and Signal Processing (ICASSP), 2017, pp. 131–135.

[26] Geoffrey Hinton, Nitish Srivastava, Kevin Swersky, Neural networks for machine learning, lecture 6a: overview of mini-batch gradient descent, 2012.

[27] Geoffrey E. Hinton, Nitish Srivastava, Alex Krizhevsky, Ilya Sutskever, Ruslan R. Salakhutdinov, Improving neural networks by preventing co-adaptation of feature detectors, arXiv preprint, arXiv:1207.0580, 2012.

[28] Andrew G. Howard, Menglong Zhu, Bo Chen, Dmitry Kalenichenko, Weijun Wang, Tobias Weyand, Marco Andreetto, Hartwig Adam Mobilenets, Efficient convolutional neural networks for mobile vision applications, arXiv preprint, arXiv:1704.04861, 2017.

[29] Toke T. Høye, Johanna Ärje, Kim Bjerge, Oskar L.P. Hansen, Alexandros Iosifidis, Florian Leese, Hjalte M.R. Mann, Kristian Meissner, Claus Melvad, Jenni Raitoharju, Deep learning and computer vision will transform entomology, Proceedings of the National Academy of Sciences (ISSN 0027-8424) 118 (2) (2021).

[30] Gao Huang, Zhuang Liu, Laurens van der Maaten, Kilian Q. Weinberger, Densely connected convolutional networks, in: IEEE Conference on Computer Vision and Pattern Recognition (CVPR), July 2017.

[31] Sergey Ioffe, Christian Szegedy, Batch normalization: accelerating deep network training by reducing internal covariate shift, in: International Conference on Machine Learning (ICML), in: Proceedings of Machine Learning Research, vol. 37, 2015, pp. 448–456.

[32] A. Kamilaris, F.X. Prenafeta-Boldú, A review of the use of convolutional neural networks in agriculture, Journal of Agricultural Science 156 (3) (2018) 312–322.

[33] Kenji Kawaguchi, Deep learning without poor local minima, in: D. Lee, M. Sugiyama, U. Luxburg, I. Guyon, R. Garnett (Eds.), Advances in Neural Information Processing Systems (NIPS), vol. 29, 2016, pp. 586–594.

[34] H. Kim, H. Nam, W. Jung, J. Lee, Performance analysis of cnn frameworks for gpus, in: IEEE International Symposium on Performance Analysis of Systems and Software (ISPASS), 2017, pp. 55–64.

[35] Diederik P. Kingma, Jimmy Ba Adam, A method for stochastic optimization, in: International Conference on Learning Representations (ICLR), 2015.

[36] Alex Krizhevsky, Ilya Sutskever, Geoffrey E. Hinton, Imagenet classification with deep convolutional neural networks, in: Advances in Neural Information Processing Systems (NIPS), 2012, pp. 1106–1114.

[37] Y. LeCun, B. Boser, J.S. Denker, D. Henderson, R.E. Howard, W. Hubbard, L.D. Jackel, Backpropagation applied to handwritten zip code recognition, Neural Computation 1 (4) (1989) 541–551.

[38] Y. LeCun, L. Bottou, Y. Bengio, P. Haffner, Gradient-based learning applied to document recognition, Proceedings of the IEEE 86 (11) (1998) 2278–2324.

[39] Y. LeCun, Y. Bengio, G. Hinton, Deep learning, Nature 521 (2015) 436–444.

[40] Jimmy Lei Ba, Jamie Ryan Kiros, Geoffrey E. Hinton, Layer normalization, arXiv preprint, arXiv:1607.06450, 2016.

[41] Hao Li, Zheng Xu, Gavin Taylor, Christoph Studer, Tom Goldstein, Visualizing the loss landscape of neural nets, in: S. Bengio, H. Wallach, H. Larochelle, K. Grauman, N. Cesa-Bianchi, R. Garnett (Eds.), Advances in Neural Information Processing Systems (NeurIPS), vol. 31, 2018, pp. 6389–6399.

[42] Qingbiao Li, Fernando Gama, Alejandro Ribeiro, Amanda Prorok, Graph neural networks for decentralized multi-robot path planning, in: IEEE/RSJ International Conference on Intelligent Robots and Systems (IROS), 2020.

[43] Zhiyuan Li, Sanjeev Arora, An exponential learning rate schedule for deep learning, in: International Conference on Learning Representations (ICLR), 2020.

[44] Min Lin, Qiang Chen, Shuicheng Yan, Network in network, in: International Conference on Learning Representations (ICLR), 2014.

[45] Tsung-Yi Lin, Priya Goyal, Ross Girshick, Kaiming He, Piotr Dollar, Focal loss for dense object detection, in: Proceedings of the IEEE International Conference on Computer Vision (ICCV), Oct 2017.

[46] Ilya Loshchilov, Frank Hutter, Decoupled weight decay regularization, in: International Conference on Learning Representations (ICLR), 2019.

[47] Lei Ma, Yu Liu, Xueliang Zhang, Yuanxin Ye, Gaofei Yin, Brian Alan Johnson, Deep learning in remote sensing applications: a meta-analysis and review, ISPRS Journal of Photogrammetry and Remote Sensing 152 (2019) 166–177.

[48] Andrew L. Maas, Awni Y. Hannun, Andrew Y. Ng, Rectifier nonlinearities improve neural network acoustic models, in: International Conference on Machine Learning (ICML), vol. 30, 2013, p. 3.

[49] Razvan Pascanu, Tomas Mikolov, Yoshua Bengio, On the difficulty of training recurrent neural networks, in: International Conference on Machine Learning, 2013, pp. 1310–1318.

[50] Ning Qian, On the momentum term in gradient descent learning algorithms, Neural Networks 12 (1) (1999) 145–151.

[51] Prajit Ramachandran, Barret Zoph, Quoc Le, Searching for activation functions, in: Workshop Track of International Conference on Learning Representations, 2018.

[52] Olga Russakovsky, Jia Deng, Hao Su, Jonathan Krause, Sanjeev Satheesh, Sean Ma, Zhiheng Huang, Andrej Karpathy, Aditya Khosla, Michael Bernstein, Alexa Berg, Li Fei-Fei, ImageNet large scale visual recognition challenge, International Journal of Computer Vision 115 (3) (2015) 211–252.

[53] Karen Simonyan, Andrew Zisserman, Very deep convolutional networks for large-scale image recognition, in: International Conference on Learning Representation (ICLR), 2015.

[54] Leslie.N. Smith, Cyclical learning rates for training neural networks, in: IEEE Winter Conference on Applications of Computer Vision (WACV), 2017, pp. 464–472.

[55] Leslie N. Smith, A disciplined approach to neural network hyper-parameters: Part 1 – learning rate, batch size, momentum, and weight decay, US Naval Research Laboratory Technical Report 5510-026, arXiv:1803.09820, 2018.

[56] Samuel L. Smith, Pieter-Jan Kindermans, Quoc V. Le, Don't decay the learning rate, increase the batch size, in: International Conference on Learning Representations (ICLR), 2018.

[57] Nitish Srivastava, Geoffrey Hinton, Alex Krizhevsky, Ilya Sutskever, Ruslan Salakhutdinov, Dropout: a simple way to prevent neural networks from overfitting, Journal of Machine Learning Research 15 (1) (2014) 1929–1958.

[58] Christian Szegedy, Wei Liu, Yangqing Jia, Pierre Sermanet, Scott Reed, Dragomir Anguelov, Dumitru Erhan, Vincent Vanhoucke, Andrew Rabinovich, Going deeper with convolutions, in: IEEE Conference on Computer Vision and Pattern Recognition (CVPR), June 2015.

[59] Christian Szegedy, Vincent Name, Sergey Ioffe, Jon Shlens, Zbigniew Wojna, Rethinking the inception architecture for computer vision, in: IEEE Conference on Computer Vision and Pattern Recognition (CVPR), June 2016.

[60] Mingxing Tan, Quoc Le, EfficientNet: rethinking model scaling for convolutional neural networks, in: Kamalika Chaudhuri, Ruslan Salakhutdinov (Eds.), Proceedings of the 36th International Conference on Machine Learning, in: Proceedings of Machine Learning Research, vol. 97, 2019, pp. 6105–6114.

[61] Yuchi Tian, Kexin Pei, Suman Jana, Baishakhi Ray Deeptest, Automated testing of deep-neural-network-driven autonomous cars, in: International Conference on Software Engineering (ICSE), 2018, pp. 303–314.

[62] Dmitry Ulyanov, Andrea Vedaldi, Victor Lempitsky, Instance normalization: the missing ingredient for fast stylization, arXiv:1607.08022, 2016.

[63] Li Wan, Matthew Zeiler, Sixin Zhang, Yann Le Cun, Rob Fergus, Regularization of neural networks using dropconnect, in: Sanjoy Dasgupta, David McAllester (Eds.), International Conference on Machine Learning (ICML), vol. 28, 2013, pp. 1058–1066.

[64] Weitao Wan, Yuanyi Zhong, Tianpeng Li, Jiansheng Chen, Rethinking feature distribution for loss functions in image classification, in: IEEE Conference on Computer Vision and Pattern Recognition (CVPR), June 2018.

[65] Ashia C. Wilson, Rebecca Roelofs, Mitchell Stern, Nathan Srebro, Benjamin Recht, The marginal value of adaptive gradient methods in machine learning, in: International Conference on Neural Information Processing Systems (NIPS), 2017, pp. 4151–4161.

[66] M. Wu, W. Su, L. Chen, Z. Liu, W. Cao, K. Hirota, Weight-adapted convolution neural network for facial expression recognition in human-robot interaction, IEEE Transactions on Systems, Man, and Cybernetics: Systems (2019) 1–12.

[67] Yuxin Wu, Kaiming He, Group normalization, in: Proceedings of the European Conference on Computer Vision (ECCV), 2018, pp. 3–19.

[68] Rikiya Yamashita, Mizuho Nishio, Richard Kinh Do Gian, Kaori Togashi, Convolutional neural networks: an overview and application in radiology, Insights Imaging 9 (2018) 611–629.

[69] Fisher Yu, Vladlen Koltun, Multi-scale context aggregation by dilated convolutions, arXiv:1511.07122, 2016.

[70] Jingzhao Zhang, Tianxing He, Suvrit Sra, Ali Jadbabaie, Why gradient clipping accelerates training: a theoretical justification for adaptivity, in: International Conference on Learning Representations (ICLR), 2020.

[71] Xiangyu Zhang, Xinyu Zhou, Mengxiao Lin, Jian Sun Shufflenet, An extremely efficient convolutional neural network for mobile devices, in: IEEE/CVF Conference on Computer Vision and Pattern Recognition (CVPR), 2018, pp. 6848–6856.

[72] Zhifen Zhang, Guangrui Wen, Shanben Chen, Weld image deep learning-based on-line defects detection using convolutional neural networks for al alloy in robotic arc welding, Journal of Manufacturing Processes 45 (2019) 208–216.

[73] Bolei Zhou, Aditya Khosla, Agata Lapedriza, Aude Oliva, Antonio Torralba, Learning deep features for discriminative localization, in: IEEE Conference on Computer Vision and Pattern Recognition (CVPR), June 2016.

CHAPTER

Graph convolutional networks

4

Negar Heidari, Lukas Hedegaard, and Alexandros Iosifidis
Department of Electrical and Computer Engineering, Aarhus University, Aarhus, Denmark

4.1 Introduction

With the rise of internet and social media, a huge amount of data has been created by the users so that graph structured data can be seen everywhere around us. For instance, the data created by social networks, citation networks, and knowledge bases are modeled as graphs of enormous sizes, as the corresponding graphs are formed by billions of nodes and edges. Accordingly, it is important to develop learning methods that are able to capture the patterns appearing in the graph-structured data and infer the information encoded by them. Deep neural networks have been very successful in processing Euclidean data structures, such as audio, text, images, and videos, which are formed by regular 1D, 2D, and 3D grids, in different learning tasks such as natural language processing [1,2], speech recognition [3], and computer vision [4,5]. However, neural networks used for processing Euclidean data structures are not suited for processing data that do not come in a regular grid structure.

Graph Neural Networks (GNNs) have attracted an increasing research interest and they have been very successful in processing non-Euclidean data structures, such as graphs, in different application areas. Many methods have been developed to extend RNNs, CNNs, and MLPs to process graph structured data. These methods are categorized into Recurrent GNNs (RecGNNs), Graph Convolutional Networks (GCNs), and Graph Autoencoders (GAEs), respectively. The first GNN methods were RecGNNs [6–9], which learn feature representations for the graph nodes by passing information between each node and its neighbors until reaching equilibrium. This idea of message passing was used in GCNs as well. GCNs [10–15] build a multilayer neural network, formed by layers that learn feature representations for the graph nodes by applying graph convolution. These representations are employed for classification of individual nodes or the entire graph. The graph convolution layer can be defined based on spectral graph theory or the message passing notion in RecGNNs. GAEs are neural networks trained in an unsupervised manner to reconstruct the structural information of the graph, such as its nodes, its edges, or the graph adjacency matrix for graph generation [16–19] or to learn feature representations for graph nodes and edges [20–23].

In problems where the input data forms a graph structure which is updated over regular time intervals, a spatiotemporal graph structure is created. The spatiotempo-

Deep Learning for Robot Perception and Cognition. https://doi.org/10.1016/B978-0-32-385787-1.00009-9

ral graph can be defined as a sequence of graphs having two types of graph node connections, that is, connections between the nodes in each graph which form the set of spatial edges, and connections between the nodes at different time steps, which form the set of temporal edges. The Spatiotemporal Graph Convolutional Networks (ST-GCN) are able to process spatiotemporal graphs. They are multilayer neural networks, which capture both spatial structure and temporal dynamics of the underlying spatiotemporal graph by applying spatial and temporal graph convolutions at each layer. Recently, ST-GCNs attracted a great research interest in computer vision tasks such as skeleton-based human action recognition where an action is represented by a sequence of human body poses. Each body pose is represented by the arrangement of spatially connected human body joints, and the sequence of skeletons is thus modeled as a spatiotemporal graph.

Using graph structured data sets, three main learning tasks can be formulated, which are briefly described as follows:

- **Node classification:** The goal of node classification methods is to predict a label for each graph node in a supervised or semisupervised setting. In these methods, the GCN model extracts high-level feature representation for each node in a supervised setting or propagates the label information from labeled nodes through the whole graph in a semisupervised setting. Using the extracted high-level node features, a label will be predicted for each node by a classification layer.
- **Graph classification:** The methods targeting graph classification tasks use a pooling layer before the classification layer in order to produce a feature vector aggregating information of the whole graph, which is then introduced to the classification layer predicting a label. The pooling layer can be a simple global average pooling, which produces a mean vector of all the node features.
- **Edge prediction and classification:** The goal of edge prediction and classification tasks is either to generate a graph by predicting the edges between nodes and building a new graph adjacency matrix, or to learn feature representations for the graph edges.

In this chapter, we focus on GCNs and introduce the most representative methods developed in this area to generalize the notion of convolution from grid structured data to graph structured data. GCN methods are mainly categorized into spectral methods, which are based on spectral graph theory, and spatial methods. In this section, we start with a formal definition a graph. In Section 4.2, spectral-based GCN and the most representative methods in this category are covered. Section 4.3 introduces spatial-based GCN methods and it is followed by a comparison of spatial and spectral-based GCN methods related to their potential advantages and disadvantages. In Section 4.4, the attention based GCN is discussed. In Section 4.5, our focus is on the efficiency and scalability of the GCN methods and we introduce methods, which are able to process large graphs. Finally, an overview of benchmark graph data sets used in different GNN tasks and the available software libraries for implementing GCNs are provided in Section 4.6.

4.1.1 Graph definition

A graph is formally defined as $G = (\mathcal{V}, \mathcal{E})$, where $\mathcal{V} = \{v_i\}_{i=1}^{N}$ denotes the set of N graph nodes and $\mathcal{E} = \left\{e_{ij} \in \mathcal{V} \times \mathcal{V},\ i, j = 1, \ldots, N\right\}$ denotes the set of graph edges expressing the connection between each pair of nodes (v_i, v_j). The graph adjacency matrix $\mathbf{A} \in \mathbb{R}^{N \times N}$ encodes pairwise connections between the nodes through the edges and it is defined as follows:

$$\mathbf{A}_{ij} = \begin{cases} 1, & \text{if } i \neq j \text{ and } e_{ij} \in \mathcal{E}, \\ 0, & \text{otherwise.} \end{cases} \tag{4.1}$$

When $\mathbf{A}$ is symmetric, that is, when for every edge $e_{ij} \in \mathcal{E}$ connecting nodes v_i and v_j the corresponding edge e_{ji} also belongs to $\mathcal{E}$, G is an undirected graph. When this is not true, G is a directed graph. When the connections between nodes are associated with real values encoding their strength, the graph adjacency matrix is replaced with the corresponding graph weight matrix[1] formed by real values. In that case, G is a weighted graph. In the following, we consider undirected and unweighted graphs, that is, graphs defined by symmetric adjacency matrices.

The degree of each node $deg(v_i)$ indicates the number of its neighbors, which are directly connected to it with an edge. The graph degree matrix $\mathbf{D} \in \mathbb{R}^{N \times N}$ is a diagonal matrix, which gives us information about the node degrees, and it is defined as follows:

$$\mathbf{D}_{ij} = \begin{cases} \sum_j \mathbf{A}_{ij}, & \text{if } i = j, \\ 0, & \text{otherwise.} \end{cases} \tag{4.2}$$

Based on this definition, the neighborhood of a node is defined as

$$\mathcal{N}(v_i) = \left\{v_j \in \mathcal{V} | e_{ij} \in \mathcal{E}\right\}. \tag{4.3}$$

Moreover, each graph node can be represented by a feature vector $\mathbf{x}_i \in \mathbb{R}^D$, leading to the feature matrix of the graph denoted as $\mathbf{X} \in \mathbb{R}^{N \times D}$ storing the feature vectors of all nodes. When the graph feature matrix is not available, information encoded in the graph edge set $\mathcal{E}$ can be exploited to calculate one through graph embedding methods such as DeepWalk [24], node2vec [25], and LINE [26]. The output of graph embedding methods is a set of feature vectors which can be used by GCNs. Here, we should note that there exists a variety of graph embedding methods which have been proposed for a wide range of problems, like data visualization based on neural networks [27–29], data visualization based on matrix factorization [30–32], subspace learning based on spectral analysis [33], and classification based on graph structures [34,35], which are beyond the scope of this chapter. For more information on these methods, we refer the reader to related review articles [36,37].

[1] The graph weight matrix should not be confused with the weight matrix of a neural layer used in MLPs.

4.2 Spectral graph convolutional network

Spectral GCNs are based on graph signal processing [38–40] where the signals are defined on undirected graphs. In other words, a graph function, or graph signal, is defined as a mapping from each node of the graph to a real value. Given the graph adjacency and the graph degree matrices, the graph Laplacian matrix is defined as $\mathbf{L} = \mathbf{D} - \mathbf{A}$ so that

$$\mathbf{L}_{ij} = \begin{cases} deg(v_i), & \text{if } i = j, \\ -1, & \text{if } i \neq j \text{ and } (v_i, v_j) \in \mathcal{E}, \\ 0, & \text{otherwise.} \end{cases} \tag{4.4}$$

$\mathbf{L}$ encodes the graph connectivity and determines the smoothness of the graph function. In a smooth graph function, its value does not vary a lot across connected nodes. This can be expressed formally by the Dirichlet energy of a graph function [41,42]:

$$E(\mathbf{f}) = \frac{1}{2} \sum_{i,j} \mathbf{A}_{ij} \left(\mathbf{f}(v_i) - \mathbf{f}(v_j)\right)^2, \tag{4.5}$$

where $\mathbf{f}$ denotes the graph function, which receives as input a node of the graph and provides as output a real value. A smooth graph function $\mathbf{f}$ can be obtained by minimizing the Dirichlet energy $E(\mathbf{f})$. This is equivalent to minimizing the difference of the graph function on pairs of connected nodes. By expanding Eq. (4.5), it can be seen that the graph Laplacian matrix is encountered in this definition:

$$E(\mathbf{f}) = \frac{1}{2} \sum_{i,j} \mathbf{A}_{ij} \left(\mathbf{f}(v_i) - \mathbf{f}(v_j)\right)^2 \tag{4.6}$$

$$= \frac{1}{2} \left(\sum_i \sum_j \mathbf{A}_{ij} \mathbf{f}(v_i)^2 - 2 \sum_{i,j} \mathbf{A}_{ij} \mathbf{f}(v_i) \mathbf{f}(v_j) + \sum_j \sum_i \mathbf{A}_{ij} \mathbf{f}(v_j)^2 \right) \tag{4.7}$$

$$= \sum_i \mathbf{D}_{ii} \mathbf{f}(v_i)^2 - \sum_{i,j} \mathbf{A}_{ij} \mathbf{f}(v_i) \mathbf{f}(v_j) \tag{4.8}$$

$$= \mathbf{f}^\top \mathbf{D} \mathbf{f} - \mathbf{f}^\top \mathbf{A} \mathbf{f} = \mathbf{f}^\top (\mathbf{D} - \mathbf{A}) \mathbf{f} = \mathbf{f}^\top \mathbf{L} \mathbf{f} \tag{4.9}$$

The smoothness of the graph function can also be expressed by the eigenvectors and eigenvalues of the normalized graph Laplacian matrix $\tilde{\mathbf{L}}$, which is defined as

$$\tilde{\mathbf{L}} = \mathbf{D}^{-1/2} \mathbf{L} \mathbf{D}^{-1/2} = \mathbf{I} - \mathbf{D}^{-1/2} \mathbf{A} \mathbf{D}^{-1/2} \tag{4.10}$$

and its elements are

$$\tilde{\mathbf{L}}_{ij} = \begin{cases} 1, & \text{if } i = j \text{ and } deg(v_i) \neq 0, \\ \frac{-1}{\sqrt{deg(v_i) deg(v_j)}}, & \text{if } i \neq j \text{ and } (v_i, v_j) \in \mathcal{E}, \\ 0, & \text{otherwise.} \end{cases} \tag{4.11}$$

Since $\tilde{\mathbf{L}}$ is a square matrix, we can use eigendecomposition $\tilde{\mathbf{L}} = \mathbf{U}\boldsymbol{\Lambda}\mathbf{U}^\top$, where $\boldsymbol{\Lambda}$ is a diagonal matrix of eigenvalues, that is, $\boldsymbol{\Lambda}_{ii} = \lambda_i$, sorted in an ascending order and $\mathbf{U} = [\mathbf{u}_0, \mathbf{u}_1, ..., \mathbf{u}_{N-1}]$ is the set of eigenvectors ordered according to the eigenvalues. Because $\tilde{\mathbf{L}}$ is symmetric and positive semidefinite, $0 \leq \lambda_0 < \lambda_1 ... \leq \lambda_{N-1}$. The eigenvector $\mathbf{u}_0$ associated with the smallest eigenvalue λ_0, oscillates very slowly across the graph, which means that when two nodes are connected by an edge, the values of this eigenvector corresponding to those two nodes are very similar. On the other hand, the eigenvector with larger eigenvalues, that is, $\mathbf{u}_{N-1}$, oscillates more rapidly and its values are more likely to be dissimilar in connected nodes.

Using the Fourier transform, the signal can be equivalently represented in both vertex domain and graph spectral domain. Let us assume that $\mathbf{x} \in \mathbb{R}^N$ is the graph signal in the vertex domain where $\mathbf{x}_i$ denotes the signal value in the ith node v_i. The graph signal $\mathbf{x}$ can be projected into the orthonormal space formed by the normalized graph Laplacian matrix eigenvectors $\mathbf{U}$ using the Fourier transform as $\mathcal{F}(\mathbf{x}) = \mathbf{U}^\top\mathbf{x}$. The graph convolution of the signal $\mathbf{x}$ with a filter $\mathbf{g}$ is defined as $\mathbf{x} * \mathbf{g}$. The spectral convolution in the orthonormal space can be obtained by applying the Fourier transform and its inverse on the graph convolution as follows:

$$\mathbf{x} * \mathbf{g} = \mathcal{F}^{-1}(\mathcal{F}(\mathbf{x}) \odot \mathcal{F}(\mathbf{g})) = \mathbf{U}(\mathbf{U}^\top\mathbf{x} \odot \mathbf{U}^\top\mathbf{g}), \tag{4.12}$$

where $\odot$ denotes the elementwise product. The parametrized filter in the spectral domain is denoted as $\mathbf{g}_\theta = \text{diag}(\mathbf{U}^\top\mathbf{g})$ and the spectral convolution takes the form:

$$\mathbf{x} * \mathbf{g} = \mathbf{U}\mathbf{g}_\theta\mathbf{U}^\top\mathbf{x}. \tag{4.13}$$

Considering the eigendecomposition of the normalized graph Laplacian matrix, the spectral convolution in Eq. (4.13) is represented as multiplication of signal $\mathbf{x}$ with a parameterized function of normalized graph Laplacian, $\mathbf{U}\mathbf{g}_\theta\mathbf{U}^\top$, and the filter $\mathbf{g}_\theta$ can be seen as a function of normalized Laplacian eigenvalues $\boldsymbol{\Lambda}$, that is, $\mathbf{g}_\theta(\boldsymbol{\Lambda})$. Computing the eigendecomposition of $\tilde{\mathbf{L}}$ has a complexity of $\mathcal{O}(N^3)$ and the multiplication of the signal with the eigenvector matrix $\mathbf{U}$ has a complexity of $\mathcal{O}(N^2)$. Thus computation of the spectral convolution through Eq. (4.13) is computationally expensive, especially for large graphs. To reduce the computational complexity, it is possible to approximate the filter $\mathbf{g}_\theta(\boldsymbol{\Lambda})$ with a Kth order expansion of Chebyshev polynomials as follows [43]:

$$\mathbf{g}_\theta(\boldsymbol{\Lambda}) = \sum_{k=0}^{K-1} \theta_k T_k(\hat{\boldsymbol{\Lambda}}), \tag{4.14}$$

where $\boldsymbol{\theta} \in \mathbb{R}^K$ is a vector of polynomial coefficients and $\hat{\boldsymbol{\Lambda}} = \frac{2}{\lambda_{\max}}\boldsymbol{\Lambda} - \mathbf{I}_N$ is a diagonal matrix of with the eigenvalues normalized in the range $[-1, 1]$. The Chebyshev polynomials of $\hat{\boldsymbol{\Lambda}}$ are defined recursively as $T_k(\hat{\boldsymbol{\Lambda}}) = 2\hat{\boldsymbol{\Lambda}}T_{k-1}(\hat{\boldsymbol{\Lambda}}) - T_{k-2}(\hat{\boldsymbol{\Lambda}})$ with $T_0(\hat{\boldsymbol{\Lambda}}) = 1$ and $T_1(\hat{\boldsymbol{\Lambda}}) = \hat{\boldsymbol{\Lambda}}$.

Considering that $(\mathbf{U}\boldsymbol{\Lambda}\mathbf{U}^\top)^k = \mathbf{U}\boldsymbol{\Lambda}^k\mathbf{U}^\top$, the spectral convolution in Eq. (4.13) can be defined by the truncated expansion of Chebyshev polynomials of the normalized graph Laplacian matrix as follows:

$$\mathbf{x} * \mathbf{g} = \sum_{k=0}^{K-1} \boldsymbol{\theta}_k T_k(\hat{\mathbf{L}})\mathbf{x}, \tag{4.15}$$

where $\hat{\mathbf{L}} = \frac{2}{\lambda_{max}}\tilde{\mathbf{L}} - \mathbf{I}_N$, and $T_k(\hat{\mathbf{L}}) = \mathbf{U}T_k(\hat{\mathbf{\Lambda}})\mathbf{U}^\top$. Using Eq. (4.15) instead of Eq. (4.13) the computation of the spectral convolution operation becomes more efficient, as the polynomial function is calculated recursively. It depends only on the Kth-hop neighborhood of each node (i.e., it is K-localized) and the computation has a complexity of $\mathcal{O}(|\mathcal{E}|)$, which is independent to the graph size. ChebNet [11] uses this definition for spectral convolution on graphs.

The GCN method [12] simplifies the spectral graph convolution by a first-order approximation of Chebyshev polynomials, to avoid overfitting on the local neighborhood structure of graphs. By using the values $K = 2$ and $\lambda_{max} = 2$ in Eq. (4.15), the simplified convolution is defined as follows:

$$\mathbf{x} * \mathbf{g} \approx \theta_0 \mathbf{x} - \theta_1(\tilde{\mathbf{L}} - \mathbf{I}_N)\mathbf{x} = \theta_0 \mathbf{x} - \theta_1(\mathbf{D}^{-1/2}\mathbf{A}\mathbf{D}^{-1/2})\mathbf{x}. \tag{4.16}$$

To avoid overfitting, it constrains the number of parameters by setting $\theta' = \theta_0 = -\theta_1$ leading to the following formulation for spectral convolution:

$$\mathbf{x} * \mathbf{g} \approx \theta'(\mathbf{I}_N + \mathbf{D}^{-1/2}\mathbf{A}\mathbf{D}^{-1/2})\mathbf{x}. \tag{4.17}$$

Since the eigenvalues of $\mathbf{I}_N + \mathbf{D}^{-1/2}\mathbf{A}\mathbf{D}^{-1/2}$ are in range [0, 2], employing this operation in a neural network leads to numerical instabilities. Therefore it is replaced by $\tilde{\mathbf{D}}^{-1/2}\tilde{\mathbf{A}}\tilde{\mathbf{D}}^{-1/2}$ where $\tilde{\mathbf{A}} = \mathbf{A} + \mathbf{I}_N$ is the graph Adjacency matrix obtained by introducing self-loops, that is, edges that connect each graph node to itself, and $\tilde{\mathbf{D}}$ is the graph degree matrix of $\tilde{\mathbf{A}}$.

The GCN method [12] proposed a two-layer network for semisupervised node classification by stacking two GCN layers, each of which is built using the convolution operation defined in Eq. (4.17). The first layer receives as input the input feature matrix of the graph nodes $\mathbf{X} \in \mathbb{R}^{N \times D}$ and performs the following transformation:

$$\mathbf{H} = \sigma(\tilde{\mathbf{D}}^{-\frac{1}{2}}\tilde{\mathbf{A}}\tilde{\mathbf{D}}^{-\frac{1}{2}}\mathbf{X}\mathbf{W}) \tag{4.18}$$

where $\mathbf{W}$ is the matrix of filter parameters, which should be optimized by training the GCN model, and $\mathbf{H}$ is the matrix storing the transformed features produced by the layer. The graph convolution is followed by an activation function σ, which is applied in an elementwise manner. The second layer receives as input the feature matrix $\mathbf{H}$ and performs another convolution operation, followed by an elementwise activation function. Because the goal of the method is to perform classification, the activation function used for the second layer is the softmax function. The activation function used for the first layer is the Rectified Linear Unit (ReLU) function. The two layer GCN model leads to

$$\mathbf{Y} = \text{softmax}\left(\hat{\mathbf{A}}\ \text{ReLU}\left(\hat{\mathbf{A}}\mathbf{X}\mathbf{W}^{(1)}\right)\mathbf{W}^{(2)}\right), \tag{4.19}$$

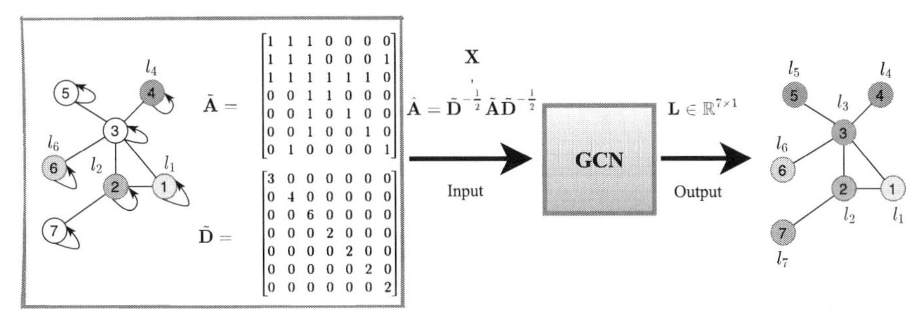

FIGURE 4.1

Illustration of a simple undirected graph with its corresponding graph adjacency and graph degree matrices. Only a subset of nodes in the input graphs are labeled with l_i and the remaining nodes are unlabeled. The normalized graph adjacency matrix and feature matrix are introduced into a GCN model. The model is trained in a semisupervised manner and propagates the information of the labeled nodes through the whole graph. The output of the model is a vector of labels $L = [l_0 \mid i = 1, ..., 7]$ containing the predicted label l_i for each graph node.

where $\hat{\mathbf{A}} = \tilde{\mathbf{D}}^{-\frac{1}{2}}\tilde{\mathbf{A}}\tilde{\mathbf{D}}^{-\frac{1}{2}}$ and $\mathbf{W}^{(1)}$, $\mathbf{W}^{(2)}$ are the learnable filter parameters of the first and second layers, respectively. The parameters of the GCN model are trained in an end-to-end manner using the backpropagation algorithm, or its variants.

Accordingly, the spectral convolution formulation proposed by the GCN method [12] bridged the gap between spectral and spatial convolution by utilizing the graph adjacency matrix for feature aggregation. In other words, the convolution operation in Eq. (4.18) can be seen as a propagation rule, which updates the features of each node and it can be used for semisupervised node classification by propagating the information from labeled nodes to all other nodes in the graph. To make this more clear, we explain it with a simple example. Let us assume that the graph illustrated in Fig. 4.1 is the graph introduced to the GCN model. In this graph, only a subset of nodes are labeled and the goal is to predict labels for all the nodes in the graph. In order to update the features of each node, we need to aggregate the features of the node itself and the features of its neighbors. Therefore, in the illustrated graph each node has a self-connection, which explains the use of $\tilde{\mathbf{A}} = \mathbf{A} + \mathbf{I}_N$ instead of $\mathbf{A}$. As can be seen in Fig. 4.1, the nodes in the graph may have different degrees. Using the matrix $\tilde{\mathbf{A}}$ for feature propagation would lead to nodes with large degrees having larger values in their feature vectors compared to nodes with smaller degrees. This issue can be solved by normalizing the graph adjacency matrix, explaining the use of $\hat{\mathbf{A}} = \tilde{\mathbf{D}}^{-\frac{1}{2}}\tilde{\mathbf{A}}\tilde{\mathbf{D}}^{-\frac{1}{2}}$. In this way, the connection between each pair of nodes is normalized by the degrees of the two nodes. The normalized graph adjacency matrix and the feature matrix of the input graph are introduced to a two-layer GCN model, which updates the features of each node by performing the propagation rule in Eq. (4.18). As a result, the extracted features are transformed into a C-dimensional space for a C-class classification problem and a label is predicted for each node based on the pseudo-probability values obtained at the output of the second layer. Here, we should note

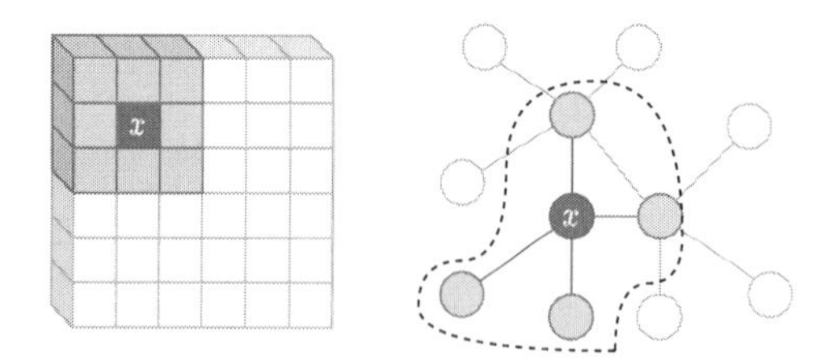

FIGURE 4.2

Left: 2D convolution on a grid where a grid-shaped filter is placed on the 2-dimensional grid data to update the features of the node x by aggregating the features of its ordered neighbors. Right: spatial graph convolution, where the feature representation of the node x is updated by aggregating the features of its neighbors.

that there is no restriction in the number of layers used to build the GCN model, that is, one can use multiple layers, stacked in a hierarchical manner, to form a multilayer GCN model. Recently, the PGCN method [44] has been proposed to automatically design a problem-specific GCN topology by jointly growing the network's structure both in width and depth in a progressive manner, and optimizing its parameters.

One of the main drawbacks of the spectral graph convolutions is that the learned filters in these methods are dependent to the eigenbasis of the underlying graph. Therefore, the learned parameters for a graph cannot be applied to another graph with a different structure. Besides, since these methods need to utilize the whole graph structure (as graph adjacency matrix) in the convolution operation, they are not scalable to large graphs.

Several methods have been proposed recently such as Adaptive Graph Convolutional Network (AGCN) [45] and Dual Graph Convolutional Network (DGCN) [46] to improve the performance of GCN by making the feature learning independent of input graph structure, and training two graph convolutional layers in parallel, respectively.

4.3 Spatial graph convolutional network

The spatial convolution operation is directly defined on the graph and it can be easily explained in the context of conventional CNNs in which the spatial structure of the images is considered. As illustrated in Fig. 4.2, the convolution operation in grid-structured data is a process of employing a weighted kernel to update the features of each node in the grid by aggregating information from its local neighbors. Its generalization to graph data is defined as a message passing function, which updates the features of each graph node by the means of message passing through the graph edges. Spatial GCN is a combination of such message passing function with a data transformation process. It is trained in an end-to-end manner to optimize an objective function, which can be defined on the individual graph nodes or the whole graph.

Diffusion Convolutional Neural Network (DCNN) [13] is a spatial GCN method, which defines the convolution operation as a diffusion process using transition matrices. A probability transition matrix $\mathbf{P} \in \mathbb{R}^{N \times N}$ is defined as the degree-normalized graph Adjacency matrix, $\mathbf{P} = \mathbf{D}^{-1}\mathbf{A}$, and $\mathbf{P}_{ij}$ indicates the probability of information transition from node v_i to node v_j. The information is propagated from each node to its neighboring nodes with a transition probability until the information distribution reaches an equilibrium. This method has several advantages in terms of classification performance, flexibility, and computational efficiency. It is flexible in the sense that the underlying graph in DCNN can be directed or undirected, weighted, or unweighted. This method can be used for different tasks including node classification and graph classification in a semisupervised or supervised manner, respectively. Moreover, DCNN represents graph diffusion as a matrix power series to encode the structural information of the graph. Hereby, we describe in detail the diffusion operation definition in this method.

The DCNN model takes the graph transition matrix and the node feature matrix $\mathbf{X} \in \mathbb{R}^{N \times D}$ as input and outputs a prediction for either nodes, edges, or the whole graph. For node classification, the diffusion convolution operation through the k-hop neighbors is defined as follows:

$$\mathbf{H}^{(k)} = \sigma((\mathbf{1}_N.\mathbf{W}^{(k)^\top}) \odot (\mathbf{P}^k\mathbf{X})), \tag{4.20}$$

where $\odot$ denotes the elementwise multiplication and σ is the nonlinear activation function. $\mathbf{W} \in \mathbb{R}^{K \times D}$ is the transformation matrix and $\mathbf{W}^{(k)} \in \mathbb{R}^{1 \times D}$ is the transformation matrix to the k-hop neighbors. $\mathbf{1}_N \in \mathbb{R}^N$ is a vector of ones, which is multiplied to $\mathbf{W}^{(k)}$ to make N copies of that. $\mathbf{P}^k$ denotes the power k of the transition matrix, which is used to aggregate features from hop k neighborhood, and $\mathbf{H}^{(k)} \in \mathbb{R}^{N \times D}$ contains the aggregated features obtained by using the k hop neighborhood. This convolution operation is performed for multiple hop values, that is, from 1 to K, and the transformed feature matrices $\mathbf{H}^{(1)}, \mathbf{H}^{(2)}, \ldots, \mathbf{H}^{(K)}$ are concatenated to produce $\mathbf{H} \in \mathbb{R}^{N \times K \times D}$.

The diffusion can also be expressed using tensor notation. For node classification, it can be defined as follows:

$$\mathbf{H} = \sigma(\mathbf{W}^\star \odot (\mathbf{P}^\star \times_3 \mathbf{X})), \tag{4.21}$$

where $\mathbf{P}^\star$ is a tensor of size $N \times K \times N$, which contains the power series $\{\mathbf{P}^1, \mathbf{P}^2, \ldots, \mathbf{P}^K\}$ of transition matrix $\mathbf{P}$, and $\mathbf{W}^\star \in \mathbb{R}^{N \times K \times D}$ is the transformation tensor to the 1 to K neighbors. This tensor is obtained by making N copies of $\mathbf{W}$ so that it has the same size as $(\mathbf{P}^\star \times_3 \mathbf{X})$, and it contains $K \times D$ learnable parameters. $\mathbf{H} \in \mathbb{R}^{N \times K \times D}$ is the diffusion convolutional representation computed by K-hops graph diffusion over the nodes' features. $\times_3$ in Eq. (4.21) denotes the 3-mode product of the tensor $\mathbf{P}^\star$ to the matrix $\mathbf{X}$. It should be noted that the diffusion convolution does not change the features' dimension.

For graph classification, $\mathbf{H}$ is of size $\mathbb{R}^{K \times D}$ and is computed by taking average over the nodes as follows:

$$\mathbf{H} = \sigma(\mathbf{W} \odot ((\mathbf{P}^{\star} \times_3 \mathbf{X}) \times_1 \mathbf{1}_N^{\top})/N). \tag{4.22}$$

The series of tensor operations in this method can be computed in polynomial time and implemented efficiently on a GPU.

Message Passing Neural Network (MPNN) [15] has generated a general framework to formulate all the existing spatial graph convolutional networks. Similar to DCNN, the convolutional operation in MPNN is defined as a K-hop message passing and the information is propagated from one node to its neighboring nodes through the graph edges. Therefore the graph convolution operation in this framework is defined using message passing and update functions as follows:

$$\mathbf{h}_{\mathcal{N}(v_i)}^{(k)} = \sum_{v_j \in \mathcal{N}(v_i)} M^{(k)}\left(\mathbf{h}_i^{(k-1)}, \mathbf{h}_j^{(k-1)}, \mathbf{x}_{ij}^{e}\right), \tag{4.23}$$

$$\mathbf{h}_i^{(k)} = U^{(k)}\left(\mathbf{h}_i^{(k-1)}, \mathbf{h}_{\mathcal{N}(v_i)}^{(k)}\right), \tag{4.24}$$

where $\mathbf{h}_i^{(k)}$ is the hidden representation of node v_i obtained by message passing through its k-hop neighbors and $\mathbf{h}_i^{(0)} = \mathbf{x}_i$. $M^{(k)}(\cdot)$ is the message passing function, which propagates the features and $U^{(k)}(\cdot)$ is the updating function which updates the features of the node. $\mathbf{h}_{\mathcal{N}(v_i)}^{(k)}$ denotes the aggregated feature vector of the neighbors of v_i, which is obtained by the message passing function $M^{(k)}(\cdot)$. $\mathbf{x}_{ij}^{e}$ denotes the feature vector of the edge connecting the neighboring nodes v_i and v_j. In simple graphs, the edges can be represented by single binary values expressing the graph node connections. The structure of a single layer of the MPNN framework is illustrated in Fig. 4.3.

This convolution operation is performed iteratively for K steps and in each step k the node features are updated using its own features and the features of its direct 1-hop neighboring nodes in step $k-1$. In this way, at the end the information has been propagated through K-hop neighbors and $\mathbf{h}_i^{(K)}$ can be introduced to a classification layer for predicting the labels for each node. For the graph classification task, a readout function is needed to generate a feature vector representing the graph as follows:

$$\mathbf{h}_G = R\left(\mathbf{h}_i^{(K)} | i = 1, ..., N\right), \tag{4.25}$$

where $R(\cdot)$ is the readout function and $\mathbf{h}_G$ is the feature vector of the graph. By specifying $M(\cdot)$, $U(\cdot)$, $R(\cdot)$, which are differentiable functions with learnable parameters in different ways, several spatial graph convolutional network methods such as convolutional networks for learning molecular fingerprints [47], Gated Graph Neural Networks (GG-NN) [48], interaction networks [49], molecular graph convolutions

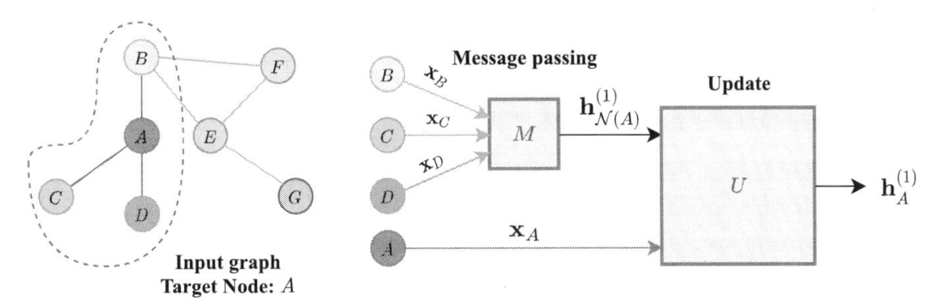

FIGURE 4.3

Illustration of convolution operation in the first layer of MPNN framework. The input graph is shown in the left side. The goal is to learn the hidden representation of the node A, so the first hop neighborhood of this node is considered. On the right side, the layer structure containing the message passing and updating functions is illustrated. The layer receives the feature vectors of node A and its first hop neighbors as input. It learns the representation of the node by performing message passing through its neighbors and applying an updating function to its own input features and the aggregated feature vector of its neighborhood.

[50], deep tensor neural networks [51], and spectral-based GCN methods [10,52,12] can be defined in this framework.

Unlike the regular GCN methods that generalize the concept of convolution from CNNs to graph structured data, some methods such as PATCHY-SAN [14] and LGCN [53] transform the graph data into grid structured data in order to apply the convolutional filter directly on the locally connected regions of the graph, similar to image-based convolutional networks that perform convolution on locally connected pixels of an image. PATCHY-SAN method [14] first orders the neighbors of each node according to their score, which can be defined based on their degree. Then it determines the number of ordered neighbors of all the nodes by selecting a fixed number of top neighbors for each node. Therefore similar to grid structured data, each graph node now has a fixed number of ordered neighbors, which serves as the convolution receptive field and a standard convolution operation is applied on each node to aggregate its neighborhood features.

LGCN [53] orders the neighbors of each node based on their feature values. For each node, it constructs a feature matrix of size $(k+1) \times D$ containing the D dimensional feature vectors of the node itself and its k neighbors. Then the feature columns of the neighbors are sorted based on their value and a fixed number of top rows are selected as the ordered neighborhood features of that node. In other words, it performs max pooling on the feature matrix of each node's neighbors followed by a standard 1D convolution, which is applied on the selected top rows of the feature matrix to generate the transformed features for the node.

4.4 Graph attention network (GAT)

Consider the graph convolutional network transformation in Eq. (4.18), which defines the feature transformation for all N graph nodes:

$$\mathbf{H} = \sigma\left(\hat{\mathbf{A}}\mathbf{X}\mathbf{W}\right).$$

The matrix multiplication by $\hat{\mathbf{A}}$ produces a weighted summation over the neighbor set for each node. For node i, the computation can be expressed equivalently as

$$\mathbf{h}_i = \sigma\left(\sum_{j \in \mathcal{N}(v_i)} \alpha_{ij}\mathbf{x}_i\mathbf{W}\right) \tag{4.26}$$

where $\mathbf{x}_i \in \mathbb{R}^D$ is the feature vector of node v_i, and $\alpha_{ij} = \hat{A}_{ij}$ is the importance weighting factor for a node v_i to its neighbor v_j in the set of neighbors $\mathcal{N}(v_i)$.

For a GCN, the importance weights α_{ij} are defined explicitly from the graph Adjacency matrix. This works well for tasks on a fixed graph, that is, in a *transductive* setting. However, the learned parameters in $\mathbf{W}$ may not generalize to unseen graphs (the *inductive* setting) because they were learned under the specific graph structure with fixed node importance weights.

This limitation can be overcome by the use of a Graph Attention Network (GAT) [54]. Inspired by the attention mechanism in [55], the importance weights α_{ij} are not defined explicitly from the graph structure. Instead, they are computed as a self-attention for neighboring nodes:

$$a_{ij} = \text{attention}(\mathbf{x}_i, \mathbf{x}_j) \tag{4.27}$$

where $a_{ij} \in \mathbb{R}$ is an unnormalized attention coefficient for input feature vectors $\mathbf{x}_i$ and $\mathbf{x}_j$. The attention mechanism could be any function of two input features, but the choice in the original paper [54] was a single-layer feedforward neural network over concatenated transformed features with LeakyReLU and softmax activations:

$$\text{attention}(\mathbf{x}_i, \mathbf{x}_i) = \text{softmax}_j\left(\text{LeakyReLU}\left(\mathbf{v}^\top\left[\mathbf{x}_i\mathbf{W} \,||\, \mathbf{x}_j\mathbf{W}\right]\right)\right). \tag{4.28}$$

Here, $||$ denotes concatenation, and $\text{LeakyReLU}(\mathbf{x}) = \max(\beta\mathbf{x}, \mathbf{x})$ for $1 > \beta > 0$. $\mathbf{W} \in \mathbb{R}^{D' \times D}$ is the weight matrix, which transforms the feature vectors, and $\mathbf{v} \in \mathbb{R}^{2D'}$ is a parameter vector for the attention layer. Notice that the attention weight vector $\mathbf{v}$ is learned from node correspondences and does not depend on the global graph structure. This lets GAT generalize better to unseen graphs than a regular GCN. Finally, the unnormalized attention coefficient normalized using a softmax over the neighborhood of the node to produce the final importance weight:

$$\alpha_{ij} = \text{softmax}_j(a_{ij}) = \frac{e^{a_{ij}}}{\sum_{k \in \mathcal{N}(v_i)} e^{a_{ik}}} \tag{4.29}$$

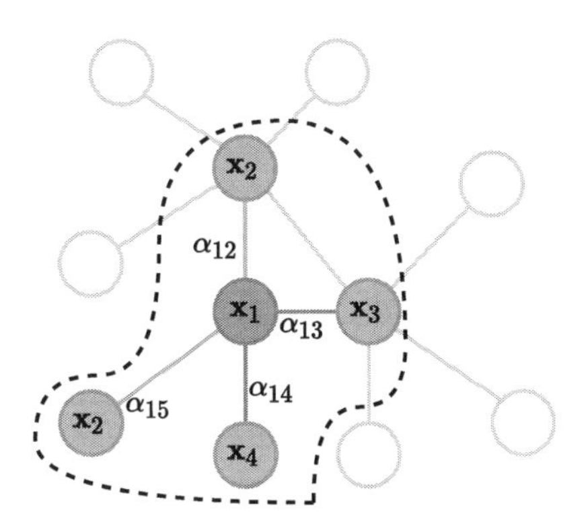

FIGURE 4.4

Illustration of edges weighted by attention coefficients α_{ij} in the neighborhood of node 1. The color intensities on the edges $\mathbf{x}_1$ denote the magnitude of the corresponding attention values α_{1j}. (For interpretation of the colors in the figure, the reader is referred to the web version of this chapter.)

The softmax operation yields positive edges for all nodes in the neighborhood of v_i that sum up to $\sum_{\sum_{j \in \mathcal{N}(v_i)}} \alpha_{ij} = 1$. An illustration of this is shown in Fig. 4.4. As with other attention-based networks [55], GAT can be extended to use *multihead attention* where K independent attention mechanisms are aggregated (typically by concatenation) to produce the new feature:

$$\mathbf{h}_i = \Big\Vert_{k=1}^{K} \sigma \left(\sum_{j \in \mathcal{N}(v_i)} \alpha_{ij}^k \mathbf{x}_i \mathbf{W}^k \right) \tag{4.30}$$

Here, k denotes the attention index for a layer. Using concatenation to aggregate the outputs, $\mathbf{h}_i$ will consist of KD features.

Now that we have covered a method of generalizing to unseen graphs, we proceed with describing how to leverage and learn from large graphs that cannot fit in computational memory all at once.

4.5 Graph convolutional networks for large graphs

Many of the existing GCN methods [10,52,47,12,14] perform the training process in a transductive setting. These methods typically need to load the whole graph data and node features in memory, and introduce them to the model at once, while the training process updates the parameters of the GCN model using full batch iterative optimization. For example, in the GCN method [12], which performs under semisupervised transductive learning, the features of each node are updated using the features of its neighbors in the previous layer. Similarly, each of these neighbors are also updated by aggregating features from their corresponding neighbors in the previous layer, and this backtracking continues until it reaches k-hops neighbors. Therefore the whole graph Laplacian and feature matrices are required during the training process. Besides, as we go deeper in the GCN, the total number of neighboring nodes, which are used for feature aggregation grows exponentially, which leads to neighbor explosion. Methods with full-batch training only target small-sized graph data sets and their application on large-scale graph data is infeasible due to their high computational complexity and memory consumption. Moreover, in Laplacian based methods like GCN [12], the node representations are dependent on the graph structure. Since they only focus on a fixed graph structure during the training process, they cannot be generalized to unseen nodes or graphs.

Recently, GCN methods which can perform on an inductive setting have been proposed. These methods are scalable to large graph data sets with millions of nodes and they are able to generate feature representations for unseen nodes or graphs efficiently. In order to alleviate the above-mentioned problems and scale GCN methods to large scale graphs, *layer sampling* and *graph sampling* methods have been proposed, which can be trained by following mini-batch optimization. In the following, we will introduce these two groups of methods.

4.5.1 Layer sampling methods

The layer sampling methods [56–59,53,60] form mini-batches by sampling a small number of nodes/edges at each layer of the model and then perform feature propagation among the sampled nodes/edges. These two steps proceed iteratively and the model weights are updated using Stochastic Gradient Descent (SGD) [61].

The GraphSAGE method [56] mitigates the neighbor explosion problem kf GCNs [12] by performing sampling on the neighboring nodes using a distinct sampling function in each layer in order to constrain the batch-size to a pre-defined fixed size. The algorithm receives as input the whole graph $G = (\mathcal{V}, \mathcal{E})$, the nodes feature matrix $\mathbf{X}$, and the mini-batch set $\mathcal{B}$ formed by the target nodes, that is, the nodes whose features will be updated. The model is composed of K convolutional layers in order to aggregate features from 1 to K-hops neighbors and it samples all the neighboring nodes needed for feature aggregation at first. Sampling K-hop neighbors for each node in $\mathcal{B}$ is like expanding a tree rooted at that node at layer K of the model recursively.

The sampling mechanism starts with the root nodes at layer K, $\mathcal{B}^K = \mathcal{B}$, and samples S_K nodes from their neighbors at layer $K - 1$ using a neighborhood sampling function F_K so that $\mathcal{B}^{K-1} = \mathcal{B}^K \cup F_K(\nu)$ for $\forall \nu \in \mathcal{B}^K$. This process continues for all the layers $k = \{K, K - 1, ..., 1\}$ with fixed sample sizes S_k and sampling functions $F_k : \nu \rightarrow 2^{\mathcal{V}}$, respectively. The neighborhood samplers F_k used in GraphSAGE are uniform random sampling functions generating independent samples. In cases where the sample size S_k is larger than the node degree, the sampling is performed with replacement. In this way, each set $\mathcal{B}^k$ contains all the needed nodes for feature aggregation in the next layer $k + 1$ to compute the feature representation nodes in $\mathcal{B}^{k+1}$.

Therefore, $\mathcal{B}^0$ contains all the sampled nodes. As an example, let us assume that we have a 2-layer network and we want to generate feature representations for the nodes in $\mathcal{B}$ using their 2-hop neighbors. For each node, we first sample S_2 nodes from its immediate neighbors and then sample $S_2 \times S_1$ nodes from their 2-hop neighbors.

Instead of learning a discrete feature representation for each node, this method trains a set of aggregation functions, one for each layer, which learn to utilize the local structure of the graph to aggregate the features from each node's neighborhood. Therefore these trained aggregation functions are able to generate feature representations for completely unseen nodes at inference time. In the forward pass, in order to generate the feature representations $\mathbf{h}_v^{(k)}$ for $\forall v \in \mathcal{B}^k$, the kth layer aggregation function $Agg^{(k)}$ aggregates the feature representations of their immediate neighborhood $\left\{\mathbf{h}_{v'}^{(k-1)}, \forall v' \in \mathcal{N}(v)\right\}$ into a single vector $\mathbf{h}_{\mathcal{N}(v)}^{(k-1)}$, while $\mathbf{h}_{v'}^{(k-1)}$ is generated at layer $k-1$. It should be noted that at $k=0$ the feature representations are the input feature vectors.

In the next step, the aggregated feature vector $\mathbf{h}_{\mathcal{N}(v)}^{(k-1)}$ is concatenated with the current representation vector of the node $\mathbf{h}_v^{(k-1)}$ and the resulting vector is introduced to a fully connected layer which transforms the features by some learnable parameters and is followed by a nonlinear activation function. Concatenation of the node features with its aggregated neighborhood feature vectors works like skip connections between different layers of the model. Experimental evaluation of GraphSAGE has shown the effectiveness of the concatenation mechanism, as it leads to significant performance improvement. At the last layer K, the final feature representations $\left\{\mathbf{h}_v^{(K)}, \forall v \in \mathcal{B}\right\}$ for the target nodes are generated, which are then introduced to the classification layer, which predicts the labels for the nodes. The sampling process and feature aggregation of the GraphSAGE method are shown in Fig. 4.5, Fig. 4.6, respectively.

GraphSAGE cannot only be trained in inductive setting with mini-batches, it can also be trained in both an unsupervised or a supervised manner and it can generalize to unseen data. Another interesting advantage of GraphSAGE is that at different levels of feature extraction, an appropriate aggregation function can be employed to capture the local structure of the graph. The aggregation functions should be differentiable and invariant to node permutations. For instance, mean aggregator, max pooling, and mean pooling aggregators are all symmetric and trainable functions, which can be employed in GraphSAGE. Mean aggregator computes the elementwise average of vectors in $\left\{\mathbf{h}_{v'}^{(k-1)}, \forall v' \in \mathcal{N}(v)\right\}$. For employing pooling aggregators, the feature vector of each neighboring node is first transformed, and then the elementwise pooling is performed on the transformed features.

An extension of GCN method with an inductive training setting can be obtained by a special form of GraphSAGE with two layers and mean aggregation function in both layers. The convolution operation in the inductive GCN is defined as follows:

$$\mathbf{h}_v^{(k)} \leftarrow \sigma\left(\mathbf{W}^{(k)} \bar{\mathbf{h}}_v^{\prime(k)}\right), \tag{4.31}$$

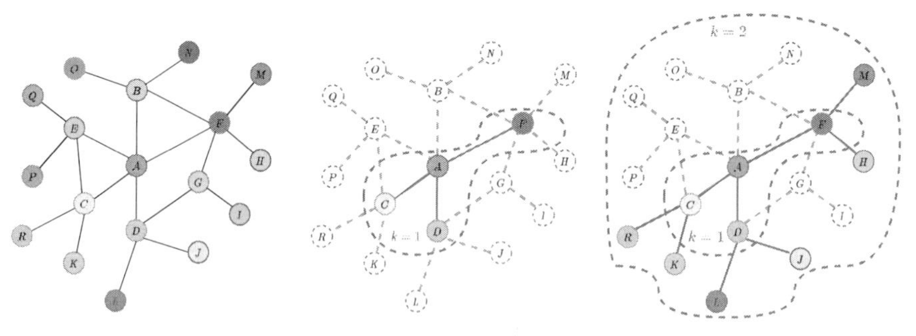

FIGURE 4.5

Visual illustration of the sampling process for a 2-layer network in the GraphSAGE method. Here, the mini-batch contains a single node A, that is, $\mathcal{B} = A$, and the number of nodes for the first and second layers (or first and second hop neighborhood of node A) are $S_1 = 3$, $S_2 = 2$, respectively. The sampling function first samples S_2 nodes from the immediate neighbors of A, and then it samples S_1 neighbors for each of the S_2 sampled nodes in the previous step. Therefore, it samples $S_2 \times S_1$ nodes in total from its 2-hop neighbors. This sampling process is repeated for each node in the batch set $\mathcal{B}$.

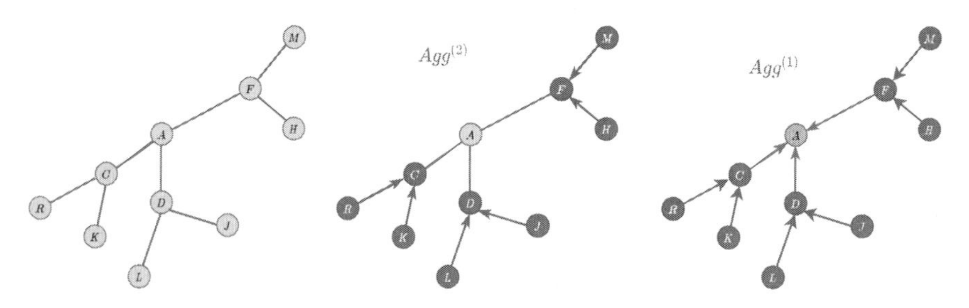

FIGURE 4.6

Illustration of GraphSAGE feature aggregation in a 2-layer network. The node A and its neighbors (which are sampled as in Fig. 4.5) are shown in the left side. In the middle, the first hop neighbors of node A are updated by aggregating features from their neighbors. Arrows indicate the feature aggregation. In the final step, illustrated on the right side, the features of node A are updated by aggregating its first-hop neighbors features. It should be noted that different aggregation functions can be used for each layer of the network (or at each step of feature aggregation process). "Agg" denotes the aggregation function.

$$\mathbf{h}_v'^{(k)} = \left\{\mathbf{h}_v^{(k-1)}\right\} \cup \left\{\mathbf{h}_{v'}^{(k-1)}, \forall v' \in \mathcal{N}(v)\right\}. \tag{4.32}$$

$\bar{\mathbf{h}}_v'^{(k)}$ in Eq. (4.31) is the mean of the node feature vector in the current state and its neighbors aggregated representation vector.

Regardless of the advantages of GraphSAGE method in generalizing GCNs to unseen data by inductive learning and mini-batch training, it still has limitations in terms of scalability to very large graphs with billions of nodes, and it cannot be employed in related real world applications. This limitation is due to the fact that it still needs to store the whole graph in the memory although it performs localized convolution by sampling a fixed number of neighbors around each of the graph nodes in the batch.

The PinSage method [58], addressed this limitation by assigning an importance score to each neighbor using random walk sampling. Moreover, it can employ multi-GPU training methods to improve both the performance and efficiency of the model to scale up to graphs with billions of nodes and edges. The PinSage method has been deployed for different recommendation tasks at Pinterest application which is one of the world's largest graph of images. In this application, the users interact with visual bookmarks, called pins, to the online images. In the following, we introduce the key properties of this method, which led to high scalability of GCNs.

- **Importance pooling:** PinSage developed an importance pooling approach for localized convolution on the nodes by using a random-walk-based sampling method instead of a uniform sampling for defining the neighborhood of each node. Random walks start from a node v and a count is kept to the number of times each of its neighbors has been visited. Each neighbor of the node thus gets an importance score, which is the normalized visit count. The subset of neighbors with the highest importance scores are selected and defined as its neighborhood. Therefore, the localized convolution in this method performs a weighted feature aggregation, which leads to less information loss.
- **Mini-batch construction:** Contrary to GraphSAGE, the PinSage method stores the whole graph Adjacency matrix and its nodes' feature vectors on the CPU memory instead of GPU. Since the convolution operations are performed on GPUs, the data needs to be transferred between CPU and GPU, which is not computationally efficient. PinSage has solved this issue by creating a small subgraph $G_{sub} = (\mathcal{V}_{sub}, \mathcal{E}_{sub})$ for each mini-batch using the nodes, which are involved in the convolution computation of that mini-batch and their sampled neighborhood. The feature matrix corresponding to each subgraph is also constructed using a reindexing approach. Therefore, at each mini-batch iteration, only a small subgraph with its corresponding feature matrix are transferred to the GPUs. By using such optimized data transfer between GPU and CPU, this method has greatly improved the GPU utilization. Moreover, this method proposed a producer-consumer approach for training, which reduces the training time. In this approach, while the GPUs are computing convolutions for the nodes in the current mini-batch, the CPU is constructing the subgraph and feature matrix for the next mini-batch in parallel.
- **Multi-GPU training:** In order to maximize the GPU usage during the forward and backward propagation, the graph nodes in each mini-batch are divided into equal-sized portions, each of which is processed by a separate GPU. It should be noted that all the GPUs use exactly the same neural network model with shared model parameters. For the optimization step, the computed gradients are aggregated across all the GPUs and a synchronous SGD step is performed.

The S-GCN method [59] has improved the time complexity of layer sampling methods by sampling only two arbitrary neighbors for each node in the graph while preserving the performance of the model. It has developed a control variate-based stochastic approximation algorithm for GCN, utilizing the historic activations of nodes in the previous layers, thus minimizing the redundant computations. It was also theoretically proved that S-GCN converges to a local optimum of GCN.

One of the main differences between the standard neural networks and GCNs is that in GCNs, each sample (which is assumed to be a graph node) is convolved with its neighbors, leading to lack of independence between the samples, while the optimization method SGD is designed based on the additive loss function for independent data samples. FastGCN [57] proposed to view graph convolution from a different perspective and developed the GCN as an integral transform of embedding functions under probability measures. In other words, it interprets each layer of the GCN as an embedding function of the nodes, which are independently sampled from the graph but are from the same probability distribution.

To reduce the computational footprint and the memory consumption caused by recursive neighborhood expansion, the FastGCN method samples the nodes independently for each layer. To compare FastGCN sampling with GraphSAGE, let us assume that at each layer l of the GraphSAGE model, S_l neighbors are sampled for each node in the batch. Then the total number of expanded neighborhood would be $S_0 \times S_1 \times ... \times S_{K-1}$ for K layers. In the FastGCN method, at each layer l, S_l nodes are sampled independently from the graph disregarding their neighbors. Therefore the total number of sampled nodes for K layers would be $S_0 + S_1 + ... + S_{K-1}$, which makes the convolutions more computationally efficient. The method also applies importance sampling to reduce the variance. However, since the nodes are sampled independently in each layer, the connections between the nodes forming the mini-batches might become too sparse, which leads to performance degradation.

The adaptive sampling method proposed in [60] addressed this problem and improved FastGCN by capturing the connections between the network layers. Similar to the FastGCN method, it constructs the network layer by layer by fixed-size sampling but the layers are not independent. In this method, the samples in each layer are conditioned on the samples on their later layers, and the connections between the sampled nodes in the earlier layer and their parents in the later layers can be used. However, training the sampling method proposed in [60] is computationally expensive.

A simple illustration of layer sampling methods is provided in Fig. 4.7. While the layer sampling methods significantly improved the training efficiency of GCNs, methods adopting them still face challenges to achieve high accuracy, scalability, and computation efficiency simultaneously.

4.5.2 Graph sampling methods

Graph sampling methods [53,62,63] perform sampling on the graph scale instead of node scale, which avoids neighbor explosion. The mini-batches are built from sub-

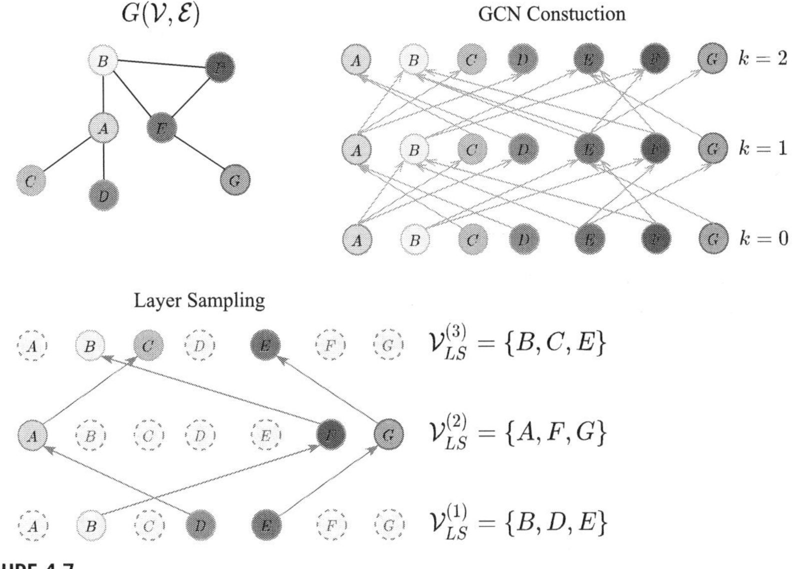

FIGURE 4.7

In layer sampling methods, the GCN model is first constructed on the whole graph. Then a mini-batch is built by a set of sampled nodes in each layer. $\mathcal{V}_{LS}^{th}$ denotes the nodes sampled by the layer sampling method in kth layer.

graphs in these methods. In the following, two of the recently proposed graph sampling methods are introduced. The graph sampling scheme is illustrated in Fig. 4.8.

The work in [62] proposed a representative graph sampling method, which assures that the connectivity characteristics of the original training graph are preserved in the sampled subgraphs and that each node in the training graph has a fair chance of being sampled. Contrary to the traditional GCN methods, where the network is directly constructed on the input graph, this graph sampling method samples a small subgraph $G_{sub} = (\mathcal{V}_{sub}, \mathcal{E}_{sub})$ for each mini-batch training iteration and constructs a complete GCN on the sampled subgraph G_{sub}. The forward propagation in layer l of this network is defined in 3 steps. In the first step, the neighbors' features are aggregated as follows:

$$\mathbf{H}_{neigh}^{(l)} = \mathbf{A}_{sub}^{(l)} \mathbf{H}_{sub}^{(l-1)} \mathbf{W}_{neigh}^{(l)}, \tag{4.33}$$

where $\mathbf{A}_{sub}^{(l)}$ denotes the graph Adjacency matrix of the sampled subgraph, and $\mathbf{H}_{sub}^{(l-1)}$ and $\mathbf{H}_{sub}^{(l)}$ are the feature matrices of nodes in layers $l-1$ and l, respectively. $\mathbf{W}_{neigh}^{(l)}$ denotes the transformation matrix which is learned for neighbors' feature aggregation. In the second step, the features of the node are transformed with another transformation matrix as follows:

$$\mathbf{H}_{self}^{(l)} = \mathbf{H}_{sub}^{(l-1)} \mathbf{W}_{self}^{(l)}, \tag{4.34}$$

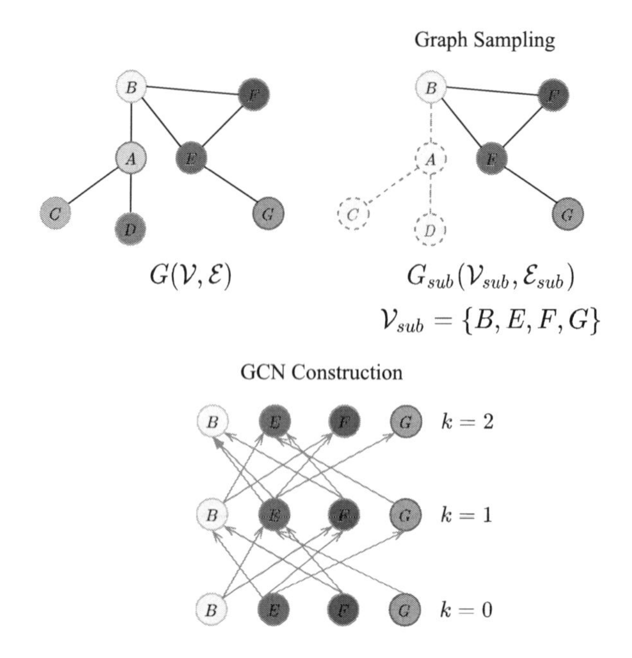

FIGURE 4.8

Illustration of graph sampling and GCN construction on the subgraph. In graph sampling methods, the subgraphs are sampled independently and all the graph nodes have sufficient chance to be sampled. The GCN model is constructed on the full subgraph and at each iteration of the training process, the full subgraph is introduced into the GCN model and the features of the nodes are updated by information propagation through their neighbors. In this figure, the model has two layers, $k = 2$, and the features are aggregated from 2-hop neighbors.

and in the final step, the features of the lth layer of the network are built by the concatenation of $\mathbf{H}^{(l)}_{neigh}$ and $\mathbf{H}^{(l)}_{self}$, which is followed by an activation function:

$$\mathbf{H}^{(l)}_{sub} = \sigma(\mathbf{H}^{(l)}_{neigh} \,||\, \mathbf{H}^{(l)}_{self}). \tag{4.35}$$

It was shown that by sampling a sufficient number representative subgraphs, the GCN model is able to capture both the training graph structure and its nodes' attributes to generate accurate feature representations. The main advantage of this method over the layer sampling methods is that it achieves both high performance and computation efficiency, simultaneously by performing convolution on a small set of nodes of the sampled subgraph. Thus neighbor explosion is avoided, while the node connections in the original graph are preserved, which leads to minimum information loss and high accuracy. Besides, the method achieves scalability with respect to graph size and number of GCN layers by developing efficient approaches for parallelizing the training process on shared-memory multiple core processing systems.

ClusterGCN method [63] employs efficient graph clustering algorithms such as Metis [64] and Graclus [65] to generate subgraphs for mini-batch training. These graph clustering methods partition the input graph into a set of dense clusters in such a way that the number of within-cluster edges is much higher than the number of between-cluster edges. Accordingly, each node and its neighbors are mostly located in the same cluster, which leads to minimized information loss. At each iteration of ClusterGCN, a set of generated clusters are selected randomly without replacement and a subgraph is constructed using these clusters. For each mini-batch, the forward pass and the optimization are performed by only using the nodes involved in the subgraph, which leads to high computational and memory efficiency.

4.6 Datasets and libraries

GNNs have applications in many different areas and many benchmark graph datasets are available, which describe different learning tasks involving graphs. TUDataset [66] contains a collection of more than 120 benchmark data sets, and experimental evaluation tools for graph learning. The statistics of a set of well-known benchmark data sets are summarized in Table 4.1. We categorize these data sets into four groups, citation networks, social networks, biochemical networks, and a group for the graph data sets used in computer vision applications. Accordingly, some of the applications of GNNs in different domains are briefly described as follows:

- **Natural language processing:** One of the graph-based methods applications in natural language processing domain is document classification [12,56,54]. Well-known data sets such as Citeseer [67], Cora [67] and Pubmed [67], include citation networks which represent documents as nodes and the cross references between documents as edges. GCN models extract high-level feature representation for each document using the node labels in a supervised setting or by propagating the label information from labeled nodes through the whole graph in a semisupervised setting. Using the extracted high-level node features, a label will be predicted for each node by a classification layer. The whole model including the classification layer and GCN layers are trained in an end-to-end manner. Other applications in this domain include neural machine translation [68], text classification [69], and sequence-to-graph learning [70].
- **Social networks and recommender systems:** A usefulness of graph-based methods can be seen in recommendation systems where we need to recommend a new item to each user while the users and items are represented as nodes and the recommendations are represented as graph edges [71,58,72]. In social networks, the users are denoted as nodes and the interaction between them or their influence on each other is modeled as graph edges [56,9,73].
- **Computer vision:** Graph-based methods have been widely used in computer vision tasks. Example computer vision problems that can be modeled as graph-based tasks are image classification [74,75], skeleton-based human action recog-

Table 4.1 Summary of benchmark graph data sets.

Category	Name	#Graphs	#Nodes(Avg.)	#Edges(Avg.)	#Features	#Classes	Year
Citation networks	Cora [67]	1	2708	5429	1433	7	2008
	Citeseer [67]	1	3327	4732	3703	6	2008
	Pubmed [67]	1	19717	44338	500	3	2008
	DBLP (v11) [86]	1	4107340	36624464	–	–	2008
	OAG v1 [86,87]	2	320,963,344	64,639,608	–	–	2017
Social networks	BlogCatalog [88]	1	10312	333983	–	39	2009
	Digg [89]	–	279.630	1,548,126	–	–	2012
	Twitter-Higgs [90]	–	456,631	14,855,875	–	–	2013
	Weibo [91]	–	1,776,950	308,489,739	–	–	2013
	COLLAB [92]	1	235,868	235,868	–	–	2015
	IMDB-Binary [92]	1000	19.8	96.5	–	2	2015
	IMDB-Multi [92]	1500	13.0	65.9	–	3	2015
	Reddit [56]	1	232965	11606919	602	41	2017
Computer vision	MNIST [93]	70000	40-75	–	–	–	1998
	CIFAR10 [94]	60000	85-150	–	–	–	2009
	Fingerprint [95]	2800	5.42	4.42	–	15	2008
	MSRC9 [85]	221	40.58	97.94	–	8	2016
	MSRC21 [85]	563	77.52	198.32	–	20	2016
	MSRC21C [85]	209	40.28	96.60	–	20	2016
	FIRSTMMDB [84,85]	41	1377.27	3074.10	–	11	2013
	NTU-RGB+D [81]	56,880	25	15000	–	60	2016
	Kinetics-skeleton [82]	300,000	18	10800	–	400	2017
Biochemical graphs	MUTAG [96]	188	17.93	19.79	7	2	1991
	PTC [97]	344	25.5	–	19	2	2003
	D&D [98]	1178	284.31	715.65	82	2	2003
	ENZYMES [99]	600	32.6	62.1	–	6	2004
	PROTEIN [100]	1113	39.06	72.81	4	2	2005
	NCI1 [101]	4110	29.87	32.30	37	2	2008
	QM9 [102]	133885	–	–	–	–	2014
	PPI [103]	24	56944	818716	50	121	2017
	ZINC [104,105]	249,456	23.1	24.9	–	12	2018
	Alchemy [106]	119487	–	–	–	–	2019

nition [76], semantic segmentation [77], human object interaction [78], and 3D point clouds segmentation and classification [79,80]. In these problems, the objects in the images are modeled by graphs representing the semantic relationships between them and the graph-based methods learn to extract features from these graphs for different purposes. The well-known data sets for skeleton-based human action recognition are NTU-RGB+D [81] and kinetics-skeletons data sets [82]. NTU-RGB+D data set is one of the largest multimodality action recognition data sets, which contains both video and 3D skeletons data that are captured by the Microsoft Kinect-v2 camera. Kinetics-skeleton data set is also a very large action recognition data set that contains YouTube video clips of different human actions. The 2D joints' coordinates on every frame are estimated using the OpenPose toolbox [83]. The widely used data set containing 3D point clouds is FIRSTMMDB [84,85] and the MSRC data sets [85] are used for semantic image processing.

- **Chemistry:** In the fields of biology and chemistry, the graphs are built by using molecules or proteins or atoms as nodes and the interaction between them as edges. In such problems, the goal is to learn molecular features, study the biochemical graph properties, and capture molecular or protein-protein interactions. GAEs can be used for solving these tasks, which are formed by an encoder and a decoder subnetworks. In the encoder part, the model employs graph convolution layers to learn graph embedding and the graph structure will be reconstructed in the decoder part by predicting edges between nodes.

There exist available open-source libraries for GNNs, which facilitate research in this area, allowing for easy model creation and reproduction of GNN results on benchmark data sets:

- **Pytorch geometric** [107] is an open-source geometric learning library implemented in Pytorch including implementations of many GNN methods.
- **Deep Graph Library (DGL)** [108] is an open-source library, which allows for fast implementation of many GNN methods in PyTorch and MXNet platforms.
- **Jraph**[2] is a lightweight open-source library implemented in JAX and provides implementation of GNN models, graph data structures, and a set of utilities for working with graphs.

4.7 Conclusion

In this chapter, we introduced graph convolutional networks and provided an overview of the most representative methods. We divided the methods in four main categories, namely spectral, spatial, attention, and sampling based methods. We started describing the concept of graph convolution in Section 4.2 by introducing the spectral graph convolution, which has a solid background in graph signal processing. Learning the feature representations of the nodes and graphs in spectral GCN methods depends heavily on the structure of the input graph, which makes the generalization of these methods to unseen data infeasible. Greatly inspired by convolutional neural networks where the spatial structure of the input grid structured data is considered, the spatial GCN methods introduced in Section 4.3 define the graph convolution operation by means of message passing through the graph edges. In Section 4.4, we described how the attention mechanisms can be used in GCN methods to improve their performance. Several methods have been proposed recently to improve the computational efficiency and scalability of the existing GCN methods. In Section 4.5, we introduced the most representative works, which are capable to train on large graph data sets efficiently and effectively. Finally in Section 4.6, a list of benchmark graph data sets used in different learning tasks was provided, along with

[2] https://github.com/deepmind/jraph.

a list open-source libraries, which facilitate research in this area by providing the implementations of methods and useful graph utility functions.

References

[1] T. Luong, H. Pham, C.D. Manning, Effective approaches to attention-based neural machine translation, in: Conference on Empirical Methods in Natural Language Processing, 2015.

[2] Y. Wu, M. Schuster, Z. Chen, Q.V. Le, M. Norouzi, W. Macherey, M. Krikun, Y. Cao, Q. Gao, K. Macherey, et al., Google's neural machine translation system: bridging the gap between human and machine translation, arXiv preprint, arXiv:1609.08144, 2016.

[3] G. Hinton, L. Deng, D. Yu, G.E. Dahl, A.-r. Mohamed, N. Jaitly, A. Senior, V. Vanhoucke, P. Nguyen, T.N. Sainath, et al., Deep neural networks for acoustic modeling in speech recognition: the shared views of four research groups, IEEE Signal Processing Magazine 29 (6) (2012) 82–97.

[4] J. Redmon, S. Divvala, R. Girshick, A. Farhadi, You only look once: unified, real-time object detection, in: Conference on Computer Vision and Pattern Recognition, 2016.

[5] S. Ren, K. He, R. Girshick, J. Sun, Faster r-cnn: towards real-time object detection with region proposal networks, IEEE Transactions on Pattern Analysis and Machine Intelligence 39 (6) (2016) 1137–1149.

[6] F. Scarselli, M. Gori, A.C. Tsoi, M. Hagenbuchner, G. Monfardini, The graph neural network model, IEEE Transactions on Neural Networks 20 (1) (2008) 61–80.

[7] C. Gallicchio, A. Micheli, Graph echo state networks, in: International Joint Conference on Neural Networks, 2010, pp. 1–8.

[8] Y. Li, D. Tarlow, M. Brockschmidt, R. Zemel, Gated graph sequence neural networks, arXiv preprint, arXiv:1511.05493, 2015.

[9] H. Dai, Z. Kozareva, B. Dai, A. Smola, L. Song, Learning steady-states of iterative algorithms over graphs, in: International Conference on Machine Learning, 2018, pp. 1106–1114.

[10] J. Bruna, W. Zaremba, A. Szlam, Y. LeCun, Spectral networks and locally connected networks on graphs, in: International Conference on Learning Representations, 2014.

[11] M. Defferrard, X. Bresson, P. Vandergheynst, Convolutional neural networks on graphs with fast localized spectral filtering, in: Advances in Neural Information Processing Systems, 2016.

[12] T.N. Kipf, M. Welling, Semi-supervised classification with graph convolutional networks, in: International Conference on Learning Representations, 2017.

[13] J. Atwood, D. Towsley, Diffusion-convolutional neural networks, in: Advances in Neural Information Processing Systems, 2016, pp. 1993–2001.

[14] M. Niepert, M. Ahmed, K. Kutzkov, Learning convolutional neural networks for graphs, in: International Conference on Machine Learning, 2016.

[15] J. Gilmer, S.S. Schoenholz, P.F. Riley, O. Vinyals, G.E. Dahl, Neural message passing for quantum chemistry, in: International Conference on Machine Learning, 2017.

[16] Y. Li, O. Vinyals, C. Dyer, R. Pascanu, P. Battaglia, Learning deep generative models of graphs, arXiv preprint, arXiv:1803.03324, 2018.

[17] J. You, R. Ying, X. Ren, W.L. Hamilton, J. Leskovec, Graphrnn: a deep generative model for graphs, arXiv preprint, arXiv:1802.08773, 2018.

[18] M. Simonovsky, N. Komodakis Graphvae, Towards generation of small graphs using variational autoencoders, in: International Conference on Artificial Neural Networks, 2018.
[19] T. Ma, J. Chen, C. Xiao, Constrained generation of semantically valid graphs via regularizing variational autoencoders, in: Advances in Neural Information Processing Systems, 2018.
[20] S. Cao, W. Lu, Q. Xu, Deep neural networks for learning graph representations, in: AAAI Conference on Artificial Intelligence, 2016.
[21] D. Wang, P. Cui, W. Zhu, Structural deep network embedding, in: International Conference on Knowledge Discovery and Data Mining, 2016.
[22] T.N. Kipf, M. Welling, Variational graph auto-encoders, arXiv preprint, arXiv:1611.07308, 2016.
[23] S. Pan, R. Hu, G. Long, J. Jiang, L. Yao, C. Zhang, Adversarially regularized graph autoencoder for graph embedding, in: International Joint Conference on Artificial Intelligence, 2018.
[24] B. Perozzi, R. Al-Rfou, S. Skiena, Deepwalk: online learning of social representations, in: International Conference on Knowledge Discovery and Data Mining, 2014.
[25] A. Grover, J. Leskovec, node2vec: scalable feature learning for networks, in: International Conference on Knowledge Discovery and Data Mining, 2016.
[26] J. Tang, M. Qu, M. Wang, M. Zhang, J. Yan, Q. Mei Line, Large-scale information network embedding, in: International Conference on World Wide Web, 2015.
[27] G. Hinton, S. Roweis, Visualizing data using t-SNE, in: Advances in Neural Information Processing Systems, 2002, pp. 833–840.
[28] L. der Maaten, G. Hinton, Visualizing data using t-SNE, Journal of Machine Learning Research 9 (11) (2008) 2579–2605.
[29] L.V.D. Maaten, Learning a parametric embedding by preserving local structure, in: Artificial Intelligence and Statistics, 2009, pp. 384–391.
[30] S.T. Roweis, L.K. Saul, Nonlinear dimensionality reduction by locally linear embedding, Science 290 (5500) (2000) 2323–2326.
[31] B. Shaw, T. Jebara, Structure preserving embedding, in: International Conference on Machine Learning, 2009.
[32] M. Ou, P. Cui, J. Pei, Z. Zhang, W. Zhu, Asymmetric transitivity preserving graph embedding, in: International Conference on Knowledge Discovery and Data Mining, 2016.
[33] S. Yan, D. Xu, B. Zhang, H. Jiang Zhang, Q. Yang, S. Lin, Graph embedding and extensions: a general framework for dimensionality reduction, IEEE Transactions on Pattern Analysis and Machine Intelligence 29 (1) (2007) 40–51.
[34] V. Mygdalis, A. Iosifidis, A. Tefas, I. Pitas, Graph embedded one-class classifiers for media data classification, Pattern Recognition 60 (2017) 585–595.
[35] A. Iosifidis, A. Tefas, I. Pitas, Graph embedded extreme learning machine, IEEE Transactions on Cybernetics 46 (1) (2016) 311–324.
[36] P. Goyal, E. Ferrara, Graph embedding techniques, applications, and performance: a survey, Knowledge-based Systems 151 (7) (2018) 78–94.
[37] H. Cai, V.W. Zheng, K.C.-C. Chang, A comprehensive survey of graph embedding: problems, techniques, and applications, IEEE Transactions on Knowledge and Data Engineering 30 (2018) 1616–1637.
[38] D.I. Shuman, S.K. Narang, P. Frossard, A. Ortega, P. Vandergheynst, The emerging field of signal processing on graphs: extending high-dimensional data analysis to networks and other irregular domains, IEEE Signal Processing Magazine 30 (3) (2013) 83–98.

[39] A. Sandryhaila, J.M. Moura, Discrete signal processing on graphs, IEEE Transactions on Signal Processing 61 (7) (2013) 1644–1656.

[40] S. Chen, R. Varma, A. Sandryhaila, J. Kovačević, Discrete signal processing on graphs: sampling theory, IEEE Transactions on Signal Processing 63 (24) (2015) 6510–6523.

[41] A. Ancona, R. Lyons, Y. Peres, Crossing estimates and convergence of Dirichlet functions along random walk and diffusion paths, Annals of Probability (1999) 970–989.

[42] T. Hutchcroft, Harmonic Dirichlet functions on planar graphs, Discrete & Computational Geometry 61 (3) (2019) 479–506.

[43] D.K. Hammond, P. Vandergheynst, R. Gribonval, Wavelets on graphs via spectral graph theory, Applied and Computational Harmonic Analysis 30 (2) (2011) 129–150.

[44] N. Heidari, A. Iosifidis, Progressive graph convolutional networks for semi-supervised node classification, IEEE Access (2021).

[45] R. Li, S. Wang, F. Zhu, J. Huang, Adaptive graph convolutional neural networks, in: AAAI Conference on Artificial Intelligence, 2018.

[46] C. Zhuang, Q. Ma, Dual graph convolutional networks for graph-based semi-supervised classification, in: International Conference on World Wide Web, 2018, pp. 499–508.

[47] D.K. Duvenaud, D. Maclaurin, J. Iparraguirre, R. Bombarell, T. Hirzel, A. Aspuru-Guzik, R.P. Adams, Convolutional networks on graphs for learning molecular fingerprints, in: Advances in Neural Information Processing Systems, 2015.

[48] Y. Li, D. Tarlow, M. Brockschmidt, R.S. Zemel, Gated graph sequence neural networks, in: Y. Bengio, Y. LeCun (Eds.), International Conference on Learning Representations, 2016.

[49] P. Battaglia, R. Pascanu, M. Lai, D. Jimenez Rezende, et al., Interaction networks for learning about objects, relations and physics, Advances in Neural Information Processing Systems 29 (2016) 4502–4510.

[50] S. Kearnes, K. McCloskey, M. Berndl, V. Pande, P. Riley, Molecular graph convolutions: moving beyond fingerprints, Journal of Computer-Aided Molecular Design 30 (8) (2016) 595–608.

[51] K.T. Schütt, F. Arbabzadah, S. Chmiela, K.R. Müller, A. Tkatchenko, Quantum-chemical insights from deep tensor neural networks, Nature Communications 8 (1) (2017) 1–8.

[52] M. Defferrard, X. Bresson, P. Vandergheynst, Convolutional neural networks on graphs with fast localized spectral filtering, Advances in Neural Information Processing Systems 29 (2016) 3844–3852.

[53] H. Gao, Z. Wang, S. Ji, Large-scale learnable graph convolutional networks, in: International Conference on Knowledge Discovery & Data Mining, 2018.

[54] P. Velickovic, G. Cucurull, A. Casanova, A. Romero, P. Liò, Y. Bengio, Graph attention networks, in: International Conference on Learning Representations, 2018.

[55] A. Vaswani, N. Shazeer, N. Parmar, J. Uszkoreit, L. Jones, A.N. Gomez, L. u. Kaiser, I. Polosukhin, Attention is all you need, in: Advances in Neural Information Processing Systems, vol. 30, 2017, pp. 5998–6008.

[56] W. Hamilton, Z. Ying, J. Leskovec, Inductive representation learning on large graphs, in: Advances in Neural Information Processing Systems, 2017.

[57] J. Chen, T. Ma, C. Xiao, Fastgcn: fast learning with graph convolutional networks via importance sampling, in: International Conference on Learning Representations, 2018.

[58] R. Ying, R. He, K. Chen, P. Eksombatchai, W.L. Hamilton, J. Leskovec, Graph convolutional neural networks for web-scale recommender systems, in: International Conference on Knowledge Discovery & Data Mining, 2018.

[59] J. Chen, J. Zhu, L. Song, Stochastic training of graph convolutional networks with variance reduction, in: International Conference on Machine Learning, 2018.
[60] W. Huang, T. Zhang, Y. Rong, J. Huang, Adaptive sampling towards fast graph representation learning, Advances in Neural Information Processing Systems 31 (2018) 4558–4567.
[61] H. Robbins, S. Monro, A stochastic approximation method, The Annals of Mathematical Statistics (1951) 400–407.
[62] H. Zeng, H. Zhou, A. Srivastava, R. Kannan, V. Prasanna, Accurate, efficient and scalable graph embedding, in: International Parallel and Distributed Processing Symposium, 2019, pp. 462–471.
[63] W.-L. Chiang, X. Liu, S. Si, Y. Li, S. Bengio, C.-J. Hsieh, Cluster-gcn: an efficient algorithm for training deep and large graph convolutional networks, in: International Conference on Knowledge Discovery & Data Mining, 2019.
[64] G. Karypis, V. Kumar, A fast and high quality multilevel scheme for partitioning irregular graphs, SIAM Journal on Scientific Computing 20 (1) (1998) 359–392.
[65] I.S. Dhillon, Y. Guan, B. Kulis, Weighted graph cuts without eigenvectors a multilevel approach, IEEE Transactions on Pattern Analysis and Machine Intelligence 29 (11) (2007) 1944–1957.
[66] C. Morris, N.M. Kriege, F. Bause, K. Kersting, P. Mutzel, M. Neumann, Tudataset: a collection of benchmark datasets for learning with graphs, arXiv preprint, arXiv:2007.08663, 2020.
[67] P. Sen, G. Namata, M. Bilgic, L. Getoor, B. Galligher, T. Eliassi-Rad, Collective classification in network data, AI Magazine 29 (3) (2008) 93.
[68] J. Bastings, I. Titov, W. Aziz, D. Marcheggiani, K. Sima'an, Graph convolutional encoders for syntax-aware neural machine translation, in: Conference on Empirical Methods in Natural Language Processing, 2017, pp. 1957–1967.
[69] L. Yao, C. Mao, Y. Luo, Graph convolutional networks for text classification, in: AAAI Conference on Artificial Intelligence, 2019.
[70] D. Beck, G. Haffari, T. Cohn, Graph-to-sequence learning using gated graph neural networks, in: Association for Computational Linguistics, 2018, pp. 273–283.
[71] R.v.d. Berg, T.N. Kipf, M. Welling, Graph convolutional matrix completion, arXiv preprint, arXiv:1706.02263, 2017.
[72] F. Monti, M.M. Bronstein, X. Bresson, Geometric matrix completion with recurrent multi-graph neural networks, in: Conference on Neural Information Processing Systems, 2017, pp. 3697–3707.
[73] J. Zhang, X. Shi, J. Xie, H. Ma, I. King, D. Yeung Gaan, Gated attention networks for learning on large and spatiotemporal graphs, in: Conference on Uncertainty in Artificial Intelligence, 2018, pp. 339–349.
[74] V.G. Satorras, J.B. Estrach, Few-shot learning with graph neural networks, in: International Conference on Learning Representations, 2018.
[75] M. Kampffmeyer, Y. Chen, X. Liang, H. Wang, Y. Zhang, E.P. Xing, Rethinking knowledge graph propagation for zero-shot learning, in: IEEE Conference on Computer Vision and Pattern Recognition, 2019.
[76] S. Yan, Y. Xiong, D. Lin, Spatial temporal graph convolutional networks for skeleton-based action recognition, in: AAAI Conference on Artificial Intelligence, 2018.
[77] X. Qi, R. Liao, J. Jia, S. Fidler, R. Urtasun, 3d graph neural networks for rgbd semantic segmentation, in: IEEE International Conference on Computer Vision, 2017, pp. 5199–5208.

[78] S. Qi, W. Wang, B. Jia, J. Shen, S.-C. Zhu, Learning human-object interactions by graph parsing neural networks, in: Proceedings of the European Conference on Computer Vision (ECCV), 2018, pp. 401–417.

[79] Y. Wang, Y. Sun, Z. Liu, S.E. Sarma, M.M. Bronstein, J.M. Solomon, Dynamic graph cnn for learning on point clouds, ACM Transactions on Graphics (ToG) 38 (5) (2019) 1–12.

[80] L. Landrieu, M. Simonovsky, Large-scale point cloud semantic segmentation with superpoint graphs, in: IEEE Conference on Computer Vision and Pattern Recognition, 2018, pp. 4558–4567.

[81] A. Shahroudy, J. Liu, T.-T. Ng, G. Wang, Ntu rgb+ d: a large scale dataset for 3d human activity analysis, in: IEEE Conference on Computer Vision and Pattern Recognition, 2016, pp. 1010–1019.

[82] W. Kay, J. Carreira, K. Simonyan, B. Zhang, C. Hillier, S. Vijayanarasimhan, F. Viola, T. Green, T. Back, P. Natsev, et al., The kinetics human action video dataset, arXiv preprint, arXiv:1705.06950, 2017.

[83] Z. Cao, T. Simon, S.-E. Wei, Y. Sheikh, Realtime multi-person 2d pose estimation using part affinity fields, in: IEEE Conference on Computer Vision and Pattern Recognition, 2017, pp. 7291–7299.

[84] M. Neumann, P. Moreno, L. Antanas, R. Garnett, K. Kersting, Graph kernels for object category prediction in task-dependent robot grasping, in: Online Proceedings of the Eleventh Workshop on Mining and Learning with Graphs, 2013, pp. 0–6.

[85] M. Neumann, R. Garnett, C. Bauckhage, K. Kersting, Propagation kernels: efficient graph kernels from propagated information, Machine Learning 102 (2) (2016) 209–245.

[86] J. Tang, J. Zhang, L. Yao, J. Li, L. Zhang, Z. Su, Arnetminer: extraction and mining of academic social networks, in: International Conference on Knowledge Discovery and Data Mining, 2008.

[87] A. Sinha, Z. Shen, Y. Song, H. Ma, D. Eide, B.-J. Hsu, K. Wang, An overview of Microsoft academic service (mas) and applications, in: International Conference on World Wide Web, 2015, pp. 243–246.

[88] L. Tang, H. Liu, Relational learning via latent social dimensions, in: International Conference on Knowledge Discovery and Data Mining, 2009.

[89] T. Hogg, K. Lerman, Social dynamics of digg, EPJ Data Science 1 (1) (2012) 5.

[90] M. De Domenico, A. Lima, P. Mougel, M. Musolesi, The anatomy of a scientific rumor, Scientific Reports 3 (2013) 2980.

[91] J. Zhang, B. Liu, J. Tang, T. Chen, J. Li, Social influence locality for modeling retweeting behaviors, in: International Joint Conference on Artificial Intelligence, 2013.

[92] P. Yanardag, S. Vishwanathan, Deep graph kernels, in: International Conference on Knowledge Discovery and Data Mining, 2015.

[93] Y. LeCun, L. Bottou, Y. Bengio, P. Haffner, Gradient-based learning applied to document recognition, Proceedings of the IEEE 86 (11) (1998) 2278–2324.

[94] A. Krizhevsky, G. Hinton, et al., Learning multiple layers of features from tiny images, 2009.

[95] K. Riesen, H. Bunke, Iam graph database repository for graph based pattern recognition and machine learning, in: Joint IAPR International Workshops on Statistical Techniques in Pattern Recognition (SPR) and Structural and Syntactic Pattern Recognition (SSPR), 2008, pp. 287–297.

[96] A.K. Debnath, R.L. Lopez de Compadre, G. Debnath, A.J. Shusterman, C. Hansch, Structure-activity relationship of mutagenic aromatic and heteroaromatic nitro com-

pounds. Correlation with molecular orbital energies and hydrophobicity, Journal of Medicinal Chemistry 34 (2) (1991) 786–797.

[97] H. Toivonen, A. Srinivasan, R.D. King, S. Kramer, C. Helma, Statistical evaluation of the predictive toxicology challenge 2000–2001, Bioinformatics 19 (10) (2003) 1183–1193.

[98] P.D. Dobson, A.J. Doig, Distinguishing enzyme structures from non-enzymes without alignments, Journal of Molecular Biology 330 (4) (2003) 771–783.

[99] I. Schomburg, A. Chang, C. Ebeling, M. Gremse, C. Heldt, G. Huhn, D. Schomburg, Brenda, the enzyme database: updates and major new developments, Nucleic Acids Research 32 (suppl_1) (2004) D431–D433.

[100] K.M. Borgwardt, C.S. Ong, S. Schönauer, S. Vishwanathan, A.J. Smola, H.-P. Kriegel, Protein function prediction via graph kernels, Bioinformatics 21 (suppl_1) (2005) i47–i56.

[101] N. Wale, I.A. Watson, G. Karypis, Comparison of descriptor spaces for chemical compound retrieval and classification, Knowledge and Information Systems 14 (3) (2008) 347–375.

[102] R. Ramakrishnan, P.O. Dral, M. Rupp, O.A. Von Lilienfeld, Quantum chemistry structures and properties of 134 kilo molecules, Scientific Data 1 (1) (2014) 1–7.

[103] M. Zitnik, J. Leskovec, Predicting multicellular function through multi-layer tissue networks, Bioinformatics 33 (14) (2017) i190–i198.

[104] V.P. Dwivedi, C.K. Joshi, T. Laurent, Y. Bengio, X. Bresson, Benchmarking graph neural networks, arXiv preprint, arXiv:2003.00982, 2020.

[105] W. Jin, R. Barzilay, T. Jaakkola, Junction tree variational autoencoder for molecular graph generation, arXiv preprint, arXiv:1802.04364, 2018.

[106] G. Chen, P. Chen, C.-Y. Hsieh, C.-K. Lee, B. Liao, R. Liao, W. Liu, J. Qiu, Q. Sun, J. Tang, et al., Alchemy: a quantum chemistry dataset for benchmarking AI models, arXiv preprint, arXiv:1906.09427, 2019.

[107] M. Fey, J. Eric Lenssen, F. Weichert, H. Müller Splinecnn, Fast geometric deep learning with continuous b-spline kernels, in: IEEE Conference on Computer Vision and Pattern Recognition, 2018, pp. 869–877.

[108] M. Wang, L. Yu, D. Zheng, Q. Gan, Y. Gai, Z. Ye, M. Li, J. Zhou, Q. Huang, C. Ma, et al., Deep graph library: towards efficient and scalable deep learning on graphs, arXiv preprint, arXiv:1909.01315, 2019.

CHAPTER

Recurrent neural networks

5

Avraam Tsantekidis, Nikolaos Passalis, and Anastasios Tefas
Department of Informatics, Aristotle University of Thessaloniki, Thessaloniki, Greece

5.1 Introduction

Melodies, written or spoken languages and stock price charts are data, which despite their apparent differences share a common attribute. Their structure is different from other data such as static images (e.g., handwritten digits [1], clothing items [2]) and multivariate measurements of objects (e.g., petal lengths of flowers [3]). That differentiating factor is that they are defined in a sequential manner. Both music and language make sense only when viewed in a certain order. From our perspective, an image can make sense when viewed as a whole or rotated by an angle, but listening to the notes of a song played all at once or with varying timing would not make sense. Many types of data, such as natural language and sound tracks, contain additional information in the way they are ordered, forming a *sequence* of data points. This type of sequence is usually referred to as a *time series*.

Deep Learning (DL) approaches such as neural networks often require a constant input size that is determined during the definition of the model and the data that are used. A naive approach to utilizing neural networks for analyzing such sequences is to process them in windows. Given a *time series* $\mathbf{X} = \{x_0, x_1, \dots, x_n\}$ where n is the time series length and x_i is the data of each time step, we can sample ranges of sequential samples, concatenate and flatten them in a deterministic way. This approach should yield 1-dimensional vectors $\mathbf{x}_{ij}$ that contain all the data points between time steps i and j.

A very strong prior of time series data is the order of the data points. When this sequence is flattened to a 1-d vector and fed all at once to a feedforward neural network, the ordering information is preserved since all sequences are flattened in the same manner, but it is not directly used in the training process. This approach, although correct, can be considered as a missed opportunity. The main intuition inferred here is that the analysis of a sequence can benefit from a hierarchical approach through time. Building upon an internal state as each step of the sequence is processed, a rich representation can be developed. Adding this prior to a Machine Learning (ML) model can be limiting in terms of "degrees of freedom" in how a sequence is processed, but also can be beneficial in matching what is a more biologically plausible way that the human brain processes data sequences as well. This idea of a recursive

Deep Learning for Robot Perception and Cognition. https://doi.org/10.1016/B978-0-32-385787-1.00010-5

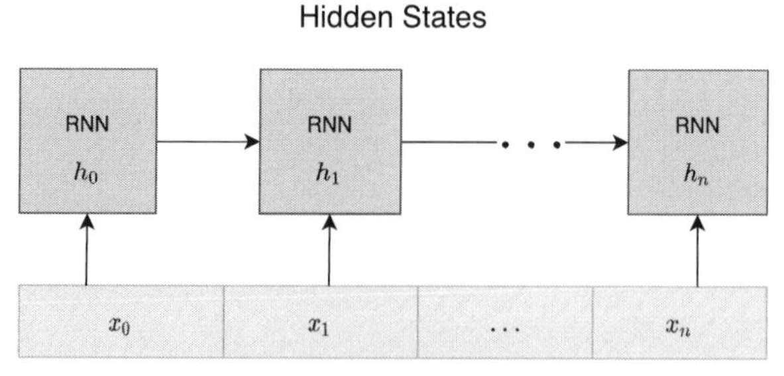

FIGURE 5.1

Basic Recurrent Neural Network (RNN) functionality. On each time step t the RNN observes an input $\mathbf{x}_t$ and the previous hidden state $\mathbf{h}_{t-1}$ and updates its hidden states to $\mathbf{h}_t$.

or recurrent approach in which artificial neurons process information can be found in multiple publications [4–6] dating back to 1943.

Another more practical reason to have a hierarchical approach to process sequences is the removal of the fixed size input limitation. Many types of sequence data, such as sounds and language, vary in length, so processing them by sampling fixed-sized $\mathbf{x}_{ij}$ vectors can present a learning and engineering challenge in itself. Being able to hierarchically process each step removes this limitation, allowing the parsing of sequences of unspecified length. By also having an internal state that is updated at each step throughout the sequence, a rich representation of the current and past states can be built as a fixed-length vector that can be separately utilized at each step by simpler models requiring such a fixed-sized input.

Indeed, the most common way to process sequence data using deep learning is by employing Recurrent Neural Networks (RNNs). As the name suggests, RNNs have a recurrent aspect in their operation, which is the repeated processing and updating of their internal state. Using their internal representation, RNNs are capable of exhibiting dynamic temporal behavior and processing sequences of arbitrary length, while being able to model the long-term relationships between distant points in a sequence. Their internal state is a vector of values $\mathbf{h}$ colloquially referred to as their hidden state. The hidden state is the RNN's only way of detecting patterns across the data sequence, so all latent information extracted at every step is included in that hidden state. In Fig. 5.1 a visualization of the basic RNN functionality is shown.

The rest of the chapter is structured as follows. In the next section, the simplest recurrent neural network structure is presented and its inner structure is examined, helping to build intuition around the functionality of neural networks working with sequences. Next, several improved versions of recurrent neural networks are presented, explaining how they overcome the problems of the vanilla RNN and their inner workings. Finally, applications of RNNs in various domains are presented and concluding remarks are drawn.

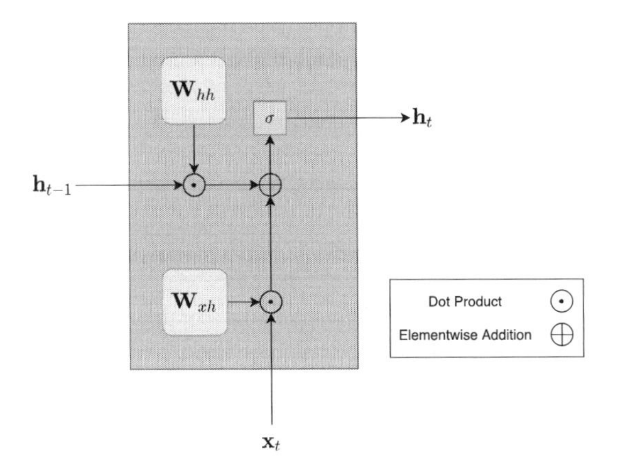

FIGURE 5.2

Vanilla RNN's internal computations. $\mathbf{W}_{hh}$ denotes the hidden-to-hidden weight matrix, $\mathbf{W}_{xh}$ denotes the input-to-hidden weight matrix and σ denotes the nonlinear activation.

5.2 Vanilla RNN

The first iteration of recurrent neural networks directly applied the logic of updating a hidden state based on the input of the current step and the hidden state of the previous step. The hidden state is usually initialized as a null vector, meaning all elements are zero.[1] The main trainable parameters of a "vanilla" RNN are the two weight matrices $\mathbf{W}_{hh} \in \mathbb{R}^{m_h \times m_h}$ and $\mathbf{W}_{xh} \in \mathbb{R}^{m_x \times m_h}$, where m_h is the number of hidden neurons of the RNN and m_x is the number of features of the input, that is, the dimensionality of the time series measurements. These parameter matrices are used to transform the previous hidden activation and the input, respectively.

The input sequence matrix is denoted $\mathbf{X} \in \mathbb{R}^{n \times m_x}$, where n is the number of steps in the sequence. At each processing step of the sequence a nonlinear transformation is performed on the current input step $\mathbf{x}_t \in \mathbb{R}^{m_x}$ using $\mathbf{W}_{xh}$ and another one on $\mathbf{h}_{t-1}$ using $\mathbf{W}_{hh}$. The result of both transformations has the same size m_h, so the two results are combined using an elementwise summation as shown in Fig. 5.2. This can also be described with notation as

$$\mathbf{h}_t = \sigma(\mathbf{h}_{t-1}\mathbf{W}_{hh} + \mathbf{x}_t\mathbf{W}_{xh}), \tag{5.1}$$

where we denote the dot product with the dot symbol $\odot$ and the elementwise summation as a simple sum, since their matching size makes the operation broadcastable. The symbol σ denotes a nonlinear activation function, which is usually the *tanh* or the *sigmoid* function.

[1] The initial states of an RNN can be treated as parameters and trained as every other parameter in a neural network, but the improvement potential is not significant, thus it has not gained traction as a technique.

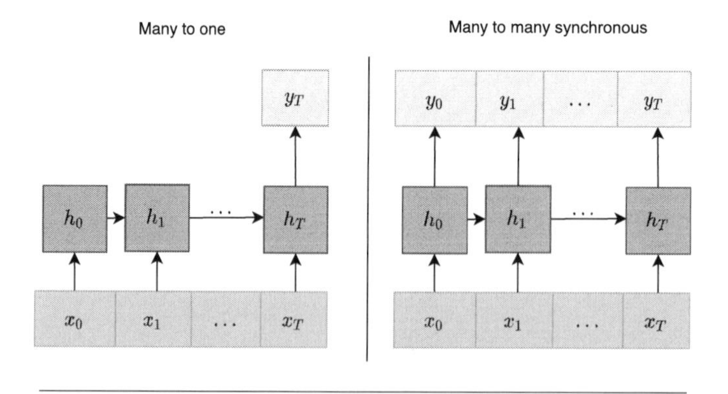

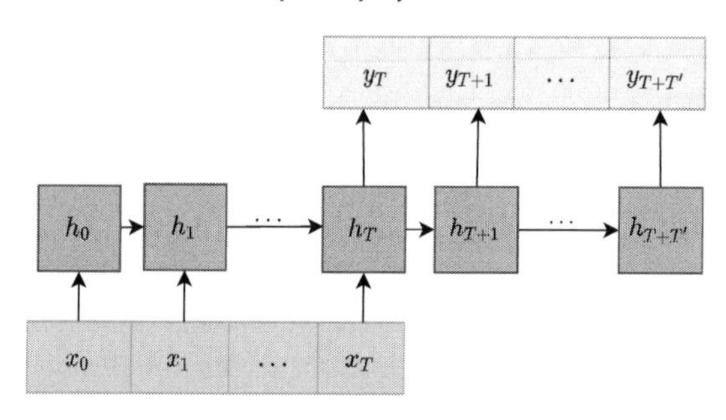

FIGURE 5.3

Modes of operation for an RNN. Green boxes (mid gray in print version) are the input sequence, red boxes (gray in print version) are the hidden activations, yellow boxes (light gray in print version) are the predictions made by the networks. "Many-to-one" parses a sequence from the beginning to the end, and at the final step makes a single prediction. "Many-to-many synchronous" makes a prediction for every step of the sequence as they are processed by the RNN. It is also referred to as sequence-to-sequence. "Many-to-many asynchronous" first processes the whole sequence and then starts producing a sequence for its output.

There are multiple avenues to utilize RNNs in the hierarchical processing of sequential data. In some cases, the requirement of inference is a prediction of some label or value for every step while others require one prediction at the end of a sequence. Some examples of the methods to utilize an RNN are shown in Fig. 5.3. The potential uses for RNNs vary from simply parsing a sequence and predicting something after processing it completely (e.g., classifying music genres, determining a comment's sentiment), to making a prediction for every step of the sequence processed (e.g., detecting anomalies throughout a time series, autoregressive prediction of a sequence), and generating a new sequence at the end of the current sequence (e.g., generating text after an initial prompt).

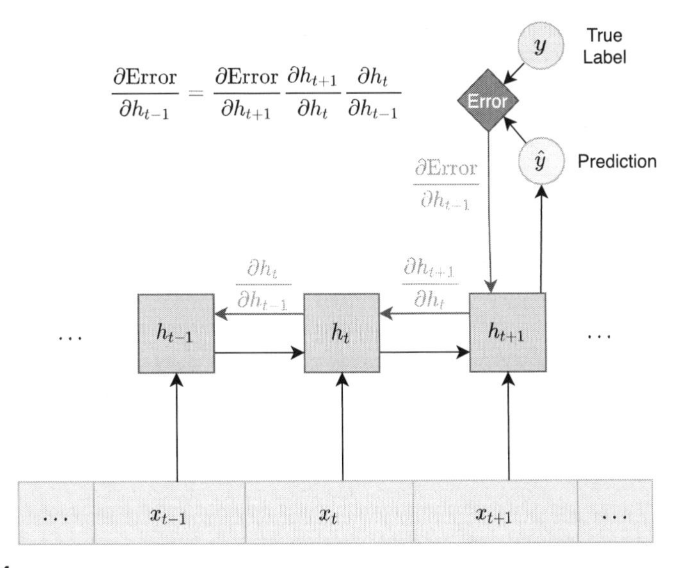

FIGURE 5.4

Backpropagation Through Time (BPTT).

The diagrams of the RNN operation presented here are unrolled, but it should not confuse the reader as the RNN functions are modeled as a directed cycle graph and are capable to process an arbitrary number of steps. Unrolling helps to better understand their operations and how their training works. To train an RNN, the backpropagation Through Time (BPTT) is utilized, which is first proposed in [7,8]. Similar to back-propagation, BPTT utilizes the chain rule to efficiently propagate the error gradient to the trainable weight to apply gradient descent. Using the unrolled RNN computation graph, the gradients are propagated through the computation steps of the hidden state as shown in Fig. 5.4. Formally, the gradient of the activation at step n w.r.t the error (loss) function $\mathcal{L}$ is defined as

$$\frac{\partial \mathcal{L}(T)}{\partial h_t} = \frac{\partial h_{t+1}}{\partial h_t} \cdots \frac{\partial h_T}{\partial h_{T-1}} \frac{\partial \mathcal{L}(T)}{\partial h_T} \tag{5.2}$$

where $\mathcal{L}(T)$ is the loss value at step T. A partial derivative matrix can also be referred to as the *Jacobian*. To utilize the error propagation and train the network by executing gradient descent, the gradients of the weight parameters $\mathbf{W}_{hh}$ are needed. The accumulated gradients that the weights of RNN incur across all time steps (considering a

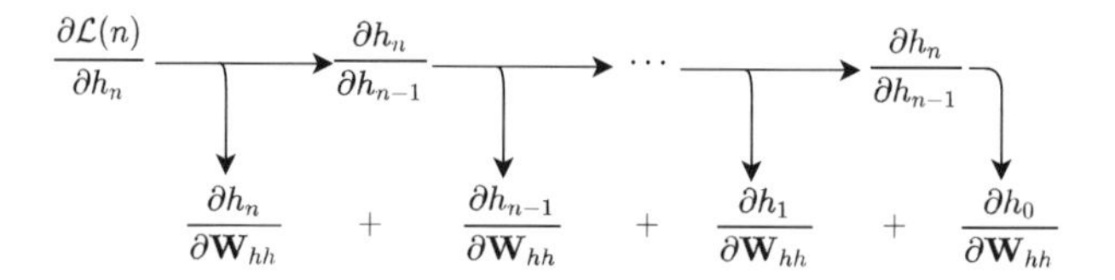

FIGURE 5.5

Gradient accumulation of RNN parameters in backpropagation through time of the loss from a single step.

total of T time steps) are calculated as

$$\begin{aligned}\frac{\partial \sum_{i=0}^{n} \mathcal{L}(n)}{\partial \mathbf{W}_{hh}} = & \frac{\partial \mathcal{L}(0)}{\partial h_0} \frac{\partial h_0}{\partial \mathbf{W}_{hh}} + \\ & \frac{\partial \mathcal{L}(1)}{\partial h_1} (\frac{\partial h_0}{\partial \mathbf{W}_{hh}} + \frac{\partial h_1}{\partial h_0} \frac{\partial h_0}{\partial \mathbf{W}_{hh}}) + \\ & \vdots \\ & \frac{\partial \mathcal{L}(T)}{\partial h_T} (\frac{\partial h_T}{\partial \mathbf{W}_{hh}} + \frac{\partial h_T}{\partial h_{T-1}} (\frac{\partial h_{T-1}}{\partial \mathbf{W}_{hh}} + \frac{\partial h_{T-1}}{\partial h_{T-2}} (\cdots \frac{\partial h_0}{\partial \mathbf{W}_{hh}}))),\end{aligned} \tag{5.3}$$

where we consider that at every time step the RNN produces a prediction, which is used to calculate some loss $\mathcal{L}$. The calculation might look dense, but it can be made easier by observing two components. One component is that there is a loss for each time step $1, 2, \ldots, n$ and the gradient each one accumulates is added together, or alternatively can be averaged to make the gradient distribution more invariant to the number of time steps. The second component is that each loss is affected by all previous hidden activations, and as a result produces multiple gradient values, which is shown in the last line of Eq. (5.3). To accommodate the calculation of gradients across multiple time steps from the loss of a single time step, we use the chain rule to calculate the gradient at each step as the error propagates backwards as shown in Fig. 5.5. This propagation of errors to every previous step of the RNN's computation graph is the reason that this type of network can manage to learn dependencies across multiple time steps and take advantage of the sequential nature of the input data.

An issue that arises for the simple RNN presented here is the problem of vanishing and exploding gradients [9,10]. The exploding gradient issue occurs when the long-term components of the gradients of an RNN become exponentially larger for longer sequences compared to shorter ones. Similarly, the vanishing gradient issue occurs when the long-term components of gradients become exponentially closer to 0 over longer sequences compared to shorter ones.

Intuitively, the weight matrix $\mathbf{W}_{hh}$ is multiplied by the hidden activations in the forward pass and by their gradients in the backward pass. Long term dependency learning signals need to be propagated through multiple steps of the hidden activations, which may be difficult depending on the properties of the $\mathbf{W}_{hh}$ matrix. In a naively simplistic scenario, with a matrix size of 1×1 and linear activation function, the long term dependencies propagated backwards shrink exponentially if $||\mathbf{W}_{hh}||_2 < 1$ or exponentially grow if $||\mathbf{W}_{hh}||_2 > 1$. Since the RNN backpropagation step recursively calculates a function $\mathbf{W}_{hh}\sigma'(\cdot)$ depending on $\mathbf{W}_{hh}$, the hidden values will exponentially increase or decrease. The intuition here is that given a value $0 < \eta < 1$, as $t \to \infty$, then $\eta^t \to 0$, which results in the vanishing gradients phenomenon. Similarly, if $\eta > 1$, as $t \to \infty$, then $\prod_{i=0}^{t} \eta = \eta^t \to \infty$, which gives rise to exploding gradients as the number of steps increases. The further from 1 this multiplication value lies, the faster the divergence or convergence.

To generalize this intuition, we first utilize the gradient equation for calculating the error propagated from the hidden activations across a single step:

$$\frac{\partial h_{t+1}}{\partial h_t} = \mathbf{W}_{hh}\, \text{diag}(\sigma'(h_t)), \tag{5.4}$$

where σ' is the elementwise derivative of the activation function applied on the RNN hidden activations h_t. Since the derivatives of the activation functions are bounded and $\left\|\text{diag}(\sigma'(h_t))\right\|$, the 2-norm of the Jacobian matrix of the hidden activations can be bounded, then

$$\forall t, \left\| \frac{\partial h_{t+1}}{\partial h_t} \right\|_2 \leq \left\| \mathbf{W}_{hh}^T \right\|_2 \left\|\text{diag}(\sigma'(h_t))\right\|_2 < \frac{1}{\gamma}, \gamma < 1. \tag{5.5}$$

The γ value is specific to the activation function ($\gamma = 1$ for *tanh* and $\gamma = \frac{1}{4}$ for *sigmoid*). From Eq. (5.5), we derive that an $\eta \in \mathbb{R}$ exists such that $\forall t \left\| \frac{\partial h_{t+1}}{\partial h_t} \right\| \leq \eta < 1$, and

$$\left\| \frac{\partial \mathcal{L}_t}{\partial h_t} \prod_{i=k}^{t-1} \left(\frac{\partial h_{i+1}}{\partial h_i} \right) \right\|_2 \leq \eta^{t-k} \left\| \frac{\partial \mathcal{L}_t}{\partial h_t} \right\|_2 \tag{5.6}$$

which resembles the naive example presented before. As the gradient is propagated backwards and is multiplied by a value $\eta < 1$, $t - k$ times, the resulting gradient becomes exponentially smaller, contributing insignificantly to the learning of long-term dependencies. The inverse of the above inequality equations demonstrates the exploding gradient case. A solution proposed for the exploding gradient problem is called "gradient clipping." To avoid the rapid increase of the scale of gradient values, the norm of the gradient is used to clip the values to be within a threshold:

$$\hat{g} \leftarrow \frac{\hat{g}}{\|\hat{g}\|} \cdot \text{threshold, if } \hat{g} > \text{ threshold, or } \hat{g} \text{ otherwise,} \tag{5.7}$$

where $\hat{g}$ is the gradient $\frac{\partial \mathcal{L}}{\partial h_t}$. In the case of the vanishing gradient, [10] suggests a regularizer that preserves the error norm as it travels backwards in the sequence.

5.3 Long-short term memory

Long-Short Term Memory (LSTM) was first proposed by [11] to solve the problem of vanishing and exploding gradients by proposing a different architectural approach to RNNs. This is achieved by protecting its hidden activation using gates between each of its transaction points with the rest of its layer. The hidden activation that is protected is called the cell state. The protection of the cell state is undertaken by the three LSTM gates, namely, the forget, input, and output gates.

During the forward pass, the first gate that acts upon the cell is the forget gate. It determines which of the cell's activations are forgotten and by how much. It achieves this by multiplying all the cell's elements with a vector $\mathbf{f}_t \in (0, 1)^{m_h}$. If the forget gate emits a value close to zero, the corresponding element in the cell state will be wiped and set to zero, whereas if it outputs a value close to one, then the cell will fully retain the value of that element.

The next gate to act upon the cell is the input gate, which determines what portion of new information will be added to the protected state. This happens in combination with the calculation of a new candidate cell state $\mathbf{c}'_t$. Similar to the forget gate, the input gate $\mathbf{i}_t \in (0, 1)^{m_h}$ is multiplied by the candidate state $\mathbf{c}'_t$ and added to the cell state. This prevents unnecessary additions to the cell state.

Finally, there is the output gate, which is important for the backwards pass (backpropagation). It determines which parts of the cell state need to be propagated forward and be included in the output of the network. The forward flow of information across the described gates is shown in Fig. 5.6.

The following equations describe the behavior of the LSTM model [11]:

$$\mathbf{f}_t = \sigma(\mathbf{W}_{xf}\mathbf{x}_t + \mathbf{W}_{hf}\mathbf{h}_{t-1} + \mathbf{b}_f) \tag{5.8}$$

$$\mathbf{i}_t = \sigma(\mathbf{W}_{xi}\mathbf{x}_t + \mathbf{W}_{hi}\mathbf{h}_{t-1} + \mathbf{b}_i) \tag{5.9}$$

$$\mathbf{c}'_t = \tanh(\mathbf{W}_{hc}\mathbf{h}_{t-1} + \mathbf{W}_{xc}\mathbf{x}_t + \mathbf{b}_c) \tag{5.10}$$

$$\mathbf{c}_t = \mathbf{f}_t\mathbf{c}_{t-1} + \mathbf{i}_t\mathbf{c}'_t \tag{5.11}$$

$$\mathbf{o}_t = \sigma(\mathbf{W}_{xo}\mathbf{x}_t + \mathbf{W}_{ho}\mathbf{h}_{t-1} + \mathbf{b}_o) \tag{5.12}$$

$$\mathbf{h}_t = \mathbf{o}_t \tanh(\mathbf{c}_t) \tag{5.13}$$

where $\mathbf{f}_t$, $\mathbf{i}_t$, and $\mathbf{o}_t$ are the activations of the input, forget, and output gates at time-step t, which control how much of the input and the previous state will be considered, and how much of the cell state will be included in the hidden activation of the network. The protected cell activation at time step t is denoted by $\mathbf{c}_t$, whereas $\mathbf{h}_t$ is the activation that will be given to the next layer. The notations of the weight matrices $\mathbf{W}_{(..)}$ are explicitly denoted with subscripts of their transform. $\mathbf{b}_{(.)}$ denotes the bias vector of each gate. $\mathbf{W}_{xf}$ and $\mathbf{W}_{hf}$ are the matrices that transform the input $\mathbf{x}_t$ and hidden state $\mathbf{h}_{t-1}$, respectively, to the forget gate dimension, $\mathbf{W}_{xi}$ and $\mathbf{W}_{hi}$ transform the input and hidden state to the input gate dimension and so on. There are simplified notation combining the input and hidden state into a single matrix, thus having a single matrix per gate $\mathbf{W}_f$, $\mathbf{W}_i$, $\mathbf{W}_c$, $\mathbf{W}_o$. For this chapter, we keep the more detailed notation $\mathbf{W}_{(..)}$ for completeness.

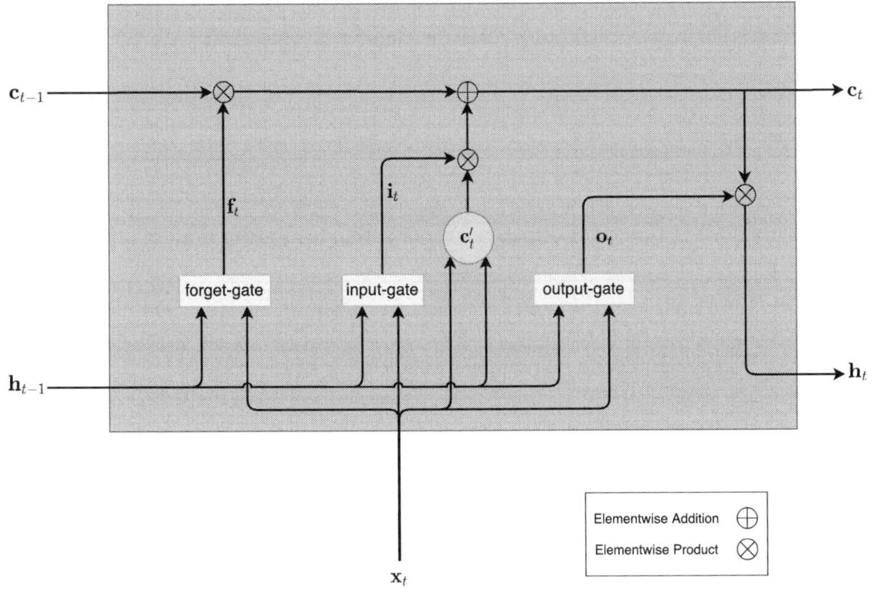

FIGURE 5.6

LSTM computation graph. Check Eq. (5.8) for detail about how input x_t and hidden state h_{t-1} are combined in each gate.

Although not obvious, the additive nature of the gates to modify the cell's state has a great impact on the effects of vanishing and exploding gradients. To understand why this happens, we must expand the gradient calculations of the LSTM's cell state. According to Eq. (5.11) the gradient can be calculated using the chain rule as

$$\begin{aligned}
\frac{\partial \mathbf{c}_t}{\partial \mathbf{c}_{t-1}} &= \frac{\partial \mathbf{c}_t}{\partial \mathbf{f}_t}\frac{\partial \mathbf{f}_t}{\partial \mathbf{h}_{t-1}}\frac{\partial \mathbf{h}_{t-1}}{\partial \mathbf{c}_t} + \frac{\partial \mathbf{c}_t}{\partial \mathbf{i}_t}\frac{\partial \mathbf{i}_t}{\partial \mathbf{h}_{t-1}}\frac{\partial \mathbf{h}_{t-1}}{\partial \mathbf{c}_t} + \frac{\partial \mathbf{c}_t}{\partial \mathbf{c}'_t}\frac{\partial \mathbf{c}'_t}{\partial \mathbf{h}_{t-1}}\frac{\partial \mathbf{h}_{t-1}}{\partial \mathbf{c}_t} + \frac{\partial \mathbf{c}_{t-1}}{\partial \mathbf{c}_{t-1}}\mathbf{f}_t \\
&= \mathbf{c}_{t-1}\,\sigma'(\mathbf{W}_{xf}\mathbf{x}_t + \mathbf{W}_{hf}\mathbf{h}_{t-1})\mathbf{W}_{hf}\;\mathbf{o}_{t-1}\tanh'(\mathbf{c}_{t-1}) \\
&\quad + \mathbf{c}'_t\sigma'(\mathbf{W}_{if}\mathbf{x}_t + \mathbf{W}_{hi}\mathbf{h}_{t-1})\mathbf{W}_{hi}\mathbf{o}_{t-1}\tanh'(\mathbf{c}_{t-1}) \\
&\quad + \mathbf{i}_t\tanh'(\mathbf{W}_{xc}\mathbf{x}_t + \mathbf{W}_{hc}\mathbf{h}_{t-1})\mathbf{o}_{t-1}\tanh'(\mathbf{c}_{t-1}) \\
&\quad + f_t
\end{aligned} \tag{5.14}$$

The intuitive difference between the LSTM gradient and the vanilla RNN gradient (Eq. (5.4)) are the additional terms of the LSTM gates that lack the multiplication of the same weight matrix along all backward steps. The LSTM's gradient has four terms that are added together and some of them are not multiplied directly with a weight matrix. This allows for each term to have a different gradient factors at each step, either increasing or decreasing the gradient step norm, thus avoiding exponentially increasing or decreasing the gradient values across steps. The vanishing/exploding gradient problem is prevented in most cases, but it might still show up in instances where the circumstances line up in just the right way, so that it may still show up. In most cases, the gradients are much better behaved than the vanilla RNN and using an LSTM is preferred in most cases.

One variant of LSTM proposed in [12] adds "peepholes" that allows the LSTM to consider the protected state when deciding what to forget, input, and output. This is achieved by modifying the gates as follows:

$$\mathbf{f}_t = \sigma(\mathbf{W}_{cf}\mathbf{c}_{t-1} + \mathbf{W}_{xf}\mathbf{x} + \mathbf{W}_{hf}\mathbf{h}_{t-1} + \mathbf{b}_f) \tag{5.15}$$

$$\mathbf{i}_t = \sigma(\mathbf{W}_{ci}\mathbf{c}_{t-1} + \mathbf{W}_{xi}\mathbf{x} + \mathbf{W}_{hi}\mathbf{h}_{t-1} + \mathbf{b}_i) \tag{5.16}$$

$$\mathbf{o}_t = \sigma(\mathbf{W}_{co}\mathbf{c}_{t-1} + \mathbf{W}_{xo}\mathbf{x}_t + \mathbf{W}_{ho}\mathbf{h}_{t-1} + \mathbf{b}_o), \tag{5.17}$$

which also adds cell state weight matrices $\mathbf{W}_{cf}$, $\mathbf{W}_{ci}$, $\mathbf{W}_{co}$.

5.4 Gated recurrent unit

The Gated Recurrent Unit (GRU) proposed in [13] is similar to the LSTM but has a much simpler gating mechanism. It utilizes two gates, namely, the reset and update gates. No protected cell state is used, hence the gates are applied directly to the hidden state. This difference makes the GRU more computationally efficient than the LSTM. The computation steps of a GRU are the following:

$$\mathbf{r}_t = \sigma(\mathbf{W}_{xz}\mathbf{x}_t + \mathbf{W}_{hz}\mathbf{h}_{t-1} + \mathbf{b}_z) \tag{5.18}$$

$$\mathbf{z}_t = \sigma(\mathbf{W}_{xz}\mathbf{x}_t + \mathbf{W}_{hz}\mathbf{h}_{t-1} + \mathbf{b}_z) \tag{5.19}$$

$$\mathbf{h}'_t = \tanh(\mathbf{W}_{hh}\mathbf{r}_t\mathbf{h}_{t-1} + \mathbf{W}_{xh}\mathbf{x}_t + \mathbf{b}_h) \tag{5.20}$$

$$\mathbf{h}_t = (1 - \mathbf{z}_t)\mathbf{h}_{t-1} + \mathbf{z}_t\mathbf{h}'_t, \tag{5.21}$$

where $\mathbf{r}_t$ is the reset gate value, $\mathbf{z}_t$ is the update gate, and $\mathbf{h}'_t$ is the candidate hidden state that is incorporated proportionally to the hidden state as shown in Eq. (5.21).

The reset gate works in the same way as the forget gate of the LSTM, whereby $\mathbf{r} \in (0, 1)^{m_h}$ using the sigmoid function and the resulting reset vector is multiplied with the last hidden state, thus allowing the hidden state to dispose of unneeded information. The update gate, which also utilizes a sigmoid function, is then used to proportionally update the state $\mathbf{h}_t$ using a newly calculated state $\mathbf{h}'$. The computation computational graph of the GRU is shown in Fig. 5.7.

5.5 Other RNN variants

There have been studies into many different variants that attempt to improve upon existing architectures. One such example is the Depth-Gated RNNs [14], where an additional connection between the protected cell states of LSTM is used with a depth gate. Another interesting approach is the clockwork RNN [15], where separate RNN

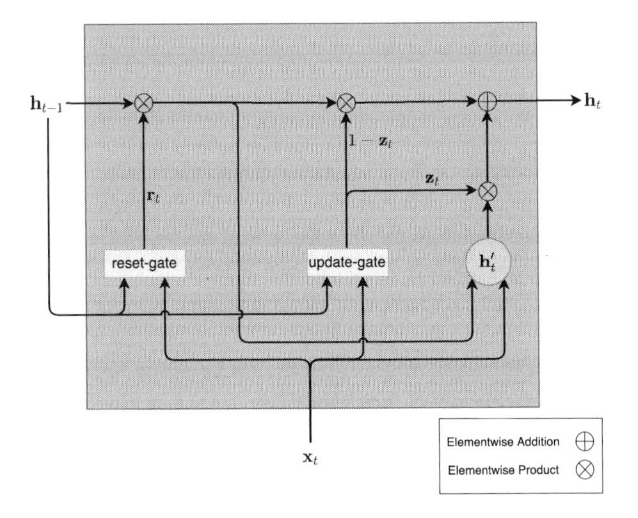

FIGURE 5.7

GRU computation graph.

layers are activated with different intervals allowing for different temporal granularity while processing the input. In [16], LSTM nodes are arranged in a multidimensional grid and connected in a spatiotemporal manner.

Studies have been conducted comparing the performance of various RNN configurations [17,18] on multiple tasks, and the main result has been that the performance does not differ in a wider manner. Some tasks are better suited for some networks than others but not in a way that a specific network type such as an LSTM or a GRU can be considered better. Although many of these studies present some improvements on the simpler versions of LSTMs, they have not been widely adopted. When selecting a model architecture to apply to a problem, it helps to begin with simpler models, where the results can be interpreted more easily and solutions to common problems are more widely available.

5.6 Applications

Many applications were afforded by the representational capacity of the RNNs. When the data points that must be analyzed can be described as a sequence, either explicitly, as in the case of time series, or implicitly, in some other applications, RNNs provide a robust tool to use. Even when a sequence is not clearly present in the data, such as an image, it can emerge from the way we choose to process it (e.g., in [19]).

A simpler example of using RNNs is to classify time series. RNNs have been used to classify human activity using mobile and wearable sensors [20], which can be used to monitor individuals in assisting living or elderly care, detecting sports injuries and security applications in smart homes. In [21], images (or video frames) have features extracted from them using a Convolutional Neural Network. Then, using the extracted features of a video frame sequentially, an LSTM processes them and predicts the activity taking place in the video. Another area of sequence classification is in medical data [22] to predict inpatient mortality soon after admission, unexpected readmissions after discharges, and long length stays.

Although the simple classification use-cases improve upon previous results, the most interesting achievements of RNN applications are in the sequence-to-sequence domain, which can be thought as many-to-many prediction as shown in Fig. 5.3. In [21], which is also discussed in the last paragraph for the classification, the authors also used a global feature set of videos to produce a sequence of words that can be used to accurately caption a video given as input. Similarly, [23] employed a CNN to extract features from an image and then used an LSTM to generate a caption for the images. In these captioning applications, the features are produced once for the whole video or image and then the LSTM is constantly presented with the same features on each step of the sequence along with the word vector produced on the previous time step. In this way, the RNN can determine which features to use based on both what the image contains and what is already emitted. RNN have also been used with attention to sequentially process images and focus on their different parts, improving the quality of generated captions [24].

An even greater advancement on the image captioning front is the question answering application, where RNN-based models are given an image and a question about it, and they can predict the correct answer [25]. The computational power of the RNN gave rise to the visual question-answering applications, which have continued to improve [26,27]. A creative use of adapting tasks to sequential processing is presented in [19], where LSTMs are used with the ability to change their view of an image using Gaussian filters, allowing them to focus on different parts of the image without changing their receptive field size. Using the same technique, they can generate images by selectively writing on patches of an output image. Direct transcription from voice waveform to text is also implemented using RNNs [28], where a bidirectional LSTM and a Connectionist Temporal Classification (CTC) scheme that removes the need to align the input temporally with the label sequence are employed.

Another set of applications, which has widely adopted RNNs concerns the text generation and translation tasks. In text generation, a prompt is provided as input to an RNN based model and the model can indefinitely generate text from that prompt. That can encompass context such as chatbots but also sentence completion and correction. In [29] an LSTM is trained to perform translation of sentences from English to French, also presenting potential uses for multilanguage translation by keeping a separate encoder for projecting an input sentence into a semantic space and then transforming it back into the intended language thus translating it.

Finally, a suitable domain to apply the capabilities of the RNN for processing time series is prediction tasks in the context of financial markets. In [30], LSTMs are used to predict the price movement of the price having limit order book data as input, framed as a supervised learning task. For financial tasks, Reinforcement Learning

(RL) is more suitable and in [31] LSTMs agents are trained to optimize their trading behavior in order to maximize profits in the simulated FOREX markets.

5.7 Concluding remarks

In the past, as with other types of deep neural networks, RNNs did not achieve impressive performance, but once problems, such as the vanishing and exploding gradients, were mostly solved, and computational power was more readily available, along with massive amounts of data, which became accessible to researchers, RNNs managed to reach their true potential. Taking advantage of the prior hidden in the sequential format of data and modeling a neural network architecture after that type of processing has given great advantages. In recent years an advanced form of attentions and smart querying of data has given rise to transformers [32] that in many cases have surpassed RNNs in many tasks. RNNs are still a worthy tool in a machine learning toolbox and are still readily used in many applications that need to keep a state in a sequential context.

The research of the advances in RNNs for the writing of this chapter was assisted by the write-ups from [33,34].

References

[1] Y. LeCun, C. Cortes, C. Burges, Mnist handwritten digit database, ATT Labs [Online]. Available: http://yann.lecun.com/exdb/mnist, 2 2010.

[2] H. Xiao, K. Rasul, R. Vollgraf, Fashion-mnist: a novel image dataset for benchmarking machine learning algorithms, arXiv preprint, arXiv:1708.07747, 2017.

[3] R.A. Fisher, The use of multiple measurements in taxonomic problems, Annals of Eugenics 7 (2) (1936) 179–188.

[4] W.S. McCulloch, W. Pitts, A logical calculus of the ideas immanent in nervous activity, The Bulletin of Mathematical Biophysics 5 (4) (1943) 115–133.

[5] D.E. Rumelhart, G.E. Hinton, R.J. Williams, Learning internal representations by error propagation, Tech. Rep., California Univ San Diego La Jolla Inst for Cognitive Science, 1985.

[6] M.I. Jordan, Serial order: a parallel distributed processing approach, in: Advances in Psychology, vol. 121, Elsevier, 1997, pp. 471–495.

[7] P.J. Werbos, Generalization of backpropagation with application to a recurrent gas market model, Neural Networks 1 (4) (1988) 339–356.

[8] M.C. Mozer, A focused backpropagation algorithm for temporal, Backpropagation: Theory, Architectures, and Applications 137 (1995).

[9] Y. Bengio, P. Simard, P. Frasconi, Learning long-term dependencies with gradient descent is difficult, IEEE Transactions on Neural Networks 5 (2) (1994) 157–166.

[10] R. Pascanu, T. Mikolov, Y. Bengio, On the difficulty of training recurrent neural networks, in: International Conference on Machine Learning, 2013, pp. 1310–1318, URL http://www.jmlr.org/proceedings/papers/v28/pascanu13.pdf.

[11] S. Hochreiter, J. Schmidhuber, Long short-term memory, Neural Computation 9 (8) (1997) 1735–1780.
[12] F.A. Gers, J. Schmidhuber, Recurrent nets that time and count, in: Proceedings of the IEEE-INNS-ENNS International Joint Conference on Neural Networks. IJCNN 2000. Neural Computing: New Challenges and Perspectives for the New Millennium, vol. 3, IEEE, 2000, pp. 189–194.
[13] K. Cho, B. Van Merriënboer, C. Gulcehre, D. Bahdanau, F. Bougares, H. Schwenk, Y. Bengio, Learning phrase representations using rnn encoder-decoder for statistical machine translation, arXiv preprint, arXiv:1406.1078, 2014.
[14] K. Yao, T. Cohn, K. Vylomova, K. Duh, C. Dyer, Depth-gated recurrent neural networks, arXiv preprint, arXiv:1508.03790, 9 2015.
[15] J. Koutnik, K. Greff, F. Gomez, J. Schmidhuber, A clockwork rnn, arXiv preprint, arXiv: 1402.3511, 2014.
[16] N. Kalchbrenner, I. Danihelka, A. Graves, Grid long short-term memory, arXiv preprint, arXiv:1507.01526, 2015.
[17] R. Jozefowicz, W. Zaremba, I. Sutskever, An empirical exploration of recurrent network architectures, in: International Conference on Machine Learning, 2015, pp. 2342–2350.
[18] K. Greff, R.K. Srivastava, J. Koutník, B.R. Steunebrink, J. Schmidhuber, Lstm: a search space odyssey, IEEE Transactions on Neural Networks and Learning Systems 28 (10) (2016) 2222–2232.
[19] K. Gregor, I. Danihelka, A. Graves, D.J. Rezende, D. Wierstra, Draw: a recurrent neural network for image generation, arXiv preprint, arXiv:1502.04623, 2015.
[20] H.F. Nweke, Y.W. Teh, M.A. Al-Garadi, U.R. Alo, Deep learning algorithms for human activity recognition using mobile and wearable sensor networks: state of the art and research challenges, Expert Systems with Applications 105 (2018) 233–261.
[21] J. Donahue, L. Anne Hendricks, S. Guadarrama, M. Rohrbach, S. Venugopalan, K. Saenko, T. Darrell, Long-term recurrent convolutional networks for visual recognition and description, in: Proceedings of the IEEE Conference on Computer Vision and Pattern Recognition, 2015, pp. 2625–2634.
[22] A. Rajkomar, E. Oren, K. Chen, A.M. Dai, N. Hajaj, M. Hardt, P.J. Liu, X. Liu, J. Marcus, M. Sun, et al., Scalable and accurate deep learning with electronic health records, NPJ Digital Medicine 1 (1) (2018) 18.
[23] O. Vinyals, A. Toshev, S. Bengio, D. Erhan, Show and tell: a neural image caption generator, in: Proceedings of the IEEE Conference on Computer Vision and Pattern Recognition, 2015, pp. 3156–3164.
[24] K. Xu, J. Ba, R. Kiros, K. Cho, A. Courville, R. Salakhudinov, R. Zemel, Y. Bengio, Show, attend and tell: neural image caption generation with visual attention, in: International Conference on Machine Learning, 2015, pp. 2048–2057.
[25] M. Ren, R. Kiros, R. Zemel, Exploring models and data for image question answering, Advances in Neural Information Processing Systems 28 (2015) 2953–2961.
[26] J. Lu, J. Yang, D. Batra, D. Parikh, Hierarchical question-image co-attention for visual question answering, Advances in Neural Information Processing Systems 29 (2016) 289–297.
[27] P. Anderson, X. He, C. Buehler, D. Teney, M. Johnson, S. Gould, L. Zhang, Bottom-up and top-down attention for image captioning and visual question answering, in: Proceedings of the IEEE Conference on Computer Vision and Pattern Recognition, 2018, pp. 6077–6086.
[28] A. Graves, N. Jaitly, Towards end-to-end speech recognition with recurrent neural networks, in: International Conference on Machine Learning, 2014, pp. 1764–1772.

[29] I. Sutskever, O. Vinyals, Q.V. Le, Sequence to sequence learning with neural networks, Advances in Neural Information Processing Systems 27 (2014) 3104–3112.

[30] A. Tsantekidis, N. Passalis, A. Tefas, J. Kanniainen, M. Gabbouj, A. Iosifidis, Using deep learning for price prediction by exploiting stationary limit order book features, Applied Soft Computing (2020) 106401.

[31] A. Tsantekidis, N. Passalis, A.-S. Toufa, K. Saitas-Zarkias, S. Chairistanidis, A. Tefas, Price trailing for financial trading using deep reinforcement learning, IEEE Transactions on Neural Networks and Learning Systems (2020).

[32] A. Vaswani, N. Shazeer, N. Parmar, J. Uszkoreit, L. Jones, A.N. Gomez, Ł. Kaiser, I. Polosukhin, Attention is all you need, in: Advances in Neural Information Processing Systems, 2017, pp. 5998–6008.

[33] C. Olah, Understanding lstm networks, https://colah.github.io/posts/2015-08-Understanding-LSTMs/, 2015.

[34] A. Karpathy, The unreasonable effectiveness of recurrent neural networks, Andrej Karpathy blog 21 (2015) 23, http://karpathy.github.io/2015/05/21/rnn-effectiveness/.

CHAPTER

Deep reinforcement learning

6

Avraam Tsantekidis, Nikolaos Passalis, and Anastasios Tefas
Department of Informatics, Aristotle University of Thessaloniki, Thessaloniki, Greece

6.1 Introduction

Most of the Deep Learning (DL) models discussed until now came to prominence by improving their performance on two fundamental machine learning tasks, namely, supervised and unsupervised learning. As [1] notes, although the terms supervised and unsupervised learning seem to exclusively cover the whole spectrum of ML methods, that is not the case. Indeed, the most succinct and simple explanation of learning is the *maximization of a reward signal through trial-and-error*. This simple concept that exists throughout nature as a mechanism that helps humans and animals to survive is systematically explored in the field of Reinforcement Learning (RL). The interaction with the environment and the feedback received, shapes our behavior depending on the reward signal we receive. When a positive reward is received, such as finding food and enjoying eating, the behavior (or policy) leading up to that feedback is reinforced. Similarly, when a negative feedback is experienced, such as falling and hurting ourselves, the behavior that leads to it is diminished or avoided in the future.

In some cases trial-and-error is used to describe how supervised learning works, where a model learns from its mistake in predicting the desired output. This is very different from RL, since the error signal in supervised learning is generated from prior knowledge that must be previously extracted and exist in the labels used. Although supervised learning from labeled data can achieve generalization to previously unseen data, in the RL case, an agent can continue learning from any new and unknown situations based on the reward signal alone. Unsupervised learning, which aims at discovering structure in unlabeled data, is also different from RL for a similar reason. Although in RL learning structures might benefit an agent, the end goal is simply to maximize the reward. Along the way some structures of the observable state might be identified by the employed model, but this is merely a side effect of the learning process and not the end goal, as in unsupervised learning.

Reinforcement learning is comprised of the following basic components: the *policy*, *reward signal*, and *value function*. Some methods also include a model of the observable environment and are called *model-based* approaches. These methods allow RL agents to improve its policy without interacting with a real environment, thus

Deep Learning for Robot Perception and Cognition. https://doi.org/10.1016/B978-0-32-385787-1.00011-7

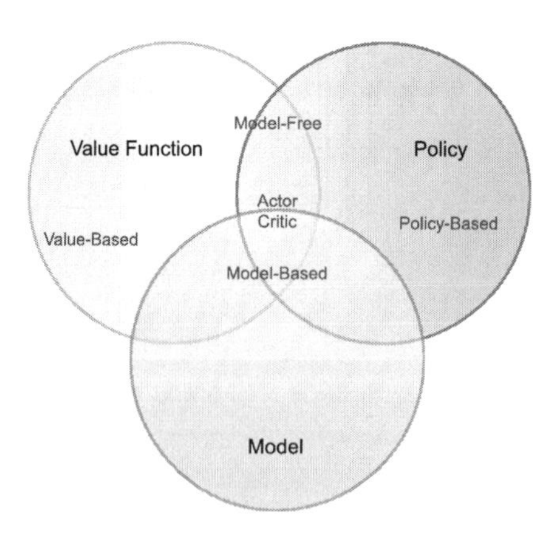

FIGURE 6.1

Components utilized in reinforcement learning and where known methods land among them. Diagram recreated from [2].

enabling a form of planning based on the understanding of the environment. When an environment can be simulated adequately without concern about the number of interaction from which an agent can learn from, then a *model-free* approach can be used. In Fig. 6.1 a diagram is drawn of how RL components are used by different methods. In this chapter, we will cover *model-free* approaches that utilize value-based and policy-based learning.

Both value and policy-based methods assign a value of the state of the environment, but value-based methods aim to select the actions that will lead to higher value states, while policy-based methods direct learn a policy utilizing state values in an indirect way. Value-based methods have been shown to be less sample-efficient and slower to train, while policy-based methods can be trapped in local minima because the policy itself is used to sample the environment. Both methods have been developed and augmented with techniques to improve upon their shortcoming, which we discuss later in this section.

In recent years, after the unprecedented success of deep learning in several tasks of machine learning [3–8], both supervised and unsupervised, the existing RL methods started to be adapted to the deep learning paradigm. The newly developed frameworks such as Tensorflow [9] and Pytorch [10], and the ever growing hardware acceleration capabilities of Graphics Processing Units (GPUs) and Application-Specific Integrated Circuits (ASICs) have allowed for significant performance improvements in RL tasks. In this chapter, we explore the conventional RL methods and how deep learning was applied to augment them and remarkably improve their performance, solving tasks previously thought impossible to be solved by machine learning and even surpassing human performance in many of them, such as Atari games, chess [11], go [12], and others.

6.2 Value-based methods

When mentioning the term value in a RL setting, we usually refer to the state value. That is an arbitrary value that can be assigned to a specific state we may find ourselves in. This value is usually based or tries to estimate the long term discounted return G_t of a specific state which is defined as

$$\begin{aligned} G_t &= r_{t+1} + \gamma r_{t+2} + \gamma^2 r_{t+3} \dots \\ &= \sum_{i=0}^{\infty} \gamma^i r_{t+i+1}, \end{aligned} \tag{6.1}$$

where r_t is the reward received at time step t, γ is the exponential decaying factor of future rewards. Discounting is used for several reasons. A reward far into the future is less valuable than a reward that is closer to becoming realized. Moreover, in some cases an RL environment might evolve over infinite time steps and summing all rewards will make the return value diverge to infinity. Having an accurate estimate of the expected returns would allow a policy to select the optimal action for any given state s_t. The optimal value function is defined as

$$V_\pi^*(s) = \mathbb{E}[G_t |\, s_t] = \mathbb{E}\left[\sum_{i=0}^{\infty} \gamma^i r_{t+i+1} \middle|\, s_t\right], \tag{6.2}$$

where $V_\pi^*(s)$ is the state value function for policy π, and s denotes the state. This value estimation of each state is needed in RL because there is no explicit label on each state that informs us whether we are in a good position or not. To find the optimal value of each state the Bellman optimality equation can be used:

$$V(s_t) = \mathbb{E}[r_{t+1} + \gamma V(s_{t+1})] \tag{6.3}$$

Since our estimates $V(s)$ at the beginning of training are not accurate, we introduce a learning rate or "step" denoted as $\alpha \in [0, 1]$ to gradually improve the estimates

$$V(s_t) \leftarrow V(s_t) + \alpha[r_{t+1} + \gamma V(s_{t+1}) - V(s_t)] \tag{6.4}$$

This is part of the Time Difference (TD) learning methods, where in this case $r_{t+1} + \gamma V(s_{t+1})$ is an unbiased estimate of $V(s)$. As a shorthand, we define

$$\delta_t = r_{t+1} + \gamma V(s_{t+1}) - V(s_t), \tag{6.5}$$

as the temporal difference error δ_t. By utilizing the value of the estimate of the next step to train the current step, we are *bootstraping* the value estimator. This allows us to start training an estimator before knowing exactly how an episode concludes.

One might ask, how does having a good state value estimator help an RL agent perform better, since the rewards are based on the action an agent takes. In simple terms, knowing which states are expected to have good return values allows for a policy to be trained that tries to transition more often toward those high return states. There are multiple ways to go about training a policy using value state estimates and we will explore some of them in the following sections.

6.2.1 Q-learning

A simple approach to utilize value functions is the direct approach taken by the Q-leaning method [13,14]. In this approach, we try to learn multiple values for each state, one for each possible action. This value is called the Q-value of a state action and is denoted as $Q(s, a)$. If an agent has access to the true Q-values of every state-action pair in an environment then an optimal policy can be built without actually needing to learn a policy explicitly. This would be done by simply looking up all state-action values and selecting the action that has the highest value:

$$\max_{a} Q(s, a)$$

For state-action values the Bellman expectation equation is utilized to determine the value of the state-action tuple:

$$Q(s_t, a_t) = \mathbb{E}[r_{t+1} + \gamma Q(s_{t+1}, a_{t+1}) | \, s_{t+1}, a_{t+1}] \tag{6.6}$$

Similar to the state value training, we can gradually train a Q-value estimator using the update rule:

$$Q(s_t, a_t) \leftarrow Q(s_t, a_t) + \alpha[r_{t+1} + \gamma Q(s_{t+1}, a_{t+1}) - Q(s_t, a_t)] \tag{6.7}$$

In its most basic of Q-learning a "Q-table" is employed, where each row of the table represents a state and each column represents an action. As state-action tuples are sampled from an environment the corresponding cells of the matrix are update using Eq. (6.7). The structure of the Q-Table is shown in Table 6.1.

This is obviously very costly if the distinct action-state space is large and quickly becomes infeasible to manage. Constructing a table is also only possible when actions

Table 6.1 Visual representation of a Q-Table with the values of $m \times n$ state-action combinations. m is the total number of states and there are a total of n possible discrete actions.

		Actions			
		$a^{(0)}$	$a^{(1)}$	...	$a^{(n-1)}$
States	s_0	$Q(s_0, a^{(0)})$	$Q(s_0, a^{(1)})$	...	$Q(s_0, a^{(n-1)})$
	s_1	$Q(s_1, a^{(0)})$	$Q(s_1, a^{(1)})$	...	$Q(s_1, a^{(n-1)})$
	$\vdots$	$\vdots$	$\vdots$	$\ddots$	$\vdots$
	s_1	$Q(s_{m-1}, a^{(0)})$	$Q(s_{m-1}, a^{(1)})$	...	$Q(s_{m-1}, a^{(n-1)})$

are discrete. To remedy this problem, we can replace the Q-table with a parametrizable approximator that can be trained using an update rule similar to Eq. (6.7).

An interesting problem that also arises for training RL agents is the decision between exploring an environment to improve the accuracy of the Q-table and the maximization of rewards by exploiting the information of the Q-table. If exploration is given higher priority then some highly rewarding paths that require many correct consecutive actions may never be discovered, whereas if a greedy approach is taken and the action with the highest state-action value is chosen, then some rewarding paths may never come to be explored. Although heuristics such the ϵ-greedy policy for combining exploration and exploitation during RL training have been proposed [13], there does not exist a satisfying or definite answer to this problem.

6.2.2 Deep Q-learning

The simplest improvement that Deep Neural Networks (DNNs) can offer to Q-learning is to act as the Q-value approximators in-place of the costly Q-table. A DNN can observe the current state and a possible action and directly approximate the Q-value. This has the obvious benefit that a neural network is able to generalize over the available state-action pairs. By being able to generalize, an agent using a deep Q-learning might be able to accurately determine the state-action values of rare or previously unseen state-action combinations.

In past works utilizing neural networks to learn a Q-table approximator required multiple passes to calculate all the possible state-action values for a single state [15,16]. Both the state and action were part of the input to the approximator. An improvement in computation was proposed in [17], where the Deep Q-network (DQN) agent calculates all the action values for a state in a single pass as shown in Fig. 6.2. The neural network used to predict the Q-values of each state action can then be used in combination with an ϵ-greedy policy to explore both randomly and greedily an environment and train a DQN with the ability to predict the state-action values.

To train the DQN an environment is first sampled to gather "experiences." These sampled experiences are referred to as *episodes* and are appended to a *replay memory,* which draws inspiration from psychology research [18,19]. To update the Q-value estimate, episodes are retrieved from the replay memory and a loss function derived from (6.7) is calculated as

$$L_{s,a,r,s'} = \left(r + \gamma \max_{a'} Q(s', a') - Q(s, a)\right)^2, \tag{6.8}$$

where s' is the state that the agent arrived after taking action a at state s. In the loss, we assume that the Q-value estimation is based on selecting the optimal actions so the maximum state-action value of the possible actions is used as the value of the next state. The gradient of the loss function is backpropagated through the DQN and its weights are updated, gradually improving the predictions of the Q-value. As is usual

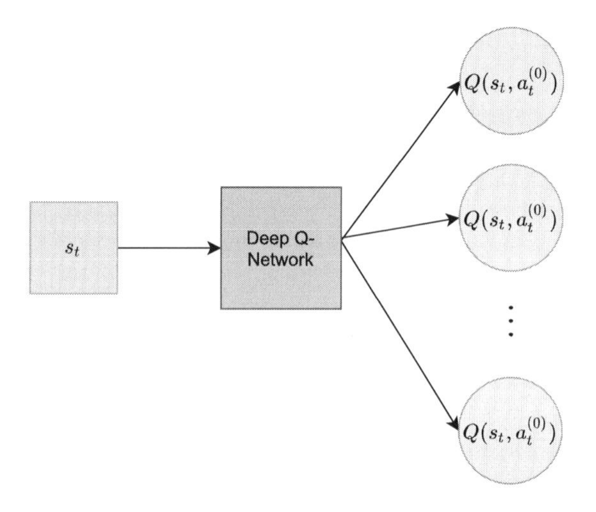

FIGURE 6.2

Computation of state-action values for all possible actions of a single state in a single pass.

for neural networks the loss function is applied over a batch of multiple transitions and then averaged to improve the gradients' accuracy.

This basic version of DQN has been shown to be problematic because of instabilities introduced by the temporal differencing of the loss, since the same parametrization of the DQN is used for the calculation of the target value $r_t + \gamma \max_{a'} Q(s', a')$ as the one that produced the initial state-action value $Q(s, a)$. Because the maximum values are selected as the target, the resulting training steps cause an overestimation of the Q-values by the DQN. This type of problem occurs when we combine the use of functional approximation, temporal differencing with bootstrapping and off-policy methods and is usually referred to as the "deadly triad" [1].

A solution partly alleviating this issue is presented in [20], which introduces Double-DQN, where the network that samples the environment and has its parameters updated during training is decoupled from the network used to calculate the target value. The two networks are usually referred to as the value network, which is the one being optimized actively, and the target network. The target network weights are updated, either by directly copying the weights of the value network at some slow interval (hard update), or by slowly shifting the weights using an exponential moving average between the current target and value weights.

The replay memory during training is constantly receiving new sampled experiences as the agent continues to evaluate the network and the existing memories become older and less useful. To ensure only relevant and newer memories exist in memory for the agent to be trained from, the memory is modeled as a first-in-first-out queue, where the oldest episodes are ejected after the memory becomes filled. Some improvements also include prioritized experience replay [21] where transitions are not sampled randomly during training for the replay memory but are prioritized based on some measure of their importance, like their temporal difference error.

More sophisticated methods, such as dueling DQN [22], noisy networks for exploration [23], and others [24], further improved deep Q-learning. These improvements have also shown their individual contributions in works such as [24] that conducts an ablation study across them, while also showing the impressive improvements when all of them are applied simultaneously.

6.3 Policy-based methods

A policy can be described as a mapping from states to actions or to action probability distributions, which depends on whether the policy is deterministic or stochastic. In general, to ensure exploration, the policy function is always stochastic, emitting a probability distribution denoted as $\pi(a|s)$. A policy usually outputs a vector of preferences over actions and is passed through a softmax activation function to bound it to a probability distribution:

$$\pi(a|s) = \frac{e^{h(s,a)}}{\sum_i^n e^{h(s,a^{(i)})}}, \tag{6.9}$$

where h is the preference function, which in the case of a neural network, can be the raw activations of the output neurons before the application of the softmax function. Since policy π is an trainable approximator with tunable parameters $\boldsymbol{\theta}$ it can also be written as $\pi_{\boldsymbol{\theta}}$.

In contrast with the state-action value learning, which might be slow to converge, a policy can converge much faster to a good performance [25]. In some cases, state-action values are simple to approximate and learn but in many cases it can be quite complicated. By avoiding to build a policy indirectly from an action-state value approximator and directly building it from the observation of the state can yield better results in many situations.

6.3.1 Policy gradient

To learn a policy when using a neural network, an objective function must be defined and gradient ascent to be employed to improve said policy. Although this might seem different than the loss function and gradient descent explained in earlier chapters, it is exactly equivalent. We denote this objective function as $J(\boldsymbol{\theta})$ and define it as

$$J(\boldsymbol{\theta}) = \sum_s \mu(s) \sum_a \pi(a|s) \sum_{s',r} p(s',r|s,a)r, \tag{6.10}$$

where $p(s',r|s,a)$ is the probability to transition from state s with action a to state s' and receiving reward r. The term μ is the stationary distribution given policy π [1], defined as

$$\mu(s) = \lim_{t\to\infty} Pr\{s_t | a_{0:t} \sim \pi\}.$$

This is true under the assumption that the stationary distribution does not change as long as the actions are selected under said policy $\pi_{\boldsymbol{\theta}}$ with parameters $\boldsymbol{\theta}$:

$$\mu(s') = \sum_s \mu(s) \sum_a \pi(a|s, \boldsymbol{\theta}) p(s'|s, a), \text{ for all } s' \in S. \tag{6.11}$$

In concrete terms, what the objective J tries to capture is that to maximize a reward signal, we must increase the probability of selecting the actions that produce state transitions that yield the highest rewards, while also considering the probability of those state transitions occurring in the first place. One simple example of the gradient ascent path we need to follow come from the REINFORCE Monte Carlo policy gradient on-policy algorithm [26] based on episodic samples:

$$\nabla J(\boldsymbol{\theta}) = \mathbb{E}\left[G_t \frac{\nabla \pi(a_t | s_t, \boldsymbol{\theta})}{\pi(a_t | s_t, \boldsymbol{\theta})} \right], \tag{6.12}$$

where G_t denotes the returns defined in Eq. (6.1). The return of the whole episode is needed so sampled trajectories must be completed before being able to perform the gradient computation, which is one reason it is considered a Monte Carlo method. To account for reward schedules that might be consistently higher or lower than 0 (among other configurations) a simple kind of "normalization" can be performed on the factor G_t of our gradient step. By subtracting an appropriate baseline $b(s)$, we can ensure that better rewards contribute more in the gradient step compared to smaller positive rewards. The only requirement for this variable is that it must be independent of action a. The resulting update function of the parameters with this baseline is

$$\boldsymbol{\theta}_{t+1} = \boldsymbol{\theta}_t + \alpha(G_t - b(s_t)) \frac{\nabla \pi(a_t | s_t, \boldsymbol{\theta})}{\pi(a_t | s_t, \boldsymbol{\theta})}, \tag{6.13}$$

where α is the step size parameter (learning rate). A simple baseline that can be used is a state value approximator $\hat{v}(s_t, \boldsymbol{w})$ trained to predict G_t.

6.3.2 Actor-critic methods

Although Monte-Carlo methods, such as REINFORCE, can successfully train an agent to maximize a reward signal, they suffer from high variance because their usage of the completed episode returns G_t, and thus are much slower to converge than bootstrapping methods that directly utilize the state values to train a policy. This is where actor-critic methods appear, where the actor is the policy and the critic is the state value approximator. The state value function tries learns to approximate the optimal value function as shown in Eq. (6.2), considering that the returns are sampled under policy π. By utilizing a learned state value V_π based on their policy the update rule for policy π becomes

$$\begin{aligned}\boldsymbol{\theta}_{t+1} &= \boldsymbol{\theta}_t + \alpha(r_{t+1} - \gamma V_\pi(s_{t+1}))\frac{\nabla_{\boldsymbol{\theta}}\pi_{\boldsymbol{\theta}}(a_t \mid s_t, \boldsymbol{\theta})}{\pi_{\boldsymbol{\theta}}(a_t \mid s_t, \boldsymbol{\theta})} \\ &= \boldsymbol{\theta}_t + \alpha\delta_t\frac{\nabla_{\boldsymbol{\theta}}\pi_{\boldsymbol{\theta}}(a_t \mid s_t, \boldsymbol{\theta})}{\pi_{\boldsymbol{\theta}}(a_t \mid s_t, \boldsymbol{\theta})} \\ &= \boldsymbol{\theta}_t + \alpha\delta_t\nabla_{\boldsymbol{\theta}}\log\pi_{\boldsymbol{\theta}}(a_t \mid s_t, \boldsymbol{\theta}),\end{aligned} \tag{6.14}$$

where δ_t is the temporal difference residual or error defined in Eq. (6.5). By utilizing the state value function the policy is able to learn in an online manner through the one-step bootstrapping objective. The training of both actor and critic is done simultaneously since the critic itself can be trained using the temporal difference as the objective to improve its prediction accuracy of the discounted returns.

There exist many approaches for determining how strong each step must be considered during training, the one presented in Eq. (6.14) uses the temporal difference residual, but many options are available. In [27], some different options are considered, for example,

$$\mathrm{g} = \mathbb{E}\left[\sum_{t=0}^{\infty}\Psi_t\nabla_{\boldsymbol{\theta}}\log\pi_{\boldsymbol{\theta}}(a_t \mid s_t, \boldsymbol{\theta})\right], \tag{6.15}$$

where g is the gradient used in 6.14 to train the policy parameters and Ψ can take the value of the temporal difference $\Psi_t = \delta_t$. Some other options include the immediate reward following the action at time step t, the reward minus some baseline (similar to Eq. (6.13)), the state-action value or the total reward of the trajectory. The most recently used instantiation though is the advantage function:

$$A_\pi(s_t, a_t) = Q_\pi(s_t, a_t) - V_\pi(s_t), \tag{6.16}$$

where A_π denotes the advantage under policy π of the transition from state s_t with action a_t. By using the appropriate γ-just advantage estimator, we can avoid training both a state-action value and a state value approximator since

$$\begin{aligned}\mathbb{E}_{s_{t+1}}\left[\delta_t^{V_{\pi,\gamma}}\right] &= \mathbb{E}_{s_{t+1}}[r_t + \gamma V_{\pi,\gamma}(s_{t+1}) - V_{\pi,\gamma}(s_t)] \\ &= \mathbb{E}_{s_{t+1}}[Q_{\pi,\gamma}(s_t, a_t) - V_{\pi,\gamma}(s_t)] = A_{\pi,\gamma}(s_t, a_t),\end{aligned} \tag{6.17}$$

where $V_{\pi,\gamma}(s_t)$ is the γ-just state value function and the advantage is an unbiased estimator of $A(s_t, a_t)$.

6.3.3 Deep policy gradient-based methods

Extending policy gradient methods with deep learning augmentation has shown great promise. The obvious choice for actor-critic methods is to implement both the state value approximator and the policy as deep neural networks that can process the state information in more complex ways.

6.3.3.1 Actor-critic

The approach for applying actor-critic explained in Section 6.3.2. The actor network processes a given state and produces a distribution of probabilities over actions, usually with the same number of actions as neurons with the softmax activations applied on the network's output [28]. The critic network is similarly a neural network that predicts the value of the state usually having a linear activation at the last layer to allow flexibility in the ranges it can to the state value.

6.3.3.2 Trust region policy optimization

One critical problem when using policy gradient algorithms is training stability, where a policy changes too quickly causing problems for both the state value approximation and for the exploration and convergence in general. To alleviate this problem [29] introduced a constraint on the parameter updates of the policy such that the Kullback–Leibler divergence between the new and the old policy stays within a limit:

$$\mathbb{E}_{s \sim \rho^{\pi_{\theta_{\text{old}}}}}[D_{\text{KL}} \pi_{\theta_{\text{old}}}(\cdot|s) || \pi_{\theta}(\cdot|s)] \leq \epsilon, \tag{6.18}$$

where $\pi_{\theta_{\text{old}}}$ is the policy as it was parametrized when the states s where sampled, while π_θ is the latest parametrization of the policy at the time of the update step and D_{KL} denotes the Kullback–Leibler divergence. The resulting effect is a smoother exploration of the environment that avoids sudden changes in policy and allows the critic to catch up while also leading to more stable training.

Despite the aforementioned advantages, Trust Region Policy Optimization (TRPO) is complicated to apply during training, so a new approach was proposed in [30], namely the Proximal Policy Optimization (PPO) where a simpler constraint was proposed that can be directly incorporated into the training loss as

$$J^{\text{CLIP}}(\boldsymbol{\theta}) = \mathbb{E}\left[\min(\text{r}(\boldsymbol{\theta}) A_{\pi_{\theta_{\text{old}}}}(s,a), \text{clip}(\text{r}(\boldsymbol{\theta}), 1-\epsilon, 1+\epsilon) A_{\pi_{\theta_{\text{old}}}}(s,a)\right], \tag{6.19}$$

where $\text{r}(\boldsymbol{\theta})$ is defined as the ratio of the current action probabilities over the past action probabilities:

$$\text{r}(\boldsymbol{\theta}) = \frac{\pi_\theta(a|s)}{\pi_{\theta_{\text{old}}}(a|s)}. \tag{6.20}$$

The loss J^{CLIP} allows the negatively advantaged probabilities to apply without any clipping on the parameters of the policy. At the same time, probabilities that make the ratio exceed the range $(1-\epsilon, 1+\epsilon)$ are clipped if they have positive advantages. Although much easier to implement and faster to train than TRPO, the Proximal Policy Optimization (PPO) has managed similar results with its predecessor.

Another interesting development also concerns the use of the same base neural network module that shares parameters across the actor and the critic. This has been shown to be unstable and to require careful tuning of the loss weighting between the actor and critic loss. On the other hand, sharing parameters also helps by allowing the shared parameters to learn better features of the state, thus improving the performance and generalization of an agent. One suggested combination is the phasic policy gradient [31], where the training is done in two interleaved phases. First, the policy is trained using the previously discussed PPO algorithm. Then on the second phase the critic is trained along with any other potential auxiliary loss, while also an additional "behavioral cloning" loss is added for the policy:

$$L^{\text{joint}} = L^{\text{aux}} + \beta_{\text{clone}} \cdot \hat{\mathbb{E}}_t \left[KL \left[\pi_{\theta_{\text{old}}} (\cdot \mid s_t), \pi_\theta (\cdot \mid s_t) \right] \right], \tag{6.21}$$

where hyperparameter β_{clone} controls how strongly the policy needs to be constraint to its original distribution at the end of the first phase. This improvement has shown impressive performance improvements.

6.4 Concluding remarks

In this chapter, we explored a number of reinforcement learning methods and how they have been augmented with deep learning showing promising results for many difficult problems. Reinforcement learning has been brought to the forefront of research in recent years, due to its achievement and the research continues to further improve its capabilities. Its connection to how our brain works with its own reward signal via dopamine and the potential that it may carry a reward prediction error. Many improvements introduced in the early days of reinforcement learning were inspired from psychology research on humans and animals, while at the same time its results have in turn inspired research directions in psychology. Since living organisms also seem to respond to reward signals, reinforcement learning is looking like an important piece of the puzzle of the manifestation of intelligence and a milestone toward general artificial intelligence.

References

[1] R.S. Sutton, A.G. Barto, Reinforcement Learning: An Introduction, MIT Press, 2018.

[2] D. Silver, Rl course by David Silver - Lecture 1: introduction to reinforcement learning, https://www.youtube.com/watch?v=2pWv7GOvuf0, 2015.

[3] A. Krizhevsky, I. Sutskever, G.E. Hinton, Imagenet classification with deep convolutional neural networks, in: F. Pereira, C.J.C. Burges, L. Bottou, K.Q. Weinberger (Eds.), Advances in Neural Information Processing Systems, vol. 25, Curran Associates, Inc., 2012, pp. 1097–1105.

[4] K. He, X. Zhang, S. Ren, J. Sun, Deep residual learning for image recognition, in: Proceedings of the IEEE Conference on Computer Vision and Pattern Recognition, 2016, pp. 770–778.

[5] I. Goodfellow, J. Pouget-Abadie, M. Mirza, B. Xu, D. Warde-Farley, S. Ozair, A. Courville, Y. Bengio, Generative adversarial nets, Advances in Neural Information Processing Systems 27 (2014) 2672–2680.

[6] A. Karpathy, L. Fei-Fei, Deep visual-semantic alignments for generating image descriptions, in: Proceedings of the IEEE Conference on Computer Vision and Pattern Recognition, 2015, pp. 3128–3137.
[7] K. Simonyan, A. Zisserman, Very deep convolutional networks for large-scale image recognition, arXiv preprint, arXiv:1409.1556, 2014.
[8] C. Szegedy, W. Liu, Y. Jia, P. Sermanet, S. Reed, D. Anguelov, D. Erhan, V. Vanhoucke, A. Rabinovich, Going deeper with convolutions, in: Proceedings of the IEEE Conference on Computer Vision and Pattern Recognition, 2015, pp. 1–9.
[9] M. Abadi, A. Agarwal, P. Barham, E. Brevdo, Z. Chen, C. Citro, G.S. Corrado, A. Davis, J. Dean, M. Devin, S. Ghemawat, I. Goodfellow, A. Harp, G. Irving, M. Isard, Y. Jia, R. Jozefowicz, L. Kaiser, M. Kudlur, J. Levenberg, D. Mané, R. Monga, S. Moore, D. Murray, C. Olah, M. Schuster, J. Shlens, B. Steiner, I. Sutskever, K. Talwar, P. Tucker, V. Vanhoucke, V. Vasudevan, F. Viégas, O. Vinyals, P. Warden, M. Wattenberg, M. Wicke, Y. Yu, X. Zheng, TensorFlow: large-scale machine learning on heterogeneous systems, software available from https://www.tensorflow.org/, 2015.
[10] A. Paszke, S. Gross, F. Massa, A. Lerer, J. Bradbury, G. Chanan, T. Killeen, Z. Lin, N. Gimelshein, L. Antiga, A. Desmaison, A. Kopf, E. Yang, Z. DeVito, M. Raison, A. Tejani, S. Chilamkurthy, B. Steiner, L. Fang, J. Bai, S. Chintala, Pytorch: an imperative style, high-performance deep learning library, in: H. Wallach, H. Larochelle, A. Beygelzimer, F. d'Alché-Buc, E. Fox, R. Garnett (Eds.), Advances in Neural Information Processing Systems, vol. 32, Curran Associates, Inc., 2019, pp. 8024–8035.
[11] D. Silver, T. Hubert, J. Schrittwieser, I. Antonoglou, M. Lai, A. Guez, M. Lanctot, L. Sifre, D. Kumaran, T. Graepel, et al., Mastering chess and shogi by self-play with a general reinforcement learning algorithm, arXiv preprint, arXiv:1712.01815, 2017.
[12] D. Silver, A. Huang, C.J. Maddison, A. Guez, L. Sifre, G. Van Den Driessche, J. Schrittwieser, I. Antonoglou, V. Panneershelvam, M. Lanctot, et al., Mastering the game of go with deep neural networks and tree search, Nature 529 (7587) (2016) 484–489.
[13] C.J.C.H. Watkins, Learning from delayed rewards, 1989.
[14] C.J. Watkins, P. Dayan, Q-learning, Machine Learning 8 (3–4) (1992) 279–292.
[15] M. Riedmiller, Neural fitted q iteration–first experiences with a data efficient neural reinforcement learning method, in: European Conference on Machine Learning, Springer, 2005, pp. 317–328.
[16] S. Lange, M. Riedmiller, Deep auto-encoder neural networks in reinforcement learning, in: The 2010 International Joint Conference on Neural Networks (IJCNN), IEEE, 2010, pp. 1–8.
[17] V. Mnih, K. Kavukcuoglu, D. Silver, A.A. Rusu, J. Veness, M.G. Bellemare, A. Graves, M. Riedmiller, A.K. Fidjeland, G. Ostrovski, et al., Human-level control through deep reinforcement learning, Nature 518 (7540) (2015) 529–533.
[18] E.L. Thorndike, Animal Intelligence: Experimental Studies, Macmillan, 1911.
[19] W. Schultz, P. Dayan, P.R. Montague, A neural substrate of prediction and reward, Science 275 (5306) (1997) 1593–1599.
[20] H. van Hasselt, A. Guez, D. Silver, Deep reinforcement learning with double q-learning, 2015.
[21] T. Schaul, J. Quan, I. Antonoglou, D. Silver, Prioritized experience replay, 2016.
[22] Z. Wang, T. Schaul, M. Hessel, H. Hasselt, M. Lanctot, N. Freitas, Dueling network architectures for deep reinforcement learning, in: International Conference on Machine Learning, PMLR, 2016, pp. 1995–2003.

[23] M. Fortunato, M.G. Azar, B. Piot, J. Menick, I. Osband, A. Graves, V. Mnih, R. Munos, D. Hassabis, O. Pietquin, et al., Noisy networks for exploration, arXiv preprint, arXiv: 1706.10295, 2017.
[24] M. Hessel, J. Modayil, H. Van Hasselt, T. Schaul, G. Ostrovski, W. Dabney, D. Horgan, B. Piot, M. Azar, D. Silver Rainbow, Combining improvements in deep reinforcement learning, in: Proceedings of the AAAI Conference on Artificial Intelligence, vol. 32, 2018.
[25] O. Simsek, S. Algorta, A. Kothiyal, Why most decisions are easy in tetris—and perhaps in other sequential decision problems, as well, in: International Conference on Machine Learning, 2016, pp. 1757–1765.
[26] R.J. Williams, Simple statistical gradient-following algorithms for connectionist reinforcement learning, Machine Learning 8 (3–4) (1992) 229–256.
[27] J. Schulman, P. Moritz, S. Levine, M. Jordan, P. Abbeel, High-dimensional continuous control using generalized advantage estimation, 2018.
[28] T.P. Lillicrap, J.J. Hunt, A. Pritzel, N. Heess, T. Erez, Y. Tassa, D. Silver, D. Wierstra, Continuous control with deep reinforcement learning, arXiv preprint, arXiv:1509.02971, 2015.
[29] J. Schulman, S. Levine, P. Abbeel, M. Jordan, P. Moritz, Trust region policy optimization, in: International Conference on Machine Learning, 2015, pp. 1889–1897.
[30] J. Schulman, F. Wolski, P. Dhariwal, A. Radford, O. Klimov, Proximal policy optimization algorithms, arXiv preprint, arXiv:1707.06347, 2017.
[31] K. Cobbe, J. Hilton, O. Klimov, J. Schulman, Phasic policy gradient, arXiv preprint, arXiv:2009.04416, 2020.

CHAPTER

Lightweight deep learning

7

Paraskevi Nousi, Maria Tzelepi, Nikolaos Passalis, and Anastasios Tefas
Department of Informatics, Aristotle University of Thessaloniki, Thessaloniki, Greece

7.1 Introduction

Deep convolutional neural networks have been excelling continuously on various challenging visual analysis tasks and competitions, including the ILSVRC object recognition and detection challenges [1], and the PASCAL VOC challenges [2]. Deep models with parameter-heavy architectures have been successfully trained and deployed on such tasks, partly due to the availability of large collections of annotated datasets, such as the ImageNet or COCO data sets [3], and partly due to the continuous development of increasingly more powerful Graphical Processing Units (GPUs). However, the power consumption and sheer size of such models inhibit their use on mobile and embedded systems, for example, on autonomous robots. The GPUs available for deployment on such systems are inadequate in terms of computational power, thus severely slowing down the performance of large models and rendering real-time deployment almost impossible. Moreover, memory constraints prohibit the direct deployment of large models even when real-time requirements can be relaxed. On the other hand, applications related to visual analysis tasks have become progressively more popular, increasing the demand of deployment of large Convolutional Neural Networks (CNNs) on embedded devices.

This has led recent research toward the optimization of heavyweight CNN architectures for deployment on devices with limited resources. Methods developed for the purpose of allowing large models to be utilized in mobile devices may be divided into two categories based on whether the optimization occurs during or after the training phase, for example, training small networks or compressing a large network after training has converged.

Algorithmic approaches, such as depthwise separable convolutions [4], can also be leveraged to reduce the computational burden. MobileNets [5] exploit both of the aforementioned approaches and were proposed specifically for feature extraction for the purpose of deployment on mobile devices, as their name suggests. SqueezeNet [6] is another lightweight architecture, which achieves AlexNet-level accuracy [7] on the ImageNet classification dataset, while requiring 50× fewer parameters. This is achieved by modifying the standard convolutional layers into fire modules and using small (1×1) and fewer filters in the traditional layers, while maintaining the

Deep Learning for Robot Perception and Cognition. https://doi.org/10.1016/B978-0-32-385787-1.00012-9

dimensions of the input image until the later layers. Other small networks include flattened nets [8] and factorized nets [9], which feature factorized convolutions.

Post-training optimizations effectively separate the training and deployment phases of large models. For example, knowledge distillation methods [10] belong to this category. Furthermore, weight quantization methods may be applied to reduce the memory requirements of a network while attempting to maintain the information contained in the original parameters [11]. Quantization of a deep learning model refers to the process of converting the network parameters to bitwidths lower than single-precision floating point format (FP32), typically to half-precision format (FP16) or low-precision 8-bit integer format (INT8). The benefits of quantization include compressed model sizes, as well as improved inference speed and power usage depending on the hardware platform used [11,12]. Among the typical drawbacks of quantization is the loss of inference accuracy, as well as a dependency on software libraries to efficiently operate on the quantized tensors, the throughput of which heavily relies on the corresponding hardware. It should be noted that quantization can also be used during training to produce quantized models. NVIDIA's TensorRT[1] library offers weight quantization techniques among other optimizations, including layer/tensor fusion schemes, leading to significant speed gains in some cases.

In the context of robotics, the real-time requirements of robotic vision tasks directly clash with the low computational capabilities of embedded systems. Furthermore, deep models typically take up a lot of memory to be stored, which may simply be unavailable. Thus arises the need for reducing the size and computational requirements of deep models to increase their speed while maintaining their accuracy.

This chapter aims to provide insight into various methods geared toward improving the performance of deep lightweight neural networks in a multitude of tasks, including classification and regression, as well as object detection and tracking. The rest of the chapter is structured as follows. Section 7.2 presents lightweight architectures and how they can be effectively used in classification and detection tasks in robotics applications, running at real-time even on embedded devices. Section 7.3 presents various regularization techniques and their importance in training lightweight architectures, which might be prone to converging to local minima more often than their heavier counterparts as a result of their reduced number of parameters. Section 7.4 presents a learnable bag-of-features module, which can improve the performance of deep networks, facilitating the use of lighter architectures on tasks like regression and visual object tracking. Finally, Section 7.5 presents an early-exit architecture for deep networks, which offers adaptive inference capabilities, and which can be tuned to run at real-time for a specific scenario.

[1] https://developer.nvidia.com/tensorrt.

7.2 Lightweight convolutional neural network architectures

7.2.1 Lightweight CNNs for classification

Real-time deployment for high resolution input is of pivotal importance for robotics applications. Consider, for example, the task of person detection in drone-captured images. Objects in such scenarios are of extremely small size, and the practice of resizing the input image in order to meet the limits of on-board deployment at sufficient speed, that is followed by the most state-of-the-art object detectors, results to further shrinkage of the object of interest, rendering the detection even impossible. Furthermore, in [13] where a computation-efficient CNN model is proposed for mobile devices with limited computing power, it is demonstrated that in order to achieve real-time deployment, one has to reduce the input frame resolution to 224×224, sacrificing also the accuracy. In addition, several works have emerged in the recent literature toward designing lightweight models. For example, a recent work proposes to replace 3×3 convolutions with 1×1 convolutions to create a very small network capable of reducing the number of parameters by a factor of 50 while obtaining high accuracy in [6]. Motivated by these observations, a methodology for developing lightweight models capable of effectively operating in real-time (i.e., about 25 frames per second) on devices with limited computational resources and also for high resolution input, for generic classification problems, [14,15] is presented. That is, lightweight fully convolutional models with the design requirement of real-time deployment on low-power GPU for input resolutions of 720p and 1080p have been developed.

The target of this approach is to provide semantic heatmaps by predicting the probability of the object's presence for each location within the captured high-resolution scene. That is, models are trained with RGB input of size $H \times W$, and then test images are propagated to the network, and for every window $H \times W$, the output of the network at the last convolutional layer is computed. The above procedure can also assist toward active perception tasks in robotics applications. That is, the heatmaps generated by the lightweight models can provide a control signal, which indicates to the robot where to look to get a better view of the object under consideration. For example, considering the task of person detection on drone captured input, a real-time model for person detection can run on board so as to produce heatmaps of person presence that will indicate to the drone to move and rotate so as to get a better view of the person.

Considering binary classification problems (e.g., crowd, noncrowd) two architectures consisting of only five convolutional layers are proposed, by discarding the deepest layers and pruning filters of the widely used VGG-16 model [16]. That is, the first four convolutional layers of the VGG-16 model are used with pruned filters, while the last convolutional layer consists of two channels. Based on the attribute that the first model can run in real-time using an NVIDIA Jetson TX2 system for 720p (1280×720) resolution image and the second one can run in real-time for 1080p (1920×1080) resolution image; the models are abbreviated as VGG-720p and VGG-1080p, respectively. Details on kernel sizes, channels, utilized stride, and

Table 7.1 Architecture of VGG-720p network – Input 32 × 32 / Input 64 × 64.

Layer	Kernel	Stride	Pad	Max pooling	Channels
conv1_1	3 × 3	1 / 1	1 / 1	– / –	16
conv1_2	3 × 3	1 / 1	1 / 1	✓/-	16
conv2_1	3 × 3	1 / 1	1 / 1	– / –	24
conv2_2	3 × 3	1 / 4	1 / 1	✓/ ✓	16
conv_last	8 × 8	1 / 1	0 / 0	– / –	2

Table 7.2 Architecture of VGG-1080p network – Input 32 × 32 / Input 64 × 64.

Layer	Kernel	Stride	Pad	Max pooling	Channels
conv1_1	3 × 3	2 / 1	0 / 0	– / –	8
conv1_2	3 × 3	1 / 2	0 / 0	✓/ –	8
conv2_1	3 × 3	1 / 1	0 / 0	– / –	6
conv2_2	3 × 3	1 / 2	0 / 0	– / –	6
conv_last	8 × 8	1 / 1	0 / 0	– / –	2

pad of each layer of the two architectures can be found in Table 7.1 and Table 7.2. The above mentioned models can operate in real-time on low-power GPU, with the assistance of the TensorRT deep learning inference optimizer.

7.2.2 Lightweight object detection

7.2.2.1 Real-time generic object detection on embedded devices

In [17], several object classification and detection algorithms are studied and their performance on various mobile devices, including a Jetson TX2 platform, is compared in terms of speed. The latency versus throughput trade-off is evaluated for different batch sizes and it is shown that using batch sizes larger than 1 is more computationally efficient. In real-time applications, however, images must be processed one-by-one sequentially as they are captured.

In [18], MobileNets are pitted against other popular feature extractors, including the Inception V2 model [19], in the context of feature extraction for object detection, and the effect of altering the input size on the detection precision is examined, among other factors. Larger input sizes lead to larger heatmaps and denser object detection, but impose heavy memory and computational constraints. In contrast, smaller input sizes are processed faster but lead to coarser, less accurate predictions.

For the purpose of object detection, single-stage detectors such as the Single Shot Detector (SSD) [20] and the You Only Look Once (YOLO) detector [21], are significantly faster than their two-stage, region-based counterparts. Although region-based detectors, such as Faster R-CNN [22], are more accurate, they tend to be slower than single-stage detectors as demonstrated in [18].

SSD [20] is a single stage multiobject detector, meaning that a single feed forward pass of an image suffices for the extraction of multiple bounding boxes with coordinate and class information and no Region of Interest (ROI) pooling occurs internally.[2] In its original form, the detector relied on the VGG16 [16] architecture for feature extraction, and added a number of layers upon it to extract better defined boxes. Two versions were proposed, one running at 300×300 input size and one at 500×500, with the latter producing the best results in terms of detection precision while being significantly slower than the first.

In [18], SSD was used as a metaarchitecture for single stage object detection and compared against region-based detectors. Among the findings of that work, was that SSD with MobileNets and Inception V2 for the feature extraction step provided the best time performance at the cost of lower detection precision, as evaluated on the challenging COCO data set.

Another factor which affects the speed of detection is the number of classes to be recognized, since recognizing more classes require more parameters increasing the size of a model and decreasing its speed. Thus in applications were only one or a few classes of objects are to be detected, the detector should be trained to only detect those, to save time. Furthermore, since the last step of detection typically involves using Nonmaximum Suppression (NMS) to collapse overlapping bounding boxes, which depends on the number of detected boxes, this approach can lead to significant speed gains when applicable.

Similar in nature to SSD, YOLO [21] is a widely used object detector, whose popularity may be attributed to its simplicity stemming from its ability to detect multiple objects with a single forward pass of an image, in combination with its speed which surpasses that of SSD. YOLO relies on a custom architecture for its feature extraction step, which is pretrained on ImageNet and publicly available. Its fully-convolutional architecture allows the network to be trained and deployed at any resolution, although odd multiples of 32—the network's total subsampling factor—are preferred to produce a final heatmap, which effectively separates the image into equally sized, overlapping regions. Each such region is responsible for detecting any object whose center lies in it by fitting precomputed anchor boxes to the groundtruth bounding boxes. Thus the input size affects not only the size of the heatmaps, thus the speed of the classifier, but also the maximum number of boxes that can be detected.

A smaller version, named Tiny YOLO, is also available and performs object detection based on the same principles. Using half the convolutional layers, Tiny YOLO sacrifices precision for the sake of speed. The tiny version is also fully convolutional and subsamples the input image by a factor of 32. Thus for an input of 416×416, a 13×13 final heatmap is produced, whose depth depends on the number of classes to be detected and the number of precomputed anchors. Note that only this final output depends on the number of classes. However, NMS is once again used as a post-processing step, and is affected by the total number of detected boxes which naturally grows as the number of classes to detect increases.

[2] This is described in more detail in Chapter 11.

Table 7.3 Lightweight fully-convolutional architecture used for real-time face detection.

Layer	Kernel	Channels	Input	Output	Parameters
conv1	3×3	24	$32 \times 32 \times 3$	$30 \times 30 \times 24$	648
prelu1		24	$30 \times 30 \times 24$	$30 \times 30 \times 24$	24
conv2	4×4	24	$30 \times 30 \times 24$	$14 \times 14 \times 24$	9216
prelu2		24	$14 \times 14 \times 24$	$14 \times 14 \times 24$	24
conv3	4×4	32	$14 \times 14 \times 24$	$11 \times 11 \times 32$	12288
prelu3		32	$11 \times 11 \times 24$	$32 \times 32 \times 3$	32
conv4	4×4	48	$11 \times 11 \times 32$	$8 \times 8 \times 48$	24576
prelu4		48	$8 \times 8 \times 48$	$8 \times 8 \times 48$	48
conv5	4×4	32	$8 \times 8 \times 48$	$5 \times 5 \times 32$	24576
prelu5		32	$5 \times 5 \times 32$	$5 \times 5 \times 32$	32
conv6	3×3	16	$5 \times 5 \times 32$	$3 \times 3 \times 16$	4608
prelu6		16	$3 \times 3 \times 16$	$3 \times 3 \times 16$	16
conv7	3×3	2	$3 \times 3 \times 16$	$1 \times 1 \times 2$	288

7.2.2.2 Real-time face detection

In recent years, commercial and professional robotic units and mobile devices rely on face detection for various applications, such as intelligent video shooting, privacy preservation, security, or affective computing. In robotics, specifically, face detection acts as the first step toward a multitude of applications, such as facial expression recognition, person classification, reality augmentation, but also as a standalone task, for example, as a safety precaution in human-centric tasks.

Deep learning methods have been deployed for the task of face detection with impressive results [23,24], although many approaches use complex networks with hundreds of thousands of parameters, making them inappropriate for real-time face detection with constrained computational power. In [25], a lightweight method for face detection is proposed, trained on input images of size 32×32, using a small neural network with only a little over 76 thousand parameters.

Specifically, a fully convolutional neural network with seven layers is trained for the binary face/nonface classification task on 32×32 image patches, and deployed on full size images during the testing phase, producing heatmaps of face existence probabilities. The architecture is summarized in Table 7.3. Note that zero padding and a stride value of 1 is used for all layers, while the PReLU [26] nonlinearity is used after all convolutional layers except for the last one.

Such lightweight architectures require careful tuning during the training phase, in comparison to heavier alternatives, as they are more prone to false positive detections, having less capacity for recognizing background objects correctly (i.e., not as faces). One way to address this issue, is via progressively adapting the training set, which consists of positive (face) and negative (nonface) samples, to facilitate the training process, similar to curriculum learning methods [27]. Intuitively, easier positive examples should be preferred for the early stages of training, so as to steer the network toward learning the general representation corresponding to the desired class. The balance between positive and negative examples should also be taken into consideration, for example, by maintaining a preset ratio for the number of samples belonging to the two classes. This is also used in generic object detectors, like SSD [20]. As the network converges and learns to correctly identify easy samples, the training set is augmented with slightly harder samples using a progressive positive example mining scheme.

A sample's difficulty can be determined in a number of ways, but perhaps the most relevant method is to use the network's classification confidence for a random set of samples to score the samples. Samples for which the network outputs a high probability are considered as easy samples, whether negative or positive, whereas samples for which the network outputs a low probability are considered hard.

As for negative samples, these are collected in conjunction to the positive ones to maintain the balance, but in order of difficulty instead. Harder negative samples are progressively added to the training set to guide the training process, using a hard negative sample mining procedure. Let $\mathcal{N}_t$ be a collection of images that serve as a pool of negative (nonface) samples, and $\mathcal{P}_t$ be a pool of positive (face) samples, where t denotes that there are multiple steps to the data set augmentation procedure, and thus different pools of negative and positive samples. Let $\mathcal{D}_0$ denote the original training set, consisting of a set of positive samples $\mathcal{P}_0$ and negative samples $\mathcal{N}_0$:

$$\mathcal{D}_0 = \mathcal{P}_0 \cup \mathcal{N}_0 \tag{7.1}$$

on which the network is trained until convergence. A set of images is then passed through the network and the samples are ranked based on the network's classification confidence. The hardest false positive samples $\mathcal{F}_1$ are added to the original set of negative samples, creating $\mathcal{N}_1 = \mathcal{N}_0 \cup \mathcal{F}_1$. The set of positive samples is also enhanced using a preset number of true positive samples $\mathcal{T}_1$, creating $\mathcal{P}_1 = \mathcal{P}_0 \cup \mathcal{T}_1$. The new training data set is formulated as $\mathcal{D}_1 = \mathcal{N}_1 \cup \mathcal{P}_1$. This process can be repeated for a preset number of steps, creating a new training set at each step:

$$\mathcal{N}_{t+1} = \mathcal{N}_t \cup \mathcal{F}_{t+1} \tag{7.2}$$

$$\mathcal{P}_{t+1} = \mathcal{P}_t \cup \mathcal{T}_{t+1} \tag{7.3}$$

$$\mathcal{D}_{t+1} = \mathcal{P}_{t+1} \cup \mathcal{N}_{t+1} \tag{7.4}$$

The training process begins again, starting from the previous convergence point, and the entire process can be repeated T times until the network has reached its learning capacity, that is, when the classification accuracy no longer increases. The training and deployment phases of this network are summarized in Figs. 7.1 and 7.2, respectively.

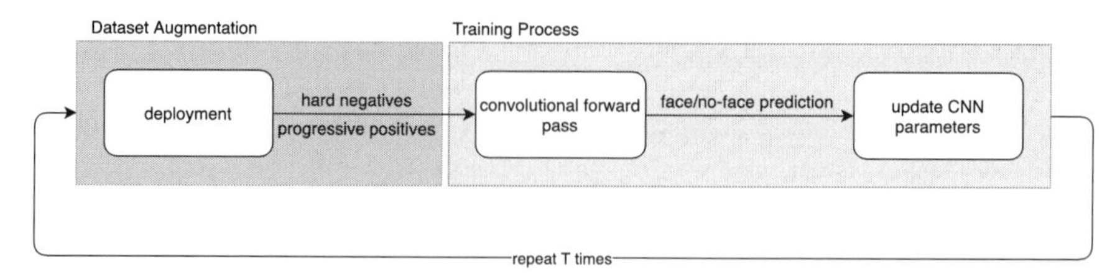

FIGURE 7.1

Training procedure for lightweight face detection with multiple steps of progressive dataset augmentation.

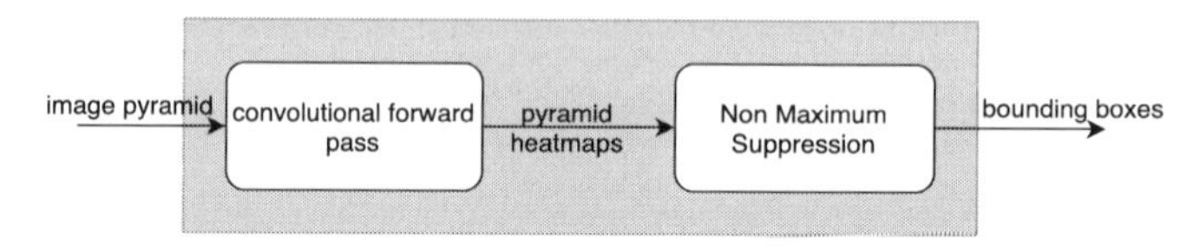

FIGURE 7.2

Deployment phase of lightweight face detection network after the training phase has converged.

7.3 Regularization of lightweight convolutional neural networks

The lightweight models are usually less accurate as compared to the more complex ones. However, this can constitute a major flaw in applications where the improved speed, attained by the lightweight nature of the models at the expense of accuracy, is as important as the accuracy, such as the problem of crowd detection toward crowd avoidance. Therefore, several regularization methods have been proposed in order to control overfitting and improve the performance of the lightweight deep learning models. Regularization, in general, constitutes a crucial element in deep learning, since neural networks are prone to overfitting due to their high capacity. The probably most common regularization technique is the $L2$ regularization, which considering a model that learns a function $f(\mathbf{x};\mathbf{w})$, where $\mathbf{x}$ corresponds to the input vector and $\mathbf{w}$ to the weights, introduces a penalty $\Omega(\mathbf{w}) = \frac{1}{2}\|\mathbf{w}\|_2^2$ to the loss function. Correspondingly, $L1$ regularization adds a penalty $\Omega(\mathbf{w}) = \|\mathbf{w}\|_1$. Dropout [28] has been also established as a successful regularization technique, where for each training sample forward propagated through the network, a randomly selected subset of the activations is set to zero, and then error is backpropagated through the remaining activations in each epoch. Another regularization method consists in early stopping [29] where training is stopped where the performance on a validation set degrades. Besides, multitask learning [30,31] has been employed as a way of enhancing the generalization ability of a model. In the multitask learning concept, the model is modified to predict targets for multiple different tasks at once. Several regularization techniques employing multitask learning have been proposed. In this section, three regularization methods for improving the performance of lightweight deep learning models are presented, based on the concept of multi-task learning. The first one is based on Graph Embedding (GE) framework [32], the second considers binary classification problems with a genuine versus impostor class, and the third one is based on the Quadratic Mutual Information (QMI) criterion [33].

7.3.1 Graph embedded-based regularizer

The graph embedded-based regularization method [34] works by introducing the ideas described in the Graph-Embedded (GE) framework in deep learning utilizing a multiple-loss architecture. Considering a deep neural architecture for classification purposes with a conventional supervised loss, for example, softmax loss (i.e., cross entropy loss), which is widely used in deep neural networks, the graph embedded-based regularization works by attaching one or multiple additional losses which impose certain constraints, motivated by the GE algorithms. The Euclidean loss (sum of squares) can be easily employed to implement them.

Consider the set of N feature representations $\mathcal{Y}^L = \{\mathbf{y}_i^L, i = 1, \ldots, N\}$ of the feature space generated by a specific deep neural layer, L, for the corresponding input of N input images $\mathcal{X} = \{\mathbf{X}_i, i = 1, \ldots, N\}$ as the corresponding projection of training data in the GE concept. Then the softmax loss is defined as

$$L_s = -\frac{1}{N}\sum_{i=1}^{N}\sum_{k=1}^{K} l_{i,k}\log(p_{i,k}), \tag{7.5}$$

where K is the number of classes, $l_{i,k} \in \{0, 1\}$ is a binary indicator that takes the value 1, if the class label k is the correct classification for the sample i, and $p_{i,k}$ is the predicted softmax probability the sample i to belong to the class k.

The Euclidean loss for a target representation $\mathbf{t}_i$, which is determined by a specific GE algorithm is defined as

$$L_e = \frac{1}{2N}\sum_{i=1}^{N} \|\mathbf{y}_i^L - \mathbf{t}_i\|_2^2 \tag{7.6}$$

The Euclidean loss can be attached to all the deep neural layers. Thus for a deep model of N_L layers, the total loss in the regularized training scheme is computed by summing the above losses:

$$L = L_s + \sum_{i=1}^{N_L} \eta_i L_e, \tag{7.7}$$

where the parameter $\eta_i \in [0, 1]$ controls the relative importance of the regularizer of each deep layer, and $\eta_i = 0$ means that no regularizer is attached to the ith layer.

In the following, the Discriminant Analysis (DA) regularization approach, the Minimum Enclosed Ball (MEB) regularization, and the Locally Linear Embedding

(LLE)-inspired regularization are introduced and their objectives are accordingly defined using appropriate target representations in the Euclidean loss. Note that the presented regularization method is straightforward, and can materialize all the GE algorithms (e.g., the marginal Fisher analysis [32], etc.).

7.3.1.1 Discriminant analysis regularization

Inspired by the Linear Discriminant Analysis (LDA) [35] method that aims at best separating training samples belonging to different classes, by projecting them into a new low-dimensional space, so as the between-class separability is maximized, while the within-class variability is minimized, the DA regularized training method, in addition to the softmax loss that preserves the between class separability, includes the Euclidean loss that aims to bring the training samples of the same class closer to the class centroid. That is, considering a labeled representation $(\mathbf{y}_i^L, l_i)$, where $\mathbf{y}_i^L$ corresponds to the image representation, while l_i corresponds to the image label, the additional objective is to minimize the squared distance between $\mathbf{y}_i^L$ and the mean representation of its class.

Let us denote as $\mathcal{X} = \{\mathbf{X}_i, i = 1, \dots, N\}$ the training set consisting of N images, $\mathcal{Y}^L = \{\mathbf{y}_i^L, i = 1, \dots, N\}$ the set of N corresponding feature representations emerged in the L layer of a deep neural model, and $\mathcal{C}^i = \{\mathbf{c}_k, k = 1, \dots, K^i\}$ the set of K^i representations of the ith image, belonging to the same class. The mean vector of the K^i representations of $\mathcal{C}^i$ to the certain image representation $\mathbf{y}_i^L$ is then computed and denoted as $\boldsymbol{\mu}_c^i$. That is, $\boldsymbol{\mu}_c^i = \frac{1}{K^i} \sum_k \mathbf{c}_k$.

Then the additional goal is defined by the following optimization problem:

$$\min_{\mathbf{y}_i^L \in \mathcal{Y}^{\mathcal{L}}} \mathcal{J}_{DA} = \min_{\mathbf{y}_i^L \in \mathcal{Y}^{\mathcal{L}}} \sum_{i=1}^{N} \|\mathbf{y}_i^L - \boldsymbol{\mu}_c^i\|_2^2, \tag{7.8}$$

7.3.1.2 Minimum enclosing ball regularization

In this regularization approach, an objective which aims at finding a projection that minimizes the variance among the training samples is applied. The rationale behind this idea is rooted in the radius-margin based Support Vector Machines (SVM) [36–38]. More specifically, in [39], it is stated that the generalization error bound of the max-margin SVMs depends on not only the squared separating margin, γ^2, of the positive/negative training samples, but on the radius-margin ratio, R^2/γ^2, where R corresponds to the radius of the minimum enclosing ball of all the training samples. For a fixed feature space, the dependency of the error bound on the radius can be ignored in the optimization procedure, since the radius, R, is constant. However, when R is defined by the minimum enclosing ball of the training data, the model has the risk that the margin can be increased by simply expanding the minimum enclosing ball of the training data in the feature space. In order to remedy this problem, an algorithm that optimizes the error bound, taking account of both the margin and the radius, in the context of multiple kernel learning, is proposed in [37].

Toward this end, as the softmax loss aims to distinguish the training samples' feature representations belonging to different classes, since feature representation especially of the negative class may be extremely expanded in the feature space generated by the neural layer, the MEB regularizer aims at shrinking the radius of the MEB of the training samples.

Let us denote by $\mathcal{X} = \{\mathbf{X}_i, i = 1, \dots, N\}$ the set of N training images, and by $\mathcal{Y}^L = \{\mathbf{y}_i^L, i = 1, \dots, N\}$ the set of N feature representations of the deep neural layer, L. The radius of the minimum enclosing ball of all the training samples is abbreviated as R_{MEB}. The squared radius is formally expressed by the following equation:

$$R_{MEB}{}^2 = \min_{R, \mathbf{y}_0^L} R^2, \quad s.t. \quad \|\mathbf{y}_i^L - \mathbf{y}_0^L\|_2^2 \leq R^2, \quad \forall i, \tag{7.9}$$

where $\mathbf{y}_0^L$ is the centroid of all the training samples $\mathbf{y}_i^L$.

However, this definition suffers from a major shortcoming. That is, it can not be applied in terms of mini-batch training, since it requires the centroid of all the training data. In order to tackle this issue, an approximation of the above definition is utilized. The radius of the minimum enclosing ball of the training data is expressed using the maximum pairwise distance over all pairs of training samples. That is,

$$\tilde{R}^2_{MEB} = \max_{i,j} \|\mathbf{y}_i^L - \mathbf{y}_j^L\|_2^2 \tag{7.10}$$

In [38], the authors proved that the radius R_{MEB} is well approximated by $\tilde{R}_{MEB}$ with the following inequality:

$$\tilde{R}_{MEB} \leqslant R_{MEB} \leqslant \frac{1 + \sqrt{3}}{2} \tilde{R}_{MEB} \tag{7.11}$$

Thus, instead of minimizing the squared radius of the smallest sphere enclosing all the training samples, for simplicity the squared diameter that is defined by the maximum pairwise distance over all pairs of the training samples is minimized, since this does not affect the solution of the minimization problem, and following also the work in [38].

Subsequently, since the approximated radius is defined over all the pairs of training samples, the following minimization problem utilizing the softmax function over the max operator, which is nonsmooth, as it is shown in [40] is first formulated and then the approximated radius is further relaxed to make it appropriate for mini-batch training:

$$\min_{\mathbf{y}_i^L \in \mathcal{Y}^\mathcal{L}} \mathcal{J}_{MEB} = \min_{\mathbf{y}_i^L \in \mathcal{Y}^\mathcal{L}} \sum_{i,j}^{N} k_{ij} \|\mathbf{y}_i^L - \mathbf{y}_j^L\|_2^2, \tag{7.12}$$

where

$$k_{ij} = \frac{\exp(a\|\mathbf{y}_i^L - \mathbf{y}_j^L\|_2^2)}{\sum_{i,j}^{N} \exp(a\|\mathbf{y}_i^L - \mathbf{y}_j^L\|_2^2)} \tag{7.13}$$

measures the correlation of the two samples, while the parameter a controls the approximation degree to max operator. When a is infinite, the approximation is identical to the max operator, while when $a = 0$, $k_{ij} = \frac{1}{N^2}$. The relaxed definition of Eq. (7.12) allows for defining the minimization objective in terms of mini-batch training, instead of the whole data set. That is, for a set $\mathcal{B}$ of training samples' feature representations of a batch, Eq. (7.12) becomes

$$\min_{\mathbf{y}_i^L \in \mathcal{Y}^{\mathcal{L}}} \mathcal{J}_{MEB} = \min_{\mathbf{y}_i^L \in \mathcal{Y}^{\mathcal{L}}} \sum_{\mathbf{y}_i^L, \mathbf{y}_j^L \in \mathcal{B}} k_{ij} \|\mathbf{y}_i^L - \mathbf{y}_j^L\|_2^2, \tag{7.14}$$

Thus for $a = 0$, $k_{ij} = \frac{1}{|\mathcal{B}|^2}$, where $|\mathcal{B}|$ is the cardinality of set $\mathcal{B}$, the minimization problem can be formulated as follows in terms of mini-batch training:

$$\min_{\mathbf{y}_i^L \in \mathcal{Y}^{\mathcal{L}}} \mathcal{J}_{MEB} = \min_{\mathbf{y}_i^L \in \mathcal{Y}^{\mathcal{L}}} \sum_{\mathbf{y}_i^L \in \mathcal{B}} \|\mathbf{y}_i^L - \boldsymbol{\mu}\|_2^2, \tag{7.15}$$

where $\boldsymbol{\mu} = \frac{1}{|\mathcal{B}|} \sum_{\mathbf{y}_j^L \in \mathcal{B}} \mathbf{y}_j^L$.

7.3.1.3 LLE-inspired regularization

Motivated by the LLE algorithm, which maps the input data to a new lower dimensional space so as the relationship between the neighboring samples are preserved, a corresponding regularizer can be implemented. That is, let $\mathcal{X} = \{\mathbf{X}_i, i = 1, \ldots, N\}$ be a set of N images of the training set, and $\mathcal{Y}^L = \{\mathbf{y}_i^L, i = 1, \ldots, N\}$ the set of the corresponding feature representations emerged in the L layer of a deep neural model. For each feature representation $\mathbf{y}_i^L$, the set $\mathcal{N}_{(i)}$ that contains its nearest neighbors is also considered. Then the goal is to minimize the Euclidean distance between each feature representation $\mathbf{y}_i^L$ and the mean vector $\boldsymbol{\mu}_{nn}^i$ of its nearest representations.

Thus the following optimization problem is defined:

$$\min_{\mathbf{y}_i^L \in \mathcal{Y}^{\mathcal{L}}} \mathcal{J}_{LLE} = \min_{\mathbf{y}_i^L \in \mathcal{Y}^{\mathcal{L}}} \sum_{i=1}^{N} \|\mathbf{y}_i^L - \boldsymbol{\mu}_{nn}^i\|_2^2, \tag{7.16}$$

where $\boldsymbol{\mu}_{nn}^i = \frac{1}{|N_{(i)}|} \sum_{\mathbf{y}_j^L \in N_{(i)}} \mathbf{y}_j^L$, and $|N_{(i)}|$ denotes the cardinality of set $\mathcal{N}_{(i)}$. However, note that finding the nearest representations for each training sample, comes with additional computational cost.

7.3.1.4 Clustering-based DA regularization

Inspired by the Clustering-based Discriminant Analysis (CDA) [41], which assumes a multimodal data distribution inside classes where each class consists of several subclasses, and aims to enhance the between class discrimination by minimizing the scatter within each subclass, while separating subclasses belonging to different classes, the CDA regularizer is proposed.

That is, let $\mathcal{X} = \{\mathbf{X}_i, i = 1, \ldots, N\}$ be the set of N images of the training set, $\mathcal{Y}^L = \{\mathbf{y}_i^L, i = 1, \ldots, N\}$ the set of N feature representations emerged in the L layer of a deep neural model. Let also $\mathcal{S}^{ck}$ be the set of all image representations $\mathbf{y}_i^L$ that belong to the kth subclass of the cth class.

Then the following optimization problem is defined:

$$\min_{\mathbf{y}_i^L \in \mathcal{Y}^\mathcal{L}} \mathcal{J}_{CDA} = \min_{\mathbf{y}_i^L \in \mathcal{Y}^\mathcal{L}} \sum_{i=1}^{N} \|\mathbf{y}_i^L - \boldsymbol{\mu}_{ck}^i\|_2^2, \tag{7.17}$$

where $\boldsymbol{\mu}_{ck} = \frac{1}{|\mathcal{S}^{ck}|} \sum_{\mathbf{y}_j^L \in \mathcal{S}^{ck}} \mathbf{y}_j^L$, $|\mathcal{S}^{ck}|$ is the cardinality of set $\mathcal{S}^{ck}$, and $\boldsymbol{\mu}_{ck}^i = \boldsymbol{\mu}_{ck}$, if $\mathbf{y}_i^L \in \mathcal{S}^{ck}$. However, this approach also comes with additional computational cost of using a clustering algorithm in order to define subclasses in each class.

7.3.2 Class-specific discriminant regularizer

This regularization method, [14], aims to exploit the nature of binary classification problems, where the one class describes a specific concept, while the other class describes anything but this concept (i.e., genuine versus impostor class). The proposed regularizer traces its origins to LDA based methods [42]; however, based on the extremely wide variation of the impostor class, class-specific concepts [43] are exploited. Specifically, while the classifier aims at distinguishing samples belonging to different classes, it is proposed to enhance the genuine class discrimination, by demanding the representations of the feature space generated by a specific deep neural layer belonging to the genuine class, to come closer to the class centroid. The L_2 norm can be used as similarity measure. The additional CSD criterion acts as regularizer to the classification objective. It could also be demanded for the remaining class to be away from the genuine class centroid; however, the classification objective is considered to preserve the between class separability.

Let us denote as $\mathcal{X} = \{\mathbf{X}_i, i = 1, \ldots, N\}$ the training set consisting of N images, $\mathcal{Y}^L = \{\mathbf{y}_i^L, i = 1, \ldots, N\}$ the set of N corresponding feature representations emerged in the L layer of a deep neural model. Each sample is associated with a class label $c_i = \{0, 1\}$, where the class 1 corresponds to the genuine class. Consider also as $\mathcal{S} = \{(\mathbf{y}_i^L, c_i) : c_i = 1\}$ the set of samples belonging to the genuine class. Then the following objective is defined:

$$\min_{\mathbf{y}_i^L \in \mathcal{S}} \mathcal{J}_{CSD} = \min_{\mathbf{y}_i^L \in \mathcal{S}} \sum_{\mathbf{y}_i^L \in \mathcal{S}} \|\mathbf{y}_i^L - \boldsymbol{\mu}_c\|_2^2, \tag{7.18}$$

where $\boldsymbol{\mu}_c = \frac{1}{|\mathcal{S}|} \sum_{\mathbf{y}_j^L \in \mathcal{S}} \mathbf{y}_j^L$. Optimizing objective (7.18) lets the network learn parameters such that data samples of the genuine class are closely mapped to their class centroid. The Euclidean loss (sum of Ssquares) is used to implement the regularizer. The regularizer can be attached to one or multiple neural layers. Thus for a deep neural model of N_L layers, the total loss in the regularized training scheme is computed by summing the above losses:

$$L_{total} = L_{classification} + \sum_{i=1}^{N_L} \eta_i L_e, \tag{7.19}$$

where the parameter $\eta_i \in [0, 1]$ controls the relative importance of the Euclidean loss, L_e, of each deep layer. It should be finally noted that it is straightforward to show that the optimization problem in Eq. (7.18) can be reformulated as an accumulation of minimization of pairwise distances, [44], that is,

$$\min_{\mathbf{y}_i^L \in \mathcal{S}} \mathcal{J}_{CSD} = \min_{\mathbf{y}_i^L \in \mathcal{S}} \sum_{\mathbf{y}_i^L, \mathbf{y}_j^L \in \mathcal{S}} \|\mathbf{y}_i^L - \mathbf{y}_j^L\|_2^2, \tag{7.20}$$

and thus the proposed regularizer can also be implemented in terms of mini-batch training.

7.3.3 Mutual information regularizer

Finally, the third regularizer, [15], is motivated by the Quadratic Mutual Information (QMI) [33] criterion. Specifically, apart from the classification loss, a regularization loss derived from the so-called *information potentials* of the QMI is introduced. The additional objective is to maximize the QMI, expressed utilizing the information potentials, between the data representations and the corresponding class labels.

Thus in the following, a brief description of the mutual information measure and its quadratic variant is provided, and then the MI regularizer is presented.

Assume random variable Y representing the image representations of the feature space generated by a specific deep neural layer, and a discrete-valued variable C that represents the class labels. For each feature representation $\mathbf{y}$ there is a class label c. The MI measures dependence between random variables, first introduced by Shannon, [45]. That is, the MI measures how much the uncertainty for the class label c is reduced by observing the feature vector $\mathbf{y}$. Let $p(c)$ be the probability of observing the class label c, and $p(\mathbf{y}, c)$ the probability density function of the corresponding joint distribution.

The MI between the two random variables is defined as

$$MI(Y, C) = \sum_c \int_{\mathbf{y}} p(\mathbf{y}, c) \log \frac{p(\mathbf{y}, c)}{p(\mathbf{y})P(c)} d\mathbf{y}, \tag{7.21}$$

where $P(c) = \int_{\mathbf{y}} p(\mathbf{y}, c) d\mathbf{y}$. MI can also be interpreted as a Kullback–Leibler divergence between the joint probability density $p(\mathbf{y}, c)$ and the product of marginal probabilities $p(\mathbf{y})$ and $P(c)$.

QMI is derived by replacing the Kullback–Leibler divergence by the quadratic divergence measure [33]. That is,

$$QMI(Y,C)=\sum_{c}\int_{\mathbf{y}}\left(p(\mathbf{y},c)-p(\mathbf{y})P(c)\right)^2 d\mathbf{y}. \tag{7.22}$$

And thus, by expanding Eq. (7.22) one can arrive at the following equation:

$$\begin{aligned}QMI(Y,C)=&\sum_{c}\int_{\mathbf{y}}p(\mathbf{y},c)^2 d\mathbf{y}+\sum_{c}\int_{\mathbf{y}}p(\mathbf{y})^2P(c)^2 d\mathbf{y}\\&-2\sum_{c}\int_{\mathbf{y}}p(\mathbf{y},c)p(\mathbf{y})P(c)d\mathbf{y}.\end{aligned} \tag{7.23}$$

The quantities appearing in Eq. (7.23) are called *information potentials* and they are defined as follows: $V_{IN}=\sum_c\int_{\mathbf{y}}p(\mathbf{y},c)^2d\mathbf{y}$, $V_{ALL}=\sum_c\int_{\mathbf{y}}p(\mathbf{y})^2P(c)^2d\mathbf{y}$, $V_{BTW}=\sum_c\int_{\mathbf{y}}p(\mathbf{y},c)p(\mathbf{y})P(c)d\mathbf{y}$, and thus the quadratic mutual information between the data samples and the corresponding class labels can be expressed as follows in terms of the information potentials:

$$QMI=V_{IN}+V_{ALL}-2V_{BTW}. \tag{7.24}$$

Assuming that there are N_c different classes, each of them consisting of J_p samples, the class prior probability for the c_p class is given as $P(c_p)=\frac{J_p}{N}$, where N corresponds to the total number of samples. Kernel density estimation [46] can be used to estimate the joint density probability: $p(\mathbf{y},c_p)=\frac{1}{N}\sum_{j=1}^{J_p}K(\mathbf{y},\mathbf{y}_{pj};\sigma^2)$, for a symmetric kernel K, with width σ, where the notation $\mathbf{y}_{pj}$ is used to refer to the j-th sample of the pth class, as well as the probability density of Y as $p(\mathbf{y})=\sum_{p=1}^{J_p}p(\mathbf{y},c_p)=\frac{1}{N}\sum_{j=1}^{N}K(\mathbf{y},\mathbf{y}_j;\sigma^2)$.

Thus, Eq. (7.24) is formulated as follows:

$$V_{IN}=\frac{1}{N^2}\sum_{p=1}^{N_c}\sum_{k=1}^{J_p}\sum_{l=1}^{J_p}K(\mathbf{y}_{pk},\mathbf{y}_{pl};2\sigma^2), \tag{7.25}$$

$$V_{ALL}=\frac{1}{N^2}\left(\sum_{p=1}^{N_c}\left(\frac{J_p}{N}\right)^2\right)\sum_{k=1}^{N}\sum_{l=1}^{N}K(\mathbf{y}_k,\mathbf{y}_l;2\sigma^2), \tag{7.26}$$

$$V_{BTW}=\frac{1}{N^2}\sum_{p=1}^{N_c}\frac{J_p}{N}\sum_{j=1}^{J_p}\sum_{k=1}^{N}K(\mathbf{y}_{pj},\mathbf{y}_k;2\sigma^2). \tag{7.27}$$

The kernel function $K(\mathbf{y}_i,\mathbf{y}_j;\sigma^2)$ expresses the similarity between two samples i and j. There are several choices for the kernel function, [46]. In this work a Euclidean based similarity is used as kernel metric, in order to avoid defining the width of the kernel, in order to ensure that a meaningful probability estimation is obtained, since fine tuning the width is not a straightforward task [47]. Given two vectors $\mathbf{y}_i, \mathbf{y}_j$, the Euclidean-based similarity is defined as $K_{ED} = \frac{1}{1+\|\mathbf{y}_i - \mathbf{y}_j\|_2^2}$.

The pairwise interactions described above between the samples can be interpreted as follows:

- V_{IN} expresses the interactions between pairs of samples inside each class
- V_{ALL} expresses the interactions between all pairs of samples, regardless of the class membership
- V_{BTW} expresses the interactions between samples of each class against all other samples

Thus in the MI regularization, apart from the optimization criterion defined by the conventional supervised loss function, which aims at separating the samples belonging to different classes, an additional optimization criterion is introduced utilizing the information potential defined in Eq. (7.24). It is assumed that the supervised loss preserves the V_{BTW} information potential, which aims to separate samples belonging to different classes. Then the additional objective is to maximize pairwise interactions between the samples described by the $V_{IN} + V_{ALL}$ quantities. The derived joint optimization criterion defines an additional loss function, which is attached to the penultimate convolutional layer and acts as regularizer to the main classification objective:

$$L_{MI} = -(V_{IN} + V_{ALL}), \tag{7.28}$$

where

$$V_{IN} = \frac{1}{N^2} \sum_{p=1}^{N_c} \sum_{k=1}^{J_p} \sum_{l=1}^{J_p} K_{ED}(\mathbf{y}_{pk}, \mathbf{y}_{pl}), \tag{7.29}$$

and

$$V_{ALL} = \frac{1}{N^2} \left(\sum_{p=1}^{N_c} \left(\frac{J_p}{N}\right)^2 \right) \sum_{k=1}^{N} \sum_{l=1}^{N} K_{ED}(\mathbf{y}_k, \mathbf{y}_l). \tag{7.30}$$

The total loss for the network training is defined as

$$L_{total} = L_{classification} + \eta L_{MI}, \tag{7.31}$$

where the parameter $\eta \in [0, 1]$ controls the relative importance of L_{MI}. Gradient descent is utilized to solve the above optimization problem.

7.4 Bag-of-features for improved representation learning

Convolutional Neural Networks (CNNs) require using global pooling layers in order to handle arbitrary-sized images [48,49] since without them the output of the last convolutional layer would depend on size of the input image. Therefore without using global pooling, it would be very difficult to adjust the complexity of the feedforward process simply by resizing the input image, rendering the application of these architectures on embedded and mobile devices, such as real-time systems on robots, especially challenging. On the other hand, global pooling methods allow for significantly reducing the dimensionality of the vector extracted from the last convolutional layer [48,49], while also providing a representation that has constant size and which does not depend on the size of the network's input. However, despite the ability of such pooling layers to handle images of any size, they come with several drawbacks, for example, they rely on the previous layers of the network to provide scale invariance, they lead to loss of valuable information regarding the distribution of the feature vectors [50], etc.

These limitations were addressed to a great extent by a powerful pooling layer inspired by the Bag-of-Features (BoF) model (also known as Bag-of-Visual-Words (BoVW) model) [51,52]. Bag-of-features-based pooling acts a trainable quantization-based layer and enables us to overcome many of the aforementioned limitations. It is worth noting that the BoF model was also originally proposed to solve a similar problem, that is, to extract a small, constant length summary representation over a set that contains feature vectors extracted from a visual object. To better understand how BoF works, one can consider the steps involved when using this model:

1. A feature extractor, such as SIFT extractor [53], is used to process an input image and extract feature vectors that describe various of its regions. Note that the number of feature vectors that are extracted depend on various parameters (e.g., the size of the image, the type of interest point detector that is used, etc.) and in many cases this leads to a different number of feature vectors per image.
2. Then the dictionary learning process follows, where a set of prototype feature vectors is learned, typically called *codewords*. The set of all codewords is also called *codebook* or *dictionary*.
3. Finally, in order to extract a constant length representation for each image each of the extracted feature vectors is quantized using the codewords. Then the final histogram representation of each input image can be extracted simply by counting the number of feature vectors that were assigned to each codeword.

Note that even though this process was originally designed for handcrafted feature extractors, it can be also readily applied for feature maps extracted out of convolutional layers, as shown in Fig. 7.3.

Neural BoF approaches, such as [54], extend the classical BoF model by allowing to learn the codebook in a way that accommodates the task at hand. For DL models, this means that the codebook can be learned simply by using the regular back-propagation algorithm. In the rest of this section a neural formulation of the BoF pooling layer [54] is presented. However, it is worth noting that such a formulation can be also extended to handle many different architectures and tasks, such as time-series forecasting [55] and color-constancy applications [56].

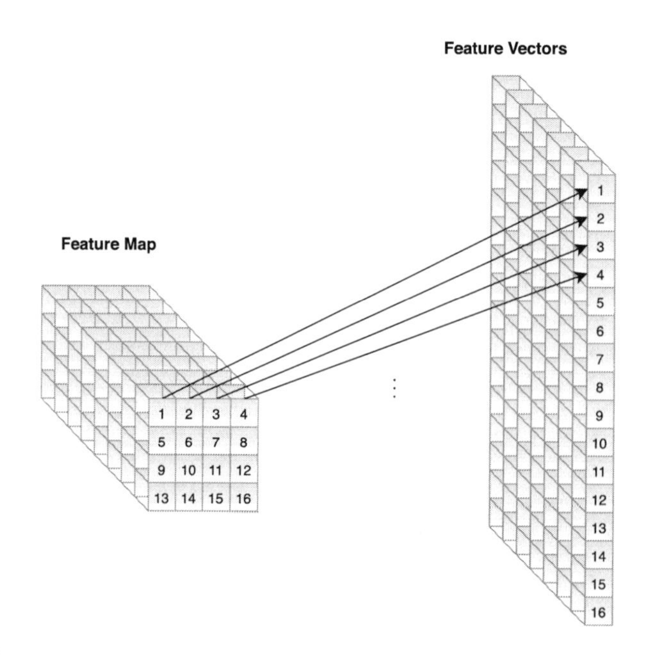

FIGURE 7.3

A feature map extracted from a convolutional layer can be readily converted into a set of feature vectors that can be fed to the BoF model.

Let $\mathbf{x}_j \in \mathbb{R}^{N_F}$ denote the jth feature vector extracted for an image x fed into a CNN, where N_F refers to the number of channels of the Lth convolutional layer that is used for the feature extraction process. The size of the feature map, that is, its width and height, defines the number of feature vectors that will be eventually available to the BoF layer. For example, if a 12×12 feature map is extracted, then 144 feature vectors will be used by the BoF layer. To simplify the presentation of the BoF, the notation N is used to refer to the number of feature vectors extracted from an image.

Each image can be represented using the set of N extracted feature vectors (extracted from the corresponding feature map, as shown in Fig. 7.3):

$$\mathbf{x}_j \in \mathbb{R}^{N_F} (j = 1, \ldots, N). \tag{7.32}$$

BoF first quantizes these feature vectors and then extracts a histogram vector for each image. Note that this process, as it will become apparent later in this section, completely decouples the length of the histogram vector from the number of the extracted feature vectors and their dimensionality. Instead, the dimensionality of the histogram vector is merely controlled by the number of codewords that are used during the quantization process.

Usually, BoF, when used in DL models, is formulated using two neural layers. The first one measures the similarity of the input feature vectors to the codewords and quantizes the input feature vectors. The second one performs the accumulation of these similarity vectors and creates the final histogram representation. These two layers compose a unified architecture that is differentiable and can be used with any DL model. Regarding the first layer note that any differentiable similarity metric can be used to perform the quantization process. Usually a soft BoF formulation is used [54], which employs a Radial Basis Function (RBF) kernel. Therefore, the output of kth codeword neuron $[\boldsymbol{\phi}(\mathbf{x})]_k$, that is, the neuron that measures the similarity with the kth codeword, is defined as

$$[\boldsymbol{\phi}(\mathbf{x})]_k = \exp(-||\mathbf{x} - \mathbf{v}_k||_2/\sigma_k) \in \mathbb{R}, \tag{7.33}$$

where $\mathbf{x}$ is a feature vector and $\mathbf{v}_k$ is the corresponding codeword (center of the RBF neuron). Note that typically a scaling factor σ_k is used to adjust the RBF kernel to the actual distribution of the feature vectors. A total of N_K neurons are used in this layer, which also implicitly control the size of the extracted histogram vector. Also, note that usually the output of this layer is normalized in order to provide a distribution over the similarity to the N_K codewords as

$$[\boldsymbol{\phi}(\mathbf{x})]_k = \frac{\exp(-||\mathbf{x} - \mathbf{v}_k||_2/\sigma_k)}{\sum_{m=1}^{N_K} \exp(-||\mathbf{x} - \mathbf{v}_m||_2/\sigma_m)} \in \mathbb{R}. \tag{7.34}$$

Note that the first layer extracts one *membership vector* for each input feature vector. These vectors are then accumulated in the second layer of the BoF model simply by aggregating them as

$$\mathbf{s} = \frac{1}{N_i} \sum_{j=1}^{N_i} \boldsymbol{\phi}(\mathbf{x}_j) \in \mathbb{R}^{N_K}, \tag{7.35}$$

where $\boldsymbol{\phi}(\mathbf{x}) = ([\boldsymbol{\phi}(\mathbf{x})]_1, \ldots, [\boldsymbol{\phi}(\mathbf{x})]_{N_K})^T \in \mathbb{R}^{N_K}$ denotes the output vector of the RBF layer. The histogram $\mathbf{s}$ provides a distribution over the codewords, each of which corresponds to a different semantic prototype [50], and as a result, describes the semantic content of each image. This vector can be then fed to the fully connected layers of a DL architecture or directly used for other representation learning tasks.

The aforementioned process extracts a histogram that only models the distribution of the extracted feature vectors, throwing away most of the spatial information (which is implicitly provided by the position from which they were extracted). To overcome this limitation, spatial segmentation schemes can be used, such as the spatial pyramid matching method [51]. To this end the image is divided into a number of regions and a different histogram vector is compiled for each region. Then these histograms are fused together to form the final histogram representation, as shown in Fig. 7.4. Note that when such schemes are used, the size of the extracted representation increases by a multiplicative factor N_S, where N_S is the number of spatial regions used.

Finally, the parameters of the model, that is, the centers of the RBF neurons, along with their scaling factors, can be directly learned using gradient descent along with

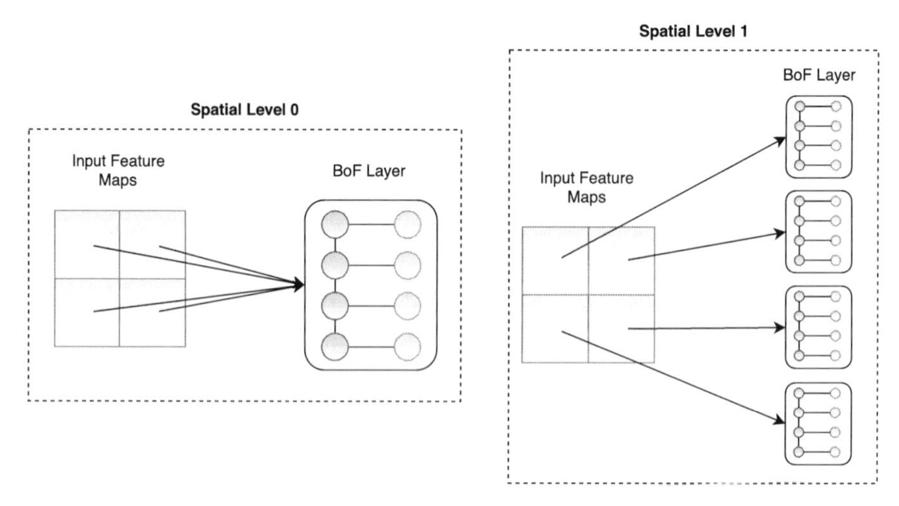

FIGURE 7.4

Using spatial segmentation multiple histograms can be extracted using the same BoF layers, allowing for capturing more spatial information and creating more discriminative representations.

the rest of the parameters of the model:

$$\Delta(\mathbf{V}, \boldsymbol{\sigma}) = -(\eta_V \frac{\partial \mathcal{L}}{\partial \mathbf{V}}, \eta_\sigma \frac{\partial \mathcal{L}}{\partial \boldsymbol{\sigma}}), \tag{7.36}$$

where $\mathcal{L}$ is the loss function used for training the network and η_V and η_σ are the learning rates for the codebook and scaling factors. Note that in many cases setting these two learning rates to two different values, for example, using a larger value for the scaling factors, can accelerate the learning process. Also, pretraining the codebook by first performing k-means on a set of extracted feature vectors, as proposed in [57], can further improve the performance of the BoF layer, especially when pretrained feature extractors are employed.

7.4.1 Convolutional feature histograms for real-time tracking

In visual object tracking, histogram-based representations and representation learning has been widely used with impressive results, as shown in the mean shift tracking method [58], as well as in KCF [59], among others. In these approaches, as well as in similar ones, it is typical to extract histogram representations from the visual target's color representation, for example, in RGB or HSV space, or on engineered features, such as Histogram of Oriented Gradients (HOG) features. In general, histogram representations are more robust to appearance changes, such as those related to object pose and viewpoint [54]. Thus trackers which rely on histogram representations (e.g.,

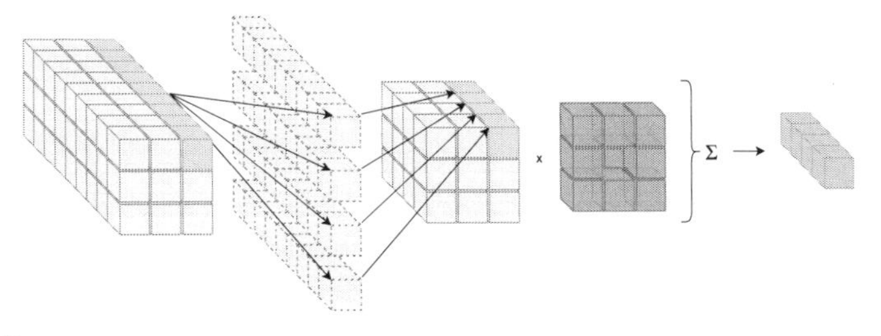

FIGURE 7.5

Single histogram extraction process. Each feature vector in the original input volume of size $3 \times 3 \times 9$ is compared against each of the 4×9 codewords, and memberships of size $3 \times 3 \times 4$ are extracted. After an average pooling operation of size 3×3, a single histogram of size $1 \times 1 \times 4$ is extracted.

[60–62]) are better equipped to deal with challenges related to interframe appearance variations of the tracked target.

In [63], BoF layers [54] were used in a fully convolutional neural architecture, taking advantage of the discriminative ability of convolutional features as well as of the aforementioned benefits of histogram representations. Let $\mathbf{F} \in \mathbb{R}^{H \times W \times C}$ be the feature vector representation of the target, where H, W are the representation's spatial dimensions and the number of channels is denoted as C (e.g., target size in pixels and color channels). To extract histogram representations from these vectors, a set of M codewords $\mathbf{B} \in \mathbb{R}^{M \times C}$ needs to be extracted first. Typically, these can be extracted by clustering the data samples into M clusters, and the resulting cluster centers constitute the codewords. Then each feature vector $\mathbf{f}_{ij} \in \mathbb{R}^C$ in $\mathbf{F}$, with i, j indexing its spatial dimensions, is compared against each of the codewords $\mathbf{b}_m \in \mathbb{R}^C$, $m = 1, \ldots, M$ from $\mathbf{B}$, extracting $H \times W$ membership vectors $\mathbf{G} \in \mathbb{R}^{H \times W \times M}$, by using a similarity metric, such as the normalized Euclidean similarity:

$$g_{ijm} = \frac{\exp(-\|\mathbf{f}_{ij} - \mathbf{b}_m\|_2^2)}{\sum_{n=1}^{M} \exp(-\|\mathbf{f}_{ij} - \mathbf{b}_n\|_2^2)} \tag{7.37}$$

where g_{ijm} is the membership of the $\{i, j\}$-th feature vector to the mth codeword and the notation $\| \cdot \|_2$ refers to the l_2 norm.

After all memberships have been calculated, the histogram representation can be extracted by aggregating the membership vectors spatially. Formally, a single histogram $\mathbf{q} \in \mathbb{R}^M$ may be extracted as

$$\mathbf{q} = \frac{1}{H \cdot W} \sum_{i=1}^{H} \sum_{j=1}^{W} \mathbf{g}_{ij} \tag{7.38}$$

The above process is illustrated in Fig. 7.5. Each of the feature vectors in the original feature volume is compared against each of the codewords, and a new volume of memberships is computed. The memberships are weighted spatially so that the final histogram representation is extracted.

The use of Euclidean similarity in the aforementioned process is quite slow, and thus inappropriate for use in visual object tracking applications. The LBoF [50] module, formulated as a normalized convolutional layer, is a better match. In this case, the memberships are computed as the absolute value of the inner product between each feature vector and each codeword, and subsequently normalized by the sum of all memberships by spatial location:

$$g_{ijm} = \frac{|\mathbf{b}_m^T \cdot \mathbf{f}_{ij}|}{\sum_{n=1}^{N} |\mathbf{b}_n^T \cdot \mathbf{f}_{ij}|}. \tag{7.39}$$

The histogram representation can then be obtained normally, that is, by aggregating the resulting memberships as in Eq. (7.38). In practice, Eq. (7.39) can be implemented as a convolutional layer followed by an absolute value activation function and a sum normalization operation, making it significantly faster.

Note that Eq. (7.38) obliterates spatial information by aggregating an input volume into a single histogram. This is often undesirable and can be counteracted by extracting multiple histogram representations by using a spatial sliding window method to aggregate the memberships. Formally, using a sliding window of size $p \times p$ with stride 1, multiple histogram representations $\mathbf{Q}^{H_q \times W_q \times M}$ can be obtained as

$$\mathbf{q}_{kl} = \frac{1}{p \cdot p} \sum_{i=k-p/\!/2+1}^{k+p/\!/2+1} \sum_{j=l-p/\!/2+1}^{l+p/\!/2+1} \mathbf{g}_{ij} \tag{7.40}$$

where $/\!/$ denotes integer division, $k = 1, \ldots, H_q$ and $l = 1, \ldots, W_q$. This can be translated into a standard average pooling layer. Stride can be introduced into Eq. (7.40) to produce less or even nonoverlapping histograms.

Given a target region $\mathcal{T}$, corresponding to the bounding box of the desired target to be tracked, a target model $\mathbf{Q} \in \mathbb{R}^{H_q \times W_q \times M}$ can be computed using the process described above. A candidate model $\mathbf{P} \in \mathbb{R}^{H_p \times W_p \times M}$ can be extracted from a search region $\mathcal{S}$. Then each histogram corresponding to the search region can be compared against the corresponding histogram from the target. The comparison can be made using any histogram comparison method, such as the Bhattacharyya coefficient for two histograms $\mathbf{p}$, $\mathbf{q}$:

$$s = \sum_{m=1}^{M} \sqrt{p_m q_m} \tag{7.41}$$

such that a map $\mathbf{S} \in [0, 1]^{U \times V}$ of similarities is produced for all possible locations of the target, where U, V depend on the sizes of the compared histogram volumes. For multiple target histograms, a single similarity score is produced by averaging the similarity of each individual histogram with the corresponding search region histogram.

Table 7.4 Tracker network architecture, based on AlexNet [7].

Layer	Kernel size	Filters	Stride
conv_block1	11×11	96	2
max_pool1	3×3		2
conv_block2	5×5	256	1
⊢ lbof1	$n = 5,\ p = 5$	256	1
max_pool2	3×3		2
conv_block3	3×3	384	1
conv_block4	3×3	384	1
└ lbof2	$n = 3,\ p = 3$	256	1
lbof1 ⊕ lbof2		512	

The spatial location at which the similarity is maximum is chosen as the new location of the target in the search region.

Multiple BoF modules can be used in the same network to incorporate information from various stages of the network, for example, to combine low-level features with high-level ones, as shown in Table 7.4. Each of the BoF modules is parameterized by the size of the codewords (i.e., filters) n, the number of codewords M, and the size and stride of the pooling operation p, s, respectively. By opting to maintain the receptive field of the histograms about the same as that of the respective convolutional layer, only the number of codewords M must be defined for each LBoF module. Let $\{\mathbf{Q}^d \in \mathbb{R}^{H_q \times W_q \times M^d}\}_{d=1}^{D}$ represent the histograms extracted at multiple levels $d = 1, \ldots, D$ of a neural network. The final representation is obtained as

$$\mathbf{Q} = \mathbf{Q}^1 \oplus \cdots \oplus \mathbf{Q}^D \tag{7.42}$$

where the $\oplus$ denotes the concatenation operation along the channel dimension. Note the histogram representations are only used at the end of the layer hierarchy and they are not the input of any convolutional layer. This allows the natural flow of gradients that a standard convolutional network would have to still subsist, allowing the network to be trained more effectively. Furthermore, a learnable, per-codeword weight vector $\mathbf{W} \in \mathbb{R}^{M_1 + \cdots + M_d}$ is applied on the final representation in an elementwise multiplication, to weigh the importance of each codeword, and thus each histogram representation:

$$\mathbf{Q}_w = \mathbf{W} \odot \mathbf{Q} \tag{7.43}$$

where $\odot$ symbolizes elementwise multiplication. By optimizing these weights during training, the network learns to attend to the various levels of histograms as needed.

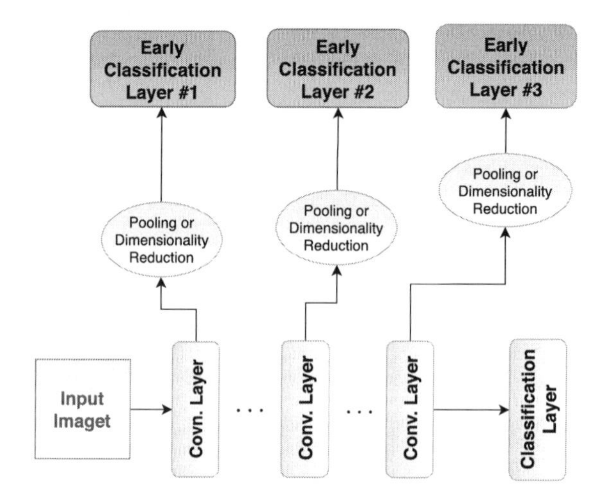

FIGURE 7.6

Early exits can be used directly (by employing pooling or dimensionality reduction layers) to estimate the final output of the network at various points of its computational graph.

7.5 Early exits for adaptive inference

Many different applications, such as real-time robots, require models that are able to dynamically adapt to the available resources and load, for example, by providing faster (yet less accurate) inference when the system is loaded and more refined and accurate (yet slower) predictions when the load is lower. For example, for a face recognition system, a higher load is expected when many people appear in a frame as opposed to a lower load when only a few people appear in a frame. In order to meet the constant inference requirements, which are imposed by many real-time systems, DL models that can adaptively alter their computational graph on-the-fly in order to keep a constant inference time can be used. These models, known as models with *adaptive computational graphs* [64–66], enable the dynamical adaptation of the inference process to the available computational resources by altering the path that an input samples follows on the model's graph.

Despite the effectiveness of this approach, developing models with such adaptive computational graphs is not always straightforward. Perhaps the easiest way to achieve this is by including additional early exits layers on top of specific layer of the network [64,66,67], as shown in Fig. 7.6. Using early exits enable us to estimate the output of the model without having to feedforward the whole network, providing an effective way to adapt the model on-the-fly to the available resources. This process can be also used to allow the network to adapt to the difficulty of a sample, for example, using earlier exists for easier samples and later exits for more difficult ones [68]. This also allows for reducing the energy consumption of an embedded system, as well as the average inference speed, given an appropriate and effective way to select which exit should be used based on the difficulty on each sample.

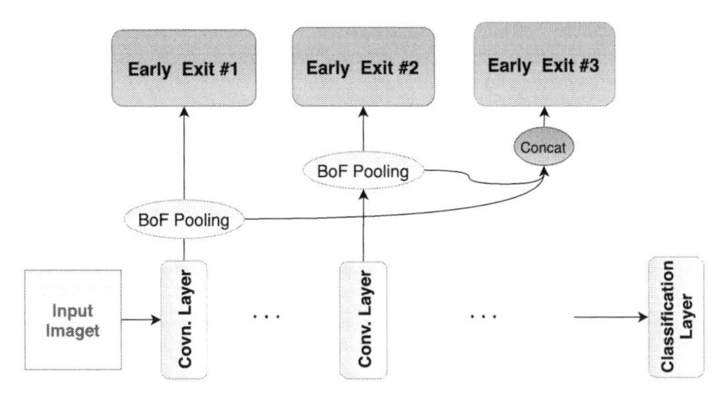

FIGURE 7.7

Replacing pooling or dimensionality reduction layers with the bag-of-features model allows for having more efficient early exits formulations that can be also used in a hierarchical fashion.

Unfortunately, using early exits, especially on earlier layers of a network, is not always straightforward, due to the enormous dimensionality of the respective feature maps. Some methods employ aggressive subsampling methods, for example, global average pooling [69], which, however, ignore both the spatial information and the distribution of the extracted feature vectors, reducing the accuracy of early exits. This problem is partly addressed in [67], where a series of densely connected structures is employed. However, this requires a significant number of structural changes in the architecture of a network and cannot be easily used with existing neural networks. On the other hand, in [64], this problem is addressed using additional dimensionality reduction layers, which can alleviate this issue. However, these extra layers can also increase the computational complexity of using early exits.

In this section, another recently proposed approach [70] is presented, which resolves these issues by using a bag-of-features-based [50] representation, as presented in Section 7.4, extracted from each convolutional layer where an early exit is used. In this way, a compact summary representation is created, enabling the efficient use of early exits, as shown in Fig. 7.7. Also, this allows to efficiently reuse the information extracted by the previous layers, which can further increase the prediction accuracy for the latter exit layers, at almost no extra cost. Therefore, bag-of-features-based early exits allow us to estimate the final output of the network at various points of the inference process, without having to feedforward through the whole network, while also keeping more information regarding the distribution of the features extracted from each layer, as well as encoding more spatial information, compared to previous approaches. Furthermore, this approach can be also used with efficient hierarchical aggregation schemes that allow for constructing common histogram representation spaces, which can gradually refine the estimation of the network using information extracted from previous layers, as also shown in Fig. 7.8. The rest of this section is structured as follows. In Section 7.5.1 the bag-of-features-based early exits method is

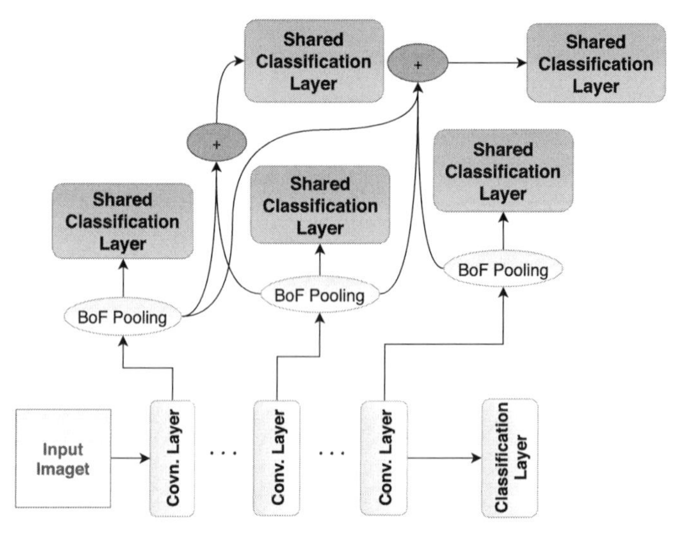

FIGURE 7.8

The efficiency of early exits can be further improved by reusing the classification layer among different early exits, as well as using additive histogram representations. This can reduce the number of required parameters by forming implicit common representations spaces, where the predictions are gradually refined as more early exits are used.

presented, while in Section 7.5.2 different ways to use adaptive inference with early exits to speedup DL models are discussed.

7.5.1 Early exits using bag-of-features

The notation $f_{\mathbf{W}}(\mathbf{x}, i)$ is used to refer to the response of the ith layer of a DL model that has a total of m layers (trained to solve a classification problem). Also, the notation $\mathbf{W}$ is used to refer to learnable parameters of the model, while the notation $\mathbf{x}$ is used to refer to the input to the neural network. For this section, consider the case of convolutional DL models that operate on images. However, this is without loss of generality, since the proposed method can be also readily applied for other kinds of input data as well. Therefore an input image is denoted as $\mathbf{x} \in \mathbb{R}^{W \times H \times C}$, where W is its width, H is its height, and C is the number of channels. Also, let $\mathbf{y}^{(i)} = f_{\mathbf{W}}(\mathbf{x}, i) \in \mathbb{R}^{W_i \times H_i \times C_i}$ be the output feature map of the ith convolutional layer. Again, W_i, H_i, and C_i refer to the width, height, and number of channels of the corresponding output. Also, the final output of the DL model, which is used to estimate the probability of each sample to belong to a different class (out of a total of N_C classes), is denoted by $\mathbf{y} = f_{\mathbf{W}}(\mathbf{x}, m) \in \mathbb{R}^{N_C}$.

DL models are typically trained using back-propagation to minimize a loss function $\mathcal{L}$:

$$\mathbf{W}' = \mathbf{W} - \eta \sum_{j=1}^{N} \frac{\partial \mathcal{L}(f_{\mathbf{W}}(\mathbf{x}_j, m), \mathbf{t}_j)}{\partial \mathbf{W}}, \tag{7.44}$$

where $\mathcal{X} = \{\mathbf{x}_1, \mathbf{x}_2, \ldots, \mathbf{x}_N\}$ is a training set of N images and each image is accompanied by a target (ground truth annotation) vector $\mathbf{t}_i \in \mathbb{R}^{N_C}$. The updated parameters of the network, after one optimization step, are denoted by $\mathbf{W}'$. For the rest of this section, assume that the cross-entropy loss function is used, since the model is trained for classification tasks:

$$\mathcal{L}(\mathbf{y}, \mathbf{t}) = -\sum_{i=1}^{N_C} [\mathbf{t}]_i \log([\mathbf{y}]_i), \tag{7.45}$$

where $[\mathbf{y}]_i$ refers to the ith element of a vector $\mathbf{y}$.

Early exits propose using an additional estimator $g^{(i)}_{\mathbf{W}_i}(\cdot)$ on the feature maps extracted at the ith layer of a model as

$$g^{(i)}_{\mathbf{W}_i}\left(\mathbf{y}^{(i)}\right) = g^{(i)}_{\mathbf{W}_i}(f_{\mathbf{W}}(\mathbf{x}, i)), \tag{7.46}$$

where $\mathbf{W}_i$ refers to the parameters of the corresponding estimator. This formulation allows for estimating the output of the network at different levels, without feedforwarding the whole network. Therefore the total inference time can be adjusted by selecting a different estimator and stopping the inference process there.

Since the dimensionality of the extracted feature map is typically too large to be directly handled by a lightweight estimator, for example, a linear layer, the bag-of-features model is used on each early exit. Therefore first each feature vector that is extracted from a different spatial location of a feature map is quantized by employing a set of N_K codewords. Each codeword is denoted by $\mathbf{v}_{ij}$, where i refers to the layer on which the BoF layer is used and j to the codewords. As explained in [57], each prototypical concept is represented by a codeword, which allows for efficiently constructing compact summary representations of the *concepts* that appear on each feature map. Also, note that a different set of codewords is used for each early exits, since each convolutional layer captures concepts of different abstraction. Therefore, for each feature vector $\mathbf{y}^{(i)}_{kl}$ a membership vector can be extracted, that is calculated as

$$[\mathbf{u}_{ikl}]_j = \frac{K(\mathbf{y}^{(i)}_{kl}, \mathbf{v}_{ij})}{\sum_{m=1}^{N_K} K(\mathbf{y}^{(i)}_{kl}, \mathbf{v}_{im})} \in [0, 1], \tag{7.47}$$

where (k, l) is the location from which the feature vector was extracted and $K(\cdot)$ is a kernel function that is used to measure the similarity between a feature vector and a codeword. In other words the membership vectors measure how similar is each feature vector to each prototypical concept (note that $||\mathbf{u}_{ikl}||_1 = 1$, where $|| \cdot ||_1$ denotes the l_1 norm of a vector).

To compile the final constant length summary histogram for each image the membership vectors can simply be aggregated as

$$\mathbf{s}^{(i)} = \frac{1}{W_i H_i} \sum_{k=1}^{W_i} \sum_{l=1}^{H_i} \mathbf{u}_{ikl} \in \mathbb{R}^{N_K}. \tag{7.48}$$

This histogram vector provides a semantic summary of the features extracted from each layer. As a result, this compact vector can be used to fit the early exit estimator instead of performing naive pooling operations (e.g., average or sum pooling) that lead to loosing valuable discriminative information or using more complex convolutional dimensionality reduction layers. Note that spatial segmentation schemes [57] can be also used to keep more spatial information into the extracted representation, if needed.

To further improve the performance of models that use early exits a hierarchical inference structure can be built, as shown in Fig. 7.7, by incrementally concatenating the current BoF representation with the ones extracted from previous layers. As a result, the final representation $\mathbf{s}^{(i,h)}$ can be extracted from the ith layer as

$$\mathbf{s}^{(i,h)} = \begin{cases} \mathbf{s}^{(i)} & \text{if } i = 1 \\ \mathbf{s}^{(i)} \frown \mathbf{s}^{(i-1,h)} & \text{if } i > 1 \end{cases}, \tag{7.49}$$

where the operator $\mathbf{a} \frown \mathbf{b}$ is used to refer to the concatenation of vectors $\mathbf{a}$ and $\mathbf{b}$. By caching and reusing the representations extracted from the previous layers the classification accuracy can increase at virtually no additional cost, apart from the use estimators with more parameters. The number of parameters can be further reduced by using an additive formulation for building the histogram representations (Fig. 7.8) as

$$\mathbf{s}^{(i,h)} = \begin{cases} \mathbf{s}^{(i)} & \text{if } i = 1 \\ (1-\alpha)\mathbf{s}^{(i)} + \alpha \mathbf{s}^{(i-1,h)} & \text{if } i > 1 \end{cases}, \tag{7.50}$$

where α is a decay factor, used to perform exponential averaging over the histograms. This also enables us to reuse the same estimator [68], further reducing the number of parameters required for early exits by building an implicit common histogram representation space, which is incrementally refined as more early exits are used.

7.5.2 Adaptive inference with early exits

Apart from using early exits for adapting the computational graph to the available resources, early exits can be also used to stop the inference process early for easier input samples. However, in order to do so, an appropriate stopping criterion should be defined. Perhaps the most straightforward solution is to estimate the difficulty of each sample based on the uncertainty of the network $r(\mathbf{x}, i)$ for a given sample $\mathbf{x}$ at the ith exit based on the confidence of the network for the class with the highest activation. This uncertainty can be estimated as

$$r(\mathbf{x}, i) = [g^{(i)}_{\mathbf{W}_i}(\mathbf{y}^{(i)})]_k, \tag{7.51}$$

where

$$k = \arg_{k'} \max[g^{(i)}_{\mathbf{W}_i}(\mathbf{y}^{(i)})]_{k'}. \tag{7.52}$$

Then the threshold for stopping the inference process at the ith early exit can be calculated based on the mean activation of the winning neurons:

$$\mu_i = \frac{1}{N}\sum_{j=1}^{N}[g^{(i)}_{\mathbf{W}_i}(\mathbf{y}_j)]_k, \tag{7.53}$$

where $g^{(i)}_{\mathbf{W}_i}$ denotes the output of the ith exit and $k = \arg\max g^{(i)}_{\mathbf{W}_i}(\mathbf{y}_j)$. Therefore the computation graph is adapted to the difficulty of the input sample by stopping the inference process on an early exit when the confidence on a class is higher than the corresponding threshold, that is,

$$r(\mathbf{x}, i) > \alpha\mu_i, \tag{7.54}$$

where α is an inference hyperparameter. That is, using larger values for this parameter leads to more accurate results, but more layers of the network will be used, increasing the inference time. On the other hand, using smaller values for the α parameter can increase the speed, but can also lead to less accurate classification results, since the inference process will use only a few early layers of the network.

Even though the aforementioned process is easy to implement, the value of the inference hyperparameter α is not bounded, requiring a significant amount of experimentation in order to select the most appropriate value. More recent early exit formulations [71] can overcome this issue by calculating this threshold while taking into account both the mean activation of a layer μ_i (to provide a lower bound for the confidence), as well as the output of the model μ_C (to provide an upper bound). As a result, the threshold can be estimated as

$$r(\mathbf{x}, i) > (1-\alpha)\mu_i + \alpha\mu_C, \tag{7.55}$$

for $\alpha \in [0, 1]$. This allows for having more interpretable values for α, since this hyperparameter is now bounded between 0 and 1.

7.6 Concluding remarks

Deep learning methods have provided significant improvements in performance terms in multiple visual analysis tasks, such as classification, object detection, and tracking. Such tasks have a multitude of applications in robotics, and are thus worthy of investigation. One key aspect of this investigation, however, is heavily influenced by the constraints imposed by the use of low-power embedded devices, both in terms of computation speed and memory restrictions. Deep learning models typically contain hundreds of thousands or even million of trainable parameters, which give them their edge in terms of performance, but also hinder their direct use in robotics applications, which suffer the above limitations.

Thus recent research has focused on introducing lightweight neural architectures and developing methods, which improve the effectiveness of such models. Specifically, regularization techniques can be utilized during the training phase of lightweight models, which have been shown to improve the performance of multiple architectures in classification tasks. Furthermore, bag-of-features learnable modules, which create histograms of convolutional features and can be trained in an end-to-end fashion, offer significantly improved models in tasks like regression and visual object tracking, allowing for the use of more lightweight models. Finally, an early-exit strategy was presented, which can be tuned to facilitate its use in applications with limited resources by effectively adapting their speed-accuracy trade-off depending on the application. Despite such advances, the research field of lightweight neural architectures remains challenging as embedded devices impose significant constraints on the models that can be successfully deployed on them.

References

[1] O. Russakovsky, J. Deng, H. Su, J. Krause, S. Satheesh, S. Ma, Z. Huang, A. Karpathy, A. Khosla, M. Bernstein, A.C. Berg, L. Fei-Fei, ImageNet large scale visual recognition challenge, International Journal of Computer Vision (IJCV) 115 (3) (2015) 211–252.

[2] M. Everingham, L. Van Gool, C.K.I. Williams, J. Winn, A. Zisserman, The PASCAL Visual Object Classes Challenge 2012 (VOC2012) Results.

[3] T.-Y. Lin, M. Maire, S. Belongie, J. Hays, P. Perona, D. Ramanan, P. Dollár, C.L. Zitnick, Microsoft coco: common objects in context, in: European Conference on Computer Vision, Springer, 2014, pp. 740–755.

[4] F. Chollet, Xception: deep learning with depthwise separable convolutions, in: The IEEE Conference on Computer Vision and Pattern Recognition (CVPR), 2017.

[5] A.G. Howard, M. Zhu, B. Chen, D. Kalenichenko, W. Wang, T. Weyand, M. Andreetto, H. Adam, Mobilenets: efficient convolutional neural networks for mobile vision applications, arXiv preprint, arXiv:1704.04861, 2017.

[6] F.N. Iandola, S. Han, M.W. Moskewicz, K. Ashraf, W.J. Dally, K. Keutzer, Squeezenet: Alexnet-level accuracy with 50x fewer parameters and <0.5mb model size, arXiv:1602.07360, 2016.

[7] A. Krizhevsky, I. Sutskever, G.E. Hinton, Imagenet classification with deep convolutional neural networks, in: Advances in Neural Information Processing Systems, 2012, pp. 1097–1105.

[8] J. Jin, A. Dundar, E. Culurciello, Flattened convolutional neural networks for feedforward acceleration, arXiv preprint, arXiv:1412.5474, 2014.

[9] M. Wang, B. Liu, H. Foroosh, Factorized convolutional neural networks, in: ICCV Workshops, 2017, pp. 545–553.
[10] G. Hinton, O. Vinyals, J. Dean, Distilling the knowledge in a neural network, arXiv preprint, arXiv:1503.02531, 2015.
[11] S. Han, H. Mao, W.J. Dally, Deep compression: compressing deep neural networks with pruning, trained quantization and Huffman coding, arXiv preprint, arXiv:1510.00149, 2015.
[12] H. Wu, P. Judd, X. Zhang, M. Isaev, P. Micikevicius, Integer quantization for deep learning inference: principles and empirical evaluation, arXiv preprint, arXiv:2004.09602, 2020.
[13] X. Zhang, X. Zhou, M. Lin, J. Sun, Shufflenet: an extremely efficient convolutional neural network for mobile devices, in: Proceedings of the IEEE Conference on Computer Vision and Pattern Recognition, 2018, pp. 6848–6856.
[14] M. Tzelepi, A. Tefas, Class-specific discriminant regularization in real-time deep cnn models for binary classification problems, Neural Processing Letters 51 (2) (2020) 1989–2005.
[15] M. Tzelepi, A. Tefas, Improving the performance of lightweight cnns for binary classification using quadratic mutual information regularization, Pattern Recognition (2020) 107407.
[16] K. Simonyan, A. Zisserman, Very deep convolutional networks for large-scale image recognition, arXiv preprint, arXiv:1409.1556, 2014.
[17] J. Hanhirova, T. Kämäräinen, S. Seppälä, M. Siekkinen, V. Hirvisalo, A. Ylä-Jääski, Latency and throughput characterization of convolutional neural networks for mobile computer vision, arXiv preprint, arXiv:1803.09492, 2018.
[18] J. Huang, V. Rathod, C. Sun, M. Zhu, A. Korattikara, A. Fathi, I. Fischer, Z. Wojna, Y. Song, S. Guadarrama, et al., Speed/accuracy trade-offs for modern convolutional object detectors, in: IEEE CVPR, 2017.
[19] S. Ioffe, C. Szegedy, Batch normalization: accelerating deep network training by reducing internal covariate shift, in: International Conference on Machine Learning, 2015, pp. 448–456.
[20] W. Liu, D. Anguelov, D. Erhan, C. Szegedy, S. Reed, C.-Y. Fu, A.C. Berg, Ssd: single shot multibox detector, in: European Conference on Computer Vision, Springer, 2016, pp. 21–37.
[21] J. Redmon, A. Farhadi, Yolo9000: better, faster, stronger, arXiv preprint, arXiv:1612.08242, 2016.
[22] S. Ren, K. He, R. Girshick, J. Sun, Faster r-cnn: towards real-time object detection with region proposal networks, in: Advances in Neural Information Processing Systems, 2015, pp. 91–99.
[23] P. Hu, D. Ramanan, Finding tiny faces, in: Proceedings of the IEEE Conference on Computer Vision and Pattern Recognition, 2017, pp. 951–959.
[24] Y. Bai, Y. Zhang, M. Ding, B. Ghanem, Finding tiny faces in the wild with generative adversarial network, in: Proceedings of the IEEE Conference on Computer Vision and Pattern Recognition, 2018, pp. 21–30.
[25] D. Triantafyllidou, P. Nousi, A. Tefas, Fast deep convolutional face detection in the wild exploiting hard sample mining, Big Data Research 11 (2018) 65–76.
[26] K. He, X. Zhang, S. Ren, J. Sun, Delving deep into rectifiers: surpassing human-level performance on imagenet classification, in: Proceedings of the IEEE International Conference on Computer Vision, 2015, pp. 1026–1034.

[27] L. Jiang, Z. Zhou, T. Leung, L.-J. Li, L. Fei-Fei, Mentornet: learning data-driven curriculum for very deep neural networks on corrupted labels, in: International Conference on Machine Learning, PMLR, 2018, pp. 2304–2313.
[28] G.E. Hinton, N. Srivastava, A. Krizhevsky, I. Sutskever, R.R. Salakhutdinov, Improving neural networks by preventing co-adaptation of feature detectors, arXiv preprint, arXiv: 1207.0580, 2012.
[29] R. Collobert, S. Bengio, Links between perceptrons, mlps and svms, in: Proceedings of the Twenty-First International Conference on Machine Learning, 2004, p. 23.
[30] R. Caruana, Multitask learning, Machine Learning 28 (1) (1997) 41–75.
[31] S. Ruder, An overview of multi-task learning in deep neural networks, arXiv preprint, arXiv:1706.05098, 2017.
[32] S. Yan, D. Xu, B. Zhang, H.-J. Zhang, Q. Yang, S. Lin, Graph embedding and extensions: a general framework for dimensionality reduction, IEEE Transactions on Pattern Analysis and Machine Intelligence 29 (1) (2007) 40–51.
[33] K. Torkkola, Feature extraction by non-parametric mutual information maximization, Journal of Machine Learning Research 3 (Mar) (2003) 1415–1438.
[34] M. Tzelepi, A. Tefas, Graph embedded convolutional neural networks in human crowd detection for drone flight safety, IEEE Transactions on Emerging Topics in Computational Intelligence (2019).
[35] R.A. Fisher, The use of multiple measurements in taxonomic problems, Annals of Eugenics 7 (2) (1936) 179–188.
[36] H. Do, A. Kalousis, M. Hilario, Feature weighting using margin and radius based error bound optimization in svms, Machine Learning and Knowledge Discovery in Databases (2009) 315–329.
[37] H. Do, A. Kalousis, A. Woznica, M. Hilario, Margin and radius based multiple kernel learning, Machine Learning and Knowledge Discovery in Databases (2009) 330–343.
[38] H. Do, A. Kalousis, Convex formulations of radius-margin based support vector machines, in: International Conference on Machine Learning, 2013, pp. 169–177.
[39] V.N. Vapnik, V. Vapnik, Statistical Learning Theory, Vol. 1, Wiley, New York, 1998.
[40] L. Lin, K. Wang, W. Zuo, M. Wang, J. Luo, L. Zhang, A deep structured model with radius-margin bound for 3d human activity recognition, arXiv preprint, arXiv: 1512.01642, 2015.
[41] A. Iosifidis, A. Tefas, I. Pitas, Representative class vector clustering-based discriminant analysis, in: 2013 Ninth International Conference on Intelligent Information Hiding and Multimedia Signal Processing, IEEE, 2013, pp. 526–529.
[42] M. Tzelepi, A. Tefas, Human crowd detection for drone flight safety using convolutional neural networks, in: European Signal Processing Conference (EUSIPCO), Kos, Greece, 2017.
[43] G. Goudelis, S. Zafeiriou, A. Tefas, I. Pitas, Class-specific kernel-discriminant analysis for face verification, IEEE Transactions on Information Forensics and Security 2 (3) (2007) 570–587.
[44] M. Kyperountas, A. Tefas, I. Pitas, Salient feature and reliable classifier selection for facial expression classification, Pattern Recognition 43 (3) (2010) 972–986.
[45] C.E. Shannon, A mathematical theory of communication, ACM SIGMOBILE Mobile Computing and Communications Review 5 (1) (2001) 3–55.
[46] D.W. Scott, Multivariate Density Estimation: Theory, Practice, and Visualization, John Wiley & Sons, 2015.
[47] S.-T. Chiu, Bandwidth selection for kernel density estimation, The Annals of Statistics (1991) 1883–1905.

[48] K. He, X. Zhang, S. Ren, J. Sun, Spatial pyramid pooling in deep convolutional networks for visual recognition, IEEE Transactions on Pattern Analysis and Machine Intelligence 37 (9) (2015) 1904–1916.

[49] M. Oquab, L. Bottou, I. Laptev, J. Sivic, Learning and transferring mid-level image representations using convolutional neural networks, in: Proceedings of the IEEE Conference on Computer Vision and Pattern Recognition, 2014, pp. 1717–1724.

[50] N. Passalis, A. Tefas, Learning neural bag-of-features for large-scale image retrieval, IEEE Transactions on Systems, Man, and Cybernetics: Systems 47 (10) (2017) 2641–2652.

[51] S. Lazebnik, C. Schmid, J. Ponce, Beyond bags of features: spatial pyramid matching for recognizing natural scene categories, in: Proceedings of the IEEE Conference on Computer Vision and Pattern Recognition, vol. 2, 2006, pp. 2169–2178.

[52] J. Sivic, A. Zisserman, Video Google: a text retrieval approach to object matching in videos, in: Proceedings of the IEEE International Conference on Computer Vision, 2003, pp. 1470–1477.

[53] D.G. Lowe, Object recognition from local scale-invariant features, in: Proceedings of the IEEE International Conference on Computer Vision, vol. 2, 1999, pp. 1150–1157.

[54] N. Passalis, A. Tefas, Neural bag-of-features learning, Pattern Recognition 64 (2017) 277–294.

[55] N. Passalis, A. Tefas, J. Kanniainen, M. Gabbouj, A. Iosifidis, Temporal logistic neural bag-of-features for financial time series forecasting leveraging limit order book data, Pattern Recognition Letters (2020).

[56] F. Laakom, N. Passalis, J. Raitoharju, J. Nikkanen, A. Tefas, A. Iosifidis, M. Gabbouj, Bag of color features for color constancy, IEEE Transactions on Image Processing 29 (2020) 7722–7734.

[57] N. Passalis, A. Tefas, Training lightweight deep convolutional neural networks using bag-of-features pooling, IEEE Transactions on Neural Networks and Learning Systems 30 (6) (2018) 1705–1715.

[58] D. Comaniciu, V. Ramesh, P. Meer, Real-Time Tracking of Non-rigid Objects Using Mean Shift, in: Proceedings IEEE Conference on Computer Vision and Pattern Recognition. CVPR 2000 (Cat. No. PR00662), Vol. 2, IEEE, 2000, pp. 142–149.

[59] J.F. Henriques, R. Caseiro, P. Martins, J. Batista, High-speed tracking with kernelized correlation filters, IEEE Transactions on Pattern Analysis and Machine Intelligence 37 (3) (2014) 583–596.

[60] O. Zoidi, A. Tefas, I. Pitas, Visual object tracking based on local steering kernels and color histograms, IEEE Transactions on Circuits and Systems for Video Technology 23 (5) (2013) 870–882.

[61] W. Zhong, H. Lu, M.-H. Yang, Robust object tracking via sparsity-based collaborative model, in: 2012 IEEE Conference on Computer Vision and Pattern Recognition, IEEE, 2012, pp. 1838–1845.

[62] S. He, Q. Yang, R.W. Lau, J. Wang, M.-H. Yang, Visual tracking via locality sensitive histograms, in: Proceedings of the IEEE Conference on Computer Vision and Pattern Recognition, 2013, pp. 2427–2434.

[63] P. Nousi, A. Tefas, I. Pitas, Dense convolutional feature histograms for robust visual object tracking, Image and Vision Computing (2020) 103933.

[64] S. Teerapittayanon, B. McDanel, H.-T. Kung, Branchynet: fast inference via early exiting from deep neural networks, in: Proceedings of the International Conference on Pattern Recognition, 2016, pp. 2464–2469.

[65] A. Veit, S. Belongie, Convolutional networks with adaptive inference graphs, in: Proceedings of the European Conference on Computer Vision, 2018, pp. 3–18.
[66] Y. Bai, S.S. Bhattacharyya, A.P. Happonen, H. Huttunen, Elastic neural networks: a scalable framework for embedded computer vision, in: Proceedings of the European Signal Processing Conference, 2018, pp. 1472–1476.
[67] G. Huang, D. Chen, T. Li, F. Wu, L. van der Maaten, K.Q. Weinberger, Multi-scale dense networks for resource efficient image classification, in: Proceedings of the International Conference on Learning Representations, 2018.
[68] N. Passalis, J. Raitoharju, A. Tefas, M. Gabbouj, Efficient adaptive inference for deep convolutional neural networks using hierarchical early exits, Pattern Recognition 105 (2020) 107346.
[69] Y. Zhou, Y. Bai, S.S. Bhattacharyya, H. Huttunen, Elastic neural networks for classification, in: Proceedings of the IEEE International Conference on Artificial Intelligence Circuits and Systems, 2019, pp. 251–255.
[70] N. Passalis, J. Raitoharju, A. Tefas, M. Gabbouj, Adaptive inference using hierarchical convolutional bag-of-features for low-power embedded platforms, in: Proceedings of the IEEE International Conference on Image Processing, 2019, pp. 3048–3052.
[71] N. Passalis, J. Raitoharju, A. Tefas, M. Gabbouj, Efficient adaptive inference leveraging bag-of-features-based early exits, in: Proceedings of the IEEE Workshop on Multimedia Signal Processing, 2020.

CHAPTER

Knowledge distillation

Nikolaos Passalis, Maria Tzelepi, and Anastasios Tefas
Department of Informatics, Aristotle University of Thessaloniki, Thessaloniki, Greece

8.1 Introduction

The increasing complexity of Deep Learning (DL) models led to the development of a wide variety of methods for having faster, more lightweight, and flexible models, ranging from applying quantization methods [1,2] and pruning techniques [3,4] to architectures that are lightweight by design, such as MobileNets [5]. All of these methods allowed for developing smaller and faster models. However, this came with some shortcomings. These smaller models are typically less accurate, leading to significant challenges for many critical applications where accuracy is as important as speed, for example, for developing crowd detection and avoidance systems for autonomous unmanned aerial vehicles [6]. This phenomenon might sound like a reasonable trade-off, since it is expected that decreasing the number of parameters in a neural network would lead to a lower accuracy. However, most (if not all) state-of-the-art DL architectures are largely overparameterized and the additional parameters/complexity in many cases just help to discover better solutions, instead of merely increasing its representational capacity [7,8]. This means that in many cases the over-parametrized DL models could be potentially "compressed" into smaller ones, if we had the appropriate tools for training them [7].

These observations fueled the interest for better understanding the learning dynamics of neural networks, as well as for developing more effective training methods that could mitigate the aforementioned phenomena. Among the most well-known methods for improving the effectiveness of the training process for DL models is *knowledge distillation* [9], also known as *knowledge transfer* [10] or *neural network distillation* [11]. These methods are capable of improving the effectiveness of the training process by *transferring* the knowledge encoded in a large and complex neural network into a smaller one. Typically, the larger model is called the *teacher* model, while the smaller one is called the *student* model, to highlight the similarity with the anthropocentric training approaches, where a teacher *transfers* its knowledge to a student using the most appropriate teaching techniques. In the case of neural networks, the knowledge encoded by the teacher model, which is also known as *dark knowledge* [12,13], describes various aspects regarding the inner workings of the larger models, which the student model should mimic.

Deep Learning for Robot Perception and Cognition. https://doi.org/10.1016/B978-0-32-385787-1.00013-0

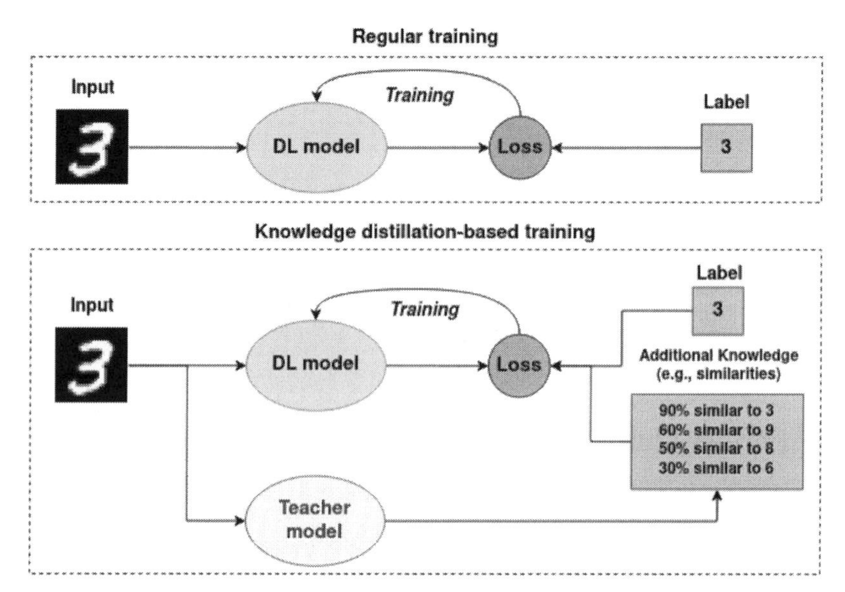

FIGURE 8.1

The process of knowledge distillation. Instead of directly training a DL model with the ground-truth labels/annotations, we employ another model, which acts as a teacher, and provides additional implicit / *dark* knowledge (e.g., similarities between the training samples and the labels) to the student model. An image from the well-known MNIST data set was used in this figure [14].

Even though a solid theoretical underpinning of why and when knowledge distillation methods work has not been fully established yet [15], there are several explanations on why this process is so effective. First, knowledge distillation acts as a regularizer, encoding our prior knowledge and beliefs regarding the regions of the hypothesis space that should be considered. Therefore the teacher model *steers* the student toward a potentially better region of the optimization landscape, often leading to a better solution and improved generalization. Note that, as we mentioned earlier, the overparameterization of the teacher model sometimes only helps to discover a better solution [7,8], thus improving the effective capacity of the teacher, even through the correct hypotheses can be described by a model with lower representational capacity.

Furthermore, knowledge distillation also has information-theoretic consequences, since the student model is trained by using more information compared with regular training using only the ground truth information. This can be better understood by an example. Consider the case of a neural network trained to classify handwritten digits, as shown in Fig. 8.1. Regular supervised learning algorithms typically aim to maximize the response of the neuron that corresponds to the correct class, ignoring that the similarities of each digit with the rest of the classes might also provide useful information to the model. On the other hand, knowledge distillation methods, such as distillation between the last (decision) layers of the networks [11], extract such information from the teacher model and then trains the student models using both the ground truth labels, as well as the implicit (dark) knowledge extracted from the teacher. Even though this dark knowledge can be provided in various forms, most methods employ either similarities or relations between samples and/or classes, as shown in Fig. 8.1.

The effectiveness of knowledge distillation methods led to a very rich literature on the topic. This chapter aims to provide a brief introduction to knowledge distillation, as well as present some of the representative methods that will equip the reader with the necessary knowledge and tools to apply these methods in practice, as well as follow this rapidly advancing field. The interested reader is referred to [16], for an extensive review of knowledge distillation methods, as well as applications that span to other areas, such as defending neural networks against adversarial attacks and application in natural language processing (NLP). The rest of this chapter is structured as follows. In Section 8.2, we present the seminal neural network distillation approach [11], which kick-started the field. Then we present a generalization of this approach (Section 8.3), which provides a general probabilistic view on knowledge distillation, allowing for going beyond classification tasks and overcoming some significant limitations that existed in earlier methods [10]. Note that both of these methods employ only one layer for transferring the knowledge. However, other methods also demonstrated that employing multiple layers of the teacher model allow for even more effectively training the student networks. Thus in Section 8.4, we present multilayer knowledge distillation methods that employ multiple layers for the process of knowledge distillation, further improving its effectiveness. Finally, in Section 8.5 we present some more advanced ways to train the teacher model, allowing for deriving more effective distillation methods, such as online distillation methods, which allow for simultaneously training both the teacher and student models, as well as, self-distillation methods that do not even use a different teacher model and are capable of reusing the knowledge extracted from the student model.

8.2 Neural network distillation

Neural network distillation, or simply *distillation*, was among the seminal methods that fueled the interest in knowledge distillation [11]. Distillation was motivated by the observation that training a single model that mimics the behavior of a larger and more complicated ensemble of models typically leads to better accuracy than directly training the same single model for the task at hand [17]. Later it was demonstrated that this process is also effective when transferring the knowledge from a single larger teacher model to a smaller student.

Distillation targets DL models trained for classification tasks and works by using the teacher model to extract implicit similarities between the training samples and the categories, as shown in Fig. 8.2. These similarities, which are also called *soft targets* to distinguish them from the *hard* ground truth targets, reveal more information both

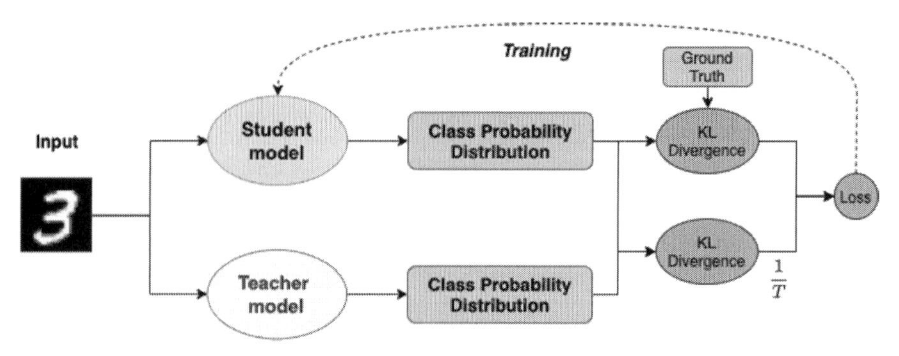

FIGURE 8.2

Neural network distillation works by first using a teacher model to produce a softer label distribution for each input sample. Then the student is trained to mimic this distribution, as well as correctly classifying the input samples according to the ground truth labels.

regarding the way the teacher models the knowledge, as well as regarding the structure of the data. However, extracting this kind of information from DL models is not always straightforward, since they are typically structured and trained to suppress the similarities with the labels apart from the correct class. For example, note that DL models trained for classification tasks typically use the *softmax* activations for producing a probability distribution over the classes, which is designed to suppress all the activation apart from the highest one. Neural network distillation overcomes this issue by introducing a *temperature* hyperparameter T that allows for producing a softer distribution over the candidate categories, as also shown in Fig. 8.3 ($T = 1$ refers to the regular output of the network).

An overview of the distillation process is provided in Fig. 8.2. First, let $\mathbf{x}$ denote an input sample that is fed into the teacher DL model $f(\cdot)$. Also, let $f^{logits}(\cdot)$ denote the *logits* of the model, that is, the activations of the final layer *before* feeding them to a softmax layer. The output of the teacher is calculated as

$$[f(\mathbf{x})]_i = \frac{\exp([f^{logits}(\mathbf{x})]_i)}{\sum_{j=1}^{N_C} \exp([f^{logits}(\mathbf{x})]_j)}, \tag{8.1}$$

where the notation $[\mathbf{x}]_i$ is used to refer to the ith element of the vector $\mathbf{x}$ and N_C is the number of categories. Also, note that $f(\mathbf{x}) \in \mathbb{R}^{N_C}$ and $f^{logits}(\mathbf{x}) \in \mathbb{R}^{N_C}$. Distillation introduces the temperature hyper-parameter T that allows for controlling the smoothness of the output distribution of the teacher p_i as

$$p_i^{(T)} = \frac{\exp([f_{\mathbf{W}}^{logits}(\mathbf{x})]_i/T)}{\sum_{j=1}^{N_C} \exp([f^{logits}(\mathbf{x})]_j/T)}, \tag{8.2}$$

where $T = 1$ is set to 1 during regular operation. During the distillation process we can fine tune the value of T in order to obtain a distribution with the appropriate

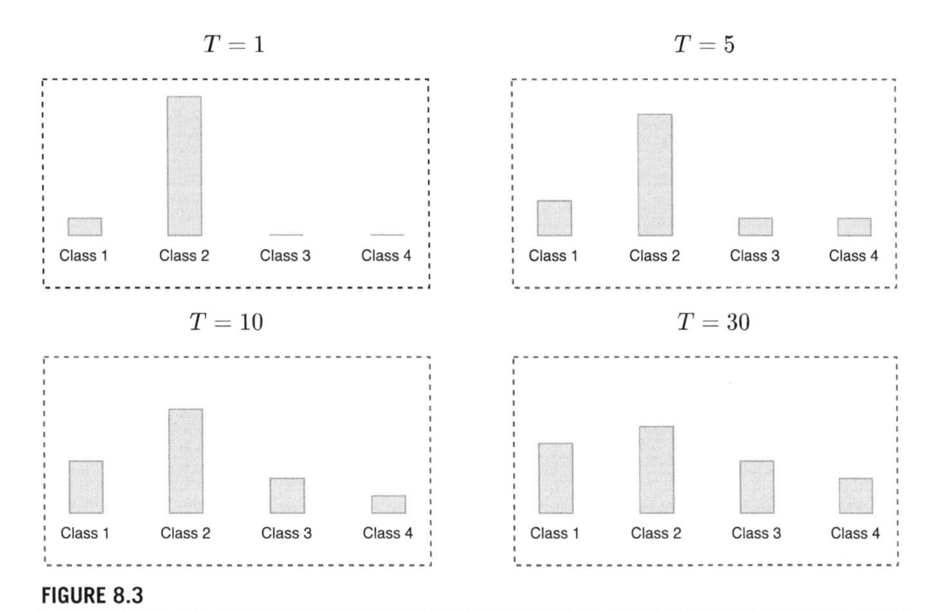

FIGURE 8.3

Effect of using different softmax temperatures (T) on the output distribution of a DL model. Note that using a higher temperature leads to a softer distribution without altering the logits. Four toy classes were used for the illustration.

entropy. More specifically, higher values of T make the distribution softer, as shown in Fig. 8.3, while smaller values make the distribution more peaked.

Similarly, the softened output of the student model $g_{\mathbf{W}}(\cdot)$ is defined as

$$q_i^{(T)} = \frac{\exp([g_{\mathbf{W}}^{logits}(\mathbf{x})]_i / T)}{\sum_{j=1}^{N_C} \exp([g_{\mathbf{W}}^{logits}(\mathbf{x})]_j / T)}, \tag{8.3}$$

where $g_{\mathbf{W}}^{logits}$ is similarly defined and $\mathbf{W}$ refers to the trainable parameters of the model. Then the student model can be trained to minimize the KL divergence between the output of the student and the ground truth labels, as well as the softened output of the student and the softened output of the teacher. Therefore, the loss function used in distillation is defined as

$$\mathcal{L}(\mathbf{x}, \mathbf{t}) = D_{KL}\left(\mathbf{t}, \mathbf{q}^{(1)}\right) + \alpha \frac{1}{T^2} D_{KL}\left(\mathbf{p}^{(T)}, \mathbf{q}^{(T)}\right), \tag{8.4}$$

where $\mathbf{q} = [q_1, \ldots, q_{N_C}]^T$, $\mathbf{p} = [p_1, \ldots, p_{N_C}]^T$, and $\mathbf{t}$ refers to the one-hot encoded ground truth vector for each sample, while KL divergence D_{KL} is defined as

$$D_{KL}(\mathbf{q}, \mathbf{p}) = \sum_{i=1}^{N_C} p_i \log\left(\frac{p_i}{q_i}\right). \tag{8.5}$$

Note that the gradients for the KL divergence term for the soft-targets scale by T^2. Therefore this term in (8.4) is multiplied by $\frac{1}{T^2}$ to ensure that both KL terms have about the same weight during the optimization. If needed, the weight of the soft-targets can be increased or decreased by altering the value of hyperparameter α in (8.4).

The student model is trained by minimizing the loss defined in (8.4):

$$\mathbf{W} = \arg\min_{\mathbf{W}} \sum_{i=1}^{N} \mathcal{L}(\mathbf{x}_i, \mathbf{t}_i), \tag{8.6}$$

where $\mathbf{x}_i$ and $\mathbf{t}_i$ refer to the samples and targets of the training set and N is the number of samples in the training set. Note that a different *transfer* set can be also used for transferring the knowledge to the student model, apart from the labeled training set. This allows for effectively exploiting large sources of unlabeled data. Also, note that the teacher model is pretrained and fixed during this process. As we will discuss in Section 8.5, there are also other options that lead to more advanced distillation approaches that can overcome some limitations that arise from keeping the teacher fixed during the whole training process.

Selecting the most appropriate temperature T can be critical for obtaining significant improvements when applying the distillation process. Unfortunately, there is no rule for selecting the most appropriate value for T, except for the observation that typically values between 2 and 10 work the best for most practical problems. However, it is worth noting that there is experimental evidence that connects the value of T to the difficulty of the problem. More specifically, in [18] it is argued that for more difficult problems, smaller values of T that are closer to 1, sometimes work better. Finally, Hinton et al. also discussed some interesting connections of distillation loss with minimizing the squared loss between the logits of the models when a high enough temperature is used [11]. More specifically, it has been shown that if logits have been zero-meaned and the temperature is high enough, then minimizing $D_{KL}(\mathbf{p}^{(T)}, \mathbf{q}^{(T)})$ is equivalent to minimizing $\sum_{i=1}^{N_C}([g_{\mathbf{W}}(\mathbf{x})]_i - [f(\mathbf{x})]_i)^2$ (up to a scaling factor).

8.3 Probabilistic knowledge transfer

Even though neural network distillation was capable of improving the performance of networks trained for classification tasks, it was not directly applicable for DL models that performed other tasks, such as representation learning. This limitation was addressed by methods designed to perform representation-based knowledge distillation, such as [10,19,20]. This section focuses on Probabilistic Knowledge Transfer (PKT), which first models the probability distribution of the samples in the teacher space and then trains a student which mimics this distribution [10]. More specifically, PKT works by modeling the pairwise relations between samples, that is, the distribution that expresses how probable is to select the sample i if we have already selected

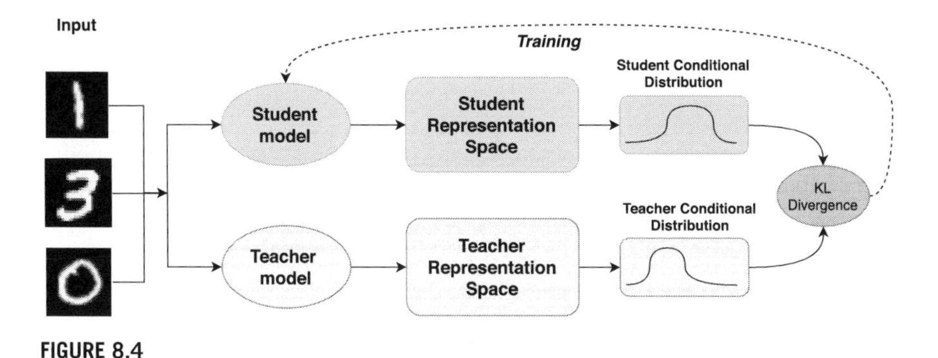

FIGURE 8.4

Probabilistic Knowledge Transfer (PKT) allows for transferring the knowledge without using any ground-truth labels. PKT works by training a student model that mimics the conditional probability distribution of the teacher model. In this way, the student models learn to recreate the geometry of the teacher into a potentially lower-dimensional space.

the sample j. Therefore, PKT models the probability $p_{i|j}$ instead of the probability $P_{c|i}$, that is, the probability that the i-th sample belongs to class c, which is used in the regular neural network distillation process [11]. In this way, PKT removes the dependence on class labels and can be applied to any kind of problems, ranging from metric learning to regression problems. Then any divergence metric can be used to measure the distance between the teacher and student distributions. Note that PKT is a fully unsupervised method, even though it can be combined with any other supervised loss. In this chapter, we will first analytically derive the PKT method and discuss how the student model can be trained in order to mimic the behavior of the student model. Then we provide further insight into the way the PKT method works by discussing connections of PKT with an information-theoretic metric, the mutual information.

The way the PKT method works is summarized in Fig. 8.4. First, we employ a teacher network $f(\cdot) \in \mathbb{R}^{N_t}$ to extract a representation for each input sample. The student model is denoted by $g_{\mathbf{W}}(\cdot) \in \mathbb{R}^{N_s}$. The dimensionality of the embedding space of the student (N_s) and teacher model (N_t) can be different. This prohibits the use of simpler methods for transferring the knowledge from the teacher model into the student, for example, by directly matching the representation of the teacher to the representation of the student, highlighting the need for more advanced methods that can transfer the knowledge between spaces of different dimensionality. During the process of knowledge distillation the parameters $\mathbf{W}$ of the student model are optimized to "mimic" the behavior of the teacher model. It is also worth noting that there is actually no constraint on what the functions $f(\cdot)$ and $g_{\mathbf{W}}(\cdot)$ are. We only need the output value of the teacher $f(\cdot)$ for every sample fed to the student $g_{\mathbf{W}}(\cdot)$ and a way to optimize the parameters of $g_{\mathbf{W}}(\cdot)$ to minimize the given objective function.

To simplify the notation used in this section, we let $\mathbf{x}_i = f(\mathbf{t}_i)$ denote the representation extracted from the teacher model and $\mathbf{y}_i = g_{\mathbf{W}}(\mathbf{t}_i)$ denote the representation extracted from student model, where $\mathbf{t}_i$ is a transfer sample contained in the transfer set $\mathcal{T} = \{\mathbf{t}_1, \mathbf{t}_2, \ldots, \mathbf{t}_N\}$ used for knowledge distillation. Note that, as before, N denotes the number of samples used for knowledge distillation. Also, we use two continuous random variables X and Y to model the distribution of the teacher and student networks, respectively.

It is well known that pairwise interactions between data samples can be used to model and describe the geometry of feature spaces, a property that has been successfully used in many different areas [21,22]. To this end, we typically employ the joint probability density of the representation of two data samples. These joint densities in the embedding space of the student and teacher can be estimated using Kernel Density Estimation (KDE) [23] as

$$p_{ij} = p_{i|j} p_j = \frac{1}{N} K(\mathbf{x}_i, \mathbf{x}_j; 2\sigma_t^2), \tag{8.7}$$

and

$$q_{ij} = q_{i|j} q_j = \frac{1}{N} K(\mathbf{y}_i, \mathbf{y}_j; 2\sigma_s^2). \tag{8.8}$$

Note that a kernel function $K(\mathbf{a}, \mathbf{b}; \sigma_t^2)$, with width σ_t, that measures the similarity between two vectors **a** and **b** is needed for estimating these probabilities. Minimizing the divergence between the probability distributions defined by the representation of the samples in the student and teacher space allows for transferring the knowledge from the teacher to the student model, since it allows for learning a feature space that maintains the same relations that appear in the teacher model. In other words, minimizing the divergence between the distribution of the teacher $\mathcal{P}$ and the distribution of the student $\mathcal{Q}$ aims to learn a space where each sample will have the same neighbors in both the teacher and student spaces. Therefore this process implicitly models the manifold of the data in the teacher space and tries to recreate it in the (potentially lower dimensional) student space. Also, note that class labels are not required during this process, effectively allowing for performing this process with fully unlabeled transfer sets.

Even though the joint probability distribution can be used to effectively model the geometry of the data and perform knowledge transfer, this might not be enough, since training a more lightweight student model that mimics the exact geometry of the more complicated teacher model is often not possible. To this end, PKT employs the conditional probability distribution instead of the joint one. To understand why this makes the optimization easier, despite that in both of these cases the divergence between the probability distributions is minimized when the kernel similarities are equal for both models, we need to consider the following. The conditional probability distribution models how probable it is for each sample to select each of its neighbors [22], which in turn allows for more precisely describing the local regions between the samples, instead of the global ones. This is also the reason why the conditional probabilities have been employed for dimensionality reduction methods, such as the t-SNE algorithm [22].

Therefore, the teacher's conditional probability distribution can be estimated as

$$p_{i|j} = \frac{K(\mathbf{x}_i, \mathbf{x}_j; 2\sigma_t^2)}{\sum_{k=1,k\neq j}^{N} K(\mathbf{x}_k, \mathbf{x}_j; 2\sigma_t^2)} \in [0, 1], \tag{8.9}$$

and the student's distribution as

$$q_{i|j} = \frac{K(\mathbf{y}_i, \mathbf{y}_j; 2\sigma_s^2)}{\sum_{k=1,k\neq j}^{N} K(\mathbf{y}_k, \mathbf{y}_j; 2\sigma_s^2)} \in [0, 1]. \tag{8.10}$$

Note that these probabilities are limited to the range [0, 1] and they sum to 1, that is, $\sum_{i=0,i\neq j}^{N} p_{i|j} = 1$ and $\sum_{i=0,i\neq j}^{N} q_{i|j} = 1$.

Selecting the most appropriate kernel for estimating these probabilities is critical for the effectiveness of the process of knowledge transfer [24,25]. Typically, the Gaussian kernel is used to this end, which is defined as

$$K_{Gaussian}(\mathbf{a}, \mathbf{b}; \sigma) = exp\left(-\frac{||\mathbf{a} - \mathbf{b}||_2^2}{\sigma}\right), \tag{8.11}$$

where the notation $||\cdot||_2$ is used for the l^2 norm of a vector, while the width (scaling factor) of the kernel is denoted by σ. This formulation leads to the regular Kernel Density Estimation (KDE) method for estimating the conditional probabilities [23], which, unfortunately, is not very effective in high-dimensional spaces, requiring the tedious and careful tuning of the widths of the kernels. Several heuristics have been proposed to overcome this issue [26]. PKT mitigates this issue by employing a more robust formulation that does not require extensive tuning. Therefore, instead of using a Gaussian kernel, a cosine-based affinity metric is used:

$$K_{cosine}(\mathbf{a}, \mathbf{b}) = \frac{1}{2}(\frac{\mathbf{a}^T\mathbf{b}}{||\mathbf{a}||_2||\mathbf{b}||_2} + 1) \in [0, 1]. \tag{8.12}$$

This formulation allows for avoiding finetuning the kernel's bandwidth, while providing a more robust affinity estimation, since it is established that for many tasks Euclidean metrics are inferior to cosine-based ones [27,28].

Furthermore, there are several ways to calculate the divergence between the teacher distribution $\mathcal{P}$ and student distribution $\mathcal{Q}$, which is required to effectively transfer the knowledge between the models. PKT employs the Kullback–Leibler (KL) divergence:

$$\mathcal{KL}(\mathcal{P}||\mathcal{Q}) = \int_{\mathbf{t}} \mathcal{P}(\mathbf{t}) \log \frac{\mathcal{P}(\mathbf{t})}{\mathcal{Q}(\mathbf{t})} d\mathbf{t}. \tag{8.13}$$

Note that in practice, we employ the transfer set to approximate the distributions $\mathcal{P}$ and $\mathcal{Q}$. Therefore the loss function that can be used for optimizing the student is defined as

$$\mathcal{L} = \sum_{i=1}^{N} \sum_{j=1, i \neq j}^{N} p_{j|i} \log\left(\frac{p_{j|i}}{q_{j|i}}\right). \tag{8.14}$$

Note that minimizing the divergence for pairs of points that are closer in the teacher space is more important that the distant ones, as shown in (8.14), since KL is an asymmetric distance metric. Therefore, keeping the local geometry of the embedding space is more important in this formulation than accurately mimicking the relations between all the samples (including very distant ones). This provides more flexibility during the training process, which is especially important when more lightweight student models, which are not capable of fully recreating the global geometry of the teacher, are used. As discussed in [10], if the whole geometry should be maintained, then symmetric distance metrics could be used, for example, the quadratic divergence measure:

$$D_Q(\mathcal{P}, \mathcal{Q}) = \int_{\mathbf{x}} (\mathcal{P}(\mathbf{t}) - \mathcal{Q}(\mathbf{t}))^2 d\mathbf{t}. \tag{8.15}$$

The metrics can also be further adapted to the needs of the each application, as demonstrated in [25].

Again, similar to the neural network distillation methods presented in Section 8.2, the student model is trained by minimizing the loss defined in (8.14):

$$\mathbf{W} = \arg\min_{\mathbf{W}} \mathcal{L} = \arg\min_{\mathbf{W}} \sum_{i=1}^{N} \sum_{j=1, i \neq j}^{N} p_{j|i} \log\left(\frac{p_{j|i}}{q_{j|i}}\right). \tag{8.16}$$

Typically, gradient descent is used to this end. Therefore the derivative of the loss function with respect to the parameters can be calculated as

$$\frac{\partial \mathcal{L}}{\partial \mathbf{W}} = \sum_{i=1}^{N} \sum_{j=1, i \neq j}^{N} \frac{\partial \mathcal{L}}{\partial q_{j|i}} \sum_{l=1}^{N} \frac{\partial q_{j|i}}{\partial \mathbf{y}_l} \frac{\mathbf{y}_l}{\partial \mathbf{W}}, \tag{8.17}$$

where $\frac{\mathbf{y}_l}{\partial \mathbf{W}}$ is the partial derivative of the student model with respect to its parameters. In practice, calculating the loss by employing the whole training set is infeasible, since it has quadratic complexity and requires quadratic space with respect to the number of training samples. Therefore PKT calculates the loss function described in (8.14) in small batches (typically of 64–256 samples). This process is similar to Nyström-based methods that approximate the full similarity matrix of the data [29], which can indeed speed up convergence, without negatively affecting the optimization. However, note that it is crucial to shuffle the training set during each epoch in order to ensure that different pairs of samples are used for each optimization epoch when calculating the loss function.

There are also interesting information-theoretic consequences for the PKT method, as discussed in [10]. More specifically, PKT also aims to maintain the same

the information potentials for the student network,

$$V_{IN}^{(s)} = \frac{1}{N^2} \sum_{p=1}^{N_c} \sum_{k=1}^{J_p} \sum_{l=1}^{J_p} K(\mathbf{y}_{pk}, \mathbf{y}_{pl}; 2\sigma_s^2), \tag{8.31}$$

$$V_{ALL}^{(s)} = \frac{1}{N^2} \left(\sum_{p=1}^{N_c} (\frac{J_p}{N})^2 \right) \sum_{k=1}^{N} \sum_{l=1}^{N} K(\mathbf{y}_k, \mathbf{y}_l; 2\sigma_s^2), \tag{8.32}$$

and

$$V_{BTW}^{(s)} = \frac{1}{N^2} \sum_{p=1}^{N_c} \frac{J_p}{N} \sum_{j=1}^{J_p} \sum_{k=1}^{N} K(\mathbf{y}_{pj}, \mathbf{y}_k; 2\sigma_s^2). \tag{8.33}$$

Of course, in the case of regular KDE, different widths σ_t and σ_s should be used for the teacher and the student model, after appropriately tuning them.

The knowledge can be transferred between the teacher and student models by maintaining the same amount of MI between the two random variables X and Y with respect to their class labels, that is, $I(X, C) = I(Y, C)$. If we employ the aforementioned QMI-based formulation, then the respective information potentials must be equal between the two models. This in turn implies that the joint densities provided in (8.7) and (8.8) must also be equal to each other, which can be achieved by solving the same minimization problem as in (8.14). It worth noting that this is not the only configuration that maintains the same MI between the models. Other configurations might exist as well. However, it is enough to use one of them to transfer the knowledge.

There is a number of other related representation-learning based distillation methods that have been proposed following PKT, which model different kinds of interactions between the transfer samples. For example, in [19] instead of measuring the conditional probabilities, the proposed *relative teacher* approach employs the distance between the training samples. Then the pairwise distances between the training samples are matched using a divergence measure. Therefore the following loss function is used:

$$\mathcal{L} = \sum_{i=1}^{N} \sum_{j=1, i \neq j}^{N} \left| \left\| \mathbf{x}_i - \mathbf{x}_j \right\|_2 - \left\| \mathbf{y}_i - \mathbf{y}_j \right\|_2 \right|. \tag{8.34}$$

A similar approach was also proposed in [20], where the cosine similarity was used instead of the Euclidean distance, and the Huber loss was used for the similarity matching, as well as in [32], where the Frobenious norm was used instead.

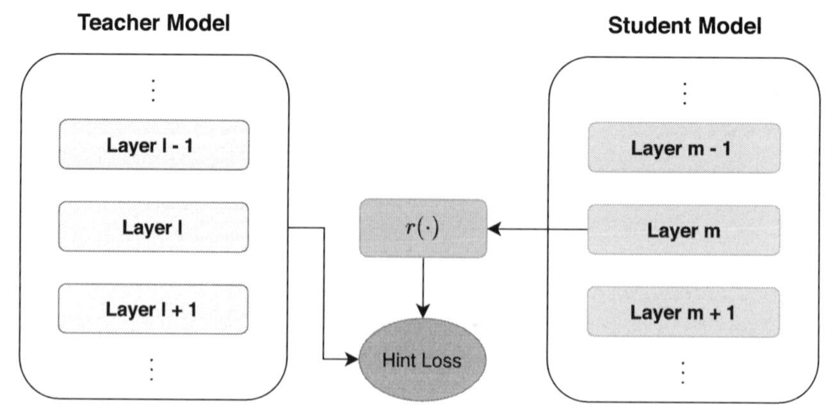

FIGURE 8.5

Hint-based distillation process. Note that the knowledge is transferred between two layers of the teacher and student, after appropriate alerting the dimensionality of the student model to match the dimensionality of the teacher using a regressor $r(\cdot)$.

8.4 Multilayer knowledge distillation

The two methods presented in the previous two sections focus on transferring the knowledge using only the output layers of the teacher and student networks. However, DL models encode useful knowledge in their intermediate layers as well. This has led to the development of a number of methods that focus on employing these intermediate layers to further improve the effectiveness of knowledge distillation. In this section, we present different methods that can perform this kind of *multilayer* distillation.

8.4.1 Hint-based distillation

Hint-based distillation was among the first methods capable of performing knowledge distillation by employing several intermediate layers of a network [33]. Hint-based distillation is summarized in Fig. 8.5 and works by (a) selecting a number of intermediate layers from the teacher and student and then (b) imposing an appropriate loss in order to distill the knowledge encoded in this intermediate layer of the student to the corresponding layer of the student. This is achieved by introducing the following loss term:

$$\mathcal{L}_{hint}^{(l,m)} = \sum_{i=1}^{N} \left\| \mathbf{x}_i^{(l)} - r(\mathbf{y}_i^{(m)}) \right\|_2^2, \tag{8.35}$$

where the notation $\mathbf{x}_i^{(l)}$ is used to refer to the lth layer of the teacher network, $\mathbf{y}_i^{(m)}$ is used to refer to the mth layer of the student network and $r(\cdot)$ is a regressor that allows for matching the dimensionality between $\mathbf{x}_i^{(l)}$ and $\mathbf{y}_i^{(m)}$. For example, when transferring the knowledge from a fully connected layer with N_l dimensions to a layer with N_m dimensions, the regressor $r(\cdot)$ can be defined as

$$r(\mathbf{x}) = \mathbf{W}_r \mathbf{x}, \tag{8.36}$$

where the projection matrix of the regressor is defined as $\mathbf{W}_r \in \mathbb{R}^{N_m \times N_l}$. The matrix $\mathbf{W}_r$ is optimized along with the rest of the parameters of the student model. Note that the additional parameters introduced by $\mathbf{W}_r$ to the student model are not part of it, since they merely exist to *guide* (or provide *hints* to) the learning process. After the student network has been trained, these parameters can be removed. The loss provided by (8.35) is typically combined with the regular neural network distillation loss.

Selecting the most appropriate sets of layers l and m is crucial for improving the performance of the student network when hint-based distillation is used. If an early layer of the teacher model is matched with a too late layer of the student network, then the student network might suffer from *overregularization*. On the other hand, if a layer of the teacher is matched with a too early layer of the student, then the student might collapse the representation in this layer [34], removing useful information too early, and thus reducing the accuracy of the model. Multiple pairs of intermediate layers can also be used, to further improve the performance of the student. However, it should be again stressed that using too many pairs of layers imposes a strong regularization effect of the network, which can eventually lead to decreasing the performance of the network [33]. Finally, using fully connected regressors can be prohibitively expensive in the case of convolutional neural networks. In this case, a convolutional regressor is used aiming to match the number of channels and size of the features maps extracted from both the teacher and student. Afterwards, (8.35) can be directly applied on the extracted feature maps.

8.4.2 Flow of solution procedure distillation

Another approach to introduce knowledge regarding the way the teacher model works to the student is by employing the *Flow of Solution Procedure* (FSP) matrix [9]. The FSP matrix describes how a network transforms its representation across two different layers. Typically, the FSP matrix is calculated using the output of two convolutional layers. The FSP matrix between the lth and the mth layers of the teacher network is defined as

$$[\mathbf{G}^{(T)}]_{p,k} = \frac{1}{HW} \sum_{i=1}^{H} \sum_{j=1}^{W} [\mathbf{x}^{(l)}]_{ijp} [\mathbf{x}^{(m)}]_{ijk}, \tag{8.37}$$

where $\mathbf{x}^{(l)} \in \mathbb{R}^{H \times W \times N_l}$ is the output feature map of the lth layer of the teacher model and $\mathbf{x}^{(m)} \in \mathbb{R}^{H \times W \times N_m}$ is the output feature map of the mth layer of the teacher model. Note that in contrast with the hint-based approach, where the notation l and m was used to refer to the layers of the teacher and student, respectively, in this section we

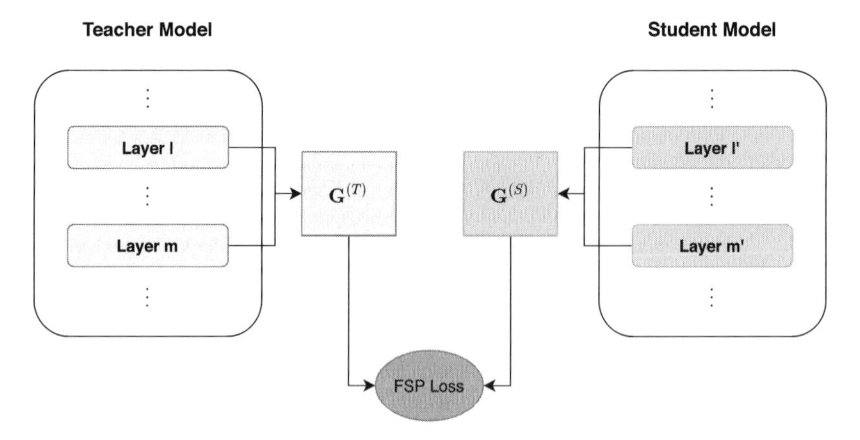

FIGURE 8.6

FSP-based distillation process. The knowledge is transferred between l-m pair of the teacher model to the l'-m' pair of the student model. Note that number of channels in the lth layer of the teacher must match the number of channels in the l'th layer of the student. The same must be true for the mth and m'th layers.

use this notation to refer to layers of the *same* (teacher) network. The dimensionality of the FSP matrix depends on the number of channels in both feature maps, that is, $\mathbf{G}^{(T)} \in \mathbb{R}^{N_l \times N_m}$. Note that in this way FSP matrix provides information regarding the way the teacher transforms the channels across intermediate layers, encoding in this way knowledge regarding the way the teacher model works.

Similarly, we can calculate the FSP matrix $\mathbf{G}^{(S)}$ for the student model. Then the FSP loss is defined as

$$\mathcal{L}_{FSP} = \sum_{i=1}^{N} \left\| \mathbf{G}^{(T)}(\mathbf{t}_i) - \mathbf{G}^{(S)}(\mathbf{t}_i) \right\|_F^2, \tag{8.38}$$

where the notation $\mathbf{G}^{(T)}(\mathbf{t}_i)$ and $\mathbf{G}^{(S)}(\mathbf{t}_i)$ is used to refer to the FSP matrix calculated when the ith transfer sample is fed to the network and $\| \cdot \|_F$ refers to the Frobenius norm of a matrix.[1] This process is summarized in Fig. 8.6. Similar to (8.35), this loss can be defined for multiple l-m pairs of the teacher network and l'-m' pairs of the student network.

A limitation of this approach is that it requires the pairs of layers between which the knowledge is transferred to have the same number of channels between the teacher and student. That is, the lth layer of the teacher must have the same dimensionality (number of channels) as the l'th layer of the teacher (as well as the mth layer of the teacher and m'th layer of the student), since the loss provided in (8.38) can only

[1] Note that the original FSP paper refers to using the l^2 norm without specifying whether this refers to pairs of matrices or vectors [9].

be calculated between the FSP matrices with the same dimensions. In practice, this often limits the application of the FSP approach to residual neural networks [35], where we employ the FSP loss between sets of residual modules. For example, five residual modules could be used for calculating the FSP matrix for the teacher and two residual modules could be used for the student, as further discussed in [9], in order to ensure that the FSP matrix of the student and the teacher will have the same dimensions. Furthermore, it is usually suggested to apply the FSP procedure in two steps: (a) first optimize the networks using the FSP loss and then (b) apply the loss that corresponds to the final task [9].

8.4.3 Other multilayer distillation methods

Apart from hint-based distillation and FSP-based distillation, several other methods have been proposed to exploit the knowledge provided by the intermediate layers of the teacher model. For example, *attention transfer* distills the knowledge of intermediate layers by training a student model that mimics the *attention* maps of the teacher model [36]. Another approach includes the so-called *Factor Transfer* (FT), which employs a *paraphrazer* module to extract the knowledge factors from any layer of the teacher that are then transferred through an additional translator module [37]. Also, in [38] it has been proposed to employ the activation boundaries, that is, the separating hyperplanes formed by the neurons, to transfer the knowledge from the teacher to student. Furthermore, in [39], the PKT method has been extended by allowing for effectively matching the distribution of intermediate layers.

For all the aforementioned methods, it is crucial to correctly match the layer of the teacher layer, which is used to mine the knowledge, to the layer of the student layer used to transfer the knowledge to. This can be trivial for compatible architectures, that is, ResNets with the same number of blocks [35], but challenging for more heterogeneous distillation setups. Indeed, as it was demonstrated in [34], an incorrect matching between these layers can either over-regularize the student or lead to an early representation collapse, negatively impacting the performance of the student model. To overcome this issue, an intermediate network, that has an architecture compatible with the student, but it is more powerful than the student, is employed in [34], allowing for reducing the *gap* between the layers.

8.5 Teacher training strategies

The previous sections focused on deriving more effective distillation approaches for extracting the knowledge from the teacher model and transferring it to the student model. However, the teacher model can also significantly affect the effectiveness of the distillation process. For example, recent evidence suggests that the *teacher-student gap* can indeed affect the distillation efficiency [40]. In other words, using a very powerful teacher is not the optimal choice when training a very lightweight student. In these cases, a smaller teacher, which is possibly less accurate, yet simpler, is usually a better choice.

There are three main categories of distillation methods based on how the teacher model is trained: (a) *offline distillation*, (b) *online distillation,* and (c) *self-distillation*. These approaches are depicted in Fig. 8.7. Offline distillation refers to the standard setup that we have considered until now, that is, the teacher model is trained beforehand, its weights are fixed and then the knowledge is transferred to the student model, while in online distillation the teacher is trained along with the student model. A special case of offline distillation has been proposed in [40], where several teachers are trained one-by-one using distillation, aiming at gradually arriving to a smaller student model, reducing in this way the teacher-student gap. Furthermore, ensembles of multiple teacher models have also been considered for improving the effectiveness of distillation, by increasing the diversity of the knowledge encoded by the teacher [41].

Finally, in self-distillation the student itself is used to extract useful knowledge for improving its accuracy, that is, the student also acts as a teacher, as shown in Fig. 8.7. There are several recent self-distillation methods proposed in the literature. In its simplest form, self-distillation methods transfer the knowledge from a previous version of the student to a newer generation [42], while this process can be repeated multiple times, potentially further increasing the accuracy of the later generations of the network. Also, in [43], the knowledge is transferred from latter layers a DL model into its earlier layers, allowing for learning more discriminative early layers. A similar approach is also followed in [44], where it is combined with an early exit strategy [45], which enables dynamic adaptations of the computational graph of the model to the available computational resources [46].

8.6 Concluding remarks

Knowledge distillation is an especially useful tool not only for developing more accurate lightweight models for various robotics tasks, but also for developing DL models in general. For example, distillation methods have been successfully used to tackle domain adaptation tasks [47–49]. In this chapter, we presented the most important tools that DL practitioners and researchers can employ to this end. That is, we presented the neural network distillation approach, which can be applied for DL models trained for classification tasks, as well as the PKT method, which can be applied for generic representation learning tasks. Then we demonstrated how multilayer knowledge distillation methods can be used to further improve the efficiency of distillation methods. However, such methods should be used with care, especially when applied on heterogeneous architectures, where an incorrect layer matching between the student and teacher can lead to decreasing the performance of the student model. Finally, we discussed three different approaches that can be used for training the teacher model (offline, online, and self-distillation). The most appropriate one can be selected according to the scenario and available computational resources for training the models. Though knowledge distillation is a rapidly advancing field, we believe

OFFLINE DISTILLATION

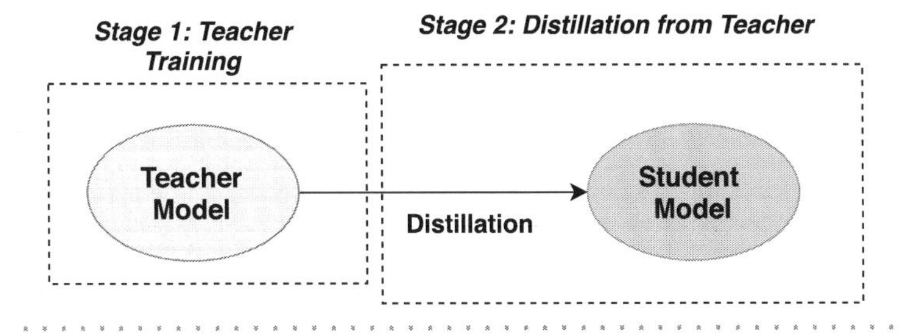

ONLINE DISTILLATION

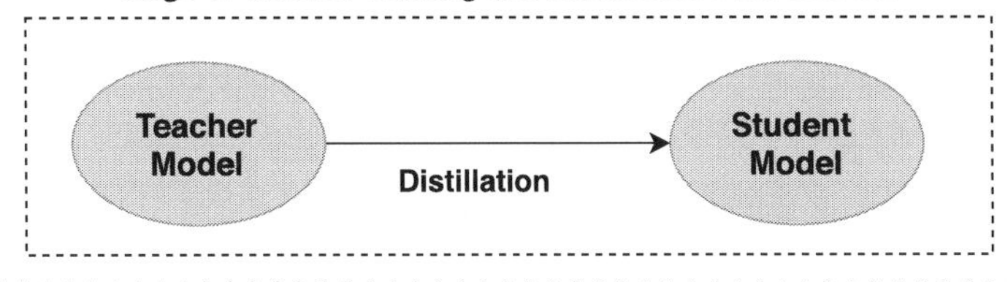

SELF-DISTILLATION

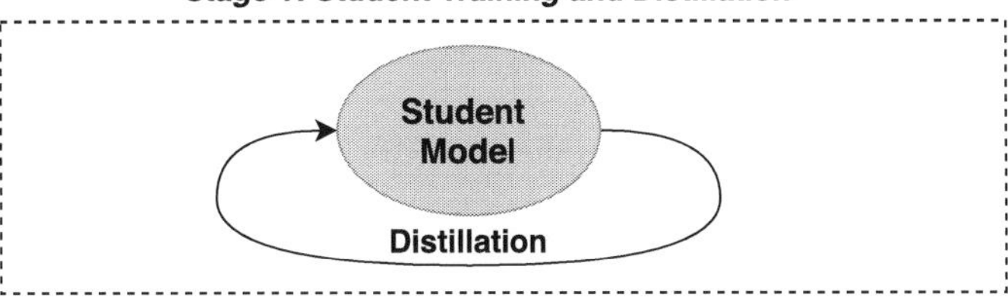

FIGURE 8.7

Comparing three different distillation approaches: (top) offline distillation, (middle) online distillation and (bottom) self-distillation. In offline distillation the teacher is pretrained and fixed, while in online distillation the teacher is trained along with the student model. Finally, in self-distillation, various aspects of the student model (e.g., earlier snapshots, layers, etc.) are used to regularize its training process, acting as a teacher for itself.

that this chapter provided the necessary knowledge to allow the reader to follow the recent advances in the field and use any of the existing distillation approaches.

References

[1] Z. Cai, X. He, J. Sun, N. Vasconcelos, Deep learning with low precision by half-wave gaussian quantization, in: Proceedings of the IEEE Conference on Computer Vision and Pattern Recognition, 2017, pp. 5918–5926.

[2] N. Floropoulos, A. Tefas, Complete vector quantization of feedforward neural networks, Neurocomputing 367 (2019) 55–63.
[3] S. Han, H. Mao, W.J. Dally, Deep compression: compressing deep neural networks with pruning, trained quantization and Huffman coding, in: Proceedings of the International Conference on Learning Representations, 2016.
[4] Y. He, X. Zhang, J. Sun, Channel pruning for accelerating very deep neural networks, in: Proceedings of the IEEE International Conference on Computer Vision, 2017, pp. 1389–1397.
[5] A.G. Howard, M. Zhu, B. Chen, D. Kalenichenko, W. Wang, T. Weyand, M. Andreetto, H. Adam, Mobilenets: efficient convolutional neural networks for mobile vision applications, arXiv preprint, arXiv:1704.04861, 2017.
[6] M. Tzelepi, A. Tefas, Human crowd detection for drone flight safety using convolutional neural networks, in: Proceedings of the European Signal Processing Conference, 2017, pp. 743–747.
[7] S. Arora, R. Ge, B. Neyshabur, Y. Zhang, Stronger generalization bounds for deep nets via a compression approach, in: Proceedings of the International Conference on Machine Learning, 2018, pp. 254–263.
[8] Z. Allen-Zhu, Y. Li, Y. Liang, Learning and generalization in overparameterized neural networks, going beyond two layers, in: Proceedings of the Advances in Neural Information Processing Systems, 2019, pp. 6158–6169.
[9] J. Yim, D. Joo, J. Bae, J. Kim, A gift from knowledge distillation: fast optimization, network minimization and transfer learning, in: Proceedings of the IEEE Conference on Computer Vision and Pattern Recognition, 2017, pp. 4133–4141.
[10] N. Passalis, A. Tefas, Learning deep representations with probabilistic knowledge transfer, in: Proceedings of the European Conference on Computer Vision (ECCV), 2018, pp. 268–284.
[11] G. Hinton, O. Vinyals, J. Dean, Distilling the knowledge in a neural network, in: NIPS Deep Learning Workshop, 2014.
[12] A.K. Balan, V. Rathod, K.P. Murphy, M. Welling, Bayesian dark knowledge, in: Proceedings of the Advances in Neural Information Processing Systems, 2015, pp. 3438–3446.
[13] Z. Tang, D. Wang, Z. Zhang, Recurrent neural network training with dark knowledge transfer, in: Proceedings of the International Conference on Acoustics, Speech and Signal Processing, 2016, pp. 5900–5904.
[14] Y. LeCun, L. Bottou, Y. Bengio, P. Haffner, Gradient-based learning applied to document recognition, Proceedings of the IEEE 86 (11) (1998) 2278–2324.
[15] M. Phuong, C. Lampert, Towards understanding knowledge distillation, in: Proceedings of the International Conference on Machine Learning, 2019, pp. 5142–5151.
[16] J. Gou, B. Yu, S.J. Maybank, D. Tao, Knowledge distillation: a survey, arXiv preprint, arXiv:2006.05525, 2020.
[17] C. Buciluă, R. Caruana, A. Niculescu-Mizil, Model compression, in: Proceedings of the ACM SIGKDD International Conference on Knowledge Discovery and Data Mining, 2006, pp. 535–541.
[18] G. Chen, W. Choi, X. Yu, T. Han, M. Chandraker, Learning efficient object detection models with knowledge distillation, in: Proceedings of the Advances in Neural Information Processing Systems, 2017, pp. 742–751.
[19] L. Yu, V.O. Yazici, X. Liu, J.v.d. Weijer, Y. Cheng, A. Ramisa, Learning metrics from teachers: compact networks for image embedding, in: Proceedings of the IEEE Conference on Computer Vision and Pattern Recognition, 2019, pp. 2907–2916.

[20] W. Park, D. Kim, Y. Lu, M. Cho, Relational knowledge distillation, in: Proceedings of the IEEE Conference on Computer Vision and Pattern Recognition, 2019, pp. 3967–3976.
[21] G.E. Hinton, S.T. Roweis, Stochastic neighbor embedding, in: Proceedings of the Advances in Neural Information Processing Systems, 2003, pp. 857–864.
[22] L.v.d. Maaten, G. Hinton, Visualizing data using t-sne, Journal of Machine Learning Research 9 (Nov) (2008) 2579–2605.
[23] D.W. Scott, Multivariate Density Estimation: Theory, Practice, and Visualization, John Wiley & Sons, 2015.
[24] N. Passalis, A. Tefas, Unsupervised knowledge transfer using similarity embeddings, IEEE Transactions on Neural Networks and Learning Systems 30 (3) (2018) 946–950.
[25] N. Passalis, M. Tzelepi, A. Tefas, Probabilistic knowledge transfer for lightweight deep representation learning, IEEE Transactions on Neural Networks and Learning Systems (2020).
[26] B.A. Turlach, Bandwidth selection in kernel density estimation: a review, in: CORE and Institut de Statistique, 1993.
[27] C.D. Manning, P. Raghavan, H. Schütze, Introduction to Information Retrieval, 2008.
[28] D. Wang, H. Lu, C. Bo, Visual tracking via weighted local cosine similarity, IEEE Transactions on Cybernetics 45 (9) (2015) 1838–1850.
[29] P. Drineas, M.W. Mahoney, On the Nyström method for approximating a Gram matrix for improved kernel-based learning, Journal of Machine Learning Research 6 (Dec) (2005) 2153–2175.
[30] T.M. Cover, J.A. Thomas, Elements of Information Theory, John Wiley & Sons, 2012.
[31] K. Torkkola, Feature extraction by non-parametric mutual information maximization, Journal of Machine Learning Research 3 (Mar) (2003) 1415–1438.
[32] F. Tung, G. Mori, Similarity-preserving knowledge distillation, in: Proceedings of the IEEE International Conference on Computer Vision, 2019, pp. 1365–1374.
[33] A. Romero, N. Ballas, S.E. Kahou, A. Chassang, C. Gatta, Y. Bengio, Fitnets: hints for thin deep nets, arXiv preprint, arXiv:1412.6550, 2014.
[34] N. Passalis, M. Tzelepi, A. Tefas, Heterogeneous knowledge distillation using information flow modeling, in: Proceedings of the IEEE/CVF Conference on Computer Vision and Pattern Recognition, 2020, pp. 2339–2348.
[35] K. He, X. Zhang, S. Ren, J. Sun, Identity mappings in deep residual networks, in: Proceedings of the European Conference on Computer Vision, Springer, 2016, pp. 630–645.
[36] S. Zagoruyko, N. Komodakis, Paying more attention to attention: improving the performance of convolutional neural networks via attention transfer, in: Proceedings of the International Conference on Learning Representations 2017, 2017.
[37] J. Kim, S. Park, N. Kwak, Paraphrasing complex network: network compression via factor transfer, in: Proceedings of the Advances in Neural Information Processing Systems, 2018, pp. 2760–2769.
[38] B. Heo, M. Lee, S. Yun, J.Y. Choi, Knowledge transfer via distillation of activation boundaries formed by hidden neurons, in: Proceedings of the AAAI Conference on Artificial Intelligence, vol. 33, 2019, pp. 3779–3787.
[39] N. Passalis, M. Tzelepi, A. Tefas, Multilayer probabilistic knowledge transfer for learning image representations, in: Proceedings of the IEEE International Symposium on Circuits and Systems, 2020, pp. 1–5.
[40] S.I. Mirzadeh, M. Farajtabar, A. Li, N. Levine, A. Matsukawa, H. Ghasemzadeh, Improved knowledge distillation via teacher assistant, in: Proceedings of the AAAI Conference on Artificial Intelligence, vol. 34, 2020, pp. 5191–5198.

[41] G. Panagiotatos, N. Passalis, A. Iosifidis, M. Gabbouj, A. Tefas, Curriculum-based teacher ensemble for robust neural network distillation, in: Proceedings of the European Signal Processing Conference, 2019, pp. 1–5.
[42] T. Furlanello, Z. Lipton, M. Tschannen, L. Itti, A. Anandkumar, Born again neural networks, in: Proceedings of the International Conference on Machine Learning, 2018, pp. 1607–1616.
[43] L. Zhang, J. Song, A. Gao, J. Chen, C. Bao, K. Ma, Be your own teacher: improve the performance of convolutional neural networks via self distillation, in: Proceedings of the IEEE International Conference on Computer Vision, 2019, pp. 3713–3722.
[44] M. Phuong, C.H. Lampert, Distillation-based training for multi-exit architectures, in: Proceedings of the IEEE International Conference on Computer Vision, 2019, pp. 1355–1364.
[45] N. Passalis, J. Raitoharju, A. Tefas, M. Gabbouj, Efficient adaptive inference for deep convolutional neural networks using hierarchical early exits, Pattern Recognition 105 (2020) 107346.
[46] N. Passalis, J. Raitoharju, A. Tefas, M. Gabbouj, Adaptive inference using hierarchical convolutional bag-of-features for low-power embedded platforms, in: Proceedings of the IEEE International Conference on Image Processing, 2019, pp. 3048–3052.
[47] T. Asami, R. Masumura, Y. Yamaguchi, H. Masataki, Y. Aono, Domain adaptation of dnn acoustic models using knowledge distillation, in: Proceedings of the IEEE International Conference on Acoustics, Speech and Signal Processing, IEEE, 2017, pp. 5185–5189.
[48] S. Ao, X. Li, C.X. Ling, Fast generalized distillation for semi-supervised domain adaptation, in: Proceedings of the AAAI Conference on Artificial Intelligence, 2017, pp. 1719–1725.
[49] L. Hedegaard, O.A. Sheikh-Omar, A. Iosifidis, Supervised domain adaptation using graph embedding, in: Proceedings of the International Conference on Pattern Recognition, 2018.

CHAPTER

Progressive and compressive learning

Dat Thanh Tran[a], Moncef Gabbouj[a], and Alexandros Iosifidis[b]
[a]*Faculty of Information Technology and Communication Sciences, Tampere University, Tampere, Finland*
[b]*Department of Electrical and Computer Engineering, Aarhus University, Aarhus, Denmark*

9.1 Introduction

During the last decade, deep neural networks have become the primary backbone in many computer vision [1–4], natural language processing [5–7], or financial forecasting [8,9] systems. Nowadays, names such as ResNet [10], DenseNet [11], or BERT [6] are encountered in almost any paper in the deep learning community. Although the idea of using artificial neurons dates back to 1960s, its large scale adoption only gained momentum during the past decade. The emergence of deep neural networks started with the landmark event in 2012 when a neural network called AlexNet [12] won the ImageNet Large Scale Visual Recognition Challenge (ILSVRC) by a large margin compared to the runner up. The primary discovery from the technical report of AlexNet was that the number of hidden layers, also known as the *depth* of the neural network, is crucial for reaching its high performance. However, stacking more hidden layers to a neural network and optimizing the parameters of such deep networks was not an easy task due to several problems such as bad initialization, and gradient vanishing or exploding phenomena. In 2015, the proposal of skip-connection in ResNet [10], also known as residual connection, has enabled us to successfully train very deep networks with hundreds or even thousands of hidden layers. Two years later, the idea of skip-connection was extended into dense-connection in DenseNet [11], leading to a new generation of state-of-the-art networks achieving better performances while having lower complexity. Although the error rates in ILSVRC have been gradually narrowed down over the years, it took the collective efforts of the entire research community several years to refine the manual network architectural designs and discover better design patterns improving performance.

In order to exempt practitioners from the time-consuming network architecture design process, a new area of research called Neural Architecture Search (NAS) has risen recently [13]. With neural architecture search, the focus has shifted from designing a single network architecture to designing a space of network architectures and the accompanying algorithm that searches this space to find the best architectures for a particular learning task. The structure of the space of potential network architectures

Deep Learning for Robot Perception and Cognition. https://doi.org/10.1016/B978-0-32-385787-1.00014-2

highly affects the computational complexity of a given NAS algorithm. Progressive neural network learning is a class of NAS algorithms that have highly structured search spaces: each algorithm progressively constructs the final network architecture by extending the current network topology with more neurons and neural connections at each step. For this reason, algorithms that fall into this category often have a much low training complexity compared to those operating in an unstructured search space. Besides, since candidate architectures are evaluated sequentially from having a lower complexity to higher complexity until convergence, a progressive neural network learning algorithm is capable of generating compact network architectures. The progressive neural network learning topic will be covered in Section 9.2.

While neural architecture search can solve the tedious problem of designing the architecture of a neural network, deep neural networks as machine learning solutions often confront with hardware barriers when being deployed to mobile or embedded devices. To make the deep learning paradigm more applicable and practical in real-world scenarios, effort have been dedicated to devising methodologies that can reduce the computational and memory complexities of pretrained deep neural networks. These methodologies are often referred to as *model compression*, which includes post-training parameter pruning [14], parameter quantization [15,16], that is, using low-precision representation for floating point numbers, low-rank approximation of the network's parameters [17–19], or a combination of these techniques [20]. Moreover, hardware-aware run-time complexity has also been incorporated as a minimization objective in some NAS algorithms [21,22]. While significant progress has been made in these areas, solutions provided from model compression methodologies or hardware-aware NAS are mainly developed and demonstrated on highly powerful mobile devices with multiple-core processors. It is also worth noting that in many applications, such powerful and expensive units are expected to perform multiple tasks at the same time. For example, sensor data acquisition, inference about the environment and navigation are often seen as common requirements in robotics. In cases where a complete on-device processing solution is unattainable, hybrid solutions can serve as a good and practical alternative, especially in the area of Internet-of-Things (IoT). That is, some preprocessing steps are implemented on-device and intermediate data is transferred to a centralized computing server for major computation steps. Compressive learning is a learning paradigm that can provide such a hybrid solution. In compressive learning signals are simultaneously acquired and compressed at the sensor level, and the machine learning model performs inference directly based on the compressed measurements, bypassing the signal reconstruction step. The compressive learning topic will be covered in Section 9.3.

9.2 Progressive neural network learning

Progressive Neural Network Learning (PNNL) is a class of algorithms that aim to build the network architecture in a data-driven and incremental manner until the learning performance converges. A PNNL algorithm starts with a minimal network topology, which is often a single hidden layer network with a few hidden neurons, then it iteratively increases the learning capacity by adding new blocks of neurons to the current network topology and optimizing the corresponding new connections. For forming new synaptic connections to the existing architecture several algorithms were proposed, each following a different expansion rule. For example, in broad learning system [23] the network topology progression strategy only expands the width of the neural network which is restricted to have two hidden layers. On the other hand, progressive operational perceptron [24] operates by incrementally expanding the depth of the network using a predefined maximal topology template with each layer having a fixed number of neurons. In addition to the approaches that solely enlarge a network only in terms of width or depth, there exist more flexible algorithms such as those in [25] and [26] in which both the width and the depth of the network is progressively increased.

Besides the network progression strategy, different algorithms also adopt different optimization strategies. The majority of them can be divided into two main categories, namely those following a convex optimization approach and those following a stochastic gradient-based optimization approach. For those algorithms that employ convex optimization techniques, such as the broad learning system [23] or the progressive learning network [25], randomized hidden neurons are often used and analytical solutions are only derived for the output layer. Nowadays, with the increasing availability of large datasets and affordable computing hardware, stochastic gradient-based optimization has also been a popular choice for building progressively neural networks [24,26,27]. Compared to the convex optimization approach, employing stochastic gradient-based optimization is less memory-intensive, however, at the cost of longer training times. Since algorithms that employ convex optimization also utilize randomized neurons, the final network architectures obtained by such algorithms are often larger than those generated by algorithms based on stochastic gradient-based optimization. There are also algorithms that combine both optimization techniques, such as the heterogeneous multilayer generalized operational perceptron [26], which takes advantage of convex optimization and randomized neurons to speed up its architecture evaluation procedure and stochastic gradient descent to fine-tune the best candidate architectures. This combination does not only enable efficient training, but also the finding of compact network architectures.

In the next sections, we will present two representative algorithms that utilize convex optimization and randomization, namely the broad learning system [23] in Section 9.2.1 and the progressive learning network [25] in Section 9.2.2. Then we will cover representative algorithms that adopt stochastic gradient-based optimization, namely the progressive operational perceptron and its variants in Section 9.2.3. In Section 9.2.4, we will cover the heterogeneous multilayer generalized operational perceptron algorithm that takes advantage of both optimization approaches. Finally, in Section 9.2.5, we will cover a method that was proposed to speed up and enhance the progressive learning procedure.

9.2.1 Broad learning system

Broad Learning System (BLS) is a two-hidden layer network architecture, which is a generalization of the random vector functional link neural network [28,29]. Given a training set of N samples, let us denote the inputs and the labels as $\mathbf{X} \in \mathbb{R}^{N \times D_{\text{in}}}$ and $\mathbf{Y} \in \mathbb{R}^{N \times D_{\text{out}}}$, respectively. The first hidden layer, also called the feature layer, is formed by the concatenation of *feature nodes*. Feature nodes represent nonlinear features extracted from the input via the following transformation:

$$\mathbf{Z}_i = \phi(\mathbf{X}\mathbf{W}_{f_i} + \mathbf{1}_N \mathbf{B}_{f_i}), \tag{9.1}$$

where $\mathbf{Z}_i \in \mathbb{R}^{N \times D_{f_i}}$ denotes the output of the ith feature node, $\mathbf{W}_{f_i} \in \mathbb{R}^{D_{\text{in}} \times D_{f_i}}$ denotes the random weights and $\mathbf{B}_{f_i} \in \mathbb{R}^{1 \times D_{f_i}}$ denotes the bias term and $\mathbf{1}_N \in \mathbb{R}^N$ denotes a constant vector of ones, and $\phi(\cdot)$ is a nonlinear elementwise activation function such as the sigmoid or the Rectified linear Unit (ReLU) functions. Multiple feature nodes can be used for the first hidden layer, and the concatenation of n feature nodes is denoted by $\mathbf{Z}^{(n)} = [\mathbf{Z}_1, \mathbf{Z}_2, \ldots, \mathbf{Z}_n] \in \mathbb{R}^{N \times (D_{f_1} + \cdots + D_{f_n})}$. Here, we use $[\mathbf{A}, \mathbf{B}]$ to express the horizontal concatenation of matrices $\mathbf{A} \in \mathbb{R}^{P \times Q_1}$ and $\mathbf{B} \in \mathbb{R}^{P \times Q_2}$, resulting in a matrix of size $P \times (Q_1 + Q2)$.

The second hidden layer, also known as the enhancement layer, is formed by the concatenation of *enhancement nodes*. Enhancement nodes represent highly nonlinear features, which are extracted from the feature nodes $\mathbf{Z}^{(n)}$ via the following transformation:

$$\mathbf{H}_j = \zeta\left(\mathbf{Z}^{(n)}\mathbf{W}_{e_j} + \mathbf{1}_N \mathbf{B}_{e_j}\right), \tag{9.2}$$

where $\mathbf{H}_j \in \mathbb{R}^{N \times D_{e_j}}$ denotes the output of the jth enhancement node, $\mathbf{W}_{e_j} \in \mathbb{R}^{(D_{f_1} + \cdots + D_{f_n}) \times D_{e_j}}$ denotes the random weights connecting all feature nodes and the jth enhancement node and $\mathbf{B}_{e_j} \in \mathbb{R}^{1 \times D_{e_j}}$ denotes the corresponding bias term and $\mathbf{1}_N$ denotes a constant vector of ones, and $\zeta(\cdot)$ is a nonlinear element-wise activation function, such as sigmoid or rectified linear unit (ReLU). The choice of $\phi(\cdot)$ and $\zeta(\cdot)$ is determined by the user.

BLS generates predictions by learning a linear classifier or regressor based on the *feature nodes* and *enhancement nodes*. Specifically, the predictions generated by a BLS network are computed as follows:

$$\tilde{\mathbf{Y}} = [\mathbf{Z}^{(n)}, \mathbf{H}^{(m)}]\mathbf{W}_n^m, \tag{9.3}$$

where $\mathbf{H}^{(m)} = [\mathbf{H}_1, \mathbf{H}_2, \ldots, \mathbf{H}_m] \in \mathbb{R}^{N \times (D_{e_1} + \cdots + D_{e_m})}$ denotes the concatenation of m enhancement nodes; $\mathbf{W}_n^m \in \mathbb{R}^{(D_{f_1} + \cdots + D_{f_n} + D_{e_1} + \cdots + D_{e_m}) \times D_{\text{out}}}$ denotes the output weight matrix that connects the output nodes with the n feature nodes and m enhancement nodes.

An illustration of the BLS model is shown in Fig. 9.1. This network formulation is widely adopted in the follow-up works that extend the BLS model, for example, [30,31]. However, we should note that in the original work, the authors also described

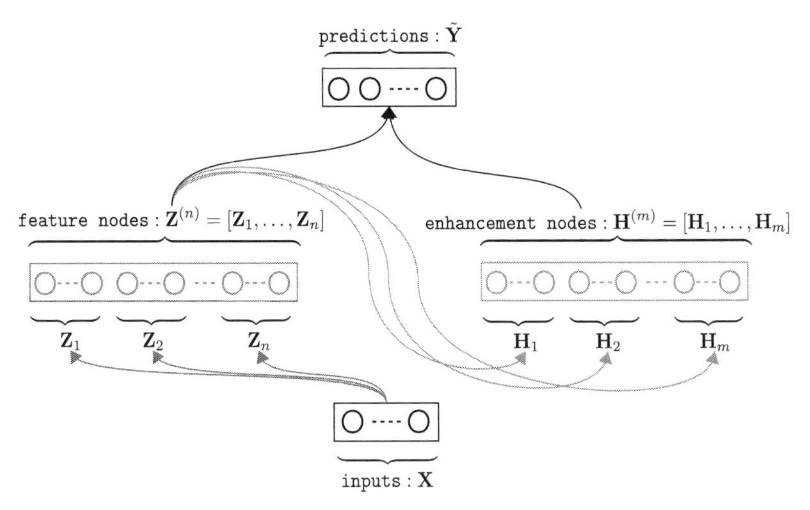

FIGURE 9.1

Broad learning system.

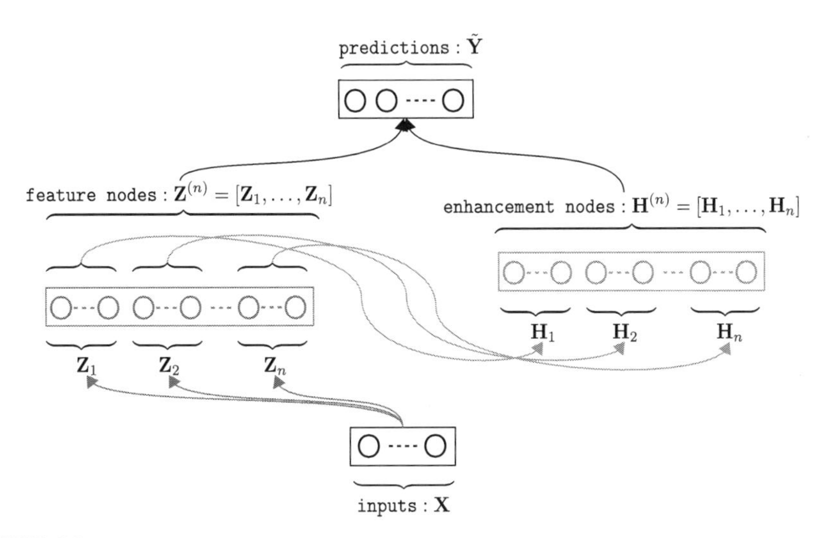

FIGURE 9.2

Broad learning system with alternative enhancement node construction.

another formulation variant, which is called BLS with alternative enhancement node construction, which is shown in Fig. 9.2. As can be seen from Figs. 9.1 and 9.2, the only difference between the two formulations is on how an enhancement node is computed. In the standard formulation described in Eq. (9.2), each enhancement node is connected to all feature nodes while in the alternative construction, one enhancement node is connected to a corresponding feature node. As a result, the number of feature nodes and enhancement nodes must be the same in the alternative formulation. When this property holds for the standard formulation, under certain assumptions, the authors in [23] have proven that the two formulations are equivalent. For the rest of this section, we will cover the standard BLS formulation. Interested readers can consult the original article [23] for further details regarding the alternative formulation.

Since BLS is a progressive learning model, the optimization procedure consists of two steps: optimizing the initial network topology and progressive network expansion until convergence. BLS belongs to the class of models that utilize randomization and convex optimization technique. During both the initial network optimization phase and the progressive expansion phase, BLS assigns random weights drawn from a uniform distribution to the feature nodes and the enhancement nodes, and only provides analytical solution for the output weights. Given an initial network topology having n feature nodes and m enhancement nodes, the solution of the output weight matrix is calculated by solving the least-squares objective and is given by

$$\mathbf{W}_n^m = \left([\mathbf{Z}^{(n)}, \mathbf{H}^{(m)}]\right)^{\dagger} \mathbf{Y}, \tag{9.4}$$

where $(\cdot)^{\dagger}$ denotes the pseudo-inverse operator.

After determining the parameters of the initial network architecture, there are two options to expand this network, namely adding new feature nodes or adding new enhancement nodes. Without the loss of generality, we consider the case when one enhancement node having a dimension of $D_{e_{m+1}}$ is added to the initial network topology. This expansion step leads to the addition of the weight matrix $\mathbf{W}_{e_{m+1}} \in \mathbb{R}^{(D_{f_1}+\cdots+D_{f_n})\times D_{e_{m+1}}}$, connecting n feature nodes to the new enhancement node. The values of $\mathbf{W}_{e_{m+1}}$ are again obtained by randomly sampling from a uniform distribution. In addition to $\mathbf{W}_{e_{m+1}}$, the expansion of a new enhancement node also leads to the expansion of the output weight matrix from $\mathbf{W}_n^m \in \mathbb{R}^{(D_{f_1}+\cdots+D_{f_n}+D_{e_1}+\cdots+D_{e_m})\times D_{\text{out}}}$ to $\mathbf{W}_n^{m+1} \in \mathbb{R}^{(D_{f_1}+\cdots+D_{f_n}+D_{e_1}+\cdots+D_{e_{m+1}})\times D_{\text{out}}}$. Following Eq. (9.4), the values of $\mathbf{W}_n^{m+1}$ can be computed as follows:

$$\mathbf{W}_n^{m+1} = \left([\mathbf{Z}^{(n)}, \mathbf{H}^{(m+1)}]\right)^{\dagger} \mathbf{Y}, \tag{9.5}$$

where $\mathbf{H}^{(m+1)} = [\mathbf{H}^{(m)}, \mathbf{H}_{m+1}]$.

We define $\mathbf{A}_n^m = [\mathbf{Z}^{(n)}, \mathbf{H}^{(m)}]$ and $\mathbf{A}_n^{m+1} = [\mathbf{Z}^{(n)}, \mathbf{H}^{(m+1)}]$. A straightforward way to compute the pseudo-inverse of $\mathbf{A}_n^{m+1}$ is to perform singular value decomposition on $\mathbf{A}_n^{m+1}$. However, as the number of enhancement nodes increases, repeating this step leads to a significant amount of computation. In order to reduce the computational complexity when computing the pseudoinverse of $\mathbf{A}_n^{m+1}$, BLS adopts an incremental approach to compute $(\mathbf{A}_n^{m+1})^{\dagger}$ based on $(\mathbf{A}_n^m)^{\dagger}$ as follows:

$$(\mathbf{A}_n^{m+1})^{\dagger} = \begin{bmatrix} (\mathbf{A}_n^m)^{\dagger} - \mathbf{D}\mathbf{B}^T \\ \mathbf{B}^T \end{bmatrix}, \tag{9.6}$$

where $\mathbf{D} = (\mathbf{A}_n^m)^\dagger \mathbf{H}_{m+1}$ and

$$\mathbf{B}^T = \begin{cases} \mathbf{C}^\dagger, & \text{if } \mathbf{C} \neq 0, \\ (\mathbf{1} + \mathbf{D}^T\mathbf{D})^{-1}\mathbf{B}^T(\mathbf{A}_n^m)^\dagger, & \text{if } \mathbf{C} = 0, \end{cases} \tag{9.7}$$

with $\mathbf{C} = \mathbf{H}_{m+1} - \mathbf{A}_n^m\mathbf{D}$. Combining Eq. (9.5) and Eq. (9.6) leads to the following equation that can be used to calculate the new output weight matrix $\mathbf{W}_n^{m+1}$ based on the initial output weight matrix $\mathbf{W}_n^m$:

$$\mathbf{W}_n^{m+1} = \begin{bmatrix} \mathbf{W}_n^m - \mathbf{D}\mathbf{B}^T\mathbf{Y} \\ \mathbf{B}^T\mathbf{Y} \end{bmatrix}. \tag{9.8}$$

A similar approach can also be derived for incrementally updating the output weight matrix when new feature nodes are added. Specifically, let us consider the case when one feature node ($\mathbf{Z}_{n+1}$) is added to the current network architecture of n feature nodes and m enhancement node. This new feature node leads to the generation of the following enhancement nodes:

$$\mathbf{H}_{ex_m} = [\zeta(\mathbf{Z}_{n+1}\mathbf{W}_{ex_1} + \mathbf{B}_{ex_1}), \ldots, \zeta(\mathbf{Z}_{n+1}\mathbf{W}_{ex_m} + \mathbf{B}_{ex_m})], \tag{9.9}$$

where $\mathbf{W}_{ex_1}, \ldots, \mathbf{W}_{ex_m}, \mathbf{B}_{ex_1}, \ldots \mathbf{B}_{ex_m}$ are randomly generated.

The output layer now not only takes $\mathbf{A}_n^m$ as inputs but also $\mathbf{Z}_{n+1}$ and $\mathbf{H}_{ex_m}$. That is, $\mathbf{A}_{n+1}^m = [\mathbf{A}_n^m, \mathbf{Z}_{n+1}, \mathbf{H}_{ex_m}]$. The update equation for $(\mathbf{A}_{n+1}^m)^\dagger$ based on $(\mathbf{A}_n^m)^\dagger$ is the following:

$$(\mathbf{A}_{n+1}^m)^\dagger = \begin{bmatrix} (\mathbf{A}_n^m)^\dagger - \mathbf{D}\mathbf{B}^T \\ \mathbf{B}^T \end{bmatrix}, \tag{9.10}$$

where $\mathbf{D} = (\mathbf{A}_n^m)^\dagger [\mathbf{Z}_{n+1}, \mathbf{H}_{ex_m}]$ and

$$\mathbf{B}^T = \begin{cases} \mathbf{C}^\dagger, & \text{if } \mathbf{C} \neq 0, \\ (\mathbf{1} + \mathbf{D}^T\mathbf{D})^{-1}\mathbf{B}^T(\mathbf{A}_n^m)^\dagger, & \text{if } \mathbf{C} = 0, \end{cases} \tag{9.11}$$

with $\mathbf{C} = [\mathbf{Z}_{n+1}, \mathbf{H}_{ex_m}] - \mathbf{A}_n^m\mathbf{D}$.

While providing a fast solution to update the parameters of the network when new enhancement nodes or feature nodes are added, no particular progression strategy is proposed in [23]. Since the enhancement nodes take the feature nodes as inputs, we empirically found that adding the feature nodes until convergence before adding the enhancement nodes is a good progression strategy for the BLS model. Besides, the empirical results in [23] suggest that the total dimension of the enhancement nodes or the feature nodes should be on the order of thousands in order to obtain good performance.

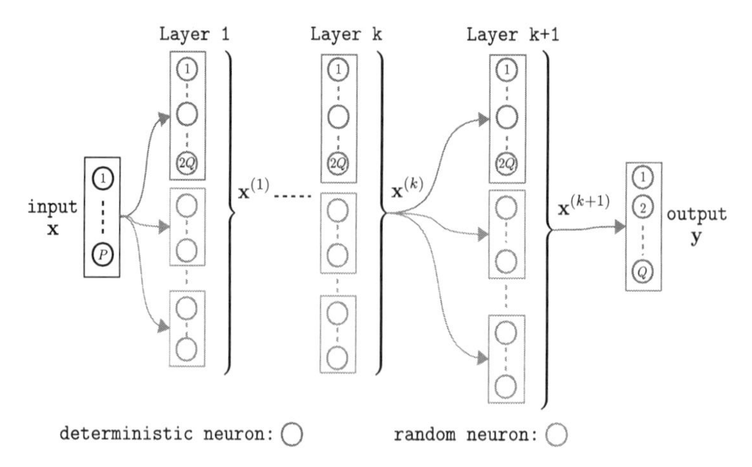

FIGURE 9.3

Progressive Learning Network (PLN) with $k+1$ hidden layers. (For interpretation of the colors in the figure, the reader is referred to the web version of this chapter.)

9.2.2 Progressive learning network

Progressive Learning Network (PLN) [25] is also an algorithm that adopts convex optimization and randomization approaches. Different from broad learning system, the progression strategy in PLN allows us to construct multilayer fully-connected networks, instead of only two hidden layers as in BLS. More specifically, the progression strategy in PLN constructs the network architecture in a layer-by-layer manner until convergence, starting from an one hidden layer network. Each hidden layer in PLN comprises of both deterministic and randomized neurons. The weight values of deterministic neurons are computed by solving a linear optimization problem while the weight values of the randomized neurons are computed by sampling from a random uniform distribution. To construct a new hidden layer, the algorithm first computes the deterministic neurons, and then incrementally adds neurons with random parameters until the learning performance saturates.

To demonstrate how a hidden layer is optimized in the PLN model, we assume that the current network architecture has k hidden layers and the PLN algorithm proceeds by adding the $(k+1)$-th hidden layer to the current network architecture. At this step, the weights of the previous k hidden layers are fixed and the PLN algorithm only optimizes the $(k+1)$-th hidden layer as well as the output layer. We illustrate a $(k+1)$-hidden layer PLN network in Fig. 9.3.

Let us denote the ith sample and its target vector by $\mathbf{x}_i \in \mathbb{R}^P$ and $\mathbf{y}_i \in \mathbb{R}^Q$, respectively. In addition, given $\mathbf{x}_i$ as the input to the network, let us denote $\mathbf{x}_i^{(k)} \in \mathbb{R}^{P_k}$ as the output of the kth hidden layer. As mentioned previously, a hidden layer in PLN consists of both deterministic neurons and randomized neurons. Thus $\mathbf{x}_i^{(k+1)}$ is formed by the concatenation of deterministic and randomized neurons as follows:

$$\mathbf{x}_i^{(k+1)} = \phi\left(\begin{bmatrix} , \mathbf{d}_i^{(k+1)} \\ \mathbf{r}_i^{(k+1)} \end{bmatrix}\right) \tag{9.12}$$

where $\mathbf{d}_i^{(k+1)}$ and $\mathbf{r}_i^{(k+1)}$ denote the outputs of deterministic and randomized neurons of the $(k+1)$-th hidden layer, respectively. $\phi(\cdot)$ is an elementwise activation function that satisfies the Progression Property (PP). According to [25], an elementwise nonlinear function $\phi(\cdot)$ holds progression property if there are two known linear transformations $\mathbf{V}_N \in \mathbb{R}^{M\times N}$ and $\mathbf{U}_N \in \mathbb{R}^{N\times M}$ such that $\mathbf{U}_N\phi(\mathbf{V}_N\mathbf{x}) = \mathbf{x}$, $\forall \mathbf{x} \in \mathbb{R}^N$. It can be easily verified that the PP holds for ReLU activation function by setting:

$$\mathbf{V}_N = \begin{bmatrix} \mathbf{I}_N \\ -\mathbf{I}_N \end{bmatrix} \in \mathbb{R}^{2N\times N}, \tag{9.13}$$

$$\mathbf{U}_N = [\mathbf{I}_N | -\mathbf{I_N}] \in \mathbb{R}^{N\times 2N}, \tag{9.14}$$

where $\mathbf{I}_N$ is the identity matrix of N dimensions.

The deterministic neurons are computed using the following relation:

$$\mathbf{d}_i^{(k+1)} = \mathbf{V}_Q\mathbf{W}_{k+1}^*\mathbf{x}_i^{(k)}, \tag{9.15}$$

where $\mathbf{V}_Q$ is the PP holding matrix as defined in Eq. (9.13). $\mathbf{W}_{k+1}^*$ is obtained by solving the regularized least-squares optimization problem:

$$\mathbf{W}_{k+1}^* = \underset{\mathbf{W}_{k+1}}{\operatorname{argmin}}\left(\sum_i \|\mathbf{y}_i - \mathbf{W}_{k+1}\mathbf{x}_i\|_2^2 + \lambda\|\mathbf{W}_{k+1}\|_2^2\right), \tag{9.16}$$

where λ is a predefined regularization coefficient.

The randomized neurons are generated by applying a linear transformation:

$$\mathbf{r}_i^{(k+1)} = \mathbf{R}_{k+1}\mathbf{x}_i, \tag{9.17}$$

where $\mathbf{R}_{k+1}$ is a random matrix with values drawn from a uniform distribution.

As we can see from Eq. (9.15), for any hidden layer, the dimension of the deterministic neurons is always $2Q$, which is twice the number of dimensions of the target vectors. The dimension of the randomized neurons is progressively increased until the learning performance converges.

Given the output of the $(k+1)$-th hidden layer $\mathbf{x}_i^{(k+1)}$, the PLN model makes prediction by learning a linear classifier/regressor that minimizes the reconstruction error:

$$\mathbf{O}^* = \underset{\mathbf{O}}{\operatorname{argmin}} \sum_i \|\mathbf{y}_i - \mathbf{O}\mathbf{x}_i^{(k+1)}\|_2^2 \tag{9.18}$$

$$\text{such that: } \|\mathbf{O}\|_2^2 \leq \alpha\|\mathbf{U}_Q\|_2^2, \tag{9.19}$$

where α is a predefined regularization coefficient and $\mathbf{U}_Q$ is the PP holding property matrix defined in Eq. (9.14).

The optimization problem in Eq. (9.18) is a convex optimization problem, which can be solved by existing convex solvers, like the Alternating Direction Method of Multipliers (ADMM) [32].

9.2.3 Progressive operational perceptron and its variants

Progressive Operational Perceptron [24] (POP) is an algorithm that heavily takes advantage of the gradient-based optimization to not only learn the network topology but also the assignment of different nonlinear transformations to different layers following a data-driven process. Before diving into the details of the progression strategy and the optimization routine in POP, it is important to become familiar with a generalized neuron model that serves as a core component in POP, as well as in other PNNL algorithms that we will cover next in this chapter.

The neuron model is a critical component that defines the modeling ability of a neural network. The most common neuron model used nowadays is based on the design of McCulloch–Pitts perceptron [33] (hereafter simply referred to as perceptron), which performs a linear transformation followed by a nonlinear activation function. There have been several attempts [34–36] to design novel neuron models that can better capture the nonlinear relationships often existing between the inputs and the targets. Generalized Operational Perceptron (GOP) [24] is such a neuron model that can express a wide range of nonlinear transformations, including the one modeled by the perceptron. GOP is designed to better simulate biological neurons in the mammalian nervous system. Specifically, a GOP performs three distinct operations, namely the nodal, the pooling and the activation operations that correspond to the operations observed in a biological neuron.

Let us consider the ith GOP of the kth layer in a fully-connected GOP network. A GOP takes a vector as input, which represents the output vector of the previous layer, and outputs a scalar. In the following, we use the notation $x_j^{(k-1)} \in \mathbb{R}$ for $j = 1, \ldots, N_{k-1}$ to denote the jth output of the $(k-1)$-th layer. Thus the output of the GOP under consideration is denoted by $x_i^{(k)} \in \mathbb{R}$. The first operation performed by a GOP is the *nodal* operation, which modifies each individual output of the previous layer using a synaptic weight:

$$z_{ji}^{(k)} = \psi_i^k \left(x_j^{(k-1)}, w_{ji}^{(k)} \right), \quad \forall j = 1, \ldots, N_{k-1}, \tag{9.20}$$

where $\psi_i^{(k)}(\cdot)$ is the *nodal operator* of the ith GOP in the kth layer; $x_j^{(k-1)} \in \mathbb{R}$ is the jth output of the $(k-1)$th layer, and $w_{ji}^{(k)} \in \mathbb{R}$ is the synaptic weight connecting the jth neuron of the $(k-1)$th layer to the ith neuron of the kth layer.

The second operation performed by a GOP is the *pooling* operation, which summarizes the modified input signals generated by the nodal operation into a scalar:

$$t_i^{(k)} = \rho_i^{(k)} \left(z_{1i}^{(k)}, \ldots, z_{N_{k-1}i}^{(k)} \right) + b_i^{(k)}, \tag{9.21}$$

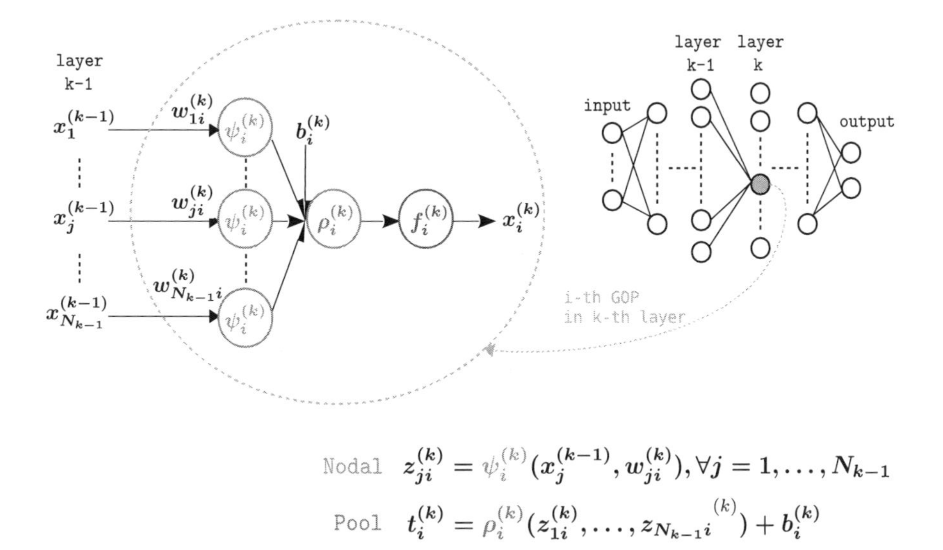

FIGURE 9.4

Generalized operational perceptron. (For interpretation of the colors in the figure, the reader is referred to the web version of this chapter.)

where $\rho_i^{(k)}(\cdot)$ denotes the *pooling operator*, and $z_{ji}^{(k)}$, $j = 1, \ldots, N_{k-1}$ denote the outputs of the nodal operation in Eq. (9.20). $b_i^{(k)} \in \mathbb{R}$ denotes the learnable bias term, which is used to offset the pooled result.

Similar to other neuron models, the final operation of a GOP is the activation operation:

$$x_i^{(k)} = f_i^{(k)}\left(t_i^{(k)}\right), \tag{9.22}$$

where $f_i^{(k)}(\cdot)$ denotes the *activation operator* and $t_i^{(k)}$ denotes the output of the pooling operation.

Fig. 9.4 illustrates the working mechanism of generalized operational perceptron. As can be seen from this figure, a GOP is characterized by the set of synaptic weights including the bias and the set of operators (nodal, pooling and activation). The functional form of an operator is selected, usually through optimization, from a library of predefined functions such as the ones provided in Table 9.1. Hereafter, we use the term *operator set* to refer to a specific choice of one nodal, one pooling and one activation operators. Each GOP is equipped with an operator set, which can be different from the operator sets of other GOPs.

Although the abstraction through operator set in GOP allows a higher degree of flexibility and diversity, optimizing neural networks that are formed by GOPs is a challenging problem since we need to optimize not only the synaptic weights but also

Table 9.1 Operator library. An operator set is formed by selecting one nodal operator, one pooling operator and one activation operator.

Nodal Ψ	$z_{ji}^{(k)} = \psi_i^{(k)}(x_j^{(k-1)}, w_{ji}^{(k)})$
Multiplication	$w_{ji}^{(k)} x_j^{(k-1)}$
Exponential	$\exp(w_{ji}^{(k)} x_j^{(k-1)}) - 1$
Harmonic	$\sin(w_{ji}^{(k)} x_j^{(k-1)})$
Quadratic	$w_{ji}^{(k)} (x_j^{(k-1)})^2$
Gaussian	$w_{ji}^{(k)} \exp(-w_{ji}^{(k)} (x_j^{(k-1)})^2)$
DoG	$w_{ji}^{(k)} x_j^{(k-1)} \exp(-w_{ji}^{(k)} (x_j^{(k-1)})^2)$
Pool (P)	$t_i^{(k)} = \rho_i^{(k)}(z_{1i}^{(k)}, \ldots, z_{N_{k-1}i}^{(k)})$
Summation	$\sum_{j=1}^{N_{k-1}} z_{ji}^{(k)}$
1-Correlation	$\sum_{j=1}^{N_{k-1}-1} z_{ji}^{(k)} z_{(j+1)i}^{(k)}$
2-Correlation	$\sum_{j=1}^{N_{k-1}-2} z_{ji}^{(k)} z_{(j+1)i}^{(k)} z_{(j+2)i}^{(k)}$
Maximum	$\max_j (z_{ji}^{(k)})$
Activation (F)	$x_i^{(k)} = f_i^{(k)}(t_i^{(k)})$
Sigmoid	$1/(1+\exp(-t_i^{(k)}))$
Tanh	$\sinh(t_i^{(k)})/\cosh(t_i^{(k)})$
ReLU	$\max(0, t_i^{(k)})$

the assignment of operator set to each specific GOP. Given a predefined library of L operator sets and a GOP network of M neurons, there are L^M different ways to make the operator set assignment, corresponding to L^M different neural networks having different functional forms. That is, the space of nonlinear functions encapsulated by a GOP network scales exponentially with respect to the number of neurons.

Progressive operational perceptron is an algorithm that trains an instantiation of GOP networks in a layer-by-layer manner, given a predefined network template, which specifies the maximum number of layers and the number of GOPs in each layer. To make the optimization problem computationally tractable, POP assigns the same operator set to all GOPs within the same layer. This constraint significantly reduces the number of candidate assignments, which are evaluated by a gradient descent-based optimization process.

Specifically, given a target loss value α and a maximal network template of k hidden layers $T = [D_{\text{in}}, N_1, \ldots, N_K, D_{\text{out}}]$ with the kth hidden layer having N_k GOPs, POP iteratively learns one hidden layer at each step and terminates when the target loss value is achieved or all hidden layers have been learned. At the kth step, POP forms a neural network with k hidden layers having the following network template $T_k = [D_{\text{in}}, N_1, \ldots, N_k, D_{\text{out}}]$. At this point, all previous $k - 1$ hidden layers have been optimized and fixed. Since all GOPs within a layer share the same operator set, the optimization problem at the kth step is to find the best operator set assignment and the weight values for the kth hidden layer and the output layer. POP tackles this problem by evaluating a reduced set of possible assignments via a two-pass Greedy Iterative Search (GIS) procedure.

Let ϕ_k and ϕ_o be the operator sets of the kth hidden layer and output layer, respectively. In the first pass of GIS, ϕ_k is selected randomly from the operator library and the best performing ϕ_o^* is chosen by evaluating all possible operator sets. That is, for every operator set that is assigned to ϕ_o we randomly initialize the weights of the kth hidden layer and the output layer and train the network instance for E epochs. Once ϕ_o^* is found, the algorithm continues by fixing ϕ_o^* and optimizing ϕ_k^* through the same evaluation procedure, that is, now the selected ϕ_o^* is used and the best operator set ϕ_k^* is found by using all operator sets. All the steps conducted in the first pass are repeated for the second pass of GIS, except the selection of the initial ϕ_k, which is set to the one selected by the first pass of GIS.

After determining best operator sets for the kth hidden layer and the output layer through the two-pass GIS, the assignment ϕ_k^* and the synaptic weights of the kth hidden layer are fixed. The algorithm continues to build the next hidden layer from the maximal network template if the current loss value is larger than the predefined target value α, otherwise it terminates the process. In the first case the output layer is discarded and a new hidden and output layers are learned, while in the latter case the learned output layer is kept and the final network architecture is formed by all the hidden layers determined by the successive GIS optimization processes and the output layer of the final last layer.

Using the network construction strategy of POP, there are P^2 different ways to assign the operator set for the kth hidden layer and the output layer at step k, resulting in P^2 different network candidates for evaluation. Based on the two-pass GIS procedure, optimizing a hidden layer in POP requires four iterations over all operator set choices, which only evaluates $4P$ candidate assignments. Although this approach reduces the computational complexity significantly, it is suboptimal since only a small subset of the search space is evaluated.

In order to speed up and enhance the optimization procedure in POP, a faster variant of POP called POPfast was proposed in [27]. The key idea in POPfast that enables improvements over POP in computational complexity is the relaxation of the output layer as a linear layer followed by an appropriate activation function, that is, the softmax function for classification problems and the identity function for regression problems. This relaxation reduces the search space from having P^2 to P candidate candidates when learning a new hidden layer. This is because with the relaxation, we no longer need to optimize the operator set of a hidden layer in conjunction with that of the output layer. At the kth step, POPfast constructs a k hidden layer network similar to POP. The algorithm finds the best performing operator set ϕ_k^* for the kth hidden layer by iterating through all operator sets. For each candidate operator set, it randomly initializes the weight values of the kth hidden layer and the output layer, and it optimizes them for few epochs. Compared to POP, POPfast only requires one iteration (versus four in POP) over all the operator set while evaluating the entire search space (versus a small subset of search space as in POP).

A memory augmented variant of POPfast, known as POPmem, is proposed in [27], which builds the network architecture so that the output layer and the newly constructed hidden layer have direct access to lower-level features extracted by previous layers. This direct access is enabled via *memory paths*, which are information-preserving linear transformations.

Let $\mathbf{x}^{(k-1)}$ be the input to the kth hidden layer, with $\mathbf{x}^{(0)}$ being the input data (the input to the network). In addition, let us denote the transformation performed by the GOPs in the kth hidden layer as $\mathcal{F}_k$. The kth hidden layer only observes $\mathbf{x}^{(k-1)}$ as its input and does not have direct access to $\mathbf{x}^{(k-2)}, \mathbf{x}^{(k-3)}, \ldots, \mathbf{x}^{(0)}$, which are the shallower features extracted by the previous hidden layers and the input. Similarly, the output layer only observes $\mathcal{F}_k(\mathbf{x}^{(k-1)})$ as its input and does not have direct access to information from previous hidden layers. POPmem provides a remedy for this problem by incorporating a memory path in parallel to the kth hidden layer formed by GOPs. Let us denote by $\mathcal{G}$ a linear projection that preserves information of the input data. The type of information retained depends on how the weight matrix of $\mathcal{G}$ is obtained. For example, using Principal Component Analysis (PCA) $\mathcal{G}$ can preserve the energy of the input data while using Linear Discriminant Analysis (LDA) $\mathcal{G}$ can preserve the linear class separability between different classes.

At the kth step, POPmem constructs a k hidden layer network. The kth hidden layer is formed by concatenating N_k GOPs, that is, $\mathcal{F}^{(k)}(\mathbf{x}^{(k-1)})$ and the information preserved from the $(k-1)$-th hidden layer, that is, $\mathcal{G}^{(k)}(\mathbf{x}^{(k-1)})$. $\mathcal{G}^{(k)}$ denotes the memory path of the kth hidden layer, which is estimated from $\mathbf{x}_i^{(k-1)}$, with i denoting the index of the training sample. For example, to preserve the energy of the data, $\mathcal{G}^{(k)}(\mathbf{x}) = \mathbf{W}_{\mathcal{G}_k}^T \mathbf{x}^{(k-1)}$ is obtained by solving the signal reconstruction problem:

$$\mathbf{W}_{\mathcal{G}_k} = \underset{\mathbf{W}}{\operatorname{argmin}} \sum_i \left\| \mathbf{x}_i^{(k-1)} - \mathbf{W}\mathbf{W}^T \mathbf{x}_i^{(k-1)} \right\|_2^2 \qquad (9.23)$$
$$\text{subject to: } \mathbf{W}^T\mathbf{W} = \mathbf{I},$$

where $\mathbf{W}^T$ denotes the transpose of $\mathbf{W}$ and $\mathbf{I}$ denotes the identity matrix of appropriate size.

By construction, we have the following recursive relation in the network architectures that are learned by POPmem:

$$\mathbf{x}^{(k)} = \begin{bmatrix} \mathcal{F}^{(k)}(\mathbf{x}^{(k-1)}) \\ \mathcal{G}^{(k)}(\mathbf{x}^{(k-1)}) \end{bmatrix}, \forall k = 1, \ldots, K. \qquad (9.24)$$

Here we should note that at the kth step in POPmem, $\mathcal{G}^{(k)}$ is determined by solving the corresponding optimization problem that defines the information-preserving

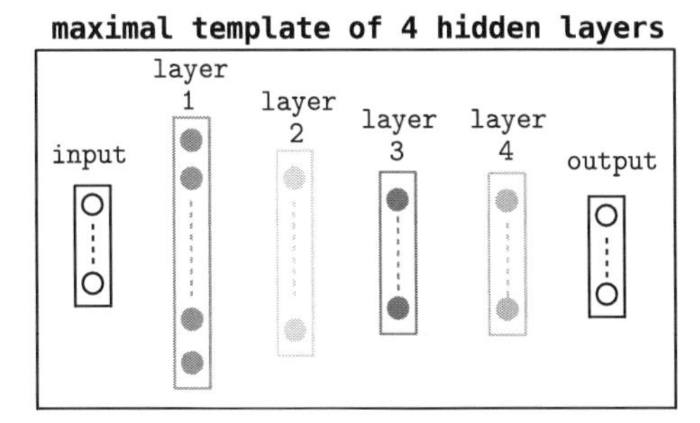

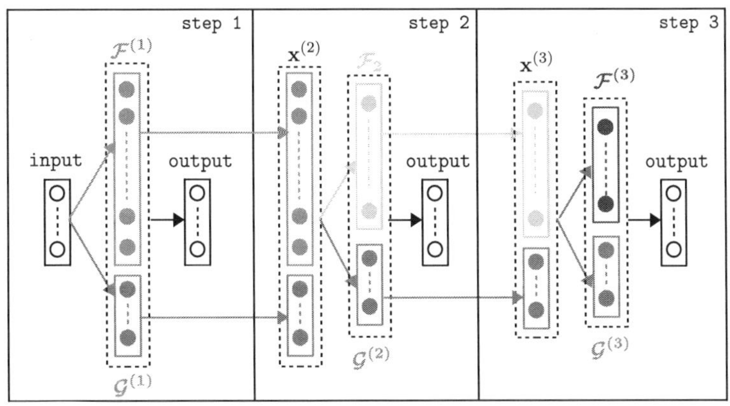

FIGURE 9.5

The progression of POPmem until the third hidden layer. (For interpretation of the colors in the figure, the reader is referred to the web version of this chapter.)

property. After that, the parameters of $\mathcal{G}^{(k)}$, that is, $\mathbf{W}_{\mathcal{G}_k}$, are fixed during the optimization of $\mathcal{F}^{(k)}$. In POPmem, the GOPs of the kth hidden layer ($\mathcal{F}^{(k)}$) and the output layer are optimized in the same manner as in POPfast. In Fig. 9.5, we illustrate the network construction until the third hidden layer by POPmem.

For POP and its variants, once the final network architecture has been determined, the parameters of the entire network can be further finetuned using stochastic gradient descent with a small learning rate.

9.2.4 Heterogeneous multilayer generalized operational perceptron

As we have seen from the previous section, POP and its variants tackle the exponential search space in GOP networks by imposing a homogeneity constraint, that is, GOPs within a layer share the same operator set assignment. Although this decision reduces the computational complexity when evaluating operator sets and constructing the network architecture using stochastic gradient descent algorithm, it significantly reduces the space of potential network architectures. In addition, using POP and its variants, the design of the final network architecture is only semiautomatic since the

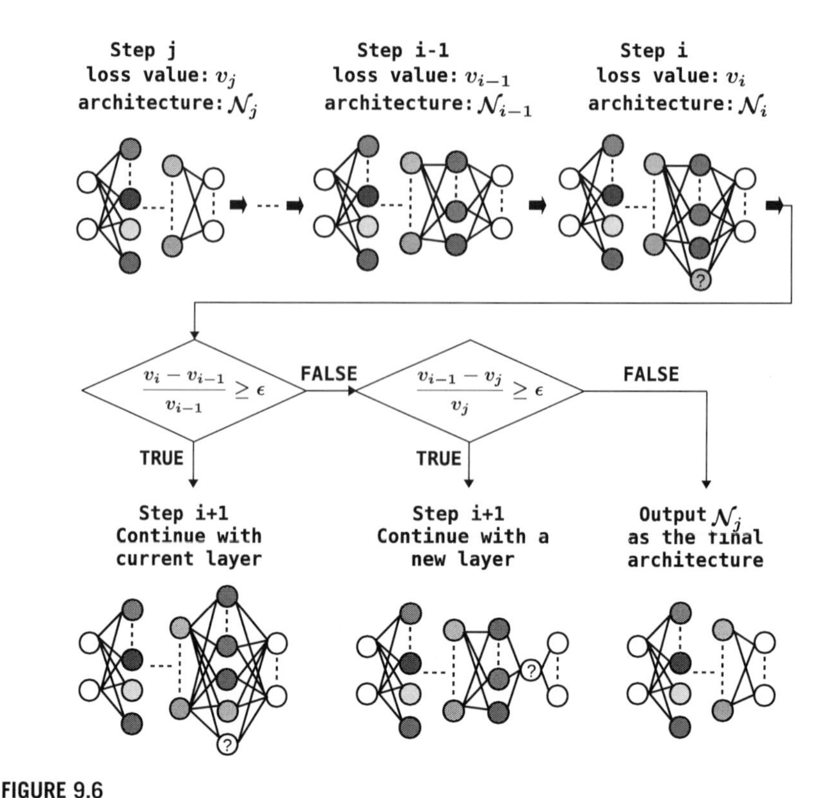

FIGURE 9.6

The progression strategy in HeMLGOP.

users still need to define the maximum number of layers and the number of neurons for each layer.

Heterogeneous Multilayer Generalized Operational Perceptron (HeMLGOP) [26] overcomes the aforementioned limitations in POP and its variants. Similar to BLS and PLN algorithms that have been covered in Sections 9.2.1 and 9.2.2, respectively, HeMLGOP also takes advantage of randomization and convex optimization to speedily evaluate the operator set assignment. This addresses the computational bottleneck encountered in POP and its variants when evaluating potential operator sets. Besides, HeMLGOP increases heterogeneity characteristic into the learned network architecture by allowing different GOPs within the same hidden layer to have different operator sets, enabling the neural network under construction the capability to model a wider range of functions. Finally, HeMLGOP is an algorithm that learns a neural network architecture in a fully automatic manner, being capable of constructing network architectures of any width and depth size in a data-driven manner, depending on the difficulty of the problem at hand.

Fig. 9.6 illustrates the progression strategy adopted by HeMLGOP. The algorithm builds the network architecture in a block-by-block manner, with S_b being the number of neurons of a block, given as a hyperparameter. At the first step, HeMLGOP forms a single hidden layer network with one block of GOPs. Once this block of neurons is optimized in conjunction with the output layer, the algorithm moves to the next step. At each step, the algorithm expands the current network architecture by adding a new block of neurons to the last hidden layer, hereafter also referred to as the current hidden layer. A hidden layer continues to expand until the performance obtained is saturated, which is measured by checking the rate of improvement:

$$\frac{v_i - v_{i-1}}{v_{i-1}} \leq \epsilon, \tag{9.25}$$

where ϵ is a predefined threshold, and v_i denotes the loss value obtained at step i. Here, we assume that $\mathcal{N}_{i-1}$ and $\mathcal{N}_i$ have the same number of hidden layers, with $\mathcal{N}_i$ denoting the network architecture obtained at step i. That is, at the end of the i-th step, if the relation in Eq. (9.25) does not hold, the progression in the current hidden layer continues. Otherwise, the new block of neurons that is added at step i is removed from the network architecture. This is because the addition of this new block only leads to a marginal relative performance improvement (less than a given threshold ϵ). In this scenario, HeMLGOP further evaluates whether a new hidden layer should be formed or the algorithm should terminate by evaluating the following relation:

$$\frac{v_{i-1} - v_j}{v_j} \leq \zeta, \tag{9.26}$$

where j denotes the step index such that $D_j = D_i - 1$ and $D_{j+1} = D_i$, with D_i denoting the number of hidden layers of the network architecture obtained at the end of step i. Here, we use the loss value v_{i-1} representing the performance of the current architecture because the new block of neurons added at step i has been removed. ζ is a predefined threshold value to control the progression between layers.

If the relation in Eq. (9.26) holds, it means that the inclusion of the last hidden layer only produces a small relative performance improvement. In this case, the algorithm terminates and outputs $\mathcal{N}_j$, which has $D_j = D_i - 1$ hidden layers. On the other hand, if Eq. (9.26) does not hold, including the last hidden layer to the network produces a noticeable improvement and the algorithm continues by building a new hidden layer at the next step.

To enable a computationally tractable search space when optimizing a block of neurons, HeMLGOP adopts two architectural constraints. The first constraint forces GOPs within the same block to share the same operator set assignment. Although this constraint might seem to promote homogeneity as in POP and its variants, it is not the case for networks constructed by HeMLGOP since each layer consists of multiple blocks, which can utilize different operator sets. The second constraint is the fixed operator set assignment for the output layer. Similar to POPfast, the output layer in HeMLGOP only performs linear transformation, followed by a task-specific activation function, such as the softmax for classification tasks and the identity function for regression tasks. This constraint eliminates the need to optimize operator set assignment of the output layer in conjunction with the hidden layer as is done in POP.

When HeMLGOP adds a new block of neurons to the current layer, all previous blocks are fixed (both the operator set assignments and synaptic weights), and the algorithm searches for the best performing operator set for the new block and optimizes its weights along with the weights of the output layer. To select a suitable operator set for the new block, the algorithm relies on randomization to evaluate the fitness of a given assignment. Specifically, for each operator set, HeMLGOP assigns the operator set to the new block and its parameters are randomly initialized from a uniform distribution. The parameters of the output layer are determined by solving the linear regression problem using the representation of the last hidden layer as input, that is,

$$\mathbf{W} = \mathbf{H}^{\dagger}\mathbf{Y}, \tag{9.27}$$

where $\mathbf{H}$ denotes the outputs of the current hidden layer and $\mathbf{Y}$ denotes the targets. After the parameters of the output layer are computed, the performance is measured for this operator set assignment. After iterating through all operator sets, HeMLGOP selects the operator set leading to the best performance, and the parameters of the new block and the output layer are optimized using stochastic gradient descent for few epochs. This intermediate finetuning process is essential in order to fully take advantage of the modeling capacity of the new block of GOPs.

Similar to POP and its variants, once the final network architecture has been determined, all parameters of the network can be jointly optimized using stochastic gradient descent with a small learning rate to further improving performance. Here, we should note that this final finetuning step is only beneficial when sufficient training data is available. On the other hand, when the training data is scarce, jointly finetuning all parameters can easily lead to overfitting.

Due to its computational efficiency and the ability to produce very compact network architectures, HeMLGOP has been applied to many applications such as face verification [37], learning to rank [38], and financial forecasting with imbalanced data [39]. The implementations of HeMLGOP and other related algorithms are publicly available as a standard Python package [40], which allows easy installation and further development.

9.2.5 Subset sampling and online hyperparameter search for training enhancement

Up until now, we have covered different progressive neural network learning algorithms with different network construction and optimization strategies. A common characteristic in progressive neural network learning algorithms is that they require repetitive computations using the training data, regardless of the optimization approach adopted. This characteristic can lead to a large amount of computation and long training time, especially for those that rely on stochastic gradient-based optimization. It was recently demonstrated in [41] that such repetitive computations over the training data can be reduced while achieving similar or better performance. Two key ideas were proposed in [41], which are the use of subsets of training data instead of the whole training set and the use of different hyperparameter values at different learning steps.

The advantage of using only a small subset of the training data optimizing the new connections at each incremental learning step is two-fold. In addition to the obvious computational cost reduction per incremental step, the use of different samples to train different groups of neurons also promotes neurons of different groups to capture different patterns in the data. In fact the idea of using a subset of available data to optimize a learning model is not new. In the field of active learning where the algorithms need to pick the most representative samples among a big pool of samples to be labeled, subset selection procedure plays a central role [42–44]. Different from active learning, the sample selection strategy for progressive learning can take advantage of the label information. The following three subset selection strategies were used in [41]:

- *Random uniform selection*: at each progressive learning step, select randomly M samples from the training set of N samples ($M < N$).
- *Top-M selection based on training error*: at each progression learning step, select the M samples that induce the highest loss values, given the current network architecture.
- *Top-M selection based on diverse training error*: K-means clustering is applied to the predicted vectors and for each cluster, select $\lfloor M/C \rfloor$ samples with C denoting the number of clusters and $\lfloor \cdot \rfloor$ is the floor operator.

The key observation when using a subset of training data in [41] is that the number of unique samples that an algorithm uses during progressive learning is an important factor for the final performance. For this reason, despite being unsupervised (without using the label information) and simple, random sampling achieves very good performance. The other two subset selection strategies can lead to a strong bias toward difficult samples, thus resulting in less diverse subsets of data being presented to the networks.

The second idea for enhancing performance of progressive network learning methods proposed by [41] is the incorporation of online hyperparameter search. In all algorithms described above the values of hyperparameters are fixed throughout the whole network construction phase. The best hyperparameter combination is often determined by running the method multiple times, measuring the performance of all runs on the validation set, and choosing the hyperparameter values leading to the best validation performance, which are then shared for the construction of all layers of the network. The process is both time-consuming and suboptimal. Since PNNL algorithms gradually increase the capacity of the neural network at each step, it is intuitive that the network might require different amount of regularization at different steps. To search for a suitable hyperparameter setting, let $\mathcal{H}$ denote the set of all hyperparameter value combinations, and Q be the cardinality of $\mathcal{H}$. At step k, after determining the subset $\mathcal{S}_k$ of the training data, we solve Q optimization problems using $\mathcal{S}_k$ and Q different configurations of hyperparameters. The algorithm then selects the solution that achieves the best performance on the validation set for the current step k.

9.3 Compressive learning

The classical signal processing pipeline often involves separate steps of signal acquisition, compression, storage, or transmission, followed by reconstruction and analysis. In order to ensure high-fidelity reconstruction, an analog signal should be sampled at a rate that is twice the highest frequency component in the signal, the so-called Nyquist rate. However, for a certain class of signals possessing sparse or compressible representation in some domain, the Compressive Sensing (CS) theory guarantees that near perfect reconstruction is possible even when the signals are sampled at sub-Nyquist rates [45,46]. In fact, many signals that we encounter in the real-world often possess this property. For example, smooth signals are compressible in the Fourier domain or some piecewise smooth signals such as consecutive frames in a video have sparse coefficients in wavelet domain.

Different from the classical signal acquisition approach, CS combines the signal sampling and compression steps at the sensor level. What we obtain from a CS device is a few compressed measurements of the signal, rather than the full amount of samples that has been obtained by sensing and discretization at a rate higher than the Nyquist rate. The compression step in CS is simply a linear projection of the samples to a lower dimension, making it highly efficient to be implemented on the sensor. Although this approach can provide a great solution for those scenarios where storing or transmitting the full signal is infeasible, working with the compressed measurements is often a daunting task. For example, a considerable amount of effort has been dedicated to develop signal recovery algorithms from compressed measurements, providing insights and theoretical conditions for a perfect reconstruction [45–47].

Although signal recovery is necessary and important in some applications such as the reconstruction of Magnetic Resonance Imaging (MRI) images for visual diagnosis by medical experts, there are many applications that do not explicitly require signal recovery. For example, object detection is often the primary task in many radar applications. Furthermore, signal reconstruction can be undesirable in sensitive scenarios since this step can potentially reveal private information, leading to the infringement of data protection legislation. Because of these reasons, it is natural to consider the problem of learning from compressed measurements without reconstruction, which is referred to as Compressive Learning (CL) in literature. While CS has been extensively studied and covered, literature on CL is rather scattered despite its high potential for robotic and IoT applications. In the following, we will cover CL methodologies, which can be categorized into two groups, namely vector-based CL and tensor-based CL. The main difference between the two categories lies in the way the compressed measurements are obtained. For vector-based CL methods, the signal acquisition model contains a single CS operator, considering the input signal as a vector while in tensor-based CL methods, the signal acquisition model adopts multiple separable CS operators, considering the input as a multidimensional signal.

In the following, we denote scalar values by either lowercase or uppercase letters ($x, y, X, Y, \dots$), vectors by lowercase boldface letters ($\mathbf{x}, \mathbf{y}, \dots$), matrices (2D tensor) by uppercase boldface letters ($\mathbf{X}, \mathbf{Y}, \mathbf{\Phi}, \dots$), and tensor as calligraphic capital letters ($\mathcal{X}, \mathcal{Y}, \dots$).

9.3.1 Vector-based compressive learning

One of the main components of vector-based CL (hereafter we simply refer to as CL) methods is the CS component, which takes as input a signal representing by vector $\mathbf{y} \in \mathbb{R}^I$. As we mentioned before, the signal of interest in CS is assumed to have a sparse or compressible representation $\mathbf{x} \in \mathbb{R}^J$ in some basis (or dictionary) $\mathbf{\Psi} \in \mathbb{R}^{I \times J}$. That is, $\mathbf{y}$ can be expressed as a linear combination of atoms (columns) of the dictionary $\mathbf{\Psi}$, with $\mathbf{x}$ being the combination coefficients:

$$\mathbf{y} = \mathbf{\Psi}\mathbf{x} \quad \text{such that } \|\mathbf{x}\|_0 \leq K \text{ and } K \ll J, \tag{9.28}$$

where $\|\mathbf{x}\|_0$ denotes the number of non-zero elements in $\mathbf{x}$, which reflects the degree of sparsity in the signal. When $J > I$, which is the case frequently adopted in CS literature, $\mathbf{\Psi}$ is called an overcomplete dictionary.

In CS, the signal $\mathbf{y}$ is compressed by a linear projection to a lower-dimensional space:

$$\mathbf{z} = \mathbf{\Phi}\mathbf{y}, \tag{9.29}$$

where $\mathbf{z} \in \mathbb{R}^M$, $M < I$ denotes the compressed measurements, which correspond to the output of a CS device. $\mathbf{\Phi} \in \mathbb{R}^{M \times I}$ denotes the sensing operator, which is the matrix representing the linear projection performed by the CS device.

In literature, Eqs. (9.28) and (9.29) are often combined into a single equation, which is referred to as the CS model:

$$\mathbf{z} = \mathbf{\Phi}\mathbf{\Psi}\mathbf{x} \quad \text{such that } \|\mathbf{x}\|_0 \leq K \text{ and } K \ll J. \tag{9.30}$$

The main problem considered in CS is the recovery of $\mathbf{y}$ from the compressed measurement $\mathbf{z}$. As such, a standard approach when building a learning model with data obtained from a CS device is to simply apply methods that have been developed for uncompressed signal and learn a decision function $f(\tilde{\mathbf{y}})$, where $\tilde{\mathbf{y}}$ denotes the reconstructed signal. While being straightforward, this approach includes the signal reconstruction step, which is computational expensive.

The idea of learning a decision function based on the compressed measurements, that is, $f(\mathbf{z})$, was introduced by [48]. In this work, the problem of learning in the compressed domain from the maximum likelihood perspective was investigated. Using the maximum likelihood principle, it was suggested that a given classification accuracy is only dependent on the noise level, which is independent of the underlying sparsity of the images. In addition, the generalized maximum likelihood framework was employed to investigate the effects of geometric transformations with respect to the classification performance, and it was found out that the number of measurements required grows linearly with the dimensionality of the manifold formed by the set of transformed images, such that a certain accuracy level of the generalized maximum likelihood classifier is maintained. Finally, the matched filter classifier (a special case of the generalized maximum likelihood classifier) was applied in the compressed domain, leading to the so-called *smashed filter* algorithm. The smashed filter algorithm was further extended to the classification of human actions [49], where 3D correlation filters along the spatial and temporal dimension were applied to extract features from compressed measurements of video frames. In addition to action recognition, the smashed filter concept was also applied to compressively sensed facial images for face recognition [50].

In several works in CS, a random sensing operator is often considered. Given a random sensing operator, theoretical work on random linear projections of smooth manifolds provides an important result for learning a linear classifier on the compressed domain [51]. This analysis suggests that a small number of random linear projections can faithfully preserve geometric information of a signal lying on a smooth manifold. More specifically, it was shown that there exists a sufficient number of projections M such that, with high probability, the pairwise Euclidean and geodesic distances between points on the manifold are preserved under the given random projections. This result implies that the performance of a linear classifier in the compressed domain can match the performance of the same classifier in the original signal domain.

While the aforementioned theoretical work was only developed in the general context of random projections of smooth manifolds, without any empirical results demonstrated in CL, the work in [52] (which was also the first to introduce the term compressive learning) provided a similar theoretical result for the Support Vector Machine (SVM) classifier with empirical verification, focusing on the problem of learning from compressed measurements. Specifically, the analytical bounds for training a linear SVM using compressed measurements $\mathbf{z}$ were derived, indicating that when the sensing matrix satisfies the distance-preserving property, the performance of a linear SVM operating on $\mathbf{z}$ is equivalent to the performance of a best linear threshold classifier operating on the original signal $\mathbf{y}$. This theoretical result was also supported by experiments in image texture classification.

The work in [53] focusing on the class of signals that are described by the Gaussian Mixture Model (GMM) also contributed to theoretical analysis of CL. In this work, the source signal is assumed to follow the GMM model, which is motivated from the fact that natural image patches can be successfully modeled by GMMs. The GMM that describes the source signal is also assumed to be known in the analysis. An upper bound to the probability of misclassification of the maximum-a-posteriori classifier was derived, given that the sensing matrix is drawn from a zero-mean Gaus-

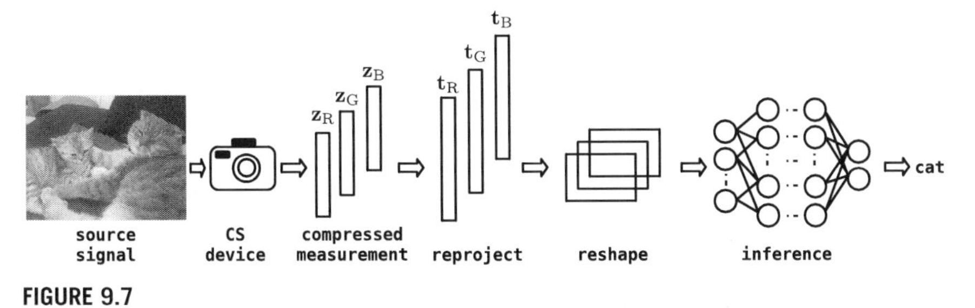

FIGURE 9.7

Compressive Learning system proposed in [56].

sian distribution and the compressed measurements were contaminated with standard white Gaussian noise.

While early works in CL focus on providing theoretical insights of simple models operating on compressed domains, such as linear models, later works shifted toward the practicality of CL for real-world problems by utilizing nonlinear learning models and focusing more on empirical validation. Since the emergence in the ImageNet Large Scale Visual Recognition Challenge 2012, deep neural networks have become a de facto choice for learning nonlinear relationships. They soon made their entrance into the CL literature in [54], where a Convolutional Neural Network (CNN) was used to classify compressed measurements of handwritten character images obtained from a random sensing matrix. The empirical results in [54] showed that a nonlinear classifier such as CNN can yield superior performance compared to smashed filters. Furthermore, the experimental results in this work also suggested that the CL paradigm can produce relatively good classification performance even when the compression rate is up to $100\times$.

The idea of combining CS and deep learning was further refined by the works in [55,56], where the idea of end-to-end CL learning was first introduced. That is, instead of utilizing a random sensing operator as predominantly done in prior works, the algorithm jointly optimizes both the machine learning model and the sensing operator via stochastic gradient descent. Since the linear projections in CS are optimized in conjunction with the classifier toward minimizing the learning objective, they are better at capturing the relevant information for the learning problem from the source signal. As a result, the approach proposed in [55,56] achieved far better performance when compared to the case where random sensing matrices were applied. In addition, later works have soon adopted and extended this idea to different applications [57–59].

The end-to-end CL architecture for RGB image recognition proposed by [56] is illustrated in Fig. 9.7. As we can see from this figure, after obtaining the compressed measurements $\mathbf{z}_R$, $\mathbf{z}_G$, $\mathbf{z}_B$ from the R, G, B color channels respectively, a reprojection step is applied to the compressed measurements by linearly up-sampling them using the transpose of the corresponding sensing operator, that is, $\mathbf{t}_R = \mathbf{\Phi}_R^T \mathbf{z}_R$, $\mathbf{t}_G = \mathbf{\Phi}_G^T \mathbf{z}_G$, $\mathbf{t}_B = \mathbf{\Phi}_B^T \mathbf{z}_B$. After performing an appropriate reshaping step, the up-sampled features are introduced to a CNN. As parameter initialization is an important step in stochastic optimization, the sensing matrices $\mathbf{\Phi}_{\mathrm{R}}$, $\mathbf{\Phi}_{\mathrm{G}}$, $\mathbf{\Phi}_{\mathrm{B}}$ are initialized such that the mean squared reconstruction error is minimized. The CNN classifier is initialized with weights learned from uncompressed data. After the initialization step, all parameters are updated with stochastic gradient descent.

9.3.2 Tensor-based compressive learning

While the majority of signals that we encounter appear naturally in tensor form, the CL paradigm described in the previous section only considers a sensing model for signals represented as vectors. The multidimensional representation of a signal often exhibits natural semantic differences inherent in different dimensions of the signal. For this reason, it is desirable to design learning models that are capable of exploiting the tensor structure of the data. Regarding the problem of learning from compressed measurements, Multilinear Compressive Learning (MCL) which is a CL framework that takes into account the multidimensional structure of signals naturally represented in the tensor form was proposed in [60].

Before diving into the details of MCL, it is necessary to review some important notions and properties in multilinear algebra. The term *mode* is often used to refer to the dimension of a tensor. For example, a tensor $\mathcal{X} \in \mathbb{R}^{I_1 \times \cdots \times I_k \times \cdots \times I_K}$ is said to have K modes with the dimension of the kth mode being equal to I_k. One of most important operations in multilinear algebra is the mode-k product, which defines the product between a tensor and a matrix. The mode-k product between a tensor $\mathcal{X} = [x_{i_1}, \ldots, x_{i_K}] \in \mathbb{R}^{I_1 \times \ldots I_k \times \cdots \times I_K}$ and a matrix $\mathbf{W} \in \mathbb{R}^{J_k \times I_k}$ is another tensor of size $I_1 \times \cdots \times J_k \times \cdots \times I_K$ and denoted by $\mathcal{X} \times_k \mathbf{W}$. The elements of $\mathcal{X} \times_k \mathbf{W}$ are defined as

$$[\mathcal{X} \times_k \mathbf{W}]_{i_1,\ldots,i_{k-1},j_k,i_{k+1},\ldots,i_K} = \sum_{i_k=1}^{I_K} [\mathcal{X}]_{i_1,\ldots,i_{k-1},i_k,\ldots,i_K} [\mathbf{W}]_{j_k,i_k}. \tag{9.31}$$

When multiplying a tensor with multiple matrices along different modes, we often chain together the notation, that is,

$$\mathcal{X} \times_1 \mathbf{W}_1 \times_2 \mathbf{W}_2 \times \cdots \times_k \mathbf{W}_K \tag{9.32}$$

denotes the multiplication of tensor $\mathcal{X}$ with $\mathbf{W}_1$ along the first mode, with $\mathbf{W}_2$ along the second mode and so on.

For a series of multiplications in distinct modes, the order of computation does not affect the result. That is,

$$\mathcal{X} \times_i \mathbf{W}_i \times_j \mathbf{W}_j = \mathcal{X} \times_j \mathbf{W}_j \times_i \mathbf{W}_i \qquad (i \neq j). \tag{9.33}$$

In addition to mode-k product, another operation that involves two matrices exists, which is the Kronecker product. The Kronecker product between two matrices $\mathbf{A} \in$

$\mathbb{R}^{M\times N}$ and $\mathbf{B}\in\mathbb{R}^{P\times Q}$ is denoted as $\mathbf{A}\otimes\mathbf{B}$. The result is a matrix which has $MP\times NQ$ elements and is calculated by

$$\mathbf{A}\otimes\mathbf{B}=\begin{bmatrix}\mathbf{A}_{11}\mathbf{B} & \dots & \mathbf{A}_{1N}\mathbf{B}\\ \vdots & \ddots & \vdots\\ \mathbf{A}_{M1}\mathbf{B} & \dots & \mathbf{A}_{MN}\mathbf{B}\end{bmatrix}. \tag{9.34}$$

Using the Kronecker product definition, we can express a series of mode-k products by a vector-matrix multiplication. This means that

$$\mathcal{Y}=\mathcal{X}\times_1\mathbf{W}_1\times\dots\times_N\mathbf{W}_N \tag{9.35}$$

can be implemented as follows:

$$\mathbf{y}=(\mathbf{W}_1\otimes\dots\otimes\mathbf{W}_N)\mathbf{x}, \tag{9.36}$$

where $\mathbf{y}=\mathrm{vec}(\mathcal{Y})$ and $\mathbf{x}=\mathrm{vec}(\mathcal{X})$, with $\mathrm{vec}(\cdot)$ denoting the vectorization operation.

In order to retain the multidimensional structure of the source signal, it is necessary to acquire and compress the source signal so that the structural information is captured in the compressed measurements. MCL uses the Multilinear Compressive Sensing (MCS) model for this purpose. Let us denote a source signal having multidimensional structure as $\mathcal{Y}\in\mathbb{R}^{I_1\times\dots\times I_K}$. Instead of using the assumption that $\mathrm{vec}(\mathcal{Y})$ is sparse with respect to some dictionary as in CS, the MCS model assumes that $\mathcal{Y}$ is sparse with respect to a set of dictionaries $(\mathbf{\Psi}_1,\dots,\mathbf{\Psi}_K)$, and $\mathcal{Y}$ can be represented by a sparse Tucker model [61]:

$$\mathcal{Y}=\mathcal{X}\times_1\mathbf{\Psi}_1\times\dots\times_K\mathbf{\Psi}_K. \tag{9.37}$$

The source signal $\mathcal{Y}$ can be acquired in a compressed manner while retaining its tensor structure by using a set of linear sensing operators along each mode:

$$\mathcal{Z}=\mathcal{Y}\times_1\mathbf{\Phi}_1\times\dots\times_K\mathbf{\Phi}_K, \tag{9.38}$$

where $\mathbf{\Phi}_k\in\mathbb{R}^{M_k\times I_k}$, $(M_k<I_k)$ denotes the sensing operator in the kth mode and $\mathcal{Z}\in\mathbb{R}^{M_1\times\dots\times M_K}$ denotes the compressed measurement obtained from the sensor.

Here, we should note that the MCS model can be easily implemented using the standard CS hardware by using the property mentioned in Eqs. (9.35) and (9.36). That is, the set of separable sensing operators $(\mathbf{\Phi}_1,\dots,\mathbf{\Phi}_K)$ in MCS can be combined into a single sensing operator $(\mathbf{\Phi}\in\mathbb{R}^{(M_1\dots M_K)\times(I_1\dots I_K)})$ using the Kronecker product $(\mathbf{\Phi}=\mathbf{\Phi}_1\otimes\dots\otimes\mathbf{\Phi}_K)$, and $\mathcal{Z}$ is obtained, after appropriate reshaping, via

$$\mathrm{vec}(\mathcal{Z})=\mathbf{\Phi}\mathrm{vec}(\mathcal{Y}). \tag{9.39}$$

Beside the fact that multidimensional structures of the source signal are preserved in the MCS model, this sensing model significantly reduces the memory and computational complexity of the sensing step. In order to observe this, let us take as an

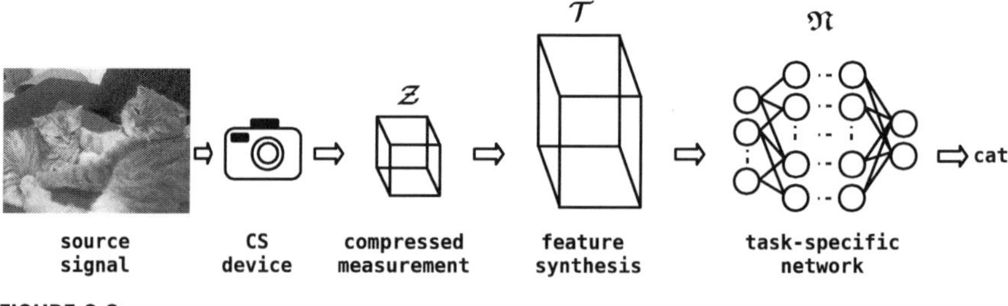

FIGURE 9.8

Multilinear compressive learning.

example the task of sensing and compressing grayscale images of size 1080 × 1920 with a compression rate of 100×. Using the CS model in Eq. (9.29), we need to store a sensing matrix of size (1080 ∗ 1920) × (108 ∗ 192), having approximately 42.9 billion parameters, while with the same compression rate, we only need to store two sensing matrices of size 1080 × 108 and 1920 × 192, having approximately 0.5 million parameters.

There are two other components in MCL, which are illustrated in Fig. 9.8, the Feature Synthesis (FS) component and the task-specific neural network ($\mathfrak{N}$). After obtaining the compressed measurement $\mathcal{Z}$ from a CS device, the MCL model synthesizes features $\mathcal{T}$ from $\mathcal{Z}$ through the FS component, which contains multilinear operations that preserve the tensor structure in $\mathcal{Z}$:

$$\mathcal{T} = \mathcal{Z} \times_1 \mathbf{\Theta}_1 \times \cdots \times_K \mathbf{\Theta}_K, \tag{9.40}$$

where $\mathbf{\Theta}_k \in \mathbb{R}^{P_k \times M_k}$, $(k = 1, \ldots, K)$ denotes the parameters of the FS component.

Here, we should note that the dimensions $(P_1 \times \cdots \times P_K)$ of the synthesized feature $\mathcal{T}$ depend on the given learning task and the input specification of the task-specific neural network. This is because the synthesized feature $\mathcal{T}$ is introduced to the third component of the MCL model, which is the task-specific neural network $\mathfrak{N}$, to generate predictions. $\mathfrak{N}$ is a neural network that is often used to solve the same learning problem using uncompressed data. Thus the architecture of $\mathfrak{N}$ is selected depending on the learning task. For example, when the task is to detect objects from an image, $\mathfrak{N}$ can be a YOLO object detector [3].

The three components of the MCL model are optimized in an end-to-end manner using stochastic gradient descent. For the initialization of their parameters, the following strategy was proposed in [60].

Parameters of the task-specific neural network are initialized by training it with the set of uncompressed data:

$$\underset{\mathbf{\Theta}_{\mathfrak{N}}}{\operatorname{argmin}} \sum_{i}^{N} L\big(c_i, f_{\mathfrak{N}}(\tilde{\mathcal{Y}}_i)\big), \tag{9.41}$$

where $\boldsymbol{\Theta}_{\mathfrak{N}}$ and $f_{\mathfrak{N}}$ denote the parameters and the function representing $\mathfrak{N}$, respectively, $L(\cdot)$ denotes the learning objective function, and $\tilde{\mathcal{Y}}_i$ and c_i denote the available uncompressed data sample and the corresponding label. Here, we should note that the uncompressed data ($\tilde{\mathcal{Y}}$) can be different from the source signal ($\mathcal{Y}$) that we wish to capture with the CS device. For example, to learn an MCL model for object detection, we might want to use available data sets with annotations such as the PASCAL data set or MS COCO data set at resolution $480 \times 480 \times 3$, while our CS device resolution is at 1080×1920. In this case, the dimensions of $\tilde{\mathcal{Y}}$ are $480 \times 480 \times 3$ while the source signal $\mathcal{Y}$ has dimensions of 1080×1920.

Parameters of the sensing and FS component are initialized with values that minimize the following reconstruction error:

$$\underset{\substack{\boldsymbol{\Phi}_1,\ldots,\boldsymbol{\Phi}_K \\ \boldsymbol{\Theta}_1,\ldots,\boldsymbol{\Theta}_K}}{\operatorname{argmin}} \sum_{i}^{N} \|\tilde{\mathcal{Y}}_i - f_{\mathrm{FS}}(f_{\mathrm{S}}(\mathcal{Y}_i))\|_{\mathrm{F}}^2, \tag{9.42}$$

where f_{S} and f_{FS} denote the functions representing the sensing and FS component, respectively, and $\|\cdot\|_{\mathrm{F}}$ denotes the Frobenius norm. Here, we should note that when we do not have access to the source signal $\mathcal{Y}_i$ that corresponds to the uncompressed data $\tilde{\mathcal{Y}}_i$, we can imitate $\mathcal{Y}_i$ by applying an upsampling or downsampling method to $\tilde{\mathcal{Y}}_i$ to generate $\mathcal{Y}_i$ with the desired resolution.

Experimental results in object classification and face recognition problems in [60] showed that by considering the multidimensional structure of images, the MCL model cannot only produce better classification performance but is also more efficient in terms of memory and computational cost, compared to the end-to-end CL framework proposed in [56] that considers the source signals in the form of vectors.

While the FS component proposed in [60] only consists of multilinear operations, the operations that extract features from the compressed measurements in general do not need to be linear or multilinear, as long as they are capable of preserving the multidimensional structure in $\mathcal{Z}$. To take advantage of the modeling capability of nonlinear transformations, an extension of the MCL model that synthesizes nonlinear features from the compressed measurements was proposed in [62]. This extended MCL model is dubbed as multilinear compressive learning with prior knowledge (MCLwP) since it comes with a novel training method that can incorporate prior knowledge of a high-quality compressed domain and feature space into the target MCL model.

In MCLwP a high-quality compressed domain for $\mathcal{Z}$ and feature space for $\mathcal{T}$ are discovered by training a prior model. This prior model also comprises of three components that correspond to the three components of an MCL model, that is, a nonlinear sensing component (g_{S}), a nonlinear FS component (g_{FS}), and a task-specific neural network ($g_{\mathfrak{N}}$). The nonlinear sensing component of the prior model has several convolution layers with nonlinear activation and max-pooling layers to reduce dimensions of the source signal to target dimensions of the compressed measurements $\mathcal{Z}$. On the other hand, the nonlinear FS component consists of several convolution layers

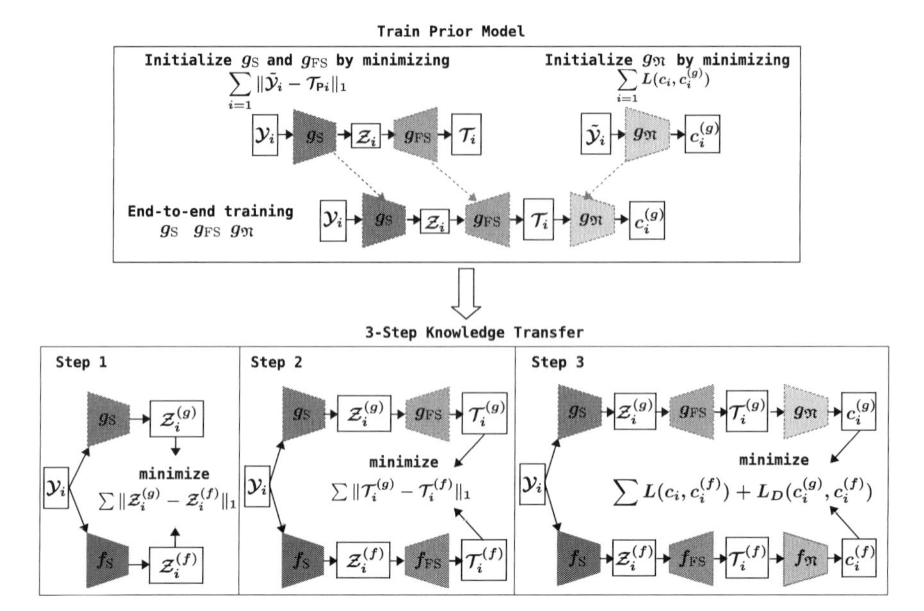

FIGURE 9.9

Multilinear compressive learning trained with the prior model.

with nonlinear activation and upsampling layers to transform $\mathcal{Z}$ to $\mathcal{T}$. The structure of the task-specific neural network is dependent on the learning task, thus similar to the MCL model. The prior model is optimized in the same manner as the MCL model. That is, before the end-to-end optimization step, the sensing and synthesis components are initialized with values that minimize the reconstruction error of the uncompressed data while the task-specific neural network is initialized by training on the uncompressed data.

After the prior model is optimized, its knowledge about the learning task from the compressed measurements is transferred to the target MCL model. Here, we should note that since there is no structural constraint on the FS component in a MCL model, it was proposed in [62] to use the same nonlinear structure that is used in the prior model for the FS component in the target MCL model. The new optimization procedure for the target MCL using the prior model consists of three steps, which are illustrated in Fig. 9.9. In the first step, the MCS component is optimized to mimic the output of the nonlinear sensing component as follows:

$$\underset{\Phi_1,\ldots,\Phi_K}{\operatorname{argmin}} \sum_i^N \|g_S(\mathcal{Y}_i) - f_S(\mathcal{Y}_i)\|_1. \tag{9.43}$$

In the second step, both the sensing and the FS components in the target MCL model are optimized to mimic the features synthesized by the prior model as follows:

$$\underset{\substack{\boldsymbol{\Phi}_1,\ldots,\boldsymbol{\Phi}_K\\ \boldsymbol{\Theta}_1,\ldots,\boldsymbol{\Theta}_K}}{\operatorname{argmin}} \sum_{i}^{N} \| g_{\text{FS}}(g_{\text{S}}(\mathcal{Y}_i)) - f_{\text{FS}}(f_{\text{S}}(\mathcal{Y}_i)) \|_1. \tag{9.44}$$

In the last step, all components of the target MCL model are optimized to mimic the predictions generated by the prior model, as well as to minimize the learning objective using the labeled data as follows:

$$\underset{\substack{\boldsymbol{\Phi}_1,\ldots,\boldsymbol{\Phi}_K\\ \boldsymbol{\Theta}_1,\ldots,\boldsymbol{\Theta}_K\\ \boldsymbol{\Theta}_{\mathfrak{N}}}}{\operatorname{argmin}} \sum_{i}^{N} \left(L_D(c_i^{(g)}, c_i^{(f)}) + L(c_i, c_i^{(f)}) \right), \tag{9.45}$$

where c_i, $c_i^{(g)}$, and $c_i^{(f)}$ denote the true label, the prediction produced by the prior model and the target MCL model, respectively, $L_D()$ denotes a distillation loss, such as the one proposed in [63], and $L(\cdot)$ denotes the learning objective function.

While a highly nonlinear prior model is capable of discovering high-quality compressed domain and feature space, it is only effective when sufficient labeled data is available. When labeled data is scarce, a prior model can easily fall into the overfitting regime, exhibiting poor generalization. As a consequence, the resulting MCL model trained with bad prior knowledge will not generalize to unseen data. While labeled data can be expensive to obtain, in many cases, we can easily obtain data coming from the same distribution without any label. In this scenario, the work in [62] also includes a semisupervised extension of the aforementioned training technique with a prior model. The key idea of this semi-supervised training method is on how the prior model can be trained to provide suggestive annotations for the unlabeled data, which is then used together with the labeled data to perform knowledge transfer.

A straightforward way to generate suggestive labels for the unlabeled data is to simply use the prior model that has been trained with a limited number of labeled samples to make predictions on the unlabeled samples. This approach, however, is defective since the prior model trained with a small number of samples without any further modification is likely to be overfitting, thus generating many false labels. To overcome this issue, [62] proposed a self-labeling process that incrementally updates the prior model and generates suggestive labels using a small number of unlabeled data at a time.

Let us denote the set of labeled data as $\mathfrak{L} = \{(\mathcal{Y}_i, \tilde{\mathcal{Y}}, c_i) | i = 1, \ldots, N\}$ and the set of unlabeled data as $\mathfrak{U} = \{(\mathcal{Y}_j, \tilde{\mathcal{Y}}_j) | j = N+1, \ldots, M\}$. At first, the nonlinear sensing and FS components of the prior model are initialized with values that minimize the reconstruction error of the uncompressed data. Since the label information is not required in this step, both labeled and unlabeled data can be used. Similar to the supervised version, the task-specific network is initialized with uncompressed data from the labeled set. After initialization, all components of the prior model are trained in an end-to-end manner to optimize the learning objective using the incrementally enlarged label set $\tilde{\mathfrak{L}}$. Initially, the enlarged label set only contains samples from the labeled set, that is, $\tilde{\mathfrak{L}} = \mathfrak{L}$. After every T back-propagation epochs, $\tilde{\mathfrak{L}}$ is expanded with those samples (and their predicted labels) in $\mathfrak{U}$ that have the most confident predictions from the current prior model, given a probability threshold ρ. That is, $\tilde{\mathfrak{L}} = \tilde{\mathfrak{L}} \cup \mathfrak{C}$, with $\mathfrak{C}$ defined as

$$\mathfrak{C} = \Big\{(\mathcal{Y}, c) \mid \mathcal{Y} \in \mathfrak{U} \wedge g_{\mathfrak{N}}(g_{\text{FS}}(g_{\text{S}}(\mathcal{Y})))_{\max} \geq \rho, \quad c = \text{argmax}(g_{\mathfrak{N}}(g_{\text{FS}}(g_{\text{S}}(\mathcal{Y}))))\Big\}. \tag{9.46}$$

After each iteration of this process, the samples added to $\tilde{\mathfrak{L}}$ are removed from the unlabeled set, that is, $\mathfrak{U} = \mathfrak{U} \setminus \mathfrak{C}$. The optimization procedure terminates when $\mathfrak{C} = \varnothing$.

The final set $\tilde{\mathfrak{L}}$ and the prior model are used to train the target MCL model in a similar manner as in the supervised algorithm. That is, the optimization steps are the same as in Eqs. (9.43), (9.44), and (9.45), except that the sum is computed over the samples in $\tilde{\mathfrak{L}}$. It has been shown in [62] that using a prior model, an MCL model can obtain noticeable performance gains. In addition, the semisupervised training method described above can indeed alleviate the overfitting problem when only few samples have labels.

While the computational advantage of the MCL over vector-based CL is obvious, it is unclear how the dimensions of the compressed measurement $\mathcal{Z}$ should be selected, given a target compression rate. For example, when the dimensions of the source signal are 100×100 and the compression rate is $100\times$, there are multiple ways to specify the dimensions of $\mathcal{Z}$, such as 10×10 or 20×5 or 25×4 and so on. It would require a large number of experiments to determine the best configuration for $\mathcal{Z}$. In addition to $\mathcal{Z}$, in a real-world problem there is also the need to determine optimal values for the CS device resolution ($\mathcal{Y}$) as well as the dimensions of uncompressed data ($\tilde{\mathcal{Y}}$). The work in [64] addressed this problem through an empirical study and suggested a performance indicator that can help us determine such information without conducting the entire optimization process of the MCL model. Specifically, it was found that the mean-squared errors obtained during the initialization of the sensing and FS components in Eq. (9.42) highly correlates with the final classification errors. This finding suggests that in order to produce an estimate ranking between different dimension configurations, conducting only the initialization steps in the MCL model is sufficient, thus the overall computational cost of training multiple MCL models for model selection can be highly reduced.

9.4 Conclusions

In this chapter, we have covered the two learning paradigms of Progressive Neural Network Learning (PNNL) and Compressive Learning (CL), which have been extensively developed in recent years to enhance the training and deployment of deep neural networks. The presented algorithms are especially suitable for the development of perception and cognition applications in computationally constrained environments, such as the ones we often encounter in robotics.

The first paradigm concerns with algorithms that automatically construct and optimize neural network architectures in a progressive manner, starting from a very small network topology and progressing toward ones with higher capacity until the learning performance converges. We covered different topology construction strategies and optimization techniques that are popular in this area. More specifically, the BLS and PLN algorithms, which are representative examples following convex optimization and randomization approaches, were described in detail in Section 9.2.1 and Section 9.2.2, respectively. We covered a series of algorithms based on gradient-based optimization that utilize a generalized neuron model (GOP) such as POP and its variants in Section 9.2.3. Our treatise on progressive neural network learning concludes with HeMLGOP in Section 9.2.4, an efficient algorithm that combines randomization, convex optimization, and gradient-based optimization to learn compact network architectures with competitive performance.

While the first learning paradigm provides us potentially compact network architectures by means of completely data-driven solutions, the second paradigm covered in this chapter brings forth hybrid solutions for data acquisition and inference in computationally constrained environments, via the combination of compressive sensing and machine learning. Data acquired by a compressive sensing device, which is highly compressed with minimal computations, can be easily transmitted to a centralized server to be directly processed without the need for reconstruction. In Section 9.3.1, we summarized different theoretical results that were developed for linear models and focused more on recent developments using deep neural networks. Section 9.3.2, which covers the class of compressive learning algorithms that preserves the intrinsic nature of multidimensional signals, concludes our treatise on compressive learning.

References

[1] S. Ren, K. He, R. Girshick, J. Sun, Faster r-cnn: towards real-time object detection with region proposal networks, IEEE Transactions on Pattern Analysis and Machine Intelligence 39 (6) (2016) 1137–1149.

[2] K. He, G. Gkioxari, P. Dollár, R. Girshick, Mask r-cnn, in: Proceedings of the IEEE International Conference on Computer Vision, 2017, pp. 2961–2969.

[3] J. Redmon, A. Farhadi, Yolov3: an incremental improvement, arXiv preprint, arXiv:1804.02767, 2018.

[4] Z. Cao, G. Hidalgo, T. Simon, S.-E. Wei, Y. Sheikh, Openpose: realtime multi-person 2d pose estimation using part affinity fields, IEEE Transactions on Pattern Analysis and Machine Intelligence 43 (1) (2019) 172–186.

[5] A. Vaswani, N. Shazeer, N. Parmar, J. Uszkoreit, L. Jones, A.N. Gomez, Ł. Kaiser, I. Polosukhin, Attention is all you need, Advances in Neural Information Processing Systems 30 (2017) 5998–6008.

[6] J. Devlin, M.-W. Chang, K. Lee, K. Toutanova Bert, Pre-training of deep bidirectional transformers for language understanding, arXiv preprint, arXiv:1810.04805, 2018.
[7] Z. Dai, Z. Yang, Y. Yang, J. Carbonell, Q.V. Le, R. Salakhutdinov, Transformer-xl: attentive language models beyond a fixed-length context, arXiv preprint, arXiv:1901.02860, 2019.
[8] D.T. Tran, A. Iosifidis, J. Kanniainen, M. Gabbouj, Temporal attention-augmented bilinear network for financial time-series data analysis, IEEE Transactions on Neural Networks and Learning Systems 30 (5) (2018) 1407–1418.
[9] Z. Zhang, S. Zohren, S. Roberts, Deeplob: deep convolutional neural networks for limit order books, IEEE Transactions on Signal Processing 67 (11) (2019) 3001–3012.
[10] K. He, X. Zhang, S. Ren, J. Sun, Deep residual learning for image recognition, in: Proceedings of the IEEE Conference on Computer Vision and Pattern Recognition, 2016, pp. 770–778.
[11] G. Huang, Z. Liu, L. Van Der Maaten, K.Q. Weinberger, Densely connected convolutional networks, in: Proceedings of the IEEE Conference on Computer Vision and Pattern Recognition, 2017, pp. 4700–4708.
[12] A. Krizhevsky, I. Sutskever, G.E. Hinton, Imagenet classification with deep convolutional neural networks, Communications of the ACM 60 (6) (2017) 84–90.
[13] T. Elsken, J.H. Metzen, F. Hutter, Neural architecture search: a survey, arXiv preprint, arXiv:1808.05377, 2018.
[14] D. Blalock, J.J.G. Ortiz, J. Frankle, J. Guttag, What is the state of neural network pruning?, arXiv preprint, arXiv:2003.03033, 2020.
[15] I. Hubara, M. Courbariaux, D. Soudry, R. El-Yaniv, Y. Bengio, Quantized neural networks: training neural networks with low precision weights and activations, Journal of Machine Learning Research 18 (1) (2017) 6869–6898.
[16] Y. Choukroun, E. Kravchik, F. Yang, P. Kisilev, Low-bit quantization of neural networks for efficient inference, in: 2019 IEEE/CVF International Conference on Computer Vision Workshop (ICCVW), IEEE, 2019, pp. 3009–3018.
[17] M. Jaderberg, A. Vedaldi, A. Zisserman, Speeding up convolutional neural networks with low rank expansions, arXiv preprint, arXiv:1405.3866, 2014.
[18] D.T. Tran, A. Iosifidis, M. Gabbouj, Improving efficiency in convolutional neural networks with multilinear filters, Neural Networks 105 (2018) 328–339.
[19] X. Yu, T. Liu, X. Wang, D. Tao, On compressing deep models by low rank and sparse decomposition, in: Proceedings of the IEEE Conference on Computer Vision and Pattern Recognition, 2017, pp. 7370–7379.
[20] Y. Cheng, D. Wang, P. Zhou, T. Zhang, A survey of model compression and acceleration for deep neural networks, arXiv preprint, arXiv:1710.09282, 2017.
[21] B. Wu, X. Dai, P. Zhang, Y. Wang, F. Sun, Y. Wu, Y. Tian, P. Vajda, Y. Jia, K. Keutzer, Fbnet: hardware-aware efficient convnet design via differentiable neural architecture search, in: Proceedings of the IEEE Conference on Computer Vision and Pattern Recognition, 2019, pp. 10734–10742.
[22] F. Scheidegger, L. Benini, C. Bekas, A.C.I. Malossi, Constrained deep neural network architecture search for iot devices accounting for hardware calibration, in: Advances in Neural Information Processing Systems, 2019, pp. 6056–6066.
[23] C.P. Chen, Z. Liu, Broad learning system: an effective and efficient incremental learning system without the need for deep architecture, IEEE Transactions on Neural Networks and Learning Systems 29 (1) (2017) 10–24.
[24] S. Kiranyaz, T. Ince, A. Iosifidis, M. Gabbouj, Progressive operational perceptrons, Neurocomputing 224 (2017) 142–154.

[25] S. Chatterjee, A.M. Javid, M. Sadeghi, P.P. Mitra, M. Skoglund, Progressive learning for systematic design of large neural networks, arXiv preprint, arXiv:1710.08177, 2017.
[26] D.T. Tran, S. Kiranyaz, M. Gabbouj, A. Iosifidis, Heterogeneous multilayer generalized operational perceptron, IEEE Transactions on Neural Networks and Learning Systems 31 (3) (2019) 710–724.
[27] D.T. Tran, S. Kiranyaz, M. Gabbouj, A. Iosifidis, Progressive operational perceptrons with memory, Neurocomputing 379 (2020) 172–181.
[28] Y.-H. Pao, Y. Takefuji, Functional-link net computing: theory, system architecture, and functionalities, Computer 25 (5) (1992) 76–79.
[29] Y.-H. Pao, G.-H. Park, D.J. Sobajic, Learning and generalization characteristics of the random vector functional-link net, Neurocomputing 6 (2) (1994) 163–180.
[30] S. Feng, C.P. Chen, Fuzzy broad learning system: a novel neuro-fuzzy model for regression and classification, IEEE Transactions on Cybernetics 50 (2) (2018) 414–424.
[31] M. Han, S. Feng, C.P. Chen, M. Xu, T. Qiu, Structured manifold broad learning system: a manifold perspective for large-scale chaotic time series analysis and prediction, IEEE Transactions on Knowledge and Data Engineering 31 (9) (2018) 1809–1821.
[32] S. Boyd, N. Parikh, E. Chu, Distributed Optimization and Statistical Learning via the Alternating Direction Method of Multipliers, Now Publishers Inc., 2011.
[33] W.S. McCulloch, W. Pitts, A logical calculus of the ideas immanent in nervous activity, The Bulletin of Mathematical Biophysics 5 (4) (1943) 115–133.
[34] S. Qian, H. Liu, C. Liu, S. Wu, H. San Wong, Adaptive activation functions in convolutional neural networks, Neurocomputing 272 (2018) 204–212.
[35] X. Jiang, Y. Pang, X. Li, J. Pan, Y. Xie, Deep neural networks with elastic rectified linear units for object recognition, Neurocomputing 275 (2018) 1132–1139.
[36] F. Fan, J. Xiong, G. Wang, Universal approximation with quadratic deep networks, Neural Networks 124 (2020) 383–392.
[37] D.T. Tran, S. Kiranyaz, M. Gabbouj, A. Iosifidis, Knowledge transfer for face verification using heterogeneous generalized operational perceptrons, in: 2019 IEEE International Conference on Image Processing (ICIP), IEEE, 2019, pp. 1168–1172.
[38] D.T. Tran, A. Iosifidis, Learning to rank: a progressive neural network learning approach, in: ICASSP 2019-2019 IEEE International Conference on Acoustics, Speech and Signal Processing (ICASSP), IEEE, 2019, pp. 8355–8359.
[39] D. Thanh Tran, J. Kanniainen, M. Gabbouj, A. Iosifidis, Data-driven neural architecture learning for financial time-series forecasting, arXiv preprint, arXiv:1903.06751, 2019.
[40] D.T. Tran, S. Kiranyaz, M. Gabbouj, A. Iosifidis, Pygop: a Python library for generalized operational perceptron algorithms, Knowledge-Based Systems 182 (2019) 104801.
[41] D.T. Tran, M. Gabbouj, A. Iosifidis, Subset sampling for progressive neural network learning, in: Proceedings of the IEEE International Conference on Image Processing, 2020.
[42] B. Settles, Active learning literature survey, Tech. Rep., University of Wisconsin–Madison Department of Computer Sciences, 2009.
[43] V. Birodkar, H. Mobahi, S. Bengio, Semantic redundancies in image-classification datasets: the 10% you don't need, arXiv preprint, arXiv:1901.11409, 2019.
[44] H. Shokri-Ghadikolaei, H. Ghauch, C. Fischione, M. Skoglund, Learning and data selection in big datasets, in: 36th International Conference on Machine Learning, Long Beach, California, PMLR 97, 2019, 2019.
[45] E.J. Candes, J.K. Romberg, T. Tao, Stable signal recovery from incomplete and inaccurate measurements, Communications on Pure and Applied Mathematics: A Journal Issued by the Courant Institute of Mathematical Sciences 59 (8) (2006) 1207–1223.

[46] D.L. Donoho, Compressed sensing, IEEE Transactions on Information Theory 52 (4) (2006) 1289–1306.
[47] E.J. Candès, M.B. Wakin, An introduction to compressive sampling, IEEE Signal Processing Magazine 25 (2) (2008) 21–30.
[48] M.A. Davenport, M.F. Duarte, M.B. Wakin, J.N. Laska, D. Takhar, K.F. Kelly, R.G. Baraniuk, The smashed filter for compressive classification and target recognition, in: Computational Imaging V, vol. 6498, International Society for Optics and Photonics, 2007, p. 64980H.
[49] K. Kulkarni, P. Turaga, Reconstruction-free action inference from compressive imagers, IEEE Transactions on Pattern Analysis and Machine Intelligence 38 (4) (2015) 772–784.
[50] S. Lohit, K. Kulkarni, P. Turaga, J. Wang, A.C. Sankaranarayanan, Reconstruction-free inference on compressive measurements, in: Proceedings of the IEEE Conference on Computer Vision and Pattern Recognition Workshops, 2015, pp. 16–24.
[51] R.G. Baraniuk, M.B. Wakin, Random projections of smooth manifolds, Foundations of Computational Mathematics 9 (1) (2009) 51–77.
[52] R. Calderbank, S. Jafarpour, Finding needles in compressed haystacks, in: 2012 IEEE International Conference on Acoustics, Speech and Signal Processing (ICASSP), IEEE, 2012, pp. 3441–3444.
[53] H. Reboredo, F. Renna, R. Calderbank, M.R. Rodrigues, Compressive classification, in: 2013 IEEE International Symposium on Information Theory, IEEE, 2013, pp. 674–678.
[54] S. Lohit, K. Kulkarni, P. Turaga, Direct inference on compressive measurements using convolutional neural networks, in: 2016 IEEE International Conference on Image Processing (ICIP), IEEE, 2016, pp. 1913–1917.
[55] A. Adler, M. Elad, M. Zibulevsky, Compressed learning: a deep neural network approach, arXiv preprint, arXiv:1610.09615, 2016.
[56] E. Zisselman, A. Adler, M. Elad, Compressed learning for image classification: a deep neural network approach, in: Handbook of Numerical Analysis, vol. 19, Elsevier, 2018, pp. 3–17.
[57] B. Hollis, S. Patterson, J. Trinkle, Compressed learning for tactile object recognition, IEEE Robotics and Automation Letters 3 (3) (2018) 1616–1623.
[58] A. Değerli, S. Aslan, M. Yamac, B. Sankur, M. Gabbouj, Compressively sensed image recognition, in: 2018 7th European Workshop on Visual Information Processing (EUVIP), IEEE, 2018, pp. 1–6.
[59] Y. Xu, K.F. Kelly, Compressed domain image classification using a multi-rate neural network, arXiv preprint, arXiv:1901.09983, 2019.
[60] D.T. Tran, M. Yamaç, A. Degerli, M. Gabbouj, A. Iosifidis, Multilinear compressive learning, IEEE Transactions on Neural Networks and Learning Systems (2020).
[61] L. De Lathauwer, B. De Moor, J. Vandewalle, A multilinear singular value decomposition, SIAM Journal on Matrix Analysis and Applications 21 (4) (2000) 1253–1278.
[62] D.T. Tran, M. Gabbouj, A. Iosifidis, Multilinear compressive learning with prior knowledge, arXiv preprint, arXiv:2002.07203, 2020.
[63] G. Hinton, O. Vinyals, J. Dean, Distilling the knowledge in a neural network, arXiv preprint, arXiv:1503.02531, 2015.
[64] D.T. Tran, M. Gabbouj, A. Iosifidis, Performance indicator in multilinear compressive learning, in: Proceedings of the IEEE Symposium Series on Computational Intelligence, 2020.

CHAPTER

Representation learning and retrieval

10

Maria Tzelepi, Paraskevi Nousi, Nikolaos Passalis, and Anastasios Tefas
Department of Informatics, Aristotle University of Thessaloniki, Thessaloniki, Greece

10.1 Introduction

Representation learning refers to the process of learning a representation $\mathbf{y}_i = f(\mathbf{x}_i)$ from an input object $\mathbf{x}_i$ toward a specific task, for example, classification, retrieval, clustering, and others. Recent advances in deep learning, that inherently incorporates representation learning following a hierarchy of representations where the low level representations define the higher level ones [1], have altered the landscape of existing representation learning understanding. In this chapter, we first focus on autoencoders, which constitute a significant tool for representation learning tasks due to their wide range of application. Subsequently, we present representative representation learning methods utilizing autoencoders for classification and clustering tasks [2].

Subsequently, we focus on a natural representation learning task that is essential for Content Based Image Retrieval (CBIR). CBIR refers to the process of obtaining images that are relevant to a query image from a large collection based on their visual content [3]. This can be more formally articulated as follows: Given the feature representations $\mathcal{X} = \{\mathbf{x}_i, i = 1, \dots, N\}$ of the images to be searched and the feature representation $\boldsymbol{q}$ of the query image, obtained using a feature extraction method, the CBIR output is a ranked set a ranked set of images in terms of a similarity measure to the query representation $\boldsymbol{q}$. Thus apart from the similarity measure, a critical issue associated with the CBIR performance is the feature extraction. That is, it is of pivotal importance to extract meaningful information from raw data in order to eliminate the so-called semantic-gap [4] between the low level representations of images (e.g., color) and their higher level semantic concepts.

Research on representation learning for CBIR includes works that employ reconstruction-based objectives (i.e., autoencoders) [5] for learning representations in an unsupervised manner, and also works that employ supervised objectives [6]. The latter ones provide superior performance over existing unsupervised approaches by generating discriminative representations. However, these methods usually work with triplet-based losses which utilize triplets of similar and dissimilar samples to a specific sample. This usually comes with some shortcomings. For example, triplet sample learning is difficult to be implemented in large scale, and usually active learning is used in order to select meaningful triplets that can indeed contribute to learning [16].

Deep Learning for Robot Perception and Cognition. https://doi.org/10.1016/B978-0-32-385787-1.00015-4

Furthermore, it has been demonstrated that learning highly discriminative representations renders them inappropriate for retrieving images that belong to classes that were not seen during the training [7].

Thus we first focus on a more generic representation learning method toward the retrieval task that is capable of exploiting not only the class category of the data producing more discriminative representations like most of the existing approaches, but also the geometric structure of the data in an unsupervised manner, as well as the user's feedback using relevance feedback [8]. As it is demonstrated in the following, more efficient retrieval-oriented representations can be learned using model retraining with respect to different kinds of information both in terms of retrieval performance and memory requirements. Subsequently, we present a representation learning method suitable for retrieving objects that belong to classes that were not seen during the training process [7]. As it is demonstrated, efficient representations can be learned by encoding information arising from the in-class variance allowing us to efficiently represent images belonging to unseen classes.

This chapter aims to provide a brief introduction to deep representation learning techniques mainly for the retrieval task and present some of the most representative methods that will equip the reader with the necessary knowledge in order to apply these methods in practice. The rest of the chapter is structured as follows. Section 10.2 provides an introduction to autoencoders and their application in representation learning tasks. Section 10.3 presents the seminal approaches that introduced deep learning, and particularly deep Convolutional Neural Networks (CNN), to the retrieval domain. Section 10.4 presents model retraining approaches for producing efficient representations for the CBIR task. Section 10.5 presents a variance preserving supervised method for CBIR.

10.2 Discriminative and self-supervised autoencoders

Autoencoders (AE) are a class of unsupervised neural networks, which are trained to reconstruct their input, by first mapping it onto an intermediate representation, typically of lower dimension [9], naturally offered for representation learning tasks. In general, an AE consists of an encoding part, which maps the input to an intermediate representation, and a decoding part which reconstructs the input and typically has an architecture that is symmetrical to the encoder. The encoder and decoder can consist of multiple layers, each of which is accompanied by learnable parameters, such as the weights and biases of fully connected layers. Let $\mathbf{x} \in \mathbb{R}^D$ denote a D-dimensional input vector. Then an autoencoder can be formally defined by its two parts as

$$\hat{\mathbf{x}} = f(g(\mathbf{x})) \tag{10.1}$$

where $g(\cdot)$, $f(\cdot)$ are the encoding and decoding functions, respectively, and $\hat{\mathbf{x}} \in \mathbb{R}^D$ is the network's output, which is trained to approximate the input. The encoding and decoding functions can have symmetrical or asymmetrical architectures, and typically consist of multiple layers of, for example, fully connected layers, convolutional

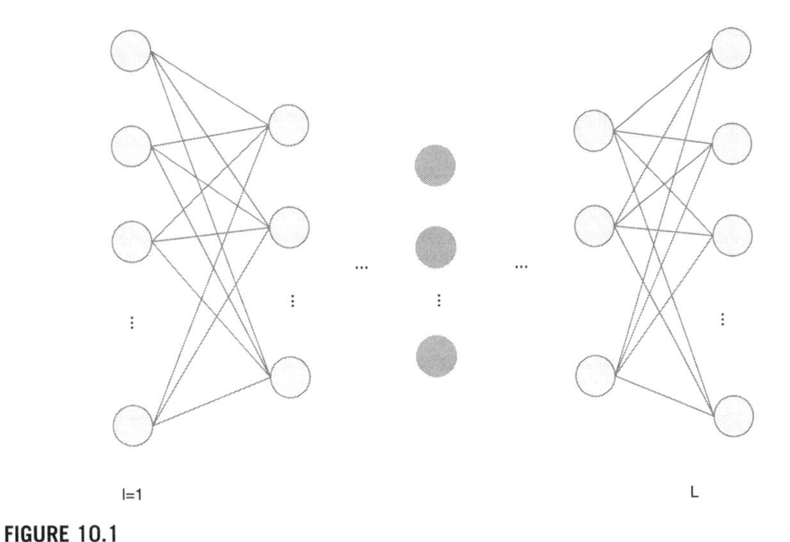

FIGURE 10.1

Typical architecture of a fully connected autoencoder.

layers and even recurrent modules. A useful operation for the decoder in particular is that of transposed convolution, of fractionally-strided convolution, used in practice to increase the spatial dimension of feature maps.

Although they are not limited in this scope, typically, AEs are used for dimensionality reduction. In this case, let $\mathbf{y} \in \mathbb{R}^d$, $d < D$, denote the output of the encoder, that is, $\mathbf{y} = g(\mathbf{x})$, then $\mathbf{y}$ may be regarded as a compressed version of $\mathbf{x}$. An AE of this form can be trained by minimizing the Mean Square Error (MSE) between the network's input and output, which corresponds to the reconstruction error:

$$\sum_{i=1}^{N} \|\hat{\mathbf{x}}_i - \mathbf{x}_i\|_2^2 \tag{10.2}$$

over all training samples, with respect to the network's parameters. As data labels are not taken into consideration, an AE trained using the object described here is fully unsupervised. Fig. 10.1 presents a typical architecture for an AE consisting of L fully connected layers. The input and output layer consist of the same number of neurons D. Multiple nonlinear layers lead to the intermediate representation. The decoder then tries to reconstruct the input via multiple nonlinear layers.

Due to their ability to extract semantically meaningful representations without the use of labels, AEs have been widely studied for a variety of tasks, including clustering [10,2], classification [11,12], and image retrieval [13,14]. Furthermore, when label information is available, it can be incorporated into the training scheme of AEs to produce representations, which are more discriminative, as shown in [15,16].

In [16] in particular, discriminative criteria were deployed to force pairs of representations corresponding to the same class to be closer together in the latent Euclidean subspace than to other representations corresponding to different classes, using a triplet loss method. In face recognition, [17] proposed the use of pose variations at given degrees of yaw rotation of the same face and mapping the variations back to the neutral pose progressively, by manipulating the loss functions for the variations. In [18], discriminative criteria are applied to AEs, assuming a Gaussian process prior between the hidden representation of the AE and the label subspace, in a nonparametric fashion. In [19], a method was proposed to discriminate between features that are relevant to the facial expression recognition and features that are irrelevant to this task.

In [15] the loss function of an AE has been augmented by making use of the samples' labels, thus incorporating supervised classification into the training of the AE. The samples' labels can also be incorporated into the AE, in order to produce hidden representations that make use of class information to better discriminate between samples of different classes. For a two-class problem, samples belonging to one class might be forced to have better AE reconstructions than those belonging to the other class, as suggested in [20]. Then the two classes may be distinguished by the error of their reconstruction. In the following paragraphs, we focus on discriminative autoencoders in both supervised and unsupervised scenarios, for learning representations toward classification and clustering tasks, following [11,12,2].

The latent representation produced by an AE is learned via minimizing the network's reconstruction error. Intuitively, if the target to be learned for each sample is a modified version of itself, such that it is closer to other samples of the same class, the network will learn to reconstruct samples that belong to classes, which are more easily separable. This modification should be reflected by the intermediate representation, thus producing well-separated low-dimensional representations of the network's input data.

Let $\tilde{\mathbf{x}}_i$ be the modified target reconstruction of sample $\mathbf{x}_i$, $i = \{1, \ldots, N\}$. The loss function to be minimized for these modified autoencoders becomes

$$\sum_{i=1}^{N} \|\hat{\mathbf{x}}_i - \tilde{\mathbf{x}}_i\|_2^2 \tag{10.3}$$

summed over all data samples. Thus the AE is trained to learn to reconstruct shifted versions of its input samples. This shift is reflected in the learned representation, which may then be used in clustering and classification tasks. Depending on the subsequent task, various shifting methods can be used to produce the shifted targets, to extract representations which are more properly separated. In general, the shifting process can be formulated as

$$\tilde{\mathbf{x}}_i = (1 - \alpha) \cdot \mathbf{x} + \alpha \cdot \mathbf{m}_i \tag{10.4}$$

where $\mathbf{m}_i$ is a linear combination of all samples in the training data set:

$$\mathbf{m}_i = \sum_{j=1}^{N} W_{ij} \mathbf{x}_j \tag{10.5}$$

and the hyperparameter α controls the intensity of the shift, that is, larger values correspond to larger steps toward $\mathbf{m}_i$. Thus different types of shifts can be achieved by properly formulating the shifting weights matrix $\mathbf{W}$.

For clustering applications, better cluster separability can be achieved by increasing the intracluster compactness, that is, how close together samples of the same cluster lie in the low-dimensional space, or the intercluster separability, that is, how far away different clusters are from each other. Let $\mathcal{V}_k$ be the kth cluster containing similar samples in the low-dimensional representation space, and $\mathbf{v}_k \in \mathbb{R}^d$, $k = 1, \ldots, K$ be the center of this cluster, that is, the mean vector of all samples belonging to that cluster. Then weight matrix elements W_{ij} for a given sample $\mathbf{x}_i$ can be defined as $\frac{1}{|\mathcal{V}_k|}$ for all samples $\mathbf{x}_j$ belonging to the same cluster as $\mathbf{x}_i$, or, formally

$$W_{ij} = \begin{cases} \frac{1}{|\mathcal{V}_k|} & \text{if } i, j \in \mathcal{V}_k \\ 0 & \text{otherwise.} \end{cases} \tag{10.6}$$

Intuitively, neighboring samples should affect each sample $\mathbf{x}_i$ more than other same-cluster samples. If $\mathcal{N}_i$ is the set of $\mathbf{x}_i$'s n closest neighbors within their cluster $\mathcal{V}_k$, then the shifting weights may also be defined as

$$W_{ij} = \begin{cases} \frac{1}{|\mathcal{N}_i|} & \text{if } j \in \mathcal{N}_i \\ 0 & \text{otherwise.} \end{cases} \tag{10.7}$$

Furthermore, repulsion weights can be defined, which combine training samples that the input should be shifted away from

$$\tilde{\mathbf{x}}_i = (1 + \beta)\mathbf{x} - \beta\mathbf{m}_i \tag{10.8}$$

where β is again a hyperparameter, which controls the intensity of the shift, and the vector $\mathbf{m}_i$ is a linear combination of the training samples with weights given by a matrix $\mathbf{W}$.

As an example, let $\mathcal{V}_{-k}$ denote the union of all clusters but the kth, for example, $\mathcal{V}_{-3} = \mathcal{V}_1 \cup \mathcal{V}_2 \cup \mathcal{V}_4 \cup \cdots \cup \mathcal{V}_K$. The shift of a sample $\mathbf{x}_i$ away from all rivaling samples can then be formulated by defining $\mathbf{W}$ as

$$W_{ij} = \begin{cases} \frac{-1}{|\mathcal{V}_{-k}|} & \text{if } j \in \mathcal{V}_{-k} \\ 0 & \text{otherwise.} \end{cases} \tag{10.9}$$

The set $\mathcal{R}_i$ of the r closest samples to $\mathbf{x}_i$ which belong to rival clusters can also define penalty weights, formulated as

$$W_{ij} = \begin{cases} \frac{-1}{|\mathcal{R}_i|} & \text{if } j \in \mathcal{R}_i \\ 0 & \text{otherwise.} \end{cases} . \tag{10.10}$$

In the supervised case, the new targets may be shifted so that the samples are moved towards their class centroids with weights given by

$$W_{ij} = \begin{cases} \frac{1}{|\mathcal{C}_k|} & \text{if } \{i, j\} \in \mathcal{C}_k \\ 0 & \text{otherwise.} \end{cases} \tag{10.11}$$

applied to Eq. (10.4), where $\mathcal{C}_k$ is set of samples belonging to the kth class.

Respectively, the distances between samples and centroids of rival classes could be enlarged by moving each sample away from the mean of all samples belonging to other classes by defining weights:

$$W_{ij} = \begin{cases} \frac{-1}{|\mathcal{Q}_i|} & \text{if } j \in \mathcal{Q}_i \\ 0 & \text{otherwise,} \end{cases} \tag{10.12}$$

to be applied in Eq. (10.8), where $\mathcal{Q}_i$ is the set of samples belonging to a class different than the class $\mathbf{x}_i$ belongs to. In practice, only the neighbors of rival classes within a given range of each sample could be taken into consideration instead of the mean of all rival class samples.

10.3 Deep representation learning for content based image retrieval

The recent advent and evolution of Deep Learning (DL) [21] provided models that achieved outstanding performance in a plethora of computer vision tasks [22]. Motivated by the exceptional performance of CNNs in various recognition tasks, such as image classification [23], and also by the observation that features extracted from the activation of a CNN trained in a fully supervised fashion on a large, fixed set of object recognition tasks can be repurposed to novel generic recognition tasks, [24], deep CNNs were introduced in the vivid research area of CBIR [25]. The seminal approach of applying deep CNNs in the retrieval domain is to extract the feature representations from a pretrained model by feeding images in the input layer of the model and taking activation values drawn for a specific neural layer. This can be more formally articulated as follows. We consider a deep CNN model, $\phi(\cdot\,; \mathcal{W})$, with $N_L \in \mathbb{N}$ layers, and set of parameters $\mathcal{W} = \{\mathbf{W}_1, \ldots, \mathbf{W}_{N_L}\}$, where $\mathbf{W}_L$ are the weights of a specific layer L, trained for distinguishing between c class categories. Then an input image $\mathbf{x}$ is propagated to the network, and the output at a certain layer L: $\phi(\mathbf{x}\,; \mathbf{W_L})$ can be used as image representation for the retrieval task. Subsequently, later works directed on utilization of the convolutional layers, so as to exploit the spatial information of these layers. Since the representations extracted from these layers are high-dimensional, pooling techniques are also investigated for producing representations of up to a few hundred dimensions. These works use either sum-pooling [26,27] or max-pooling [28] techniques.

10.4 Model retraining methods for image retrieval

Utilizing a CNN model trained on a large dataset, for example, ImageNet [23], for a classification task in order to derive the representations either from the fully connected or the convolutional layers, usually achieves considerable performance, as compared to the existing shallow approaches. This is attributed to the fact that the models trained on such data sets have learned representations that sufficiently describe the physical world. Therefore, the aforementioned semantic-gap between the low level representations of images and their higher level concepts is eliminated to some extent. However, a key target of deep representation learning, is to produce efficient representations toward a specific learning task. Thus, model retraining methods in order to learn more efficient retrieval-oriented feature representations have also been proposed in the recent literature [6,29]. These methods typically work by utilizing triplet-based losses, for example, a deep CNN is retrained with a similarity learning objective function, considering triplets of relevant and irrelevant instances obtained from the fully connected layers of the pretrained model is proposed in [6].

In this section, we focus on a more generic representation learning method towards the retrieval task that is capable of exploiting any kind of available information [8,30]. That is, in an image retrieval task, may exist information regarding the labels of the data set, information acquired from users' feedback, and no other information rather than the data set to be searched. Thus a model pretrained for a generic classification task is utilized for extracting the image representations from a specific layer, subsequently new targets are computed with respect to the available information, a regression task is formulated, and the model's weights are fine tuned toward the retrieval task. Then more efficient retrieval-oriented image representations are produced from the retrained model. This practice is accompanied with several advantages as compared to the triplet-based retraining approaches. Considering multiple relevant and multiple irrelevant samples in the training procedure for each training sample, absolve us from finding meaningful triplets that can indeed contribute to learning. Additionally, defining representation targets for the training samples and simple regression, instead of defining more complex loss functions that need three samples for each training step results to a significant boost in the training speed.

The method includes the following steps:

1. A model pretrained for a generic classification task is employed in order to derive the feature representations from the activations of a certain layer.
2. The model is properly adapted by discarding the layers following the layer utilized for feature extraction.
3. The new targets for the model retraining are formulated based on the available information.
4. The model is retrained, exploiting the idea that a deep neural architecture can nonlinearly distort the feature space in order to modify the feature representations, with respect to the available information.

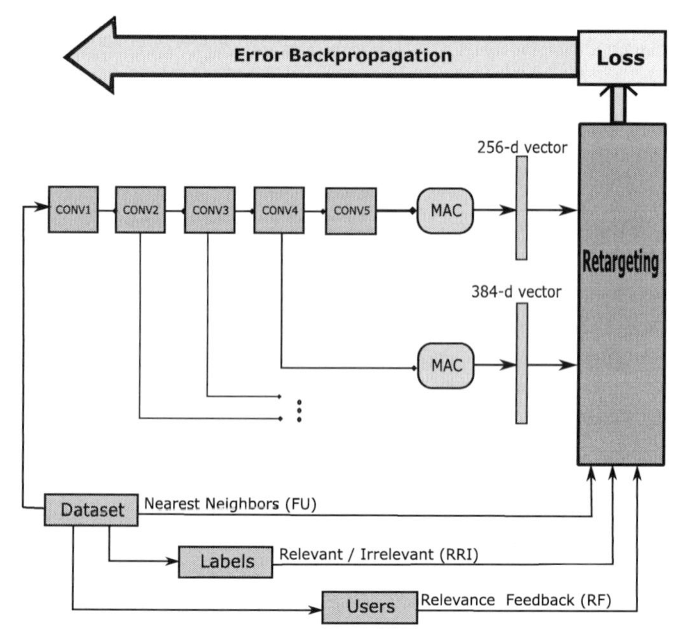

FIGURE 10.2

The FU, RRI, and RF retraining approaches applied on the convolutional layers of the pretrained model. The feature representations are extracted from a certain convolutional layer, accompanied with a max-pooling technique, named MAC [28]. Then the new targets are formulated with respect to the available information and the error is computed and propagated backwards through the network.

The retraining approaches with respect to the available information consist in: the Fully Unsupervised (FU) retraining, if no information is available except for the data set to be searched, the Retraining with Relevance Information (RRI), in the case that the labels of the data set or of part of the data set are available, and the Relevance-Feedback (RF) based retraining, if feedback from users is available. Each of retraining approach can be applied independently or in a pipeline, where each approach operates as a pretraining step to another retraining process. It is also noted that the approaches presented in this section can be applied to several layers, while also using different similarity metrics (e.g., cosine similarity, Euclidean distance). The convolutional layers were used in the presented approaches, since they provide more compact feature representations conveying spatial information, allowing for producing more efficient representations both in terms of retrieval performance and speed. The retraining methodology is schematically described in Fig. 10.2. These approaches are also applied on the fully connected layers, utilizing the Euclidean distance as similarity metric in [30].

Finally, these approaches have also been combined with other representation methods for CBIR, providing improved performance. For example, a work introduc-

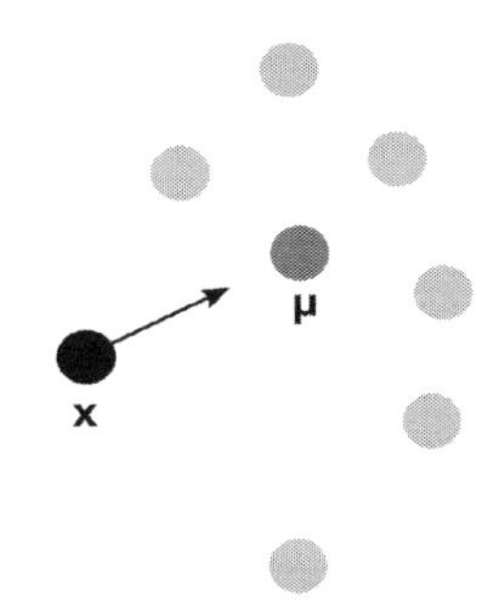

FIGURE 10.3

Schematic description of the FU retraining approach. The goal is to maximize the cosine similarity of each image representation $\boldsymbol{x}$ with the mean vector $\boldsymbol{\mu}$ of its n nearest representations.

ing the widely used term-weighting technique in document-retrieval, term frequency-inverse document frequency (tf-idf), into deep CNNs for CBIR has been proposed in [31]. In this work, the learned filters of the convolutional layers of a pretrained CNN model are treated as the detectors of the visual words, since they have been trained to be activated in different visual patterns. Thus since the activations of each filter provide information about the degree of presence of the visual pattern that the filter has learned during the training procedure, are considered as the tf part. The work also investigates different approaches for computing the idf part. Finally, the tf-idf method is applied in combination with the aforementioned retraining methodologies providing improved retrieval performance. In the rest of this section, we introduce the three retraining approaches and we explain how they work.

10.4.1 Fully unsupervised retraining

The idea behind the FU retraining approach is to amplify the primary retrieval assumption that the relevant image representations are closer to a certain query representation in the feature space. The rationale behind this approach is rooted to the cluster hypothesis, which states that documents in the same cluster are likely to satisfy the same information need [32]. That is, the pretrained CNN model is retrained on the given dataset, aiming at maximizing the cosine similarity between each image representation and its n nearest representations, in terms of cosine distance. A schematic description of the FU retraining approach is illustrated in Fig. 10.3.

Let $\mathcal{I} = \{\mathbf{I}_i, i = 1, \dots, N\}$ be the set of N images to be searched, and $\mathcal{X} = \{\mathbf{x}_i, i = 1, \dots, N\}$ their corresponding feature representations emerged in the L layer. Let also $\boldsymbol{\mu}^i$ be the mean vector of the $n \in \{1, \dots, N-1\}$ nearest representations to $\mathbf{x}_i$, denoted as $\mathcal{X}^i = \{\mathbf{x}^i_l, l = 1, \dots, N-1\}$. That is,

$$\boldsymbol{\mu}^i = \frac{1}{n}\sum_{l=1}^{n}\mathbf{x}^i_l. \tag{10.13}$$

The new target representations for the images of $\mathcal{I}$ can be determined by solving the following optimization problem:

$$\max_{\mathbf{x}_i \in \mathcal{X}} \mathcal{J} = \max_{\mathbf{x}_i \in \mathcal{X}} \sum_{i=1}^{N} \frac{\mathbf{x}_i^{\mathsf{T}} \boldsymbol{\mu}^i}{\|\mathbf{x}_i\| \, \|\boldsymbol{\mu}^i\|}. \tag{10.14}$$

The above optimization problem is solved using gradient descent. The first-order gradient of the objective function $\mathcal{J}$ is given by

$$\frac{\partial \mathcal{J}}{\partial \mathbf{x}_i} = \frac{\partial}{\partial \mathbf{x}_i} \left(\sum_{i=1}^{N} \frac{\mathbf{x}_i^{\mathsf{T}} \boldsymbol{\mu}^i}{\|\mathbf{x}_i\| \, \|\boldsymbol{\mu}^i\|} \right) = \frac{\boldsymbol{\mu}^i}{\|\mathbf{x}_i\| \, \|\boldsymbol{\mu}^i\|} - \frac{\mathbf{x}_i^{\mathsf{T}} \boldsymbol{\mu}^i}{\|\mathbf{x}_i\|^3 \, \|\boldsymbol{\mu}^i\|} \mathbf{x}_i. \tag{10.15}$$

The update rule for the vth iteration for each image can be formulated as

$$\mathbf{x}_i^{(v+1)} = \mathbf{x}_i^{(v)} + \eta \left(\frac{\boldsymbol{\mu}^i}{\|\mathbf{x}_i^{(v)}\| \, \|\boldsymbol{\mu}^i\|} - \frac{\mathbf{x}_i^{(v)\mathsf{T}} \boldsymbol{\mu}^i}{\|\mathbf{x}_i^{(v)}\|^3 \, \|\boldsymbol{\mu}^i\|} \mathbf{x}_i^{(v)} \right), \quad \mathbf{x}_i \in \mathcal{X}. \tag{10.16}$$

Finally, a normalization step is introduced, in order to better control the learning rate, as follows:

$$\mathbf{x}_i^{(v+1)} = \mathbf{x}_i^{(v)} + \eta \|\mathbf{x}_i^{(v)}\| \, \|\boldsymbol{\mu}^i\| \left(\frac{\boldsymbol{\mu}^i}{\|\mathbf{x}_i^{(v)}\| \, \|\boldsymbol{\mu}^i\|} - \frac{\mathbf{x}_i^{(v)\mathsf{T}} \boldsymbol{\mu}^i}{\|\mathbf{x}_i^{(v)}\|^3 \, \|\boldsymbol{\mu}^i\|} \mathbf{x}_i^{(v)} \right), \quad \mathbf{x}_i \in \mathcal{X}. \tag{10.17}$$

Using the above representations as targets in the layer of interest, a regression task is formulated for the neural network, which is initialized on the pretrained model's weights and is trained on the utilized data set, using back-propagation. The Euclidean loss is used during training for the regression task. Thus the procedure is integrated by feeding the entire data set into the input layer of the retrained adapted model and obtaining the new representations.

10.4.2 Retraining with relevance information

This approach aims to enhance the performance of the deep CNN descriptors exploiting the relevance information deriving from the available class labels. To achieve this goal, considering a labeled representation $(\mathbf{x}_i, y_i)$, where $\mathbf{x}_i$ is the image representation and y_i is the corresponding image label, the convolutional neural layers of the CNN model used for the feature extraction are adapted, aiming to maximize the cosine similarity between $\mathbf{x}_i$ and the m nearest relevant representations, and simultaneously to minimize the cosine similarity between $\mathbf{x}_i$ and the l nearest irrelevant representations, in terms of cosine distance. As relevant are defined the images belonging to same class, while as irrelevant the images belonging to different classes. A schematic description of the RRI retraining approach is illustrated in Fig. 10.4.

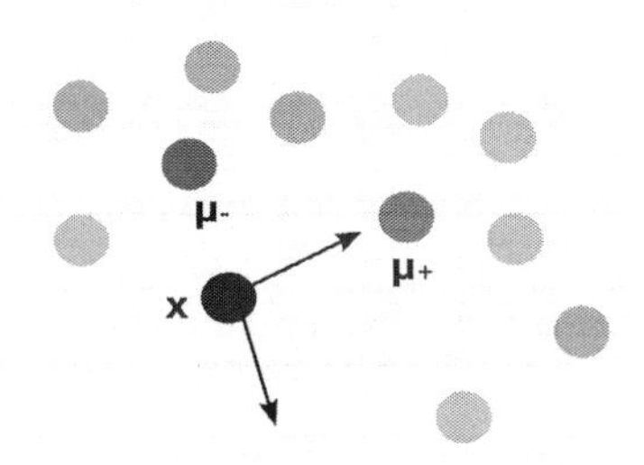

FIGURE 10.4

Schematic description of the RRI approach. The goal is to maximize the cosine similarity of each image representation $\boldsymbol{x}$ with the mean vector $\boldsymbol{\mu}+$ of its m nearest relevant representations and simultaneously to minimize the cosine similarity between $\boldsymbol{x}$ and the mean vector $\boldsymbol{\mu}-$ of its l nearest irrelevant representations.

Let $\mathcal{I} = \{\mathbf{I}_i, i = 1, \ldots, N\}$ be a set of N images of the search set provided with relevance information, and $\mathbf{x} = F_L(\mathbf{I})$ the output of the L layer of the pretrained CNN model on an input image $\mathbf{I}$. Let also $\mathcal{X} = \{\mathbf{x}_i, i = 1, \ldots, N\}$ be the set of N feature representations emerged in the L layer, $\mathcal{R}^i = \{\mathbf{r}_k, k = 1, \ldots, K^i\}$ the set of K^i relevant representations of the ith image and by $\mathcal{C}^i = \{\mathbf{c}_j, j = 1, \ldots, L^i\}$ the set of L^i irrelevant representations. The mean vector of the m nearest representations of R^i to the certain image representation $\mathbf{x}_i$, and the mean vector of the l nearest representations of C^i to $\mathbf{x}_i$ are computed, and be denoted by $\boldsymbol{\mu}^i_+$ and $\boldsymbol{\mu}^i_-$, respectively. Then the new target representations for the images of $\mathcal{I}$ can be determined by solving the following optimization problems:

$$\max_{\mathbf{x}_i \in \mathcal{X}} \mathcal{J}^+ = \max_{\mathbf{x}_i \in \mathcal{X}} \sum_{i=1}^{N} \frac{\mathbf{x}_i^\mathsf{T} \boldsymbol{\mu}^i_+}{\|\mathbf{x}_i\| \, \|\boldsymbol{\mu}^i_+\|}, \tag{10.18}$$

$$\min_{\mathbf{x}_i \in \mathcal{X}} \mathcal{J}^- = \min_{\mathbf{x}_i \in \mathcal{X}} \sum_{i=1}^{N} \frac{\mathbf{x}_i^\mathsf{T} \boldsymbol{\mu}^i_-}{\|\mathbf{x}_i\| \, \|\boldsymbol{\mu}^i_-\|}. \tag{10.19}$$

The normalized update rules for the υth iteration can be formulated as

$$\mathbf{x}_i^{(\upsilon+1)} = \mathbf{x}_i^{(\upsilon)} + \zeta \|\mathbf{x}_i^{(\upsilon)}\| \, \|\boldsymbol{\mu}^i_+\| \left(\frac{\boldsymbol{\mu}^i_+}{\|\mathbf{x}_i^{(\upsilon)}\| \, \|\boldsymbol{\mu}^i_+\|} - \frac{\mathbf{x}_i^{(\upsilon)\mathsf{T}} \boldsymbol{\mu}^i_+}{\|\mathbf{x}_i^{(\upsilon)}\|^3 \, \|\boldsymbol{\mu}^i_+\|} \mathbf{x}_i^{(\upsilon)} \right), \quad \mathbf{x}_i \in \mathcal{X} \tag{10.20}$$

and

$$\mathbf{x}_i^{(v+1)} = \mathbf{x}_i^{(v)} - \beta \|\mathbf{x}_i^{(v)}\| \, \|\boldsymbol{\mu}_-^i\| \left(\frac{\boldsymbol{\mu}_-^i}{\|\mathbf{x}_i^{(v)}\| \, \|\boldsymbol{\mu}_-^i\|} - \frac{\mathbf{x}_i^{(v)\top} \boldsymbol{\mu}_-^i}{\|\mathbf{x}_i^{(v)}\|^3 \, \|\boldsymbol{\mu}_-^i\|} \mathbf{x}_i^{(v)} \right), \quad \mathbf{x}_i \in \mathcal{X}. \tag{10.21}$$

Consequently, the combinatory normalized update rule, deriving by adding the Eqs. (10.20) and (10.21) can be formulated as

$$\begin{aligned} \mathbf{x}_i^{(v+1)} = \mathbf{x}_i^{(v)} &+ \zeta \|\mathbf{x}_i^{(v)}\| \, \|\boldsymbol{\mu}_+^i\| \left(\frac{\boldsymbol{\mu}_+^i}{\|\mathbf{x}_i^{(v)}\| \, \|\boldsymbol{\mu}_+^i\|} - \frac{\mathbf{x}_i^{(v)\top} \boldsymbol{\mu}_+^i}{\|\mathbf{x}_i^{(v)}\|^3 \, \|\boldsymbol{\mu}_+^i\|} \mathbf{x}_i^{(v)} \right) \\ &- \beta \|\mathbf{x}_i^{(v)}\| \, \|\boldsymbol{\mu}_-^i\| \left(\frac{\boldsymbol{\mu}_-^i}{\|\mathbf{x}_i^{(v)}\| \, \|\boldsymbol{\mu}_-^i\|} - \frac{\mathbf{x}_i^{(v)\top} \boldsymbol{\mu}_-^i}{\|\mathbf{x}_i^{(v)}\|^3 \, \|\boldsymbol{\mu}_-^i\|} \mathbf{x}_i^{(v)} \right), \quad \mathbf{x}_i \in \mathcal{X} \end{aligned}. \tag{10.22}$$

Thus as in the previous approach, using the above target representations the neural network is retrained on the images provided with relevance information using back-propagation.

10.4.3 Relevance feedback based retraining

The idea of this approach is rooted in the relevance feedback philosophy. Relevance feedback refers to the ability of users to impart their judgment regarding the relevance of search results to the retrieval procedure in order to ameliorate its performance [33]. In this retraining approach, information from different users' feedback is available. This information consists of queries and relevant and irrelevant images to these queries. This approach works by modifying the model parameters in order to maximize the cosine similarity between a specific query and its relevant images and minimize the cosine similarity between it and its irrelevant ones. A schematic description of the RF-based retraining approach is illustrated in Fig. 10.5.

Let $\mathcal{Q} = \{\mathbf{Q}_k, k = 1, \ldots, K\}$ be a set of queries, $\mathcal{I}_+^k = \{\mathbf{I}_i, i = 1, \ldots, Z\}$ a set of relevant images to a certain query, $\mathcal{I}_-^k = \{\mathbf{I}_j, j = 1, \ldots, O\}$ a set of irrelevant images, $\mathbf{x} = F_L(\mathbf{I})$ the output of the L layer of the pretrained CNN model on an input image $\mathbf{I}$, and $\mathbf{q} = F_L(\mathbf{Q})$ the output of the L layer on a query. Then $\mathcal{X}_+^k = \{\mathbf{x}_i, i = 1, \ldots, Z\}$ denotes the set of feature representations emerged in L layer of Z images that have been qualified as relevant by a user, and $\mathcal{X}_-^k = \{\mathbf{x}_j, j = 1, \ldots, O\}$ denotes the set of O irrelevant feature representations.

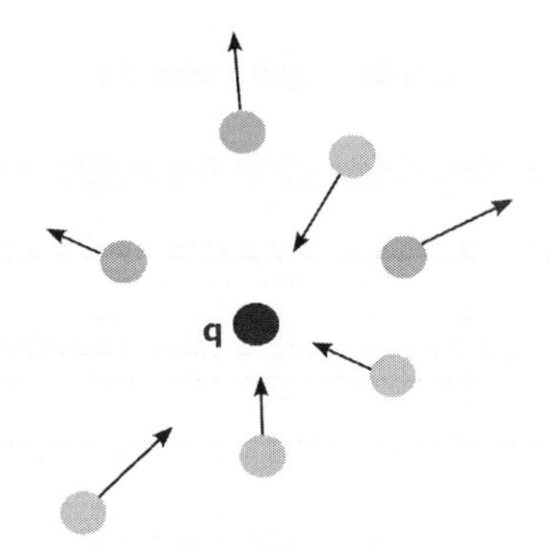

FIGURE 10.5

Schematic description of the RF-based retraining approach. The goal is to maximize the cosine similarity between a specific query q and its relevant images and minimize the cosine similarity between it and its irrelevant ones.

The new target representations for the relevant and irrelevant images can be respectively determined by solving the following optimization problems:

$$\max_{\mathbf{x}_i \in \mathcal{X}_+^k} \mathcal{J}^+ = \max_{\mathbf{x}_i \in \mathcal{X}_+^k} \sum_{i=1}^{Z} \frac{\mathbf{x}_i^\mathsf{T} \boldsymbol{q}^k}{\|\mathbf{x}_i\| \, \|\boldsymbol{q}^k\|}, \tag{10.23}$$

$$\min_{\mathbf{x}_j \in \mathcal{X}_-^k} \mathcal{J}^- = \min_{\mathbf{x}_j \in \mathcal{X}_-^k} \sum_{j=1}^{O} \frac{\mathbf{x}_j^\mathsf{T} \boldsymbol{q}^k}{\|\mathbf{x}_j\| \, \|\boldsymbol{q}^k\|}. \tag{10.24}$$

The normalized update rules for the vth iteration can be formulated as

$$\mathbf{x}_i^{(v+1)} = \mathbf{x}_i^{(v)} + \alpha \|\mathbf{x}_i^{(v)}\| \, \|\boldsymbol{q}^k\| \left(\frac{\boldsymbol{q}^k}{\|\mathbf{x}_i^{(v)}\| \, \|\boldsymbol{q}^k\|} - \frac{\mathbf{x}_i^{(v)\mathsf{T}} \boldsymbol{q}^k}{\|\mathbf{x}_i^{(v)}\|^3 \, \|\boldsymbol{q}^k\|} \mathbf{x}_i^{(v)} \right), \quad \mathbf{x}_i \in \mathcal{X}_+^k \tag{10.25}$$

and

$$\mathbf{x}_j^{(v+1)} = \mathbf{x}_j^{(v)} - \alpha \|\mathbf{x}_j^{(v)}\| \, \|\boldsymbol{q}^k\| \left(\frac{\boldsymbol{q}^k}{\|\mathbf{x}_j^{(v)}\| \, \|\boldsymbol{q}^k\|} - \frac{\mathbf{x}_j^{(v)\mathsf{T}} \boldsymbol{q}^k}{\|\mathbf{x}_j^{(v)}\|^3 \, \|\boldsymbol{q}^k\|} \mathbf{x}_j^{(v)} \right), \quad \mathbf{x}_j \in \mathcal{X}_-^k. \tag{10.26}$$

Similar to the other approaches, using the above representations as targets in the layer of interest, the neural network is retrained on the set of relevant and irrelevant images.

10.5 Variance preserving supervised representation learning

There is a rich literature on supervised representation learning approaches that range from simpler and seminal methods, such as using the contrastive [34] and triplet [35] losses, to more refined and efficient methods, such as the N-pair loss [36], lifted structural embedding loss [37], and multisimilarity loss [38]. These methods typically work by learning discriminative spaces that increase the separability of objects that are not similar, while bringing closer the representation of objects that are similar.

Despite the effectiveness of the aforementioned approaches, there are still critical limitations regarding the way they work, since they tend to wrap the representation space in an *unnatural way*, collapsing the intrinsic manifold of the data, and promoting overfitting phenomena, as explained in [7]. To better understand these limitations, consider the case of the contrastive loss [34,39], which is widely used in many representation learning applications:

$$\frac{1}{2}\delta_{ij}||\mathbf{x}_i - \mathbf{x}_j||_2^2 + \frac{1}{2}(1-\delta_{ij})\max(0, k - ||\mathbf{x}_i - \mathbf{x}_j||_2^2), \tag{10.27}$$

where $\mathbf{x}_i$ denotes the vector representation of the ith training sample (typically extracted from a DL model), k denotes the minimum distance between dissimilar classes and

$$\delta_{ij} = \begin{cases} 1, & \text{if the } i\text{th and } j\text{th samples are similar,} \\ 0, & \text{otherwise.} \end{cases} \tag{10.28}$$

By examining (10.27), it is easy to see that the contrastive loss forces the samples of the same class to collapse into one point, while pulling apart points that are not similar, but are within a radius of k of each other.

Even though in many cases the similar samples are not collapsed into a singular point, since the limited learning capacity of the used DL model, along with the use of regularization methods can reduce such phenomena, there is an intrinsic flaw in the way contrastive loss as well as other losses that promote this in-class collapse work. Reducing the in-class variance, by collapsing the in-class representation, can lead to improved in-domain accuracy, since it leads to more discriminative representations. However, at the same time, restricting in-class variance can lead to removing useful information regarding the similarities between the in-class samples. As a result, it would be more difficult to identify possible subclasses of the data. This can be better understood by the following example. Suppose that there are two classes for which a representation has been optimized using the contrastive loss, that is, "car" and "bicycle." If we optimize the representation using the contrastive loss we will force the representation to suppress any attributes that are irrelevant to the class, for example, color, type of vehicle, etc. As a result, the learned representation would be excellent for discriminating between "cars" and "bicycles," but probably useless for retrieving subclasses of these categories, for example, "sedans," "red cars," etc. Indeed, recent evidence suggests that taking into account these variations, for example, through observing the similarities between different classes [40], can indeed lead to improved performance for many DL models.

Also, promoting this kind of in-class collapse can harm the generalization ability for classes that are not seen during the training. Indeed, there are several works that highlight that overfitted representations indeed lead to reduced out-of-domain precision [41,42]. It is also worth noting that these phenomena are not limited to the contrastive loss, since many of the existing supervised representation learning objectives, for example, triplet-based objectives [43,16], are also affected by these.

Even through these behaviors can have a significant impact on the performance of various retrieval systems, there are only a few methods proposed to tackle these issues. These phenomena were studied in [41], where the effect of retrieving classes that were not seen during the training of supervised hashing methods was evaluated. Furthermore in [42,44] entropy-based objectives were used to allow for optimizing representations in a more regularized way. In this section, we focus on a recently proposed method that works by employing an auxiliary task during representation learning, that is, to maintain the in-class variance instead of suppressing it. As demonstrated in [7], this approach can indeed significantly improve the performance of the learned representation for various tasks. In the rest of this section, we introduce this method and explain how it works.

Let $f_{\mathbf{W}}(\mathbf{x}_i) \in \mathbb{R}^m$ denote a feature extraction model, where the notation $\mathbf{W}$ is used to refer to the trainable parameters of the model. This model is used for representation learning, that is, for extracting a representation $\mathbf{z}_i = f_{\mathbf{W}}(\mathbf{x}_i)$ from an input object $\mathbf{x}_i$. In this section, we will focus on image inputs. However, this is without loss of generality, since the proposed method can be readily applied for any kind of inputs. Furthermore, we assume that the model will operate in the typical content-based image retrieval setup, that is, a query image $\mathbf{q}$ will be provided and the system should return a ranking of the images in a database $\mathcal{X} = \{\mathbf{x}_1, \mathbf{x}_2, ..., \mathbf{x}_N\}$ according to their relevance to $\mathbf{q}$.

Learning feature extractors that maintain as much information (variance) for each class, while also ensuring that objects from different classes will be separated is not straightforward, since maximizing (or even just maintaining) the intraclass variance typically conflicts with most discriminative objectives that are employed by representation learning methods that aim to minimize the intra-class variance. To overcome this limitation, recent methods [7], aim at learning a discriminative representation space, while also learning the latent generative factors for each class, as shown in Fig. 10.6. To this end, a different Gaussian distribution is used to model the distribution of each class. Then each Gaussian is pushed away from the other ones, ensuring that the learned space will be discriminative enough, while the latent generative factors of the data are encoded in each Gaussian (repulsive loss in Fig. 10.6), allowing for maintaining the in-class variance (reconstruction + KL loss in Fig. 10.6).

Let us first consider the case of having only one class. The goal is to model the intra-class variations between them. To do so, a variational learning approach can be employed to learn the latent generative factors of the class [45–47]. As a result,

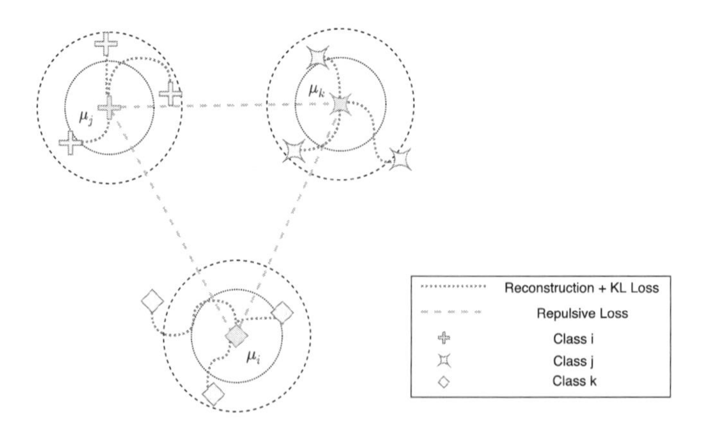

FIGURE 10.6

Variance-preserving supervised representation learning: Each training sample is affected by two different *forces*. First, a reconstruction loss is employed to ensure that the generative factors of each class will be modeled, while KL divergence attracts each sample to its distribution. Then the Gaussians used to model the generative distribution are pushed apart (repulsive loss) to increase the discriminative ability of the learned representation.

sampling a vector $\mathbf{z}$ from the latent space and then using an image decoder $f_{dec}(\mathbf{z})$ allows for reconstructing each image. Therefore, in order to model each class as good as possible the probability of correctly reconstructing each image $\mathbf{x}$ when sampling the latent space should be maximized:

$$P(\mathbf{x}) = \int_{\mathbf{z}} P(\mathbf{x}|\mathbf{z}) P(\mathbf{z}) d\mathbf{z}, \tag{10.29}$$

where $P(\mathbf{x}|\mathbf{z})$ denotes a Gaussian distribution centered around the reconstructed sample, after using the decoder $f_{dec}(\cdot)$, with a diagonal covariance matrix multiplied by σ:

$$P(\mathbf{x}|\mathbf{z}) = \mathcal{N}(\mathbf{x}; f_{dec}(\mathbf{z}), \sigma \mathbf{I}). \tag{10.30}$$

Furthermore, the notation $P(\mathbf{z})$ is used to refer to the density function of the latent space. Typically, this is set to be a Gaussian distribution with zero-mean and unit variance [45]:

$$P(\mathbf{z}) = \mathcal{N}(\mathbf{z}; \mathbf{0}, \mathbf{I}), \tag{10.31}$$

where

$$\mathcal{N}(\mathbf{x}; \boldsymbol{\mu}, \boldsymbol{\Sigma}) = (2\pi)^{-\frac{k}{2}} \det(\boldsymbol{\Sigma})^{\frac{1}{2}} \exp\left(-\frac{1}{2}(\mathbf{x} - \boldsymbol{\mu})^T \boldsymbol{\Sigma}^{-1} (\mathbf{x} - \boldsymbol{\mu})\right), \tag{10.32}$$

and k is the number of dimensions of $\mathbf{x}$.

Directly optimizing (10.29) is typically not tractable. Instead, we can employ the evidence lower bound (ELBO) to this end as

$$E_{z \sim Q(\mathbf{z}|\mathbf{x})}[\log P(\mathbf{x}|\mathbf{z}) - \mathcal{D}_{KL}(Q(\mathbf{z}|\mathbf{x})||P(\mathbf{z}))], \tag{10.33}$$

where $Q(\mathbf{z}|\mathbf{x})$ denotes a tractable model that is fitted to approximate $P(\mathbf{z}|\mathbf{x})$, and $\mathcal{D}_{KL}$ denotes the Kullback–Leibler (KL) divergence. By using Monte Carlo sampling, along with the reparametrization trick, this bound can be optimized [45]. Usually $Q(\mathbf{z}|\mathbf{x})$ is modeled using a neural network that predicts the mean $\mu_Q(\mathbf{x})$ and the covariance matrix $\Sigma_Q(\mathbf{x})$ (which is constrained to be diagonal) for an input image $\mathbf{x}$. Then the distribution Q can be modeled using a Gaussian as

$$Q(\mathbf{z}|\mathbf{x}) = \mathcal{N}(\mathbf{z}; \mu_Q(\mathbf{x}), \Sigma_Q(\mathbf{x})). \tag{10.34}$$

Therefore the representation of each sample is derived as

$$f_{\mathbf{W}}(\mathbf{x}) = \mu_Q(\mathbf{x}), \tag{10.35}$$

allowing for implicitly learning the representation space through this process.

This variational approach enables to learn the generative factors for each class. However, the learned representation should be also discriminative. To this end, separate Gaussians, each with separate mean $\boldsymbol{\mu}_i$ and unit variance, can be used for generating images of different classes. Therefore $P(\mathbf{z}|c = i)$ is defined to be the probability of reconstructing an image of the ith class, when drawing a sample from the ith Gaussian. As before, this probability should be maximized and can be calculated as

$$P(\mathbf{x}|c = i) = \int_{\mathbf{z}} P(\mathbf{x}|\mathbf{z}) P(\mathbf{z}|c = i) d\mathbf{z}, \tag{10.36}$$

where $P(\mathbf{z}|c = i) = \mathcal{N}(\boldsymbol{\mu}_i, 1)$. Using a different Gaussian for each class allows for separately modeling the in-class variance for each class. This problem can be solved as before by optimizing the ELBO leading to

$$\mathcal{L}_{var}(\mathbf{x}) = [\log P(\mathbf{x}|\mathbf{z}) - \alpha_{KL}\mathcal{D}_{KL}(Q(\mathbf{z}|\mathbf{x})||P(\mathbf{z}|c = i))], \tag{10.37}$$

where α_{KL} is used to control the importance of minimizing the KL-divergence:

$$\begin{aligned} &\mathcal{D}_{KL}(Q(\mathbf{z}|\mathbf{x})||P(\mathbf{z}|c = i))] \\ &= \frac{1}{2}\Big(\mathrm{tr}(\Sigma_Q(\mathbf{x})) + (\mu_Q(\mathbf{x}) - \boldsymbol{\mu}_i)^T(\mu_Q(\mathbf{x}) - \boldsymbol{\mu}_i) \\ &\quad - m - \log\det\left(\Sigma_Q(\mathbf{x})\right)\Big), \end{aligned} \tag{10.38}$$

where m denotes the dimensionality of the latent space. Furthermore, it should be noted that $\log P(\mathbf{x}|\mathbf{z})$ is proportional to $-||f_{dec}(\mathbf{z}) - \mathbf{x}||_2^2$ (after appropriately scaling it by σ).

Even though (10.37) ensures that the in-class variance will be modeled, it does not provide any guarantee regarding the discrimination ability of the latent space. This issue can be trivially addressed by requiring each Gaussian to have a distance of at least ρ from the other ones. In this way, the different classes can be separated, while also encoding their generative factor. To this end, the centers of the Gaussian $\boldsymbol{\mu}_i$ are constrained to have a distance of at least ρ by using a simple margin-based loss:

$$\mathcal{L}_{sup} = \frac{1}{\rho} \sum_{i=1}^{N_C} \sum_{j=1, j \neq i}^{N_C} \max(0, \rho - ||\boldsymbol{\mu}_i - \boldsymbol{\mu}_j||_2^2)^2 \tag{10.39}$$

where ρ refers to that minimum distance between the used Gaussians and N_C is the number of them.

The final loss function that can be used for training DL models is formulated as

$$\mathcal{L} = \mathcal{L}_{var} + \mathcal{L}_{sup}, \tag{10.40}$$

which allows for capturing the in-class variations, by using a reconstruction-based objective, while also ensuring that different classes will be separated in the latent space. The hyperparameter α_{KL} enables us to control the trade-off between modeling the in-class variation and increasing the discrimination ability of the representation. Also, to ensure that the initial representation will be discriminative enough, the Gaussian centers are initialized using orthogonal initialization scaled by the desired distance ρ, which controls how far apart the Gaussians should be. Finally, note that after training the DL model $f_{\mathbf{W}}(\mathbf{x}) = \mu_Q(\mathbf{x})$, which is used to provide the representation for each input sample, the decoder $f_{dec}(\cdot)$ is no longer needed and can be discarded.

10.6 Concluding remarks

Recent advances in Deep Learning (DL), which inherently incorporate representation learning, have altered the landscape of representation learning tasks, such as image retrieval. Autoencoders are an especially useful tool for representation learning towards various tasks, for example, [5,2]. In this chapter, we presented how autoencoders work, while we also provided some representative works so as the DL researches to apply them for several tasks in practice. Subsequently, we focused on an extensively studied representation learning task, which is essential for Content Based Image Retrieval (CBIR). We first provided the basic concepts of CBIR utilizing DL techniques. Then we presented some representative representation learning methods that aim to learn more efficient representations toward the CBIR task, either in unsupervised or in supervised manner. Finally, we presented a representation learning method suitable for retrieving objects that belong to classes that were not seen during the training process. Even though deep representation learning is a rapidly progressing area with a wide spectrum of applications, we believe that this chapter provided the necessary knowledge for allowing the reader to follow the recent advances in deep representation learning.

References

[1] L. Deng, D. Yu, Deep learning: methods and applications, Foundations and Trends in Signal Processing 7 (3–4) (2014) 197–387.

[2] P. Nousi, A. Tefas, Self-supervised autoencoders for clustering and classification, Evolving Systems (2018) 1–14.

[3] R. Datta, J. Li, J.Z. Wang, Content-based image retrieval: approaches and trends of the new age, in: Proceedings of the 7th ACM SIGMM International Workshop on Multimedia Information Retrieval, ACM, 2005, pp. 253–262.

[4] A.W. Smeulders, M. Worring, S. Santini, A. Gupta, R. Jain, Content-based image retrieval at the end of the early years, IEEE Transactions on Pattern Analysis and Machine Intelligence 22 (12) (2000) 1349–1380.

[5] A. Krizhevsky, G.E. Hinton, Using very deep autoencoders for content-based image retrieval, in: ESANN, Vol. 1, Citeseer, 2011, p. 2.

[6] J. Wan, D. Wang, S.C.H. Hoi, P. Wu, J. Zhu, Y. Zhang, J. Li, Deep learning for content-based image retrieval: a comprehensive study, in: Proceedings of the ACM International Conference on Multimedia, ACM, 2014, pp. 157–166.

[7] N. Passalis, A. Iosifidis, M. Gabbouj, A. Tefas, Variance-preserving deep metric learning for content-based image retrieval, Pattern Recognition Letters 131 (2020) 8–14.

[8] M. Tzelepi, A. Tefas, Deep convolutional learning for content based image retrieval, Neurocomputing 275 (2018) 2467–2478.

[9] P. Vincent, H. Larochelle, Y. Bengio, P.-A. Manzagol, Extracting and composing robust features with denoising autoencoders, in: Proceedings of the 25th International Conference on Machine Learning, 2008, pp. 1096–1103.

[10] J. Xie, R. Girshick, A. Farhadi, Unsupervised deep embedding for clustering analysis, in: International Conference on Machine Learning, 2016, pp. 478–487.

[11] P. Nousi, A. Tefas, Deep learning algorithms for discriminant autoencoding, Neurocomputing 266 (2017) 325–335.

[12] P. Nousi, A. Tefas, Discriminatively trained autoencoders for fast and accurate face recognition, in: International Conference on Engineering Applications of Neural Networks, Springer, 2017, pp. 205–215.

[13] P. Wu, S.C. Hoi, H. Xia, P. Zhao, D. Wang, C. Miao, Online multimodal deep similarity learning with application to image retrieval, in: Proceedings of the 21st ACM International Conference on Multimedia, 2013, pp. 153–162.

[14] M.A. Carreira-Perpinán, R. Raziperchikolaei, Hashing with binary autoencoders, in: Proceedings of the IEEE Conference on Computer Vision and Pattern Recognition, 2015, pp. 557–566.

[15] J.T. Rolfe, Y. LeCun, Discriminative recurrent sparse auto-encoders, in: 1st International Conference on Learning Representations, ICLR 2013, 2013.

[16] F. Schroff, D. Kalenichenko, J. Philbin, Facenet: a unified embedding for face recognition and clustering, in: Proceedings of the IEEE Conference on Computer Vision and Pattern Recognition, 2015, pp. 815–823.

[17] M. Kan, S. Shan, H. Chang, X. Chen, Stacked progressive auto-encoders (spae) for face recognition across poses, in: Proceedings of the IEEE Conference on Computer Vision and Pattern Recognition, 2014, pp. 1883–1890.

[18] J. Snoek, R.P. Adams, H. Larochelle, Nonparametric guidance of autoencoder representations using label information, Journal of Machine Learning Research 13 (1) (2012) 2567–2588.

[19] S. Rifai, Y. Bengio, A. Courville, P. Vincent, M. Mirza, Disentangling factors of variation for facial expression recognition, in: European Conference on Computer Vision, Springer, 2012, pp. 808–822.

[20] S. Razakarivony, F. Jurie, Discriminative autoencoders for small targets detection, in: 2014 22nd International Conference on Pattern Recognition, IEEE, 2014, pp. 3528–3533.

[21] L. Deng, A tutorial survey of architectures, algorithms, and applications for deep learning, APSIPA Transactions on Signal and Information Processing 3 (2014).

[22] W. Liu, Z. Wang, X. Liu, N. Zeng, Y. Liu, F.E. Alsaadi, A survey of deep neural network architectures and their applications, Neurocomputing 234 (2017) 11–26.

[23] A. Krizhevsky, I. Sutskever, G.E. Hinton, Imagenet classification with deep convolutional neural networks, in: Advances in Neural Information Processing Systems, 2012, pp. 1097–1105.

[24] J. Donahue, Y. Jia, O. Vinyals, J. Hoffman, N. Zhang, E. Tzeng, T. Darrell, Decaf: a deep convolutional activation feature for generic visual recognition, in: ICML, 2014, pp. 647–655.

[25] A. Babenko, A. Slesarev, A. Chigorin, V. Lempitsky, Neural codes for image retrieval, in: European Conference on Computer Vision (ECCV), Springer, 2014, pp. 584–599.

[26] A. Babenko, V. Lempitsky, Aggregating local deep features for image retrieval, in: Proceedings of the IEEE International Conference on Computer Vision, 2015, pp. 1269–1277.

[27] Y. Kalantidis, C. Mellina, S. Osindero, Cross-dimensional weighting for aggregated deep convolutional features, in: European Conference on Computer Vision (ECCV) Workshops, Springer, 2015, pp. 685–701.

[28] G. Tolias, R. Sicre, H. Jégou, Particular object retrieval with integral max-pooling of cnn activations, CoRR, arXiv:1511.05879 [abs], 2015.

[29] A. Gordo, J. Almazán, J. Revaud, D. Larlus, Deep image retrieval: learning global representations for image search, in: European Conference on Computer Vision, Springer, 2016, pp. 241–257.

[30] M. Tzelepi, A. Tefas, Deep convolutional image retrieval: a general framework, Signal Processing Image Communication 63 (2018) 30–43.

[31] N. Kondylidis, M. Tzelepi, A. Tefas, Exploiting tf-idf in deep convolutional neural networks for content based image retrieval, Multimedia Tools and Applications 77 (23) (2018) 30729–30748.

[32] E.M. Voorhees, The cluster hypothesis revisited, in: Proceedings of the 8th Annual International ACM SIGIR Conference on Research and Development in Information Retrieval, ACM, 1985, pp. 188–196.

[33] Y. Rui, T.S. Huang, M. Ortega, S. Mehrotra, Relevance feedback: a power tool for interactive content-based image retrieval, IEEE Transactions on Circuits and Systems for Video Technology 8 (5) (1998) 644–655.

[34] R. Hadsell, S. Chopra, Y. LeCun, Dimensionality reduction by learning an invariant mapping, in: Proceedings of the IEEE Computer Society Conference on Computer Vision and Pattern Recognition, vol. 2, 2006, pp. 1735–1742.

[35] K.Q. Weinberger, L.K. Saul, Distance metric learning for large margin nearest neighbor classification, Journal of Machine Learning Research 10 (Feb) (2009) 207–244.
[36] K. Sohn, Improved deep metric learning with multi-class n-pair loss objective, in: Proceedings of the Advances in Neural Information Processing Systems, 2016, pp. 1857–1865.
[37] H. Oh Song, Y. Xiang, S. Jegelka, S. Savarese, Deep metric learning via lifted structured feature embedding, in: Proceedings of the IEEE Conference on Computer Vision and Pattern Recognition, 2016, pp. 4004–4012.
[38] X. Wang, X. Han, W. Huang, D. Dong, M.R. Scott, Multi-similarity loss with general pair weighting for deep metric learning, in: Proceedings of the IEEE Conference on Computer Vision and Pattern Recognition, 2019, pp. 5022–5030.
[39] E. Simo-Serra, E. Trulls, L. Ferraz, I. Kokkinos, P. Fua, F. Moreno-Noguer, Discriminative learning of deep convolutional feature point descriptors, in: Proceedings of the IEEE International Conference on Computer Vision, 2015, pp. 118–126.
[40] R. Müller, S. Kornblith, G. Hinton, Subclass distillation, arXiv preprint, arXiv:2002.03936, 2020.
[41] A. Sablayrolles, M. Douze, N. Usunier, H. Jégou, How should we evaluate supervised hashing?, in: Proceedings of the IEEE International Conference on Acoustics, Speech and Signal Processing, 2017, pp. 1732–1736.
[42] N. Passalis, A. Tefas, Entropy optimized feature-based bag-of-words representation for information retrieval, IEEE Transactions on Knowledge and Data Engineering 28 (7) (2016) 1664–1677.
[43] S.J. Pan, Q. Yang, A survey on transfer learning, IEEE Transactions on Knowledge and Data Engineering 10 (22) (2010) 1345–1359.
[44] N. Passalis, A. Tefas, Learning bag-of-embedded-words representations for textual information retrieval, Pattern Recognition 81 (2018) 254–267.
[45] D.P. Kingma, M. Welling, Auto-encoding variational Bayes, in: Proceedings of the International Conference on Learning Representations, 2014.
[46] G. Lu, X. Zhao, J. Yin, W. Yang, B. Li, Multi-task learning using variational auto-encoder for sentiment classification, Pattern Recognition Letters (2018).
[47] J. Lim, Y. Yoo, B. Heo, J.Y. Choi, Pose transforming network: learning to disentangle human posture in variational auto-encoded latent space, Pattern Recognition Letters 112 (2018) 91–97.

CHAPTER

Object detection and tracking

11

Kateryna Chumachenko[a], Moncef Gabbouj[a], and Alexandros Iosifidis[b]

[a]*Faculty of Information Technology and Communication Sciences, Tampere University, Tampere, Finland*

[b]*Department of Electrical and Computer Engineering, Aarhus University, Aarhus, Denmark*

11.1 Object detection

Object detection is one of the common computer vision tasks, finding its applications in various areas, including autonomous driving, robotics, and visual surveillance. By nature, object detection is similar to the task of image classification, but with an additional localization component, hence making it a significantly harder task. That is, rather than classifying the given image into a set of predefined classes as done in image classification, the goal of an object detection algorithm is to find all the objects of interest in the image and determine their locations. Moreover, the number of objects in the image is unknown beforehand, and objects might correspond to different classes, which need to be identified by the algorithm as well. Therefore, the detection of a single object in an image consists of the location of the object defined by a *bounding box*, the class it belongs to, and (optionally) a confidence score. Bounding box is a box enclosing the object as tightly as possible, and often defined by four parameters, namely, the center coordinate (alternatively, top-left coordinate), width, and height. Conventionally, bounding boxes are axis-aligned, but other definitions have been proposed as well, including rotated bounding boxes [1] and pixel-level segmentation masks [2]. An example of an object detection method output can be seen in Fig. 11.1.

Common challenges associated with the task of object detection include partial object occlusions, variations in object scale as well as in shape and aspect ratio. In addition, as the number of objects in the image is unknown beforehand, object detection models should be able to handle variable-size output. Multiple object detection methods have been proposed, ranging from sliding window based classifiers with handcrafted features to fast deep learning based detectors. The recent models achieving top performance are primarily based on convolutional neural networks, which are trained on large-scale data sets such as PASCAL VOC [3] or MS COCO [4], consisting of thousands annotated bounding box images.

Deep Learning for Robot Perception and Cognition. https://doi.org/10.1016/B978-0-32-385787-1.00016-6

FIGURE 11.1

Example of successful object detection.

11.1.1 Object detection essentials

Most object detectors operate by selecting a rich set of candidate regions from the image and classifying them into the ones containing objects of interest and the ones containing nonobject regions (commonly referred to as background). Essentially, such an approach allows to reduce an object detection task to a simpler task of image classification. The candidate regions to be classified can be simply obtained in a brute-force manner by sliding a $k \times k$ window over the image, or several such windows of different sizes to account for multiple potential object scales. However, such a brute-force approach results in enormous amount of candidate regions to be classified, hence it is impractical. As an alternative, various region proposal methods have emerged, predicting the regions within an image that are likely to contain objects [5–7]. Some methods rely on bounding box scoring. Following an initial sliding window based candidate extraction, each region is assigned an *objectness score* based on certain features, which may include saliency, edge density, size, and location, among others. The objectness score essentially denotes the likelihood of a candidate region to contain an object. The candidate regions with low objectness score are further filtered out from the candidate pool. One example of such methods is EdgeBoxes [6] that performs candidate region selection based on edges enclosed within it. Other methods follow a different approach, first segmenting an image into a set of regions and further iteratively combining them based on similarity. One example of such methods is selective search [5] that essentially applies a graph-based segmentation to the image and hierarchically groups the obtained regions by their similarity resulting in around 2000 region proposals per image. Methods which use information in the intermediate layers of a CNN have been used for reranking the regions provided by object proposal methods [7]. As we will see in the latter sections, algorithms such as Selective Search and EdgeBoxes have been completely superseded by deep learning models, so we will not discuss them in detail here. Following candidate region extrac-

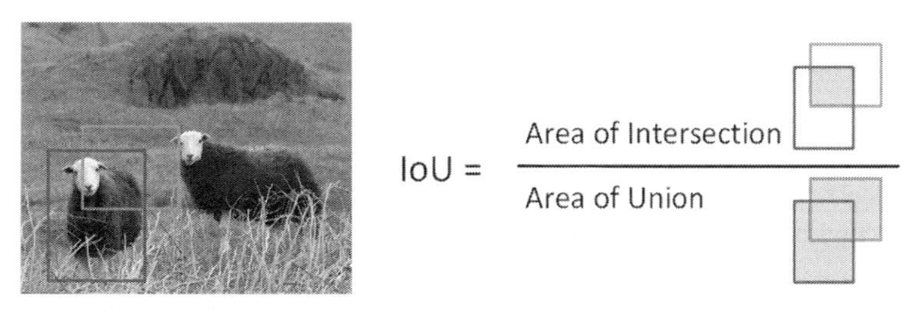

FIGURE 11.2

Intersection over union.

tion, classifiers are further applied to determine whether an object is present, taking as input features extracted from the corresponding regions. The features can range from handcrafted ones in earlier methods to those extracted by deep neural networks in more recent methods. Nowadays, most object detectors incorporate the two main steps of candidate region selection and region classification into a single model.

11.1.1.1 Nonmaximum suppression

Following the above-described approach, obtaining good detection results requires a good set of bounding box candidates. A candidate set covering as many locations in the image as possible ensures that no object will be missed given that classification is successful. On the other hand, a large set of bounding box candidates results in a high chance of multiple candidate regions lying close to the true bounding box and partially containing the same object. These regions are likely to be (correctly) classified as containing objects, and thus result in duplicate detections of the same objects. At the same, the bounding box candidates lying close to the true bounding box might localize the object equally well to the level that the human eye will not be able to distinguish the true bounding box. In order to suppress these duplicate detections and only select one bounding box per object as the final output, a process referred to as *nonmaximum suppression* is applied. Nonmaximum suppression takes a set of bounding boxes predicted by a detector as input, finds the ones corresponding to the same objects and selects only one of the bounding boxes per object, thus filtering out the rest of the bounding boxes.

For finding which bounding boxes correspond to the same object, it is necessary to define a metric that would reflect the spatial similarity of two bounding boxes. *Intersection over Union (IoU)* is a metric commonly used to this end. It is defined as the ratio between the area of intersection of the two bounding boxes and the area of their union. An example can be seen from Fig. 11.2. As can be seen in that figure, an IoU value of one indicates perfect alignment of two bounding boxes. For this reason, this metric is also used for evaluating the object localization performance of a detector by considering as one of the bounding boxes the ground truth box to be matched with a detection result. Generally, an IoU value above 0.5 is considered to be a good result.

Given a metric reflecting spatial similarity of two bounding boxes, we can proceed with defining nonmaximum suppression (NMS) algorithm. NMS performs bounding box selection according to the confidence score of each bounding box provided by the detector. At first, all predictions are sorted according to their confidence scores. The most confident bounding box is selected as the true one and added to the list of final outputs. IoU values between the selected box and all other bounding boxes in the list are calculated and predictions scoring above a given threshold θ (usually, $\theta = 0.5$) are considered as belonging to the same object, and hence removed from the list. The process continues iteratively until there are no unprocessed predictions left. Nonmaximum suppression is applied as a post-processing step to the vast majority of object detection methods.

11.1.1.2 Performance evaluation

For evaluating the performance of object detection methods, most works use the *Mean Average Precision (MAP)* metric. Recall that precision is defined as the ratio of true positive predictions to the sum of true positives and false positives, and recall is defined as the ratio of true positive predictions to all the positive samples in the data set. In the context of object detection, a prediction is said to be true positive if its IoU with the ground truth bounding box of the same class as the predicted one is larger than a given threshold θ. Depending on the adopted experimental protocol, θ can be defined as 0.5 or 0.75 (these results are referred to as $MAP_{0.5}$, $MAP_{0.75}$) or an average between MAP at multiple values of θ can be calculated. For example, COCO data set challenge calculates an average of MAP at IoUs between 0.5 to 0.95 with a step size 0.05 [4].

Average precision is defined by the area under the precision-recall curve, and the Mean average precision is simply the mean of this metric over all object classes. More specifically, for calculating the average precision, all detections in a data set within a certain class are sorted according to their confidence scores in descending order. Naturally, recall will increase as one traverses the list in the reverse way, with precision changing depending on the amount of true positive predictions in the list. That is to say, going down in the list of predictions, precision and recall are calculated for every rank, with recall increasing after every true positive detection. This way, the precision-recall curve is created, essentially showing how many of the predictions detected at a given confidence level were true positives versus how many of the true objects were found. An average precision is then defined as the area under the precision-recall curve. Different baselines consider slightly different ways of calculating this area, but these varieties are rather minor and their discussion is omitted in this chapter for the sake of simplicity. In a multiclass setting, an average between the average precisions of all classes is calculated constituting the final mean average precision score.

11.1.1.3 Traditional object detection methods

Early methods for object detection mainly applied classifiers image patch representations based on handcrafted features, and adopted the sliding window approach. Early image patch representations were mainly designed based on handcrafted features, including SIFT [8], SURF [9], HOG [10] features. One of the notable examples of an object detector that is performed in a sliding window manner based on handcrafted features is the **Viola Jones** [11] detector developed in 2001 primarily as a solution for the face detection problem. Although being a simple method with rather poor performance in unconstrained environments compared to state-of-the-art deep learning based methods, it still finds application in a variety of problems, primarily targeting embedded devices where fast inference is crucial. The Viola Jones algorithm utilizes a set of Haar-like features for identifying potential face regions from the image. The features are obtained by using a set of square and rectangular binary filters, which calculate the difference between the regions of the image windows where they are applied, hence allowing to identify edges and line features. Using these features, a set of classifiers are applied in a cascaded manner to obtain the final decision on the presence of face in candidate regions. Another notable detector represents image patches using **Histogram of Oriented Gradients (HoG)** features [10,12]. Given an image, it calculates two matrices representing the gradients of the image in the directions of the two image axes. An image is further divided into a cell grid and the gradient magnitudes are calculated at each pixel. Inside each cell, a histogram is created by assigning magnitudes to the bins of the corresponding orientation. Further, histograms are combined in a sliding window by vectorizing and concatenating them to obtain features, which are subsequently used by a classifier for determining the presence of an object.

Object detectors based on hand-crafted features generally provide benefits in terms of lightweight models, fast inference speed, and good interpretability. On the other hand, these models are only capable to successfully detect objects of specific scales, and successful detection of objects across multiple scales requires utilization of image pyramids, hence limiting the speed benefits of such methods. However, the deep learning revolution did not omit the field of object detection, leading to the appearance of a large number of high-performing deep learning based object detectors, and leaving the methods such as those described above far behind in terms of detection performance. Modern deep learning based object detection methods are primarily based on convolutional neural networks that allow to extract richer image patch representations and lead to better performance. Most of the methods follow the principle of candidate region selection followed by classification. In fact, early deep learning based object detectors combine CNNs-based feature extraction with conventional methods for candidate region selection and classifications. Therefore they are referred to as two-stage methods and will be discussed in Section 11.1.2. With the development of deep learning, more recent methods shifted toward end-to-end deep learning models. At the same time, one limitation of deep learning based object detection methods lies in their high computational cost, and a line of research has been devoted to parameter sharing and speeding up the inference time. The main breakthrough in this area was achieved by one-stage detectors discussed in Section 11.1.3. Further, some novel ideas, such as anchors and anchor-free detectors were introduced, as discussed in Section 11.1.4. In the next sections, we will discuss the most prominent and historically representative methods within each category.

11.1.2 Two-stage object detectors

Two-stage object detectors directly follow the principle of candidate region selection followed by classification, hence their name. Although previously mentioned handcrafted feature based methods also follow this two-stage approach, the term "two-stage detector" generally refers specifically to deep learning based detectors. As previously stated, the first stage (also referred to as region proposal stage) selects the regions potentially having an object present, and the second stage performs the foreground-background classification of the region.

One of the first deep learning methods following the two-stage approach is **R-CNN** [13]. In R-CNN, the selective search algorithm [5] extracts the region proposals from the image, resulting in a set of candidate bounding boxes. Having obtained the region proposals, a convolutional neural network (specifically, AlexNet [14]) is used as a feature extractor on top of each extracted proposal, producing a 4096-dimensional feature vector from each proposal. The CNN is pretrained on ImageNet data set for image classification, and further fine-tuned on warped region proposal windows obtained by the training set of the data set describing the object detection problem at hand. Specifically, for each image, proposals overlapping with the ground truth bounding box of a object with an IoU exceeding a threshold value (0.5 is commonly used) are considered as positive examples for the corresponding object class, while region proposals overlapping with a ground truth bounding box with an IoU lower than the threshold (0.5 is also commonly used) or depicting background regions are considered as negative examples for all object classes. Further, a One-versus-Rest multiclass classification scheme based linear support vector machine (binary) classifiers is used, classifying the features obtained from each of the regions to one of the predefined object categories. To improve the localization precision, an additional step of bounding box regression is performed. At this step, the regions classified as containing objects are regressed to the final bounding box coordinates via regularized least squares. The least squares weight matrix is obtained by training the regression of positive proposals (selected by high IoU with the ground truth bounding boxes) to true ground truth boxes.

A scheme showing the structure of R-CNN [13] as well as the two-stage detectors to be further discussed is shown in Fig. 11.3. As can be seen, one major limitation comes from the fact that each extracted region proposal is processed by CNN separately, resulting in significant computational overhead. This means that the CNN needs to be used multiple times, one per region proposal to detect objects for a single image. Besides, the model is not trainable in an end-to-end fashion, resulting in additional overhead. A remedy for these limitations was proposed in **Fast R-CNN** [15]. There, instead of extracting features of each proposal separately, the image is processed by the CNN as a whole (specifically, VGG16 [16] is used). Since the network is fully convolutional, and considering the receptive fields of the convolution filters at

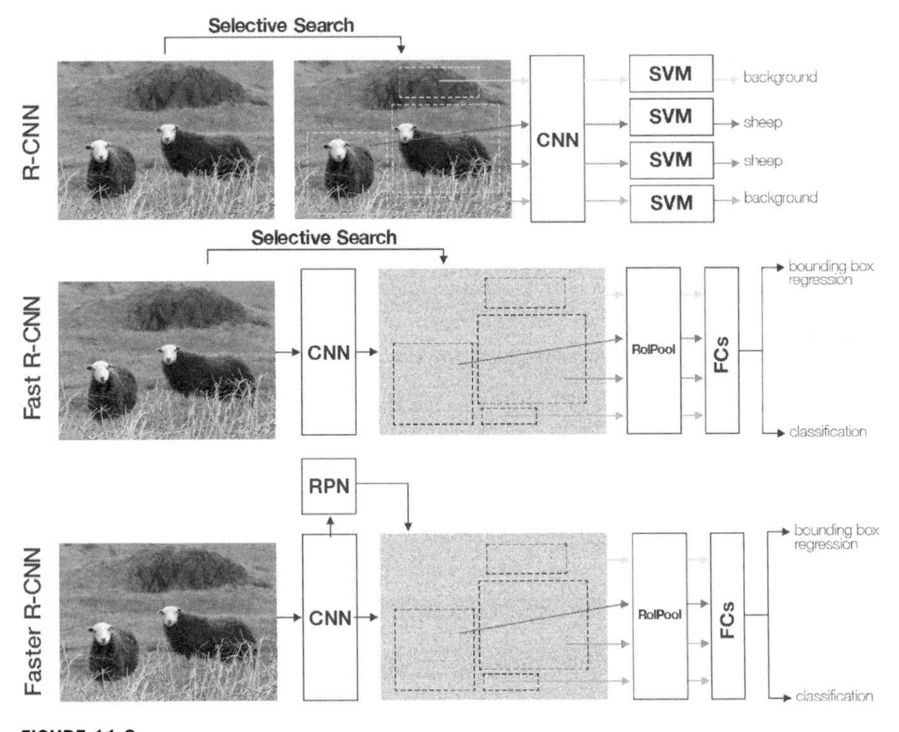

FIGURE 11.3

Two-stage Object Detection methods.

each layer of the network, correspondences between the region proposals in the input image and regions in the feature map can be determined. Thus the region proposal features can be extracted directly from the feature representation of the image at the last convolutional layer. That is to say, Fast R-CNN takes as input an image and a set of bounding box region proposals (obtained via selective search applied on the input image). The image is processed by several convolutional and pooling layers, after which a fixed-length feature vector is extracted from the feature map according to each region proposal obtained via selective search. Extraction is performed using a layer referred to as *Region of Interest (RoI) pooling* layer. The goal of RoI pooling layer is to obtain a fixed-size feature representation from variable-size input representation. An arbitrary sized $h \times w$ region is pooled to a fixed-size $H \times W$ feature map by dividing the region into an $H \times W$ grid of $\frac{h}{H} \times \frac{w}{W}$ and pooling the maximal value within each region. This way, all features obtained from each region proposal have the same size.

Having extracted the features of each region proposal via RoI pooling, each feature is fed into a sequence of fully-connected layers, further ending with two output layers. One of the output layers aims at regressing the bounding box coordinates producing 4 outputs $(t_x^k, t_y^k, t_w^k, t_h^k)$ for each of the object classes. These values correspond to the bounding box position and scale offsets relative to the region proposal. The second output layer predicts the softmax probability of object presence for each object class. The model is trained in a multitask manner with two losses corresponding to bounding box regression (L_{los}) and classification (L_{cls}). The regression loss is the smooth L1 loss, and classification loss is the log loss, defined as follows:

$$L(p, u, t^u, t^*) = L_{cls}(p, u) + \lambda e^u L_{loc}(t^u, t^*) \tag{11.1}$$

$$L_{cls}(p, u) = -\log p_u \tag{11.2}$$

$$L_{loc}(t^u, t^*) = \sum_{i \in \{x, y, w, h\}} smooth_{L_1}(t_i^u - t_i^*), \tag{11.3}$$

$$smooth_{L_1}(x) = \begin{cases} 0.5x^2, & \text{if } |x| < 1 \\ |x| - 0.5, & \text{otherwise,} \end{cases} \tag{11.4}$$

where u is the ground truth class label (with $u = 0$ being the background class), e^u is an indicator variable $e^u = 1$ if $u \geq 1$ and $e^u = 0$, otherwise, p is the predicted class probability distribution, t^u is the predicted bounding box for class u, and t^* is the ground truth bounding box.

As can be noticed, Fast R-CNN combines the feature extraction and classification stages into a single model, making a step toward an end-to-end trainable detector. During testing, the detection is conducted by running a selective search on the input image and performing a single forward pass with the trained network. This way, Fast R-CNN becomes approximately $9\times$ faster than R-CNN while achieving better detection performance. However, the creation of region proposals is still left out as a separate task. The third incarnation of R-CNN networks, **Faster R-CNN** [17], combines all of these steps into one model.

In Faster R-CNN, the functionality of the selective search algorithm producing region proposals is incorporated in the network architecture by introducing a separate module referred to as *Region Proposal Network (RPN)*. RPN is applied on top of the feature representation obtained from the CNN producing a set of region proposals, each accompanied with an objectness score. RPN is applied in a sliding window, taking as input a $n \times n$ patch of the feature map, and applying a convolutional layer, followed by two 1×1 convolutional layers for bounding box regression and classification. Note that these are not yet the final predictions of object classes and bounding boxes, but predictions of regions where objects of any class might be present.

Instead of directly predicting the final bounding box coordinates at each sliding window location, the model predicts the offsets relative to a set of k reference boxes, referred to as *anchor boxes*. Each anchor is essentially a bounding box shape defined by a scale and aspect ratio, and designed in a way that would incorporate some prior knowledge of the common object shapes in the training set. An example is depicted in Fig. 11.4. Given a set of k anchor boxes (in practice, Faster R-CNN uses $k = 9$), each of them is centered at each sliding window location of the image, and predictions are done relative to each anchor box at each position. That is, instead of predicting one offset per sliding window location, k distinct offsets relative to distinct

FIGURE 11.4

Anchor boxes are the reference boxes with predefined scale and aspect ratio and evaluated on different locations in the image (each sliding window location in Faster R-CNN, or each grid cell in YOLOv2/YOLOv3).

anchors are predicted. Following this, the regression layer produces $4k$ outputs, and classification layer produces $2k$ outputs. The use of anchor boxes helps incorporate prior knowledge on common shapes of objects, achieve scale invariance, as well as simplify the learning objective by providing a reference to the model. As we will see in the following sections, anchor boxes became an integral part of many object detection methods.

RPN is trained in a multitask manner, with a loss comprising of classification and regression losses:

$$L(p_i, t_i) = \frac{1}{N_{cls}} \sum_i L_{cls}(p_i, p_i^*) + \lambda \frac{1}{N_{reg}} \sum_i p_i^* L_{reg}(t_i, t_i^*) \tag{11.5}$$

where i denotes the anchor index, t_i and t_i^* denote the coordinates of the predicted and ground truth bounding box, respectively, and p_i and p_i^* denote the predicted and ground truth probability of an anchor. N_{reg} and N_{cls} are normalization parameters defined by the number of anchor locations and samples in the mini-batch, respectively. Positive and negative anchors are determined by assigning positive or negative binary labels to the anchor boxes depending on their IoU with the ground truth. The regression and classification losses are similar to those of fast RCNN.

So far, we described the RPN module of Faster R-CNN, which can produce region proposals given an image. In order to construct a full detection model, Fast R-CNN can be employed for performing classification and bounding box regression based on regions proposed by RPN. Both networks rely on a CNN as a feature extractor. For efficient and fast inference, it is desired that these networks share the parameters. However, training of the detection head relies on region proposals, and fine-tuning of it would break the parameter sharing property. As can be seen, the model cannot be trained end-to-end. Instead, training is done iteratively, generally following a 4-step process:

1. RPN is trained (initialized from ImageNet) and the obtained proposals are used to train Fast R-CNN (initialized from ImageNet).
2. The obtained backbone is then used to reinitialize RPN.
3. The weights of shared layers are usually frozen at this point, and only RPN layers are updated.
4. The detection head layers are fine-tuned using the same shared backbone weights and the new RPN proposals.

Faster R-CNN was able to combine the whole object detection pipeline within a single model. However, an end-to-end trainable model is still not achieved. Nevertheless, the accuracy and speed obtained by Faster R-CNN made it a particularly notorious object detector serving as a base for many subsequent methods. Examples include feature pyramid networks [18] that perform detection at multiple scales and mask R-CNN [2] that improves detection performance by jointly training for bounding box prediction and semantic segmentation.

11.1.3 One-stage detectors

Faster R-CNN performs detection in two stages, the first being the prediction of region proposals, and the second the classification of these regions into object categories along with a regression to more precise bounding boxes. A natural desired step for achieving an end-to-end trainable model is the combination of these two stages into one, that is, rather than first finding "object regions" and further classifying them, we would like to directly predict the regions accompanied with class labels. The methods following this approach are generally referred to as one-stage or single-stage detectors, taking an image as input and directly regressing it to the output bounding boxes of the objects depicted in it. Due to this combination of the two stages, these methods generally result in significantly higher speed.

The first attempts to one-stage detection include **Single Shot Detector (SSD)** [19] and **You Only Look Once (YOLO)** [20], which are rather similar in their principle.

YOLO utilizes a 24-layer convolutional network for feature extraction, followed by 2 fully connected layers. The first 20 convolutional layers are pretrained on ImageNet for classification, and after adding the remaining layers, the model is further trained for object detection. YOLO splits an image into an $S \times S$ grid cell (specifically, 7×7) and predicts B bounding boxes along with their confidence scores, as well as a single class probability distribution for each grid cell. Each bounding box is parameterized by the central point of the box relative to the grid cell, and the width and height relative to the whole image. The confidence score of a bounding box is formally defined as the IoU between the predicted bounding box and the corresponding ground truth box. As a result, the model produces an output of size $S \times S \times (B * 5 + C)$. Note that unlike Faster R-CNN that predicts offsets from anchor boxes, YOLO predicts the final bounding box coordinates directly.

YOLO is trained with the following sum of squared error based loss function jointly optimizing for localization and classification:

$$L = \lambda_{coord} \sum_{i=0}^{S^2} \sum_{j=0}^{B} e_{ij}^{obj} [(x_i - x_i^*)^2 + (y_i - y_i^*)^2]$$

$$+ \lambda_{coord} \sum_{i=0}^{S^2} \sum_{j=0}^{B} e_{ij}^{obj} [(\sqrt{w_i} - \sqrt{w_i^*})^2 + (\sqrt{h_i} - \sqrt{h_i^*})^2]$$

$$+ \sum_{i=0}^{S^2} \sum_{j=0}^{B} e_{ij}^{obj} (g_i - g_i^*)^2$$

$$+ \lambda_{noobj} \sum_{i=0}^{S^2} \sum_{j=0}^{B} e_{ij}^{noobj} (g_i - g_i^*)^2 + \sum_{i=0}^{S^2} e_i^{obj} \sum_{c} (p_i(c) - p_i^*(c))^2, \tag{11.6}$$

where g_i corresponds to the confidence score of the bounding box at the corresponding location, and $p_i(c)$ is the probability of class c at the corresponding location. Here, $e_i^{obj} = 1$ if an object appears in cell i and 0 otherwise, and $e_{ij}^{obj} = 1$ if the jth predictor is responsible for that prediction. x and y denote the coordinates of the bounding box center relative to the grid cell, w and h denote the width and height of the corresponding predicted box. λ_{noobj} and λ_{coord} are scaling factors controlling the loss from confidence predictions of bounding boxes not containing objects, and from coordinate predictions, respectively. These scaling factors are introduced as a way to account for imbalance between bounding boxes containing and not containing objects by downscaling the loss of the latter ones, as well as to put more importance on localization rather than classification. In practice, values of $\lambda_{coord} = 5$ and $\lambda_{noobj} = 0.5$ are used.

In turn, **SSD** utilizes ImageNet pretrained VGG16 network as a backbone, on top of which only convolutional layers are added, with their size decreasing with depth. This allows to perform multiscale detection by utilizing separate predictors on top of several feature scales (i.e., convolutional layers at different depths). Prediction is done relative to sets of anchor boxes (defined separately for each feature scale), applied to each feature map location. Therefore, each predictor is represented by a $3 \times 3 \times (c + 4)k$ convolutional layer, where k denotes the number of anchor boxes, and $c + 4$ corresponds to a class distribution along with 4 spatial parameters that are predicted for each anchor at each feature scale. Training is done with multitask loss, where localization is optimized with a smooth L1 loss similar to Faster R-CNN, and classification is optimized with a softmax loss. As can be seen, the main ideas behind SSD and YOLO are rather similar in their one-shot operation, while major differences between them lie in the use of anchor boxes and prediction over multiple scales in SSD, as well as prediction performed with convolutional kernels rather than fully-connected layers. As a result, SSD is able to achieve better speed and performance as compared to YOLO.

Nevertheless, YOLO method has been subsequently improved a number of times, resulting in multiple "versions" of this algorithm (with only the first 3 developed by the original authors of YOLO). The discussion of detailed architectural and training routine changes between them is omitted here, and we only consider some of the essential ideas. **YOLOv2** [21] steps back from direct prediction of bounding box coordinates and, following ideas of Faster R-CNN and SSD, predicts offsets relative to the anchor boxes instead, hence simplifying the learning task. This also helps to resolve the issue of one grid cell containing multiple objects, as class probability distribution prediction is now changed to "per-anchor" rather than "per-grid cell." Anchors are determined by applying K-Means based clustering on the training set with the distance between the box and the cluster centroid defined as $d(b, c) = 1 - IoU(b, c)$. Other modifications include certain architectural changes as well as utilization of lighter (but as powerful) backbone classification network consisting of 19 convolutional layers, hence speeding up the prediction. That is, YOLOv2 is able to achieve improved performance at a higher speed, outperforming Faster R-CNN on both measures.

YOLOv3 [22] incorporates several improvements that further push the performance of the algorithm. The most important novelty of YOLOv3 is detection across multiple scales that is achieved by adding detection heads consisting of 1×1 detection kernels with a number of channels matching the desired output dimensionality (i.e., $B(5 + C)$) to three different locations in the network. Notably, this idea was already proven to be useful in SSD. In addition, the 19-layer backbone of YOLOv2 is replaced by a deeper 53-layer network, resulting in an additional overhead but a richer feature representation leading to improved performance. In addition, instead of predicting the class of each bounding box via softmax, multiple logistic classifiers are used and trained with binary cross-entropy loss.

Although providing significant speed improvements, one-stage detectors such as YOLO(v1) and SSD still fail to match the state-of-the-art performance of improved two-stage detectors in terms of detection performance. One reason for this is the huge class imbalance between easy-to-classify background candidate regions and true objects. Class imbalance results in inefficient training as the majority of locations are easy negatives providing no useful information to the model, while overwhelming the training. In two-stage detectors this problem is not as significant, as region proposals narrow down the candidate object locations and filter a significant amount of background locations. In contrast, one-stage detectors evaluate a significantly larger set of candidate object locations defined by a grid and/or anchor boxes, as in the cases of SSD and YOLO. This problem is often addressed by incorporating *hard negative mining* during training samples selection process, that is, rather than selecting all of the negative examples, only the ones resulting in highest confidence loss are selected. This approach, however, does not always eliminate the problem. As a remedy, an object detector referred to as **RetinaNet** [23] adopted a novel loss function that puts more attention on hard-to-classify examples, while attending less to the easy examples. The loss function is referred to as **focal loss** and is derived from the cross-entropy loss. Here, we consider the binary case, while extension to multiple categories is straightforward:

$$FL(p_t) = -\alpha_t(1 - p_t)^\gamma \log(p_t), \tag{11.7}$$

$$p_t = \begin{cases} p, & \text{if } y = 1 \\ 1 - p, & \text{otherwise,} \end{cases} \tag{11.8}$$

where y is the ground truth label, p is the predicted probability for $y = 1$, α_t is a weighting factor equal to α when $y = 1$ and $1 - \alpha$, otherwise, for a user-defined parameter $0 \leq \alpha \leq 1$. α_t is responsible for balancing the positive/negative examples, and γ is a focusing parameter responsible for weighting the hard/easy examples, $\gamma \geq 0$.

RetinaNet follows a common object detection architecture with an off-the-shelf network and two branches, with one for object classification, and another for bounding box regression. To this end, a *feature pyramid network* is used. FPNs are commonly used in object detection among other tasks as being efficient in creating multiscale representations of data, and FPN-based variants of Faster R-CNN have been proven to be useful [18]. By augmenting a standard CNN, FPN creates a top-down path with lateral connections between the layers resulting in several "levels" of representations with different scales from a single input image. In RetinaNet, similar to other detectors, localization is predicted relative to a set of anchor boxes. In particular, 9 anchors are used for each FPN level. Under this formulation and the novel focal loss, RetinaNet was able to achieve state-of-the-art performance in terms of object detection, outperforming the two-stage detectors while preserving the fast speed of one-stage object detectors. Focal Loss later became a widely-utilized loss function in many computer vision tasks.

As can be seen, SSD and RetinaNet explore the ideas of multiscale predictions and anchor boxes, and the more recent versions of YOLO converged to the ideas of anchor box utilization and multiscale predictions as well; these ideas became essential in many of the object detection frameworks. However, the selection of meaningful anchor boxes requires careful design choices and parameter tuning. SSD develops a selection process for anchor boxes at different feature scales, and YOLO determines them based on K-Means clustering with IoU-based distance metric. To step away from the necessity of making these design choices, some methods refrain from using anchor boxes and rely on different bounding box representations.

11.1.4 Anchor-free detectors

The later single-stage detectors primarily rely on incorporation of prior knowledge in the detection process in the form of anchor boxes. Not only this requires careful design choices during anchor box creation related to their size, aspect ratio, and amount, but also adds the necessity to rely on subsequent nonmaximum suppression to remove the excessive detections. In turn, nonmaximum suppression requires careful selection of threshold, hence adding overhead to the detection process. These facts led to emergence of methods not relying on anchor boxes, referred to as anchor-free detectors.

One notable anchor-free object detector is CornerNet [24], which instead of relying on common representation of bounding boxes by a central point and width-height, it parameterizes bounding boxes using top-left and bottom-right corner points pairs. The model predicts the corners directly, hence eliminating the need of anchor box creation. Note that this approach is different from direct prediction of bounding box coordinates, as corner points are predicted independently from each other and only matched to form a bounding box at a later stage.

The CornerNet model is based on hourglass network [25], on top of which two branches are added with one branch responsible for predicting a set of top-left corners, and the other one predicting bottom-right corners. Each branch predicts an $H \times W \times C$ heatmap of corner locations with C being the number of classes. During training, the true ground truth corner is considered as a positive example, and among the negative examples the predicted locations lying closer to the true corner are penalized less than the ones lying further. This is due to the fact that precise corner locations are often unnecessary, as the points lying close enough can still produce good bounding boxes. The amount of penalization is defined by fitting a Gaussian to the true corner point with $\sigma = \frac{r}{3}$, where r defines the radius around the true corner point that includes all the points resulting in a bounding box producing at least IoU of 0.3 with true ground truth bounding box. The training is subsequently done with a modified version of the focal loss.

In the vast majority of convolutional architectures, images are progressively downsampled during processing. When predicting the heatmaps and upsampling them back to obtain the estimated locations in the image coordinates, a certain loss of precision can occur, being especially significant for small objects. To account for this, in addition to heatmaps, each branch predicts a set of offsets for top-left and bottom-right points. An offset is defined as

$$\mathbf{o}_k = \left(\frac{x_k}{n} - \lfloor\frac{x_k}{n}\rfloor, \frac{y_k}{n} - \lfloor\frac{y_k}{n}\rfloor\right), \tag{11.9}$$

where x_k and y_k denote the coordinates of corner k, and $\lfloor . \rfloor$ denotes the floor operation. Training of offset predictor is done with smooth L1 loss.

Assuming that each image can contain multiple objects, after prediction of a number of corner points, they need to be matched with each other to produce bounding boxes. To achieve this, an embedding is predicted for each detected corner, and pairs of top-left bottom-right corners are formed by finding closest points in the embedding space. This is achieved by training to keep the embeddings of corners of the same object close to each other, while pushing them away from corners of other objects:

$$L_{pull} = \frac{1}{N}\sum_{k=1}^{N}\left[(e_{t_k} - e_k)^2 + (e_{b_k} - e_k)^2\right] \tag{11.10}$$

$$L_{push} = \frac{1}{N(N-1)} \sum_{k=1}^{N} \sum_{j=1, j \neq k}^{N} \max(0, 1 - |e_k - e_j|), \tag{11.11}$$

where e_{t_k} and e_{b_k} are the embeddings of top-left and bottom-right corners of object k, respectively, and e_k is their average.

Usually, bounding box corners do not lie directly on the object, but in some proximity of it, making their direct prediction difficult as no visual cues can show the presence of a bounding box corner on the image. CornerNet proposes a novel pooling layer that assists the prediction of corner points by incorporating prior information. It is safe to assume that the top-left corner lies somewhere on the top-left direction from the object, as well as bottom-right corner lies below on the right side. The corner pooling layer takes two feature maps as input (produced by the output of two hourglass modules in the network). For a top-left corner candidate point (i, j) on the feature map, corner pooling layer max pools the vectors between (i, j) and (i, H) from one feature map, and all the vectors between (i, j) and (W, j) for the other feature map. The pooled results are subsequently summed up. Similarly, in a bottom-right branch, pooling is done in the $(0, j)$ to (i, j) and $(i, 0)$ to (i, j), respectively. Corner pooling is applied directly to the output of hourglass network in each branch separately, after which a set of convolutional layers are applied and the result is added up via skip-connection with the network output. The three prediction branches corresponding to heatmaps, offsets, and embeddings are added on top.

A similar idea is used in the detector referred to as **CenterNet** [26], where objects are modeled as single points denoting the center of the corresponding bounding boxes. Since there is only one point responsible for the detection of an object, embedding-based matching of points belonging to the same objects similar to CornerNet is unnecessary. Otherwise, the keypoint predictor is trained similar to CornerNet. This means that, using the hourglass network, points are predicted based on a modified focal loss with penalization defined by fitting a Gaussian at the true ground truth point, and an offset prediction is trained with L1 loss. In order to obtain the bounding box from its central point, an additional branch is introduced, which is responsible for prediction of the bounding box width and height, and trained with L1 loss.

11.2 Object tracking

Object tracking is another fundamental task in computer vision and robotics. The goal of an object tracking algorithm is to locate the *arbitrary* object(s) of interest in a sequence of frames, based on the location of the object(s) in the first frame. Similar to object detection, the location of an object is defined by a bounding box (conventionally, axis-aligned). Unlike object detection, the algorithms are (generally) agnostic of the specific object classes, and are expected to be able to track various types of objects, as defined by the provided initial bounding box.

Generally, two separate classes of problems are distinguished within the field of object tracking, namely, single object tracking (sometimes referred to as visual object

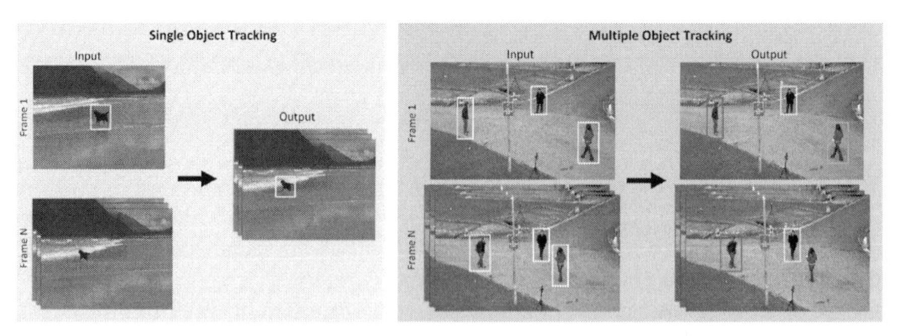

FIGURE 11.5

Single vs. multiple object tracking tasks. *Source:* MOT challenge dataset [28].

tracking) [27] and multiple object tracking [28]. In single object tracking, the objective is to determine the location of the object of interest in a sequence of frames, given its location in the initial frame of the sequence. In multiple object tracking, the difference lies not only in the presence of multiple objects to be tracked, but also in the problem formulation itself. In multiple object tracking, the locations of objects in all frames are assumed to be known, defined by a set of bounding boxes, but their association between each other in different frames is to be determined. In practical applications, for example, pedestrian tracking, these bounding boxes are generally obtained by applying an object detection algorithm. The task of the tracker therefore lies in determining associations of given bounding boxes with objects and finding their trajectories. The difference between single and multiple object tracking is shown in Fig. 11.5. In this section, we will consider several approaches for addressing the single object tracking and multiple object tracking problems.

11.2.1 Single object tracking

The problem of single object tracking is formulated as locating the object of interest (target), defined by a bounding box in the first frame of a video, in all subsequent frames. A bounding box is generally defined either as $b = [c_x, c_y, w, h]$, where (c_x, c_y) is a central coordinate of the bounding box, and w and h correspond to the width and height of the object, respectively; or as $b = [x_1, y_1, x_2, y_2]$, where (x_1, y_1) and (x_2, y_2) correspond to the top-left and bottom-right coordinates of the bounding box.

For a period of time, the conventional approach to single object tracking relied on training online discriminative foreground-background (target–nontarget) classifiers based on the information provided in the first frame, and subsequently updating them online. The classifier is applied to different regions of the frame (or certain visual features extracted from them), and the region classified as a target with the highest confidence is selected as the optimal target location at the current frame. This approach was further outperformed by correlation filter based trackers that learn a filter based on the target given in the first frame [29,30]. The filter is correlated with patches from the search region in the subsequent frames, and the patch corresponding to the highest response is selected as a new position of the target. Based on the features extracted from the new location, the filter is updated. The benefit of this approach lies in the fact that cross-correlation can be computed efficiently in Fourier domain using fast Fourier transform, resulting in higher tracking speed. With the evolution of deep learning models, correlation based trackers started to employ features extracted from deep networks in their filter learning process. At the same time, new trackers relying solely on deep learning, as well as a new family of trackers based on similarity learning with Siamese networks trained fully offline started to evolve. Moreover, some of the modern state-of-the-art tracking methods, such as ATOM [31] and DIMP [32], successfully combine online and offline training in their learning process.

Challenges generally associated with the task of single object tracking include variations in scale, shape, illumination transformation, occlusion, or other deformations. Many algorithms rely on online training, with the tracker being updated online with visual cues obtained from predicted locations of the target in previous frames. Therefore when visual appearance of the target changes significantly, or the object is occluded, trackers are likely to fail and subsequently "*drift*" as a result of error accumulation. Due to this fact, in recent years, a more clear segregation between short-term and long-term object tracking methods is emerging. Unlike short-term trackers, long-term tracking algorithms are expected to be able to detect their own failures and redetect lost objects even after they reappear in the scene. Generally, this is achieved by incorporating prediction of a confidence score to the location prediction, and applying stronger redetection models across the whole frame when the confidence is low, while continuing tracking with conventional short-term trackers otherwise.

11.2.1.1 Tracking with correlation filters

Correlation filters applied to tracking started to gain increased attention as an idea with the **Minimum Output Sum of Squared Error tracker (MOSSE)** tracker [29]. The general idea of correlation filter based trackers is similar to that of discriminative classifiers, that is, multiple candidate regions are evaluated at each frame, with the best one being selected as an optimal target location. Instead of performing foreground-background classification, the best candidate region is selected by correlating a filter learnt from the initial frame with all the candidate regions and selecting the one producing the highest response. The filter is subsequently updated online. However, unlike discriminative classifier based trackers, correlation filters have proven to be more efficient, utilizing only a fraction of the computational power compared to the most complex (at the time) trackers while achieving competitive performance. This efficiency is achieved by taking advantage of performing cross-correlation in the Fourier domain, where it is equivalent to performing elementwise multiplication. This way, using fast Fourier transform and its inverse transform, fast tracking is achieved.

MOSSE tracker [29] learns the filter from a set of images $\mathbf{X}_i$ and the corresponding ground truths $\mathbf{Y}_i$, with each ground truth defined such that it has a compact 2D

Gaussian shaped peak centered at the center of the object of interest in the training image (frame) $\mathbf{X}_i$. The initial filter is learned by sampling 8 training samples from the initial frame, with training samples obtained by applying affine transformations to the ground truth.

Taking advantage of the fact that cross-correlation is equivalent to elementwise multiplication in the Fourier domain, training of the filter $\mathbf{H}_i$ is performed as follows:

$$\hat{\mathbf{H}}_i^* = \frac{\hat{\mathbf{Y}}_i}{\hat{\mathbf{X}}_i}, \tag{11.12}$$

where the $\hat{\mathbf{H}}_i$, $\hat{\mathbf{Y}}_i$, and $\hat{\mathbf{X}}_i$ correspond to the Fourier domain representations of the corresponding filter, the ground truth, and the image. Hereafter, $\hat{\cdot}$ denotes the Fourier representation, and $\cdot^*$ corresponds to the complex conjugate. To find the optimal filter $\hat{\mathbf{H}}_i^*$ MOSSE minimizes the sum of squared errors between the ground truth and the predicted response in the Fourier domain as follows:

$$\min_{\hat{\mathbf{H}}^*} \sum_i |\hat{\mathbf{H}}_i \odot \hat{\mathbf{H}}^* - \hat{\mathbf{Y}}_i|^2, \tag{11.13}$$

where $\odot$ corresponds to elementwise multiplication. The closed-form solution for the optimal filter can be obtained as

$$\hat{\mathbf{H}}^* = \frac{\sum_i \hat{\mathbf{Y}}_i \odot \hat{\mathbf{X}}_i^*}{\sum_i \hat{\mathbf{X}}_i \odot \hat{\mathbf{X}}_i^*}. \tag{11.14}$$

Further, to account for appearance changes during tracking, online updates are performed as follows:

$$\hat{\mathbf{H}}^* = \frac{\hat{\mathbf{A}}_i}{\hat{\mathbf{B}}_i}, \tag{11.15}$$

$$\hat{\mathbf{A}}_i = \eta \hat{\mathbf{Y}}_i \odot \hat{\mathbf{X}}_i^* + (1-\eta)\hat{\mathbf{A}}_{i-1}, \tag{11.16}$$

$$\hat{\mathbf{B}}_i = \eta \hat{\mathbf{X}}_i \odot \hat{\mathbf{X}}_i^* + (1-\eta)\hat{\mathbf{B}}_{i-1}, \tag{11.17}$$

where η is the learning rate parameter. A value of $\eta = 0.125$ is used.

Although relying on rather simple objective and simple formulation, MOSSE resulted in the establishment of the area of correlation filtering based trackers, resulting in outstanding (at the time) performance and achieving up to $\times 20$ increased speed compared to competing methods.

Further, it was shown that the linear correlation filter solution in the Fourier domain is equivalent to applying linear regression to the circulant matrix of the target sample in the spatial domain, that is, a matrix containing all possible cyclic image shifts of the target. This insight explains the robustness of MOSSE tracker, but also provides the potential for further correlation filter based tracking extensions, including the use of nonlinear formulations. One notable algorithm proposing such a nonlinear approach is **KCF (Kernelized Correlation Filter)** [30]. There the solution is obtained in a nonlinear feature space exploiting the kernel trick. Recall that the kernelized regression is given by

$$\boldsymbol{\alpha} = (\mathbf{K} + \lambda \mathbf{I})^{-1}\mathbf{y}, \tag{11.18}$$

where $\mathbf{K}$ is the kernel matrix consisting of dot products between pairs of samples in the feature space, $\boldsymbol{\alpha}$ is the solution of the regression problem in the kernel space [33], $\mathbf{y}$ is the groundtruth, and $\mathbf{I}$ is an identity matrix. It can be proven that the kernel matrix $\mathbf{K}$ obtained from circulant data remains circulant for the majority of popular kernel functions, such as RBF and polynomial kernels [30]. Therefore a closed-form solution can subsequently be obtained as

$$\hat{\boldsymbol{\alpha}} = \frac{\hat{\mathbf{y}}}{\hat{\mathbf{k}}^{\mathbf{xx}} + \lambda}, \tag{11.19}$$

where $\mathbf{k}^{\mathbf{xx}}$ denotes the first row of a circulant matrix $\mathbf{K}$ and is also referred to as kernel autocorrelation. In a more general setting, kernel correlation between two vectors $\mathbf{x}$ and $\mathbf{x}'$ is defined as

$$\mathbf{k}_i^{\mathbf{xx}'} = (\mathbf{x}', P^{i-1}\mathbf{x}), \tag{11.20}$$

where P^{i-1} is a cyclic shift operator resulting in $i-1$ shifts of $\mathbf{x}$. That is, for each candidate patch, a response can be calculated by obtaining a kernel correlation vector $\mathbf{k}^{xz}$ between the target patch $\mathbf{x}$ and candidate patch $\mathbf{z}$, and subsequently performing correlation, efficiently computed in the Fourier domain:

$$\hat{\mathbf{f}}(\mathbf{z}) = \hat{\mathbf{k}}^{\mathbf{xz}} \odot \hat{\boldsymbol{\alpha}}. \tag{11.21}$$

Note that this way $\mathbf{f}(\mathbf{z})$ obtained by applying the Inverse DFT to $\hat{\mathbf{f}}(\mathbf{z})$ contains responses for all cyclic shifts of $\mathbf{z}$. Besides, solutions for efficient calculation of the kernel correlation for different kernel functions exist. As an example, consider a Gaussian kernel given as

$$\mathbf{k}^{\mathbf{xx}'} = \exp\left(-\frac{1}{\sigma^2}\left(||\mathbf{x}||^2 + ||\mathbf{x}'||^2 - 2\mathcal{F}^{-1}(\hat{\mathbf{x}}^* \odot \hat{\mathbf{x}}')\right)\right), \tag{11.22}$$

where $\mathcal{F}^{-1}$ denotes the Inverse DFT. Another benefit provided by such kernelized formulations based on circulant matrices is the ability to integrate more complex features with multiple channels (in practice, HOG features are generally used in KCF), as multiple channels can be simply summed up in the Fourier domain. That is, considering the Gaussian kernel formulation from Eq. (11.22), this is achieved by simply substituting $\hat{\mathbf{x}}^* \odot \hat{\mathbf{x}}'$ with $\sum_c \hat{\mathbf{x}}_c^* \odot \hat{\mathbf{x}}_c'$, where c denotes the channel. Similarly, a multichannel extension can be incorporated into the linear solution using a linear kernel.

Tracking by correlation filtering became an important paradigm in the object tracking field due its efficiency and good performance. The methods described above gave rise to many subsequent algorithms, including those using more complex representations than simply raw pixels or low-level handcrafted features. Benefits associated with more complex feature representations include more robustness to changes in illumination, occlusions, and other deformations. On the other hand, naturally, using more complex feature representations results in additional computational overhead when online training is performed.

11.2.1.2 Deep learning based tracking

With the evolution of deep learning and specifically convolutional neural networks, object trackers using deep feature representations started to merge. Previously described methods relying on correlation filters are trained predominantly online. Although this approach allows to account for real-time changes in visual appearance of the target, it limits the methods to using low-level handcrafted features. On the other hand, computationally intensive deep features introduce a speed limitation for online training of neural networks. Note that although multiple discriminative correlation filter based tracking algorithms using deep features have been proposed to date, their speed is still rather limited compared to real-time requirements in many applications.

These facts led to the introduction of offline pretrained deep learning methods. A tracker can be asked to perform tracking on arbitrary objects, thus eliminating the possibility of pretraining with the task of detecting specific targets. On the other hand, pretraining on existing videos can serve the goal of learning scale-, light-, and rotation-invariant feature representations of arbitrary objects that are agnostic of specific object classes. The offline pretraining objective is generally formulated as a similarity learning problem as discussed in Section 11.2.1.3 or as a regression problem on image patches extracted from previous frames [29]. Alternatively, some methods take advantage of both offline pretraining for robust feature extraction and online updates intended to account for the presence of distractors during testing [34,31,32].

Tracking with offline pretraining

One example of an early deep learning based object tracker trained fully offline is the **GOTURN** (Generic Object Tracking Using Regression Networks) tracker [35]. Due to extensive fully offline training, it is able to achieve faster speed compared to existing trackers at the time while ensuring robustness. GOTURN relies on the assumption that the target does not move significantly between adjacent frames, and thus its location at frame t can be determined by analyzing the corresponding region of frame $t-1$. That is, GOTURN operates on pairs of image crops extracted from the current and preceding frames, and outputs a directly predicted bounding box location of the target in the current frame.

The crop in the initial frame of the video is extracted so that the center of the target (c_x, c_y) corresponds to the center of the crop, and the crop has a width of w and h, enclosing the target fully, while including also contextual information (i.e., background). At each subsequent frame $t+1$, a crop centered around the central point $(c_{x,t}, c_{y,t})$ of the previous frame t is extracted for further analysis, assuming that the

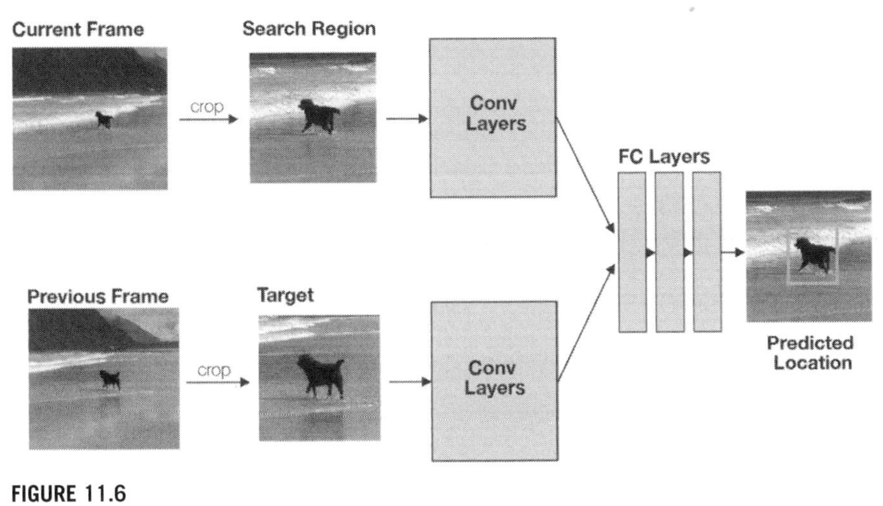

FIGURE 11.6

Structure of GOTURN tracker.

object did not move significantly and is only slightly shifted within the extracted crop. The crop at frame $t+1$ has a width of $k_1 w$ and height $k_2 h$, where the parameters k_1 and k_2 control the size of the search region in the frame (generally values of $k_1 = k_2 = 2$ are used). The two input crops are then introduced to a neural network that directly regresses the input to the bounding box of a target object in the frame, which is used to extract the crop of the subsequent frame, and the process continues for all frames in the video. The method utilizes CaffeNet architecture [36] for feature extraction, and the two crops obtained at frames $t-1$ and t are introduced to a sequence of five convolutional layers, after which they are concatenated into a single vector, which is processed with 3 fully connected layers. The output of the network is a dense layer with four neurons, representing a bounding box enclosing the target object at frame t. Training is performed offline on pairs of crops obtained from the video and image data and the L_1 loss is minimized between the predicted bounding box and the ground truth bounding box. The structure of GOTURN is shown in Fig. 11.6.

Tracking with online training

Being trained fully offline, GOTURN achieved unprecedented speed compared to the trackers developed at the time. On the other hand, when applied to environments with rapid movement and significant visual appearance changes of the objects, GOTURN does not result in enough robustness to account for these changes, as no online updates to the model are performed. Another notable tracking algorithm relying on deep features is **MDNet** (Multidomain Network) [34], that incorporates both offline and online training in the tracking procedure. The model takes advantage of multidomain learning, where the layers in the network are segregated into the domain-agnostic ones trained fully offline and the domain-specific ones updated online.

MDNet performs tracking by foreground-background classification of candidate regions in video sequences. Although the notion of background and foreground can vary significantly from sequence to sequence, some desired feature properties are shared, such as invariance to illumination, scale invariance, and tolerance to motion blur. To learn filters which are capable of producing such domain-independent visual features, MDNet takes advantage of multidomain learning, and multiple domains (i.e., multiple video sequences) are considered simultaneously during offline pretraining. The model employs a network architecture formed by three convolutional and two fully-connected layers, whose parameters are shared across domains. On top of these layers, domain-specific classification layers are added, that is, in a scenario involving K domains, the network has K domain-specific fully-connected binary classification layers trained with cross-entropy loss. During training, the mini-batches are constructed separately for each domain, and the mini-batch corresponding to kth domain updates only the kth domain-specific layers along with the shared layers. Hard negative mining is performed during mini-batch construction to improve robustness of the model by exposing it to a significant number of distractors.

During testing, an optimal target location is determined by performing foreground-background classification of 256 candidate locations sampled around the target location in the previous frame with different translation and scale parameters, and selecting the candidate with the highest response. For improving the localization precision, bounding box regression is applied on top of the proposed target location, with the bounding box regression model trained similar to R-CNN described in Section 11.1.2. The last domain-specific layer is initialized randomly, during testing, for each new sequence. Online updates are performed only on three fully-connected layers of the network, while convolutional layers are frozen. Online updates are performed with 50 positive and 200 negative training samples extracted at each frame (with an exception for the first frame, from which 500 positive and 5000 negative samples are selected). Two types of online updates are performed, that is, long-term updates intended to improve robustness, and short-term updates intended to ensure adaptiveness of the model. Long-term updates are performed using positive samples collected over a longer time span (100 frames) and negative samples collected over a short-term period (20 frames). Positive samples correspond to predicted locations of the object, and negative samples are sampled from the background. Short-term updates are performed every 20 frames, or whenever the tracker fails, using the positive and negative samples collected over this period.

Using the multidomain formulation, MDNet was able to ensure robustness of the learnt classifier and achieve state-of-the-art results in VOC 2015 challenge. At the same time, the proposed segregation of layers into shared and domain-specific ones, as well as the online updates of only a small portion of the network, allowed to reduce the processing time generally needed by other online trackers. However, the speed of this tracker is still rather far from real-time.

11.2.1.3 Tracking by similarity learning with Siamese networks

Considering the GOTURN tracker described previously, one of its limitations is its underlying assumption of low object movement between frames, which can lead to loss of objects in rapidly changing environments. In such situations, instead of searching for the object locally, we would like to be able to evaluate multiple candidate regions and ensure a global search of the object in the current frame. However, directly evaluating multiple crops results in significant overhead to the model. A remedy was proposed with a family of methods performing similarity learning based on Siamese networks [37–40]. Rather than learning online a foreground–background classifier, the goal of these methods is to learn a similarity function offline that, given two images, produces a high similarity score if they depict the same object and a low similarity score, otherwise.

One of the first methods following this approach is **SiamFC** [37] that proposed to use Siamese architectures for similarity learning. Consider a patch of an image z depicting the target object defined by the user in the first frame or predicted by the tracker at frame t. This patch is hereafter referred to as an exemplar or template patch. Considering a candidate patch x extracted from a subsequent frame $t+1$, the goal of the model is to learn a function $f(z,x)$ that is able to compare two images and return a high value if they are similar. Following this, object tracking in a video can be achieved by comparing the similarity of image patches extracted at multiple locations of an image to the exemplar image depicting the target, and selecting the one resulting in the highest value.

SiamFC relies on a fully-convolutional network φ allowing efficient evaluation of all candidate patches in frame t. Since the architecture is fully-convolutional, the necessity of providing multiple candidate patches to the network separately is removed and a much larger search area can be used as input to the network, producing a heatmap of similarity values at different candidate locations. The similarity values are obtained via a cross-correlation layer given convolutional embeddings $\varphi(\cdot)$ of the feature maps extracted from the exemplar image x and search image z:

$$f(z,x) = \varphi(z) * \varphi(x) + e_b, \tag{11.23}$$

where e_b denotes a signal which takes value $b \in R$ in every location.

The network architecture is based on the convolutional part of AlexNet, and is trained on exemplar-search image pairs with logistic loss. During tracking, no online fine-tuning is performed, and the target location is determined by the highest response at a given search image area. The search area has a fixed size and is selected at each frame by using the object location in the previous frame as a central point. Location estimation is done at multiple scales by using features of several scaled images. The structure of SiamFC is illustrated in Fig. 11.7.

SiamFC gave birth to a family of tracking algorithms based on Siamese networks. Its next instantiation, **SiamRPN** [38], significantly improved the performance of SiamFC by employing a region proposal network and formulating tracking as local one-shot detection. That is, after obtaining feature representations $\varphi(x)$ and

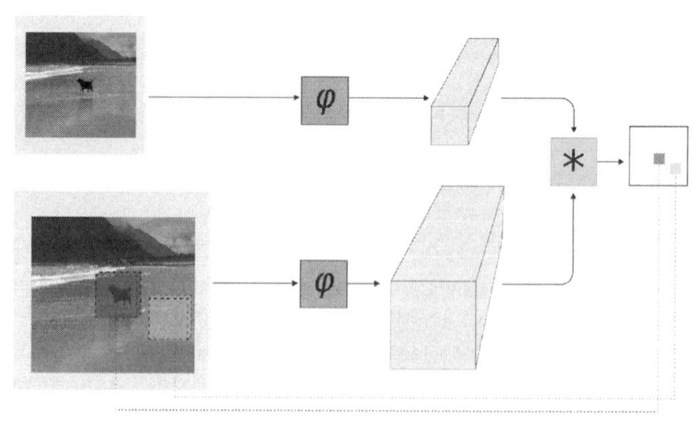

FIGURE 11.7

Structure of SiamFC tracker.

$\varphi(z)$ from backbone network branches, instead of computing their cross-correlation via a cross-correlation layer and generating a heatmap of scores similar to SiamFC, SiamRPN employs an end-to-end trainable Faster R-CNN like region proposal network to directly generate bounding box predictions. RPN consists of classification and regression branches, each taking as input both the template and the search area feature representations. These feature representations are processed with a sequence of convolutional layers followed by cross-correlation (in each branch separately) and output layers corresponding to foreground-background classification and bounding box regression. The network is trained offline on pairs of template-search area images in an end-to-end fashion.

During testing, the template branch is dropped and only the search area branch is utilized. This is achieved by passing the template patch from the initial frame through the template branch of the network, and utilizing the obtained region proposal network kernel for the search area branch in later frames. The obtained region proposals are reranked according to their displacement from the previous frame, penalizing large displacements, and the final bounding box is selected via nonmaximum suppression. This way, SiamRPN is able to achieve roughly twice the speed of SiamFC, as no multiscale testing is needed, while resulting in improved performance.

A number of tracking algorithms tried to improve performance by learning richer feature representations with deeper backbone networks. However, the latter did not meet the goal. Furthermore, it was proven that previous deep trackers were unable to improve the performance when resorting to deeper architectures due to breach of strict translation invariance in deep networks. To overcome this restriction, **SiamRPN++** [39] introduced a spatial aware sampling strategy during training that relies on shifting the target within a certain range sampled from a uniform distribution in the search image, and thus preventing the model from learning a strong center bias. This way it was able to successfully incorporate a modified ResNet architecture in the model and learn a better feature representation, hence improving the results by up to 2% on accuracy metric.

In addition, the use of a deeper network allows to extract multilevel features for multiscale prediction. Specifically, in SiamRPN++, features from the last three residual blocks of a modified ResNet architecture are extracted. These features are fed into RPN separately, and a weighted fusion layer is employed for their combination. Another modification introduced in SiamRPN++ is the use of depthwise cross-correlation instead of grouped cross-correlation of SiamRPN, resulting in a significant reduction in the number of parameters in the RPN part of the architecture, while stabilizing the optimization procedure.

Another notable tracker from a family of Siamese networks is **SiamMask** [40] that simultaneously addresses the tasks of object tracking and visual object segmentation. Similar to its predecessors, SiamMask is trained fully offline and is initialized solely from the bounding box in the initial frame. The goal is, however, extended to the prediction of a pixelwise segmentation mask. The multitask nature of the method forces it to better learn the object granularity thus improving the tracking performance as well. To achieve pixelwise object segmentation, SiamMask extends previous architectures with an additional segmentation (mask) branch and loss. Specifically, two architectures are proposed that correspond to SiamFC and SiamRPN architectures augmented with the segmentation branch and loss, with modified ResNet-50 as a backbone. Similar to SiamFC and SiamRPN, training is done on pairs of exemplar and search image patches. During training, ground truth pixel-level segmentation mask is provided for each target candidate on the grid enclosing the object. Depending on the architecture (two- or three-branch), the total loss is calculated as a weighted sum of segmentation and similarity losses in the two-branch variant, or a weighted sum of segmentation, bounding box regression, and classification losses in the three-branch variant. Segmentation loss is the binary logistic regression loss over all candidates associated with the true target (i.e., it is not considered for negative candidates). During testing, the mask is predicted per candidate region, and the output mask is selected by the region attaining the maximum similarity score. In the two-branch variant, the final bounding box is selected by axis-aligning a rectangle containing the object mask, while in the three-branch variant, it is selected by the output of the RPN branch.

11.2.2 Multiple object tracking

Another problem defined within the field of object tracking is referred to as Multiple Object Tracking (MOT). In this formulation, it is (generally) assumed that the locations of objects of interest in the form of bounding boxes are known within the video; however, the bounding boxes of different frames are not associated with their corresponding objects. The task is, hence, to associate the bounding boxes with their corresponding objects through the video, and thus determine objects' trajectories. The methods can be separated into online and offline ones, where online methods perform tracking frame-by-frame without using future information, while offline trackers operate on batches of data, assuming that the whole video sequence is available at any step.

Besides suffering from the same challenges as single object tracking, such as illumination, scale, and shape changes, multiple object tracking is generally associated with crowded environments, hence making objects occluded not only by distractors, but also by other objects being tracked. This inevitably results in identity switches and losses of objects. Such challenges require multiple object tracking to not only be able to track the objects present in the initial frame, but also to determine when new identities should be initiated and old ones should be discarded.

The majority of the methods in the field of multiple object tracking rely on a two-stage approach, where the bounding boxes are assumed to be known, for example, generated with some object detector. This approach is known as tracking-by-detection and remains the primary approach to date. As the detection process is separated from tracking, tracking essentially reduces to a data association problem defined over given bounding boxes. Association is generally performed by calculating some cost/similarity function defined over the feature representations of pairs of tracklets from frames t and $t-1$, and associating the pairs corresponding to the same objects. Some methods focus on learning better cost functions/feature representations for cost calculation, while others focus on learning how to perform better associations. Feature representations are generally based on deep features extracted with a CNN or various motion features, such as tracklet velocity, acceleration, etc. In terms of association algorithms, online tracking methods generally rely on simple association methods, such as Hungarian algorithm [41], while offline methods utilize more complex graph optimization based approaches.

Early works in multiple object tracking largely focused on motion/location information for tracking, and payed less attention to visual features of tracklets. Examples include the IOU Tracker [42] that defines the similarity between tracklets and detections by intersection over union, and performs association by selecting detections resulting in maximal IoU for each tracklet. Another example is the Simple Online Realtime Tracker (SORT) [43] that describes each tracklet by the spatial characteristics of the bounding box, and their velocity, defining the similarity between tracklets and new detections accordingly.

Although these algorithms do not provide high levels of robustness due to the fact that they disregard visual information, they provide good results in environments without many occlusions and result in significantly high tracking speed.

11.2.2.1 Tracking with deep visual representations

Once again, the deep learning revolution did not spare the field of multiple object tracking and a large amount of tracking methods utilize deep learning in various ways. The primary deep learning task within the field of MOT has notably been the learning of deep visual feature representations for tracklets. Generally, a visual model is trained with a re-identification objective on a public data set. The goal of a reidentification model is to learn to reidentify objects of the same class (e.g., people) across different camera views or the same camera but in different time instants. Generally, this is achieved by comparing pairs of images and producing their similarity score or classification label inferring whether or not they depict the same identity. Therefore model learnt with such objective results in feature representations that are able to capture the differences between objects of the same class, hence being useful in a MOT task. Here, features can be extracted from next-to-the-last layer of a reidentification model and used as embeddings for similarity calculation between tracklets and detections. Often, cosine distance is employed as the similarity metric.

One example of a multiple object tracking method using deep features for similarity matching is **Deep SORT** [44] that extends previously mentioned SORT algorithm with deep association metric. Similar to SORT, each tracklet is described by an 8-dimensional representation of $(u, v, \gamma, h, \dot{x}, \dot{y}, \dot{\gamma}, \dot{h})$, where (u, v) denotes the bounding box center, γ denotes its aspect ratio, h denotes the height, and the last four parameters denote the corresponding velocities in the image coordinates. A Kalman filter [45] with constant velocity model is used for modeling the motion, that is, for predicting the tracklet state in each subsequent frame. Each tracklet is associated with an age value initialized from zero and incremented at each frame the tracklet is not matched with any detections, and reset to zero when a match occurs.

In each frame, already known objects from previous frames are matched with new detections. This is achieved by constructing a pairwise cost matrix between the known tracklets and the new detections, and matching them in a cascade manner to be further described next. A cost matrix is defined based on two measures, that is, the visual appearance distance between the detection and tracklet, and their motion distance. Motion distance is obtained by calculating the Mahalanobis distance between predicted Kalman filter states of known tracklets with new detections:

$$d_{mot}(i, j) = (\mathbf{d}_j - \mathbf{y}_j)^T \mathbf{S}_i^{-1} (\mathbf{d}_j - \mathbf{y}_i), \tag{11.24}$$

where $(\mathbf{y}_i, \mathbf{S}_i)$ denote the projection of ith track distribution into measurement space, and $\mathbf{d}_j$ denotes jth detection. Although providing some level of certainty, this distance measure is vulnerable to occlusions, and hence, a second distance metric based on visual information is employed. To achieve this, each tracklet and each detection are accompanied with appearance descriptors $\mathbf{e}$ obtained by a convolutional neural network trained for a reidentification task, where 128-dimensional features extracted from the second last layer are utilized as descriptors. While each detection is associated with only one descriptor, each tracklet that has already existed for a number of frames is associated with a set of descriptors (limited to 100) obtained from the corresponding detections in the previous frames. This allows to achieve certain level of robustness to sudden appearance changes.

During matching, comparison between visual descriptors is performed via the cosine distance

$$d_{vis}(i, j) = \min\{1 - \mathbf{e}_j^T \mathbf{e}_k^{(i)} | \mathbf{e}_k^{(i)} \in R_i\}, \tag{11.25}$$

where R_i denotes the gallery of descriptors of ith tracklet. The total association cost is then calculated as a weighted sum of appearance and motion cost:

$$c_{i,j} = \lambda d_{mot}(i, j) + (1 - \lambda) d_{vis}(i, j). \tag{11.26}$$

In addition, a match is directly disregarded if one of the distances falls beyond a certain empirically-established threshold.

Pairwise matching is performed sequentially. Starting from the tracklets with age 1, they are matched with the new detections using the described cost matrix and associations optimized with the Hungarian algorithm. The matched pairs are removed from the pool of unmatched tracklets and detections, respectively, and matching continues for the tracklets of age 2 and so on, until all possible matches have been made. Furthermore, tracklets reaching a certain age A_{max} are killed, the matched tracklets are updated with bounding box and appearance information of the corresponding detections, and unmatched detections are used for initialization of new tracklets.

Overall, Deep SORT and other multiple object tracking methods relying on deep learning for visual feature extraction use relatively simple data association procedures, making them well suited for online real-time applications. In turn, some other methods tend to focus on learning to perform data association in a more meaningful manner instead, as will be discussed further.

11.2.2.2 Tracking as a graph optimization problem

Another set of methods in the field of multiple object tracking follow a graph approach. These methods are generally offline, that is, the bounding boxes containing the objects are known for all the frames in an image sequence, and both past and future information can be exploited for tracking. Adopting such an offline process has the benefit of exploiting the global information for object tracking, rather than only local information in successive frame pairs.

The general formulation of a graph-based multiple object tracking algorithm is as follows. Given a set of detections, we consider a directed graph $G(V, E)$ with vertices V and edges E, where the vertices correspond to the detections. Each detection is connected to all the detections of the subsequent frame via the edges, where each edge is associated with a certain cost $C(i, j)$. The goal is thus to find true trajectories of the objects in a frame sequence by identifying which edges connect the same objects in different frames. An example of such representation is shown in Fig. 11.8. Formally, the objective can be defined as a minimum-cost flow problem [46] with the objective function:

$$T^* = \text{argmin} \sum_{(i,j) \in E} C(i, j) y(i, j), \tag{11.27}$$

where T is a trajectory, $y(i, j) \in \{0, 1\}$ is an indicator function, denoting the activation of edges, i.e. $y(i, j) = 1$ denotes that the vertices connected by the corresponding edge correspond to the same object.

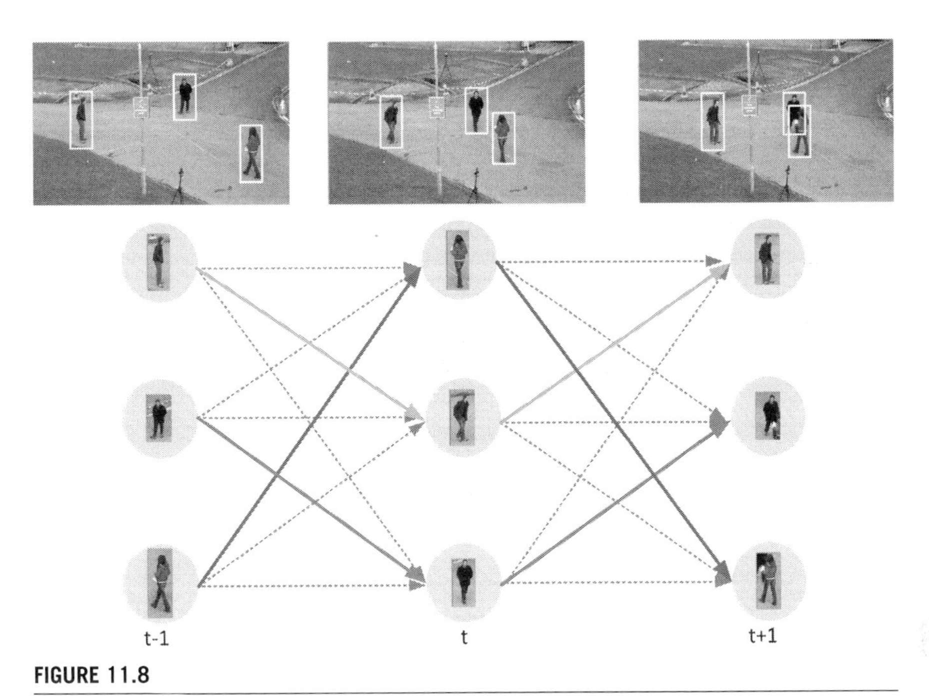

FIGURE 11.8

Graph-based representation of a MOT task, where each detection is represented with a node and an edge connecting nodes i and j is associated to cost value $C(i, j)$, with the goal of finding a set of edges resulting in minimal total cost [28].

Based on this formulation, two different approaches can generally be adopted. The first approach seeks to improve the model by learning better cost functions, while the second proposes new graph models. The former approach includes methods with appearance-based costs obtained via Siamese networks, or distance-based costs, while the latter develop more complex graph optimization strategies or formulations.

Recent methods explore different forms of graph neural networks for multiple object tracking on graphs. One notable example achieving state-of-the-art performance is **MPNTrack** [47] based on message passing network with learnt feature embeddings. Following the graph-based definition, each detection is represented by a vertex of a directed graph, with edges connecting the vertices of different frames. Each detection o_i with a corresponding image patch α_i is described by a feature embedding $N_v^{enc}(\alpha_i)$ obtained by a convolutional neural network pretrained for reidentification on public datasets. In addition, edge embedding corresponding to the edge between detections o_i and o_j are created based on the relative distance and size of the detections defined as

$$\left(\frac{2(x_j - x_i)}{h_i + h_j}, \frac{2(y_j - y_i)}{h_i + h_j}, \log \frac{h_i}{h_j}, \log \frac{w_i}{w_j} \right), \tag{11.28}$$

where x, y correspond to the top-left coordinates, and h, w correspond to the height and width. The relative distance-size vector is concatenated with the relative appearance change $||N_v^{enc}(\alpha_j) - N_v^{enc}(\alpha_i)||_2$ and the time difference $t_j - t_i$ and introduced to an MLP denoted as N_e^{enc}.

After the feature calculation and graph construction steps, several message passing steps are performed. Recall that a message passing step in message passing networks refers to the update on nodes/edges with information from adjacent nodes/edges performed sequentially for a number of steps, generally defined as

$$e_{(i,j)}^{(l)} = N_e\left([e_i^{(l-1)}, e_j^{(l-1)}, e_{(i,j)}^{(l-1)}]\right), \tag{11.29}$$

$$e_{(i)}^{(l)} = \Phi\left(\{m_{(i,j)}^{(l)}\}_{j \in N_i}\right), \tag{11.30}$$

$$m_{(i,j)}^{(l)} = N_v\left([e_i^{(l-1)}, e_{(i,j)}^{(l)}]\right) \tag{11.31}$$

where $e_{(i,j)}^{(l)}$ denotes the feature embedding of the edge connecting nodes i and j at step (l), $e_i^{(l)}$ denotes the feature embedding of node i, N_i are the set of nodes adjacent to i, Φ denotes some order-invariant operation (e.g., summation), and N_v and N_e correspond to learnable functions, for example, MLPs.

Instead of relying on general formulation of MPN updates, MPNTrack uses the structure of multiple object tracking graph and proposes to perform time-aware updates with different models used for the nodes corresponding to future and past frames. In other words, instead of directly relying on a single model N_v for updating a certain node, first separate N_v^f and N_v^p are used for future and past nodes, respectively. The node updates are therefore performed as follows:

$$e_i^{(l)} = N_v\left([e_{i,p}^{(l)}, e_{i,f}^{(l)}]\right), \tag{11.32}$$

$$e_{i,p}^{(l)} = \sum_{j \in N_i^p} m_{(i,j)}^{(l)}, \tag{11.33}$$

$$e_{i,f}^{(l)} = \sum_{j \in N_i^f} m_{(i,j)}^{(l)}, \tag{11.34}$$

$$m_{(i,j)}^{(l)} = \hat{N}_v\left([e_i^{(l-1)}, e_{(i,j)}^{(l)}, e_{(i)}^{0}]\right), \tag{11.35}$$

where $\hat{N}_v$ is replaced by N_v^p if $j \in N_i^p$, and by N_v^f if $j \in N_i^f$, with N_i^p and N_i^f corresponding to sets of nodes adjacent to i in the past and future frames, respectively.

Finally, the problem is formulated as an edge classification performed with an MLP taking as input the edge embedding at a given message passing step and outputting a single sigmoid-valued output. The model is trained with binary cross-entropy loss over all edges and message passing steps:

$$L = \frac{-1}{|E|}\sum_{l=l_0}^{l-L}\sum_{(i,j)\in E} w\, y_{(i,j)} \log\left(\hat{y}^{(l)}_{(i,j)}\right) + (1 - y_{(i,j)}) \log\left(1 - \hat{y}^{(l)}_{(i,j)}\right), \qquad (11.36)$$

where $l_0 \in 1, \ldots, L$, and w is a scalar weight to account for imbalance between active and inactive edges.

11.2.2.3 Detection-driven tracking

All the methods discussed so far rely on a two-stage approach, where detections are generated first, followed by a data association step, where bounding boxes are assigned identities. Such an approach results in limitations related to the running speed of the trackers especially in situations where many objects are present in the video. Since the detection and association steps are solved separately, the overall pipeline is not benefiting from shared feature representations, leading to larger number of parameters. As a remedy, methods combining the two tasks were published in the recent years [48,49].

One example of such detection-driven tracking method is the **Joint Detector and Embedding Model (JDE)** [48]. The method achieves multiple object tracking by jointly learning to localize bounding boxes and extract corresponding feature embeddings in one model, providing a significant speed-up compared to two-stage trackers. The model uses an FPN [18] network with Darknet-53 [22] architecture, and performs prediction on multiple scales. The feature representations extracted by the network are further propagated through three prediction heads (one for each of the scales). Each prediction head consists of three branches corresponding to bounding box regression, bounding box classification, and feature embedding prediction. The bounding box regression and classification branches are essentially similar to conventional RPN network and are trained with the cross-entropy loss for classification and Smooth L1 loss for regression. The feature embedding task is formulated as a metric learning problem and is trained with cross-entropy loss, defined as

$$L_{CE} = -\log \frac{\exp(f^T g^+)}{\exp(f^T g^+) + \sum_i \exp(f^T g_i^-)}, \qquad (11.37)$$

where f^T is an instance in a mini-batch, g^+ is the classwise weight of the positive class (i.e., the one of f^T), and g_i^- corresponds to weights of negative classes. Embeddings extracted from the feature map are processed through a shared fully-connected layer outputting classwise logits, which are used for calculation of the cross-entropy loss. The bounding box regression, classification, and feature embedding losses are further combined.

During testing, the model predicts detections along with corresponding feature embeddings. These are subsequently matched between frames as follows: at each frame, pairwise motion affinity and appearance affinity matrices A_m and A_e are created between all the existing tracklets and predictions of the current frame, where

A_e is calculated based on the cosine distance between embeddings, and A_m based on Mahalanobis distance. Assignment is solved via the Hungarian algorithm with cost matrix set to $C = \lambda A_e + (1 - \lambda)A_m$. For the matched tracklets, the motion state is further updated with a Kalman filter and feature descriptors are updated as $e_i^t = \alpha e_i^{t-1} + (1 - \alpha) f_i^t$, where f_i^t is the feature embedding of the matched prediction.

Another notable detection-driven multiple object tracker is the **Tracktor++** [49]. The method is based on Faster R-CNN [17] architecture and, at each frame, bounding boxes of the previous frame are regressed to the current frame. Practically, this is achieved by applying the RoI pooling step of Faster R-CNN to the feature representations of the current frame at locations of bounding boxes at the previous frame, assuming that the objects do not move significantly between frames. This way, the new locations for the tracklets are obtained with the preservation of identities. The predicted tracklets are killed if the corresponding confidence score falls below certain threshold. In addition, new bounding boxes are predicted at each frame via a separate detection branch. New tracklets are initiated from the detections unless they have a high IoU with one of the already existing tracklets predicted by the first branch.

To achieve further robustness, a camera motion compensation model is used for sequences exhibiting high motion. To account for occlusions and tracklets leaving and returning to the scene, an appearance-based re-identification model is used. Essentially, a reidentification model is a Siamese network (TriNet [50]) used for calculating the similarity between lost tracklets and new detections.

11.3 Conclusion

The most prominent methods for object detection, single object tracking, and multiple object tracking problems have been reviewed in this chapter. The majority of the recent methods predominantly rely on deep convolutional neural network architectures for feature representation, on top of which task-specific learning is performed. Within the field of object detection, most methods are based on the two-stage approach described in Section 11.1.2 performing bounding box prediction and classification on top of certain region proposals. One-stage approaches described in Section 11.1.3 directly regress the bounding boxes along with class labels from the image. Notable ideas in these methods include the use of anchor boxes, multiscale testing, and various forms of positive–negative samples balancing, for example, hard negative mining or the use of focal loss. In turn, some methods propose novel learning paradigms, not relying on direct bounding box prediction, such as anchor-free detectors described in Section 11.1.4.

Among the single object tracking methods, successful approaches include the use of correlation filters described in Section 11.2.1.1 with online training, methods relying on heavy pre-training with the goal of similarity learning based on Siamese networks described in Section 11.2.1.3, and methods combining both online updates and offline pretraining, with some examples given in Section 11.2.1.2. In multiple object tracking, the majority of methods proposed to date are based on the two-stage approach with the assumption of detections obtained by an object detector. Among these, methods operating on a frame-by-frame basis are largely based on deep visual representations with relatively simple association strategies as described in Section 11.2.2.1, while other methods focus on global offline optimization and are largely based on graph representations, as described in Section 11.2.2.2. At the same time, recent methods tend to shift from this two-step tracking paradigm and aim at jointly learning to track and detect. These methods are described in Section 11.2.2.3.

Novel methods in object detection and tracking are evolving with an unprecedented speed due to high demands from potential applications, including but not limited to robotics and autonomous systems. Thankfully, the availability of computational power, the creation of a large number of public data sets and baselines, and the transparency given by open access publications and open source implementations enabled the rapid development of the fields of object detection and object tracking. For broader insights, the reader is encouraged to refer to baseline challenges organized for the evaluation of novel methods and the establishment of new state-of-the-art results, that is, the visual object tracking challenge [27], the multiple object tracking challenge [28], and the MS COCO [4] and ILSVRC [51] challenges for object detection.

References

[1] Y. Jiang, X. Zhu, X. Wang, S. Yang, W. Li, H. Wang, P. Fu, Z. Luo, R2cnn: rotational region cnn for orientation robust scene text detection, arXiv preprint, arXiv:1706.09579, 2017.

[2] K. He, G. Gkioxari, P. Dollár, R. Girshick, Mask r-cnn, in: Proceedings of the IEEE International Conference on Computer Vision, 2017, pp. 2961–2969.

[3] M. Everingham, L. Van Gool, C.K. Williams, J. Winn, A. Zisserman, The Pascal visual object classes (voc) challenge, International Journal of Computer Vision 88 (2) (2010) 303–338.

[4] T.Y. Lin, M. Maire, S. Belongie, J. Hays, P. Perona, D. Ramanan, P. Dollár, C.L. Zitnick, Microsoft coco: common objects in context, in: Proceedings of the European Conference on Computer Vision, Springer, 2014, pp. 740–755.

[5] J.R. Uijlings, K.E. Van De Sande, T. Gevers, A.W. Smeulders, Selective search for object recognition, International Journal of Computer Vision 104 (2) (2013) 154–171.

[6] C.L. Zitnick, P. Dollár, Edge boxes: locating object proposals from edges, in: Proceedings of the European Conference on Computer Vision, Springer, 2014, pp. 391–405.

[7] M.A. Waris, A. Iosifidis, M. Gabbouj, Cnn-based edge filtering for object proposals, Neurocomputing 266 (2017) 631–640.

[8] D.G. Lowe, Object recognition from local scale-invariant features, in: Proceedings of the Seventh IEEE International Conference on Computer Vision, Vol. 2, IEEE, 1999, pp. 1150–1157.

[9] H. Bay, A. Ess, T. Tuytelaars, L. Van Gool, Speeded-up robust features (surf), Computer Vision and Image Understanding 110 (3) (2008) 346–359.

[10] R.K. McConnell, Method of and apparatus for pattern recognition, US Patent 4,567,610, 1986.

[11] P. Viola, M. Jones, Rapid object detection using a boosted cascade of simple features, in: Proceedings of the IEEE Conference on Computer Vision and Pattern Recognition, Vol. 1, IEEE, 2001, pp. I–I.
[12] N. Dalal, B. Triggs, Histograms of oriented gradients for human detection, in: Proceedings of the IEEE Conference on Computer Vision and Pattern Recognition, Vol. 1, IEEE, 2005, pp. 886–893.
[13] R. Girshick, J. Donahue, T. Darrell, J. Malik, Rich feature hierarchies for accurate object detection and semantic segmentation, in: Proceedings of the IEEE Conference on Computer Vision and Pattern Recognition, 2014, pp. 580–587.
[14] A. Krizhevsky, I. Sutskever, G.E. Hinton, Imagenet classification with deep convolutional neural networks, Communications of the ACM 60 (6) (2017) 84–90.
[15] R. Girshick, Fast r-cnn, in: Proceedings of the IEEE International Conference on Computer Vision, 2015, pp. 1440–1448.
[16] K. Simonyan, A. Zisserman, Very deep convolutional networks for large-scale image recognition, arXiv preprint, arXiv:1409.1556, 2014.
[17] S. Ren, K. He, R. Girshick, J. Sun, Faster r-cnn: towards real-time object detection with region proposal networks, in: Advances in Neural Information Processing Systems, 2015, pp. 91–99.
[18] T.Y. Lin, P. Dollár, R. Girshick, K. He, B. Hariharan, S. Belongie, Feature pyramid networks for object detection, in: Proceedings of the IEEE Conference on Computer Vision and Pattern Recognition, 2017, pp. 2117–2125.
[19] W. Liu, D. Anguelov, D. Erhan, C. Szegedy, S. Reed, C.Y. Fu, A.C. Berg, Ssd: single shot multibox detector, in: Proceedings of the European Conference on Computer Vision, Springer, 2016, pp. 21–37.
[20] J. Redmon, S. Divvala, R. Girshick, A. Farhadi, You only look once: unified, real-time object detection, in: Proceedings of the IEEE Conference on Computer Vision and Pattern Recognition, 2016, pp. 779–788.
[21] J. Redmon, A. Farhadi, Yolo9000: better, faster, stronger, in: Proceedings of the IEEE Conference on Computer Vision and Pattern Recognition, 2017, pp. 7263–7271.
[22] J. Redmon, A. Farhadi, Yolov3: an incremental improvement, arXiv preprint, arXiv:1804.02767, 2018.
[23] T.Y. Lin, P. Goyal, R. Girshick, K. He, P. Dollár, Focal loss for dense object detection, in: Proceedings of the IEEE International Conference on Computer Vision, 2017, pp. 2980–2988.
[24] H. Law, J. Deng, Cornernet: detecting objects as paired keypoints, in: Proceedings of the European Conference on Computer Vision, 2018, pp. 734–750.
[25] A. Newell, K. Yang, J. Deng, Stacked hourglass networks for human pose estimation, in: Proceedings of the European Conference on Computer Vision, Springer, 2016, pp. 483–499.
[26] X. Zhou, D. Wang, P. Krähenbühl, Objects as points, arXiv preprint, arXiv:1904.07850, 2019.
[27] M. Kristan, J. Matas, A. Leonardis, T. Vojir, R. Pflugfelder, G. Fernandez, G. Nebehay, F. Porikli, L. Čehovin, A novel performance evaluation methodology for single-target trackers, IEEE Transactions on Pattern Analysis and Machine Intelligence 38 (11) (2016) 2137–2155, https://doi.org/10.1109/TPAMI.2016.2516982.
[28] P. Dendorfer, A. Osep, A. Milan, K. Schindler, D. Cremers, I. Reid, S. Roth, L. Leal-Taixé, MOTChallenge: a benchmark for single-camera multiple target tracking, International Journal of Computer Vision 129 (4) (2021) 845–881.

[29] D.S. Bolme, J.R. Beveridge, B.A. Draper, Y.M. Lui, Visual object tracking using adaptive correlation filters, in: Proceedings of the IEEE Conference on Computer Vision and Pattern Recognition, IEEE, 2010, pp. 2544–2550.
[30] J.F. Henriques, R. Caseiro, P. Martins, J. Batista, High-speed tracking with kernelized correlation filters, IEEE Transactions on Pattern Analysis and Machine Intelligence 37 (3) (2014) 583–596.
[31] M. Danelljan, G. Bhat, F.S. Khan, M. Felsberg, Atom: accurate tracking by overlap maximization, in: Proceedings of the IEEE Conference on Computer Vision and Pattern Recognition, 2019, pp. 4660–4669.
[32] G. Bhat, M. Danelljan, L.V. Gool, R. Timofte, Learning discriminative model prediction for tracking, in: Proceedings of the IEEE International Conference on Computer Vision, 2019, pp. 6182–6191.
[33] A.J. Smola, B. Schölkopf, Learning with Kernels, Vol. 4, Citeseer, 1998.
[34] H. Nam, B. Han, Learning multi-domain convolutional neural networks for visual tracking, in: Proceedings of the IEEE Conference on Computer Vision and Pattern Recognition, 2016, pp. 4293–4302.
[35] D. Held, S. Thrun, S. Savarese, Learning to track at 100 fps with deep regression networks, in: Proceedings of the European Conference on Computer Vision, Springer, 2016, pp. 749–765.
[36] Y. Jia, E. Shelhamer, J. Donahue, S. Karayev, J. Long, R. Girshick, S. Guadarrama, T. Darrell, Caffe: convolutional architecture for fast feature embedding, in: Proceedings of the ACM International Conference on Multimedia, 2014, pp. 675–678.
[37] L. Bertinetto, J. Valmadre, J.F. Henriques, A. Vedaldi, P.H. Torr, Fully-convolutional Siamese networks for object tracking, in: Proceedings of the European Conference on Computer Vision, Springer, 2016, pp. 850–865.
[38] B. Li, J. Yan, W. Wu, Z. Zhu, X. Hu, High performance visual tracking with Siamese region proposal network, in: Proceedings of the IEEE Conference on Computer Vision and Pattern Recognition, 2018, pp. 8971–8980.
[39] B. Li, W. Wu, Q. Wang, F. Zhang, J. Xing, J. Yan, Siamrpn++: evolution of Siamese visual tracking with very deep networks, in: Proceedings of the IEEE Conference on Computer Vision and Pattern Recognition, 2019, pp. 4282–4291.
[40] Q. Wang, L. Zhang, L. Bertinetto, W. Hu, P.H. Torr, Fast online object tracking and segmentation: a unifying approach, in: Proceedings of the IEEE Conference on Computer Vision and Pattern Recognition, 2019, pp. 1328–1338.
[41] H.W. Kuhn, The Hungarian method for the assignment problem, Naval Research Logistics Quarterly 2 (1–2) (1955) 83–97.
[42] E. Bochinski, V. Eiselein, T. Sikora, High-speed tracking-by-detection without using image information, in: Proceedings of the IEEE International Conference on Advanced Video and Signal Based Surveillance, IEEE, 2017, pp. 1–6.
[43] A. Bewley, Z. Ge, L. Ott, F. Ramos, B. Upcroft, Simple online and realtime tracking, in: IEEE International Conference on Image Processing, IEEE, 2016, pp. 3464–3468.
[44] N. Wojke, A. Bewley, D. Paulus, Simple online and realtime tracking with a deep association metric, in: Proceedings of the IEEE International Conference on Image Processing, IEEE, 2017, pp. 3645–3649.
[45] R.E. Kalman, A new approach to linear filtering and prediction problems, Journal of Basic Engineering 82 (1) (1960) 35–45.
[46] L. Zhang, Y. Li, R. Nevatia, Global data association for multi-object tracking using network flows, in: Proceedings of the IEEE Conference on Computer Vision and Pattern Recognition, IEEE, 2008, pp. 1–8.

[47] G. Brasó, L. Leal-Taixé, Learning a neural solver for multiple object tracking, in: Proceedings of the IEEE Conference on Computer Vision and Pattern Recognition, 2020, pp. 6247–6257.

[48] Z. Wang, L. Zheng, Y. Liu, Y. Li, S. Wang, Towards real-time multi-object tracking, arXiv preprint, arXiv:1909.12605, 2019.

[49] P. Bergmann, T. Meinhardt, L. Leal-Taixe, Tracking without bells and whistles, in: Proceedings of the IEEE International Conference on Computer Vision, 2019, pp. 941–951.

[50] A. Hermans, L. Beyer, B. Leibe, In defense of the triplet loss for person re-identification, arXiv preprint, arXiv:1703.07737, 2017.

[51] O. Russakovsky, J. Deng, H. Su, J. Krause, S. Satheesh, S. Ma, Z. Huang, A. Karpathy, A. Khosla, M. Bernstein, A.C. Berg, L. Fei-Fei, ImageNet large scale visual recognition challenge, International Journal of Computer Vision 115 (3) (2015) 211–252, https://doi.org/10.1007/s11263-015-0816-y.

CHAPTER

Semantic scene segmentation for robotics 12

Juana Valeria Hurtado and Abhinav Valada
Department of Computer Science, University of Freiburg, Freiburg, Germany

12.1 Introduction

In order for robots to interact with the world, they should first have the ability to comprehensively understand the scene around them. The efficiency with which a robot performs a task, navigates, or interacts, strongly depends on how accurately it can comprehend its surroundings. Furthermore, the ability to understand context is crucial for safe operation in diverse environments [1]. However, accurate interpretation of the environment is extremely challenging, especially in real-world urban scenarios that are complex and dynamic. In these environments, robots are expected to perform their tasks precisely while they encounter diverse agents and objects. These environments themselves undergo appearance changes due to illumination variations and weather conditions, further increasing the difficulty of the task [2].

In robotics, scene understanding includes identifying, localizing, and describing the elements that compose the environment, their attributes, and dynamics. Research on novel techniques for automatic scene understanding has been exponentially increasing over the past decade due to the potential impact on numerous applications. The advances in deep learning as well as the availability of open-source data sets, and the increasing capacity of computation resources have facilitated the rapid improvement of scene understanding techniques [1]. These advances are most evident in image classification, where the goal is to determine what an image contains. The output of this classification task can be considered as a high-level representation of the scene, which enables the identification of the various objects that are present by assigning a class label to them. A different mid-level capability, known as object detection, further details the scene by simultaneously classifying and localizing the objects in the image with bounding boxes. Although this task represents the scene in greater detail, it is still unable to provide essential information or object attributes such as the shape of objects. A closely related task, known as object segmentation, aims to fill this gap by additionally providing the shape of the object inside the bounding box in terms of their segmented boundaries. We present an overview of these perception tasks in Fig. 12.1.

Semantic segmentation on the other hand presents a more integrated scene representation that includes classification of objects at the pixel-level, and their locations

Deep Learning for Robot Perception and Cognition. https://doi.org/10.1016/B978-0-32-385787-1.00017-8

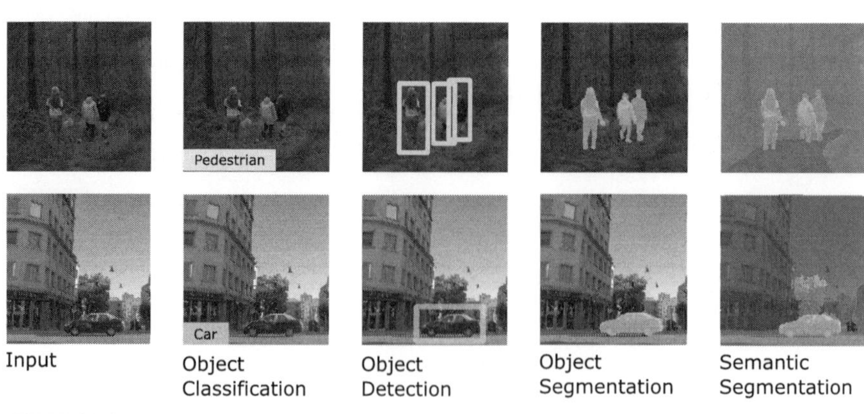

FIGURE 12.1

Each row shows an example with an input image and the corresponding output of different scene understanding tasks. Object classification identifies "what" objects compose the image, object detection predicts "where" the objects are located in the image, object segmentation outputs a mask that indicates the shape of the object. Semantic segmentation further details the input image by predicting the label of all the pixels, including the background.

in the image. Semantic segmentation is considered as a low-level task that works at the pixel resolution and unifies object detection, shape recognition, and classification. By extending the previous tasks, semantic segmentation assigns a class label to each pixel in the image. Therefore as we show in Fig. 12.2, semantic segmentation models output a full-resolution semantic prediction that contains scene contexts such as the object category, location, and shape of all the scene elements including the background. Given that this prediction is performed for each pixel of the image, it is known as a dense prediction task.

Detailed understanding of the scene by assigning a class label to each pixel facilitates various downstream robotic tasks such as mapping, navigation, and interaction. This is an important step toward building complex robotic systems for applications such as autonomous driving [3,4], robot-assisted surgery [5,6], indoor service robots [7,8], search and rescue robots [9], and mobile manipulation [10]. Therefore incorporating semantic information has strongly influenced several robotics areas such as Simultaneous Localization And Mapping (SLAM) and perception through object recognition and segmentation, highlighting the importance of semantics knowledge for robotics. Semantic information can be of special importance to tackle challenging problems such as perception under challenging environmental conditions. Robots operating in the real world encounter adverse weather, changing lightning condition from day through night. Accurately predicting the semantics in these conditions is of great importance for successful operation.

In semantic segmentation, we are not interested in identifying single instances, that is, individual detections of objects in the scene. Therefore the segmentation output does not distinguish two objects of the same semantic class as separate enti-

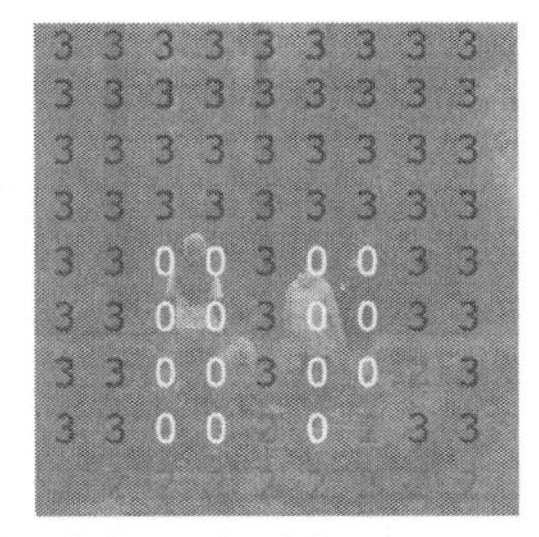

FIGURE 12.2

Illustration of the semantic segmentation output in which a semantic class label is assigned to each pixel in the image. The network predicts label indices for each pixel, which is depicted with different colors for visualization purposes. The predictions overlaid on the input image is shown on the left and the label indices overlaid on the input image is shown on the right. (For interpretation of the colors in the figures, the reader is referred to the web version of this chapter.)

ties. Another closely related perception task known as instance segmentation, allows for distinguishing instances of objects of the same class in addition to pixel-level segmentation, and Multiobject Tracking and Segmentation (MOTS) [11] tackles the problem of obtaining temporally coherent instance segmentation. Furthermore, panoptic segmentation [12] is a recently introduced task that unifies semantic and instance segmentation. We present a graphical comparison of these perception tasks in Fig. 12.3. An additional task named Multiobject Panoptic Tracking (MOPT) [13] further combines panoptic segmentation and MOTS. This elucidates the importance of developing models for a more holistic scene representation.

In this chapter, we present techniques for semantic scene segmentation, primarily using deep learning. We first discuss different algorithms and architectures for semantic segmentation, followed by the different loss functions that are typically used. We then discuss how different types of data and modalities can be used, and how video-classification models can be extended to yield temporal coherent semantic segmentation. Subsequently, we present an overview of data sets, benchmarks, and metrics that are used for semantic segmentation. Finally, we discuss different challenges and opportunities for developing advanced semantic segmentation models.

FIGURE 12.3

Comparison of semantic segmentation, instance segmentation, and panoptic segmentation tasks. Semantic segmentation assigns a class label to each pixel in the image and instance segmentation assigns an instance ID to pixels belonging to individual objects as well as semantic class label to each pixel in the image. The panoptic segmentation task unifies semantic and instance segmentation.

12.2 Algorithms and architectures for semantic segmentation

The automatic understanding of the scene semantics has been a major area of research for decades. However, unprecedented advancement in semantic segmentation methods has only been achieved recently. Deep learning has played a significant role in enabling this capability, especially after the introduction of Fully Convolutional Networks (FCNs) [14], which were proposed as a solution for semantic segmentation. In this section, we first provide a brief overview of semantic segmentation approaches used prior to Convolutional Neural Networks (CNNs) and then present important deep learning approaches, their focus on improvements and limitations. We further present challenges and proposed solutions.

12.2.1 Traditional methods

Typically, the traditional algorithms for image segmentation use clustering methodologies, contours, and edges information [15,16]. Particularly for semantic segmentation, diverse approaches initially followed the idea of obtaining a pixel-level inference by considering the relationship between spatially close pixels. To do so, various features of appearance, motion, and structure ranging in complexity were considered including pixel color, surface orientation, height above camera [17], histogram of oriented gradients (HOG) [18], Speeded Up Robust Features (SURF) [19], among others. Approaches for image semantic segmentation range from simple thresholding methods in gray images to more complex edge-based approaches [20–22] or graphical models such as Conditional Random Fields (CRF) and Markov Random Fields (MRF) [23–25]. Another group of approaches employ multiple individually pretrained object detectors with the aim of extracting the semantic information from the image [26]. In general, as traditional approaches typically rely on a priori knowledge, such as the dependency between neighboring pixels, these methods require the definition of semantic and spatial properties with respect to the application to define segmentation concept. Moreover, these methods are limited to being able to segment a specific number of object classes, which are defined by hand-selected parameters of the methods.

12.2.2 Deep learning methods

The introduction of deep learning methods and deep features presented important advances in computer vision tasks such as image classification, and led to the interest in using deep features for semantic segmentation. The initially proposed deep learning approaches for semantic segmentation used classification networks such as VGG [27] and AlexNet [28], and adapted them for the task of semantic segmentation by fine-tuning the fully connected layers [29–31]. As a result, these approaches were plagued by overfitting and required significant training time. Additionally, the semantic segmentation performance was affected by the insufficient discriminative deep features that were learned by them.

Most subsequently proposed methods suffered from low performance, and consequently several refinement strategies were incorporated such as Conditional Random Fields (CRFs) [32], Markov Random Fields (MRFs) [29], nearest neighbors [30], calibration layers [31], and super-pixels [33,34]. Refinement strategies are still used as post-processing methods to enhance the pixel classification around regions where class intersections occur [35].

Significant progress in semantic segmentation was achieved with the introduction of FCNs [14] that use the entire image to infer the dense prediction. FCN is composed of only stacked convolutional and pooling layers, without any fully connected layers, as shown in Fig. 12.4. Originally, the proposed network used a stack of convolutional layers without any downsampling to learn the mapping from pixels to pixels directly by maintaining the input resolution. To compensate for this and to learn sufficient representative features, they assemble multiple consecutive convolutional layers. This initial approach was able to generate impressive results but at a very high computational cost. Therefore the model was inefficient and its scalability was limited.

As a solution to this problem, they presented an encoder-decoder architecture. The encoder is a typical CNN pretrained for classification tasks such as AlexNet [28] or VGGNet [27]. The goal of the downsampling layers in the encoder is to capture deep contextual information that corresponds to the semantics. For its part, the decoder network is composed of deconvolutional and up-sampling layers. Its goal is to convert a low-resolution feature representation to a high-resolution image, recovering the spatial information and enabling precise localization, thereby yielding the dense classification.

Towards this direction, a deeper deconvolution network consisting of stacked deconvolution and unpooling layers is proposed in [36]. Oliveira et al. [94] employed a similar strategy called UpNet for part segmentation and tackled two problems: occluded parts and overfitting. Similarly, an encoder-decoder architecture was proposed

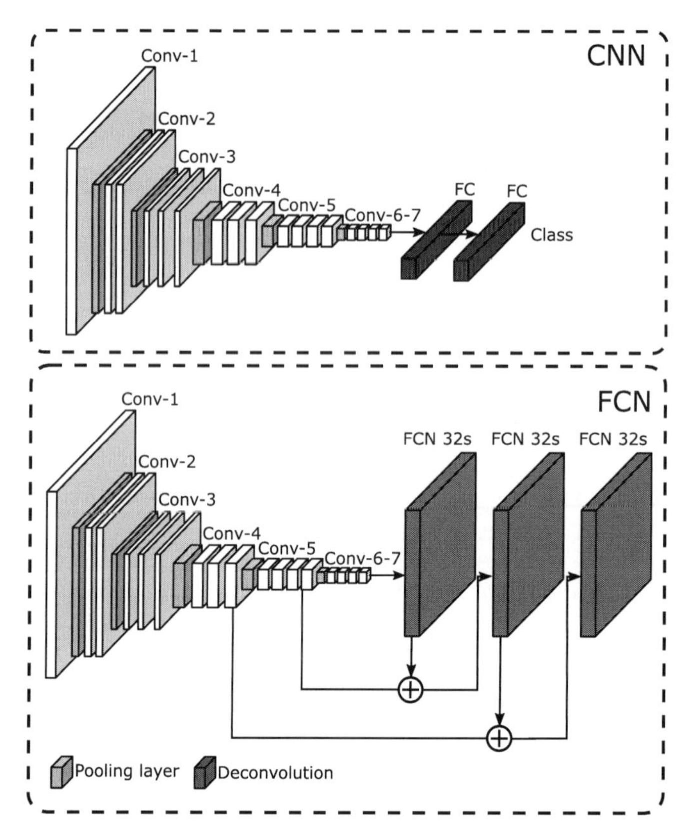

FIGURE 12.4

An example topology of a Convolutional Neural Network (CNN) used for image classification (top) and a Fully Convolutional Network (FCN) that is used for dense prediction (bottom). Note that FCNs do not contain any fully connected layers.

in [37] where the feature maps obtained with the encoder are used as input of the decoder network that up-samples the feature maps by keeping the maxpooling indices from the corresponding encoder layer. Similarly, Liu et al. [38] proposed an approach called ParseNet, which models global context directly. Such methods have demonstrated state-of-the-art results at the time of their introduction. Among these early deep learning models, UpNet and ParseNet achieve superior performance.

Specifically, to obtain deep features that enhance the performance, the convolutional layers learn feature maps with progressively coarser spatial resolutions, and the corresponding neurons have gradually larger receptive fields in the input image. This means that the earlier layers encode features of appearance and location, and the feature extracted with the later layers encodes context and high-level semantic information. As a consequence, the two main challenges of deep convolutional neural networks in semantic segmentation were identified. First, the consecutive pooling operations or convolution striding lead to smaller feature resolution. Second, the multiscale nature of the objects in the scene were difficult to be captured. Since the features at different resolutions encode context information at different scales, this information was exploited to enhance the representation of the multiscale objects. In the following subsection, we present different techniques that have been proposed to address these challenges.

12.2.3 Encoder variants

Encoders are also referred to as the backbone network in semantic segmentation architectures. The encoders used for semantic segmentation are typically based on CNNs that have been proposed for image classification. The initial semantic segmentation approaches adopted the VGG-16 [27], AlexNet [28], or GoogLeNet [39] architectures for the encoder. Each of these encoders have achieved outstanding results in the ImageNet ILSVRC14 and ILSVRC12 [40] challenges. VGG was extensively used in several semantic segmentation architectures [41,42]. A breakthrough in semantic segmentation models was achieved with the introduction of ResNet [43]. Several semantic segmentation models that employed ResNets and its variants such as Wide ResNet [44] and ResNeXt [45] achieved state-of-the-art performance on various benchmarks. Another popular encoder architecture were the new generation of GoogLeNet models such as Inception-v2, Inception-v3 [46], Inception-v4 and Inception-ResNet [47]. More recently, semantic segmentation models that employ the EfficientNet [48] family of architectures have achieved impressive results while being computationally more efficient than the previous encoders.

12.2.4 Upsampling methods

While employing multiple sequential convolution and pooling operations in the network leads to deep features and enhances the performance of perception networks, substantial information loss can occur in the downsampled representation of the input toward the end of the network. This loss in information can affect the localization of features as well as details of the scene elements, such as texture or boundary information. Diverse works in this direction have been proposed to prevent or recuperate the loss in information. As a solution to this problem, [36] introduced deconvolution networks composed of sets of deconvolution and un-pooling layers. The authors apply their proposed network on individual object proposals and combine the predicted instance-wise segmentations to generate a final semantic segmentation.

The goal of employing the upsampling operations during the decoding step is to generate the semantic segmentation output at the same resolution of the input image. Given the computational efficiency of bilinear interpolation, it was extensively used in several semantic segmentation networks [14,41]. Another common method to upsample the features maps is using deconvolution or transpose convolution layers. Transpose convolution computes the reverse of the convolution operation and it can be used to obtain the dense prediction in the decoder of semantic segmentation architectures [49–51].

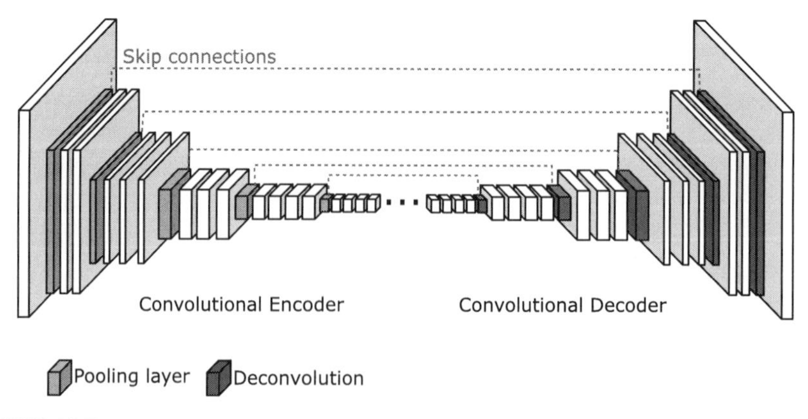

FIGURE 12.5

An example topology of an encoder-decoder architecture. Typically, the encoder is a pretrained classification CNN that uses downsampling layers to capture the contextual information. On the other hand, the decoder network is composed of up-sampling layers to recover the spatial information, yielding the pixel-level classification output with the same resolution as the input image.

12.2.5 Techniques for exploiting context

Different semantic segmentation methodologies have been proposed with the aim of exploiting semantic information, local appearance, and global context of the scene that can be extracted in early and late deep features. Several methodologies propose different strategies and architectures that fuse features in the process.

12.2.5.1 Encoder-decoder architecture

Initial semantic segmentation architectures [14,36] used deconvolution to learn the upsampling of the encoder's outputs at low resolution. In addition, SegNet [37] takes the encoder's pooling indices and later use them in the decoder to learn additional convolutional layers with the aim of densifying the feature responses. The U-Net architecture [52] implements skip connections between the encoder and the corresponding decoder layers as shown in Fig. 12.5. The skip connections connect encoder layers to the decoder, and allow directly transferring the information to deeper layers of the network. This network employs symmetry by increasing the size of the decoder to match the encoder. More recent encoder-decoder approaches [53–56] have demonstrated the effectiveness of this structure on several semantic segmentation benchmarks.

12.2.5.2 Image pyramid

The main idea with this network topology is that the same model is used to process multiscale inputs. We present the general image pyramid network in Fig. 12.6. Typically, the model weights are shared for multiple inputs, while the different size inputs have different purposes. Features corresponding to small scale inputs encode

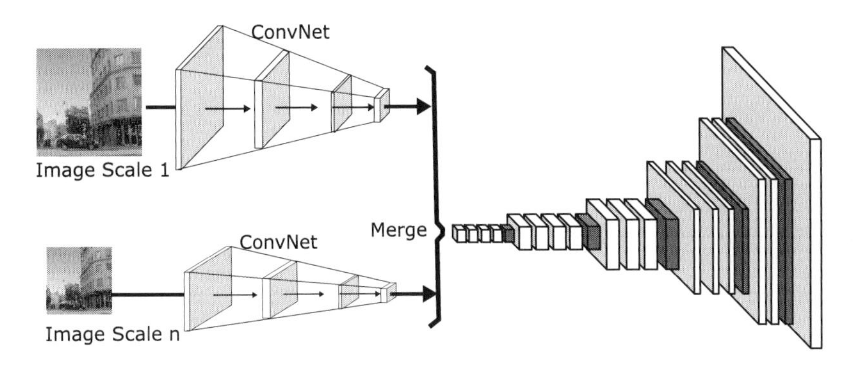

FIGURE 12.6

Topology of the image pyramid architecture. This network uses the same model to process the same input at different scales. The different scales allow the network to encode different context in the image.

the wider context, and the details from small elements are encoded and preserved by the large scale inputs. Examples of this image pyramid structure include the Laplacian pyramid used to transform the input. With this transformation, each input scale is subsequently used to feed a CNN, and finally, all feature maps scales are fused together [33]. Other methodologies directly resize the input to different scales and later fuse the obtained features [41,57,58]. However, the main restriction of this model is related to limited GPU memory, given that networks that requires deeper CNNs such as [43,44,59], which are computationally expensive.

12.2.5.3 Conditional random fields

Graphical models, especially Conditional Random Fields (CRFs) have been used as refinement layers in deep semantic segmentation architectures. The main objective is to capture low-level detail in regions where class intersections occur. These boundary regions are particularly difficult to segment with precision. This strategy includes additional modules placed consecutively to represent the longer-range context. To do so, a popular methodology is to integrate CRF into CNNs. Diverse methodologies have been presented as a refinement layer, including convolutional CRFs [60] and dense CRF [61]. Other methodologies have been proposed to train both the CRF and CNN jointly [41,62]. These methods that incorporate CRFs have demonstrated their benefits to capture contextual knowledge and exploit finer details to enhance the class label localization in the pixels.

12.2.5.4 Spatial pyramid pooling

This structure uses spatial pyramid pooling to obtain context at multiple scales, and it is graphically described in Fig. 12.7. Toward this direction, ParseNet [38] exploits image-level features. Another methodology is presented in DeepLabv2 [41], which uses atrous spatial pyramid pooling (ASPP) that includes parallel atrous convolution

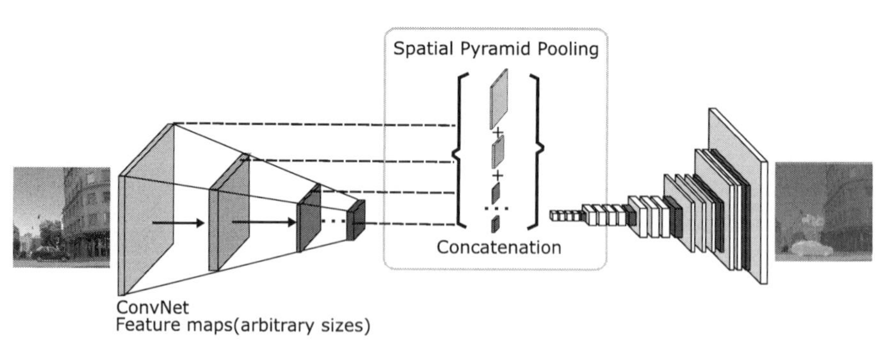

FIGURE 12.7

Topology of spatial pyramid pooling architecture used to exploit the context found at multiple scales. This network includes a new module for multilevel pooling between the convolutional and fully-connected layers. The multi-level pooling allows this network to be more robust to the variations in object scale and deformation.

layers with different rates to consider objects at different scales. This strategy improves the segmentation performance. Following this direction, a subsequent work also proposes to generate multiscale features that cover a larger scale range densely using the Densely connected Atrous Spatial Pyramid Pooling (DenseASPP) [63]. Furthermore, an efficient variant of the ASPP is proposed in Adapnet++ [2], which captures a larger effective receptive field while decreasing the required parameters by 87% using cascaded and parallel atrous convolutions. Additionally, in Pyramid Scene Parsing Network (PSPNet) [64], the authors exploit the global context information by aggregating multiple region-based contexts with a proposed pyramid pooling module. PSPNet demonstrated significant improvement on several semantic segmentation benchmarks.

12.2.5.5 Dilated convolution

Dilated convolutions, graphically presented in 12.8, are also known as atrous convolutions [41] and aim to have an effective receptive field that grows more rapidly than in contiguous convolutional filters. These convolutions are an effective strategy to preserve the feature map resolution and extract deep features without using pooling or subsampling. Nevertheless, since the feature map resolutions are not reduced with the progression of the network hierarchy, using dilated convolutions requires higher GPU storage and computation. Some of the explored models that use atrous convolutions for semantic segmentation include [65–67].

12.2.6 Real-time architectures

Many of the architectures and topologies presented so far in this chapter typically require high computational capacity and are not efficient for real-time applications. As these architectures employ large networks for their encoder such as GoogleNet [39] and ResNet [43], or use large CNN structures in different stages of the architec-

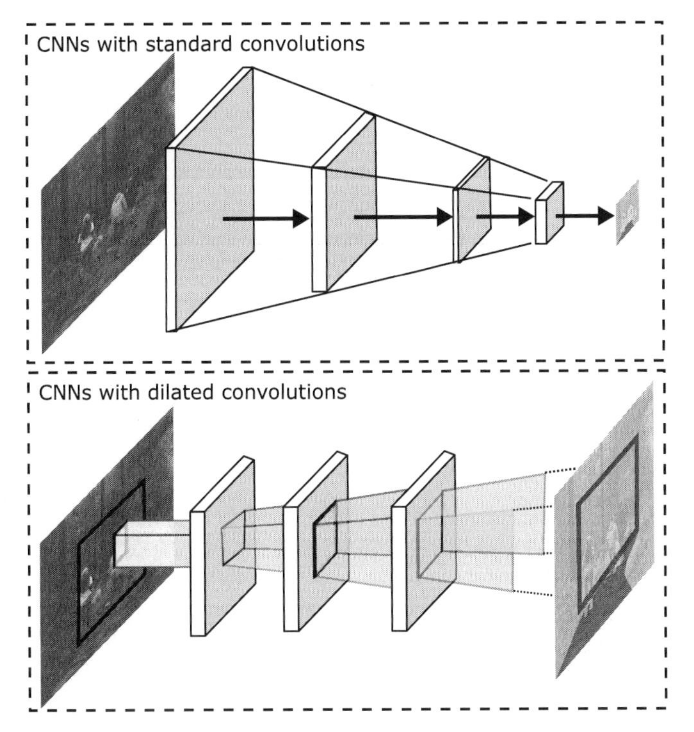

FIGURE 12.8

Illustration of CNNs with standard convolutions (top) and CNNs with dilated convolutions (bottom). The receptive fields of consecutive deep convolution layers output smaller resolution feature maps. The receptive field of dilated convolution grows more rapidly than in contiguous convolutional filters.

ture, they achieve high performance but with low efficiency in terms of computation cost and runtime. To address this problem, different works propose approaches that are more suitable for real-time applications. For example, Efficient Neural Network (Enet) [68] is a lightweight architecture in which the last stage of the model is removed to optimize the network to obtain a faster inference time. The main drawback of this architecture is that excluding the downsampling operations in the last stage of the network makes it unable to cover large objects since the receptive field is smaller.

As an alternative, in ERFNet [56] a layer is designed to use residual connections and depthwise separable convolution with the aim of increasing the receptive field. On the other hand, in the spatial pyramid pooling structure, the convolutional feature maps are resampled at the same scale before the classification layer, resulting in a computationally expensive process. To tackle this problem, ESPNet is proposed as an efficient network structure [69]. This approach aims to efficiently exploit both context and spatial features by incorporating an efficient convolutional module called ESP. ESPNet architecture is able to preserve the segmentation accuracy while being fast and small while requiring low power and low latency. ESPNet decomposes a standard convolution into two steps. First, pointwise convolutions to lessen the computation effort. Second, a spatial pyramid of dilated convolutions that re-samples the feature maps so that the network can learn the representations from a large effective receptive field.

The approaches mentioned above provide lightweight architectures but compromise the model accuracy. Other models such as BiSeNet [70] and LiteSeg [71] explore strategies to improve computational efficiency while maintaining high accuracy. In BiSeNet, the authors design two different streams. First, the spatial path generates high-resolution features by using a small stride. Second, the context path is designed to obtain an adequate receptive field with a fast downsampling approach. Later, an additional feature fusion module is employed to combine both features. In LiteSeg, the authors propose a deeper version of ASPP and use dilated convolutions and depthwise separable convolutions. As a result, LiteSeg is a faster and efficient model which provides high accuracy.

12.2.7 Object detection-based methcds

In object detection in scenes with multiple elements, the main goal is to generate a bounding box indicating each object. Nevertheless, since the input image can include a diverse number of objects or may not have any objects, the number of objects that should be detected cannot be fixed. Therefore this task cannot be solved using a standard CNNs followed by a fully connected layer with a predefined number of output classes. A straightforward approach to tackle this problem is to select different regions of interest from the image and employ a CNN on each region to classify the presence of a single object in the region. The architecture used to determine the regions out of the input image is called Region Proposal Network (RPN). RPNs are essential structures in the construction of algorithms that select a reduced group of regions of interest where objects might be located in the image. These algorithms aim to obtain an optimal number of region that allow the detection of all the elements in the scene, therefore reducing the required computation capacity. Some popular algorithms are YOLO [72], Single Shot Detector [73], and Fast-RCNN [74] and its improved version Faster-RCNN [75]. These networks facilitate the segmentation of the object inside a smaller region of the image. In this direction, the segmentation of instances is proposed in Mask-RCNN [76] and YOLACT [77]. They obtain a semantic segmentation output by drawing the segmentation masks and completing all the pixels of the input image.

12.3 Loss functions for semantic segmentation

In this section, we discuss the most commonly employed loss functions for learning the semantic segmentation task.

12.3.1 Pixelwise cross entropy loss

The pixelwise cross entropy loss function [78] inspects each pixel individually by comparing the predicted class label to the ground truth, and finally computes averages over all pixels. This loss function leads to equal learning for each pixel in the image, which can lead to problems if the image is composed of unbalanced pixel classes. With the aim of tackling class imbalance, weighting this loss for each output channel was proposed [14]. Additionally, a pixelwise weighting loss has also been proposed [52]. In this case, a larger weight is assigned to the pixels at the borders of segmented objects. The pixelwise cross entropy loss ($\mathcal{L}_{\mathcal{CE}}$) is computed as

$$\mathcal{L}_{\mathcal{CE}}(p, y) = -\sum_{c} y_{o,c} \log p_{o,c}, \tag{12.1}$$

where c is the class label, o denotes an observation, $y_{o,c} \in [0, 1]$ denotes if c is the correct classification for o, and p is the predicted probability of o belonging to c.

12.3.2 Dice loss

The Dice loss function [79] is based on the dice coefficient metric that is used to measure the overlap between two samples. This metric is also used to compute the similarity between a pair of images. The dice loss is computed for each class individually and then the average among class is computed to obtain a final loss. The dice coefficient is computed as

$$\text{Dice}(y, c) = \frac{2|X \cap Y|}{|X| + |Y|} \tag{12.2}$$

where c is the class label, X are the scores of each class and Y is a tensor with the class labels. Based on this metric, the dice loss ($\mathcal{L}_{\mathcal{D}}$) is computed as

$$\mathcal{L}_{\mathcal{D}}(y, c) = 1 - \text{Dice}(y, c) \tag{12.3}$$

12.4 Semantic segmentation using multiple inputs

Thus far in this chapter, we primarily discussed techniques for semantic segmentation that take a single image as input and yield a corresponding segmentation output for that image. In this section, we further discuss semantic segmentation methods that take multiple inputs simultaneously, namely for video semantic segmentation, point cloud semantic segmentation and multimodal semantic segmentation.

12.4.1 Video semantic segmentation

As robots move and interact with the environment, the perception is typically dynamic with changing context and scene relationships. Most often, a temporal sequence of data is available as an input to the semantic segmentation model and we can exploit this information to enforce temporal coherence in the output. For example, if an object of a particular semantic class is present in two consecutive frames, that object should be identified with the same label across the different frames, assuring the temporal coherence of the classified pixel. To do so, different works have explored various techniques to exploit temporal information to improve the overall semantic segmentation performance [80–83]. Specifically, clockwork convnets are proposed in [80] uses clock signals to control the learning rate of different layers and a LSTM-based spatio-temporal FCN is introduced in [81]. Similarly, [82] uses a spatiotemporal representation where the convolutional gated recurrent network enables learning both spatial and temporal information jointly. Furthermore, the approach presented in [83] combines convolutional gated architectures and spatial transformers for video semantic segmentation.

12.4.2 Point cloud semantic segmentation

A point cloud is a collection of points in the 3D space representing the structure of the scene. With the growing use of depth and LiDAR sensors that enable building 3D maps of the scene, methods for 3D semantic segmentation of point clouds are increasingly becoming more popular. We show in Fig. 12.9 an example of LiDAR-based semantic segmentation. We briefly discuss three main categories of techniques for 3D semantic segmentation, namely point-based methods, voxel-based methods, and projection-based methods.

Point-based methods aim to process the raw point clouds directly. PointNet [84] was the first approach to tackle semantic segmentation directly on the raw point clouds. The PointNet architecture allows working on unordered data with a point-

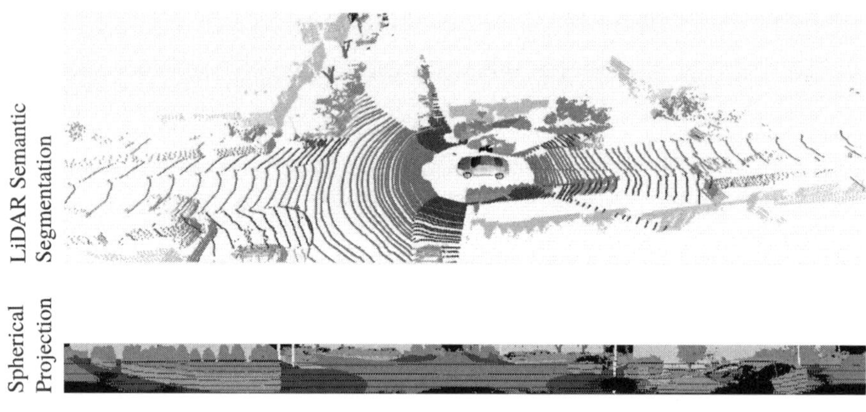

FIGURE 12.9

Semantic segmentation of point clouds. The top figure presents a class label assigned to each point in the 3D space. The bottom figure presents the point cloud projected into the 2D representation using spherical projections. The semantic segmentation prediction is obtained for each pixel in the projection.

wise learning methodology that uses shared multilayer perceptrons and subsequently symmetrical pooling functions. PoinNet++ [85] further extends PointNet and proposes to group points in a hierarchical manner. A similar approach [86] computes per-point convolutions by grouping together neighboring points into kernel cells. In contrast, [87] proposes to use a directed graph over the point cloud generating a set of interconnected superpoints to capture the structure and context information.

Voxel-based methods first convert the point cloud into a 3D voxel representation and then employ 3D CNN architectures on them. Voxels are volumetric discretizations of the 3D space. SegCloud [88] proposes to use this representation as to the input to a 3D-FCN [89]. The authors also propose a deterministic trilinear interpolation that converts the voxel predictions back to the point cloud space and subsequently employ CRFs for refinement. Nevertheless, the computation of voxels and their 3D processing consumes substantial runtime and requires very high computational cost making it unfeasible for real-time applications.

Projection-based methods project the point cloud from 3D data into 2D space to reduce the computational cost. A known transformation method is spherical projection. This approach and has been especially utilized for LIDAR semantic segmentation. The resulting 2D projection allows using 2D image methodologies for semantic segmentation, which are faster than 3D approaches. SqueezeSeg [90] and its extension SqueezeSegV2 [91] yield very efficient semantic segmentation result by utilizing spherical projections. Additionally, the authors in [92] use different 2D architectures and additional post-processing stages to refine the 3D segmentation results. A more recent approach called EfficientLPS [93] presents a model that incorporates geometric transformations while learning features in the encoder of semantic and instance segmentation networks.

12.4.3 Multimodal semantic segmentation

Semantic segmentation of RGB images has led to important advances in scene understanding. However, image-based approaches suffer from visual limitations as they are susceptible to changing weather and seasonal conditions as well as varying illumination conditions. With the availability of low-cost sensors and with the goal of improving the robustness and the granularity of segmentation for robot perception, fusion of different modalities have been explored to exploit complementary features. Alternate modalities such as thermal images and depth images have been shown to be beneficial for segmenting objects in low illumination conditions by exploiting properties of objects such as reflectance and geometry (Fig. 12.10).

There are three main categories of fusion techniques for multimodal semantic segmentation: early, late, and hybrid fusion. Early fusion, also known as data-level or input-level fusion, aims to combine the modalities before feeding them as input to the CNN. The obtained multimodal representation is later used as the input to a single model to exploit complementary features and leverage the interactions between low-level features of each modality. This technique typically requires that the utilized modalities have semantic similarities such as RGB and depth. A straightforward

FIGURE 12.10

Example of multimodal semantic segmentation using multiple modalities as input in the Freiburg Forest data set [94]. In this case, the input modalities are RGB, thermal, and depth images. The semantic segmentation output is obtained by fusing the features obtained from both modalities.

implementation of early fusion of modalities is channel stacking by concatenating multiple modalities across the channel dimension. Subsequently, a learning model can be trained end-to-end using the stacked modalities as input. In the case of RGB-Depth semantic segmentation, the first deep learning approach for multimodal early fusion proposed to concatenate the depth image with the RGB image as an additional channel [95]. Later, [96] proposed to use an encoder-decoder architecture composed of two encoder branches to extract features from RGB and depth images, and subsequently combine the obtained depth feature maps with the RGB branch. More recently, a fusion mechanism proposed in RFBNet [97] explores the interdependencies between the encoders to provide an efficient fusion strategy. Given that early fusion approaches mainly utilize a single or unified network, they are computationally more efficient than the other techniques. Nevertheless, early fusion of modalities has its own set of limitations since it always forces the network to learn fused representations from the beginning and does it enable the network to exploit cross-modal interdependencies and complementary features.

Late fusion techniques use individual CNN streams for each modality, followed by a fusion stage in the end which facilitates learning of further combined multimodal representations. In one of the early methods, an RGB and depth based geocentric embedding was proposed for object detection and segmentation [98]. To this end, the method employs two modality-specific networks to obtain feature maps that are then fed to a classifier. Other deep learning methodologies aim to map the multimodal features to a subspace [99,100]. In [99], an adaptive gating network was proposed which generates a classwise probability distribution over the modality-specific network streams. [101] exploits both multimodal and multispectral images in a late-fused convolution architecture. In a similar work, [94] proposes a fusion architecture that extracts multimodal features separately using individual network streams, followed by feeding the summation of the resulting feature maps through consecutive convolutional layers. Since the modality-specify streams are trained individually, the subsequent fusion training allows obtaining a combined prediction. As a result, the late fusion approach can potentially learn better complementary features along with fusion specific features. This strategy facilitates a more robust performance in cases when the features in each model have good classification performance, then the fu-

data set contains sequences of frames recorded with diverse sensor modalities such as high-resolution RGB, grayscale stereo cameras, and a 3D laser scanners.

12.5.1.3 Mapillary vistas

The Mapillary vistas [107] data set contains 25,000 high-resolution images annotated into 66 object categories. Mapillary vistas is one of the most extensive publicly available data sets for semantic segmentation of street scenes. Additionally, this data set presents instance-specific semantic annotations for 37 classes, and it is suitable for other scene understanding tasks such as instance segmentation and panoptic segmentation. This data set includes diverse scenes in terms of geographic extent and conditions such as weather, season, daytime, cameras, and viewpoints. The images in this data set range in resolutions from 1024×768 pixels to 4000×6000 pixels.

12.5.1.4 BDD100K: a large-scale diverse driving video database

The BDD100k [108] data set is one of the largest driving data sets consisting of diverse scenes that cover different weather conditions including sunny, overcast, rainy, as well as different times of the day. The data set consists of 100,000 videos of 4 seconds each that were captured at a resolution of 1280×720 pixels. BDD100k provides annotations for semantic segmentation, instance segmentation, multiobject segmentation and tracking, image tagging, lane detection, drivable area segmentation, multiobject detection and tracking, domain adaptation and imitation learning. It provides fine-grained pixel-level annotations for 40 object classes and the data set is split into 7000 images for training, 1000 images for validation, and 2000 images for testing.

12.5.1.5 Indian driving data set

The Indian Driving Data set (IDD) [109] data set is a collection of finely annotated images of autonomous driving scenarios. Instead of focusing on data from structured environments, this data set adds novel data from unstructured environments of scenes that do not have well-delineated infrastructures such as lanes and sidewalks. It consists of 10,000 images with resolution ranging from 720×1280 pixels to 1920×1080 pixels and finely annotated semantic segmentation labels with 34 classes. The training set contains 6993 images, while the validation and testing set have 981 and 2029, respectively.

12.5.2 Indoor data sets

These data sets contain indoor scenes such as offices and rooms. Besides semantic class labels for images, some of data sets also provide depth images and 3D models of the scenes. The class labels in these data sets include objects such as sofa, bookshelf, refrigerator, and bed. Additionally, indoor data sets present background class labels such as wall and floor. We present an example of a outdoor image in Fig. 12.12.

FIGURE 12.12

An example semantic image segmentation from the challenging ScanNet data set. The image shows an indoor scene where objects appear much closer than in outdoor scenes.

12.5.2.1 NYU-Depth V2

The NYU Depth V2 data set [110] is a collection of video sequences recorded in indoor scenarios. The data set was collected using both the RGB and Depth sensors from the Microsoft Kinect at 464 different scenes from three different cities. It contains 1449 images of nearly 900 different categories with the respective dense semantic labels and additional depth information.

12.5.2.2 SUN 3D

The SUN 3D [111] data set contains RGB-D videos of 415 sequences of 254 different indoor scenes. The sequences were captured in over 41 buildings. Only 8 sequences in this data set have been annotated with semantic segmentation labels. In addition to the RGB and depth images, the data set also provides the camera poses for each frame.

12.5.2.3 SUN RGB-D

SUN RGBD [112] is a scene understanding benchmark suite. The data set contains 10, 335 pairs of RGB-D images with pixelwise semantic annotation for both 2D and 3D, for both objects and rooms. The data set provides class labels for six important recognition tasks: semantic segmentation, object classification, object detection, context reasoning, mid-level recognition, and surface orientation and room layout estimation.

12.5.2.4 ScanNet

The ScanNet [113] data set is a large scale RGB-D video data set. It consists of 2.5 million views. Besides providing annotations for semantic segmentation, ScanNet also presents labels for 3D camera poses, surface reconstructions, and instance-level semantic segmentation.

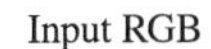

FIGURE 12.13

An example semantic image segmentation from the Microsoft common objects in context data set. The image shows a scene with a person playing sports. The class labels in the image include person, sky, and floor.

12.5.3 General purpose data sets

These data sets contain generic class labels including almost every type of object or background. Some of these data sets are the most standard benchmarks to measure progress in the semantic segmentation task as a whole. An example of this data set is presented in Fig. 12.13.

12.5.3.1 PASCAL visual object classes

PASCAL Visual Object Classes (VOC) [114] present all types of indoor and outdoor images with 20 foreground object classes and one background class with 1464 images for training, 1449 images for validation, and 1456 test image. The test set is not public and is accessible only for the challenge. Besides including pixel-level annotations for semantic segmentation, this data set also presents annotations for classification, detection, action classification, and person layout tasks.

12.5.3.2 Microsoft common objects in context

Microsoft Common Objects in Context (MS COCO) [115] is an extensive large scale data set for object detection, semantic segmentation, and captioning image set. It contains 330,000 images of complex everyday scenes with common objects. The data set contains 200,000 labeled images with 1.5 million object instances and 80 object categories. The data set is split into 82,000 images for training, 40,500 images for validation and 80,000 for testing.

12.5.3.3 ADE20K

The ADE20K [116] data set contains more than 20,000 scenes with objects and object parts annotations composing 150 semantic categories. The average image size of the samples in the data set is 1.3M pixels that can be up to 2400 $\times$ 1800 pixels.

The data set is split into 20,000 images for training and 2000 images for validation. Additionally, ADE20K also provides a public leaderboard.

12.6 Semantic segmentation metrics

Two principal criteria are usually considered during the evaluation of semantic segmentation models. The first being accuracy, which is related to the effectiveness and represents how successful the model is. The second corresponds to the computational complexity and is associated with the scalability of the model. Both criteria are essential for robots to successfully perform tasks using scene understanding models that can be deployed in resource limited systems.

12.6.1 Accuracy

Measuring the effectiveness of a semantic segmentation can be difficult, given that it requires to measure both classification and localization in the pixel space. Different metrics have been presented to measure each individual criteria or the combination of them.

12.6.1.1 ROC-AUC

ROC stands for the Receiver-Operator Characteristic curve. ROC measures a binary classification system's ability by utilizing the trade-off between the true-positive rate against the false-positive rate at various threshold settings. AUC stands for the area under the curve of this trade-off, and its maximum value is 1. This metric is useful in binary classification problems and is suitable in problems with balanced classes. Nevertheless, given that most semantic segmentation presents an unbalance between the classes, this evaluation metric is not considered in most recent challenges.

12.6.1.2 Pixel accuracy

Pixel Accuracy (PA) is a semantic segmentation metric that denotes the percent of pixels that are accurately classified in the image. This metric calculates the ratio between the amount of adequately classified pixels and the total number of pixels in the image as

$$PA = \frac{\sum_{j=1}^{k} n_{jj}}{\sum_{j=1}^{k} t_j}, \tag{12.4}$$

where n_{jj} is the total number of pixels both classified and labeled as class j. In other words, n_{jj} corresponds to the total number of true positives for class j. t_j is the total number of pixels labeled as class j.

Since there are multiple classes present in semantic segmentation, the mean pixel accuracy (mPA) represents the class average accuracy as

$$mPA = \frac{1}{k}\sum_{j=1}^{k}\frac{n_{jj}}{t_j}. \quad (12.5)$$

PA and mPA are intuitive and interpretable metrics. However, high PA does not directly imply superior segmentation performance, especially in imbalanced class data sets. In this case, when a class dominates the image, while some other classes make up only a small portion of the image, only correct classification of the dominant class will yield a high PA.

12.6.1.3 Intersection over union

In semantic segmentation, Intersection over Union (IoU) is the overlap area between the predicted segmentation and the ground truth divided by the area of union between the predicted segmentation and the ground truth. From the following equation, it is shown that IoU is the ratio of true positives (TP) to the sum of: false alarms known as false positives (FP), misses known as false negatives (FN), and hits known as true positives (TP). IoU value ranges between 0 and 1, the lower the value, the worse the semantic segmentation performance. The IoU metric can be computed as

$$IoU = \frac{TP}{TP + FP + FN}, \quad (12.6)$$

where FP_{ij} corresponds to the number of pixels which are labeled as class i, but classified as class j. Similarly, FN_{ji}, the total number of pixels labeled as class j, but classified as class i, corresponding to the misses of the class j. In the equation, the numerator extracts the overlap area between the predicted segmentation and the ground truth. The denominator represents the area of the union, by both the predicted segmentation and the ground truth bounding box.

Similar to PA, mIoU is computed to obtain the average per-class IoU as

$$mIoU = \sum_{j=1}^{k}\frac{TP_{jj}}{TP_{ij} + FP_{ji} + FN_{jj}}. \quad (12.7)$$

The IoU measure is more informative than PA given that it also punishes false alarms, whereas PA does not. The IoU is a very straightforward metric and it is very effective. Therefore, IoU and its variants are widely used as accuracy evaluation metrics in the most popular semantic segmentation challenges such as Cityscapes and VOC challenge. However, the main drawbacks of these metrics are that they only consider the labeling correctness without measuring how accurate are the boundaries of the segmentation, and the significance between false positives and misses is not measured.

12.6.1.4 Precision-recall curve-based metrics

Precision refers to the ratio of successes over the summation of successes and false alarms and recall relates the successes over the summation of successes and misses. Precision and recall is computed as

$$Precision = \frac{TP}{TP + FP}, \tag{12.8}$$

$$Recall = \frac{TP}{TP + FN}. \tag{12.9}$$

Precision-Recall Curve (PRC) is used to represent the trade-off between precision and recall for binary classification. PRC has the ability of discriminating the effects between the false positives and false negatives. Therefore, PCR-based metrics are widely used for semantic segmentation.

- F score or dice coefficient is very similar to the IoU and also ranges from 0 to 1, where 1 means the greatest similarity between the segmentation and ground truth. It is a normalized measure of similarity, and is defined as

$$F - score = \frac{Precision \times Recall}{Precision + Recall}. \tag{12.10}$$

- PRC-AuC is defined as the area under the PRC. This metric describes the precision-recall trade-off under different thresholds.
- Average Precision (AP) consists of a single value summarizing the shape and the AUC of PRC. This is one of the most used single value metric for semantic segmentation. The mean average precision (mAP) is also presented for all classes.

12.6.2 Computational complexity

Besides the model accuracy, computational complexity of the semantic segmentation model is critical factor to assess. Consequently, the following metrics have been used to measure the time to complete the task and the computational sources demanded by the model.

12.6.2.1 Runtime

It refers to the total time that the model requires to produce the output, starting from the image given as input until the dense semantic segmentation output is generated. This metric highly depends on the hardware that is used. Therefore, when this metric is reported, it also includes a description of the system used.

12.6.2.2 Memory usage

Given the ample applications of semantic segmentation, memory usage is an essential metric to report. This indicates how feasible is the deployment of the perception models on devices with limited computational resources. The goal of many semantic segmentation algorithms is to obtain the best possible accuracy with limited memory.

A commonly employed metric is the maximum memory required for the semantic segmentation model.

12.6.2.3 Floating point operations per second

The Floating Point Operations Per Second (FLOPs) refers to the number of floating-point calculations per second are required. It is used to measure the complexity of the CNN model. Assume that we use a sliding window to compute the convolution and we ignore the overhead due to nonlinear computation, then the FLOPs for the convolution kernel is given by

$$FLOPs = 2HW(C_{in}K^2 + 1)C_{out} \tag{12.11}$$

where H, W, and C_{in} are the height, width and number of channels in the input feature map respectively. K is the size of the kernel and $Cout$ is the number of channels in the output.

12.7 Conclusion

In this chapter, we discussed the semantic segmentation task for robot perception. We highlighted the essential role of deep learning techniques in the field, and we described the most popular data sets that can be used to train semantic segmentation models for robotics in different environments. We described related the algorithms, architectures, and different strategies that have been proposed to improve the semantic segmentation output. We have presented the required concepts, techniques, and general tools that comprise the topic. Semantic segmentation methods have a great potential as an important component to enhance perception systems of robot that operate and interact in real-world scenarios.

A limitation found in perception methodologies based on supervised learning is the large amount of labeled data that is required and the consequent labeling process. Given that supervised learning methods rely on a massive amount of annotated data and semantic segmentation requires dense labels for classification, the collection of data sets can be arduous, expensive, and sometimes unfeasible. Nowadays, other learning techniques such as weakly-supervised, self-supervised, and unsupervised learning have been recently explored. Transfer learning is a technique that allows to first train a general model using a large annotated data set and then fine tune the model using a reduced number of samples from the main application. Self-supervised learning aims to use details or attributes inherent to the images that can be captured without extra annotation steps. These inherent attributes can be used to initially pre-train the network and reduce the required amount of data to train the target model. These techniques represent an interesting contribution toward mitigating the dependency of training the models on a large amount of annotated data.

In the specific case of robotics, it is usually required that the employed models run in real-time and with limited computational resources. In this regard, there is still room for improvement in scene understanding. In applications where the robot is expected to rapidly react according to the conditions and situations in the environment, such as in autonomous vehicles, it is necessary to develop segmentation models with a short inference time. Developing such models without compromising on accuracy is also an interesting direction for research.

Semantic segmentation can be considered as the starting point of holistic scene representation by training models that are able to represent the complete scene, including objects and background. After semantic segmentation, other tasks that further detail the scene have been proposed. Such is the case of panoptic segmentation that combines semantic and instance segmentation. Recently, panoptic segmentation was extended to temporal sequences of frames where the instances also conserve the assigned label so that the instances are time coherent. The evolution of tasks that gradually detail the information contained in the scene highlights the importance of semantic segmentation.

References

[1] C. Premebida, R. Ambrus, Z.-C. Marton, Intelligent robotic perception systems, Applications of Mobile Robots (2018).

[2] A. Valada, R. Mohan, W. Burgard, Self-supervised model adaptation for multimodal semantic segmentation, International Journal of Computer Vision (2019) 1–47.

[3] G. Kalweit, M. Huegle, M. Werling, J. Boedecker, Interpretable multi time-scale constraints in model-free deep reinforcement learning for autonomous driving, arXiv preprint, arXiv:2003.09398, 2020.

[4] N. Radwan, W. Burgard, A. Valada, Multimodal interaction-aware motion prediction for autonomous street crossing, The International Journal of Robotics Research (IJRR) (2020).

[5] A. Tewari, J. Peabody, R. Sarle, G. Balakrishnan, A. Hemal, A. Shrivastava, M. Menon, Technique of da Vinci robot-assisted anatomic radical prostatectomy, Urology 60 (4) (2002) 569–572.

[6] Y. Qin, S. Feyzabadi, M. Allan, J.W. Burdick, M. Azizian, davincinet: joint prediction of motion and surgical state in robot-assisted surgery, arXiv preprint, arXiv:2009.11937, 2020.

[7] F. Boniardi, A. Valada, W. Burgard, G.D. Tipaldi, Autonomous indoor robot navigation using sketched maps and routes, in: Workshop on Model Learning for Human–Robot Communication at Robotics: Science and Systems (RSS), Citeseer, 2016.

[8] J.V. Hurtado, L. Londoño, A. Valada, From learning to relearning: a framework for diminishing bias in social robot navigation, arXiv preprint, arXiv:2101.02647, 2021.

[9] M. Mittal, R. Mohan, W. Burgard, A. Valada, Vision-based autonomous uav navigation and landing for urban search and rescue, arXiv preprint, arXiv:1906.01304, 2019.

[10] D. Honerkamp, T. Welschehold, A. Valada, Learning kinematic feasibility for mobile manipulation through deep reinforcement learning, arXiv preprint, arXiv:2101.05325, 2021.

[11] P. Voigtlaender, M. Krause, A. Osep, J. Luiten, B.B.G. Sekar, A. Geiger, B. Leibe, Mots: multi-object tracking and segmentation, in: Proceedings of the IEEE Conference on Computer Vision and Pattern Recognition, 2019, pp. 7942–7951.

[12] A. Kirillov, K. He, R. Girshick, C. Rother, P. Dollár, Panoptic segmentation, in: Proceedings of the IEEE Conference on Computer Vision and Pattern Recognition, 2019, pp. 9404–9413.

[13] J.V. Hurtado, R. Mohan, A. Valada, Mopt: multi-object panoptic tracking, arXiv preprint, arXiv:2004.08189, 2020.

[14] J. Long, E. Shelhamer, T. Darrell, Fully convolutional networks for semantic segmentation, in: Proceedings of the IEEE Conference on Computer Vision and Pattern Recognition, 2015, pp. 3431–3440.

[15] D. Weinland, R. Ronfard, E. Boyer, A survey of vision-based methods for action representation, segmentation and recognition, Computer Vision and Image Understanding 115 (2) (2011) 224–241.

[16] M. Sonka, V. Hlavac, R. Boyle, Image Processing, Analysis, and Machine Vision, Nelson Education, 2014.

[17] G.J. Brostow, J. Shotton, J. Fauqueur, R. Cipolla, Segmentation and recognition using structure from motion point clouds, in: European Conference on Computer Vision, Springer, 2008, pp. 44–57.

[18] N. Dalal, B. Triggs, Histograms of oriented gradients for human detection, in: 2005 IEEE Computer Society Conference on Computer Vision and Pattern Recognition (CVPR'05), Vol. 1, IEEE, 2005, pp. 886–893.

[19] H. Bay, A. Ess, T. Tuytelaars, L. Van Gool, Speeded-up robust features (surf), Computer Vision and Image Understanding 110 (3) (2008) 346–359.

[20] T. Lindeberg, M.-X. Li, Segmentation and classification of edges using minimum description length approximation and complementary junction cues, Computer Vision and Image Understanding 67 (1) (1997) 88–98.

[21] L. Barghout, Visual taxometric approach to image segmentation using fuzzy-spatial taxon cut yields contextually relevant regions, in: International Conference on Information Processing and Management of Uncertainty in Knowledge-Based Systems, Springer, 2014, pp. 163–173.

[22] S. Osher, N. Paragios, Geometric Level Set Methods in Imaging, Vision, and Graphics, Springer Science & Business Media, 2003.

[23] L. Ladický, C. Russell, P. Kohli, P.H. Torr, Associative hierarchical crfs for object class image segmentation, in: 2009 IEEE 12th International Conference on Computer Vision, IEEE, 2009, pp. 739–746.

[24] A. Montillo, J. Shotton, J. Winn, J.E. Iglesias, D. Metaxas, A. Criminisi, Entangled decision forests and their application for semantic segmentation of ct images, in: Biennial International Conference on Information Processing in Medical Imaging, Springer, 2011, pp. 184–196.

[25] J. Yao, S. Fidler, R. Urtasun, Describing the scene as a whole: joint object detection, scene classification and semantic segmentation, in: 2012 IEEE Conference on Computer Vision and Pattern Recognition, IEEE, 2012, pp. 702–709.

[26] L. Ladický, P. Sturgess, K. Alahari, C. Russell, P.H. Torr, What, where and how many? Combining object detectors and crfs, in: European Conference on Computer Vision, Springer, 2010, pp. 424–437.

[27] K. Simonyan, A. Zisserman, Very deep convolutional networks for large-scale image recognition, arXiv preprint, arXiv:1409.1556, 2014.

[28] A. Krizhevsky, I. Sutskever, G.E. Hinton, Imagenet classification with deep convolutional neural networks, Communications of the ACM 60 (6) (2017) 84–90.

[29] F. Ning, D. Delhomme, Y. LeCun, F. Piano, L. Bottou, P.E. Barbano, Toward automatic phenotyping of developing embryos from videos, IEEE Transactions on Image Processing 14 (9) (2005) 1360–1371.

[30] Y. Ganin, V. Lempitsky, n^4-fields: neural network nearest neighbor fields for image transforms, in: Asian Conference on Computer Vision, Springer, 2014, pp. 536–551.

[31] D. Ciresan, A. Giusti, L. Gambardella, J. Schmidhuber, Deep neural networks segment neuronal membranes in electron microscopy images, Advances in Neural Information Processing Systems 25 (2012) 2843–2851.

[32] J. Lafferty, A. McCallum, F.C. Pereira, Conditional random fields: probabilistic models for segmenting and labeling sequence data, in: Proc. of the Eighteenth Int. Conf. on Machine Learning, 2001.

[33] C. Farabet, C. Couprie, L. Najman, Y. LeCun, Learning hierarchical features for scene labeling, IEEE Transactions on Pattern Analysis and Machine Intelligence 35 (8) (2012) 1915–1929.

[34] B. Hariharan, P. Arbeláez, R. Girshick, J. Malik, Simultaneous detection and segmentation, in: European Conference on Computer Vision, Springer, 2014, pp. 297–312.

[35] I. Ulku, E. Akagunduz, A survey on deep learning-based architectures for semantic segmentation on 2d images, arXiv preprint, arXiv:1912.10230, 2019.

[36] H. Noh, S. Hong, B. Han, Learning deconvolution network for semantic segmentation, in: Proceedings of the IEEE International Conference on Computer Vision, 2015, pp. 1520–1528.

[37] V. Badrinarayanan, A. Kendall, R. Cipolla, Segnet: a deep convolutional encoder-decoder architecture for image segmentation, IEEE Transactions on Pattern Analysis and Machine Intelligence 39 (12) (2017) 2481–2495.

[38] W. Liu, A. Rabinovich, A.C. Berg, Parsenet: looking wider to see better, arXiv preprint, arXiv:1506.04579, 2015.

[39] C. Szegedy, W. Liu, Y. Jia, P. Sermanet, S. Reed, D. Anguelov, D. Erhan, V. Vanhoucke, A. Rabinovich, Going deeper with convolutions, in: Proceedings of the IEEE Conference on Computer Vision and Pattern Recognition, 2015, pp. 1–9.

[40] J. Deng, W. Dong, R. Socher, L.-J. Li, K. Li, L. Fei-Fei, Imagenet: a large-scale hierarchical image database, in: 2009 IEEE Conference on Computer Vision and Pattern Recognition, IEEE, 2009, pp. 248–255.

[41] L.-C. Chen, G. Papandreou, I. Kokkinos, K. Murphy, A.L. Yuille, Deeplab: semantic image segmentation with deep convolutional nets, atrous convolution, and fully connected crfs, IEEE Transactions on Pattern Analysis and Machine Intelligence 40 (4) (2017) 834–848.

[42] Z. Liu, X. Li, P. Luo, C.-C. Loy, X. Tang, Semantic image segmentation via deep parsing network, in: Proceedings of the IEEE International Conference on Computer Vision, 2015, pp. 1377–1385.

[43] K. He, X. Zhang, S. Ren, J. Sun, Deep residual learning for image recognition, in: Proceedings of the IEEE Conference on Computer Vision and Pattern Recognition, 2016, pp. 770–778.

[44] S. Zagoruyko, N. Komodakis, Wide residual networks, arXiv preprint, arXiv:1605.07146, 2016.

[45] S. Xie, R. Girshick, P. Dollár, Z. Tu, K. He, Aggregated residual transformations for deep neural networks, in: Proceedings of the IEEE Conference on Computer Vision and Pattern Recognition, 2017, pp. 1492–1500.

[46] C. Szegedy, V. Vanhoucke, S. Ioffe, J. Shlens, Z. Wojna, Rethinking the inception architecture for computer vision, in: Proceedings of the IEEE Conference on Computer Vision and Pattern Recognition, 2016, pp. 2818–2826.

[47] C. Szegedy, S. Ioffe, V. Vanhoucke, A. Alemi, Inception-v4, inception-resnet and the impact of residual connections on learning, in: Proceedings of the AAAI Conference on Artificial Intelligence, 2017.

[48] M. Tan, Q.V. Le, Efficientnet: rethinking model scaling for convolutional neural networks, arXiv preprint, arXiv:1905.11946, 2019.

[49] R. Mohan, Deep deconvolutional networks for scene parsing, arXiv preprint, arXiv: 1411.4101, 2014.

[50] J.B. de Monvel, E. Scarfone, S. Le Calvez, M. Ulfendahl, Image-adaptive deconvolution for three-dimensional deep biological imaging, Biophysical Journal 85 (6) (2003) 3991–4001.

[51] S. Saito, T. Li, H. Li, Real-time facial segmentation and performance capture from rgb input, in: European Conference on Computer Vision, Springer, 2016, pp. 244–261.

[52] O. Ronneberger, P. Fischer, T. Brox, U-net: convolutional networks for biomedical image segmentation, in: International Conference on Medical Image Computing and Computer-Assisted Intervention, Springer, 2015, pp. 234–241.

[53] C. Peng, X. Zhang, G. Yu, G. Luo, J. Sun, Large kernel matters – improve semantic segmentation by global convolutional network, in: Proceedings of the IEEE Conference on Computer Vision and Pattern Recognition, 2017, pp. 4353–4361.

[54] T. Pohlen, A. Hermans, M. Mathias, B. Leibe, Full-resolution residual networks for semantic segmentation in street scenes, in: Proceedings of the IEEE Conference on Computer Vision and Pattern Recognition, 2017, pp. 4151–4160.

[55] M. Amirul Islam, M. Rochan, N.D. Bruce, Y. Wang, Gated feedback refinement network for dense image labeling, in: Proceedings of the IEEE Conference on Computer Vision and Pattern Recognition, 2017, pp. 3751–3759.

[56] E. Romera, J.M. Alvarez, L.M. Bergasa, R. Arroyo, Erfnet: efficient residual factorized convnet for real-time semantic segmentation, IEEE Transactions on Intelligent Transportation Systems 19 (1) (2017) 263–272.

[57] L.-C. Chen, Y. Yang, J. Wang, W. Xu, A.L. Yuille, Attention to scale: scale-aware semantic image segmentation, in: Proceedings of the IEEE Conference on Computer Vision and Pattern Recognition, 2016, pp. 3640–3649.

[58] G. Lin, C. Shen, A. Van Den Hengel, I. Reid, Efficient piecewise training of deep structured models for semantic segmentation, in: Proceedings of the IEEE Conference on Computer Vision and Pattern Recognition, 2016, pp. 3194–3203.

[59] Z. Wu, C. Shen, A. Van Den Hengel, Wider or deeper: revisiting the resnet model for visual recognition, Pattern Recognition 90 (2019) 119–133.

[60] M.T. Teichmann, R. Cipolla, Convolutional crfs for semantic segmentation, arXiv preprint, arXiv:1805.04777, 2018.

[61] P. Krähenbühl, V. Koltun, Efficient inference in fully connected crfs with gaussian edge potentials, Advances in Neural Information Processing Systems 24 (2011) 109–117.

[62] L.-C. Chen, G. Papandreou, I. Kokkinos, K. Murphy, A.L. Yuille, Semantic image segmentation with deep convolutional nets and fully connected crfs, arXiv preprint, arXiv:1412.7062, 2014.

[63] M. Yang, K. Yu, C. Zhang, Z. Li, K. Yang, Denseaspp for semantic segmentation in street scenes, in: Proceedings of the IEEE Conference on Computer Vision and Pattern Recognition, 2018, pp. 3684–3692.

[64] H. Zhao, J. Shi, X. Qi, X. Wang, J. Jia, Pyramid scene parsing network, in: Proceedings of the IEEE Conference on Computer Vision and Pattern Recognition, 2017, pp. 2881–2890.
[65] J. Dai, H. Qi, Y. Xiong, Y. Li, G. Zhang, H. Hu, Y. Wei, Deformable convolutional networks, in: Proceedings of the IEEE International Conference on Computer Vision, 2017, pp. 764–773.
[66] P. Wang, P. Chen, Y. Yuan, D. Liu, Z. Huang, X. Hou, G. Cottrell, Understanding convolution for semantic segmentation, in: 2018 IEEE Winter Conference on Applications of Computer Vision (WACV), IEEE, 2018, pp. 1451–1460.
[67] Z. Wu, C. Shen, A.v.d. Hengel, Bridging category-level and instance-level semantic image segmentation, arXiv preprint, arXiv:1605.06885, 2016.
[68] A. Paszke, A. Chaurasia, S. Kim, E. Culurciello, Enet: a deep neural network architecture for real-time semantic segmentation, arXiv preprint, arXiv:1606.02147, 2016.
[69] S. Mehta, M. Rastegari, A. Caspi, L. Shapiro, H. Hajishirzi, Espnet: efficient spatial pyramid of dilated convolutions for semantic segmentation, in: Proceedings of the European Conference on Computer Vision (ECCV), 2018, pp. 552–568.
[70] C. Yu, J. Wang, C. Peng, C. Gao, G. Yu, N. Sang, Bisenet: bilateral segmentation network for real-time semantic segmentation, in: Proceedings of the European Conference on Computer Vision (ECCV), 2018, pp. 325–341.
[71] T. Emara, H.E. Abd El Munim, H.M. Abbas, Liteseg: a novel lightweight convnet for semantic segmentation, in: 2019 Digital Image Computing: Techniques and Applications (DICTA), IEEE, 2019, pp. 1–7.
[72] J. Redmon, S. Divvala, R. Girshick, A. Farhadi, You only look once: unified, real-time object detection, in: Proceedings of the IEEE Conference on Computer Vision and Pattern Recognition, 2016, pp. 779–788.
[73] W. Liu, D. Anguelov, D. Erhan, C. Szegedy, S. Reed, C.-Y. Fu, A.C. Berg, Ssd: single shot multibox detector, in: European Conference on Computer Vision, Springer, 2016, pp. 21–37.
[74] R. Girshick, J. Donahue, T. Darrell, J. Malik, Rich feature hierarchies for accurate object detection and semantic segmentation, in: Proceedings of the IEEE Conference on Computer Vision and Pattern Recognition, 2014, pp. 580–587.
[75] S. Ren, K. He, R. Girshick, J. Sun, Faster r-cnn: towards real-time object detection with region proposal networks, IEEE Transactions on Pattern Analysis and Machine Intelligence 39 (6) (2016) 1137–1149.
[76] K. He, G. Gkioxari, P. Dollár, R. Girshick, Mask r-cnn, in: Proceedings of the IEEE International Conference on Computer Vision, 2017, pp. 2961–2969.
[77] D. Bolya, C. Zhou, F. Xiao, Y.J. Lee, Yolact: real-time instance segmentation, in: Proceedings of the IEEE International Conference on Computer Vision, 2019, pp. 9157–9166.
[78] M. Yi-de, L. Qing, Q. Zhi-Bai, Automated image segmentation using improved pcnn model based on cross-entropy, in: Proceedings of 2004 International Symposium on Intelligent Multimedia, Video and Speech Processing, 2004, IEEE, 2004, pp. 743–746.
[79] F. Milletari, N. Navab, S.-A. Ahmadi V-net, Fully convolutional neural networks for volumetric medical image segmentation, in: 2016 Fourth International Conference on 3D Vision (3DV), IEEE, 2016, pp. 565–571.
[80] E. Shelhamer, K. Rakelly, J. Hoffman, T. Darrell, Clockwork convnets for video semantic segmentation, in: European Conference on Computer Vision, Springer, 2016, pp. 852–868.

[81] M. Fayyaz, M.H. Saffar, M. Sabokrou, M. Fathy, R. Klette, F. Huang, Stfcn: spatio-temporal fcn for semantic video segmentation, arXiv preprint, arXiv:1608.05971, 2016.
[82] M. Siam, S. Valipour, M. Jagersand, N. Ray, Convolutional gated recurrent networks for video segmentation, in: 2017 IEEE International Conference on Image Processing (ICIP), IEEE, 2017, pp. 3090–3094.
[83] D. Nilsson, C. Sminchisescu, Semantic video segmentation by gated recurrent flow propagation, in: Proceedings of the IEEE Conference on Computer Vision and Pattern Recognition, 2018, pp. 6819–6828.
[84] C.R. Qi, H. Su, K. Mo, L.J. Guibas, Pointnet: deep learning on point sets for 3d classification and segmentation, in: Proceedings of the IEEE Conference on Computer Vision and Pattern Recognition, 2017, pp. 652–660.
[85] C.R. Qi, L. Yi, H. Su, L.J. Guibas, Pointnet++: deep hierarchical feature learning on point sets in a metric space, in: Advances in Neural Information Processing Systems, 2017, pp. 5099–5108.
[86] B.-S. Hua, M.-K. Tran, S.-K. Yeung, Pointwise convolutional neural networks, in: Proceedings of the IEEE Conference on Computer Vision and Pattern Recognition, 2018, pp. 984–993.
[87] L. Landrieu, M. Simonovsky, Large-scale point cloud semantic segmentation with superpoint graphs, in: Proceedings of the IEEE Conference on Computer Vision and Pattern Recognition, 2018, pp. 4558–4567.
[88] L. Tchapmi, C. Choy, I. Armeni, J. Gwak, S. Savarese, Segcloud: semantic segmentation of 3d point clouds, in: 2017 International Conference on 3D Vision (3DV), IEEE, 2017, pp. 537–547.
[89] Y. Liu, B.M. Nacewicz, G. Zhao, N. Adluru, G.R. Kirk, P.A. Ferrazzano, M.A. Styner, A.L. Alexander, A 3d fully convolutional neural network with top-down attention-guided refinement for accurate and robust automatic segmentation of amygdala and its subnuclei, Frontiers in Neuroscience 14 (2020) 260.
[90] B. Wu, A. Wan, X. Yue, K. Keutzer, Squeezeseg: convolutional neural nets with recurrent crf for real-time road-object segmentation from 3d lidar point cloud, in: 2018 IEEE International Conference on Robotics and Automation (ICRA), IEEE, 2018, pp. 1887–1893.
[91] B. Wu, X. Zhou, S. Zhao, X. Yue, K. Keutzer, Squeezesegv2: improved model structure and unsupervised domain adaptation for road-object segmentation from a lidar point cloud, in: 2019 International Conference on Robotics and Automation (ICRA), IEEE, 2019, pp. 4376–4382.
[92] A. Milioto, I. Vizzo, J. Behley, C. Stachniss, Rangenet++: fast and accurate lidar semantic segmentation, in: 2019 IEEE/RSJ International Conference on Intelligent Robots and Systems (IROS), IEEE, 2019, pp. 4213–4220.
[93] K. Sirohi, R. Mohan, D. Büscher, W. Burgard, A. Valada, Efficientlps: efficient lidar panoptic segmentation, arXiv preprint, arXiv:2102.08009, 2021.
[94] A. Valada, G. Oliveira, T. Brox, W. Burgard, Deep multispectral semantic scene understanding of forested environments using multimodal fusion, in: International Symposium on Experimental Robotics (ISER), 2016.
[95] C. Couprie, C. Farabet, L. Najman, Y. LeCun, Indoor semantic segmentation using depth information, arXiv preprint, arXiv:1301.3572, 2013.
[96] C. Hazirbas, L. Ma, C. Domokos, D. Cremers, Fusenet: incorporating depth into semantic segmentation via fusion-based cnn architecture, in: Asian Conference on Computer Vision, Springer, 2016, pp. 213–228.

[97] L. Deng, M. Yang, T. Li, Y. He, C. Wang, Rfbnet: deep multimodal networks with residual fusion blocks for rgb-d semantic segmentation, arXiv preprint, arXiv:1907.00135, 2019.

[98] S. Gupta, R. Girshick, P. Arbeláez, J. Malik, Learning rich features from rgb-d images for object detection and segmentation, in: European Conference on Computer Vision, Springer, 2014, pp. 345–360.

[99] D. Eigen, M. Ranzato, I. Sutskever, Learning factored representations in a deep mixture of experts, arXiv preprint, arXiv:1312.4314, 2013.

[100] R.A. Jacobs, M.I. Jordan, S.J. Nowlan, G.E. Hinton, Adaptive mixtures of local experts, Neural Computation 3 (1) (1991) 79–87.

[101] A. Valada, G. Oliveira, T. Brox, W. Burgard, Towards robust semantic segmentation using deep fusion, in: Robotics: Science and Systems (RSS 2016) Workshop, Are the Sceptics Right? Limits and Potentials of Deep Learning in Robotics, 2016.

[102] S.-J. Park, K.-S. Hong, S. Lee, Rdfnet: rgb-d multi-level residual feature fusion for indoor semantic segmentation, in: Proceedings of the IEEE International Conference on Computer Vision, 2017, pp. 4980–4989.

[103] Y. Li, J. Zhang, Y. Cheng, K. Huang, T. Tan, Semantics-guided multi-level rgb-d feature fusion for indoor semantic segmentation, in: 2017 IEEE International Conference on Image Processing (ICIP), IEEE, 2017, pp. 1262–1266.

[104] M.H. Saffar, M. Fayyaz, M. Sabokrou, M. Fathy, Semantic video segmentation: a review on recent approaches, arXiv preprint, arXiv:1806.06172, 2018.

[105] M. Cordts, M. Omran, S. Ramos, T. Rehfeld, M. Enzweiler, R. Benenson, U. Franke, S. Roth, B. Schiele, The cityscapes dataset for semantic urban scene understanding, in: Proceedings of the IEEE Conference on Computer Vision and Pattern Recognition, 2016, pp. 3213–3223.

[106] A. Geiger, P. Lenz, C. Stiller, R. Urtasun, Vision meets robotics: the kitti dataset, The International Journal of Robotics Research (2013).

[107] G. Neuhold, T. Ollmann, S. Rota Bulo, P. Kontschieder, The mapillary vistas dataset for semantic understanding of street scenes, in: Int. Conf. on Computer Vision, 2017, pp. 4990–4999.

[108] D. Seita, Bdd100k: a large-scale diverse driving video database, The Berkeley Artificial Intelligence Research Blog. Version 511 (2018) 41.

[109] G. Varma, A. Subramanian, A. Namboodiri, M. Chandraker, C. Jawahar, Idd: a dataset for exploring problems of autonomous navigation in unconstrained environments, in: IEEE Winter Conference on Applications of Computer Vision (WACV), 2019, pp. 1743–1751.

[110] P.K. Nathan Silberman, Derek Hoiem, R. Fergus, Indoor segmentation and support inference from rgbd images, in: Proc. of the Europ. Conf. on Computer Vision, 2012.

[111] J. Xiao, A. Owens, A. Torralba, Sun3d: a database of big spaces reconstructed using sfm and object labels, in: Proceedings of the IEEE International Conference on Computer Vision, 2013, pp. 1625–1632.

[112] S. Song, S.P. Lichtenberg, J. Xiao, Sun rgb-d: a rgb-d scene understanding benchmark suite, in: Proceedings of the IEEE Conference on Computer Vision and Pattern Recognition, 2015, pp. 567–576.

[113] A. Dai, A.X. Chang, M. Savva, M. Halber, T. Funkhouser, M. Nießner, Scannet: richly-annotated 3d reconstructions of indoor scenes, in: Proc. Computer Vision and Pattern Recognition (CVPR), IEEE, 2017.

[114] M. Everingham, L. Van Gool, C.K. Williams, J. Winn, A. Zisserman, The Pascal visual object classes (voc) challenge, International Journal of Computer Vision 88 (2) (2010) 303–338.

[115] T.-Y. Lin, M. Maire, S. Belongie, J. Hays, P. Perona, D. Ramanan, P. Dollár, C.L. Zitnick, Microsoft coco: common objects in context, in: European Conference on Computer Vision, Springer, 2014, pp. 740–755.

[116] B. Zhou, H. Zhao, X. Puig, T. Xiao, S. Fidler, A. Barriuso, A. Torralba, Semantic understanding of scenes through the ade20k dataset, International Journal of Computer Vision 127 (3) (2019) 302–321.

CHAPTER

3D object detection and tracking 13

Illia Oleksiienko and Alexandros Iosifidis
Department of Electrical and Computer Engineering, Aarhus University, Aarhus, Denmark

13.1 Introduction

3D object detection and tracking are essential tasks in robotics and autonomous systems where it is crucial for a robot (or agent) to not only understand the environment, the objects inside it and how they move, but also to determine their precise locations so that it can plan its movement or actions to interact with (or avoid) other objects. Being extensions of their 2D counterparts, methods for 3D object detection and tracking share many principles and ideas with 2D object detection and tracking methods described in Chapter 11.

While both 2D and 3D object detection tasks aim at detecting the same type of objects appearing inside a scene, 3D object detection methods estimate the exact object position in the 3-dimensional scene, as opposed to the 2-dimensional position of the projection of the object on the image plane estimated by 2D object detection methods. Images captured by cameras used by 2D object detection methods can also be used for 3D object detection, but with low effectiveness due to the lack of depth information. In order to incorporate depth information for 3D object detection, several types of sensors can be used, with Lidar sensor being the most widely adopted choice to date. This sensor creates a 3D point cloud, that is, a set of 3D points, obtained from laser reflections on the surfaces of objects appearing in the scene. Point cloud is an excellent source of depth information, but it is not structured in a regular grid (like images) and it cannot be efficiently processed by conventional deep learning models, like Convolutional Neural Networks (CNNs). For this reason, preprocessing is commonly applied to point cloud data before they are introduced to deep learning models for object detection.

3D object tracking is very similar to its 2D counterpart (Chapter 11). 2D multiple object tracking is usually performed based on the association of bounding boxes enclosing the detected objects or based on detected objects' representations comparison through the consecutive time instances. These ideas can be directly used for 3D object tracking, and even many 2D object tracking methods require only minor modifications to be used for tracking objects in 3D. Due to the variety of sensors that can be used for 3D object detection and tracking, many methods try to fuse data from multiple sensors to improve detection performance, usually sacrificing processing speed.

Deep Learning for Robot Perception and Cognition. https://doi.org/10.1016/B978-0-32-385787-1.00018-X

In this chapter, we start from an overview of the 3D object detection task in Section 13.2, followed by a description of the input data that can be used by 3D object detection methods in Section 13.2.1 and the most widely used public data sets and metrics used for performance evaluation in Section 13.2.2. We group 3D object detection methods based on the type of data they use, and we describe Lidar-based methods in Section 13.2.3, methods combining image and Lidar data in Section 13.2.4, methods using monocular images in Section 13.2.5 and those using binocular images in Section 13.2.6. Further, we provide an overview of the 3D object tracking task in Section 13.3 and the most commonly used public data sets and performance evaluation metrics in Section 13.3.1. We split 3D object tracking methods into two categories, those applying tracking after detection in Section 13.3.2.1, and those combining object detection and tracking tasks in Section 13.3.2.2.

13.2 3D object detection

3D object detection is a natural extension of the 2D object detection task described in Chapter 11 to the 3-dimensional world. As opposed to 2D object detection, this task focuses on the detection of surrounding objects by estimating the 3D bounding boxes enclosing them. This requires the estimation of 3 coordinate values expressing their position in the 3D scene, 3 size dimensions expressing their width, length, and height, and one rotation angle expressing their orientation. The 3D coordinates are presented in the Cartesian coordinate system centered at the position of the sensor. Both coordinates and size dimensions are commonly expressed in meters. The rotation angle describes rotation around the vertical-axis of an object's 3D bounding box. Existing 3D object detection methods estimate only one rotation angle because the objects commonly considered in the current applications, such as cars, pedestrians, and cyclists, are parallel to the ground and their orientation can be represented with only one angle. However, with the rapid progress of the methods in this task, it is possible that object classes requiring more precise definition of objects' position and orientation will be also targeted in the future. Apart from the precise estimation of the position, the size and orientation of objects, the classification of the detected objects to a set of predefined object classes is usually targeted. Because the objective is to localize objects in the 3D world, 3D object detection methods commonly need to receive as input depth information of the scene. This can be achieved by using a variety of sensors, as it will be described in the following. An example 3D scene from the KITTI data set [1], along with the ground-truth bounding boxes of the objects appearing in it is shown in Fig. 13.1.

13.2.1 Input data for 3D object detection

There exist different types of sensors that can be used to collect information from the 3D scene and perform 3D object detection. One of the most affordable sensors is a camera that will produce monocular images providing only one point of view of

FIGURE 13.1

Example of ground-truth data in a KITTI data set. The top part of the figure corresponds to the 3D point cloud obtained by using a Lidar sensor. The points corresponding to objects are shown in purple and the 3D bounding boxes are shown in red. The bottom part of the figure shows the same object detections projected onto an image captured by a camera placed close to the Lidar sensor and synchronized with it. As can be seen, the point cloud data obtained by the Lidar sensor can capture a larger part of the scene, as the sensor collects 3D data from the surrounding of the sensor, while the image depicts a small part of it corresponding to the field of view of the camera. For instance, it can be seen that the big truck is not fully inside the camera field of view, and thus its 3D bounding box exceeds the borders of the image when it is projected to the image plane. (For interpretation of the colors in the figure, the reader is referred to the web version of this chapter.)

the 3D scene contents, or two cameras which are close to each other (stereo camera) that will produce binocular images suitable for generating disparity maps encoding depth information of the scene contents relative to the camera. Even though binocular images encode depth information of the 3D scene, existing methods exploiting such depth information do not usually lead to high performance.

Lidar (which stands for *laser imaging, detection, and ranging*) sensor is the most commonly used sensor among the state-of-the-art 3D object detection methods. It creates a 3D point cloud, which is an array of a few hundred thousand points, with the 3D position of a point expressed in the 3D coordinate space (x, y, z). Additional information such as the point reflectance level and color can also be provided. While point cloud data encode information related to the depth of objects with respect to the position of the sensor, the farther an object from the sensor is, the lower the number of points on its visible surfaces will be present in the point cloud. Moreover, the effective range of the sensor is relatively low, ranging from 50 m to 150 m [2], while the collected data cannot be easily processed by standard deep learning models, like CNNs that assume regular-grid data.

Another sensor that can capture depth information is Radar (which stands for *radio detection and ranging*). One advantage of Radar sensors compared to Lidar is that they have higher effective range (200–300 m) and can be useful for measuring the velocity of objects. However, the data collected by Radar sensors are even sparser than those collected by Lidar sensors, and currently they are not a common choice among the state-of-the-art 3D object detection methods.

13.2.2 3D object detection data sets and metrics

The KITTI data set [1] is a widely used data set for 3D object detection, containing 7481 training samples and 7518 test samples. Each of the training samples includes left and right RGB camera images, 3 temporal preceding frames for left and right cameras, Lidar point cloud data, camera calibration matrices, object bounding boxes, and class labels. The test samples include all the information provided for the training samples, excluding the object bounding boxes and class labels. Users need to generate the outputs of their model and submit them to the KITTI testing server, which will evaluate the results and provide the performance level of the method. The data set includes three object categories, namely cars, pedestrians, and cyclists. Objects are divided into three groups by their difficulty, that is, easy, moderate, and hard, based on their size and occlusion levels. The KITTI data set has some limitations: scenes are recorded during daytime, mostly under sunny weather, and it describes a class-imbalanced detection problem where 75% of the objects are cars, 4% are cyclists, and 15% are pedestrians [3].

The NuScenes data set [2] is a relatively new dataset that includes more sensors compared to the KITTI data set. It includes 1 Lidar, 5 Radars, 6 cameras, Inertial Measurement Unit (IMU), and Global Positioning System (GPS). 390,000 Lidar sweeps are available to be used for object detection.

Most of the existing 3D object detection approaches resemble the related 2D object detection ones, with the addition of a third dimension on the input data corresponding to the depth of collected 3D points. As will be described in the following sections, considerable effort has been put in transforming such 3D point clouds to data structures that resemble pseudo-images to exploit processing steps similar to those used in 2D object detection methods.

Nonmaximum suppression is used in the same way as in 2D object detection, but the Intersection over Union (IoU) metric is measured using the corresponding 3D volumes of the intersection and the union, instead of 2D area as is the case in 2D object detection. Similar to 2D object detection, the main detection performance metric is *Mean Average Precision* (MAP or mAP), which is computed as the mean of the areas under the precision-recall curve for different object classes (more details are given in Chapter 11). Calculation of average precision is based on the process used to define positive predictions, for which two strategies are followed. The first strategy uses the 3D IoU with a threshold θ (usually set to 0.5 and 0.7) to define a positive detection, while the second strategy uses the distance between the detection and the ground-truth centers on the 2D ground plane with different thresholds (usually set equal to 0.5, 1, 2, and 4 meters) and then mAP is computed by averaging the obtained results using these thresholds. These two strategies are followed by the KITTI and the nuScenes datasets, respectively.

The main inference speed metric used for the evaluation of 3D object detection methods is Frames Per Second (FPS), which indicates how many samples per second can be processed by a method. KITTI data set uses only the 3D (mean) average precision to evaluate the performance of a method. NuScenes dataset uses mAP, but also includes the following performance evaluation metrics:

- Average Translation Error (ATE), which expresses the 2D distance between centers of ground-truth and detected boxes:

$$ATE = |(x_c, y_c)_{dt} - (x_c, y_c)_{gt}|, \tag{13.1}$$

  where $(x_c, y_c)_{dt}$ represents center of detected box in top-down view, $(x_c, y_c)_{gt}$ is the center of ground-truth box in top-down view, and $|\cdot|$ is the absolute value operator;
- Average Scale Error (ASE), which is equal to $1 - IoU$ after aligning the centers and rotation angles of the ground-truth and the detected boxes;
- Average Orientation Error (AOE), which is the smallest difference between the vertical-axis rotation angles between the ground-truth and the detected boxes;
- Average Velocity Error (AVE), which is the L2-norm of the predicted and ground-truth velocities difference in top-down view;
- Average Attribute Error (AAE), which is given by $1 - acc$, where acc is the attribute classification accuracy;
- NuScenes Detection Score (NDS), which combines all the above-described metrics:

$$NDS = \frac{1}{10}\left[5mAP + \sum_{mTP \in \mathbb{TP}} (1 - \min(1, mTP))\right], \tag{13.2}$$

  where $\mathbb{TP}$ is the set of classwise mean values of ATE, ASE, AOE, AVE, and AAE.

As described before, various types of sensors can be used to collect data processed by 3D object detection methods. In the following, we provide a description of such methods categorized by the type of sensors they use.

13.2.3 Lidar-based 3D object detection methods

The most accurate 3D object detection methods rely purely on point cloud data generated by Lidar sensors. The main difficulty in using point cloud data is related to their irregular structure that makes them impossible to be used as inputs to standard deep learning models (such as CNNs) directly. To overcome this problem, VoxelNet

[4] divides point cloud data of a 3D scene into volumetric elements (voxels), which correspond to small regions of predefined size that aggregate information about all points inside them. Since the points corresponding to objects farther to the sensor are sparser, voxels located far away from the sensor in the scene are sparser compared to those located close to the sensor, which means that they commonly contain few points, or they can even be empty.

13.2.3.1 VoxelNet

VoxelNet [4] considers only a subset of the entire 3D scene. This method is evaluated on KITTI dataset that requires detections only in camera range, so all points outside the camera field of view are not considered, reducing the number of points to be processed from 100k–200k to around 20k points per scene. From such a reduced scene, VoxelNet considers a sub-scene in a cuboid shape with boundaries of $([0, x], [-y, y], [z_0, z_1])$. Here, the first dimension corresponds to depth and $[0, x]$ means that the part of the scene considered by the method starts from the position of the sensor and extends to x meters in front of it. The second dimension corresponds to width (left-right) and $[-y, y]$ means that we consider y meters in both directions. Finally, the third dimension corresponds to the height (vertical dimension) and different lower and upper limits can be selected based on the object classes considered. VoxelNet uses the ranges $([0, 70.4], [-40, 40], [-3, 1])$ for cars and $([0, 48], [-20, 20], [-3, 1])$ for pedestrians and cyclists. The range of the vertical dimension starts from a value of -3 because in KITTI data set the Lidar sensor is placed on top of a car, which means that the coordinate center is above the ground level. The large difference between the ranges of the dimensions of subscenes considered can be explained by the size of the objects belonging to each class. Cars, having a big size, can be effectively identified at higher distances, while humans who are much smaller in size are hard to detect when they are far away from the sensor as they will be represented by a small number of points. VoxelNet creates one model for the detection of cars and another one for simultaneous detection of pedestrians and cyclists, using the above-mentioned subscene limit ranges. VoxelNet is formed by three processing blocks, as shown in Fig. 13.2.

Given the input point cloud data corresponding to the subscene of a Lidar sweep, cuboid-shaped voxels with a size of $\mathbf{v} = (0.2, 0.2, 0.4)$ meters are defined. Each voxel encloses a set of points which are processed to create a feature vector by the *feature learning network*. The feature learning network performs random sampling on the points inside each voxel so that each voxel will contain at most T points. The value of T differs for different classes, that is, a value of $T = 35$ is used for cars, while $T = 45$ is used for pedestrians and cyclists. This is done to reduce the number of computations and to balance the density of voxels in different locations of the scene, as the initial point cloud can have many points inside close to the sensor areas and few points (which can lead even to empty voxels) in areas far away from the sensor.

The selected points are subsequently introduced to the so-called Voxel Feature Encoding (VFE) layers, which perform a feature transformation from R^7 to R^C space. The first layer of VFE receives as input a point representation in R^7, where the

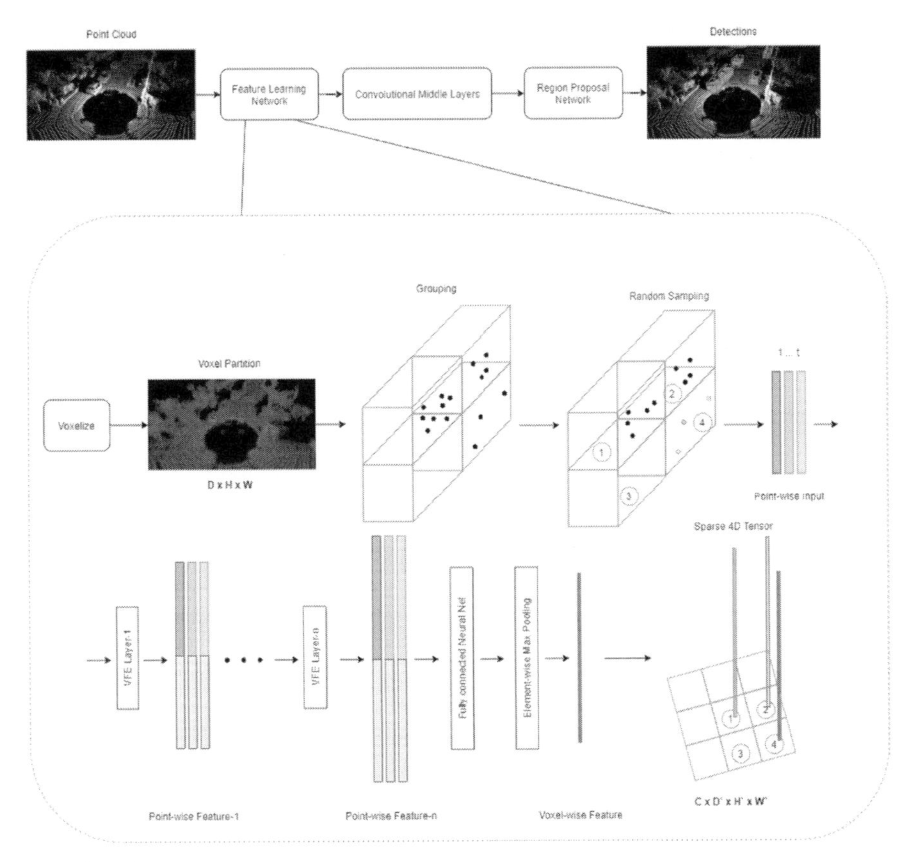

FIGURE 13.2

Structure of VoxelNet model.

first 4 dimensions (x, y, z, r) correspond to the 3D position and reflectance level of a point (which are available in KITTI point cloud data), while the remaining 3 dimensions $(\delta_x, \delta_y, \delta_z)$ correspond to offsets of the point with respect to the centroid of the voxel it belongs to. VFE layers are formed by fully-connected layers with batch normalization [5] and Rectified Linear Unit (ReLU) [6] activation function to transform pointwise features to the desired feature space. Subsequently, max-pooling is used to aggregate features from all points of a voxel into an m-dimensional vector that is concatenated to the features of the points in the voxel. This way, the representation of each point is augmented with information related to its neighbors. VoxelNet uses a value of $C = 128$ and employs two layers that perform the mappings $7 \rightarrow 32$ and $32 \rightarrow 128$, respectively, for all classes. Only nonempty voxels are processed, and then their features are placed into a 3D grid to create a 4D tensor with size (C, D', H', W'). Here, the first dimension corresponds to the C features of each voxel, and the remaining three dimensions correspond to the number of voxels that fit to the ranges $[z_0, z_1]$,

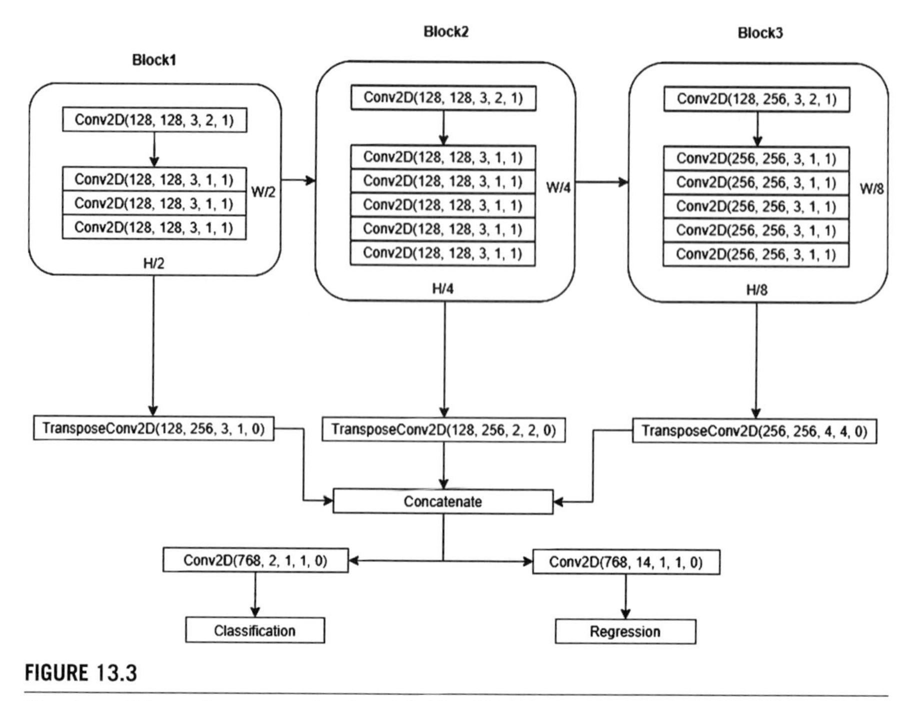

FIGURE 13.3

Structure of the region proposal network in VoxelNet.

$[-y, y]$ and $[0, x]$, respectively. For cars, the VoxelNet model creates a 4D tensor of $(128, 10, 400, 352)$ shape, while a tensor of $(128, 10, 200, 240)$ shape is created for pedestrians and cyclists. This tensor is processed by 3D convolutional middle layers to create a $(64, 2, 400, 352)$ tensor for cars or a $(64, 2, 200, 240)$ for pedestrians and cyclists. This tensor is then squeezed along the second dimension to create a 3D tensor of $(128, H, W)$ elements. Such a tensor is suitable for regular 2D Convolutional layers.

The Region Proposal Network (RPN) processes the 3D tensor obtained by the convolutional middle layers. RPN consists of 3 blocks, as can be seen in Fig. 13.3. Each of these blocks has one convolutional layer with stride 2 to reduce the width and height of the resulting tensor, and 3 layers for the first block and 5 layers for the next two blocks with stride 1. The output of each block is then up-sampled using transpose convolution to size $(256, \frac{H'}{2}, \frac{W'}{2})$. The three resulting up-sampled tensors are concatenated into one $(768, \frac{H'}{2}, \frac{W'}{2})$ block, which is used to create classification and regression outputs with sizes $(CL * A, \frac{H'}{2}, \frac{W'}{2})$ and $(7 * A, \frac{H'}{2}, \frac{W'}{2})$, respectively, where CL is the number of classes and A is the number of anchors per voxel. This means that the classification branch creates a score for each anchor to select the most suitable one and the regression branch creates bounding box predictions $(x, y, z, w, h, l, \theta)$ for each anchor, where (x, y, z) is a position of an object, (w, h, l) is a size of an object and θ is a rotation angle along a vertical axis. In case that the background is not included in the set of classes, that is, the number of object classes is equal to CL, then class scores falling below a certain threshold are considered to be background.

VoxelNet provides high detection performance with a low inference time, roughly 75% of which corresponds to the calculations conducted by the 3D convolutional middle layers. It influenced many other 3D object detection methods, such as PointPillars [7], SECOND [8], and TANet [9] which follow processing steps similar to VoxelNet.

13.2.3.2 PointPillars

As mentioned above, most of the computations in VoxelNet are performed in the 3D convolutional middle layers. Thus if the second processing block of VoxelNet can be omitted without sacrificing performance, the processing speed can be highly increased. PointPillars [7] processes the same subscene as in VoxelNet, and proposes to process the input point cloud using a different kind of voxels, the so-called pillars. The main difference between voxels and pillars is that the former divides the 3D volume of the scene along all three dimensions, while the latter divides the 3D volume of the scene along only the depth and width dimensions and aggregates information on the vertical dimension. For example, the most accurate PointPillars models use pillars of size $(0.16, 0.16, 4)$. Comparing this with the size of the 3D volume of the scene used by VoxelNet in KITTI data set, it can be seen that all voxels with the same depth and width dimensions which are stacked on top of each other are used to form a pillar. Thus since the vertical dimension is effectively discarded by the use of pillars instead of voxels, PointPillars defines a pseudo-image with (C, H, W) elements, where the first dimension corresponds to the C features of each pillar, and the remaining two dimensions correspond to the number of pillars that fit to the ranges $[-y, y]$ and $[0, x]$, respectively. Such a 3D tensor can be directly processed by regular 2D CNNs.

Processing of pillars to calculate their features is slightly different compared to the process followed by VoxelNet to process voxels. The features of each point are augmented with 5 numbers $(x_c, y_c, z_c, x_p, y_p)$. The first three of these values correspond to the distance of the point from the position of the mean point in the pillar, and the last two numbers correspond to the 2D distance of the point from the center of the pillar. Similar to VoxelNet, PointPillars randomly samples points inside a pillar when the pillar contains more than $T = 100$ points. However, in the case where the number of points inside a pillar is less than T, the resulting array is padded with zeros, leading to the same size of features at all places of the resulted pseudo-image. Finally, PointPillars uses the same region proposal network as VoxelNet, shown in Fig. 13.3.

By changing the voxelization structure of the input point cloud and removing the Cconvolutional middle layers of VoxelNet, PointPillars is more efficient compared to VoxelNet, while it can also achieve higher performance.

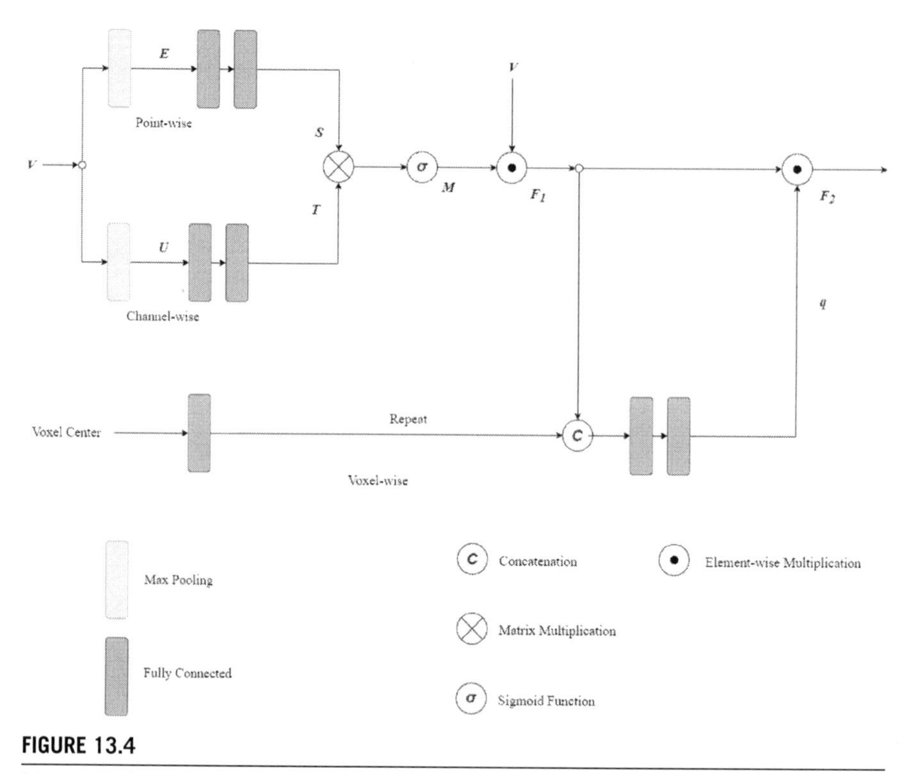

FIGURE 13.4

Structure of the triple attention module in TANet.

13.2.3.3 TANet

PointPillars became a popular method, but its performance and robustness are not ideal. TANet [9] focuses on improving these two characteristics of PointPillars by changing its first and last processing steps, that is, the voxel feature extraction and the region proposal network.

For improving the voxel feature extraction processing step, a stacked Triple Attention (TA) module is used. As can be seen in Fig. 13.4, the TA module performs three types of attention, namely pointwise, channelwise, and voxelwise, to improve the quality and the robustness of resulting voxel features. This triple attention can be used for processing both voxels and pillars (which are viewed as elongated voxels with respect to the z dimension) as will be described in the following.

Pointwise attention is designed to encode spatial correlation across the points of a voxel. It takes as input a voxel which is represented by a matrix $V \in \mathbb{R}^{T \times C}$, where T is the number of points in the voxel and C is the dimensionality of each point (which is different for voxels and pillars, as described in Section 13.2.3.1 and Section 13.2.3.2, respectively). Max-pooling across the C dimension is applied to produce pointwise responses, taking the maximum of all features for each point and creating the vector $E \in \mathbb{R}^{T \times 1}$. The resulting attention vector $S \in \mathbb{R}^{T \times 1}$ is obtained by introducing E to a two-layer fully-connected network with weights $W_1 \in \mathbb{R}^{r \times T}$ and $W_2 \in \mathbb{R}^{T \times r}$:

$$S = W_2 \delta(W_1 E), \tag{13.3}$$

where $\delta(\cdot)$ is an activation function used for the first fully-connected layer, selected to be the ReLU function.

Channelwise attention is applied in parallel to pointwise attention, but across T dimension in order to aggregate features of different points. Max-pooling is used to create channelwise responses $U \in \mathbb{R}^{1 \times C}$. Two fully-connected layers with weights $W_1' \in \mathbb{R}^{r \times C}$ and $W_2' \in \mathbb{R}^{C \times r}$ are subsequently applied to create a channelwise attention vector $J \in \mathbb{R}^{1 \times C}$:

$$J = W_2' \delta(W_1' U^T). \tag{13.4}$$

The pointwise and channelwise attention matrices are combined to create the attention matrix $M \in \mathbb{R}^{T \times C}$:

$$M = \sigma(S J^T), \tag{13.5}$$

where the sigmoid activation function $\sigma(\cdot)$ is used to normalize values to the range $[0, 1]$. Finally, the attention matrix M is combined with original voxel features by applying elementwise multiplication, leading to the attended features $F_1 \in \mathbb{R}^{T \times C}$ encoding the importance of points inside each voxel:

$$F_1 = \sigma(M \odot V), \tag{13.6}$$

where $\odot$ denotes the elementwise multiplication operator.

Voxelwise attention is used to encode the importance of voxels. All points inside a voxel are averaged to obtain the voxel center, which is transformed to a higher dimension using a fully-connected layer. The output of this layer is concatenated to F_1. Voxelwise attention weight $q \in \mathbb{R}$ is computed using two fully-connected layers that compress pointwise and channelwise attention vectors. Finally, the resulting feature matrix $F_2 \in \mathbb{R}^{T \times C}$ is computed by

$$F_2 = q F_1. \tag{13.7}$$

Several TA modules can be stacked in a hierarchical manner. TANet stacks two TA modules, where the first one operates on the original point cloud data, and the second one receives as input the outputs of the first module which have a higher dimensionality. To fuse more information, the input and the output of each TA module are concatenated (first module) or summed together (second module). After the stacked TA, fully-connected and max-pooling layers are used to create the final pseudo-image with shape (F, H, W).

TANet also employs a coarse-to-fine regression network, which processes the generated pseudo-image to create the final predictions in two steps, as shown in Fig. 13.5. First, the region proposal network from PointPillars is used to create

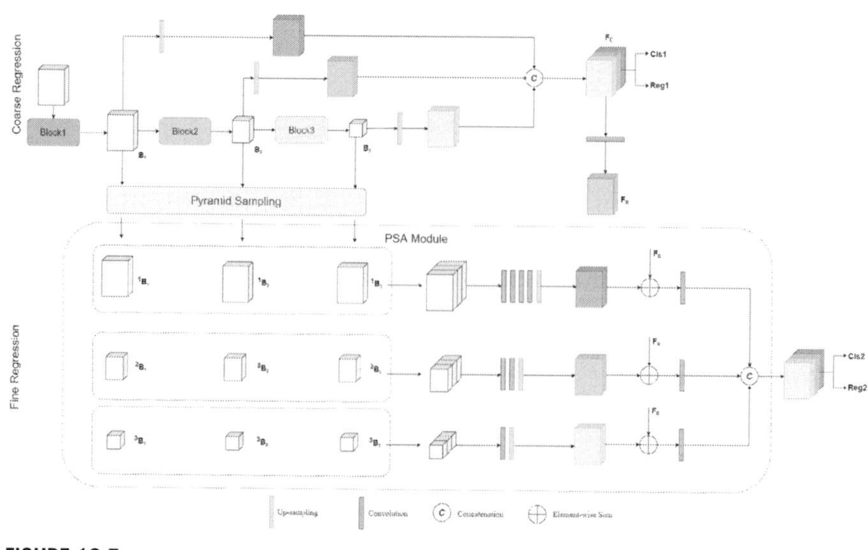

FIGURE 13.5

Coarse-to-fine regression network of TANet.

coarse detections (Cls1 and Reg1). Then the outputs of each block are up- and down-sampled (using transposed convolution and pooling, respectively) to create three sets of blocks, each of which is formed by three new blocks, which are concatenated and processed with a set of convolutional and up-sampling layers. The resulting three blocks will have the same size with the detections obtained by the coarse network, and are combined with the output of the coarse network through summation. Finally, by applying a convolutional layer and concatenating the three resulting tensors, we obtain the output of the coarse-to-fine regression network.

TANet uses the scene ranges and voxel sizes of PointPillars. It achieves higher performance and is one of the best methods for pedestrians class detection. However, its inference speed is twice lower than that of PointPillars due to complex voxel feature extraction part, but has higher overall detection accuracy and robustness in general.

13.2.3.4 HotSpotNet

All Lidar-based methods described above use anchors to create object predictions, but it is also possible to perform 3D object detection without anchors. Similar to the 2D object detection case, where related methods output a heatmap [10] in which high values indicate the center of an object and the corresponding size and offset predictions are calculated based on these central points, HotSpotNet [11] classifies voxels to two categories, namely hotspots and background voxels, and uses detection features only from hotspots for 3D object detection.

During the training stage, only a subset of voxels inside each object are marked as positive (hotspots), while all other voxels are assigned to the background class.

This allows the method to be less biased to differences in point density between close and far objects because only a subset of points for close objects will be marked as positive, while far objects will usually have a tiny number of nonempty voxels.

A backbone network, such as PointPillars [7], is used to create 3D feature maps, which are subsequently used to generate voxel-vise hotspot classifications H and bounding box predictions B. The final predictions are computed by applying the hotspot mask on the box predictions tensor using elementwise multiplication:

$$R = B \odot H. \tag{13.8}$$

HotSpotNet is slower compared to TANet, but it can achieve higher performance for all object classes.

13.2.3.5 Point-based methods

Instead of preprocessing the point cloud data to create grid-structured data, which are subsequently introduced to neural networks, it is possible to apply neural networks directly on points to generate pointwise features and use them for 3D bounding box estimation.

PointNet [12] applies order-invariant pointwise functions to address the non-fixed order of point cloud data. Summation, multiplication, and maximum operations lead to the same result for any order of their inputs, so fully-connected (FC) layers (performing summation and multiplication) and max-pooling layers (performing the maximum operator) are used to create point features. PointNet contains 2 T-Net sub-networks that are used for pose normalization at the input transformation level and the feature transformation level. T-Net consists of a set of FC layers shared between points (the same layer is applied to each point in parallel), max-pooling and matrix multiplication. PointNet uses 2 FC layers between T-Nets and 3 after the second one, followed by max-pooling to create a global feature vector with size K. This vector is concatenated to the features of all points. The resulting feature vectors are processed with FC layers to create pointwise classification scores, which lead to point cloud semantic segmentation. This method is not applied for 3D object detection, but it is heavily used as a part of other 3D object detection methods. It can be applied to the entire point cloud, part of the point cloud corresponding to a subregion of the scene, or the point cloud corresponding to an object proposal. Global features encode information about the neighborhood of points, which is useful for pointwise classification. Since, these features are obtained by applying max-pooling, at most K points can contribute to the global feature vector. These points are called critical points because removal of other points will not affect the global feature vector, resulting in identical point cloud classification and segmentation outputs.

PointNet++ [13] is an extension of PointNet that addresses the poor ability of PointNet to capture local structures. This is done by applying PointNet at different hierarchical levels, obtained by sampling and grouping points to create smaller point clouds. The lowest level of the hierarchy corresponds to the initial point cloud, while the highest level point cloud is the smallest one obtained by aggregating point information through the successive levels. A classification branch processes the higher level point cloud with a set of FC layers to create a single classification vector. Segmentation is performed by backward hierarchical pass from the smaller point cloud to the original one. Features of points at a lower level are obtained by applying interpolation to the features of the corresponding point at the higher level based on their distance to that point and processed with a unit PointNet, which is formed by shared FC layers.

PointRCNN [14] is a 2-stage 3D object detection method. At the first stage, it uses PointNet++ as a backbone for foreground points segmentation, that is, it classifies points into foreground or background classes. Ground-truth data for foreground points segmentation are obtained by using the ground-truth annotations of 3D bounding boxes. A small number of 3D box proposals is created in a bottom-up manner based on the segmentation. For each proposal, points are transformed into the canonical coordinate system centered at the middle of a proposal box. Transformed local features and stage-1 semantic features are combined and processed to refine the box proposals.

13.2.3.6 Projection-based methods

Projection-based methods try to use different types of point-cloud projections to create structured data that can be processed by standard 2D CNNs. **MVCNN** [15] and **GSL + AMTC** [16] use multiple plane projections to create separate features and combine them to perform 3D shape recognition. **VeloFCN** [17] uses cylindrical projection that creates a 2D image with size that depends on the properties of the Lidar sensor. Given the point (x, y, z) and the average horizontal and vertical angle differences $\Delta\theta$ and $\Delta\phi$, the projected position (r, c) can be obtained by

$$\begin{aligned}
\theta &= \operatorname{atan2}(y, x), \\
\phi &= \arcsin\left(\frac{z}{\sqrt{x^2+y^2+z^2}}\right), \\
r &= \lfloor \theta/\Delta\theta \rfloor, \\
c &= \lfloor \phi/\Delta\phi \rfloor.
\end{aligned} \tag{13.9}$$

The resulting projection is a 2D image that is processed by a Fully Convolutional Network (FCN) [18] that generates pointwise box and class predictions. **SqueezeSeg** [19] uses a slightly different spherical projection:

$$\begin{aligned}
\theta &= \arcsin\left(\frac{z}{\sqrt{x^2+y^2+z^2}}\right), \\
\phi &= \arcsin\left(\frac{y}{\sqrt{x^2+y^2}}\right), \\
r &= \lfloor \theta/\Delta\theta \rfloor, \\
c &= \lfloor \phi/\Delta\phi \rfloor,
\end{aligned} \tag{13.10}$$

FIGURE 13.6

Example of a spherical projection depth map calculated using a Lidar point cloud from the KITTI data set is given in the top part of a figure with the corresponding camera image in the bottom part. Each pixel represents a point in the 3D scene. Lighter colors correspond to points at a higher distance from the sensor, while darker colors correspond to points at a lower distance. Absence of a point is indicated by the darkest color and can be found in the top-center part of the projection image. (For interpretation of the colors in the figure, the reader is referred to the web version of this chapter.)

and the resulting 2D image is processed by a CNN to produce a point cloud segmentation output. An example 2D image created by using the spherical projection in Eq. (13.10) can be seen in Fig. 13.6.

Complex-YOLO [20] projects the Lidar point cloud into a Birds-eye-view (BEV) plane, creating an RGB-like top-down image with three channels corresponding to height, intensity, and density features. A similar approach is used in other methods, such as MV3D [21]. An example BEV projection can be seen in Fig. 13.7. The resulting BEV image is processed by a CNN for feature extraction. The Euler-Region-Proposal Network (E-RPN) is then applied to estimate box position, size, and a pair of real and imaginary rotation values. The complex rotation values corresponding to each prediction are then used to estimate the final rotation angle of the corresponding object.

13.2.4 Image+Lidar-based 3D object detection

Lidar-based methods lead to a good compromise between detection performance and inference speed. To improve their performance even further, some methods try to use additional information from camera images. **PointPainting** [22] applies fusion of an image captured by a camera and point cloud data obtained by a Lidar sensor to create "painted points," which correspond to 3D points augmented with their corresponding RGB color values. Not all points are painted, but only those corresponding to pixels with objects on the image. Semantic image segmentation is applied to the input image using a neural network, leading to an output map in which each pixel is assigned to a class label. Subsequently, the projection of pixels onto the point cloud allows painting points corresponding to objects. The newly generated point cloud is then passed to a Lidar-based method, leading to a slightly improved performance compared to the case where regular point cloud data is used.

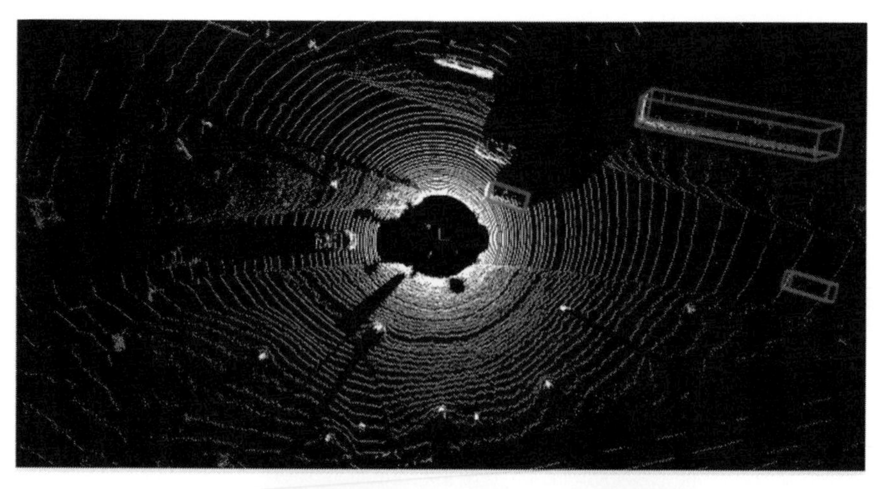

FIGURE 13.7

Example of a BEV projection formed from a KITTI input scene. The top part represents a projected Lidar scene with the ego car in the center, which moves in a left-to-right direction. Boxes represent ground-truth objects and are labeled only for the subscene inside the camera's field of view, which is shown at the bottom part of the figure. Using this projection, it can be easily seen that objects create "shadows," that is, regions in the point cloud with no points because all Lidar beams were reflected at the object's surface.

Frustum PointNets [23] applies a 2D CNN object detector on the input image to define 2D positions of objects. The 2D position of an object defines a frustum, that is, a part of the 3D scene that is projected into the selected boundaries on the camera plane. All points inside this frustum are processed with PointNet [12] to create point-wise binary classification, in which the desired class is defined by the 2D CNN and the binary classification step filters out background points. Lightweight PointNet (T-Net) with Amodal 3D Box Estimation PointNet are subsequently employed to create the final 3D bounding boxes. **Frustum ConvNet** [24] also uses 2D object proposals, but it generates a set of voxel-like frustums for each proposal. Points in frustums are processed with PointNet, which creates features for each frustum. All features are concatenated into a 2D feature map, which is processed with a fully-convolutional network similar to region proposal network of VoxelNet, but this time having four blocks.

13.2.5 Monocular 3D object detection

While Lidar-based methods can achieve good performance, the use of Lidar sensors is not so common in everyday life. For this reason, monocular 3D object detection methods aim to use color images captured by cameras to perform 3D object detection. The main challenge in this approach is related to the lack of depth information in the input data.

13.2.5.1 Prior information fusion based methods

These methods utilize prior information such as object shape or ground-plane location to fuse 2D and 3D information.

Mono3D [25] sets the assumption of a ground-plane based on the camera calibration data and uses this information to place 3D object proposals in a 3D space and to find their 2D projections on the camera plane. This approach allows avoiding multi-scale search in the image. Class semantic and instance semantic branches are used to score each 3D object proposal by shape, context, contour, and location features.

Mono3D++ [26] integrates the three tasks of depth, ground-plane, and shape prior estimation. This method is applied only for car detection. First, a 2D CNN detector creates 2D box proposals on the image plane. Based on the 2D box proposals, a two-branch network is used. The first branch generates 2D object landmarks (14 points for each vehicle), which describe the boundaries of each object with much higher precision compared to the bounding box. The second branch uses a CNN to estimate 3D orientation and scale with 3D bounding boxes and morphable shape models (3D landmarks), based on the estimated 2D landmarks. The parameters of all branches used to provide estimations for all tasks are optimized jointly. This method achieves higher performance compared to Mono3D.

13.2.5.2 Depth-estimation-based methods

MF3D [27] uses MonoDepth [28] as a submodule for monocular depth estimation. MonoDepth is an unsupervised depth estimation method that uses left and right images of a stereo camera for estimating the disparity map corresponding to these images. It uses both left and right images during training and only one image during inference. The left image is treated as reference and the disparity map is computed with respect to the right image, describing the horizontal coordinate difference of pixels that correspond to the same 3D location. Given the estimated disparity value I_d for a pixel (I_x, I_y), its corresponding 3D point is calculated by

$$\begin{aligned} x &= \frac{(I_x - C_x) \cdot z}{f}, \\ y &= \frac{(I_y - C_y) \cdot z}{f}, \\ z &= \frac{f \cdot C_b}{I_d}, \end{aligned} \tag{13.11}$$

where f is the focal length, (C_x, C_y) is the principal point, and C_b is the baseline distance. Concatenation of the monocular image and three maps, namely the depth map, the height map and the distance map, is introduced to a CNN (specifically the RPN from faster R-CNN [29]) to create region proposals. Based on region proposals and point cloud features, 2D and 3D detections are estimated.

MonoGRNet [30] consists of 4 subnetworks, which perform 2D bounding boxes estimation, instance depth estimation, 3D location estimation and local corner regression. Given 2D bounding boxes and depth estimation, the 3D bounding box center locations are predicted, which in combination with the results of local corner regression, produce final 3D bounding boxes. **OFT-Net** [31] uses ResNet-18 [32] as a feature extraction backbone network and transforms resulting features with Orthographic Feature Transform (OFT). OFT generates voxel features $g(x, y, z)$ by accumulating image features $f(u, v)$ over the area in the image corresponding to the projection of the voxel on the image plane. The top-down network processes the transformed features in BEV and predicts classification, offset, size, and angle of a 3D box at each position.

13.2.5.3 Other monocular 3D object detection methods

MonoDIS [33] and **3D bounding box estimation using deep learning and geometry** [34] use a two-stage approach in which a 2D detection network is used to create 2D object proposals and a 3D detection head uses features at the proposed locations to estimate the 3D shape and the location of the selected object. **Real-time seamless single shot 6D object pose prediction** [35] predicts object centers and 3D keypoint projection in 2D image and uses the Perspective-n-Point (PnP) algorithm in [36] to estimate the final 3D position and orientation.

13.2.6 Binocular 3D object detection

Binocular images contain information about the relative depth of objects with respect to the position of the stereo camera, and thus they can be used to address depth-related issues in monocular 3D object detection methods.

Pseudo-LiDAR [37] estimates depth from a single camera image or a pair of images by predicting a disparity map following a similar process as the one used in MF3D (Eq. (13.11)). The resulting pseudo-point cloud data can be subsequently processed by any Lidar-based 3D object detection method. Comparing the pseudo-LiDAR and MF3D methods, we can see that their difference lies in that the first one exploits existing Lidar-based methods operating on point clouds, while the latter creates 2D feature maps, which are processed by standard 2D CNNs.

Pseudo-LiDAR++ [38] uses binocular images because the quality of pseudo-point clouds created following the process in pseudo-LiDAR is influenced by the quality of depth estimation method, which works better with 2 images than with a single image [37]. Disparity estimation is achieved by using PSMNet [39], which constructs a 4D disparity cost volume $C_{disp}(u, v, d, c)$ describing differences of pixels in the left and right images with a horizontal offset d, that is, $I_l(u, v)$ and $I_r(u, v+d)$

encoded by c-dimensional feature vectors. The resulting tensor $S_{disp}(u, v, d, 1)$ is obtained through 3D convolution. Finally, the disparity map $D(u, v)$ is computed as follows:

$$D(u, v) = \sum_{d} softmax(-S_{disp}(u, v, d)) \cdot d, \tag{13.12}$$

where softmax is applied over the last dimension of S_{disp}. This method is optimized to minimize the *disparity error,* which expresses the difference between the predicted $D(u, v)$ and ground-truth $D^*(u, v)$ disparity values for all pixels, that is,

$$E_{disp} = \sum_{(u,v) \in M} |D(u, v) - D^*(u, v)|, \tag{13.13}$$

where $| \cdot |$ is the absolute value operator, and M denotes the disparity map.

In pseudo-LiDAR++, PSMNet is optimized with respect to the depth error instead of the disparity error, that is,

$$E_{depth} = \sum_{(u,v) \in M} |Z(u, v) - Z^*(u, v)|, \tag{13.14}$$

where $Z(u, v) = \frac{f_u \times b}{D(u,v)}$, $Z^*(u, v)$ are the ground-truth depth values and M denotes the depth map. This approach addresses the problem of high influence of small depth errors of nearby objects over depth errors of faraway objects.

A problem of disparity-based depth estimation is that disparity is quantized since it can only take integer values corresponding to offsets in pixel coordinates, but depth is continuous. To address this issue, Pseudo-LiDAR++ also proposed a graph-based depth correction algorithm (GDC), which uses Lidar data and a KNN graph structure to refine the predicted pseudo-points. The GDC can operate using a low-cost Lidar sensor with 4 beams (instead of the 64 usually obtained by high-cost Lidar sensors) and a small number of Lidar points and achieve good performance.

CDN [40] applies 2D semantic image segmentation on the image obtained by the left camera to segment the image in background and foreground pixels. Stereo matching of the pixels in the images obtained by the left and right cameras is performed by another network to estimate the disparity values, but this task is performed separately for background and foreground pixels. The predicted disparity map is subsequently transformed to a point cloud with an additional confidence score. Background points that are not on the ground plane are filtered out, so that only foreground and ground plane points are used to form the resulting point cloud. Then a Lidar-based method is applied to this point cloud to perform 3D object detection.

13.3 3D object tracking

3D object tracking can be considered as an extension of the 2D object tracking performed on detections of objects in a 3D scene instead of the 2D image plane. 3D object tracking methods need to detect objects in an input frame, and to associate them with their appearance in the previous and next frames of the sequence, creating the so-called tracklets, that is, lists of object positions in all frames of the sequence. The most commonly considered problem in 3D is that of multiple object tracking, where a method has to find associations between detections of more than one object in consecutive frames.

13.3.1 3D object tracking data sets and metrics

Both KITTI and nuScenes data sets contain data that can be used for the evaluation of 3D object tracking methods. KITTI tracking data set is formed by 21 training sequences and 29 testing sequences, and it additionally includes IMU/GPS data, as well as detection task data. NuScenes contains 390,000 Lidar sweeps for the object detection and 1,000 sequences of 20-second scenes for tracking. Lidar in nuScenes collects sweeps with a frequency of 20 FPS, which means that each 20-second sequence has 400 Lidar sweeps in it.

Many metrics can be used to evaluate the performance of 3D object tracking methods. NuScenes data set includes a variety of metrics, but the performance of methods is usually compared using the multiobject tracking accuracy (MOTA) and average multiobject tracking accuracy (AMOTA):

$$MOTA = 1 - \frac{\sum_t (m_t + fp_t + mme_t)}{\sum_t g_t} \tag{13.15}$$

where m_t is the number of misses, fp_t is the number of false positives, mme_t is the number of mismatches, and g_t is a number of objects at time t.

$$\begin{aligned} AMOTA &= \frac{1}{n-1} \sum_{r \in \{\frac{1}{n-1}, \frac{2}{n-1} \ldots 1\}} MOTAR, \\ MOTAR &= max\left(0, 1 - \frac{IDS_r + FP_r + FN_r - (1-r) * P}{r * P}\right), \end{aligned} \tag{13.16}$$

where r is a normalization factor, P is a number of ground-truth positives. For a given r, IDS_r is the number of identity switches (misses), FP_r is the number of false positives, FN_r is the number of false negatives.

Other performance evaluation metrics included in nuScenes data set are the following:

- MOTP, which represents misalignment between ground-truth objects and detected objects:

$$MOTP = \frac{\sum_{i,t} (d_t^i)}{\sum_t c_t}, \tag{13.17}$$

where d_t^i is a distance between object o_i and its corresponding hypothesis, c_t is the number of matches found at time t;

- MT (Mostly Tracked), which is the number of ground-truth trajectories that are covered by a predicted track for at least 80% of their respective life span;
- MML (Mostly Lost), which is the number of ground-truth trajectories that are covered by a predicted track for at most 20% of their respective life span;
- MFP (False Positives), which is the total number of predicted objects that have no corresponding ground-truth;
- MFN (False Negatives), which is the total number of ground-truth objects that were not predicted (missed);
- MIDS (Identity Switches), which is the total number of times when few tracks switched their IDs;
- MFRAG (Fragmentations), which is the total number of times a trajectory is interrupted during tracking.
- Average Multiobject tracking Precision (AMOTP), which computes the average of the MOTP metric at different recall thresholds.

The performance of methods evaluated in nuScenes is usually compared using the AMOTA metric.

KITTI data set used the MOTA metric for tracking evaluation but, starting from 2021, switched to using higher order tracking accuracy (HOTA) metrics [41]. HOTA is calculated based on a set of submetrics, which are used to evaluate strong and weak sides of a method. Similar to MOTA, detection-related concepts TP, FN, FP are used. Similar concepts are implemented for measuring association: True Positive Associations (TPA), False Negative Associations (FNA), and False Positive Associations (FPA).

For a given TP box detection c, TPA is defined as the set of TPs with the same ground truth ID (gtID) and prediction ID (prID) as c:

$$\begin{aligned} TPA(c) &= k, \\ & k \in \{TP \mid prID(k) = prID(c) \wedge gtID(k) = gtID(c)\}. \end{aligned} \tag{13.18}$$

For a given TP box detection c, FNA is defined as the set of ground-truth boxes with the same gtID as c that are either assigned to a different prID or were not detected:

$$\begin{aligned} FNA(c) &= k, \\ & k \in \{TP \mid prID(k) \neq prID(c) \wedge gtID(k) = gtID(c)\} \\ & \quad \cup \{FN \mid gtID(k) = gtID(c)\}. \end{aligned} \tag{13.19}$$

Finally, for a given TP box detection c, FPA is defined as the set of prediction boxes with the same prID as c that either have different assigned gtIDs or do not correspond to an object:

$$\begin{aligned} FPA(c) &= k, \\ & k \in \{TP \mid prID(k) = prID(c) \wedge gtID(k) \neq gtID(c)\} \\ & \quad \cup \{FP \mid prID(k) = prID(c)\}. \end{aligned} \tag{13.20}$$

For a particular localization threshold α (3D IoU in 3D object tracking), HOTA_α is defined as follows:

$$\text{HOTA}_\alpha = \sqrt{\frac{\sum_{c\in\{TP_\alpha\}} \mathcal{A}_\alpha(c)}{|TP_\alpha| + |FN_\alpha| + |FP_\alpha|}}, \tag{13.21}$$

$$\mathcal{A}_\alpha(c) = \frac{|TPA_\alpha(c)|}{|TPA_\alpha(c)| + |FNA_\alpha(c)| + |FPA_\alpha(c)|}. \tag{13.22}$$

By defining the detection accuracy score (DetA) and an association accuracy score (AssA) as follows:

$$DetA_\alpha = \frac{|TP_\alpha|}{|TP_\alpha| + |FN_\alpha| + |FP_\alpha|}, \tag{13.23}$$

$$AssA_\alpha = \frac{1}{|TP_\alpha|} \sum_{c\in\{TP_\alpha\}} \mathcal{A}_\alpha(c). \tag{13.24}$$

HOTA can be calculated as the geometric mean between detection accuracy and association accuracy:

$$HOTA_\alpha = \sqrt{DetA_\alpha \cdot AssA_\alpha}. \tag{13.25}$$

13.3.2 3D object tracking methods

There are two possible approaches to solve the 3D object tracking problem. The first one assumes that the detected object bounding boxes are given and resolves only the object association problem between two consecutive time instances. The second one performs detection and tracking simultaneously, providing the detected object bounding boxes and their associations through time at once.

13.3.2.1 Detection-based tracking

Most of 2D detection-based trackers can work using 3D bounding boxes with minor modifications, but application of specific 3D methods leads to higher performance. Similar to 2D object tracking methods, **AB3DMOT** [42] uses Hungarian algorithm [43] for data association and 3D Kalman filter [44] with a constant velocity model for motion modeling. 2D Kalman filter has the state space in the image plane, while the 3D one extends the state space of the objects to 3D, including 3D size, position, velocity, and orientation.

At each frame, the method receives a set of 3D bounding boxes and matches them with existing tracklets. This is done using the Hungarian algorithm based on 3D IoU of given 3D bounding boxes and tracklet predictions, which are created by using the 3D Kalman filter. The result of the application of the Hungarian algorithm is a set of matched pairs, unmatched tracklets and unmatched 3D bounding boxes. Matched pairs represent a pair of a tracklet and a 3D bounding box, having a high IoU between the predicted tracklet position at the current frame and the given 3D bounding box.

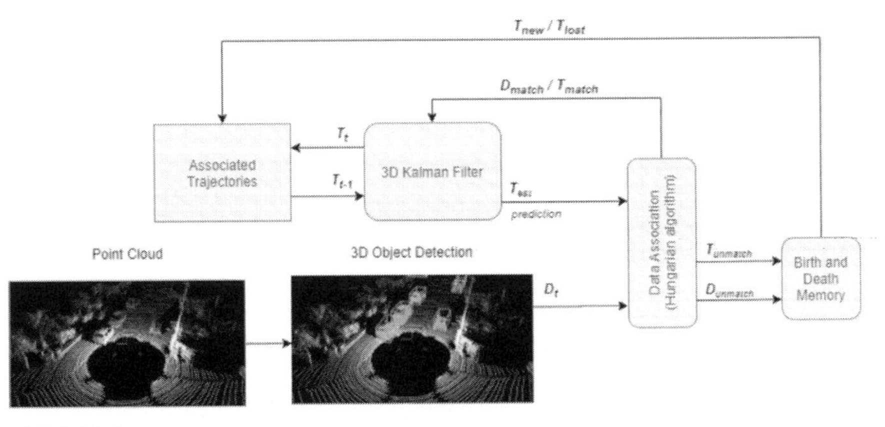

FIGURE 13.8

Processing steps of AB3DMOT method. D_t represent detections at frame t. T_t represents tracklets at frame t. D_{match}, T_{match} and $D_{unmatch}$, $T_{unmatch}$ represent matched and unmatched detections and tracklets, respectively. T_{new}, T_{lost}, represent created and deleted tracklets. T_{est} represents predicted object positions for each tracklet.

A tracklet is considered to be unmatched if its IoU with any 3D bounding box in the current frame does not exceed a predefined threshold value. Unmatched tracklets are marked as stale and when an unmatched tracklet is stale for a predefined number of frames, it is deleted. In a similar manner, a 3D bounding box is considered unmatched if its IoU with any of the tracklets in the current frame does not exceed a predefined threshold. Unmatched 3D bounding boxes are used to initialize new tracklets. The structure of the method is given in Fig. 13.8.

Leveraging shape completion for 3D Siamese tracking [45] takes as input point cloud and first frame detections to initialize tracklets with object point cloud shapes. For each future frame, the shape of a tracklet is compared with candidate shapes based on the cosine similarity of latent vectors extracted from a Siamese pointwise neural network.

Efficient tracking proposals using 2D–3D Siamese networks on LIDAR [46] uses a pair of Siamese networks, one of which is a 3D network operating on the point cloud data and the second one is a 2D network operating on BEV top-down projection. BEV region proposal network is used for fast generation of proposals, which are subsequently matched with tracklets by the 3D network. The model needs first frame predictions that are used to initialize tracklets and associated object shapes. Model shapes are constantly updated by concatenation of tracklet and predicted 3D bounding boxes points.

13.3.2.2 Simultaneous detection and tracking

CenterPoint [47] is an extension of the 2D object detection method CenterNet [48] to 3D data. CenterNet represents objects as points and estimates box size and offset based on the features at that central point. CenterPoint uses the approaches followed by SECOND [8] for voxel feature extraction and 2D feature map generation. Instead of using anchors, as in most voxel-based methods, it follows CenterNet and creates a heatmap per class to estimate center points. Features from these center points are subsequently used to estimate the 3D size and the location offset of each object. Additionally, 3D object tracking is performed, using velocity estimation. The center of an object is projected to the previous frame by applying a negative velocity estimate, and matching is performed based on the closest distance to the center of objects in the previous frame.

Mono3DT [49] uses monocular images to create 3D object predictions and a recurrent network to create object ID assignment across frames. 2D object detection is used to create proposals in an image frame and uses pixel features of those proposals to estimate a 3D location, size, and orientation predictions.

Complexer-YOLO [50] applies 3D object detection and tracking simultaneously using a combination of camera image and Lidar point cloud. ENet [51] is applied on the image for 2D semantic segmentation, which is further projected onto a voxelized point cloud with a small voxel size (0.08, 0.08, 0.19). The voxelized semantic point cloud is processed with Complex-YOLO [20] to generate 3D object detections. Based on these detections, Labeled Multi-Bernoulli Random Finite Sets (LMB RFS) [52] approach is used to assign tracking labels.

13.4 Conclusion

In this chapter, we provided an overview of the 3D object detection and tracking problems, the most popular public data sets and the metrics used for performance evaluations, along with the main ideas in recent related methods based on deep learning models. We briefly described the types of sensors that can be used to collect information from the 3D scene for 3D object detection, and we categorized the existing methods based on the type of sensors they use. Lidar-based methods provide good compromise between detection performance and inference speed. When Lidar is not available, monocular or binocular image-based methods can be implemented to either directly predict 3D bounding boxes from image features, or to create a pseudo-point cloud, which is subsequently introduced to Lidar-based methods. Lidar data can be fused with image or Radar data to improve detection performance, but at the cost of lower inference speed. 3D object tracking methods follow two different approaches. The first one, and the most popular, is to use a 3D object detection method for obtaining 3D object bounding boxes, and then associate detected boxes across frames. The second approach applies detection and tracking simultaneously.

3D object detection and tracking problems are much less explored to date, compared to their 2D counterparts, and most of the ideas used to address them are extensions of 2D object detection and tracking methods. An identified issue related to the state-of-the-art performance in these two problems is related to publicity, since many results in KITTI and nuScenes leader-boards are anonymous. For instance, the leader-board in KITTI data set has 284 results for cars detection, while only 108 of them are connected to a method described in a research paper. This phenomenon may be a result of the high demand for 3D object detection and tracking methods in real-life applications and the competition between companies for the best autonomous driving system, which highly relies on the performance of the associated 3D object detection and tracking methods.

References

[1] Andreas Geiger, Philip Lenz, Raquel Urtasun, Are we ready for autonomous driving? The KITTI vision benchmark suite, in: 2012 IEEE Conference on Computer Vision and Pattern Recognition, 2012, pp. 3354–3361.

[2] Holger Caesar, Varun Bankiti, Alex H. Lang, Sourabh Vora, Venice Erin Liong, Qiang Xu, Anush Krishnan, Yu Pan, Giancarlo Baldan, Oscar Beijbom, nuScenes: a multimodal dataset for autonomous driving, in: 2020 IEEE/CVF Conference on Computer Vision and Pattern Recognition, 2020, pp. 11618–11628.

[3] E. Arnold, O.Y. Al-Jarrah, M. Dianati, S. Fallah, D. Oxtoby, A. Mouzakitis, A survey on 3D object detection methods for autonomous driving applications, IEEE Transactions on Intelligent Transportation Systems 20 (10) (2019) 3782–3795.

[4] Yin Zhou, Oncel Tuzel, VoxelNet: end-to-end learning for point cloud based 3D object detection, in: IEEE Conference on Computer Vision and Pattern Recognition, 2018, pp. 4490–4499.

[5] Sergey Ioffe, Christian Szegedy, Batch normalization: accelerating deep network training by reducing internal covariate shift, in: 32nd International Conference on Machine Learning, 2015, pp. 448–456.

[6] Vinod Nair, Geoffrey E. Hinton, Rectified linear units improve restricted Boltzmann machines, in: 27th International Conference on Machine Learning (ICML-10), 2010, pp. 807–814.

[7] Alex H. Lang, Sourabh Vora, Holger Caesar, Lubing Zhou, Jiong Yang, Oscar Beijbom, PointPillars: fast encoders for object detection from point clouds, in: IEEE Conference on Computer Vision and Pattern Recognition, 2019, pp. 12697–12705.

[8] Yan Yan, Yuxing Mao, Bo Li, SECOND: sparsely embedded convolutional detection, Sensors 18 (10) (2018) 3337.

[9] Zhe Liu, Xin Zhao, Tengteng Huang, Ruolan Hu, Yu Zhou, Xiang Bai, TANet: robust 3D object detection from point clouds with triple attention, in: The Thirty-Fourth AAAI Conference on Artificial Intelligence, 2020, pp. 11677–11684.

[10] Xingyi Zhou, Dequan Wang, Philipp Krähenbühl, Objects as points, arXiv e-prints, 2019.

[11] Qi Chen, Lin Sun, Zhixin Wang, Kui Jia, Alan L. Yuille, Object as hotspots: an anchor-free 3D object detection approach via firing of hotspots, in: Computer Vision – ECCV 2020 – 16th European Conference, 2020, pp. 68–84.

[12] R.Q. Charles, H. Su, M. Kaichun, L.J. Guibas, PointNet: deep learning on point sets for 3D classification and segmentation, in: 2017 IEEE Conference on Computer Vision and Pattern Recognition (CVPR), 2017, pp. 77–85.

[13] Charles Ruizhongtai Qi, Li Yi, Hao Su, Leonidas J. Guibas, PointNet++: deep hierarchical feature learning on point sets in a metric space, in: Advances in Neural Information Processing Systems 30: Annual Conference on Neural Information Processing Systems, 2017, pp. 5099–5108.

[14] Shaoshuai Shi, Xiaogang Wang, Hongsheng Li, PointRCNN: 3D object proposal generation and detection from point cloud, arXiv e-prints, arXiv:1812.04244, 2018.

[15] Hang Su, Subhransu Maji, Evangelos Kalogerakis, Erik G. Learned-Miller, Multi-view convolutional neural networks for 3D shape recognition, in: 2015 IEEE International Conference on Computer Vision, 2015, pp. 945–953.

[16] Xinwei He, Song Bai, Jiajia Chu, Xiang Bai, An improved multi-view convolutional neural network for 3D object retrieval, IEEE Transactions on Image Processing 29 (2020) 7917–7930.

[17] Bo Li, Tianlei Zhang, Tian Xia, Vehicle detection from 3D Lidar using fully convolutional network, in: Robotics: Science and Systems XII, 2016.

[18] Evan Shelhamer, Jonathan Long, Trevor Darrell, Fully convolutional networks for semantic segmentation, IEEE Transactions on Pattern Analysis and Machine Intelligence 39 (4) (2017) 640–651.

[19] Bichen Wu, Alvin Wan, Xiangyu Yue, Kurt Keutzer SqueezeSeg, Convolutional neural nets with recurrent CRF for real-time road-object segmentation from 3D LiDAR point cloud, in: 2018 IEEE International Conference on Robotics and Automation, 2018, pp. 1887–1893.

[20] Martin Simon, Stefan Milz, Karl Amende, Horst-Michael Gross, Complex-YOLO: an Euler-region-proposal for real-time 3D object detection on point clouds, in: Computer Vision – ECCV 2018 Workshops, Munich, Germany, 2018, pp. 197–209.

[21] X. Chen, H. Ma, J. Wan, B. Li, T. Xia, Multi-view 3D object detection network for autonomous driving, in: 2017 IEEE Conference on Computer Vision and Pattern Recognition (CVPR), 2017, pp. 6526–6534.

[22] Sourabh Vora, Alex H. Lang, Bassam Helou, Oscar Beijbom, PointPainting: sequential fusion for 3D object detection, in: 2020 IEEE/CVF Conference on Computer Vision and Pattern Recognition, 2020, pp. 4603–4611.

[23] Charles R. Qi, Wei Liu, Chenxia Wu, Hao Su, Leonidas J. Guibas, Frustum PointNets for 3D object detection from RGB-D data, in: 2018 IEEE Conference on Computer Vision and Pattern Recognition, 2018, pp. 918–927.

[24] Zhixin Wang, Kui Jia, Frustum ConvNet: sliding frustums to aggregate local point-wise features for Amodal, in: 2019 IEEE/RSJ International Conference on Intelligent Robots and Systems, 2019, pp. 1742–1749.

[25] Xiaozhi Chen, Kaustav Kundu, Ziyu Zhang, Huimin Ma, Sanja Fidler, Raquel Urtasun, Monocular 3D object detection for autonomous driving, in: 2016 IEEE Conference on Computer Vision and Pattern Recognition, CVPR 2016, Las Vegas, NV, USA, June 27–30, 2016, 2016, pp. 2147–2156.

[26] Tong He, Stefano Soatto, Mono3D++: monocular 3D vehicle detection with two-scale 3D hypotheses and task priors, in: The Thirty-Third AAAI Conference on Artificial Intelligence, 2019, pp. 8409–8416.

[27] Bin Xu, Zhenzhong Chen, Multi-level fusion based 3D object detection from monocular images, in: 2018 IEEE Conference on Computer Vision and Pattern Recognition, 2018, pp. 2345–2353.

[28] Clément Godard, Oisin Mac Aodha, Gabriel J. Brostow, Unsupervised monocular depth estimation with left-right consistency, in: 2017 IEEE Conference on Computer Vision and Pattern Recognition, 2017, pp. 6602–6611.

[29] Shaoqing Ren, Kaiming He, Ross Girshick, Jian Sun, Faster R-CNN: towards real-time object detection with region proposal networks, arXiv e-prints, arXiv:1506.01497, 2015.

[30] Zengyi Qin, Jinglu Wang, Yan Lu, MonoGRNet: a geometric reasoning network for monocular 3D object localization, in: The Thirty-Third AAAI Conference on Artificial Intelligence, 2019, pp. 8851–8858.

[31] Thomas Roddick, Alex Kendall, Roberto Cipolla, Orthographic feature transform for monocular 3D object detection, in: 30th British Machine Vision Conference, 2019, p. 285.

[32] Abhishek Verma, Hussam Qassim, David Feinzimer, Residual squeeze CNDS deep learning CNN model for very large scale places image recognition, in: 8th IEEE Annual Ubiquitous Computing, Electronics and Mobile Communication Conference, 2017, pp. 463–469.

[33] Andrea Simonelli, Samuel Rota Bulò, Lorenzo Porzi, Manuel Lopez-Antequera, Peter Kontschieder, Disentangling monocular 3D object detection, in: 2019 IEEE/CVF International Conference on Computer Vision, 2019, pp. 1991–1999.

[34] Arsalan Mousavian, Dragomir Anguelov, John Flynn, Jana Kosecka, 3D bounding box estimation using deep learning and geometry, in: 2017 IEEE Conference on Computer Vision and Pattern Recognition, 2017, pp. 5632–5640.

[35] Bugra Tekin, Sudipta N. Sinha, Pascal Fua, Real-time seamless single shot 6D object pose prediction, in: 2018 IEEE Conference on Computer Vision and Pattern Recognition, 2018, pp. 292–301.

[36] Vincent Lepetit, Francesc Moreno-Noguer, Pascal Fua, EP*n*P: an accurate $O(n)$ solution to the P*n*P problem, International Journal of Computer Vision 81 (2) (2009) 155–166.

[37] Yan Wang, Wei-Lun Chao, Divyansh Garg, Bharath Hariharan, Mark E. Campbell, Kilian Q. Weinberger, Pseudo-LiDAR from visual depth estimation: bridging the gap in 3D object detection for autonomous driving, in: IEEE Conference on Computer Vision and Pattern Recognition, 2019, pp. 8445–8453.

[38] Yurong You, Yan Wang, Wei-Lun Chao, Divyansh Garg, Geoff Pleiss, Bharath Hariharan, Mark Campbell, Kilian Q. Weinberger, Accurate depth for 3D object detection in autonomous driving, 2019.

[39] Jia-Ren Chang, Yong-Sheng Chen, Pyramid stereo matching network, in: 2018 IEEE Conference on Computer Vision and Pattern Recognition, 2018, pp. 5410–5418.

[40] Chengyao Li, Jason Ku, Steven L. Waslander, Confidence guided stereo 3D object detection with split depth estimation, CoRR, arXiv:2003.05505, 2020.

[41] Jonathon Tyler Luiten, Aljoša Ošep, Patrick Dendorfer, Philip Torr, Andreas Geiger, Laura Leal-Taixé, Bastian Leibe, HOTA: a higher order metric for evaluating multi-object tracking, International Journal of Computer Vision 129 (2) (2021) 548–578.

[42] Xinshuo Weng, Jianren Wang, David Held, Kris Kitani, AB3DMOT: a baseline for 3D multi-object tracking and new evaluation metrics, CoRR, arXiv:2008.08063, 2020.

[43] H.W. Kuhn, The Hungarian method for the assignment problem, Naval Research Logistics Quarterly 2 (1–2) (1955) 83–97.

[44] R.E. Kalman, A new approach to linear filtering and prediction problems, Journal of Basic Engineering 82 (1) (1960) 35–45.

[45] Silvio Giancola, Jesus Zarzar, Bernard Ghanem, Leveraging shape completion for 3D Siamese tracking, in: IEEE Conference on Computer Vision and Pattern Recognition, 2019, pp. 1359–1368.

[46] Jesus Zarzar, Silvio Giancola, Bernard Ghanem, Efficient tracking proposals using 2D-3D Siamese networks on LIDAR, arXiv e-prints, 2019.

[47] Tianwei Yin, Xingyi Zhou, Philipp Krähenbühl, Center-based 3D object detection and tracking, CoRR, arXiv:2006.11275, 2020.

[48] Xingyi Zhou, Dequan Wang, Philipp Krähenbühl, Objects as points, CoRR, arXiv:1904.07850, 2019.

[49] Hou-Ning Hu, Qi-Zhi Cai, Dequan Wang, Ji Lin, Min Sun, Philipp Krähenbühl, Trevor Darrell, Fisher Yu, Joint monocular 3D vehicle detection and tracking, in: 2019 IEEE/CVF International Conference on Computer Vision, 2019, pp. 5389–5398.

[50] Martin Simon, Karl Amende, Andrea Kraus, Jens Honer, Timo Sämann, Hauke Kaulbersch, Stefan Milz, Horst-Michael Gross, Complexer-YOLO: real-time 3D object detection and tracking on semantic point clouds, in: IEEE Conference on Computer Vision and Pattern Recognition Workshops, 2019, pp. 1190–1199.

[51] Adam Paszke, Abhishek Chaurasia, Sangpil Kim, Eugenio Culurciello, ENet: a deep neural network architecture for real-time semantic segmentation, CoRR, arXiv:1606.02147, 2016.

[52] D.S. Bryant, B. Vo, B. Vo, B.A. Jones, A generalized labeled multi-Bernoulli filter with object spawning, IEEE Transactions on Signal Processing 66 (23) (2018) 6177–6189.

CHAPTER

Human activity recognition

14

Lukas Hedegaard, Negar Heidari, and Alexandros Iosifidis
Department of Electrical and Computer Engineering, Aarhus University, Aarhus, Denmark

14.1 Introduction

The explosion of multimedia on the internet and the increased usage of video surveillance monitoring has led to an increasing need to process and identify explicit and unwanted content online and in public spaces. Since it is unfeasible to assign human workers to watch through and label these massive amounts of content, it is of great interest to automate the process by means of machine learning. However, we cannot simply adopt image classification techniques to handle these problems, since the temporal information is crucial in the distinction of action classes. For instance, an image depicting a person performing the action 'standing up" would not be easily distinguished by an image depicting the person performing the action "sitting down" as they will correspond to very similar human body poses, while image sequences depicting these actions would allow for easy distinction due to the difference in the temporal succession of the human poses forming these two actions. To this end, we need methods that can exploit the temporal information in image sequences or videos.

In this chapter, we give an overview of the field of human activity recognition, which (as the name implies) is concerned with recognizing the activity of humans. Here, we should note that the terms action, activity, and movement are often used interchangeably in the computer vision literature, though there have been attempts to distinguish the problems described by these terms [1]. Many different approaches to human activity recognition have been proposed throughout the years, and in this exploration of the subject we focus exclusively on the methods that use deep learning and visual input data. First, we will define the tasks in human activity recognition more precisely and give an overview of common input modalities used by the various methods. In Section 14.2, we cover different video classification architectures and methods for trimmed action recognition (Section 14.2.1, Section 14.2.2, Section 14.2.3, and Section 14.2.4), human body pose-based methods using body skeleton data (Section 14.2.5), and multistream architectures (Section 14.2.6). Next, we consider the extensions to temporal action localization (Section 14.3) and spatiotemporal action localization (Section 14.4). Finally, we provide an overview of data sets used human activity recognition (Section 14.5).

Deep Learning for Robot Perception and Cognition. https://doi.org/10.1016/B978-0-32-385787-1.00019-1

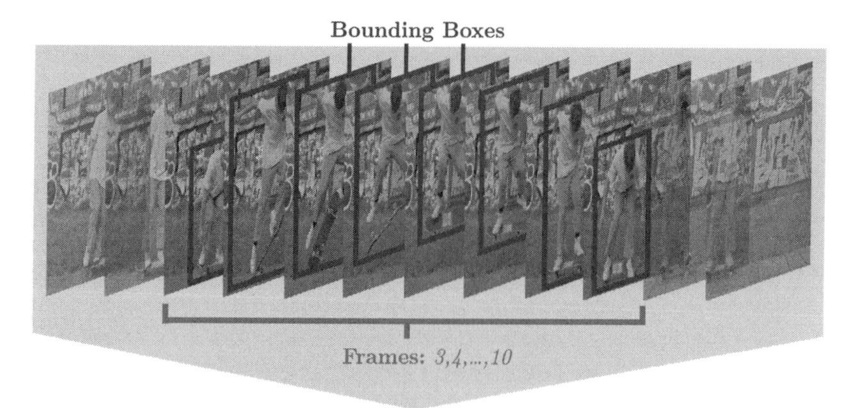

FIGURE 14.1

Illustration of the types of information involved across human activity recognition tasks.

14.1.1 Tasks in human activity recognition

Often times, "human activity recognition" is understood as the narrow setting of classifying the human action present in a trimmed video. However, the term human activity recognition also encompasses other tasks, where human activity is involved. Broadly speaking, the tasks are concerned with a mix of *what*, *when*, and *where*, expressed as an action class label, action time frames, and the spatial locations of the action in each frame (bounding boxes). These are illustrated in Fig. 14.1. There are many combinations of information to predict. What follows is an overview of the taxonomy[1] used to distinguish the most common tasks:

- **Trimmed action recognition** is only concerned with *what*, that is, what is the class of the human activity depicted in a video, and assumes that the video has been trimmed beforehand to contain only that activity. Often, this task is simply referred to as "action recognition."
- **Temporal action proposals** determination is the task of estimating *when* any action is depicted in a video, disregarding the class-label of the action itself.
- **Temporal action localization**[2] handles both *what* and *when*. The task is to predict both the class label of actions depicted in a video and the corresponding time stamps for each of the actions.
- **Spatiotemporal action localization** is concerned with *what*, *when*, and *where*. The task is to classify the action, note in which frames it is present, and to specify the spatial bounding box for the action in each frame.

[1] Taxonomy adopted from the ActivityNet challenge, http://activity-net.org.

[2] Another commonly encountered term is "Action Detection." Note, however, that action detection is often used interchangeably for temporal action localization and spatiotemporal action localization.

Table 14.1 Tasks in human activity recognition (rows) and the label information in each of them (columns).

	Class label	Time stamps	Bounding boxes
Trimmed Action Recognition	×		
Temporal Action Proposals		×	
Temporal Action Localization	×	×	
Spatiotemporal Action Localization	×	×	×

An overview of human activity recognition tasks and associated labels is shown in Table 14.1.

It should be noted that there exist settings where multiple action classes appear in the video simultaneously. This is especially prevalent in data sets for temporal action localization and spatiotemporal action localization, but can also occur in trimmed action recognition. Other tasks also exist that are human-centric, such as human pose estimation [2], dense pose estimation [3], human activity prediction [4], activity-based person identification [5] and visual voice activity detection [6], but these are beyond the scope of this chapter.

14.1.2 Input modalities for human activity recognition

Not only does the information we want to obtain vary widely between subtasks in human activity recognition; the input modalities are as just as diverse (Fig. 14.2). The most common input is RGB video,[3] which consists of a series of RGB images, sampled at a regular interval (usually in the range of 20 to 30 frames per second). Though the videos often vary in length, it is common practice to cut the video into clips containing a fixed number of frames (e.g., 32 frames per clip), sometimes with a lower frame rate (e.g., 8 frames per second) to ensure the clip covers a sufficiently large temporal extend despite the relatively few number of frames. This form of clipping is based on the assumption that human movements are slow-changing and the variations in the contents of a scene between successive video frames are small. It is a common approach for most modalities and is adopted to reduce redundant information and to accommodate network architectures that require a fixed input size, such as the 3D CNNs, which will be described in Section 14.2.2.

The plain RGB input modality can be extended with hand-crafted features such as optical flow [9] and Improved Dense Trajectories (iDT) [10]. These are calculated from the video inputs and explicitly capture information on the movements of keypoints between frames. Though this type of handcrafted features may be considered a remnant of the pipelines prior to the dominance of deep neural networks, their inclusion still yields a systematic accuracy improvement when used in multistream

[3] While under some application scenarios multiple cameras can be used, leading to the multiview human action recognition [7] and view-invariant action recognition [8] problems, the most common scenario involves a single camera.

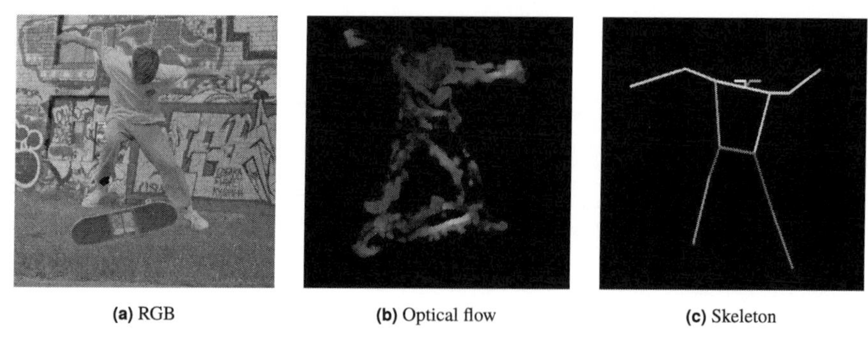

(a) RGB (b) Optical flow (c) Skeleton

FIGURE 14.2

Examples of different input modalities frequently used for human activity recognition. For the image of optical flow, the color indicates the angle of the movement, and the color intensity denotes movement magnitude. For the image of human body skeleton, different colors indicate connections of different human body joints. (For interpretation of the colors in the figure, the reader is referred to the web version of this chapter.)

network topologies alongside plain RGB video. We will explore the multistream architectures in Section 14.2.6.

In addition to RGB video, some popular devices such as Kinect and Vicon [11] have depth-sensing capabilities that let them produce a depth (D) channel. This can either be treated as a fourth channel in the (RGB+D) video input, or more commonly used as a separate stream due to the lower spatial resolution of the depth channel compared to the RGB channels.

Finally, human body skeletons can be extracted for each frame, each corresponding to the body pose of a person appearing in the scene and represented as a set of human body joints interconnected by bones. Such a human body representation can be used to define a graph structure formed by nodes (joints) and edges (bones) that is evolving through time during action execution. The computation of skeleton landmarks is a built-in functionality to both the Kinect and Vicon devices, and alternatively it can be extracted using human body pose methods, such as those in the OpenPose toolkit [12]. The Spatial Temporal Graph Convolutional Network (ST-GCN) has been successfully applied to the skeleton input modality and will be covered in Section 14.2.5.

While there are still other modalities that have been used for Human Activity Recognition, such as Electromyography (EMG), accelerometer, and wireless signal measurements, these are beyond the scope of this chapter.

14.2 Trimmed action recognition

Often, the terms human activity recognition and trimmed action recognition are used interchangeably, though the former encompasses an array of different tasks, as described in Section 14.1.1. Trimmed action recognition is the more precise term, and amounts to video classification, where the video is trimmed to the action of the human subject in the video. The action performed by this subject then constitutes the label. Though the human actions are of special interest here, the architectures described in this section (except for the skeleton-based models) are generic video-classifiers and might just as well be trained to classify the activity of a cat or other scene classification problems, for instance, the bathing conditions at a beach, provided a data set had been collected for this purpose and the corresponding methods have been trained to provide the respective classes as outputs.

In our coverage of the topic, we will start in Section 14.2.1 by describing how image classifiers can be used directly for trimmed action recognition, by using a recursive neural network for knowledge integration over time. In Section 14.2.2, the extension of 2D CNNs for image classification to 3D CNNs for video classification is described. Section 14.2.3 extends our discussion on 3D CNNs by describing how we can "inflate" pretrained 2D CNN image classifiers to improve performance in 3D models. An issue with 3D CNNs as compared to their 2D counterparts is their increased network size. To make 3D CNNs more efficient, it is possible to decompose a 3D kernel into the combination of a 2D and a 1D convolution, as we will detail in Section 14.2.4. Human body pose-based human activity recognition methods, which utilize a sequence of skeletons instead of RGB videos as input, will be described in Section 14.2.5. Most of the architectures and models described in this chapter use a multistream topology with multiple different input modalities to achieve their best results, and some methods use different streams to concurrently exploit multiple temporal extends and resolutions. The details on this are given in Section 14.2.6.

14.2.1 2D convolutional and recurrent neural network-based architectures

2D-CNNs have enjoyed almost unrivaled success in the domain of images. Specifically, the ImageNet competition and ever-improving benchmarks on the data set of the same name have led to CNN architectures that surpass human level performance on image classification tasks. In the context of human activity recognition, one can thus reuse one of these networks to extract features for each frame of the input clip, and then integrate the information along the temporal dimension using some other method. This is a case of transfer learning, where the architecture and parameters of a model trained on a classification task in the 2D image domain is transferred to the task of framewise classification in video clips. There are many possibilities of how to integrate the information across frames, including the bag-of-features approach (or its recurrent version [13]), using convolutions along the temporal dimension, and via recurrent neural networks (RNN). Specifically, the RNN-based approaches with Long Short Term Memory (LSTM) as the recursive modules have seen good success in the years 2015 to 2017 [14,15]. Their accuracy, however, was surpassed by the I3D [16] and R(2+1)D [17] architectures, which are covered in Section 14.2.3 and

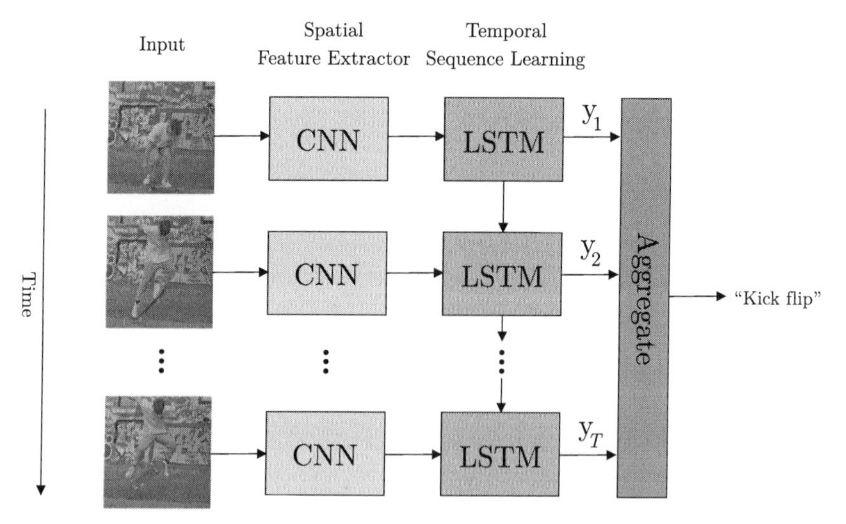

FIGURE 14.3

2D-CNN + LSTM-based architecture employed in [14] and [15]. Each frame is independently processed through a pretrained CNN, which acts a feature extractor operating on the spatial domain. These features are then introduced to a (stacked) LSTM module, which produces a prediction y_t for time step t and a hidden state $\boldsymbol{h}_t$ that is reused in the next time-step. Finally, an aggregation function produces a video-level prediction based on the predictions y_t at each time-step. Note that the exact same CNN and LSTM networks are used in each time step.

Section 14.2.4, respectively. Nevertheless, their appeal still holds today, due to the ease of training and straightforward application to online systems.

The basic 2D-CNN + LSTM architecture is depicted in Fig. 14.3. When processing a video, each frame is passed through a CNN such as AlexNet [18], that was pretrained to produce good results on the image classification task, before being fed to the recurrent module. Here, one or more (five in [15]) stacked LSTM layers process the image features and output a preliminary classification for time step t alongside a hidden state $\boldsymbol{h}_t$ that is reused in the computation on the next time-step $t + 1$. The network thus uses the same exact CNN and LSTM for processing each of the input frames forming the video, making it efficient in terms of the number of parameters. Finally, the partial classification results for all time steps are aggregated to produce the action classification result. The aggregation over time steps can take on many variants, that is, taking the final classification result; averaging over all classification results; max-pooling over time (selecting the classification result with highest certainty among all time steps); and using a weighted sum with a gain g, which increases linearly from 0 in the first output to 1 in the last [15,14]. While the linearly weighted gain was found to work slightly better for inference, the difference in performance between aggregation strategies is very small.

Often, when training a recurrent neural network, the loss is computed only for the final step, and the error is then backpropagated through time [19]. Training using this technique can lead to a slow and memory-inefficient procedure, because all intermediary results through time need to be cached in order to compute the gradients. Moreover, the error information of early time steps diminishes when long sequences are processed. In practice, training can be improved by computing a classification loss for each time step, and weighting each of the derivatives by a gain g (in a similar manner as in the weighted aggregation strategy described above) before backpropagation [15].

14.2.2 3D convolutional neural network architectures

A straightforward way of extending CNNs used for image classification tasks to be used for human activity recognition is to replace the 2D convolutions, which were suitable for processing images, with 3D convolutions. In this way, the convolution operation is extended to consider the temporal dimension in addition to the two spatial dimensions found in images. Indeed, many works have used this idea to produce well-performing models for video classification.

Like 2D convolutions, 3D convolutions[4] extract features from a local neighborhood of an input by multiplying each input position with a corresponding weight tensor, and summing all values alongside a bias term. Only, instead of having a 2D kernel that covers an area of a 2D input, 3D convolutions cover a volume on the 3D input. This is illustrated in Fig. 14.4. Given a 3D input volume **X** and a convolutional kernel **K**, the output feature map **Y** is calculated as follows:

$$Y(t,i,j) = \sum_{l}\sum_{m}\sum_{n} X(t-l, i-m, j-n)K(l,m,n) \tag{14.1}$$

In general, properties of 2D convolutions, such as the commutative property, also hold for 3D convolutions. In context of a 3D convolutional neural network, 3D convolutions are usually followed by a nonlinear activation function such as the Rectified Linear Unit (ReLU) applied on each element of the output. These are then arranged in layers of multiple convolutional kernels. Using multiple convolutional kernels in a layer gives rise to outputs with an additional channel dimension, and likewise, an input can consist of multiple channels. Denoting by C the number of input channels, T the number of frames in a video clip (i.e., temporal dimension), and H and W the spatial height and width of each input frame, a network operating on video takes an input of size $(C \times T \times H \times W)$. For example, a video formed by 16 frames, each having a resolution of 112×112 pixels and three color channels (corresponding to RGB color space) is a tensor of size $(3 \times 16 \times 112 \times 112)$.

3D convolutions have been exploited by a number of works on Trimmed Action Recognition, which either use 3D convolutions as the primary building block in a

[4] In our usage of the terms 2D and 3D convolutions, we disregard the channel dimension of the input, even though a 2D convolution (convolution over spatial dimensions) in reality uses kernels which are 3D-tensors to account for height, width and channel dimensions. Similarly, 3D convolutions (spatial dimensions + temporal dimension) have kernels that constitute a 4D tensor in practice.

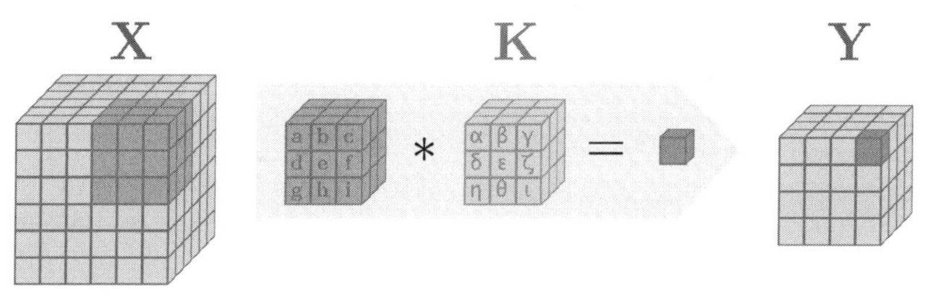

FIGURE 14.4

Illustration of 3D convolution for a kernel of size $(3 \times 3 \times 3)$ without zero padding for a single-channel input video. $(3 \times 3 \times 3)$ slices of the input tensor **X** are convolved with a kernel **K** to produce an output tensor **Y**. The convolution amounts to a multiplication of each aligned cube and a summation of all involved cubes: $a\alpha + b\beta + c\gamma + d\delta + e\epsilon + f\zeta + g\eta + h\theta + i\iota + ...$

deep CNN [20–23] or as a part of a multistage pipeline [24]. An efficient and conceptually simple model introduced in [22] is the Convolutional 3D (C3D) architecture. It consists entirely of 3D convolutional layers with kernel size of $(3 \times 3 \times 3)$, 3D pooling layers, and fully connected layers. In each 3D convolutional layer, the input is zero-padded in the spatial and temporal dimension in order to keep the dimensionality for the output feature-map. A 3D convolution with zero padding is illustrated in Fig. 14.5. To reduce the size of the feature-maps throughout the network, max pooling is used. In C3D, this is implemented as a $(2 \times 2 \times 2)$ pooling with stride 2 after all convolutional layers but the first, where a $(1 \times 2 \times 2)$ pooling is used, thus omitting pooling along the temporal dimension initially. This was done to not collapse the temporal signal too early, and to postpone temporal collapse until the last layer (an input of size 16 can be pooled by 2 at most 4 times). Finally, the network has two fully connected layers with ReLU nonlinearities before the final prediction layer with a Softmax activation function. The architecture in its entirety is summarized in Table 14.2. It is worth noting the remarkable similarity between the C3D architecture and a 2D architecture such as VGG16 [25]. They are both built with blocks of padded convolutions of kernel size 3 and ReLU activations, followed by max pooling to reduce the size of the feature map, before being passed to two fully connected layers and a prediction layer. Works proposing these two network architectures were both published in 2015, and represent clear examples of the dominant design for deep neural networks in visual tasks at the time.

While 3D-CNN architectures have shown promising results, they did not achieve state-of-the-art accuracy initially. An issue with these models is that the additional kernel dimension results in significantly more parameters per neuron compared to a 2D CNN. This, coupled with the fact that we cannot reuse parameters from ImageNet pre-trained models, makes them harder to train than their 2D-CNN counterparts. Due to difficulties related to effectively optimizing networks with such a large number of parameters, the employed architectures have been relatively shallow compared to the deep network architectures used for image classification.

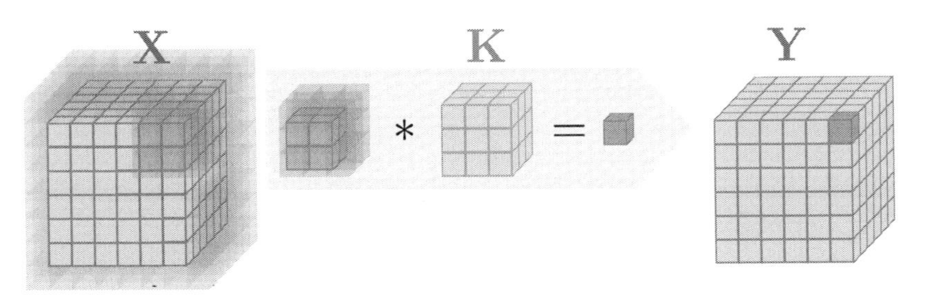

FIGURE 14.5

Illustration of 3D convolution for a kernel of size $(3 \times 3 \times 3)$ with zero padding of size $(1 \times 1 \times 1)$ for a single-channel input video. During convolution, slices of the input tensor **X** that would have gone 1 step beyond the boundaries are allowed, with values outside the boundaries set to zero. Using zero padding this way, the output tensor **Y** retains the dimensions of the input **X**.

Table 14.2 Architecture summary for C3D [22]. All convolution and pooling operators operate in 3D across the time T, height H and width W dimensions. C denotes the number of channels or neurons in the layer.

Layer name	Neurons	Kernel size $T \times H \times W$	Padding $T \times H \times W$	Feature size $C \times T \times H \times W$
Input	–	–	–	$3 \times 16 \times 112 \times 112$
Conv1a + ReLU	64	$3 \times 3 \times 3$	$1 \times 1 \times 1$	$64 \times 16 \times 112 \times 112$
Pool1	–	$1 \times 2 \times 2$	–	$64 \times 16 \times 56 \times 56$
Conv2a + ReLU	128	$3 \times 3 \times 3$	$1 \times 1 \times 1$	$128 \times 16 \times 56 \times 56$
Pool2	–	$2 \times 2 \times 2$	–	$128 \times 8 \times 28 \times 28$
Conv3a + ReLU	256	$3 \times 3 \times 3$	$1 \times 1 \times 1$	$256 \times 8 \times 28 \times 28$
Conv3b + ReLU	256	$3 \times 3 \times 3$	$1 \times 1 \times 1$	$256 \times 8 \times 28 \times 28$
Pool3	–	$2 \times 2 \times 2$	–	$256 \times 4 \times 14 \times 14$
Conv4a + ReLU	512	$3 \times 3 \times 3$	$1 \times 1 \times 1$	$512 \times 4 \times 14 \times 14$
Conv4b + ReLU	512	$3 \times 3 \times 3$	$1 \times 1 \times 1$	$512 \times 4 \times 14 \times 14$
Pool4	–	$2 \times 2 \times 2$	–	$512 \times 2 \times 7 \times 7$
Conv5a + ReLU	512	$3 \times 3 \times 3$	$1 \times 1 \times 1$	$512 \times 2 \times 7 \times 7$
Conv5b + ReLU	512	$3 \times 3 \times 3$	$1 \times 1 \times 1$	$512 \times 2 \times 7 \times 7$
Pool5	–	$2 \times 2 \times 2$	$0 \times 1 \times 1$	$512 \times 1 \times 4 \times 4$
Flatten	–	–	–	8192
FC6	4096	–	–	4096
FC7	4096	–	–	4096
Preds + Softmax	Num. classes	–	–	Num. classes

14.2.3 Inflated 3D CNN architectures

Hitherto, a major shortcoming of 3D CNNs was their inability to exploit the architectures and weights of image classification models trained on ImageNet. To solve the issue, Carriera and Zisserman [16] proposed to reuse the architectures and weights of state-of-the-art 2D CNNs for image recognition, and to *inflate* the 2D convolutional kernels to 3D. In practice, this means taking each $(K \times K)$ 2D convolutional kernel in a source model (e.g., VGG16), repeating it K times along a new 3rd dimension and finally dividing the weights by K. At the heart of the technique is the observation that we can treat an image as a "boring video" by repeating it along the temporal dimension is used. The 3D convolutional filters can then be bootstrapped by enforcing a "boring video fixed point," that is, that the output of a 3D filter applied to a boring video, should yield the same result as the corresponding 2D filter would for the image that made the boring video. For example, consider the convolution of a (3×3) patch of an image and a (3×3) kernel:

$$\begin{aligned}\mathbf{K} * \mathbf{X} &= \begin{bmatrix} 1 & 0 & -1 \\ 2 & 0 & -2 \\ 1 & 0 & -1 \end{bmatrix} * \begin{bmatrix} 0.1 & 0.2 & 0.3 \\ 0.4 & 0.5 & 0.6 \\ 0.7 & 0.8 & 0.9 \end{bmatrix} \\ &= 1 \cdot 0.1 + 2 \cdot 0.4 + 1 \cdot 0.7 - 1 \cdot 0.3 - 2 \cdot 0.6 - 1 \cdot 0.9 \\ &= -0.8\end{aligned}$$

If we naïvely repeat the kernel three times to produce a tensor kernel $\mathbf{K} \in \mathbb{R}^{3\times3\times3}$ to match a boring video patch $\mathbf{X}$, their convolution would produce three times the desired result. But if the kernel weights are divided by three, the results match

$$\mathbf{K} * \mathbf{X} = \sum_{1}^{3} \frac{1}{3} \cdot 0.1 + \frac{2}{3} \cdot 0.4 + \frac{1}{3} \cdot 0.7 - \frac{1}{3} \cdot 0.3 - \frac{2}{3} \cdot 0.6 - \frac{1}{3} \cdot 0.9 = -0.8$$

Moreover, because the response of activation functions and max-pooling are the same in 3D as for 2D, the whole network satisfies the boring-video fix-point.

Using an inflated Inception-v1 [26] network in a multistream setup with an optical flow input in the second stream, the inflated 3D (I3D) model was able to yield state-of-the-art results on multiple benchmarks in trimmed action recognition.

14.2.4 Factorized (2+1)D CNN architectures

Consider the associative property for convolutions for a feature map $\mathbf{X}$ and two kernels:

$$(\mathbf{X} * \mathbf{K}_1) * \mathbf{K}_2 = \mathbf{X} * (\mathbf{K}_1 * \mathbf{K}_2) \tag{14.2}$$

On the left-hand side of Eq. (14.2), a feature map is convolved sequentially by multiple kernels in a CNN (disregarding the activation function). Without the nonlinearities, we could have freely convolved all the kernels prior to convolving the feature map, as is done on the right-hand side of Eq. (14.2). This means that we can construct a new kernel $\mathbf{K}$ by convolving two others, $\mathbf{K}_1 * \mathbf{K}_2$, and that convolving the feature map $\mathbf{X}$ with the new kernel $\mathbf{K}$ yields the same result as first convolving the feature map with one kernel $\mathbf{K}_1$ and then the second kernel $\mathbf{K}_2$. For example, consider the decomposition of a Sobel operator [27] used for the detection of horizontal

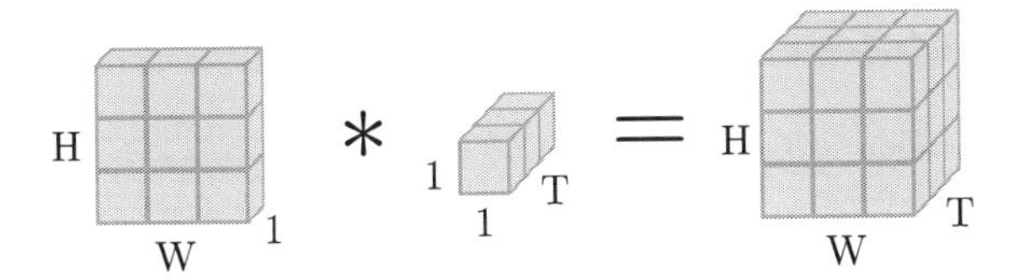

FIGURE 14.6

The convolution of a $(1 \times H \times W)$ and $(T \times 1 \times 1)$ kernel with padding yields a resulting convolutional kernel of dimension $(T \times H \times W)$.

edges:

$$\mathbf{K}_1 * \mathbf{K}_2 = \mathbf{K}$$

$$\begin{bmatrix} 1 & 0 & -1 \end{bmatrix} * \begin{bmatrix} 1 \\ 2 \\ 1 \end{bmatrix} = \begin{bmatrix} 1 & 0 & -1 \\ 2 & 0 & -2 \\ 1 & 0 & -1 \end{bmatrix}$$

Convolving the feature map $\mathbf{X}$ with the kernels $\mathbf{K}_1$ and $\mathbf{K}_2$ sequentially gives the exact same results as convolving it once with $\mathbf{K}$ (see the prior example):

$$\begin{aligned}
(\mathbf{X} * \mathbf{K}_1) * \mathbf{K}_2 &= \left(\begin{bmatrix} 0.1 & 0.2 & 0.3 \\ 0.4 & 0.5 & 0.6 \\ 0.7 & 0.8 & 0.9 \end{bmatrix} * \begin{bmatrix} 1 & 0 & -1 \end{bmatrix} \right) * \begin{bmatrix} 1 \\ 2 \\ 1 \end{bmatrix} \\
&= \begin{bmatrix} (0.1 \cdot 1 + 0.2 \cdot 0 + 0.3 \cdot (-1)) \\ (0.4 \cdot 1 + 0.5 \cdot 0 + 0.6 \cdot (-1)) \\ (0.7 \cdot 1 + 0.8 \cdot 0 + 0.9 \cdot (-1)) \end{bmatrix} * \begin{bmatrix} 1 \\ 2 \\ 1 \end{bmatrix} \\
&= \begin{bmatrix} -0.2 \\ -0.2 \\ -0.2 \end{bmatrix} * \begin{bmatrix} 1 \\ 2 \\ 1 \end{bmatrix} \\
&= -0.2 \cdot 1 - 0.2 \cdot 2 + -0.2 \cdot 1 \\
&= -0.8
\end{aligned}$$

Likewise, we can decompose a $(T \times H \times W)$ 3D kernel as the convolution of a $(1 \times H \times W)$ kernel with a $(T \times 1 \times 1)$ kernel. This is illustrated in Fig. 14.6.

Though not all 3D convolutional kernels can be represented as the convolution of a separate 2D and 1D kernel, doing so can reduce the computational complexity of convolving an input 3D tensor with the resulting kernel from $\mathcal{O}(T \cdot H \cdot W)$ to $\mathcal{O}(T + H \cdot W)$. For example, using this approach for convolution with a $(3 \times 3 \times 3)$ kernel and without using a bias term, the number of parameters and floating point operations (multiplications and additions) is reduced from $3^3 = 27$ to $3 + 3^2 = 12$. This idea was exploited in the Pseudo 3D blocks (P3D) [28] and the R(2+1)D [17] architectures, by using separate 2D and 1D convolutions to approximate 3D kernels. The "R" in the R(2+1)D alludes to the residual connection (aka skip connection) that was popularized by ResNet [29]; mechanistically, it adds the input of a block to its output. Unlike our previous example, a ReLU activation was added after each convolution of the (2+1)D block, yielding double the number of nonlinearities as compared to a regular 3D convolution with ReLU. According to the authors, this has the benefit of increasing the complexity of functions that can be represented by the network. The reduction in the number of parameters from using (2+1)D blocks instead of 3D convolutions can either be used to decrease the total network size and run-time, or it can be used to increase the number of neurons in the network by keeping the overall number of parameters unchanged. For a (2+1)D convolutional block with and input of C_{i-1} channels and output of C_i channels to have the same number of parameters as a regular 3D convolution, the first (spatial) convolution should be formed by M_i neurons:

$$M_i = \frac{T \cdot H \cdot W \cdot C_{i-1} \cdot C_i}{(H \cdot W \cdot C_{i-1}) + (T \cdot C_i)}. \tag{14.3}$$

For example, with $T = H = W = 3$ and $C_{i-1} = C_i$, the intermediary subspace has dimension $M_i = 2.25 \times C_i$. Using this approach, R(2+1)D [17] was able to surpass or match the benchmark results for I3D.

14.2.5 Skeleton-based action recognition

A fundamental way to describe a human action is to determine human body poses which appear during action execution along with their temporal succession. Given a video as a sequence of frames, the human body pose depicted in each frame can be represented by estimating the spatial arrangement of the human body joints and forming the corresponding body skeleton. Efficient and accurate body skeleton detection methods have exist and can be easily used due to their integration in popular toolboxes such as OpenPose [12]. Compared to other data modalities used for human activity recognition such as RGB videos, optical flow, and depth images, body skeletons provide several advantages: They encode high-level information of human body poses and dynamic body motions in a compact and simple structure, and they are invariant to view-point, illumination variations, body scale, human appearance, motion speed, context noise, and background complexity [30]. While these representational benefits can lead to improved performance for actions involving one person, human body skeletons cannot be easily extended to complex activities and multiple persons, where the scene context is needed to distinguish actions involving similar human body poses. For example, consider a player running during a football game. The exploitation of scene information, for example, related to the appearance of the field, the ball, and other players, can be the defining factor in distinguishing the action "play football" from the action "run."

Recently, many deep learning based methods have been proposed, which take advantage human skeleton dynamics for skeleton-based human action recognition. These methods are generally categorized into three groups. The first group consists of methods that are based on recurrent neural networks, which mostly utilize LSTM networksto model the temporal dynamics of the action in a sequence of skeletons.

Examples in this group of methods include deep LSTM [31], ST-LSTM [32], VA-LSTM [33], and STA-LSTM [34]. These methods form multiple feature vectors, each representing a body skeleton forming the action using the 2D or 3D coordinates of the body joints, and subsequently apply temporal analysis on the sequence of the feature vectors. The second group of methods are based on CNNs. Because CNNs require their inputs to follow a regular grid structure, these methods transform the body joint coordinates of each body skeleton to a pseudo-image, and the sequence of such pseudo-images is subsequently processed by the CNN. Notable works in this group are [35–39]. Methods belonging to these two groups are not fully capable of exploiting information encoded in the non-Euclidean structure of the body skeletons. Moreover, they are not able to benefit from the information encoded by the spatial connections between the body joints. The third group of methods is based on Graph Convolutional Networks (GCNs), which are able to process the structural information encoded in the body skeletons more efficiently. GCNs [40] generalize the notion of convolution from grid-structured data to graph data structures, and have been very successful when applied in skeleton-based human action recognition. Here, each skeleton is represented as a graph, which models the corresponding body pose by a set of joints connected with edges following the natural spatial connections indicated by the human body structure. The temporal dynamics in an action are modeled by connecting the graph structures representing the corresponding human body skeletons in successive time steps corresponding to different frames. The Spatial-Temporal Graph Convolution Network (ST-GCN) [41] exploited such a spatiotemporal graph structure for GCN-based action recognition, and several methods have built on top of this idea in recent years. In the remainder of this section, we shall delve further into ST-GCNs.

14.2.5.1 Spatial-temporal graph convolution network

Given a sequence of skeletons with body joint coordinates, the ST-GCN method first constructs a spatiotemporal graph by connecting each joint to its neighboring joints in the same body skeleton and the corresponding joints in the preceding and successive skeletons in the temporal sequence. Then spatiotemporal GCN layers are applied to the graph, in order to capture both spatial and temporal patterns of the skeleton's motion, and to finally recognize the human action. Formally, a spatiotemporal graph on a sequence of skeletons is defined as $\mathcal{G} = (\mathcal{V}, \mathcal{E})$, where

$$\mathcal{V} = \{v_{ti} \mid t,\ i \in \mathbb{Z},\ \ 1 \leq t \leq T,\ \ 1 \leq i \leq V\}$$

denotes the vertice/node set of V body joints in a sequence of skeletons of T time steps, and $\mathcal{E}$ is the set of spatial and temporal edges expressing the connectivity of the body joints through space and time domains. The spatial edges express the natural connections between the body joints in each skeleton and the temporal edges connect each body joint to its corresponding joint in the previous and next time steps. Fig. 14.7 (right) illustrates the spatiotemporal graph constructed on a sequence of 3 skeletons representing a person walking.

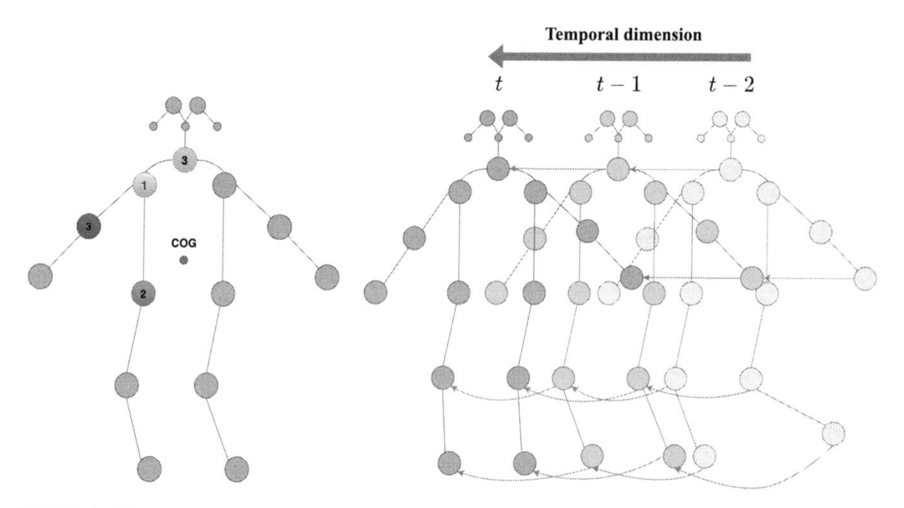

FIGURE 14.7

Left: the 3 neighboring subsets of a joint obtained by the spatial partitioning process in [41] are shown in different colors. The Center of Gravity (COG) is shown as a red dot. The joint under consideration is the one shown in red color, which forms the subset of root node ($p = 1$). Joints closer to the COG than the root node form the second subset ($p = 2$) and are shown in green color. Joints farther to the COG than the root node of the third subset ($p = 3$) are shown in orange color. Right: illustration of spatiotemporal graph constructed on a sequence of three skeletons. In temporal domain, each joint is connected to its corresponding joint in the next time step. The temporal connections for some of the nodes are shown by directed arrows. (For interpretation of the colors in the figure, the reader is referred to the web version of this chapter.)

Given the spatiotemporal graph and the node features as input, the ST-GCN model updates the features in both spatial and temporal domains by performing graph convolutions. By generalizing the definition of convolution on 2D feature maps, the spatial graph convolution on each skeleton is defined as [41]:

$$f_{out}(v_{ti}) = \sum_{v_{tj} \in \mathcal{N}(v_{ti})} \frac{1}{Z_{ti}(v_{tj})} f_{in}(v_{tj}) \mathbf{W}_{l_{ti}(v_{tj})}, \tag{14.4}$$

where $f_{out}(v_{ti})$ denotes the output features for v_{ti} which is the ith node at time step t, $f_{in}(v_{tj})$ denotes the input features of v_{tj}, and $\mathcal{N}(v_{ti})$ denotes the set of neighboring nodes to v_{ti}, which is defined as $\mathcal{N}(v_{ti}) = \{v_{tj} \mid d(v_{ti}, v_{tj}) < D\}$. $d(v_{ti}, v_{tj})$ is the shortest path distance from v_{ti} to v_{tj}. In ST-GCN, the shortest distance is set to $D = 1$ which means that, at each step, the convolution is performed on each node and its immediate neighbors. $\mathbf{W}_{l_{ti}(v_{tj})}$ denotes the transformation used to calculate how input features of node v_{tj} affect the output features of node v_{ti}.

In 2D convolution on grid data, the central pixel and its neighboring pixels which are convolved with a 2D kernel have a fixed spatial order. Therefore the kernel values are indexed based on the spatial order of pixels and all the pixels are convolved with a shared kernel. In a skeleton graph, the nodes can have different degrees and different number of neighbors, so there is no fixed spatial order around each node of the spatial graph. ST-GCN method addresses this issue by employing a spatial partitioning process to partition the neighboring set of a node v_{ti} into a fixed number of P subsets and assigning each subset a numeric label $l_{ti} \in \{1, ..., P\}$. The neighboring nodes v_{tj} assigned to a label $l_{ti}(v_{tj}) = p$ form the neighborhood $\mathcal{N}_p(v_{ti})$, where $\bigcup_{p=1}^{P} \mathcal{N}_p(v_{ti}) = \mathcal{N}(v_{ti})$, and are used to update the features of v_{ti}. The spatial partitioning process considers the structure of the skeletons so that the neighboring nodes of each body joint is divided into $P = 3$ subsets, which are defined as follows:

1. the root node itself,
2. the neighbors of the root node which are closer to the center of gravity (COG) than the root node,
3. the remaining neighbors of the root node, that is, those that are farther to the COG than the root node.

The COG is defined as the average of all the body joints' coordinates in a skeleton. Fig. 14.7 (left) shows the partitioned neighbors of a joint in a skeleton, which are illustrated in different colors. Accordingly, $l_{ti}(v_{tj})$ in Eq. (14.4) maps each neighbor v_{tj} of node v_{ti} into one of the above mentioned three subsets. Each neighboring subset is then associated with a transformation matrix $\mathbf{W}_p$, $p \in \{1, 2, 3\}$. Therefore by employing this spatial partitioning, the model learns a fixed number of $P = 3$ transformation matrices, which are shared between all the nodes and their neighboring subsets for feature extraction. The normalization factor $Z_{ti}(v_{tj}) = \left|\{v_{tk} \mid l_{ti}(v_{tk}) = l_{ti}(v_{tj})\}\right|$ denotes the number of nodes in each partition and balances the contribution of each of the neighboring subsets in the output features $f_{out}(v_{ti})$ of the target node v_{ti}.

Thus the graph structure in each skeleton is modeled by P Adjacency matrices $\mathbf{A}_p \in \mathbb{R}^{V \times V}$ having values of $\mathbf{A}_p^{(ij)} = 1$ if joints i and j exhibit a connectivity of type $p = \{1, 2, 3\}$, and $\mathbf{A}_{p,ij} = 0$ otherwise. The normalized Adjacency matrix $\hat{\mathbf{A}}_p \in \mathbb{R}^{V \times V}$ for each subset of neighbors p is defined as

$$\hat{\mathbf{A}}_p = \mathbf{D}_p^{-\frac{1}{2}} \mathbf{A}_p \mathbf{D}_p^{-\frac{1}{2}}, \tag{14.5}$$

where $\mathbf{D}_p^{(ii)} = \sum_j^V \mathbf{A}_p^{(ij)} + \varepsilon$, is the (diagonal) degree matrix, and $\varepsilon = 0.001$ is used to avoid issues related to empty rows in $\mathbf{D}_{(ii)_p}$ when calculating $\hat{\mathbf{A}}_p$. The Adjacency matrix of the root node denotes the nodes' self-connections and it is set to $\mathbf{A}_1 = \mathbf{I}_V$, where $\mathbf{I}_V \in \mathbb{R}^{V \times V}$ is the identity matrix.

In practice, the ST-GCN model receives as input the adjacency matrices of the spatiotemporal graph $\mathcal{G}$ and a sequence of T_{in} skeletons denoted as $\mathbf{X} \in \mathbb{R}^{C_{in} \times T_{in} \times V}$, where C_{in} is the number of input channels for each body joint and V denotes the number of body joints in each skeleton. The spatial convolution in each layer of ST-GCN model is performed by employing the layerwise propagation rule of GCNs [40] and it is defined as follows:

$$\mathbf{H}_s^{(l)} = ReLU\left(\sum_p \left(\hat{\mathbf{A}}_p \otimes \mathbf{M}_p^{(l)}\right)\mathbf{H}^{(l-1)}\mathbf{W}_p^{(l)}\right), \tag{14.6}$$

where $\mathbf{H}_s^{(l)}$ is the output of the spatial convolution at layer l and $\mathbf{H}^{(l-1)}$ is the output of the spatiotemporal GCN at layer $l-1$. The input of the first hidden layer of the model is defined as $\mathbf{H}^{(0)} = \mathbf{X}$. The matrix $\mathbf{W}_p^{(l)} \in \mathbb{R}^{C^{(l)} \times C^{(l-1)}}$ performs a linear transformation of the features associated to the pth partition of the graph, leading to a change of the nodes' representation from a feature space with $C^{(l-1)}$ dimensions to one with $C^{(l)}$ dimensions. $\otimes$ is the elementwise product of two matrices and $\mathbf{M}_p^{(l)} \in \mathbb{R}^{V \times V}$ is a learnable attention mask, which highlights the importance of the connections between different body joints (spatial edges) in a skeleton to capture the most important relations between the parts of the human pose at a time step of a human action. The output of the spatial convolution $\mathbf{H}_s^{(l)}$ is introduced into the temporal convolution operation. In practice, the spatial convolution operation performs $C^{(l)}$ standard 2D convolutions with filters of size $C^{(l-1)} \times 1 \times 1$ on $\mathbf{H}^{(l-1)}$ to map the features into a $C^{(l-1)}$ dimensional subspace and multiplies the resulting tensor with the attention masked adjacency matrix $(\hat{\mathbf{A}}_p \otimes \mathbf{M}_p^{(l)})$ on the last dimension V.

In order to model the temporal dynamics through the sequence of skeletons, the temporal aspect of the spatiotemporal graph, which connects the same body joint through consecutive skeletons is employed to extend the spatial convolution to the temporal domain of the data. In this regard, the neighboring set of each node v_{ti} in temporal domain is defined as

$$\mathcal{N}(v_{ti}) = \left\{v_{qj} \mid d(v_{tj}, v_{ti}) \leq K, |q-t| \leq \lfloor \Gamma/2 \rfloor\right\}, \tag{14.7}$$

where Γ is the temporal kernel size, which denotes the number of consecutive skeletons which are involved for feature aggregation in temporal domain. In practice, the temporal convolution performs $C^{(l)}$ standard 2D convolutions with kernels of size $C^{(l)} \times \Gamma \times 1$ on $\mathbf{H}_s^{(l)}$ and the output would be $\mathbf{H}^{(l)}$, which is of size $C^{(l)} \times T^{(l)} \times V$. This means that the features of a node v_{ti} and those corresponding to the same joint in a time window of Γ frames centered at time t are convolved with the filter of the temporal convolution to update the features of each node v_{ti} in the spatiotemporal graph, that is, a value of $K = 0$ is used. Depending on the temporal kernel size Γ, the stride and zero padding size in temporal convolution operation, the $T^{(l)}$ can be less than or equal to T_{in}. Fig. 14.8 shows the standard 2D convolution operation in both spatial and temporal domains.

The ST-GCN method has some drawbacks related to the graph construction process, which are addressed by more recently proposed methods, such as the 2s-AGCN [42], the DGNN [43], and the GCN-NAS [44] methods. It employs a predefined (and fixed) graph structure that represents the naturally existing physical connections between the human body joints. While it uses an attention mechanism

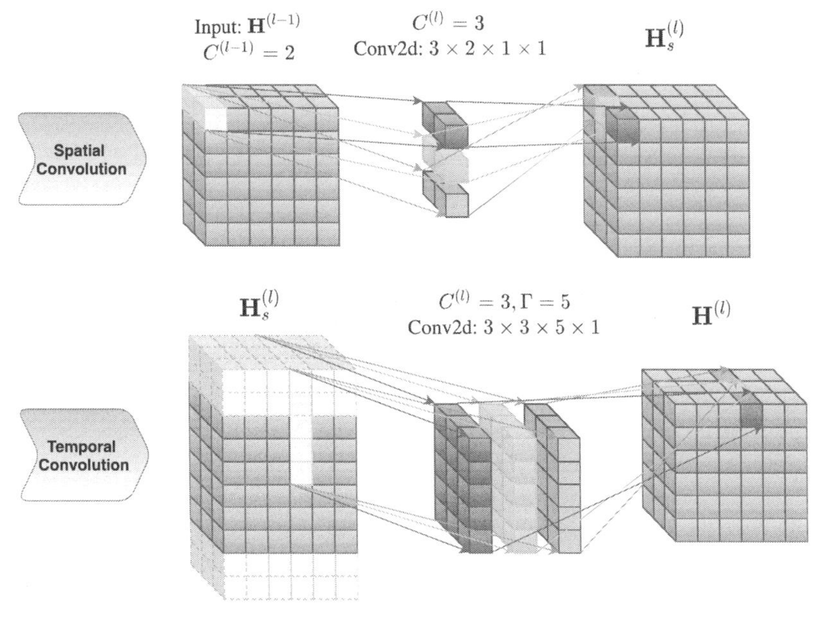

FIGURE 14.8

Illustration of spatial and temporal convolutions at lth layer of the ST-GCN network. The values of $C^{(l-1)} = 2$, $C^{(l)} = 3$ and $\Gamma = 5$ are used. In order to keep the temporal dimension unchanged so that $T^{(l)} = T^{(l-1)}$, zero padding is used in temporal convolution. Conv2d denotes the standard 2D convolutional operation.

in which the learnable attention mask can be used to highlight (or diminish) connections between the joints which are informative for an action, the fact that this attention mask is applied using elementwise multiplication does not allow the method to learn connections between joints that do not appear in the predefined Adjucency matrices $\mathbf{A}_p$. It has been shown that such a graph construction process is not optimal for the human action recognition task, since in some actions the interaction between some of the body joints, which are not naturally connected in the human body is very important to classify the action. As an example, for an action like "hugging" or "clapping" the connection between two hands is very important even though it does not exist naturally in the human body.

The 2s-AGCN [42] method which is built on top of ST-GCN method, successfully addressed this issue and improved the performance of action recognition by proposing an additive attention mechanism in which the learnable attention mask is added to the graph Adjacency matrix to both highlight (or diminish) the existing graph edges for each action and also potentially add some new edges to the graph, which connect the disconnected joints to help the model classify the action correctly. Moreover, in addition to the global graph which encodes the human body skeleton in general, the 2s-AGCN method also learns an individual data-dependent graph which represents a unique structure for each input sample and increases the model's flexibility for classifying various actions. In more details, this method utilizes the sum of natural human body structure, the learnable attention mask and a data-dependent graph, which encodes the soft connections between human body joints for each action. GCN-NAS [44] method also learns the graph structure adaptively during the training process by exploring a search space providing various dynamic graph modules. Besides, it considers multiple-hop neighbors of each node in the graph in order to increase the representational capacity of the model. DGNN [43] represents the skeletons as a directed acyclic graph to model the dependencies between joints and bones simultaneously. This method constructs the directed graph using the human body joints as graph nodes and the bones as edges while the edges are grouped into incoming edges and outgoing edges based on their direction.

In most of the ST-GCN-based methods, the model has a fixed architecture with a large number of parameters. Several methods have been proposed to reduce the computational cost and the memory requirements. SGN [45] proposes a compact model by introducing high-level semantics of the human body joints such as joint type and frame index into the model explicitly. Shift-GCN [46] also improves the computational complexity by employing lightweight pointwise convolutions and shift graph convolutions instead of regular graph convolution. DPRL+GCNN [47] and TA-GCN [48] improve the efficiency in terms of floating point operations (FLOPs) by selecting the most informative skeletons in a sequence and PST-GCN [49] method finds a compact and data-dependent network architecture for the ST-GCN in a progressive manner. ST-BLN [50] analyzes the connection of GCN layers with additive spatial attention to bilinear layers and propose the spatiotemporal bilinear network (ST-BLN), which does not require the use of predefined body skeletons and allows for more flexible and potentially more efficient design of the model.

14.2.6 Multistream architectures

So far, we have focused on the neural network design for a single input. However, most of the methods we have covered above use ensembles or topologies with multiple streams in conjunction to achieve their highest performance. The motivation for combining multiple types of inputs is that each separate input stream will contain information that other streams lack (or at least have it more readily available). In this section, we will go through some common network architectures, which use multiple streams.

14.2.6.1 Multimodal

A common combination in the multimodal setting is to use a stream of hand-crafted features, such as optical flow, to explicitly input motion information to the network in addition to the RGB video frames. This combination takes inspiration in the two-stream hypothesis in neuroscience [51], which models the two hypothesized pathways from the human visual cortex: A *ventral* stream performs recognition and detection on still objects, and a *dorsal* stream recognizes motion. The work in [52] proposed a corresponding two-stream neural network which has a stream that operates on RGB-frames, and another one which uses stacked frames of optical flow.

The architecture of the network streams were 2D CNNs as used in plain image recognition, but in general the two-stream setup can and has been used for all the architecture types discussed so far, that is, 2D CNN + RNN [14,15] (Section 14.2.1), 3D CNNs [23] (Section 14.2.2), inflated 3D CNNs [16] (Section 14.2.3), and the factorized (2+1)D CNN [17] (Section 14.2.4). The setup has also been extended beyond two streams by using Improved Dense Trajectories (iDT) in addition to [23] or as a replacement for [22] optical flow.

The multistream setup has also been used in skeleton-based human action recognition methods to improve the model's performance in terms of classification accuracy. Some methods such as AS-GCN [53], 2s-AGCN [42], GCN-NAS [44], and TA-GCN [48] improved performance by forming a two stream network and training the first and second stream using joint and bone data, respectively. In these methods, the final classification result is obtained by fusing the predicted Softmax scores of the two network streams. DGNN [43] also benefits from the motions taking place in a sequence of skeletons and utilizes the optical flow data to train the model. It forms a four-stream model and trains each network stream using a distinct data, from the set of joint, bone, joint's motion, and bone's motion data, and fuses the Softmax predictions of all the network streams to get the final outcome.

There are also multiple ways of fusing input modalities, as explored in [52] and [54]. The most common is to perform *averaging* (or equivalently summation) for each pair of channels in the two streams, when the dimensionality of both streams are equal. This was found to work well by following both the strategies of late fusion, that is, combining the modalities at the network end [52,14–17], and mid-network fusion [54]. Other options include taking the point and channelwise *max* across streams, and *concatenation* of channels for the stream, optionally followed by a pointwise 1D *convolution* to reduce the number of channels. The approach using concatenation and pointwise 1D convolutions was found to work best among the options explored in [54] for mid-network fusion. For iDT features as the additional modality, late fusion using channel concatenation in conjunction with a Support Vector Machine (SVM) for performing predictions has also been used successfully [22,23].

14.2.6.2 Multiresolution

Instead of using a different input modality, one may also use multiple variations on the same input to improve performance or make the network more efficient. One such setup is to use multiple resolutions of an input video; one covering a large spatial extend (the *context*) of the video at low resolution, and one with a higher resolution cropped to the center of the video (the *fovea*, cf. the human eye). This takes advantage of the bias present in many videos where the most important motifs are centered. Given an input video formed by frames of size 178×178 pixels, late fusion by concatenation of the input downscaled to 89×89 pixels for context, and a 89×89-pixel center crop of the input as fovea was used in [21], to achieve a $2 - 4\times$ speedup in runtime performance without loss of accuracy.

14.2.6.3 Multitemporal

A recent approach, which achieves state of the art performance, is the *SlowFast* network architecture [55]. Here, a *Slow* pathway uses a high temporal stride (low frame rate) input video to a high-capacity network stream in order to capture the semantics of slow changing objects in the scene. Similarly, a *Fast* pathway uses low temporal stride (high frame rate) input to a low capacity network stream to capture more rapidly moving objects in the video. The pathways are fused using mid-network lateral connections from the fast to the slow pathway with concatenation. In order to match the sizes before fusing, a time-strided 3D convolution with a kernel size of $T \times H \times W = 5 \times 1 \times 1$, temporal stride α and $2\beta C$ output channels is used, assuming a slow pathway of feature dimension $C \times T \times H \times W$ and fast pathway dimensions $\beta C \times \alpha T \times H \times W$.

Another approach named *Feature Aggregation for Spatio-TEmporal Redundancy* (FASTER) uses a combination of a computationally expensive network every N frames and a cheap network in the other frames, and accumulates them using a specially made FAST-GRU [56].

14.3 Temporal action localization

In temporal action localization, we go beyond the standard single-label paradigm used in classification. Here, the task is to specify the fine-grained time (video frames) in which an action is present in addition to the action-class label. Typically, this also entails recognizing multiple classes with temporal overlap. For instance, the action of performing a "Kick Flip" on a skateboard also includes a simultaneous "Jump" action (see Fig. 14.1). The data is thus labeled differently than in trimmed action recognition; it has not been trimmed to a single action. Accordingly, the input videos are often of longer duration, and multiple (or no) action labels may be present in each time instance.

In order to accommodate the more flexible labeling requirements in temporal action localization, the binary cross entropy over each time t and class c is used as loss:

$$\mathcal{L}(\mathbf{X}) = \sum_{t,c} z_{tc} \log(p(c \mid \mathbf{X}_t)) + (1 - z_{tc}) \log(1 - p(c \mid \mathbf{X}_t)) \qquad (14.8)$$

where $\mathbf{X}$ is the input video, z_{tc} the ground truth label, which is 1 if the class c is present at time t and 0 otherwise, and $p(c \mid x_t)$ is the model prediction for class c at time t. Contrary to the single-label setting, which uses a Softmax activation function to distribute the probability density over all classes, a Sigmoid activation is used here to let each class probability be computed separately.

All of the network topologies that we have discussed so far for trimmed action recognition are also applicable for temporal action localization. Networks such as the Two-stream CNN [52] or I3D [16] are often used as base networks or as part of an extended architecture. Architectures tailored for temporal action localization predominantly put their focus on exploiting the temporal sequence of actions, learning for instance that "basketball dunk" has a high likelihood of coming after a 'basketball dribble" and to cooccur with a "jump." LSTM-based methods lend themselves well to this kind of modeling and can be used to capture the dependencies on a high level [57]. The *temporal Gaussian mixture layer* [58] is another method, which captures the long-term dependencies by parameterizing a temporal convolution by the location and variance of a collection of Gaussians. In practice, these layers are used on top of the features of a trimmed activity recognition network such as I3D.

In the evaluation of temporal action localization methods, the presence or absence of each actions needs to be considered for each frame. This can be captured by the mean average precision metric.

14.4 Spatiotemporal action localization

Extending temporal action localization with an estimation of the spatial bounding boxes for the action(s) in each frame leads to the task of spatiotemporal action localization. As in temporal action localization (Section 14.3), the human subject may perform multiple simultaneous actions. Moreover, we may have multiple human subjects in a clip, the actions of whom must be located and classified separately.

Spatiotemporal action localization can be considered a mix between trimmed human activity recognition and object detection, as techniques from both fields are used for the task. Broadly speaking, object detection techniques are used for localizing the human in each frame, and an action recognition model is applied to classify the action that takes place in that location. Thus the detection and classification pipeline for spatiotemporal action localization can be adapted from object detection with few modifications. The temporal information in the input video is most important to the classification of actions, and less so for detecting the boundaries of the human subject. In [59] and [55], the pipeline was set up accordingly, using the faster R-CNN [60] metaalgorithm with minor adaptations. As usual in human activity recognition, a 3D feature extractor is used to produce a set of spatiotemporal features. Any of the previously described networks can be used to produce these, such as the two-stream CNN [52], the 2D-CNN+LSTM [14,15], the I3D [16], the R(2+1)D [17]. In parallel, a region proposal network with a deep 2D-CNN as feature extractor (considering only a single frame), evaluates candidate regions using a finite set of anchor boxes. The region(s) detecting a human constitute the bounding boxes, and are used to extract a region from the spatiotemporal features. These are then warped to a predefined size of quadratic shape using either ROI Pooling [61] or the improved ROI Align [62], before being average pooled along the temporal dimension, and finally classified using a multiclass classifier as described in Section 14.3. This pipeline, as depicted in Fig. 14.9, with a SlowFast feature extractor (see Section 14.2.6.3) was used to achieve state of the art results on the challenging AVA benchmark [59,55].

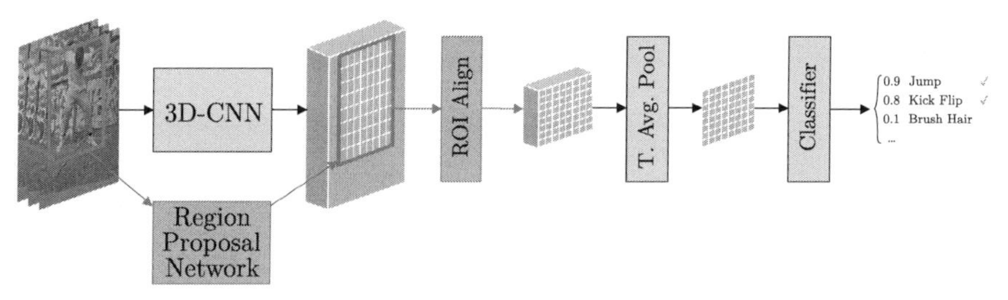

FIGURE 14.9

Faster R-CNN based pipeline for spatiotemporal action localization. The blue boxes denote components found in the usual trimmed action recognition pipeline, while the pink boxes are components found in object detection. The green 3D boxes illustrate the produced features. (For interpretation of the colors in the figure, the reader is referred to the web version of this chapter.)

The above-described pipeline is based on the Region-CNN (R-CNN) family of object detectors, which employ an *anchor-based* approach, evaluating a predefined set of anchor boxes to find a matching location. Alternatively, an *anchor-free* object detection method such as CenterNet [63] could also be adapted as was done in [64].

14.5 Data sets for human activity recognition

A central part of any task in deep learning is the availability of good data sets. In this section, we give a brief categorization of data sets in the domain, followed by a curated list of open-source data sets. Data sets in human activity recognition come in a plethora of sizes, content types, and annotation levels. In terms of annotation, a single video-level action class label is the most basic annotation available. Extending this, a video may have multiple labels associated (e.g., "jump" and 'kick flip"). While this is the level of annotation needed for the trimmed activity recognition, more densely labeled videos are required for temporal action localization. Action classes can be assigned at every frame of the video or on select frames only (e.g., every N frames). For spatiotemporal action localization, spatial bounding boxes (BB) must be supplied in addition, noting the action classes in each BB.

The contents of the videos also vary widely. In the simplest case, a video depicts a single person performing some action, in other cases multiple people are present simultaneously. Regarding the camera viewpoint, the part of the subject captured may be restricted to the upper body (*gesture recognition*), face (*facial expression recognition*), or captured from a first-person viewpoint (*egocentric activity recognition*). Also, the subjects may be situated in different environments which can be broadly

Table 14.3 List of data sets in human activity recognition. What follows is a list of abbreviations used. **Task**: Trimmed Action Recognition (TAR), Temporal Action Localization (TAL), Spatiotemporal Action Localization (STL); **Viewpoint**: full body visible (F), upper body visible (U), lower body visible (L), egocentric viewpoint 4et(E); **Environment**: controlled (C), wild (W); **Modalities**: intensity (I), red (R), green (G), blue (B), multiview (MV), depth (D), skeleton (S), infrared (IR), electromyography (EMG); **Size**: videos (v), hours (h), clips (c), frames (f); **Classes**: * action classes × object classes, † verb classes × noun classes, ‡ weakly annotated data.

Name	Task	Viewpoint	Env.	Modalities	Size	Classes	Year
ActivityNet 100 [65]	TAL	F, U, L	W	RGB	9,682v	100	2015
ActivityNet 200 [66]	TAL	F, U, L	W	RGB	19,994v (849h)	203	2016
AVA Actions [59]	STL	F, U	W	RGB	437v (392,000f)	80	2018
AVA Kinetics [67]	STL	F	W	RGB	238,474v (624,430f)	80	2020
Charades [68]	TAL	F, U, L	W	RGB	9,848v	157×46*	2016
EPIC-KITCHENS-55 [69]	TAL	E	W	RGB	55h (90,000c)	125×331†	2019
EPIC-KITCHENS-100 [69]	TAL	E	W	RGB	100h (39,594c)	97×300†	2020
EV-Actions [70]	TAR	F	C	RGB+D+S+EMG	7,000v	20	2019
HMDB-51 [71]	TAR	F, U, L	W	RGB	6,766v	51	2011
Hollywood 1 (HOHA) [72]	TAR	F, U, L	W	RGB	211v	8	2008
Hollywood 2 [73]	TAR	F, U, L	W	RGB	2,517v	12	2009
HUMAN4D [74]	TAR	F	C	RGB+D+S	50,306f	19	2020
Jester [75]	TAR	U	W	RGB	148,092v	27	2019
J-HMDB [76]	STL	F, U, L	W	RGB+S	928v	21	2013
Kinetics400 [77]	TAR	F	W	RGB	306,245v	400	2017
Kinetics600 [78]	TAR	F	W	RGB	495,547v	600	2018
Kinetics700 [79]	TAR	F	W	RGB	650,317v	700	2019
KTH [80]	TAR	F	C	I	2,391v	6	2004
MOBISERV-AIIA [81]	TAL, TAR	U	C	RGB-MV	96v (59.68h)	10 (15)	2015
Moments in Time [82]	TAR	F, U	W	RGB	903,964v	339	2019
MPII Cooking [83]	TAL	F, U	W	RGB	273v	65	2012
MultiTHUMOS [57]	TAL	F, U, L	W	RGB	413v (30h)	65	2015
Multimoments in Time [84]	TAR, STL	F, U	W	RGB	1,020,000v	313	2019
NTU RGB+D [85]	TAR	F	C	RGB+D+S+IR	56,880v	60	2016
NTU RGB+D 120 [86]	TAR	F	C	RGB+D+S+IR	114,480v	120	2016
Olympic Sports [87]	TAR	F	W	RGB	800v	16	2010
Something-Something V1 [88]	TAR	U, E	W	RGB	108,499v	174	2017
Something-Something V2 [88]	TAR	U, E	W	RGB	220,874v	174	2018
Sports-1M [21]	TAR	F, U, L	W	RGB	1,000,000v	487‡	2014
THUMOS-14 [89]	TAL	F, U, L	W	RGB	254h	20	2014
THUMOS-15 [90]	TAL	F, U, L	W	RGB	430h	20	2015
UCF-11 (YouTube Action) [91]	TAR	F	W	RGB	1,160v	11	2009
UCF-50 [92]	TAR	F, U, L	W	RGB	6,681v	50	2012
UCF-101 [86]	TAR	F, U, L	W	RGB	13,320v	101	2012

categorized as either a *controlled* setting, or *in the wild*.[5] Data sets in the controlled

[5] It should be noted that "wild" does not refer to wild nature here, and should rather be understood as a synonym for uncontrolled.

setting are predominantly created by having subjects perform the action on command in a well-lit space and capturing them with a still camera. This may be regarded as the easier of the two, because the actions themselves constitute the primary source of variation in videos. However, models trained on data from a controlled setting may not generalize well to real life problems, partly because of the low variation in the data sets, and partly because these data sets are often smaller in size than data sets from the wild. Data sets with activities in the wild are typically collected by gathering and annotating existing video clips from movies, YouTube, and the like. They thus exhibit much larger variation of contents, including background, lightning conditions, camera motion, subjects, subject visibility, and clip duration. Moreover, due to the relative ease of gathering data they can grow very large, as exemplified by the Kinetics 700 data set [79], which contains more 650,000 videos.

Finally, the modalities captured for an action also vary. In the simplest case, an action is represented by an RGB video or derived modalities such as human skeleton landmarks. In addition, depth information may be captured by a depth sensor, point clouds by Lidar, muscle activity by Electromyography (EMG) readings, kinetic forces by accelerometers, and thermal radiation by infrared sensors. A collection of data sets for human activity recognition is listed in Table 14.3.

14.6 Conclusion

This concludes our inquiry into the topic of human activity recognition, and its subtasks trimmed activity recognition, temporal action localization and spatiotemporal action localization. A plethora of architectures and methods have been devised to solve these tasks, and we have done our best to cover a wide range of them. Greatly inspired by image recognition, most methods use convolutional layers to extract features in the spatial dimension. Different approaches have been adopted in the handling of the temporal dimension, including recursive neural networks, 3D convolutions, and separated 2+1D convolutions. These were covered in Section 14.2.1, Section 14.2.2, Section 14.2.3, and Section 14.2.4, respectively. Beyond the input modality of RGB, a special class of methods use Graph Convolutional Networks (GCNs) to perform action recognition on a graph composed of the coordinates of human body joints (skeleton) that change over time (Section 14.2.5). In Section 14.2.6, we described how multiple modalities can be fused in a multistream network architecture, and how multiple resolutions (both spatially and temporally) can be utilized to make recognition more efficient. In Section 14.3 and Section 14.4, the extended tasks of temporal localization and spatiotemporal action localization, which are heavily inspired by object detection techniques, were covered. Finally, in Section 14.5 a curated list of data sets for human activity recognition was supplied. The field of human activity recognition is rapidly evolving, with new and improved methods being published on a regular basis. Luckily, there is great culture of transparency, and almost all new publications are made freely available online, just waiting for you to explore the topic further.

References

[1] A. Bobick, Movement, activity and action: the role of knowledge in the perception of motion, Philosophical Transactions of the Royal Society B 352 (1358) (1997) 1257–1265.

[2] T. Pfister, J. Charles, A. Zosserman, Flowing convnets for human pose estimation in videos, in: 2015 International Conference on Computer Vision, 2015, pp. 1913–1921.

[3] R.A. Güler, N. Neverova, I. Kokkinos, Densepose: dense human pose estimation in the wild, in: IEEE Conference on Computer Vision and Pattern Recognition, 2018, pp. 7297–7306.

[4] M. Ryoo, Human activity prediction: early recognition of ongoing activities from streaming videos, in: 2011 International Conference on Computer Vision, 2011, pp. 1036–1043.

[5] A. Iosifidis, A. Tefas, I. Pitas, Activity based person identification using fuzzy representation and discriminant learning, IEEE Transactions on Information Forensics and Security 7 (2) (2012) 530–542.

[6] F. Patrona, A. Iosifidis, A. Tefas, I. Pitas, Visual voice activity detection in the wild, IEEE Transactions on Multimedia 18 (6) (2016) 967–977.

[7] M. Holte, C. Tran, M. Trivedi, T. Moeslund, Human action recognition using multiple views: a comparative perspective on recent developments, in: 2011 Joint ACM Workshop on Human Gesture and Behavior Understanding, 2011, pp. 47–52.

[8] A. Iosifidis, A. Tefas, I. Pitas, View-invariant action recognition based on artificial neural networks, IEEE Transactions on Neural Networks and Learning Systems 23 (3) (2012) 412–434.

[9] G. Farnebäck, Two-frame motion estimation based on polynomial expansion, in: Image Analysis, Springer, Berlin Heidelberg, 2003, pp. 363–370.

[10] H. Wang, C. Schmid, Action recognition with improved trajectories, in: IEEE International Conference on Computer Vision (ICCV), 2013.

[11] A. Pfister, A. West, S. Bronner, J.A. Noah, Comparative abilities of Microsoft kinect and vicon 3d motion capture for gait analysis, Journal of Medical Engineering & Technology 38 (2014) 274–280.

[12] Z. Cao, G. Hidalgo Martinez, T. Simon, S. Wei, Y.A. Sheikh, Openpose: realtime multi-person 2d pose estimation using part affinity fields, IEEE Transactions on Pattern Analysis and Machine Intelligence (2019).

[13] M. Krestenitis, N. Passalis, A. Iosifidis, M. Gabbouj, A. Tefas, Recurrent bag-of-features for visual information analysis, Pattern Recognition 106 (107380) (2020) 1–11.

[14] J. Donahue, L.A. Hendricks, S. Guadarrama, M. Rohrbach, S. Venugopalan, T. Darrell, K. Saenko, Long-term recurrent convolutional networks for visual recognition and description, in: IEEE Conference on Computer Vision and Pattern Recognition (CVPR), 2015, pp. 2625–2634.

[15] Joe Yue-Hei Ng, M. Hausknecht, S. Vijayanarasimhan, O. Vinyals, R. Monga, G. Toderici, Beyond short snippets: deep networks for video classification, in: IEEE Conference on Computer Vision and Pattern Recognition (CVPR), 2015, pp. 4694–4702.

[16] J. Carreira, A. Zisserman, Quo vadis, action recognition? A new model and the kinetics dataset, in: 2017 IEEE Conference on Computer Vision and Pattern Recognition (CVPR), 2017, pp. 4724–4733.

[17] D. Tran, H. Wang, L. Torresani, J. Ray, Y. LeCun, M. Paluri, A closer look at spatiotemporal convolutions for action recognition, in: IEEE/CVF Conference on Computer Vision and Pattern Recognition (CVPR), 2018, pp. 6450–6459.

[18] A. Krizhevsky, I. Sutskever, G.E. Hinton, Imagenet classification with deep convolutional neural networks, in: Advances in Neural Information Processing Systems (NIPS), 2012.

[19] D.E. Rumelhart, G.E. Hinton, R.J. Williams, Learning internal representations by error propagation, in: Parallel Distributed Processing: Explorations in the Microstructure of Cognition, Volume 1: Foundations, 1986.
[20] S. Ji, W. Xu, M. Yang, K. Yu, 3d convolutional neural networks for human action recognition, IEEE Transactions on Pattern Analysis and Machine Intelligence 35 (1) (2013) 221–231.
[21] A. Karpathy, G. Toderici, S. Shetty, T. Leung, R. Sukthankar, L. Fei-Fei, Large-scale video classification with convolutional neural networks, in: IEEE Conference on Computer Vision and Pattern Recognition (CVPR), 2014, pp. 1725–1732.
[22] D. Tran, L. Bourdev, R. Fergus, L. Torresani, M. Paluri, Learning spatiotemporal features with 3d convolutional networks, in: IEEE International Conference on Computer Vision (ICCV), 2015, pp. 4489–4497.
[23] G. Varol, I. Laptev, C. Schmid, Long-term temporal convolutions for action recognition, IEEE Transactions on Pattern Analysis and Machine Intelligence 40 (6) (2018) 1510–1517.
[24] G.W. Taylor, R. Fergus, Y. LeCun, C. Bregler, Convolutional learning of spatio-temporal features, in: K. Daniilidis, P. Maragos, N. Paragios (Eds.), European Conference on Computer Vision (ECCV), 2010, pp. 140–153.
[25] K. Simonyan, A. Zisserman, Very deep convolutional networks for large-scale image recognition, in: International Conference on Learning Representations (ICLR), 2015.
[26] S. Ioffe, C. Szegedy, Batch normalization: accelerating deep network training by reducing internal covariate shift, in: Proceedings of Machine Learning Research, PMLR, 2015.
[27] N. Kanopoulos, N. Vasanthavada, R.L. Baker, Design of an image edge detection filter using the Sobel operator, IEEE Journal of Solid-State Circuits (1988).
[28] Z. Qiu, T. Yao, T. Mei, Learning spatio-temporal representation with pseudo-3d residual networks, in: International Conference on Computer Vision (ICCV), 2017.
[29] K. He, X. Zhang, S. Ren, J. Sun, Deep residual learning for image recognition, in: IEEE Conference on Computer Vision and Pattern Recognition (CVPR), 2016.
[30] F. Han, B. Reily, W. Hoff, H. Zhang, Space-time representation of people based on 3D skeletal data: a review, Computer Vision and Image Understanding 158 (2017) 85–105.
[31] A. Shahroudy, J. Liu, T.-T. Ng, G. Wang, NTU RGB+D: a large scale dataset for 3D human activity analysis, in: IEEE Conference on Computer Vision and Pattern Recognition, 2016, pp. 1010–1019.
[32] J. Liu, A. Shahroudy, D. Xu, G. Wang, Spatio-temporal LSTM with trust gates for 3D human action recognition, in: European Conference on Computer Vision, Springer, 2016, pp. 816–833.
[33] P. Zhang, C. Lan, J. Xing, W. Zeng, J. Xue, N. Zheng, View adaptive recurrent neural networks for high performance human action recognition from skeleton data, in: IEEE International Conference on Computer Vision, 2017, pp. 2117–2126.
[34] S. Song, C. Lan, J. Xing, W. Zeng, J. Liu, An end-to-end spatio-temporal attention model for human action recognition from skeleton data, in: AAAI Conference on Artificial Intelligence, 2017, pp. 4263–4270.
[35] T.S. Kim, A. Reiter, Interpretable 3D human action analysis with temporal convolutional networks, in: IEEE Conference on Computer Vision and Pattern Recognition Workshops, 2017, pp. 1623–1631.
[36] Q. Ke, M. Bennamoun, S. An, F. Sohel, F. Boussaid, A new representation of skeleton sequences for 3D action recognition, in: IEEE Conference on Computer Vision and Pattern Recognition, 2017, pp. 3288–3297.

[37] M. Liu, H. Liu, C. Chen, Enhanced skeleton visualization for view invariant human action recognition, Pattern Recognition 68 (2017) 346–362.
[38] B. Li, Y. Dai, X. Cheng, H. Chen, Y. Lin, M. He, Skeleton based action recognition using translation-scale invariant image mapping and multi-scale deep CNN, in: IEEE International Conference on Multimedia & Expo Workshops, 2017, pp. 601–604.
[39] C. Li, Q. Zhong, D. Xie, S. Pu, Skeleton-based action recognition with convolutional neural networks, in: IEEE International Conference on Multimedia & Expo Workshops, 2017, pp. 597–600.
[40] T.N. Kipf, M. Welling, Semi-supervised classification with graph convolutional networks, in: International Conference on Learning Representations, 2017.
[41] S. Yan, Y. Xiong, D. Lin, Spatial temporal graph convolutional networks for skeleton-based action recognition, in: AAAI Conference on Artificial Intelligence, 2018.
[42] L. Shi, Y. Zhang, J. Cheng, H. Lu, Two-stream adaptive graph convolutional networks for skeleton-based action recognition, in: IEEE Conference on Computer Vision and Pattern Recognition, 2019.
[43] L. Shi, Y. Zhang, J. Cheng, H. Lu, Skeleton-based action recognition with directed graph neural networks, in: IEEE Conference on Computer Vision and Pattern Recognition, 2019.
[44] W. Peng, X. Hong, H. Chen, G. Zhao, Learning graph convolutional network for skeleton-based human action recognition by neural searching, in: AAAI Conference on Artificial Intelligence, 2020.
[45] P. Zhang, C. Lan, W. Zeng, J. Xing, J. Xue, N. Zheng, Semantics-guided neural networks for efficient skeleton-based human action recognition, in: IEEE Conference on Computer Vision and Pattern Recognition, 2020.
[46] K. Cheng, Y. Zhang, X. He, W. Chen, J. Cheng, H. Lu, Skeleton-based action recognition with shift graph convolutional network, in: IEEE Conference on Computer Vision and Pattern Recognition, 2020.
[47] Y. Tang, Y. Tian, J. Lu, P. Li, J. Zhou, Deep progressive reinforcement learning for skeleton-based action recognition, in: IEEE Conference on Computer Vision and Pattern Recognition, 2018.
[48] N. Heidari, A. Iosifidis, Temporal attention-augmented graph convolutional network for efficient skeleton-based human action recognition, in: International Conference on Pattern Recognition, 2020.
[49] N. Heidari, A. Iosifidis, Progressive spatio-temporal graph convolutional network for skeleton-based human action recognition, arXiv preprint, arXiv:2011.05668, 2020.
[50] N. Heidari, A. Iosifidis, On the spatial attention in spatio-temporal graph convolutional networks for skeleton-based human action recognition, arXiv preprint, arXiv: 2011.03833, 2020.
[51] M.A. Goodale, A. Milner, Separate visual pathways for perception and action, Trends in Neurosciences (1992).
[52] K. Simonyan, A. Zisserman, Two-stream convolutional networks for action recognition in videos, in: Advances in Neural Information Processing Systems, 2014.
[53] M. Li, S. Chen, X. Chen, Y. Zhang, Y. Wang, Q. Tian, Actional-structural graph convolutional networks for skeleton-based action recognition, in: IEEE Conference on Computer Vision and Pattern Recognition, 2019.
[54] C. Feichtenhofer, A. Pinz, A. Zisserman, Convolutional two-stream network fusion for video action recognition, in: IEEE Conference on Computer Vision and Pattern Recognition (CVPR), 2016.

[55] C. Feichtenhofer, H. Fan, J. Malik, K. He, Slowfast networks for video recognition, in: IEEE/CVF International Conference on Computer Vision (ICCV), 2019.
[56] L. Zhu, D. Tran, L. Sevilla-Lara, Y. Yang, M. Feiszli, H. Wang, FASTER recurrent networks for efficient video classification, in: AAAI Conference on Artificial Intelligence, 2020, pp. 13098–13105.
[57] S. Yeung, O. Russakovsky, N. Jin, M. Andriluka, G. Mori, L. Fei-Fei, Every moment counts: dense detailed labeling of actions in complex videos, International Journal of Computer Vision (2017).
[58] A. Piergiovanni, M. Ryoo, Temporal Gaussian mixture layer for videos, in: Proceedings of Machine Learning Research (PMLR), vol. 97, 2019, pp. 5152–5161.
[59] C. Gu, C. Sun, D.A. Ross, C. Vondrick, C. Pantofaru, Y. Li, S. Vijayanarasimhan, G. Toderici, S. Ricco, R. Sukthankar, C. Schmid, J. Malik Ava, A video dataset of spatio-temporally localized atomic visual actions, in: IEEE Conference on Computer Vision and Pattern Recognition (CVPR), 2018.
[60] S. Ren, K. He, R. Girshick, J. Sun, Faster r-cnn: towards real-time object detection with region proposal networks, in: Advances in Neural Information Processing Systems 28, 2015.
[61] R. Girshick, Fast r-cnn, in: IEEE International Conference on Computer Vision (ICCV), 2015.
[62] K. He, G. Gkioxari, P. Dollár, R. Girshick, Mask r-cnn, in: IEEE International Conference on Computer Vision (ICCV), 2017.
[63] X. Zhou, D. Wang, P. Krähenbühl, Objects as points, arXiv preprint, arXiv:1904.07850, 2019.
[64] Y. Li, Z. Wang, L. Wang, G. Wu, Actions as moving points, in: European Conference on Computer Vision (ECCV), 2020.
[65] F.C. Heilbron, J.C. Niebles, Collecting and annotating human activities in web videos, in: Proceedings of International Conference on Multimedia Retrieval, 2014, p. 377.
[66] F.C. Heilbron, V. Escorcia, B. Ghanem, J.C. Niebles, Activitynet: a large-scale video benchmark for human activity understanding, in: IEEE Conference on Computer Vision and Pattern Recognition (CVPR), 2015, pp. 961–970.
[67] A. Li, M. Thotakuri, D.A. Ross, J. Carreira, A. Vostrikov, A. Zisserman, The ava-kinetics localized human actions video dataset, arXiv:2005.00214, 2020.
[68] G.A. Sigurdsson, G. Varol, X. Wang, A. Farhadi, I. Laptev, A. Gupta, Hollywood in homes: crowdsourcing data collection for activity understanding, in: European Conference on Computer Vision (ECCV), 2016.
[69] D. Damen, H. Doughty, G.M. Farinella, S. Fidler, A. Furnari, E. Kazakos, D. Moltisanti, J. Munro, T. Perrett, W. Price, M. Wray, The epic-kitchens dataset: collection, challenges and baselines, IEEE Transactions on Pattern Analysis and Machine Intelligence (2020).
[70] L. Wang, B. Sun, J. Robinson, T. Jing, Y. Fu, Ev-action: electromyography-vision multi-modal action dataset, in: IEEE International Conference on Automatic Face and Gesture Recognition, 2020.
[71] H. Kuehne, H. Jhuang, E. Garrote, T. Poggio, T. Serre, Hmdb: a large video database for human motion recognition, in: 2011 International Conference on Computer Vision, 2011, pp. 2556–2563.
[72] I. Laptev, M. Marszalek, C. Schmid, B. Rozenfeld, Learning realistic human actions from movies, in: IEEE Conference on Computer Vision and Pattern Recognition (CVPR), 2008, pp. 1–8.
[73] M. Marszałek, I. Laptev, C. Schmid, Actions in context, in: IEEE Conference on Computer Vision & Pattern Recognition (CVPR), 2009.

[74] A. Chatzitofis, L. Saroglou, P. Boutis, P. Drakoulis, N. Zioulis, S. Subramanyam, B. Kevelham, C. Charbonnier, P. Cesar, D. Zarpalas, S. Kollias, P. Daras, Human4d: a human-centric multimodal dataset for motions and immersive media, IEEE Access 8 (2020) 176241–176262.

[75] J. Materzynska, G. Berger, I. Bax, R. Memisevic, The jester dataset: a large-scale video dataset of human gestures, in: IEEE/CVF International Conference on Computer Vision Workshop (ICCVW), 2019, pp. 2874–2882.

[76] H. Jhuang, J. Gall, S. Zuffi, C. Schmid, M.J. Black, Towards understanding action recognition, in: International Conference on Computer Vision (ICCV), 2013, pp. 3192–3199.

[77] W. Kay, J. Carreira, K. Simonyan, B. Zhang, C. Hillier, S. Vijayanarasimhan, F. Viola, T. Green, T. Back, P. Natsev, M. Suleyman, A. Zisserman, The kinetics human action video dataset, arXiv:1705.06950, 2017.

[78] J. Carreira, E. Noland, A. Banki-Horvath, C.H.A. Zisserman, A short note about kinetics-600, arXiv:1808.01340, 2018.

[79] J. Carreira, E. Noland, C. Hillier, A. Zisserman, A short note on the kinetics-700 human action dataset, arXiv:1907.06987, 2019.

[80] I. Laptev, T. Lindeberg, Velocity adaptation of space-time interest points, in: International Conference on Pattern Recognition (ICPR), 2004, pp. 52–56.

[81] A. Iosifidis, E. Marami, A. Tefas, I. Pitas, The mobiserv-aiia eating and drinking multi-view database for vision-based assisted living, Journal of Information Hiding and Multimedia Signal Processing 6 (2) (2015) 254–273.

[82] M. Monfort, A. Andonian, B. Zhou, K. Ramakrishnan, S.A. Bargal, T. Yan, L. Brown, Q. Fan, D. Gutfruend, C. Vondrick, Moments in time dataset: one million videos for event understanding, IEEE Transactions on Pattern Analysis and Machine Intelligence (2019) 1–8.

[83] M. Rohrbach, S. Amin, M. Andriluka, B. Schiele, A database for fine grained activity detection of cooking activities, in: IEEE Conference on Computer Vision and Pattern Recognition (CVPR), 2012, pp. 1194–1201.

[84] M. Monfort, K. Ramakrishnan, A. Andonian, B.A. McNamara, A. Lascelles, B. Pan, Q. Fan, D. Gutfreund, R. Feris, A. Oliva, Multi-moments in time: learning and interpreting models for multi-action video understanding, arXiv:1911.00232, 2019.

[85] A. Shahroudy, J. Liu, T.-T. Ng, G. Wang, Ntu rgb+d: a large scale dataset for 3d human activity analysis, in: IEEE Conference on Computer Vision and Pattern Recognition, 2016.

[86] J. Liu, A. Shahroudy, M. Perez, G. Wang, L.-Y. Duan, A.C. Kot, Ntu rgb+d 120: a large-scale benchmark for 3d human activity understanding, IEEE Transactions on Pattern Analysis and Machine Intelligence (2019).

[87] J.C. Niebles, C.-W. Chen, L. Fei-Fei, Modeling temporal structure of decomposable motion segments for activity classification, in: European Conference on Computer Vision (ECCV), 2010, pp. 392–405.

[88] R. Goyal, S.E. Kahou, V. Michalski, J. Materzynska, S. Westphal, H. Kim, V. Haenel, I. Fruend, P. Yianilos, M. Mueller-Freitag, F. Hoppe, C. Thurau, I. Bax, R. Memisevic, The "something something" video database for learning and evaluating visual common sense, in: IEEE International Conference on Computer Vision (ICCV), 2017, pp. 5843–5851.

[89] Y.-G. Jiang, J. Liu, A.R. Zamir, G. Toderici, M. Laptev, Ivan anShah, R. Sukthankar, Thumos challenge 2014, URL http://crcv.ucf.edu/THUMOS14/, 2014.

[90] A. Gorban, H. Idrees, Y.-G. Jiang, A.R. Zamir, M. Laptev, Ivaand Shah, R. Sukthankar, Thumos challenge 2015, http://www.thumos.info, 2015.

[91] J. Liu, J. Luo, M. Shah, Recognizing realistic actions from videos "in the wild", in: IEEE International Conference on Computer Vision and Pattern Recognition (CVPR), 2009.
[92] K. Reddy, M. Shah, Recognizing 50 human action categories of web videos, Machine Vision and Applications 24 (2012) 971–981.

CHAPTER

Deep learning for vision-based navigation in autonomous drone racing

15

Huy Xuan Pham[a,c], Halil Ibrahim Ugurlu[a,c], Jonas Le Fevre[a], Deniz Bardakci[a,b], and Erdal Kayacan[a]

[a]*Artificial Intelligence in Robotics Laboratory (Air Lab), Department of Electrical and Computer Engineering, Aarhus University, Aarhus, Denmark*

[b]*Department of Electronics, Information and Bioengineering (DEIB), Polytechnic University of Milan, Milan, Italy*

15.1 Introduction

There are many real-world applications of unmanned aerial vehicles (UAVs) in which it is critical to navigate in cluttered unknown environments, including, but not limited to, search and rescue missions in underground mines [1,2], detection and monitoring of victims trapped under collapsed buildings [3,4], aerial radiation detection in nuclear power plants after an accident [5], aerial cinematography [75], aerial surveillance for anomaly detection [6,7], or extraterrestrial explorations [8,9]. In the aforementioned applications, knowledge about the environment is either not accessible or corrupted due to environmental dynamics. Hence advanced learning-based algorithms in perception, planning, and control tasks are advantageous to cope with uncertainties.

Due to limited energy storage capacity and inefficiency of rechargeable batteries, drones must execute their missions as fast as possible with the most efficient algorithms to expand their operating range and flight duration. Motivated by the aforementioned demand, agile aerial robots have gained increasing interest in the robotics community. Autonomous drone racing (ADR) is an ideal use case to evaluate various algorithms as the performance of the drone must push the boundaries in terms of perception and autonomy. ADR allows us to evaluate systems under difficult but realistic conditions, such as motion blur at high speed [10], limited onboard computer constraints [11], or aerodynamic effects in control [12]. Technologies developed by solving them in a drone racing context could be transferred to improve robot performance in other real-world problems in which drones must navigate in a cluttered environment without a need for external sensing.

[c] These authors contributed equally to this work.

Deep Learning for Robot Perception and Cognition. https://doi.org/10.1016/B978-0-32-385787-1.00020-8

The International Conference on Intelligent Robots and Systems (IROS) has offered an annual ADR contest since 2016, which has been a showcase for various research groups to validate their perception, navigation, and control algorithms in real-time. Recently, in 2019, Lockheed Martin and the Drone Racing League have launched the AlphaPilot Challenge [13], where drones try to fly as fast as possible by pushing the boundaries of artificial intelligence (AI) and autonomous flight.

Recent developments in deep learning (DL) methods bring innovative solution to conventional robotic problems, and they are already commonly deployed in the field. Deep neural networks (DNNs) are used in various ways, such as by improving controller performance [14], perception [15–20], or developing end-to-end motion planning strategies [21,22]. This chapter focuses on the use of DL methods for the navigation problem in ADR, where the ultimate objective is to find a collision-free trajectory through a set of racing gates and complete the trajectory as fast as possible without having a crash with the gates.

Autonomous robot navigation problem is conventionally divided into four subproblems, namely *state estimation*, *perception*, *planning*, and *control*. One approach, which may be interpreted as being conventional for navigation in robotics, is to find individual solutions for each aforementioned subproblem to achieve desired performance. Methods following the decomposition approach [11,15,17,19] offer clarity and transparency in the design, implementation, management, and debugging of the entire system. They allow understanding, testing, and optimizing each component independently. However, such an approach may require large computation power and proper system integration since each problem is solved individually. Maintenance is also an issue, as reintegration is needed if changes are induced for each component. Furthermore, in various applications such as ADR, where we do not have sufficient time to solve each subproblem separately, this approach may not be preferable. Alternatively, end-to-end planning methods [23] have been more popular with the development of cost-effective onboard processors and recent advances in DL, which is redefining solutions to several machine learning problems in fundamental ways. Those methods aim to solve the problem in one shot by mapping raw input sensor data (e.g., RGB camera image) to desired robot actions (e.g., drone velocity commands in x-y-z directions) [24]. This approach is generally computationally less expensive than the decomposition approach due to its compact structure, which is vital for onboard processors where the energy consumption is of great importance. On the other hand, end-to-end methods have a significant drawback: debugging particular issues in subproblems is more challenging, and sometimes it is impossible. The abstract layer makes the robot's actions and decisions less clear to human understanding. Eventually, it becomes more challenging to obtain safety or convergence guarantees in end-to-end planning methods. A rough comparison of system decomposition and end-to-end methods for ADR in given in Fig. 15.1.

The remainder of this chapter is organized as follows. Section 15.2 provides background information and methods that follow the system decomposition approach. The section presents hardware design (15.2.2), state estimation (15.2.3), control (15.2.4), and planning (15.2.5) with a special focus on agile flight. The section also shows

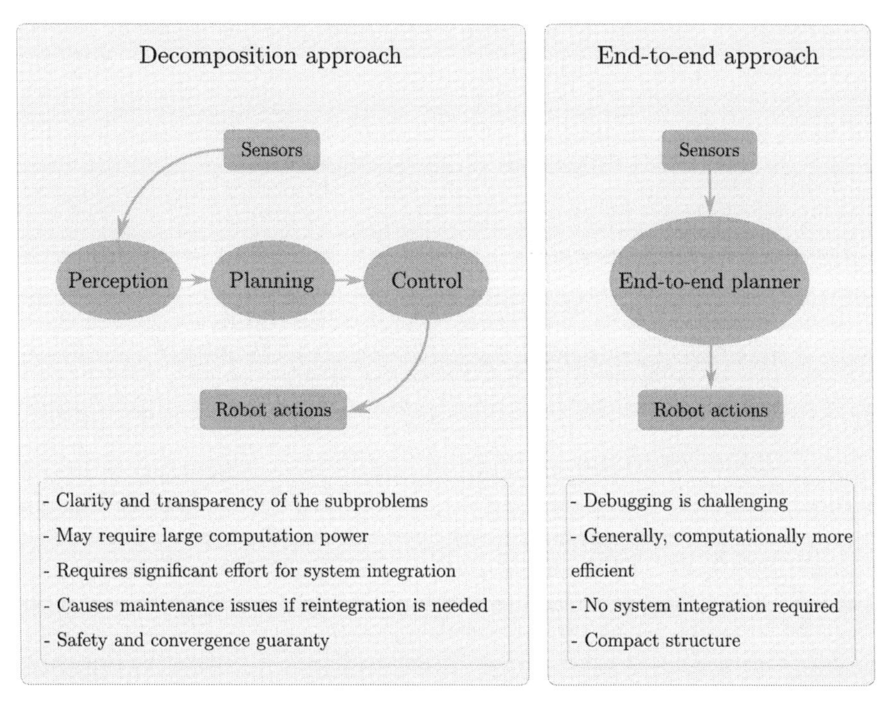

FIGURE 15.1

Comparison of system decomposition and end-to-end methods for ADR.

how convolutional neural networks (CNNs) are used for gate localization and gate pose estimation in racing drone perception, either by predicting a gate's center or building a 3D global gate map. Section 15.3 discusses alternative methods for agile planning using learning methods, where appropriate layers of abstraction can provide a direct mapping from raw images and state inputs to desired robot actions. Finally, Section 15.4 introduces a few simulation tools and data sets that would help researchers to develop, validate, and evaluate their novel algorithms for agile aerial robot navigation.

15.2 System decomposition approach in drone racing navigation

15.2.1 Related work

In the system decomposition approach for ADR, the most critical issue is to detect gates using only camera images. Conventional computer vision methods tend to fail in complex background setups with varying lighting conditions, occlusion, and motion blur [11,25]. Obviously, these methods are lacking in the generalization of the environment. DL methods, on the other hand, can successfully handle the task and generalize the information robustly. Jung et al. [15] use a DNN for gate center estimation, yet the network size is considerably large, and limited in capability due to not considering gate rotation. Kaufmann et al. [17] use two separate multilayer perceptrons to estimate the distribution of the state of a racing gate. However, it is a large-scaled DNN achieving a low inference rate on an onboard processor. Recently, Foehn et al. [19] propose a novel method that segments the four corners of the gate and identifies multiple gates' poses. Although they demonstrate remarkable performance, the method necessitates a powerful onboard computer.

In this section, we present the method which decomposes the navigation problem into separate subproblems namely control, planning, and perception. For the sake of completeness, drone hardware and state estimation methods are also provided.

15.2.2 Drone hardware

Relatively small first-person view (FPV) quadrotors are used in ADR with similar specifications to the ones that are flown by human professional pilots. A FPV drone is usually lightweight and has high-power motors to generate high thrust force for aggressive maneuverability that is critical in challenging racing tracks. Thrust-to-weight ratio is a dimensionless ratio of total thrust to takeoff mass of a drone, and it is directly proportional to the acceleration. Therefore, it has a big impact on control, as a drone with a high thrust-to-weight ratio is more responsive, capable of generating high acceleration in a short amount of time. For human-piloted small FPV drones, the ratio varies from 4 to 12 [26]. In ADR, this ratio is expected to be lower, as an onboard system, sensors, and bigger batteries might be required.

In this section, we present a prospective design of an autonomous racing quadrotor, which is prototyped at the AiRLab, Aarhus University (Fig. 15.2). The drone is built on a small carbon fiber "X" frame, which has a diagonal distance of 250 mm

(a)

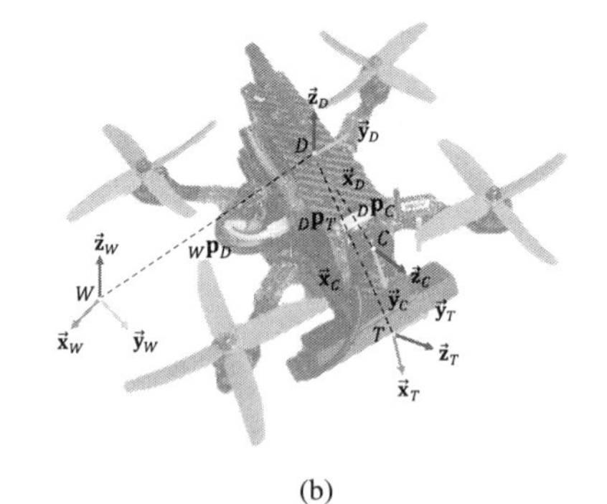

(b)

FIGURE 15.2

(a) An ADR test arena with custom-designed gates with various height and leg colors at the AiR Lab, Aarhus University. (b) Custom-designed racing drone and its coordinate frames.

between the two opposite motors. It has a thrust-to-weight ratio of 5. The drone is also equipped with an RGB camera for racing gate perception. Unlike human-piloted FPV drones, the drone has an onboard computer, a Pixhawk-based autopilot [27] module for autonomous control, and an Intel Realsense Tracking camera T265 [28] to provide state estimation. The onboard computer is an NVIDIA Jetson TX2 [29], which is one of the fastest, most power-efficient embedded AI computing devices in the market.

15.2.3 State estimation

State estimation is the process of estimating a robot's state from sensor data. The state of a robot comprises its pose (its position and orientation) with respect to the surrounding environment, its current linear and angular velocities, and other useful information. In some cases, robots can rely on the global positioning system (GPS) in outdoor environments [30] or a motion capture system in lab settings [31]. However, ADR commonly takes place in indoor, GPS-denied environments, where an external sensing system is not allowed. Also, racing drones cannot exploit 3D light detection and ranging (LiDAR) technologies [32] due to their limited payload. State estimation for agile flight is even more challenging, as jerky, aggressive motions make the problem more difficult in terms of ego-motion estimation and also vision-based estimation [33].

Simultaneous localization and mapping (SLAM) methods using visual cameras are considered in some drone racing applications [34]. However, visual SLAM is a computationally expensive algorithm when the global consistency is maintained with a single onboard computer. Furthermore, complexity increases even more as the size of the explored area becomes larger [18]. Several works in ADR opt to use more local methods found in visual-only odometry (VO) and visual inertial odometry (VIO) algorithms due to their computational efficiency on constrained onboard systems.

VO methods use input images over time to estimate local pose and motions of cameras, either by tracking and mapping a set of landmarks and features [35] or by calculating photometric errors from intensity changes in the image [36]. Mohta et al. [37] utilize a VO method namely semidirect visual odometry framework (SVO) [38] for their agile drone participated in the DARPA fast lightweight autonomy challenge. SVO jointly tracks features and minimizes the photometric error of patches around these features. The process is optimized by minimizing the reprojection error of the features using nonlinear least-squares optimization. Jung et al. [15] also deploy the same algorithm in the IROS 2017 ADR contest. However, visual inputs are not robust to low texture environments, motion blur at high speeds, and possible changing lighting conditions. To complement the camera in these conditions, VIO algorithms include inertial measurement units (IMUs) measurements to enhance motion estimation, leading to more robust performance. For instance, robust visual inertial odometry (ROVIO) [39] finds corner features and parameterizes their 3D positions with robot-centric bearing vectors and distances. It extracts multiple pyramid-level patches around these features from images to create patch-based descriptors that allow computing photometric errors. The photometric errors are used in the update step of an iterative extended Kalman filter (EKF) method based on IMU-predicted motion. ROVIO is robust in real-time and can be configured for multiple camera-IMU systems. Foehn et al. [19] use ROVIO with active drift compensations to having a robust performance in the AlphaPilot 2020.

One can also opt to use off-the-shelf solutions, such as an Intel Tracker T265 camera [40] that runs an entire visual SLAM algorithm on embedded processors and offers reasonable odometry estimation and re-localization performance even for fast vehicles. The drone in Fig. 15.2 leverages this sensor as its main state estimation component. Despite being ready-to-work and reliable, off-the-shelf devices are usually closed systems with limited options for modification and improvement. Lacking under-the-hood knowledge could lead to unaccounted behaviors that hamper the integration process of the robot system.

15.2.4 Control for agile quadrotor flight

Quadrotor is an underactuated, open-loop unstable, cross-coupling system that experiences complex aerodynamic forces during its motion [41]. Even though linear controllers are generally optimized for the trim conditions where the UAV model is linearized, ADR problem enforces UAVs to operate in their nonlinear regions due to sharp and aggressive turns along the trajectory. Therefore, model-based [12,37,41–44] and model-free [31,45–53] nonlinear controllers are needed in particular throughout agile maneuvers.

Quadrotors suffer from blade flapping and induced drag related aerodynamic effect, namely, *rotor drag*. Blade flapping and induced drag originate from flexibility and rigidity characteristics of rotors and induce forces on underactuated directions in the dynamics of a drone. Since those terms act on drone in a similar way, we can express both terms with a single term in a lumped parameter dynamical model [54]. State-of-the-art methods [55–57] exploit the differential-flatness characteristic of multirotor vehicles to simplify the controller design. Recently, Faessler et al. [12] prove that dynamics of a quadrotor are differentially flat under linear rotor drag effects and propose a cascaded control law. The method has an accurate trajectory tracking performance at high speeds. In this section, we present the dynamic model of a racing quadrotor and briefly introduce the design steps of its corresponding controller.

15.2.4.1 Dynamic model of a racing quadrotor

Based on the model with rotor drag [58], dynamics of a racing drone can be represented as

$$\dot{p} = v \tag{15.1}$$

$$\dot{R} = R\omega_\times \tag{15.2}$$

$$\dot{v} = -gz_w + cz_D - RD_{diag}R^T v \tag{15.3}$$

$$\dot{\omega} = J^{-1}(\tau - \omega \times J\omega - \tau_g - AR^T v - B\omega), \tag{15.4}$$

where p is the position of the drone's center of mass, v is the velocity, R is the transformation matrix, composed of orthonormal basis vectors of drone's body frame, ω is body rates, $\omega_\times$ is an associated skew-symmetric matrix of vector ω, c is the mass-normalized total thrust, D_{diag} is a constant diagonal matrix that associated with the mass-normalized rotor-drag coefficients, τ is the torque input vector, τ_g is the gyroscopic torque vector originated from the interaction of rotors and body, J is the inertia matrix of quadrotor, A and B are constant matrices. In other words, (15.1)–(15.2) represent translational and rotational kinematics, (15.3)–(15.4) illustrate translational and rotational dynamics of a racing drone.

We also introduce the reference thrust model [59] to compute the mass-normalized total thrust c:

$$c = c_{cmd} + k_h {v_h}^2, \tag{15.5}$$

where c_{cmd} is the commanded collective thrust input, the term k_h is an experimentally obtained positive constant, and the term v_h is the horizontal velocity vector of propellers [59].

In addition to the mentioned model choices, by using the differential flatness property of a presented dynamical model of the quadrotor, we can express its states and inputs as a function of chosen flat outputs. Faessler et al. [12] provide solid proof and adopt the corresponding feature.

15.2.4.2 Controller design

The control architecture of the racing drone adopts a cascaded control architecture which is the combination of a position controller on the high-level control loop and a body-rate controller on the low-level control loop. As a first step, we use the differential flatness property of the dynamic model of a quadrotor to compute additional reference inputs from the reference trajectory, for example, we use ω_{ref} on the computation of desired body-rates. The high level position controller also directly adopts reference trajectory terms, for example, reference acceleration a_{ref}, reference velocity v_{ref}, and reference position p_{ref} [12]. The presented high-level controller in Fig. 15.3 consider rotor drag related terms, for example, rotor drag related acceleration plays a crucial role in the computation of thrust input, c_{cmd}. From (15.5), the thrust input can be computed as

$$c_{cmd} = c - k_h {v_h}^2. \tag{15.6}$$

We compute c by projecting the desired acceleration a_{des} onto the current Z_D direction of a racing drone (see Fig. 15.2(b)), $a_{des}^T Z_D$. The desired body acceleration is obtained from

$$a_{des} = -a_{rd} + a_{fb} + a_{ref} + g z_w, \tag{15.7}$$

where a_{fb} is the proportional-derivative (PD) feedback-control term, a_{ref} is the reference acceleration, a_{rd} is the accelerations due to rotor drag, and g is the acceleration of gravity along z_w direction. Based on the corresponding formula

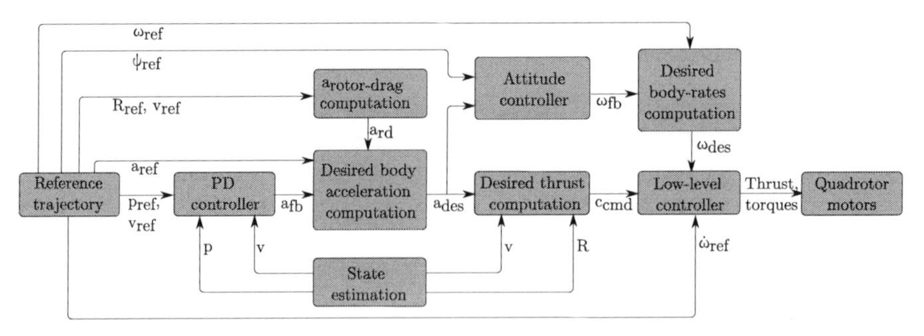

FIGURE 15.3

Control architecture of an agile quadrotor where the rotor drag effects are considered in the desired body accelerations block to obtain accurate trajectory tracking performance at high-speed flight.

$-R_{ref} D_{diag} R_{ref}^T v_{ref}$, we obtain the rotor drag related term a_{rd}. By using position and velocity control errors in PD feedback control law, we compute the a_{fb} term as

$$a_{fb} = -K_p(p - p_{ref}) - K_v(v - v_{ref}), \tag{15.8}$$

where K_p and K_v are constant diagonal gain matrices. Besides to c_{cmd}, high level controller also computes desired rotation matrix R_{des}, that is, the desired orientation of a quadrotor, the desired body rates ω_{des}, and the desired angular acceleration $\dot{\omega}_{des}$ to use them in the low-level controller. We compute the desired orientation of a quadrotor R_{des} by using the desired acceleration and reference heading angle [12]. We use R_{des} and estimation of the attitude in their quaternion representation form to make them compatible for the attitude controller [60]. The attitude controller provides the feed back terms to the desired body rates computation block as illustrated in Fig. 15.3. In addition to that, we use reference angular accelerations on the computation of the desired angular acceleration [12]. The reader is referred to [44] for the implementation steps of the low-level controller.

The racing drone in this section utilizes a Pixhawk-based autopilot [27] module to act as the low-level controller. This module takes the desired thrust and desired attitude computed by the high-level controller and generates the required rotation-per-minute value for each motor.

15.2.5 Motion planning for agile flight

Motion planning is defined as finding a feasible trajectory from an initial state to a goal state to complete a mission [61]. There are several methods for robot motion planning, such as graph search-based algorithms, artificial potential fields, and sampling-based algorithms. Although these methods are common in many robotic applications and have their advantages, they are poorly suited for agile vehicles. Graph search-based algorithms, such as the Dijkstra algorithm or D* [62], are either too slow for real-time implementation or dependent on heuristics. Artificial potential field algorithms, such as in [63], are prone to local optima and dead-locks. Sampling-based algorithms, such as RRT* [64], generate trajectories that are not continuous and, therefore, jerky.

Several research groups in ADR often choose smooth trajectory generation methods [65,66]. Unlike the aforementioned motion planning methods, these methods can generate a minimum-jerk or minimum-snap continuous trajectory for drones with low computation costs. Moreover, it allows concatenating a trajectory through multiple waypoints and attaining minimal time allocation to complete the trajectory.

A smooth trajectory $P(t)$ can be modeled as a spline, which is a function of time t defined by multiple piecewise polynomials. Let us assume that the trajectory is composed of multiple segments, where each segment connects two desired points. Each segment is modeled as a high-level polynomial $S(t)$. For quadrotors, the segment in flat output space can be decoupled into independent polynomials in each differential flatness dimension $k : x, y, z, \psi$, where $x, y, z \in \mathbb{R}$ denote the Cartesian coordinates of the drone, and $\psi \in \mathbb{R}$ is its yaw angle. Let us consider such an Nth order polynomial $S(t)$ of a segment in dimension k as

$$S_k(t) = a_0 + a_1 t + a_2 t^2 + ... + a_N t^N, \tag{15.9}$$

where $s_k = [a_0\ a_1\ a_2\ ...\ a_N]$ is a vector of the coefficients of the decoupled polynomial segment in dimension k. For simplicity, we will drop the subscript k in the following equations that optimize a decoupled trajectory in each dimension. To guarantee that the trajectory is smooth, that is, continuous at high-order derivatives (such as acceleration, jerk, and snap), we minimize a quadratic cost function as in [65]:

$$\begin{aligned} J_i(t) &= \int_0^T c_0 S(t)^2 + c_1 S'(t)^2 + c_2 S''(t)^2 + ... + c_N S^{(N)}(t)^2 \, \mathrm{d}t \\ &= s_i^T Q(T) s_i, \\ J(t) &= \begin{bmatrix} s_1 \\ \vdots \\ s_M \end{bmatrix}^T \begin{bmatrix} Q_1(T_1) & & \\ & \ddots & \\ & & Q_M(T_M) \end{bmatrix}^T \begin{bmatrix} s_1 \\ \vdots \\ s_M \end{bmatrix} + k_T \sum_{i=1}^{M} T_i, \end{aligned} \tag{15.10}$$

where $J_i(t)$ is the cost function for each segment i with a duration T, c_0, c_1, c_2, ..., c_N are the cost weights, and Q is a Hessian matrix. If we seek a minimum-snap trajectory, only c_4 is nonzero. The term $J(t)$ is the total cost function of the decoupled trajectory of M segments, concatenated from individual segment cost, plus a weighted term of duration T to allowing a trade-off between aggressiveness (i.e., short duration T) and smoothness (i.e., long duration T) of the trajectory. By tuning penalty weight k_T properly, one can achieve a time-minimal trajectory with acceptable smoothness. The cost function $J(t)$ is subject to a set of constraints of each segment. Let us assume a mapping matrix (A) between the coefficient vector s and endpoint derivatives d. For each segment i, the constraints can be modeled as in [56]:

$$\begin{aligned} A_i s_i &= d_i, \\ A_i &= [A_0 \; A_T]_i, \\ d_i &= [d_0 \; d_T]_i, \\ A_{T,i} s_i &= A_{0,i+1} s_{i+1}, \end{aligned} \tag{15.11}$$

where subscript 0 and T describe the time at the beginning and the end of each segment. While the problem above can be solved by standard quadratic programming (QP) solver, its performance degrades for a trajectory with many segments and widely varying segment times. The problem is reformulated in [65] as an unconstrained QP, by substituting the constraints into the cost function and reorganizing specified known derivatives d_F (such as zero derivatives at goal) and the unspecified derivatives d_P as

$$J = \begin{bmatrix} d_F \\ d_P \end{bmatrix}^T \begin{bmatrix} R_{FF} & R_{FP} \\ R_{PF} & R_{PP} \end{bmatrix}^T \begin{bmatrix} d_F \\ d_P \end{bmatrix}, \tag{15.12}$$

where R_{FF}, R_{FP}, R_{PF}, R_{PP} are the appropriate blocks of the augmented cost matrix constructed in [65]. Then, the optimal solution can be obtained analytically,

$$d_P^* = -R_{PP}^{-1} R_{FP}^T d_F. \tag{15.13}$$

By using (15.13), one can obtain time allocation and the optimal end-point derivatives for each segment, and the component polynomial $s_k(t)$ in (15.9), and thus the component state k at time t. One can combine the four component states $k : x, y, z, \psi$ to have the desired trajectory. The trajectory can then be sent to the quadrotor controller as a reference trajectory.

Many works, such as [19], adopt a nested motion planner architecture. It has a low-frequency global planner on top to generate a global trajectory with a long-horizon of planning through all detected gates. Then a high-frequency receding-horizon local planner [67,68] implements just a portion of the trajectory with explicit collision checks to ensure safe movements. This local planner has also replan capacity to alternate the local path to respond to changes in gate estimation.

15.2.6 Deep learning for perception

Accurate gate detection and pose estimation play a critical role on the success of the overall mission in ADR. The number of gates passed by a racer is a criterion to measure success, as it indicates how far the drone has gone in the mission. Racing gate perception is a difficult but interesting problem for computer vision. Due to agile maneuvers from the drone, gate positions, and appearance on raw input images change dramatically in short time. Due to high speed and jerky movements induced by aggressive motions, image quality also degrades from motion blur and high dynamic range effects. Recent works utilize DNN and achieve good performance. We present two DNN-based approaches for gate perception in ADR in the following subsections.

15.2.6.1 Gate center estimation

A common solution to guarantee safe gate passing is localizing the gate's center on input images, and directly minimizing the error between the drone center of mass and the detected gate center. Jung et al. propose ADRNet [15] using the single shot multiBox detector (SSD) [69]. This detector is built on top of AlexNet [70]. Fig. 15.4 provides the overview of this method, which detects bounding box offsets of the gate center on the image, and considers the center of the bounding box as the gate center. The image pixel coordinates of this pseudocenter are then extracted, and used to calculate the pixel error ϵ toward the image center (which can be aligned with the robot original heading). This pixel error is then used to calculate the lateral error E_{lat} and angular error E_{ang} in world coordinates as

$$\begin{aligned} E_{lat} &= C_1\epsilon^2 + C_2\epsilon + C_3, \\ E_{ang} &= \tan^{-1}\left(\frac{K_p E_{lat} + K_d \dot{E}_{lat}}{\lambda}\right), \\ v_{lat} &= \dot{E}_{lat}, \\ \omega &= \dot{E}_{ang}, \end{aligned} \tag{15.14}$$

where v_{lat} is the lateral velocity and ω is the angular velocity around the drone's body z-axis. They are used to control the drone to pass through the gate. The constant K_p, K_d are the control gains. The term λ is the scale factor and C_1, C_2, C_3 are the constants identified by a calibration process.

On the other hand, Morales et al. [20] use a related method, training MobileNetv2-SSDLite to detect gates in a flight arena. The method can detect multiple gates by classifying the detection result into one of the following three categories: *the immediate next gate, the second gates, and facing-backward gates*. Unlike [15], they can predict the gate distance directly by cropping a detected rectangular bounding box area and feed it to a fully connected network. The performance is also more robust due to training with semisynthetic data generated by rendering randomly multiple gates with different background and illumination conditions.

15.2.6.2 Global gate mapping

The previous planning methods using direct gate center passing could be less safe when the drone tries to pass the gate at an acute angle. It is also limited for a long-horizon motion planning for the drone to pass multiple gates. Therefore a few methods look into mapping the gate in world coordinates. These methods predict a gate center's pixel coordinates, distance, and orientation of the gate relative to the drone, and then back-project the gate from the image into the world frame. The gate pose is then incorporated into a global map, normally with the help of a probabilistic filter (such as EKF) to cope with the uncertainties of the network prediction. A motion planner, such as the one mentioned in Section 15.2.5, can rely on this map to generate a global trajectory through the perceived gates to finish a race track.

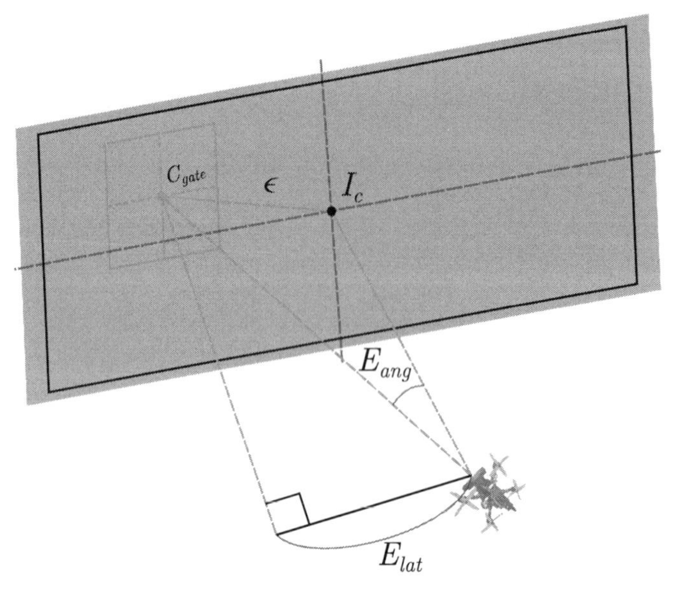

FIGURE 15.4

Outline of the method in [15] of guiding a drone to pass the gate through its center. The term ϵ is the pixel error from the perceived gate center C_{gate} to the image center I_c.

Kaufmann et al. [17] use a deep network that is a modification of the CNN architecture designed by Loquercio et al. [16] (i.e., DroNet), to first extract gate features from an RGB image stream, and then feed the extracted features into two separate multilayer perceptrons. A perceptron regresses a relative position and orientation of the gate relative to the quadrotor in spherical coordinates, while the other estimates the variance of a multivariate normal distribution that describes the current estimate of the gate's pose. The pose of the gate is updated after each prediction.

In this section, we present a related method for gate 3D mapping. Instead of using the aforementioned deep network structure in [17], we use a CNN with six hidden layers and a single fully-connected output layer. Fig. 15.5 summarizes our methodology. The method works as follows. Image features are extracted by the convolutional layers from 160×120 RGB raw images. Then the confidence values, gate center pixel coordinates, distance, and orientation of gates relative to the drone are regressed by the fully connected layer. The prediction is back-projected as a measurement of the 3D pose of the gate in the world frame. For each gate, an individual EKF is maintained to represent the probabilistic belief of the gate's pose. A consistent sparse map can be constructed. As can be seen in Fig. 15.5(b), the gates (in color) are perceived accurately compared to their ground-truth pose (green color), as they match closely.

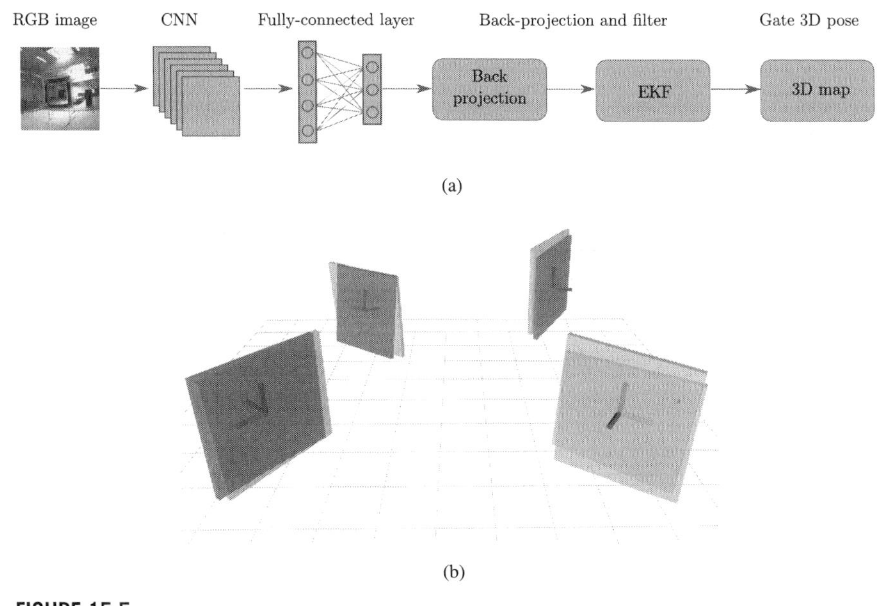

FIGURE 15.5

(a) A CNN-based perception method to obtain a global gate map from images. (b) The global 3D gate map, which shows perceived poses of multiple gates (colored) with their ground-truth (green). (For interpretation of the colors in the figure, the reader is referred to the web version of this chapter.)

15.2.7 Experimental results

This section presents the experimental results of a fully autonomous quadrotor, which combines the aforementioned modules of hardware (15.2.2), state estimation (15.2.3), controller (15.2.4), planning (15.2.5), and perception (15.2.6). Fig. 15.6 describes the software framework of the autonomous racing drone. It is implemented in the robot operating system (ROS) [71], which offers a suitable framework for system integration, as it facilitates communication between individual components working under a single system clock.

As explained in Section 15.2.3, the robot's odometry is estimated by an Intel Tracker T265. There is an independent IMU that provides attitude estimation for additional robustness. The odometry is then used by the designed controller (Section 15.2.4) to generate attitude and thrust commands to a PX4-based autopilot. The RGB image captured from the main camera is fed into the perception CNN module described in Section 15.2.6.2. The network detects a gate in the image and predicts its center location, distance, and orientation. The prediction result is then used to back-project into a 3D gate pose, with respect to the current drone's pose. A global 3D gate map is updated with the help of EKFs, to maintain global information of the racing track (Fig. 15.5(b)). The motion planner is designed as in Section 15.2.5. A low-frequency global planner generates a smooth trajectory from the drone's pose through

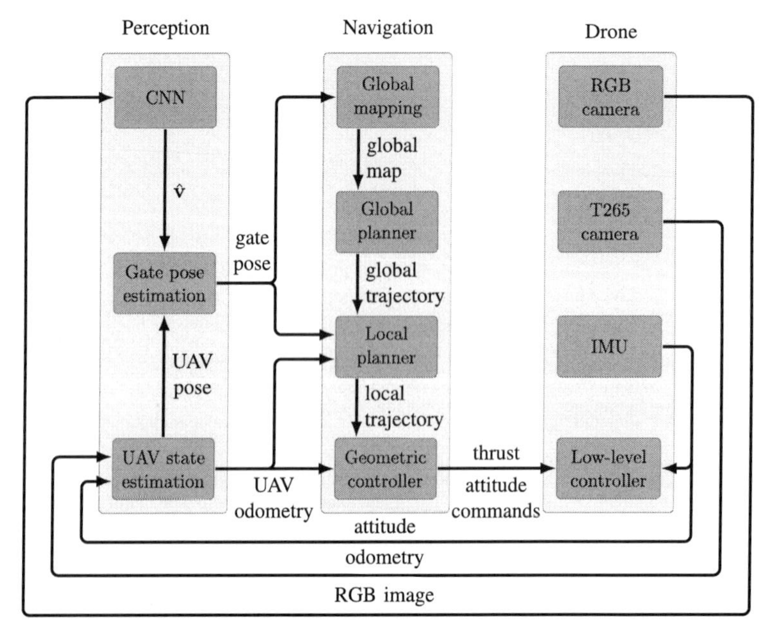

FIGURE 15.6

System architecture of the autonomous racing drone. The robot's odometry is estimated by an Intel Tracker T265 and an independent IMU. RGB images from the main camera are fed into the gate perception CNN module that can detect a gate and predict its center location, distance, and orientation. The prediction result is then back-projected and incorporated into a global 3D gate map. A low-frequency global planner generates a smooth trajectory from the drone's pose through the centers of mapped gates. A high-frequency receding-horizon local planner is deployed to implement a portion of this global path and check possible collisions along the path.

the centers of perceived gates. A high-frequency receding-horizon local planner is deployed to implement a portion of this global path and check possible collisions along the path. It can also replan if gate estimation changes or a new gate is detected.

Fig. 15.7 visualizes the drone at the current pose (depicted by the red arrow) tracking the generated trajectory, which is a combination of optimally-sampled poses (depicted by colored axes). This local path is then referenced and tracked by the designed controller. The dashed purple line in Fig. 15.7 records the actual trajectory traveled by the drone against the past reference trajectories depicted by orange continuous lines. The tracking error is mainly caused by inaccurate odometry estimation.

Fig. 15.8(a) shows the CNN's computation time (in milliseconds) over the course of the tracked trajectory. The perception module attains a high inference rate (up to 60 Hz on an NVIDIA Jetson TX2), proving its effectiveness and suitability for real-time systems. The results of the perception CNN are quite robust, allowing an estimating gate when there are changes in the location of the original gate. For orientation changes (Fig. 15.8(b)), it is less robust, only reliable up to 45°. The reason is that

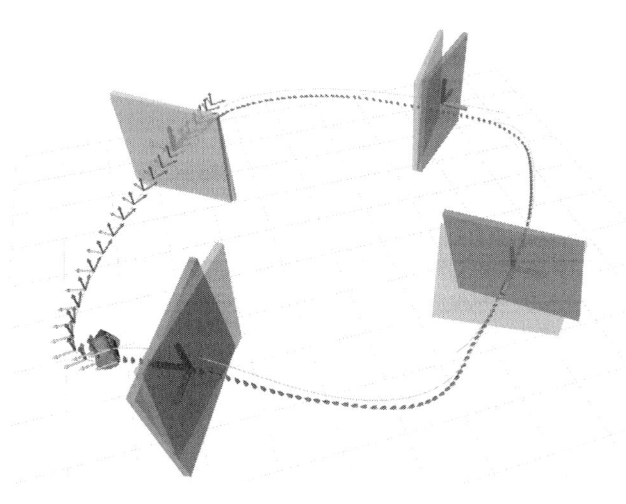

FIGURE 15.7

Trajectory of a real-time experiment. The red arrow describes the drone at the current position and heading. The perception module outputs a 3D gate map (colored, showed with their estimated poses and their ground truth in green color). A local trajectory, presented here by sampled axes, is generated by the local motion planner. The dashed purple line records the actual trajectory traveled by the drone against the past reference trajectories depicted by orange continuous lines. (For interpretation of the colors in the figure, the reader is referred to the web version of this chapter.)

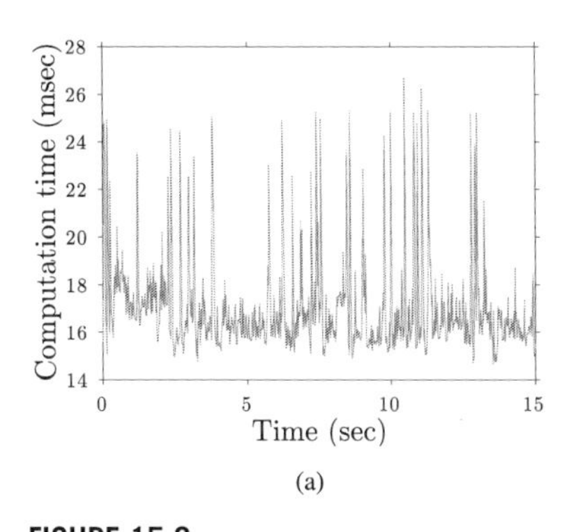

(a)

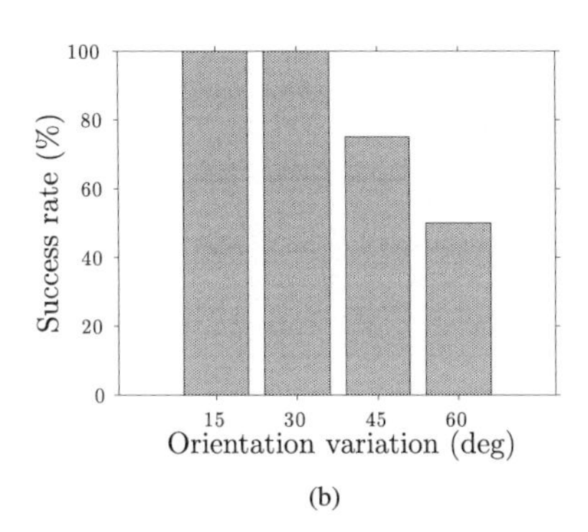

(b)

FIGURE 15.8

(a) Computation time of the perception CNN on an NVIDIA Jetson TX2. (b) Success rate with respect to orientation variations.

changing orientation affects the background appearances more, and training samples might not effectively cover the cases observed in the test flights.

15.3 Transfer learning and end-to-end planning

15.3.1 Related work

Instead of learning gate poses, Loquercio et al. [18] train an artificial neural network to generate the planner's steering function. They also show the superiority of DL with domain randomization for simulation-to-real transfer, which opens a way for DL to involve coupling localization, planning, or control blocks. For example, Bonatti et al. [75] use a variational autoencoder to learn a latent representation of gate pose estimation. Then they apply imitation learning (IL) to generate actions from the latent space. Muller et al. [21] also use IL, but they train a convolutional DNN to generate low-level actuator commands from RGB camera images. However, they ease the problem using additional guiding cones between the gates. More recently, Rojas-Peres et al. [72] use a heavier convolutional DNN to generate controller actions fed with mosaic images. Mosaic images contain consecutive camera images side-by-side to inform the network about the velocity of the drone visually. However, they apply a simple training method based on manual flight references, and finally, they achieve a smaller velocity level when compared to other methods in the literature.

15.3.2 Sim-to-real transfer with domain randomization

Learning policies from real data have a significant shortcoming, which is the high cost of generating training data in the physical world. The collection and annotation of the vast amount of data requires significant time and attention. To address this dilemma, a line of work has investigated the possibility of using simulations to train a policy and then transfer this knowledge to a real system. Training within simulations has given rise to new problems that need to be solved. The policies trained inside simulations learn to conquer problems in simulations. However, it is to be noted that simulations are only abstractions of reality and would differ from reality to some degree. As a result, when it is transferred to a real-world environment, the policies might be behaving poorly to new environments.

One way of improving the knowledge transferred is by utilizing domain randomization. The idea is to reduce overfitting by exposing the system to various scenarios by randomizing different simulation parameters, like lighting, textures, or shapes. Loquercio et al. [18] show that domain randomization can lead to an increase in closed-loop performance and improve robustness toward conditions, which are not seen at training stage. They achieve a direct sim-to-real transfer of the navigation system, and without any fine tuning on the system, it was able to navigate a real quadrotor. To train the navigation policy in simulation, they apply parameter randomization during training on illumination conditions, gate appearance, and background texture. The concept is shown in Fig. 15.9. Despite the large appearance differences between the simulated and the real environment, the policy trained in simulation via domain randomization is proven to be more robust to changes in the environment than the policy trained with data collected on the real track. Loquercio et al. [18] demon-

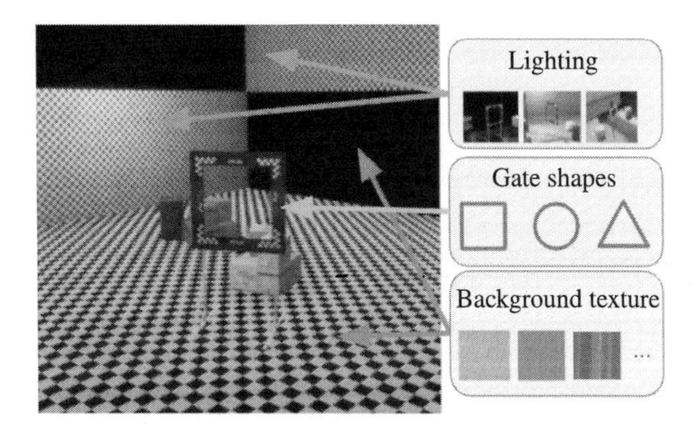

FIGURE 15.9

Domain randomization: changing lighting, shapes, and textures while training.

strate a performance increase up to 300% in simulation and up to 36% in real-world experiments when compared to that of systems not using domain randomization.

Here is some basic steps to get started with transfer learning. Step one is to create the virtual environment or use-case inside the chosen simulator. The complexity of this step very much depends on the selected simulator. Step two is to randomize the domain parameters and, as described, this is time-of-day, lighting, shapes, wind, etc. Step three is to run the simulation and gather the input material needed for the chosen model. Steps two and three will be repeated multiple times to cover a broader domain within the chosen use-case. Some simulators allow domain randomization to happen automatically and even in real-time.

15.3.3 Perceive and control with variational autoencoders

Variational autoencoders (VAE) [73] consist of two cascaded neural networks: encoder and decoder. The encoder network's input is a data point, and the output is a lower-dimensional latent representation. The decoder networks work reversely by reconstructing the data point from the latent representation of it. In other words, a VAE network regenerates the input data while passing it through a bottleneck.

Although there are various application areas for VAEs in machine learning, our focus will be on dimensionality reduction. To understand the benefit of dimensionality reduction in ADR, we can refer to perception approaches in Section 15.2. In that case, an image, a high-dimensional matrix, is used to generate the gate's pose estimation, a lower-dimensional data, which state-of-the-art motion planners can use. The methods to train end-to-end policies, such as IL or RL, are also open to similar dimensionality problems, which is called the curse of dimensionality.

As an extension to VAEs, cross-modal VAEs (CM-VAEs) [74] are using multiple data modalities. For instance, multiple encoder or decoder networks employed for the same latent representation but different data modalities. The use of CM-VAE is also enabling robust sim-to-real TL in addition to dimensionality reduction. Us-

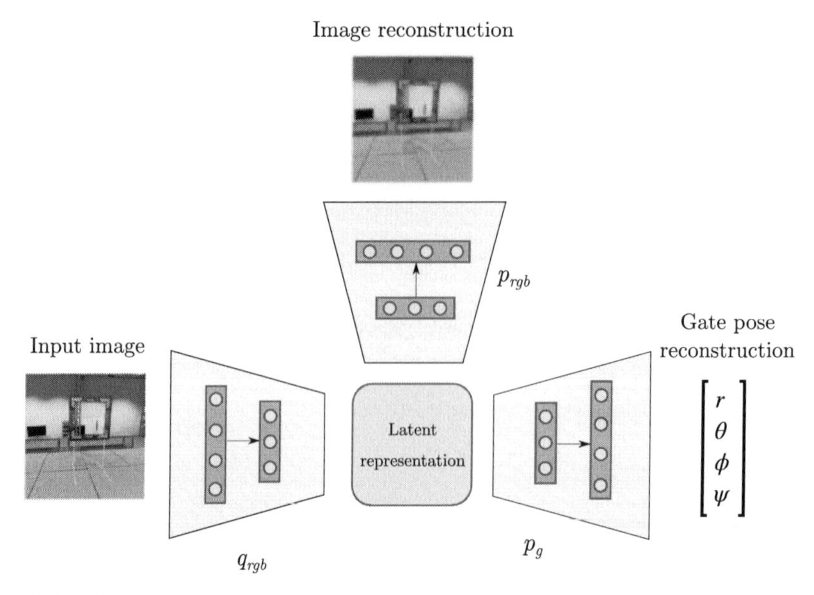

FIGURE 15.10

Cross-modal VAE architecture.

ing numerous modalities prevents the neural network from overfitting on simulation data by involving the representation's implicit regularization. In this subsection, we present this method applied in the ADR problem [75].

Two data modalities are considered to learn a cross-modal representation: RGB images from the first-person view camera and relative pose of the gate with respect to the drone. A low-dimensional latent representation of state is trained using an encoder for images and decoders for both modalities. Eventually, both labeled data and unlabeled data (e.g., real-world data) can be involved in the training process. A general CM-VAE architecture is given in Fig. 15.10.

The relative pose of the gate, i, is defined as

$$y_i = [r, \theta, \phi, \psi], \tag{15.15}$$

where (r, θ, ϕ) are the position in spherical coordinates, and ψ is the yaw angle of the gate. The gates are assumed to be not inclined, so the pitch and roll angles are zero. The encoder function is defined as

$$z_t \sim q_{rgb}(I_t), \tag{15.16}$$

where $z_t \in \mathbb{R}^N$ is sampled latent vector representation of RGB image, I_t, at time instance, t. The decoder functions are defined as

$$\hat{I}_t = p_{rgb}(z_t), \tag{15.17}$$

$$\hat{y}_t = p_g(z_t), \tag{15.18}$$

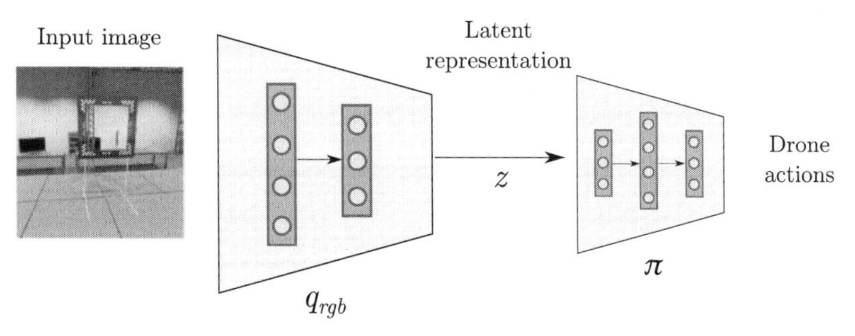

FIGURE 15.11

End-to-end neural network architecture. Input image is represented in latent space. The policy learns to generate drone actions using latent representation.

where $\hat{I}_t$ and $\hat{y}_t$ are reconstructed representations of RGB image and relative gate pose, respectively.

CM-VAE is trained based on a three-part loss function: mean squared error loss between given and predicted images, mean squared error loss between ground-truth and predicted gate pose, and Kulback–Leibler divergence loss for each sample. The decoder network for gate pose estimation is not trained for unsupervised data, but image encoder and decoder networks are still trainable.

Having a latent representation of observation, a control policy can also be built by DNNs. Fig. 15.11 shows that the encoder network connected with the policy network generates a mapping from raw observations to drone actions. The policy can be trained using IL or RL, as the former is explained in this subsection, while the latter is detailed in the following subsection.

IL is a supervised learning (SL) method based on imitating the behavior of a privileged expert. The expert is assumed to have all environmental information to label the observations with optimal actions. Then the agent needs to train a DNN having observations with correct labels. In the ADR case, an expert is usually chosen as a state-of-the-art minimum jerk trajectory tracking algorithm with ground-truth knowledge of the drone and the gate positions, as explained in Section 15.2. Later, the policy is trained with SL using images labeled with optimal actions.

15.3.4 Deep reinforcement learning

15.3.4.1 RL framework

RL [76] is a machine learning approach, along with SL and unsupervised learning. The agent discovers itself in which actions to take interacting with the environment based on a reward function. The general agent-environment interaction scheme for RL is given in Fig. 15.12. Based on its current observation of the state, s_t, the agent takes action, a_t, and receives the next state, s_{t+1}, and a reward, r_{t+1}. The agent learns a policy, $\pi(s_t)$, that outputs action given the current state to optimize the long-term sum of reward acquired from the environment.

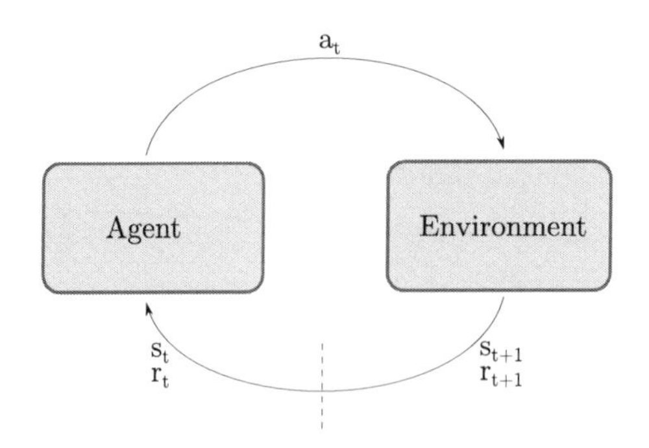

FIGURE 15.12

Agent environment interaction in RL.

A fundamental dilemma in RL is the exploration-exploitation problem. The agent should explore the state-action spaces to learn better behaviors while employing its current policy to collect rewards and fine-tune it. That makes RL hard to employ in high-dimensional or continuous state-action spaces like in the robotics field. DL is involved in RL methods using their ability to extract high-level features and generalize information, and deep RL (DRL) emerges. DL helps RL to scale the huge-size problems such as learning to play computer video games. Similarly, DRL algorithms are also used to train control policies for robots to be learned directly from sensor inputs, such as RGB cameras.

In DRL, methods are broadly classified as value-based or policy-based methods. In value-based methods, such as deep Q-learning [77], a DNN is used to approximate a value function that gives the long-term sum of reward given a state-action pair. The policy is selected to maximize the reward on value function approximation with a greedy approach. Instead of learning policies indirectly, policy-based methods [78, 79] employ a DNN to directly map states to actions. These methods are called actor-critic when they learn the policy (actor) and the value (critic) networks in parallel while the actor is updated based on the gradients in the value network [80]. Generally, a DRL problem is defined on a Markov decision process where the current state has the necessary information for the following states. Since the previous steps are irrelevant in the question, feedforward neural network implementations are preferred, such as CNN architectures for image-based observations or MLP architectures for vector-based observations. However, the observation does not contain all information of state in certain domains, which is called a partially observable Markov decision process. In that case, recursive architectures should be given preference.

15.3.4.2 Drone racing environment for DRL

A DRL agent can be assigned to learn an end-to-end neural network policy for ADR. In this context, the following assumptions are made to define ADR as a DRL problem. The drone is assumed to be forward-facing to the gate, having an RGB camera. The RGB images are considered as observations on the state, so the policy maps them to actions. The policy is trained in episodes by simulating action and observation sequences with constant discrete time steps, Δt.

An episode is randomly initiated by placing the drone with respect to the gate according to

$$p_0^x \sim \beta(1.3, 0.9) d_{max}, \tag{15.19}$$

$$p_0^y \sim \mathcal{U}(-p_0^x, p_0^x), \tag{15.20}$$

where the drone's position variables, p_0^x, p_0^y, are derived with given beta, β, and uniform, $\mathcal{U}$, distributions using d_{max} as a difficulty parameter. A number is sampled from the given Beta distribution in [0, 1] interval with higher values more probable. The action, a_t, is applied as velocity references to the controller in the x and y-axes as

$$a_t = (v_r^x, v_r^y), \tag{15.21}$$

where v_r^x and v_r^y are the reference values to the drone controller. It is to be noted that altitude is controlled separately aligned with the gate center. Finally, a sparse reward can be proposed to determine the objective only given at episode termination as

$$R = \begin{cases} 1, & \text{if drone passes gate successfully,} \\ -1, & \text{otherwise.} \end{cases} \tag{15.22}$$

Therefore an episode is terminated in two ways: success or failure. For success, there are two conditions. First, the drone should cross a virtual window from the previous time-step to the current time-step in position. Second, it should have a velocity oriented in a certain range from the forward direction. For a step to be considered as a failure, a time limit, and a boundary limit is taken into account.

15.3.4.3 Curriculum learning

Bengio et al. [81] have shown that the data provided in order during training has an effect on training performance in supervised learning. Here, the training set is divided into subsets and the training is handled from easier subset to harder subset in the problem domain. Since there is no training set in the RL domain, the curriculum is applied by the difficulty of the given task. One kind of curriculum in RL can be counted as playing against a curricular opponent or the agent itself [82]. The task becomes more difficult during training while the opponent also progresses. However, this application is only limited to two-player game scenarios. Alternatively, the initial condition or environmental parameters are shifted during RL training [83].

One of the major contributions of a curricular schedule in DRL is the exploration of sparse reward in high-dimensional state spaces. In the drone racing problem, an adaptive curriculum strategy [84] is adopted. Particularly, a difficulty parameter, $d \in [0, 1]$, is controlled by the adaptive curriculum generator. The curriculum generator is trying to keep the success rate in the range of 50% during training. The parameter

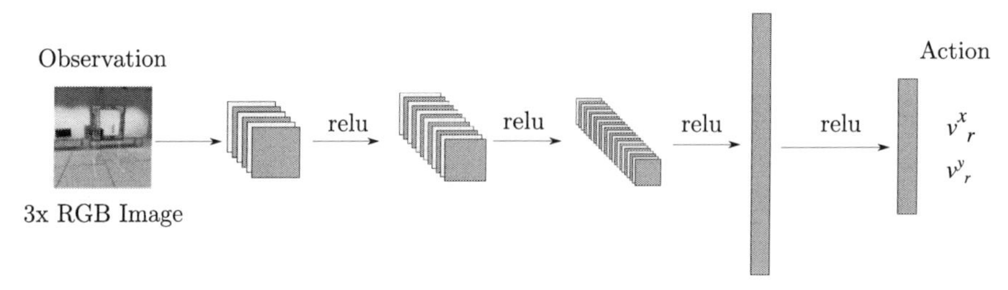

FIGURE 15.13

Policy network structure. The network input is three consecutive RGB images. There are three convolutional layers followed by two fully connected layers. The network output is the action commands to the drone.

is initialized to zero and updated according to the following equation:

$$d \longleftarrow \begin{cases} d + 0.0005, & 60\% \leq \alpha_{success}, \\ d, & 40\% < \alpha_{success} < 60\%, \\ d - 0.0005, & \alpha_{success} \leq 40\%, \end{cases} \tag{15.23}$$

where $\alpha_{success}$ is the success rate computed using the last 50 episodes. The difficulty parameter is linearly related with d_{max} in (15.19). Therefore it controls how far the drone is initialized from the gate.

15.3.4.4 Policy network architecture

We train neural network policy using proximal policy optimization algorithm [78] which is a state-of-the-art DRL algorithm. The policy network structure is illustrated in Fig. 15.13, which is a standard CNN used in DRL algorithms. We keep the images in a first-in-first-out buffer to feed our neural network policy since a single image does not infer information about the drone's velocity. The buffer size is chosen as three, considering the results shown in Table 15.1. We observe that the RL agent converges in difficulty parameter; that means it can keep the success rate above 60% while $d = 1$, if the buffer size is above two. On the other side, the computation complexity increases by the increase in buffer size. Since we do not observe a better performance with buffer size above three, we continue our experiments using three images buffered. At every layer, activation functions are rectified linear units (relu). The policy outputs the action as explained in (15.21).

Table 15.1 Effect of the size of the image buffer to the training performance.

Number of images	2	3	4
Max difficulty reached	0.8	1	1
Max success rate reached	–	90%	80%

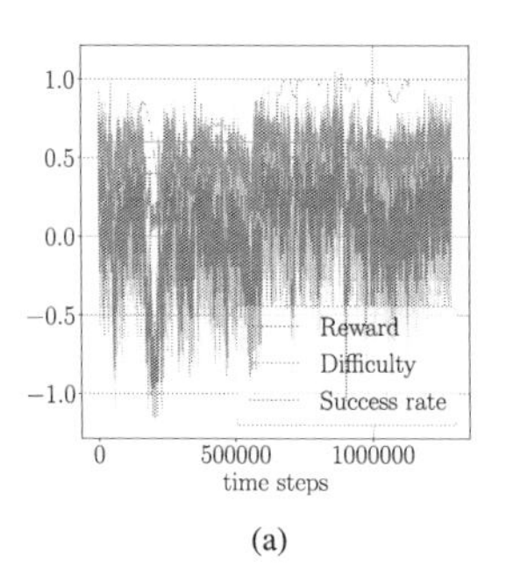

(a)

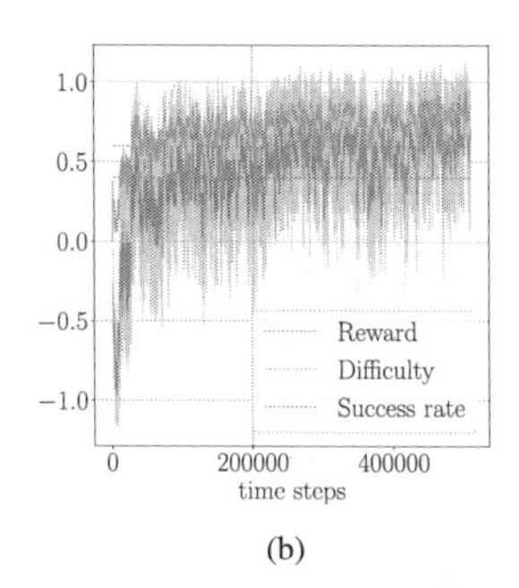

(b)

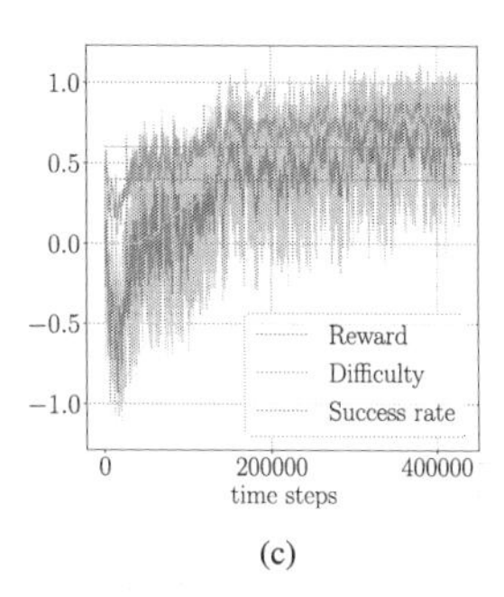

(c)

FIGURE 15.14

Sample training curves for increasing number of time-step per algorithm update: (a) 128, (b) 1024, and (c) 4096. Training performance increases both in time and variance when the hyperparameter is increased from the default value for PPO algorithm.

(a) Simulated laboratory environment in AirSim with four training gates. The gates are placed different orientation to diversify the background.

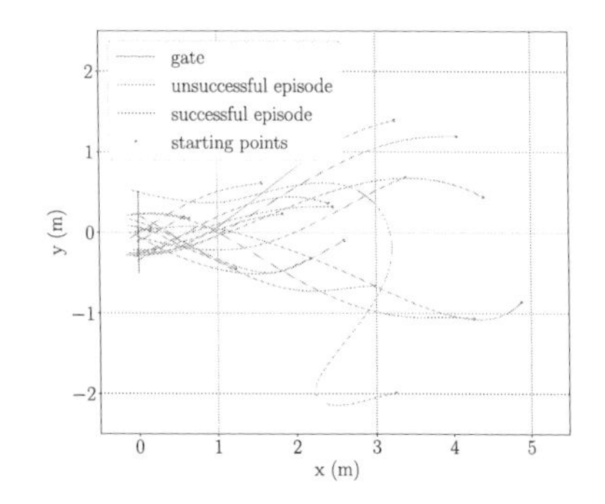

(b) Sample trajectories obtained from 20 episodes for one gate. Each trajectory is plotted in top view starting from green dot and goes through the gate denoted with blue.

FIGURE 15.15

Simulated laboratory environment and sample test trajectories of the drone. (For interpretation of the colors in the figure, the reader is referred to the web version of this chapter.)

15.3.4.5 Experimental results

We use Stable baselines [85], which is a Python package including several DRL algorithm implementation. We use the default hyperparameters except for the number of time-steps per algorithm update. In Fig. 15.14, different training curves are shown for the increasing number of the hyperparameter. The training performance increases both in time and reward variance with the increasing values of the hyperparameter.

We train and test our method in a realistic AirSim [86] simulation environment. The laboratory environment created in Unreal Engine 4 is shown in Fig. 15.15(a). We have placed four gates with different background in the environment to avoid the memorization of a single scene. Each episode is randomly initiated for one of four

Table 15.2 Success rate of the trained policy above 20 test episodes on single gate.

Gate number	1	2	3	4	Overall
Success rate	95%	95%	90%	85%	91%

gates with equal probability. We measure the speed of the networks on laptop CPU and Jetson TX2 onboard computer. We run the network 10000 times and calculate the frames per second. We get 500 Hz and 71 Hz for i7 laptop CPU and TX2, respectively.

We obtain $\sim 90\%$ success rate when we test the policy for the gates used in training. We run 20 episodes for each gate and tabulate the success rates in Table 15.2. The sample trajectories for a gate are shown in Fig. 15.15(b). The starting point of each episode is indicated by green dots. From the starting point, the policy generates and runs a velocity command according to the current observation. The trajectories are obtained using ground truth positions at each time-step.

15.4 Useful tools for data collection and training

15.4.1 Simulation environments for autonomous drone racing

Simulations have long been a valuable tool for developing robotic vehicles. It allows engineers to identify faults early in the development process and lets scientists quickly prototype and demonstrate their ideas without the risk of destroying potentially expensive hardware. While simulation systems have various benefits, many researchers see the results generated in simulation with skepticism, as any simulation system is an abstraction of reality and will vary from reality on some scale. Despite skepticism about simulation results, several trends have emerged that have driven the research community to develop better simulation systems of necessity in recent years. A major driving trend toward realistic simulators originates from the emergence of data-driven algorithmic methods in robotics, for instance, based on machine learning that requires extensive data, or RL that requires a safe learning environment to generate experiences. Simulation systems provide not only vast amounts of data but also the corresponding labels for training, which makes it ideal for these data-driven methods. This driving trend has created a critical need to develop better, more realistic simulation systems.

The ideal simulator has three main features:

1. Fast collection of large amounts of data with limited time and computations.
2. Physically accurate to represent the dynamics of the real world with high fidelity.
3. Photorealistic to minimize the discrepancy between simulations and real sensor observations.

These objectives are generally conflicting in nature: the higher accuracy in a simulation, the more computational power is needed to provide this accuracy, hence a slower

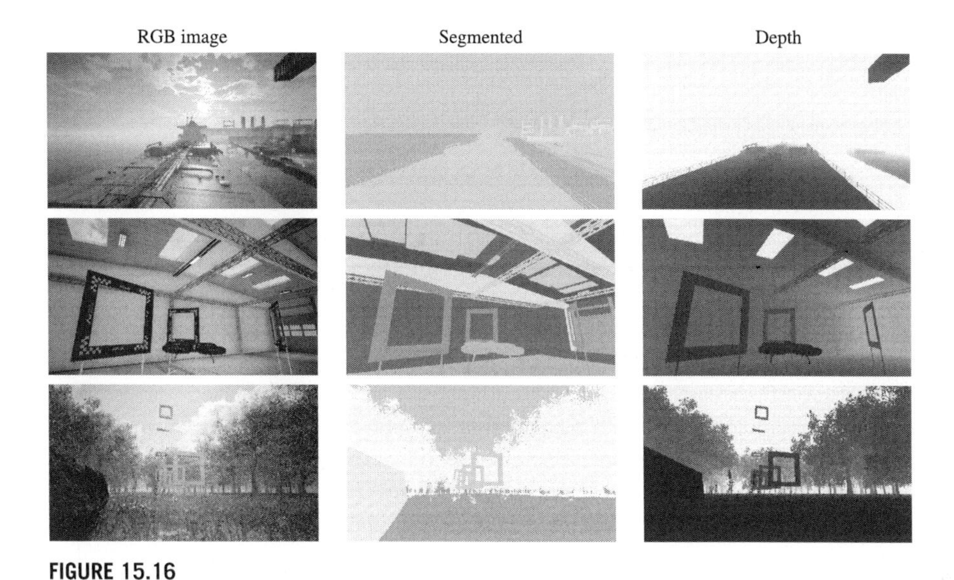

FIGURE 15.16

RGB, segmented and depth information extracted from three different environments using AirSim. The three environments are created using unreal engine and are showcasing; harbor (top row), drone lab (middle row), and outdoors (bottom row).

simulator. Therefore achieving all those objectives in a single uniform simulator is challenging. The three simulation tools that have tried to balance these objectives in a UAV simulator are FlightGoggles [87], AirSim [86], and Flightmare [88]. These simulation tools use modern 3D game engines such as the unity or unreal engine to produce high-quality, photorealistic images in real-time.

15.4.1.1 AirSim

In 2017, Microsoft released the first version of aerial informatics and robotics simulation (AirSim). AirSim is an open-source photorealistic simulator for autonomous vehicles built on unreal engine. A set of application programming interfaces (API), written in either C++ or Python, allows communication with the vehicle whether it is observing the sensor streams, collision, vehicle state, or sending commands to the vehicle. In addition, AirSim offers an interface to configure multiple vehicle models for quadrotors and supports hardware-in-the-loop as well as software-in-the-loop with flight controllers such as PX4.

AirSim differs from the other two simulators, as the dynamic model of the vehicle is simulated using NVIDIAs physics engine PhysX, a popular physics engine used by the large majority of today's video games. Despite the popularity in the gaming industry, PhysX is not specialized for quadrotors, and it is tightly coupled with the rendering engine to simulate environment dynamics. AirSim achieves some of the most realistic images, but because of this strict connection between rendering and physics simulation, AirSim can achieve only limited simulation speeds. In Fig. 15.16, examples of the photorealistic information collected via AirSim are presented, providing RGB, segmented, and depth information.

15.4.1.2 FlightGoggles

In 2019, FlightGoggles is developed at Massachusetts Institute of Technology as an open-source photorealistic sensor simulator for perception-driven robotic vehicles. The contribution from FlightGoggles consists of two separate components. First, decoupling the photorealistic rendering engine from the dynamics modeling gives the user the flexibility to choose the dynamic models' complexity. Lowering the complexity allows the rendering engine more time to generate higher quality perception data at the cost of model accuracy, whereas by increasing the complexity, more resources are allocated to accurately modeling. Second, providing an interface with real-world vehicles and actors in a motion capture system for testing vehicle-in-the-loop and human-in-the-loop. Using the motion capture system is very useful for rendering camera images given trajectories and inertial measurements from flying vehicles in the real-world, in which the collected data set is used for testing vision-based algorithms. Being able to perform vehicle-in-the-loop experiments with photorealistic sensor simulation facilitates novel research directions involving, for example, fast and agile autonomous flight in obstacle-rich environments, safe human interaction.

15.4.1.3 Flightmare

In 2020, University of Zurich robotics and perception group released the first version of Flightmare: a flexible quadrotor simulator. Flightmare shares the same motivation as FlightGoggles by decoupling the dynamics modeling from the photorealistic rendering engine to gain the flexibility to choose how much time we want to simulate an accurate model compared to gathering fast data. While not incorporating the motion capture system, Flightmare provides interfaces to the famous robotics simulator Gazebo along with different high-performance physics engines. Flightmare can simulate several hundreds of agents in parallel by decoupling the rendering module from the physics engine. This is useful for multidrone applications and enables extremely fast data collection and training, which is crucial for developing DRL applications. To facilitate RL even further, Flightmare provides a standard wrapper (OpenAI Gym), together with popular OpenAI baselines for state-of-the-art RL algorithms. Lastly, Flightmare contributes with a sizeable multimodal sensor suite, providing: IMU, RGB, segmentation, depth, and an API to extract the full 3D information of the environment in the form of a point cloud.

This section introduces some of the high-quality perception simulators on the market. Table 15.3 summarizes the main differences of the presented simulators but choosing the right one depending on the problem and what data is essential for the project. For example, if high-quality photorealistic images are the priority of the project, AirSim might be best, but if an evaluation of the quadrotor's performance is needed, maybe FlightGoggles with its vehicle-in-the-loop is the right choice. These simulation tools are being improved and extended on a daily basis, so no one can predict the right simulator to use in the future, but hopefully this section has shone

Table 15.3 Simulator feature comparison.

Simulator	Rendering	Dynamics	Sensor suite	Motion capture	RL API	Vehicles
AirSim	Unreal Engine 4	PhysX	IMU, RGB, Depth, Seg	No	No	Multiple
FlightGoggles	Unity	Flexible	IMU, RGB	Yes	No	Single
Flightmare	Unity	Flexible	IMU, RGB, Depth, Seg	No	Yes	Multiple

Table 15.4 UAV high-speed data sets.

Data sets	Top speed (m/s)	Setting	mm ground-truth	Sensor modalities
EuRoC MAV [89]	2.3	indoor	Yes	IMU, stereo camera, 3D lidar scan
Zurich urban MAV [90]	3.9	outdoor	No	IMU, RGB camera, GPS
Blackbird [91]	13.8	indoor	Yes	IMU, RGB camera, depth camera, segmentation
UPenn fast flight [92]	17.5	outdoor	No	IMU, RGB camera
UZH-FPV drone racing [10]	23.4	indoor/ outdoor	Yes	IMU, RGB camera, event camera, optical flow

some light on how simulators can be a useful tool for training, testing, and prototype and how valuable it is for autonomous vehicles.

15.4.2 Data sets

Apart from simulation tools, one can train their system with data obtained in real-world conditions. Ideally, the system used for data collection should be similar to the intended application. Careful attention must be paid to ensure the overall system is time-synchronized with ground-truth information, and all sensors are properly calibrated.

Multiple real-time data sets for studying UAVs are publicly available and can be useful for many tasks such as state estimation, perception, and even training a control policy. Delmerico et al. [10] releases a data set that includes drone racing in different environments (indoor, outdoor, different lighting conditions). The data set is helpful for studying drone state estimation, with varying difficulties. This and four other UAV data sets are listed in Table 15.4. The first two data sets are more general navigation with drones at slower speeds carrying a variety of sensors. Other data sets are at a more challenging speed. They can be used for training and evaluation of perception and state estimation algorithms. Fig. 15.17 shows sample images with different image qualities and specifications from some data sets.

A hybrid data collection method, called semisynthesis data generation [93], has caught attention. The method normally chooses a few desired parameters or objects

(a)

(b)

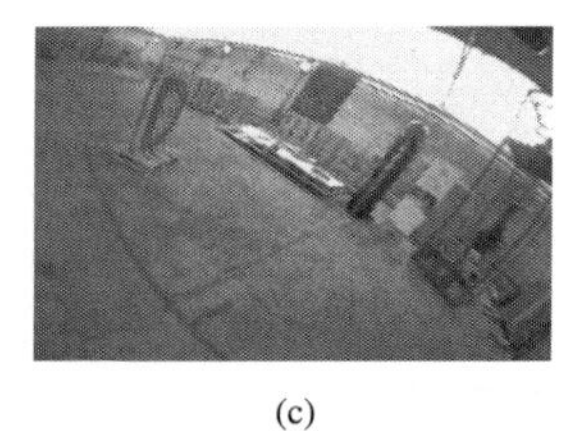
(c)

FIGURE 15.17

Sample images with different image qualities and specifications from (a) EuRoC MAV data set, (b) Zurich urban MAV data set, and (c) UZH-FPV drone racing data set.

and feeds them to a rendering mechanism that could randomly create different backgrounds, illumination conditions, or even contexts [94] of the subjects. This method can also provide the desired distribution and more content variety by controlling the rendering process. In theory, this method could provide limitless quantities of training data. For drone racing applications, Morales et al. [20] use semisynthesis data generation to obtain a robust gate detection. Loquercio et al. [18] leverage domain randomization to improve their drone's performance in gate passing even under the conditions that have not appeared in training data. By using the rendering mechanisms of photorealistic simulation tools introduced in Section 15.4.1 to generate this type of data to complement the real-time data sets, one can expect to significantly increase the robustness of their DL algorithms.

15.5 Conclusions and future work

15.5.1 Conclusions

In this chapter, we have presented two different methodologies to solve ADR navigation problem; namely *system decomposition* and *learning-based end-to-end planning*. For each approach, we highlight how DNNs can potentially improve the overall performance of the aerial robot for fast and agile flight.

First, we show that each subproblem, that is, perception, planning, and control of the system decomposition approach is well-defined, easy to formulate, and various solutions are available in the literature. DNNs have gained widespread popularity in solving these subproblems due to their ability to efficiently and accurately perform computer vision tasks, such as object detection and classification. In order to present the capabilities of this approach, a complete ADR framework is proposed, in which a DNN-based method for racing gate detection and localization is utilized in combination with conventional algorithms in state estimation, control, and motion planning. The experimental results illustrate the efficiency, efficacy, and robustness of not only the perception module but also the entire framework. However, in this methodology, each subtask must be solved separately, and it requires significant efforts to integrate separate modules, and optimize the overall performance of the system.

Second, as an alternative approach to the system decomposition method, we present end-to-end planning methods, which have caught more attention with the recent development of efficient and cost-effective onboard processors. We explain the notion of domain randomization, which helps to transfer end-to-end policies trained in simulations to the real world. A case study on end-to-end policies trained with DRL is also presented in a realistic simulation environment. A curriculum-based approach is applied to cure the exploration dilemma of the DRL problem. We qualitatively obtain an $\sim 85\%$ success rate for a gate placed in our simulated laboratory environment. We believe the accuracy level is still not sufficient for the ADR challenge. Moreover, we can assume that the success rate will drop further in real-world experiments. The presented simulations show that the method is lacking from sim-to-real transfer currently. Since DRL is subject to the exploration-exploitation dilemma, DRL algorithms are known to result in a less optimized network than supervised learning algorithms. We believe instead of depending only camera, other sensors would also be beneficial for the planner. Therefore, one open research problem is to integrate the overall system with other sensors, such as an inertial sensor and/or depth camera. The network architecture and hyperparameters of the learning algorithms are other variables that should be extensively optimized.

Since end-to-end planning methods aim to map input sensor data directly to robot actions, they are generally computationally less expensive than the system decomposition approach due to its relatively more compact structure. Therefore it is appealing for applications that demand efficiency and rapid response rate like drone racing. However, end-to-end methods have a significant drawback in debugging particular issues for a subproblem. The abstraction layers make the robot's behaviors less clear to human understanding so that a robot's action can be more challenging to verify and warrant safety. It is also difficult to extend the solutions beyond navigation. There is a research potential to alleviate these weaknesses and realize the potentials of the end-to-end methods in drone racing and future robotic applications.

The DL methods and algorithms introduced in this book chapter can be trained and improved by low-cost methods such as exploiting existing data sets, semisynthetic data generation, and sim-to-real transfer mechanism, which are also presented in his chapter. Simulation tools like: AirSim, FlightGoogles, and Flightmare have shown how engines build for the gaming industry can be used to improve the visual quality of modern simulators. Each of these tools has its own specific features and focus areas. Therefore, when choosing the most appropriate tool, one has to look into what sensors are required, the level of visual quality needed, and the expected outcome of the simulation. AirSim provides out of the box stunning visual quality, where FlightGoogles enable vehicles-in-the-loop functionality for testing real drone hardware in simulated environments. Lastly, Flightmare provides the most flexible simulator in which each of the three categories of simulation speed, model accuracy, and visual quality can be prioritized. Due to data variety and availability, DL methods are potentially more efficient and robust than the traditional computer vision approaches.

15.5.2 Future work

As indicated in the several other studies in the literature and this book chapter, the most serious problems come from the fast flight in ADR. The speed of the drone has a fundamental effect in the drone front looking camera, which is motion blur. Therefore one potential direction is to use unconventional sensor modalities, such as event-based cameras, to perceive the environment at high speeds. Event-based cameras, such as the dynamic vision sensor, are not susceptible to motion blur since they sense only the event in its image plane. Their latency in the order of microseconds makes them an ideal solution for ADR for both localization and obstacle avoidance. As a natural consequence, event-based neural networks, can be used to process event-based data. However, the challenge in using even-based cameras is that conventional computer vision methods do not work with the event-based cameras. Therefore a major contribution is needed in this direction.

Another potential research direction is the use of neuromorphic computing devices, for instance, the Loihi neuromorphic chip [95] designed by Intel Labs, which is still in its testing stage. This chip uses an asynchronous spiking neural network for high efficiency computation. Neuromorphic devices have triggered a revolution in unconventional computing technologies, but the neuromorphic computing paradigm—based on asynchronous generation and communication of spike-events—radically departs from the basic axioms of the Shannon's conventional sampling theory, which is based on synchronous sampling of analog signals. This leads to a fundamental theory gap that needs to be addressed.

References

[1] T. Dang, F. Mascarich, S. Khattak, C. Papachristos, K. Alexis, Graph-based path planning for autonomous robotic exploration in subterranean environments, in: 2019 IEEE/RSJ International Conference on Intelligent Robots and Systems (IROS), IEEE, 2019, pp. 3105–3112.

[2] M. Dharmadhikari, T. Dang, L. Solanka, J. Loje, H. Nguyen, N. Khedekar, K. Alexis, Motion primitives-based path planning for fast and agile exploration using aerial robots, in: 2020 IEEE International Conference on Robotics and Automation (ICRA), IEEE, 2020, pp. 179–185.

[3] N. Michael, S. Shen, K. Mohta, V. Kumar, K. Nagatani, Y. Okada, S. Kiribayashi, K. Otake, K. Yoshida, K. Ohno, et al., Collaborative mapping of an earthquake damaged building via ground and aerial robots, in: Field and Service Robotics, Springer, 2014, pp. 33–47.

[4] H. Nguyen, F. Mascarich, T. Dang, K. Alexis, Autonomous aerial robotic surveying and mapping with application to construction operations, arXiv preprint, arXiv:2005.04335.

[5] V. Kharchenko, A. Sachenko, V. Kochan, H. Fesenko, Reliability and survivability models of integrated drone-based systems for post emergency monitoring of npps, in: 2016 International Conference on Information and Digital Technologies (IDT), IEEE, 2016, pp. 127–132.

[6] I. Bozcan, J.L.F. Sejersen, H. Pham, E. Kayacan, Gridnet: image-agnostic conditional anomaly detection for indoor surveillance, IEEE Robotics and Automation Letters 6 (12) (2021) 1638–1645.
[7] I. Bozcan, E. Kayacan, Au-air: a multi-modal unmanned aerial vehicle dataset for low altitude traffic surveillance, in: 2020 IEEE International Conference on Robotics and Automation (ICRA), 2020, pp. 8504–8510.
[8] D.S. Bayard, D.T. Conway, R. Brockers, J.H. Delaune, L.H. Matthies, H.F. Grip, G.B. Merewether, T.L. Brown, A.M. San Martin, Vision-based navigation for the NASA Mars helicopter, in: AIAA Scitech 2019 Forum, 2019, p. 1411.
[9] N. Potter, A Mars helicopter preps for launch: the first drone to fly on another planet will hitch a ride on NASA's perseverance rover-[news], IEEE Spectrum 57 (7) (2020) 06.
[10] J. Delmerico, T. Cieslewski, H. Rebecq, M. Faessler, D. Scaramuzza, Are we ready for autonomous drone racing? The uzh-fpv drone racing dataset, in: 2019 International Conference on Robotics and Automation (ICRA), IEEE, 2019, pp. 6713–6719.
[11] S. Li, M.M. Ozo, C. De Wagter, G.C. de Croon, Autonomous drone race: a computationally efficient vision-based navigation and control strategy, Robotics and Autonomous Systems 133 (2020) 103621.
[12] M. Faessler, A. Franchi, D. Scaramuzza, Differential flatness of quadrotor dynamics subject to rotor drag for accurate tracking of high-speed trajectories, IEEE Robotics and Automation Letters 3 (2) (2017) 620–626.
[13] Alphapilot – Lockheed Martin AI drone racing innovation challenge, https://www.lockheedmartin.com/en-us/news/events/ai-innovation-challenge.html. (Accessed 13 December 2020).
[14] S. Li, E. Öztürk, C. De Wagter, G.C. de Croon, D. Izzo, Aggressive online control of a quadrotor via deep network representations of optimality principles, in: 2020 IEEE International Conference on Robotics and Automation (ICRA), IEEE, 2020, pp. 6282–6287.
[15] S. Jung, S. Hwang, H. Shin, D.H. Shim, Perception, guidance, and navigation for indoor autonomous drone racing using deep learning, IEEE Robotics and Automation Letters 3 (3) (2018) 2539–2544.
[16] A. Loquercio, A.I. Maqueda, C.R. Del-Blanco, D. Scaramuzza, Dronet: learning to fly by driving, IEEE Robotics and Automation Letters 3 (2) (2018) 1088–1095.
[17] E. Kaufmann, M. Gehrig, P. Foehn, R. Ranftl, A. Dosovitskiy, V. Koltun, D. Scaramuzza, Beauty and the beast: optimal methods meet learning for drone racing, in: 2019 International Conference on Robotics and Automation (ICRA), IEEE, 2019, pp. 690–696.
[18] A. Loquercio, E. Kaufmann, R. Ranftl, A. Dosovitskiy, V. Koltun, D. Scaramuzza, Deep drone racing: from simulation to reality with domain randomization, IEEE Transactions on Robotics 36 (1) (2019) 1–14.
[19] P. Foehn, D. Brescianini, E. Kaufmann, T. Cieslewski, M. Gehrig, M. Muglikar, D. Scaramuzza, Alphapilot: autonomous drone racing, Autonomous Robots (2021) 1–14.
[20] T. Morales, A. Sarabakha, E. Kayacan, Image generation for efficient neural network training in autonomous drone racing, in: 2020 International Joint Conference on Neural Networks (IJCNN), 2020, pp. 1–8.
[21] M. Muller, V. Casser, N. Smith, D.L. Michels, B. Ghanem, Teaching uavs to race: end-to-end regression of agile controls in simulation, in: Proceedings of the European Conference on Computer Vision (ECCV), 2018.
[22] E. Camci, E. Kayacan, End-to-end motion planning of quadrotors using deep reinforcement learning, arXiv preprint, arXiv:1909.13599.
[23] S. Levine, C. Finn, T. Darrell, P. Abbeel, End-to-end training of deep visuomotor policies, Journal of Machine Learning Research 17 (1) (2016) 1334–1373.

[24] E. Camci, D. Campolo, E. Kayacan, Deep reinforcement learning for motion planning of quadrotors using raw depth images, in: 2020 International Joint Conference on Neural Networks (IJCNN), 2020, pp. 1–7.
[25] S. Jung, S. Cho, D. Lee, H. Lee, D.H. Shim, A direct visual servoing-based framework for the 2016 iros autonomous drone racing challenge, Journal of Field Robotics 35 (1) (2018) 146–166.
[26] E. Bond, B. Crowther, B. Parslew, The rise of high-performance multi-rotor unmanned aerial vehicles-how worried should we be?, in: 2019 Workshop on Research, Education and Development of Unmanned Aerial Systems (RED UAS), IEEE, 2019, pp. 177–184.
[27] L. Meier, P. Tanskanen, F. Fraundorfer, M. Pollefeys, Pixhawk: a system for autonomous flight using onboard computer vision, in: 2011 IEEE International Conference on Robotics and Automation, IEEE, 2011, pp. 2992–2997.
[28] Intel realsense tracking camera t265, https://www.intelrealsense.com/tracking-camera-t265. (Accessed 13 December 2020), 2020.
[29] Nvidia jetson tx2 module, https://developer.nvidia.com/embedded/jetson-tx2. (Accessed 13 December 2020).
[30] I. Bozcan, E. Kayacan, Uav-adnet: unsupervised anomaly detection using deep neural networks for aerial surveillance, in: 2020 IEEE/RSJ International Conference on Intelligent Robots and Systems (IROS), 2020, pp. 1158–1164, https://doi.org/10.1109/IROS45743.2020.9341790.
[31] A. Sarabakha, E. Kayacan, Online deep learning for improved trajectory tracking of unmanned aerial vehicles using expert knowledge, in: 2019 International Conference on Robotics and Automation (ICRA), IEEE, 2019, pp. 7727–7733.
[32] T. Dang, F. Mascarich, S. Khattak, H. Nguyen, N. Khedekar, C. Papachristos, K. Alexis, Field-hardened robotic autonomy for subterranean exploration, in: Conf. on Field and Service Robot., Tokyo, Japan, 2019.
[33] D. Scaramuzza, Z. Zhang, Visual-inertial odometry of aerial robots, arXiv preprint, arXiv:1906.03289.
[34] L.O. Rojas-Perez, J. Martinez-Carranza, Metric monocular slam and colour segmentation for multiple obstacle avoidance in autonomous flight, in: 2017 Workshop on Research, Education and Development of Unmanned Aerial Systems (RED-UAS), IEEE, 2017, pp. 234–239.
[35] S. Weiss, M.W. Achtelik, S. Lynen, M.C. Achtelik, L. Kneip, M. Chli, R. Siegwart, Monocular vision for long-term micro aerial vehicle state estimation: a compendium, Journal of Field Robotics 30 (5) (2013) 803–831.
[36] C. Kerl, J. Sturm, D. Cremers, Robust odometry estimation for rgb-d cameras, in: 2013 IEEE International Conference on Robotics and Automation, IEEE, 2013, pp. 3748–3754.
[37] K. Mohta, M. Watterson, Y. Mulgaonkar, S. Liu, C. Qu, A. Makineni, K. Saulnier, K. Sun, A. Zhu, J. Delmerico, et al., Fast, autonomous flight in gps-denied and cluttered environments, Journal of Field Robotics 35 (1) (2018) 101–120.
[38] C. Forster, Z. Zhang, M. Gassner, M. Werlberger, D. Scaramuzza, Svo: semidirect visual odometry for monocular and multicamera systems, IEEE Transactions on Robotics 33 (2) (2016) 249–265.
[39] M. Bloesch, S. Omari, M. Hutter, R. Siegwart, Robust visual inertial odometry using a direct ekf-based approach, in: 2015 IEEE/RSJ International Conference on Intelligent Robots and Systems (IROS), IEEE, 2015, pp. 298–304.
[40] A. Grunnet-Jepsen, M. Harville, B. Fulkerson, D. Piro, S. Brook, Introduction to Intel realsense visual slam and the t265 tracking camera, Product Documentation.

[41] S. Bouabdallah, R. Siegwart, Full control of a quadrotor, in: 2007 IEEE/RSJ International Conference on Intelligent Robots and Systems, IEEE, 2007, pp. 153–158.
[42] M. Mehndiratta, E. Camci, E. Kayacan, Automated tuning of nonlinear model predictive controller by reinforcement learning, in: 2018 IEEE/RSJ International Conference on Intelligent Robots and Systems (IROS), IEEE, 2018, pp. 3016–3021.
[43] D. Falanga, E. Mueggler, M. Faessler, D. Scaramuzza, Aggressive quadrotor flight through narrow gaps with onboard sensing and computing using active vision, in: 2017 IEEE International Conference on Robotics and Automation (ICRA), IEEE, 2017, pp. 5774–5781.
[44] M. Faessler, D. Falanga, D. Scaramuzza, Thrust mixing, saturation, and body-rate control for accurate aggressive quadrotor flight, IEEE Robotics and Automation Letters 2 (2) (2016) 476–482.
[45] S. Patel, A. Sarabakha, D. Kircali, E. Kayacan, An intelligent hybrid artificial neural network-based approach for control of aerial robots, Journal of Intelligent & Robotic Systems 97 (2) (2020) 387–398.
[46] M. Mehndiratta, E. Kayacan, A constrained instantaneous learning approach for aerial package delivery robots: onboard implementation and experimental results, Autonomous Robots 43 (8) (2019) 2209–2228.
[47] S. Patel, A. Sarabakha, D. Kircali, G. Loianno, E. Kayacan, Artificial neural network-assisted controller for fast and agile uav flight: onboard implementation and experimental results, in: 2019 Workshop on Research, Education and Development of Unmanned Aerial Systems (RED UAS), IEEE, 2019, pp. 37–43.
[48] A. Sarabakha, N. Imanberdiyev, E. Kayacan, M.A. Khanesar, H. Hagras, Novel Levenberg–Marquardt based learning algorithm for unmanned aerial vehicles, Information Sciences 417 (2017) 361–380.
[49] A. Sarabakha, C. Fu, E. Kayacan, Intuit before tuning: type-1 and type-2 fuzzy logic controllers, Applied Soft Computing 81 (2019) 105495.
[50] E. Kayacan, R. Maslim, Type-2 fuzzy logic trajectory tracking control of quadrotor vtol aircraft with elliptic membership functions, IEEE/ASME Transactions on Mechatronics 22 (1) (2016) 339–348.
[51] E. Camci, E. Kayacan, Game of drones: uav pursuit-evasion game with type-2 fuzzy logic controllers tuned by reinforcement learning, in: 2016 IEEE International Conference on Fuzzy Systems (FUZZ-IEEE), IEEE, 2016, pp. 618–625.
[52] C. Fu, A. Sarabakha, E. Kayacan, C. Wagner, R. John, J.M. Garibaldi, Input uncertainty sensitivity enhanced nonsingleton fuzzy logic controllers for long-term navigation of quadrotor uavs, IEEE/ASME Transactions on Mechatronics 23 (2) (2018) 725–734.
[53] E. Kaufmann, A. Loquercio, R. Ranftl, M. Müller, V. Koltun, D. Scaramuzza, Deep drone acrobatics, arXiv preprint, arXiv:2006.05768.
[54] R. Mahony, V. Kumar, P. Corke, Multirotor aerial vehicles: modeling, estimation, and control of quadrotor, IEEE Robotics & Automation Magazine 19 (3) (2012) 20–32.
[55] T. Lee, M. Leok, N.H. McClamroch, Geometric tracking control of a quadrotor uav on se (3), in: 49th IEEE Conference on Decision and Control, (CDC), IEEE, 2010, pp. 5420–5425.
[56] D. Mellinger, V. Kumar, Minimum snap trajectory generation and control for quadrotors, in: 2011 IEEE International Conference on Robotics and Automation, IEEE, 2011, pp. 2520–2525.
[57] J. Ferrin, R. Leishman, R. Beard, T. McLain, Differential flatness based control of a rotorcraft for aggressive maneuvers, in: 2011 IEEE/RSJ International Conference on Intelligent Robots and Systems, IEEE, 2011, pp. 2688–2693.

[58] J.-M. Kai, G. Allibert, M.-D. Hua, T. Hamel, Nonlinear feedback control of quadrotors exploiting first-order drag effects, IFAC-PapersOnLine 50 (1) (2017) 8189–8195.
[59] J. Svacha, K. Mohta, V. Kumar, Improving quadrotor trajectory tracking by compensating for aerodynamic effects, in: 2017 International Conference on Unmanned Aircraft Systems (ICUAS), IEEE, 2017, pp. 860–866.
[60] M. Faessler, F. Fontana, C. Forster, D. Scaramuzza, Automatic re-initialization and failure recovery for aggressive flight with a monocular vision-based quadrotor, in: 2015 IEEE International Conference on Robotics and Automation (ICRA), IEEE, 2015, pp. 1722–1729.
[61] S.M. LaValle, Planning Algorithms, Cambridge University Press, 2006.
[62] A. Stentz, Optimal and efficient path planning for partially known environments, in: Intelligent Unmanned Ground Vehicles, Springer, 1997, pp. 203–220.
[63] H.X. Pham, H.M. La, D. Feil-Seifer, M. Deans, A distributed control framework for a team of unmanned aerial vehicles for dynamic wildfire tracking, in: 2017 IEEE/RSJ International Conference on Intelligent Robots and Systems (IROS), IEEE, 2017, pp. 6648–6653.
[64] S. Karaman, E. Frazzoli, Optimal kinodynamic motion planning using incremental sampling-based methods, in: 49th IEEE Conference on Decision and Control, (CDC), IEEE, 2010, pp. 7681–7687.
[65] C. Richter, A. Bry, N. Roy, Polynomial trajectory planning for aggressive quadrotor flight in dense indoor environments, in: Robotics Research, Springer, 2016, pp. 649–666.
[66] M. Burri, H. Oleynikova, M.W. Achtelik, R. Siegwart, Real-time visual-inertial mapping, re-localization and planning onboard mavs in unknown environments, in: Intelligent Robots and Systems (IROS 2015), 2015 IEEE/RSJ International Conference on, 2015.
[67] S. Liu, M. Watterson, K. Mohta, K. Sun, S. Bhattacharya, C.J. Taylor, V. Kumar, Planning dynamically feasible trajectories for quadrotors using safe flight corridors in 3-d complex environments, IEEE Robotics and Automation Letters 2 (3) (2017) 1688–1695.
[68] C. Papachristos, F. Mascarich, S. Khattak, T. Dang, K. Alexis, Localization uncertainty-aware autonomous exploration and mapping with aerial robots using receding horizon path-planning, Autonomous Robots 43 (8) (2019) 2131–2161.
[69] W. Liu, D. Anguelov, D. Erhan, C. Szegedy, S. Reed, C.-Y. Fu, A.C. Berg, Ssd: single shot multibox detector, in: European Conference on Computer Vision, Springer, 2016, pp. 21–37.
[70] A. Krizhevsky, I. Sutskever, G.E. Hinton, Imagenet classification with deep convolutional neural networks, in: Advances in Neural Information Processing Systems, 2012, pp. 1097–1105.
[71] M. Quigley, K. Conley, B. Gerkey, J. Faust, T. Foote, J. Leibs, R. Wheeler, A.Y. Ng, Ros: an open-source robot operating system, in: ICRA Workshop on Open Source Software, Vol. 3, Kobe, Japan, 2009, p. 5.
[72] L.O. Rojas-Perez, J. Martinez-Carranza, Deeppilot: a cnn for autonomous drone racing, Sensors 20 (16) (2020) 4524.
[73] D.P. Kingma, M. Welling, Auto-encoding variational Bayes, arXiv preprint, arXiv:1312.6114.
[74] A. Spurr, J. Song, S. Park, O. Hilliges, Cross-modal deep variational hand pose estimation, in: Proceedings of the IEEE Conference on Computer Vision and Pattern Recognition, 2018, pp. 89–98.
[75] R. Bonatti, R. Madaan, V. Vineet, S. Scherer, A. Kapoor, Learning visuomotor policies for aerial navigation using cross-modal representations, arXiv preprint, arXiv:1909.06993.

[76] R.S. Sutton, A.G. Barto, Reinforcement Learning: An Introduction, MIT Press, 2018.

[77] V. Mnih, K. Kavukcuoglu, D. Silver, A.A. Rusu, J. Veness, M.G. Bellemare, A. Graves, M. Riedmiller, A.K. Fidjeland, G. Ostrovski, et al., Human-level control through deep reinforcement learning, Nature 518 (7540) (2015) 529–533.

[78] J. Schulman, F. Wolski, P. Dhariwal, A. Radford, O. Klimov, Proximal policy optimization algorithms, arXiv preprint, arXiv:1707.06347.

[79] J. Schulman, S. Levine, P. Abbeel, M. Jordan, P. Moritz, Trust region policy optimization, in: International Conference on Machine Learning, PMLR, 2015, pp. 1889–1897.

[80] T.P. Lillicrap, J.J. Hunt, A. Pritzel, N. Heess, T. Erez, Y. Tassa, D. Silver, D. Wierstra, Continuous control with deep reinforcement learning, arXiv preprint, arXiv:1509.02971.

[81] Y. Bengio, J. Louradour, R. Collobert, J. Weston, Curriculum learning, in: Proceedings of the 26th Annual International Conference on Machine Learning, 2009, pp. 41–48.

[82] M. Riedmiller, R. Hafner, T. Lampe, M. Neunert, J. Degrave, T. Wiele, V. Mnih, N. Heess, J.T. Springenberg, Learning by playing solving sparse reward tasks from scratch, in: International Conference on Machine Learning, PMLR, 2018, pp. 4344–4353.

[83] C. Florensa, D. Held, M. Wulfmeier, M. Zhang, P. Abbeel, Reverse curriculum generation for reinforcement learning, in: Conference on Robot Learning, PMLR, 2017, pp. 482–495.

[84] L. Hermann, M. Argus, A. Eitel, A. Amiranashvili, W. Burgard, T. Brox, Adaptive curriculum generation from demonstrations for sim-to-real visuomotor control, in: 2020 IEEE International Conference on Robotics and Automation (ICRA), IEEE, 2020.

[85] A. Hill, A. Raffin, M. Ernestus, A. Gleave, A. Kanervisto, R. Traore, P. Dhariwal, C. Hesse, O. Klimov, A. Nichol, M. Plappert, A. Radford, J. Schulman, S. Sidor, Y. Wu, Stable baselines, https://github.com/hill-a/stable-baselines, 2018.

[86] S. Shah, D. Dey, C. Lovett, A. Kapoor, Airsim: high-fidelity visual and physical simulation for autonomous vehicles, in: Field and Service Robotics, Springer, 2018, pp. 621–635.

[87] W. Guerra, E. Tal, V. Murali, G. Ryou, S. Karaman, Flightgoggles: photorealistic sensor simulation for perception-driven robotics using photogrammetry and virtual reality, arXiv preprint, arXiv:1905.11377.

[88] Y. Song, S. Naji, E. Kaufmann, A. Loquercio, D. Scaramuzza, Flightmare: a flexible quadrotor simulator.

[89] M. Burri, J. Nikolic, P. Gohl, T. Schneider, J. Rehder, S. Omari, M.W. Achtelik, R. Siegwart, The euroc micro aerial vehicle datasets, The International Journal of Robotics Research 35 (10) (2016) 1157–1163.

[90] A.L. Majdik, C. Till, D. Scaramuzza, The Zurich urban micro aerial vehicle dataset, The International Journal of Robotics Research 36 (3) (2017) 269–273.

[91] A. Antonini, W. Guerra, V. Murali, T. Sayre-McCord, S. Karaman, The blackbird dataset: a large-scale dataset for uav perception in aggressive flight, in: International Symposium on Experimental Robotics, Springer, 2018, pp. 130–139.

[92] K. Sun, K. Mohta, B. Pfrommer, M. Watterson, S. Liu, Y. Mulgaonkar, C.J. Taylor, V. Kumar, Robust stereo visual inertial odometry for fast autonomous flight, IEEE Robotics and Automation Letters 3 (2) (2018) 965–972.

[93] A. Tsirikoglou, G. Eilertsen, J. Unger, A Survey of Image Synthesis Methods for Visual Machine Learning, Computer Graphics Forum, vol. 39, Wiley Online Library, 2020, pp. 426–451.

[94] M. Wrenninge, J. Unger, Synscapes: a photorealistic synthetic dataset for street scene parsing, arXiv preprint, arXiv:1810.08705.
[95] M. Davies, N. Srinivasa, T.-H. Lin, G. Chinya, Y. Cao, S.H. Choday, G. Dimou, P. Joshi, N. Imam, S. Jain, et al., Loihi: a neuromorphic manycore processor with on-chip learning, IEEE MICRO 38 (1) (2018) 82–99.

CHAPTER

Robotic grasping in agile production

16

Amir Mehman Sefat[a], Saad Ahmad[a], Alexandre Angleraud[a], Esa Rahtu[b], and Roel Pieters[a]

[a]*Cognitive Robotics group, Automation Technology and Mechanical Engineering, Tampere University, Tampere, Finland*

[b]*Computer Vision group, Computing Sciences, Tampere University, Tampere, Finland*

16.1 Introduction

Agile production refers to robotic production systems that operate quickly and adaptively in dynamically changing work environments. This means robots should adapt to variations in tasks, the local production environment and its contents such as work pieces. Moreover, recent advances in robotics allow for the safe interaction between humans and robots toward a common shared goal [1,2]. In such scenario, a human operator and the robot share the same workspace to complete a task in the dynamic environment. The contrast to traditional industrial production is rather evident as most commonly production environments exclude human operators and production processes have little variation in tasks and work pieces for the same production line. Besides the trend of agility in production, other developments toward smart manufacturing refer to the utilization of advanced data analytics to improve system performance and decision making. This ongoing effort to digitize industry and collect relevant data, enables the processing of data by machine and deep learning techniques [3]. As a step-by-step procedure, traditional machine learning performs feature extraction and model construction in a separated manner, as compared to deep learning, which learns features and the model in an end-to-end manner. In the context of smart manufacturing and its generation of big data [4], deep learning is therefore regarded as a great advantage, as it avoids the complex step of feature engineering. Different levels of data analytics can then extract knowledge, identify patterns, and make decisions on various processes. That is, models can act on individual production processes as well as on the factory as a whole, for example, by tracking parts and their production progress, monitoring equipment and its usage and evaluation of a process' performance and quality.

In this chapter, we narrow our focus toward one common task for robots in agile production, that is, object grasping, and its utilization of deep learning. First, we introduce in Section 16.1 robot tasks and deep learning, in context to the requirements and limitations the agile production environment sets. Following, the robot tasks of

Deep Learning for Robot Perception and Cognition. https://doi.org/10.1016/B978-0-32-385787-1.00021-X

grasping and object manipulation are described in Section 16.2, where different techniques are classified according to the sensor modality (RGB-D or point clouds) and the objects to be grasped. Evaluation of grasping and benchmarking of manipulation is discussed in Section 16.3 and Section 16.4, respectively. Finally, a brief overview of grasp data sets is given in Section 16.5.

16.1.1 Robot tasks in agile production

Robot manipulators in agile production environments are tasked with repetitive actions that require high precision. Typical examples of repetitive tasks are bin picking and packaging in warehouses and assembly for the automotive industry. Most of these industrial robots are position controlled, to ensure a high accuracy and high precision of end-effector motion [5], which has implications to their tasks and capabilities. Practically speaking, a position-controlled robot is designed to execute repetitive and predefined motions without any external disturbances. Tasks such as object pick and placement, spot welding, and painting are suitable for such robot control, as the task is well-defined and external disturbances are unlikely to occur. That is, consecutive objects are known, path motion does not change, and contact to a rigid surface is either planned or not included.

In case of robot tasks that include contact, other control modalities are common, such as force/torque and compliance control, which enables a more adaptive behavior of the robot to external disturbances and sensors [6]. Typical tasks that rely on such control methods are for example assembly (peg-in-hole, bolt screwing) and surface finishing (deburring, sanding, polishing). In these cases, external force/torque sensors provide information for the controller to act upon during task and motion execution. Drawbacks of this approach are the complexity of the control structure and how uncertain and noisy sensor measurements should be taken into account.

While in both cases the robots work independently, they are no longer standalone devices, but connected as a general effort toward Industry 4.0 [7]. This trend toward digitization offers several advantages, such as predictive maintenance and optimal factory resource deployment, which ultimately lead to a higher efficiency and productivity. Digitization also offers integration solutions, as a so-called digital twin can be utilized for the programming of robot tasks, by programming and verifying motion and task models in simulation and deploying them on the real robot [8].

Recent advances in robotics also allow for the collaboration between humans and machines in performing shared tasks in a safe way. In this context, where robot and human share the same workspace, adaptation from both the human and the robot is required such that they could work together as smoothly as possible meanwhile capturing the best capabilities from each side [2]. Interfaces to ease robot programming are an inherent part of this, by techniques such as learning from demonstration [9,10], web-based interfaces [11], augmented reality [12], and many others [13]. User experience design should then ensure efficient collaboration and acceptance from the human operator [14].

16.1.2 Deep learning in agile production

Within an industrial production environment, deep learning can be utilized in data analytics in different ways and on different processes [4]. With respect to robot manipulators and related sensor systems, deep learning has been utilized for perception, action, and control [15].

As perception tool, deep learning can detect and classify objects, and estimate their pose [16]. Robot systems can then utilize this information for grasping and manipulation tasks, toward goals such as pick and placement, assembly [9], and bin picking [17]. A broad overview and discussion on deep learning-based object pose estimation and robot grasping is given in the body of this chapter. Besides the perception of objects, visual processing is also utilized for monitoring persons in a workspace, by detection of a person and the estimation of his/her body pose. Actions of the person, by body movement and gesture recognition, can guide robot motion or serve for monitoring safety in a human-robot collaborative scenario [18] (i.e., by estimating the distance between a person and robot, according to ISO 15066 [19]). Chapter 14 in this book is dedicated to deep learning for human activity recognition.

Deep learning offers an additional approach toward robot motion and task generation, by collecting data (real or simulated), training motion and task models based on this data and deployment of the models on a real robot. In context of agile production, robot (deep) learning has been demonstrated for tasks such as bin picking [17], grasping [20], assembly [9], and peg-in-hole insertion [21]. Moreover, the interaction and collaboration between human and robot is under investigation as well, for example, by imitation learning of motion trajectories [22] and deep reinforcement learning for a collaborative packaging task [23].

16.1.3 Requirements in agile production

Development and integration of robots and robot systems in production and manufacturing environments need to follow standards on safety requirements as set by ISO 10218-1/2 [24]. In addition, if human operators are collaborating with robots or in its close vicinity, additional standard ISO 15066 [19] needs to be taken into account. Compliance to these standards can be done in several ways, for example, by vision-based safety systems [25]. As the grasping approaches in this chapter act as stand-alone systems, only the functional requirements are described and not the safety requirements as defined by the standards.

One crucial factor for the functionality of a robot grasping action in agile production is timing, for both its implementation and execution. Implementation of the grasping model implies the generation of a new grasp detection model by training a Deep Neural Network (DNN) with real or simulated data. The collection of such data can be time-consuming and difficult in situations were no accurate object models (e.g., CAD) are available or when no appropriate imaging system can be utilized. Moreover, the training of a DNN model requires computational resources that are often not present in small to medium-sized enterprises (SMEs) and external computation clusters need to be utilized. In addition, inference time of a grasp model should not be the bottleneck of the robotic system, and ideally, real-time control should be possible. This means that a grasp detection model should be lightweight (<1 GB), fast (>25 FPS) and have low latency (<100 ms), while ensuring a grasp detection or pose estimation accuracy such that parts can be picked up.

In short, as the requirements set by industrial agile production implies a fast implementation and reconfiguration of different robotic tasks, this should be reflected in the generation of robot tasks that utilize deep learning as well. A careful trade-off between inference time and input resolution should therefore be made, while considering suitable computational hardware.

16.1.4 Limitations in agile production

The industrial agile production environment sets limitations on several aspects for utilizing deep learning. Practical limitations concern the environment itself and the difficulty on capturing reliable models and data, such as sensor models with noise and accurate CAD models of objects. Following, a brief explanation is given of relevant hardware and software limitations that affect both the integration of advanced robotics in agile production and how it impacts the utilization of deep learning.

Grasping hardware

Most common industrial robot grippers have a parallel gripping mechanism that only allows one contact patch per gripper side [26]. Gripper finger design can then be utilized to customize the gripper for individual parts [27] and create a more robust grip by shape matching. Alternatively, tool change mechanisms enable multiple (custom) grippers per robot. Nevertheless, the number of parts that a robot can handle is still limited, especially when an accurate grasp pose is required for subsequent object manipulation. Multifingered grippers [28] or compliant gripping surfaces [29] are under development as well; however, in most cases, accurate sensing is still required. This gripper limitation affects deep learning, as the gripper design has to be included in the grasping model. Naturally, a more complex gripper can complicate the grasp model, while a too simple gripper design is incapable of grasping the industrial parts. Moreover, the reality that often industrial part models are unavailable hinders the learning and modeling process as well.

Grasping software

Recent state of the art in robot grasping utilizes or develops open source software that is distributed free of charge, but without warranty or liability to its use. In context to industrial applications this might pose difficulties as the understanding of the software and its integration in a production line can take considerable effort in resources and time. Moreover, support from the initial developers is not guaranteed and reliance on other software packages and hardware systems, such as sensors or processing platforms, has most likely not been tested. This affects the compliance to required safety standards even if risk assessments are carried out. Even if an industrial company would integrate a (deep learning-based) grasping model, pretrained models might not be suitable for the custom objects that a company's robot should handle. Training new models or extending existing models might be too time-consuming or simply ineffective due to an inaccessibility to computational resources (e.g., computing clusters). The need for a standard toolkit with respect to robotics and deep learning is therefore evident, effectively bridging the gap between academic research and industrial applications.

16.2 Grasping and object manipulation

Robotic object grasping and manipulation are commonplace in industrial production. For high-throughput assembly lines, fast and accurate grasping is typically established by custom-made bulk feeders that position and align the part to a specific pose [30]. A robot then grasps the object, based on a predefined grasp pose, with gripper or end-effector design that is tailored to the shape of the object. For low-volume manufacturing, this approach is however unsuitable, as it requires flexibility and reconfigurability following task changes. Ongoing research efforts toward (bin) picking therefore disregard feeders and utilize sensing to determine object and grasp pose for object handling [31], for example, by deep convolutional neural networks (CNNs) trained on large data sets of grasp attempts in a simulator or on a physical robot [17,20]. As autonomous grasping and object manipulation with robots sees many similarities to the computer vision problem of pose estimation, we first introduce the problem statement of grasp representations used in grasping literature (see Fig. 16.1 and Fig. 16.2). Following, an overview of different grasp detection methods is given, divided by their sensor modality; RGB-D or point clouds.

16.2.1 Problem statement

To a robot, the task of grasping is the successful determination of an end-effector or gripper pose $\mathbf{P}_g$ that leads to a secure and stable object grip and subsequent object lift-off from a surface. In this, the gripper pose is defined as $\mathbf{P}_g = (\mathbf{R}_g, \mathbf{t}_g) \in SE(3)$, where the rotation $\mathbf{R}_g \in SO(3)$ defines the grasp axis and direction of approach and the translation $\mathbf{t}_g \in \mathbb{R}^3$ defines the center location in between the gripper fingers (see Fig. 16.2). Sahbani et al. [32] give a detailed overview of terminology conventionally used in work related to robotic grasping, which defines the stability conditions such that the sum of all external forces and moments acting on a grasped object are zero. Furthermore, a stable grasp is defined as one that can withstand minor disturbance forces to the object or the end-effector and allows the system to restore to its original configuration. Other than stability, task-compatibility and adaptability to unseen objects are important parameters as well [32,33]. In literature, a wide variety of popular grasp representations are presented, some which combine both image and depth information [33]. Earlier works defined a grasp as 2D point coordinates $[x, y]$ in an image (see Fig. 16.1) or as 3D point coordinates $[x, y, z]$ in the robot workspace. The obvious limitation is the inability to address gripper orientation, opening width, and

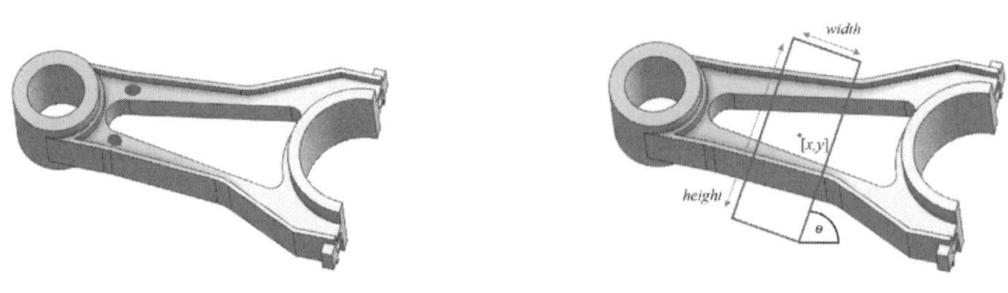

FIGURE 16.1

Grasp representation in image space: pixel coordinates $[x, y]$ (left, green circles) and oriented bounding box $[x, y, \theta, \textit{width}, \textit{height}]$ (right, green square). The yellow object is a connecting rod for a piston engine. (For interpretation of the colors in the figure, the reader is referred to the web version of this chapter.)

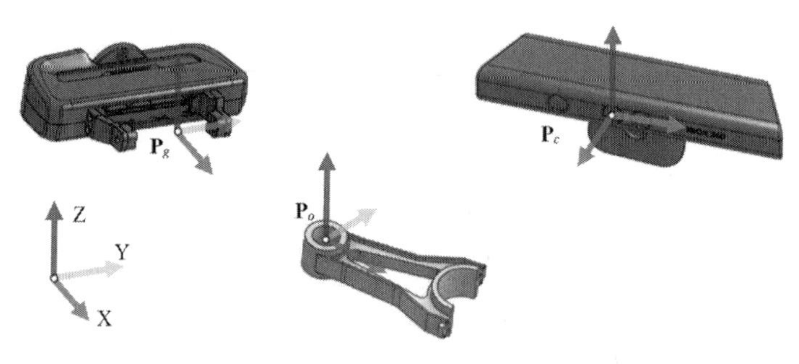

FIGURE 16.2

Grasp representation in 3D Cartesian space, with 6D poses for the gripper of the robot $\mathbf{P}_g$, the camera $\mathbf{P}_c$ and the object to be grasped $\mathbf{P}_o$.

angle of approach. Therefore, later approaches [34] used 7D oriented rectangular-box representations in robot workspace $[x, y, z, \textit{roll}, \textit{pitch}, \textit{yaw}, \textit{width}]$ and 5D on the image plane $[x, y, \theta, \textit{width}, \textit{height}]$ (see Fig. 16.1). These approaches are analogous to object detection and localization frameworks, and hence can be translated to grasping, as grasping is a detection problem in itself. Some of the later works introduced depth besides image data as input, and hence used a 5D representation $[x, y, z, \theta, \textit{width}]$ that dropped the height of the gripper [33]. In addition, other approaches regress grasps over point clouds [35–37], parametrized in object or camera coordinate frame, that is, grasps are defined as 6-DOF poses of the gripper $\mathbf{P}_g$ relative to the camera $\mathbf{P}_c = (\mathbf{R}_c, \mathbf{t}_c) \in SE(3)$, by transformation of an (estimated) object pose $\mathbf{P}_o = (\mathbf{R}_o, \mathbf{t}_o) \in SE(3)$. These approaches propose grasps that are not constrained in a single plane relative to the camera, and can be directly used as goal poses for the robot. However, these are difficult to regress directly, require high-dimensional features for learning and usually require post-refinement steps.

16.2.2 Analytical versus data-driven approaches

Early research in robotic grasping utilized analytical approaches of calculating robot kinematics and dynamics to deal with the constraints on the 3D geometry of the object to be grasped [16,33]. These techniques aim to satisfy force-closure, form-closure, or task-specific geometric constraints to find feasible contact points for a particular object and manipulator configuration. The majority of these works is concerned with finding and parameterizing the surface-normals on various flat faces of an object and then testing the force-closure condition by subjecting the angles between these normals to be within certain thresholds. Generally, a force-closed grasp is one in which the end-effector can apply required forces on the object in any direction, without letting it slip or rotate out of the gripper. Later analytical methods argued on the optimality criterion of the grasp quality [32]. This means that a metric should decide on the quality of the force-closure achieved with a certain grasp, that is, how close the grasp is to losing its force-closure, given a particular object geometry and hand configuration. Recently, it has been studied that these analytical modeling methods and metrics alone were not good measures of grasps as they do not adapt well to the challenges that are faced during execution, such as uncertainties in dynamic behavior and the unstructured environment [38]. Therefore, in the last decade, there was a general push toward machine-learning approaches, which present a more abstract, easy-to-test and indirect approach of evaluating grasp success on a large variety of objects, grippers and environmental conditions. The shift from complex mathematical modeling of the grasping itself toward the indirect mapping of perceptual features with grasp-success was made possible by the availability of high quality 3D cameras and depth sensors, increasingly powerful computational resources, and a substantial amount of invaluable research in deep neural networks and transfer learning [33]. The biggest advantage these methods provided were the vast possibilities for testing in simulation and real execution, using real and synthesized data. These methods do not explicitly guarantee equilibrium, dexterity, or stability but the premise of testing all criteria, based solely on sensor data, object representation, and carefully designed object features provides a convenient way for studying grasp synthesis [38].

16.2.3 Grasp detection with RGB-D

Known objects

Grasping known objects have been researched extensively, as it is a direct extension of object detection and object pose estimation. Different methods are used to extract information from the objects such as 2D point correspondences (e.g., SIFT, SURF, ORB), templates (e.g., HOG, surface normals, moments) or patches (e.g., image regions, superpixels) to represent objects in intermediate representations for correspondence matching [39]. Hinterstoisser et al. [40] pioneered the research in template matching by providing a complete framework for creating robust templates from existing 3D models of objects by sampling a full-view hemisphere around an object. In addition, they introduced the Linemod data set, consisting of 1100+ frame video sequences of 15 different household objects varying in shape, color, and size along with their registered meshes and CAD models. Other research includes PoseCNN [41], which combines feature and template methods, ConvPoseCNN [42] that improves PoseCNN by a fully-convolutional architecture, HybridPose [43] and PVNet [44].

Similar objects

This class of grasp-detection methods are aimed at objects that are similar, that is, a different/unseen instance of a previously seen category of objects. For example, all cups with slight intraclass variation belong to a single category of objects and all shoes belong to another. These methods learn a normalized representation of an object category and transfer grasps using sparse-dense correspondence between normalized 3D representation of the category and the partial-view object in the scene [16]. NOCS [45] presents an initial benchmark in this category by formulating a canonical-space representation per category using a vast collection of different CAD models for each class. A color-coded 2D perspective projection of this space (NOCS map) is then used to train a Mask-RCNN based network [46], in order to learn correspondences from RGB-D images of unseen instances to this NOCS map. These correspondences are later combined with a depth-map to estimate the 6D pose and size of multiple instances per class.

Novel objects

Regarding novel objects, there is no existing knowledge of object geometry and grasps are estimated directly from image and/or depth data. The majority of these methods use geometric properties inferred from input perceptual data, as a measure of grasp success [16]. Most methods are developed in an end-to-end fashion, learning from a database of grasps on a large number of different object models. These grasps are sampled exhaustively around the objects and are either evaluated on classical grasp metrics, such as the epsilon quality metric [47], manually annotated with their success measures by humans or tested with real execution [48,49]. The premise of these methods lies in the later training stage, where a deep neural network learns to produce robust grasps in general. The emphasis is on learning a robustness function that ranks a candidate grasp for various quality metrics.

Dex-Net 1.0 [48] and Dex-Net 2.0 [49] are two pioneering works that utilized this strategy and created large data sets of 3D object models for learning objective functions that minimize grasp failure, in presence of object and gripper position uncertainties and camera noise. Enforcing constraints like collision avoidance, approach-angle threshold and gripper-roll threshold, these methods provide a baseline for correlating object geometry from RGB-D images [48] or point clouds [49] with grasp robustness. Extensions of Dex-Net include suction grasping in Dex-Net 3.0 [50] and ambidextrous grasping (i.e., two or more heterogeneous grippers) in Dex-Net 4.0 [51]. Finally, [52] and [53] are excellent examples that use a fully-convolutional architecture for grasp detection.

PVN3D

To obtain a better understanding of the complexity of deep learning-based pose estimation, one network is explained in more detail. In particular, PVN3D [54] is a recent, deep pointwise 3D key-points-based voting network for 6-DOF pose estimation and utilizes single RGB-D images for inference. The network consists of the following separate blocks and their functionalities:

1. **Feature-extraction** contains two separate steps as follows:

   - PSPNet-based [55] CNN layer for feature-extraction in RGB images.
   - PointNet++-based [56] layer for geometry extraction in point clouds.

   The output features from these two are then fused by:

   - DenseFusion [57] layer for a combined RGB-D feature-embedding.

2. **3D key-point detection** comprises of shared Multilayered Perceptrons (MLP) with a semantic-segmentation block and uses the features extracted by the previous block to estimate an offset of each visible point from the target key-points in Euclidean space. The points and their offsets are then used for voting candidate key-points. The candidate key-points are then clustered using meanshift clustering [58] and cluster-centers are casted as key-point predictions.
3. **Instance semantic-segmentation** contains two modules sharing the same layers of MLPs as those of the 3D key-point detection block. These are, a semantic segmentation module that predicts per-point class label and a center-voting module to vote for different object centres in order to distinguish between object instances in the scene. The center-voting module is similar to the 3D key-point detection block in that it predicts per-point offset, which in this case votes for the candidate center of the object rather than the key-points.
4. **6 DOF pose estimation** finally executes a least-squares fitting between key-points predicted by the network and the corresponding key-points.

16.2.4 Grasp detection with point clouds

The detection of grasps with point cloud data presents a unique set of challenges uncommon to RGB and RGB-D-based methods, such as the small scale of available data sets, the high dimensionality and the unstructured nature of point clouds. Nevertheless, point clouds are a richer representation of geometry, scale, and shape as they preserve the original geometric information in 3D space without any discretization [59]. Perhaps the most widely used point cloud aggregation methods in applications of grasp detection and pose estimation are PointNet [60] and PointNet++ [56]. These two methods revolutionized geometry-encoding in point clouds by preserving permutation invariance. Point clouds are an inherently unordered data-type and any kind of global or local feature representation should not change the way they are ordered. To deal with this problem, most techniques convert point clouds into other discrete and ordered forms, for example, voxel-grids, height-maps, octomaps, surface-normals or

2D-projected gradient-maps, before aggregating into a final compact representation. PointNet, PointNet++ and their later modifications overcame this and paved way for direct usage of point clouds in order to be used with other deep learning frameworks.

Due to a better scene understanding and geometry aggregation, point clouds processed with deep neural networks have led to a whole new line of pose estimation methods. These methods filled the undesirable gap that existed in RGB/RGB-D techniques, that required data collection, training and evaluation in image coordinates in the form of 2D or 3D bounding boxes. Moreover, the need to transfer poses or grasps from image to world coordinates was removed, as these techniques could recover full 6-DOF pose of the object without having to employ any post-processing on the depth channel or without learning to estimate depth. In this regard, methods have been developed that follow the same categorization as RGB/RGB-D based pose estimation, such as correspondence-based [61–63], template-based [64], or voting-based [44,54] techniques.

Methods that generate grasps directly from point cloud data, circumvent the need for estimating the pose of an object in the scene or any prior knowledge about the shape or canonical representation of a class of objects. With point clouds, these grasps can be recovered in 6-DOF without constraining the gripper to move along the image plane, as with the RGB-D-based detectors. Noteworthy examples of such networks are [35], [65], GG-CNN [66], PointNetGDP [36], and 6-DOF GraspNet [37].

6-DOF GraspNet

To obtain a better understanding of the complexity of deep learning based grasp detection, one network is explained in more detail. In particular, 6-DOF GraspNet [37] generates grasps directly as output of the network, from 3D point clouds observed by a depth camera. The network has two main components and a third refinement step as follows:

1. **Grasp-sampling** utilizes a Variational Auto-encoder (VAE) [67] with PointNet++ encoding layers, which learns to maximize the likelihood of ground-truth positive grasps given a point cloud.
2. **Grasp-evaluation** utilizes an adversarial network to measure the probability of success of a grasp. Negative examples are included from the training data and by generating hard-negatives through random perturbation of positive grasp samples.
3. **Iterative grasp-pose refinement**: The grasps rejected by evaluation are close to success and can undergo an iterative refinement step. This transformation is found by taking a partial derivative of the success function with respect to the closest successful grasp.

16.3 Grasp evaluation

As many different approaches toward robotic grasping exist, in this section we evaluate two methods that are the state of the art in their respective approach: pose estimation with PVN3D and grasp detection with 6-DOF GraspNet. In addition, we address the metrics by which grasping should be evaluated.

16.3.1 Metrics

From an industrial perspective, the evaluation and measures for grasping can be simplified to either a successful grasp, lift and put-down, within a specified placement and pose range. With the traditional industry approach of prespecified grasp locations and object feeders, this protocol provides a suitable evaluation. However, with the novel approach of learning-based grasp generation without object-designed grippers, multiple factors influence the success of grasping. Here, we aim to address these different factors and cover the most popular metrics for grasp evaluation. This includes pose estimation metrics for grasping objects at a predefined pose, grasp candidate selection for grasp detection models, grasp evaluation metrics for grasp detection models, and the influence of the gripper on grasping.

For evaluating pose-estimation different pose-error functions can be utilized [68]. One popular metric is the Average Distance of Model Points (ADD) and the Average Closest Point Distance (ADD-S) [40]. Given a ground truth object rotation $\mathbf{R}$, translation $\mathbf{T}$, predicted rotation $\tilde{\mathbf{R}}$ and translation $\tilde{\mathbf{T}}$, ADD computes the average of the distances between pairs of corresponding 3D points $\mathbf{x}$ on the model transformed according to the ground truth pose and the estimate pose:

$$ADD = \underset{\mathbf{x} \in M}{\text{avg}} \, \|(\mathbf{R}\mathbf{x} + \mathbf{t}) - (\tilde{\mathbf{R}}\mathbf{x} + \tilde{\mathbf{T}})\|_2, \tag{16.1}$$

where M denotes the set of 3D model points and $\|\cdot\|_2$ denotes the $L2$-norm.

For objects that are symmetric in the plane along the principal axis of rotation, the rotation along this axis in the predicted pose could be ambiguous by 180 degrees since similar points would repeat after every 180 degrees and the correspondences are spurious. This rotational ambiguity is by-passed by the ADD-S metric where only the minimum of all the distances between a pair of points is considered and the average of this distance then gives the error:

$$ADD - S = \underset{\mathbf{x}_1 \in M}{\text{avg}} \, \underset{\mathbf{x}_2 \in M}{\min} \|(\mathbf{R}\mathbf{x}_1 + \mathbf{T}) - (\tilde{\mathbf{R}}\mathbf{x}_2 + \tilde{\mathbf{T}})\|_2. \tag{16.2}$$

Based on an estimated object pose, a robot grasp pose is defined to which the robot moves for grasping. This approach assumes that the predefined grasp pose is suitable for grasping the object with the available gripper. Novel learning-based approaches for grasp detection generate a grasp pose directly from object perception data, thereby bypassing the suitability check of the grasp pose. Moreover, typically a large number of grasp candidates are generated per object and only a single best grasp should be selected for execution (see Fig. 16.5). One open question in grasping research is therefore how the grasp candidates are sampled for testing and how they should be evaluated. In order to generate training data for a robot that is large enough in quantity, varied enough for generalization and provides an accurate enough representation of task constraints, some efficient heuristic measures are needed to search through a space of thousands of potentially viable grasps [69]. Even after the initial selection of these candidates, effective evaluation of the grasps and the metrics for a robot's performance on them determines the usefulness of the data and the robustness of the grasp algorithm being trained on [70]. The commonly used grasp sampling techniques are analyzed by Eppner et al. [69] and are broadly categorized into object geometry-guided sampling, (non-)uniform sampling [71], approach-based sampling, and antipodal sampling [72]. Eppner et al. [69] then devised a few intuitive metrics of comparing these sampling methods, that is, grasp coverage, robustness and precision. In simulation more than 317 billion grasps on 21 YCB-data set objects [73] are generated and the successful 1 billion grasps out of these are evaluated. It is concluded that uniform samplers have better grasp coverage because of minimal constraints, with the trade-off for efficiency. On the other hand, heuristics like approach-based sampling or antipodal sampling are efficient but might not entirely capture all possible grasps. Moreover, it was also found that antipodal grasps have higher coverage and find more robust grasps. Precision is quite low for both uniform and approach-based methods, while being significantly higher for antipodal methods. Nonuniform or geometry-based approaches, consistently perform poor on all three metrics.

In both cases, that is, grasping based on pose estimation and grasp detection, grasping is evaluated by a grasp success rate for a single object or an average grasp success rate for an object set, while including the number of attempted and failed grasps. Another common method of evaluation is an offline metric using the Cornell grasping data set [34] or Jacquard [74] data set, where a predicted grasp is successful if it has an intersection-over-union (IoU) greater than 25% and is within 30° angle, as compared to the ground-truth grasp. However, this metric is susceptible to producing large numbers of false–positive and false–negative detections, due to the sparse labeling of the data set, and low requirements for considering a match. Recent work has included resistance to disturbances to the evaluation metrics, by expected wrench resistance or the ability to resist task-specific forces and torques, such as gravity, under random perturbations [51]. In addition, it reports the mean picks per hour that the physical robot system achieves (i.e., the number of successful grasps per hour).

It is crucial to note the importance of the gripper used for the evaluation. A perfect pose estimate or grasp detection does not imply a perfect grasp, nor does it suggest which type of gripper would provide best results. Even though many works integrate the gripper as part of the grasp model, usually these are limited to only modeling the gripper width and depth [35,36,75], thereby only ensuring that the generated grasps comply to the gripper's dimensions and configuration. Compliant grippers [29] could provide an alternative and potentially improve grasp success rate, however, at the expense of a higher placement uncertainty, as, similar to many grasp detection methods, the object pose inside the gripper would be unknown.

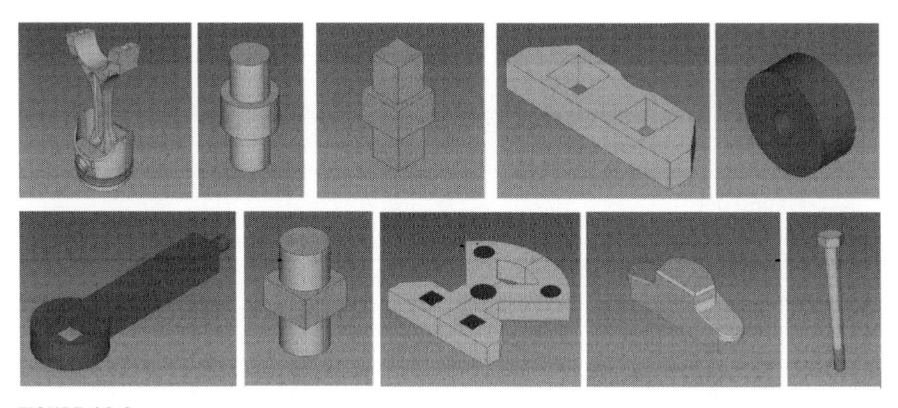

FIGURE 16.3

CAD data set consisting of parts from the (3D printed) Cranfield assembly benchmark [76] and from a Diesel engine assembly use case. From top left to top right: piston, round peg, square peg, separator, pendulum head. From bottom left to bottom right: pendulum, shaft, face plate, valve tappet, shoulder bolt.

16.3.2 Pose estimation with PVN3D

Data collection and training

Data collection utilizes simulation only, as it provides easy fine-tuning of a variety of parameters, that is, object placement and pose, lighting, etc. A set of ten objects are selected, consisting of parts from a (3D printed) Cranfield assembly benchmark [76] and from a Diesel engine assembly use case (see Fig. 16.3). A Kinect v1 camera is simulated in Gazebo and publishes RGB and depth images (i.e., 640 by 480 pixels) and all objects are simulated using their standard polygon mesh files generated from CAD models. The background is left out to be a brightly-lit plain-gray room with no walls and the objects always rest on the floor in their most stable equilibrium pose. No dense background clutter or nondata set objects are added to the environment. A moderately dense clutter of all the objects in the data set is created around the origin in the simulation environment. Hemisphere-sampling, as described in [40], is carried out around each clutter-set, with steps in yaw, pitch and scale for the simulated camera. For each sample, the data set records RGB and depth images of the scene, a gray-scale image with binary mask of each object with the respective class label and the ground truth pose of each object in camera and world coordinates.

This generates a total of 2304 collected data samples, from which a 75%–25% train-test split is applied, in order to evenly cover all possible pitches, yaws and scales in both test and training data set. A joint multitask training is carried out for 3D key-point detection and instance semantic-segmentation blocks. First, the semantic-segmentation module facilitates extracting global and local features in order to differentiate between different instances, which results in accurate localization of points and improves the key-point offset reasoning procedure. Second, learning for the prediction of key-point-offsets indirectly learns size-information as well. This helps distinguish objects with similar appearance but different size. Three key-point

Table 16.1 Area under curve (AUC) for accuracy-threshold curve for the ADD and ADD-s metric on the CAD data set.

Object	16 key-points		12 key-points		8 key-points	
	ADD	ADD-s	ADD	ADD-s	ADD	ADD-s
Piston	**97.76**	**98.16**	97.57	98.02	97.21	97.85
Round peg	97.35	97.35	**97.4**	**97.4**	97.05	97.05
Square peg	**97.12**	**97.12**	96.92	96.92	96.32	96.32
Pendulum	**96.09**	**96.09**	95.68	95.68	95.31	95.31
Pendulum-head	**97.27**	**97.27**	97.05	97.05	96.73	96.73
Separator	**96.4**	**96.4**	96.24	96.24	95.95	95.95
Shaft	**97.93**	**97.93**	97.68	97.68	97.41	97.41
Face-plate	**96.16**	**96.16**	96.15	96.15	95.88	95.88
Valve-tappet	91.7	91.7	**92.48**	**92.48**	91.58	91.58
Shoulder-bolt	**93.5**	**93.5**	93.08	93.08	91.4	91.4
Average	**96.13**	**96.17**	96.03	96.07	95.48	95.55
Inference time [sec]	1.98		1.75		1.5	
GPU memory [MB]	2775		2544		2343	

variations (i.e., 8, 12, and 16 key-points) were trained for a total of 70 epochs with batch size of 24 as recommended by the authors. The training was carried out on 4 Nvidia V100 GPUs and takes around 5–7 hours for the given batch-size, number of epochs, and training data set.

Results

Results of object pose estimation are reported in terms of area under the accuracy-threshold curve where the threshold for the ADD and ADD-S metric (see Eqs. (16.1) and (16.2)) is varied from 0 to 10 cm. These results are reported in Table 16.1.

Fig. 16.4 depicts the pose estimation results for the object set, where each object is overlaid with an estimated pose, depicted by a projected bounding box (light blue). These results demonstrate that accurate object pose estimation can be achieved with PVN3D in simulation. Unfortunately, both the inference time and model size are large (>1.5 seconds and >2 gigabytes, see Table 16.1), meaning that real-time feedback for control is not possible.

16.3.3 Grasp detection with 6-DOF GraspNet

Data collection and training

The grasp data is collected in a physics simulation, based on grasps done with a free-floating parallel-jaw gripper and objects in zero gravity. Objects retain a uniform surface density and friction coefficient and the grasping trial consists of closing the gripper in a given grasp-pose and performing a shaking motion. If the object stays enclosed in the fingers during the shaking, the grasp is labeled as positive. Grasps are sampled based on object geometry, sampling random points on object mesh surface

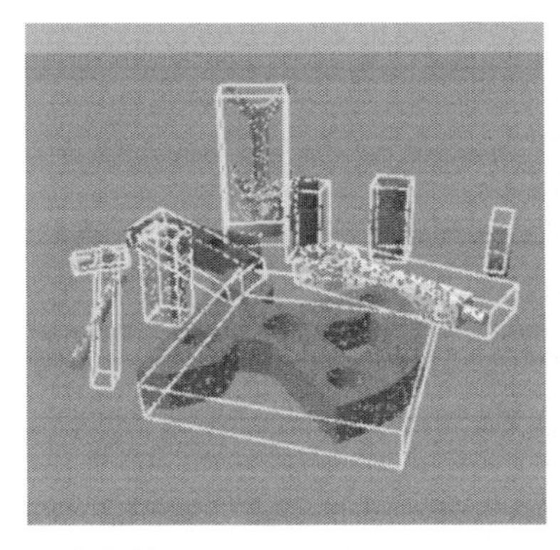
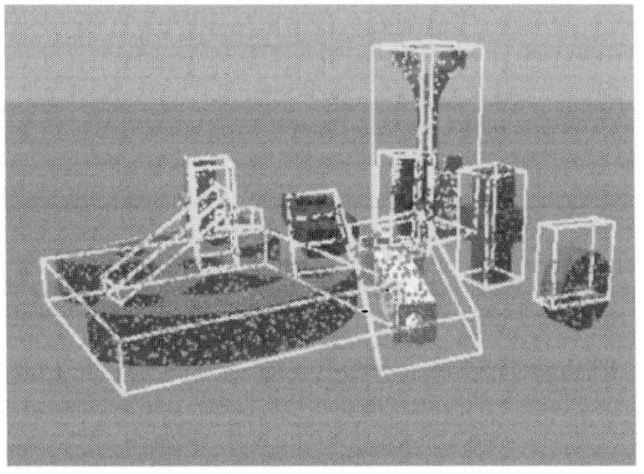

FIGURE 16.4

Images from the pose estimation result with PVN3D [54]. The 3D poses are shown as bounding boxes projected on the RGB image. (For interpretation of the colors in the figure, the reader is referred to the web version of this chapter.)

to align approach axis with the normal at each of these points. Grasps are performed on a total of 206 objects from six categories in ShapeNet [77]. From a total of over 10 million candidate grasps, 2 million successful grasps (19.4%) are generated. This data collection and training is done as part of the original work in [37], by utilizing 8 NVIDIA V100 GPUs.

Results

For evaluation, grasp detection and execution is simulated on a Franka Emika Panda[1] robot with Kinect v1 camera in Gazebo using ROS.[2] MoveIt[3] and ros_control[4] are utilized as they provide a generic and robot-agnostic framework to interface with any real or simulated robot. Results of grasp detection are reported in terms of total number of grasps generated per object and the percentage of grasps that passed the initial planning step and grasp execution test (grasp test in Table 16.2). The initial planning step excludes grasps that are infeasible, by limiting the yaw and pitch angles of grasp candidates to $[-90, 90]$ and $[-90, 45]$ degrees, respectively. Fig. 16.5 depicts the detected grasps before (middle figure) and after (right figure) grasp candidate filtering for an industrial part (piston). Table 16.2 reports results on grasp generation and grasp testing for a subset of objects. These results demonstrate that, while grasps can be generated with 6-DOF GraspNet in simulation, stable grasping is difficult to achieve. Reasons for this are likely due to the simple candidate filtering approach and the fact that simulation is a poor representation of the real world. Object and gripper

[1] https://franka.de.

[2] https://www.ros.org.

[3] https://moveit.ros.org.

[4] https://wiki.ros.org/ros_control.

Table 16.2 Results from grasp simulation experiments from grasp detection with 6-DOF GraspNet [37], reported based on the number of generated grasps and grasp success: % of grasps that ended in successful holding of the object.

Object	Generated grasps	Grasp test [%]
Piston	40	42.5
Round peg	22	77.3
Square peg	19	47.4
Pendulum	36	8.3
Separator	39	5.1
Shaft	21	42.9

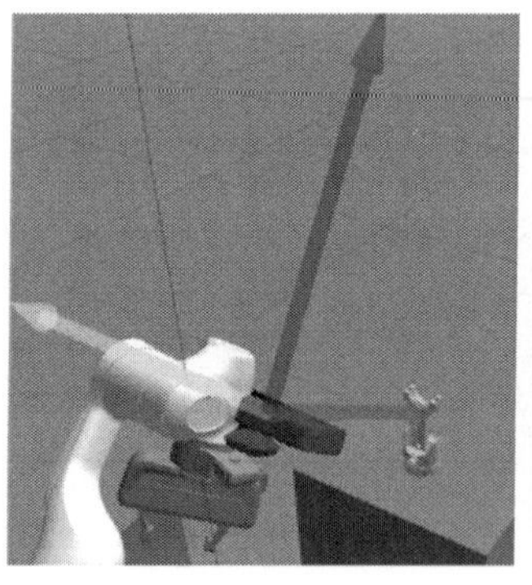
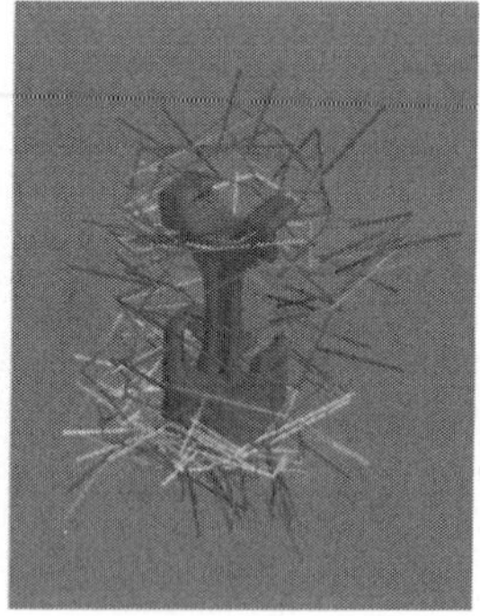
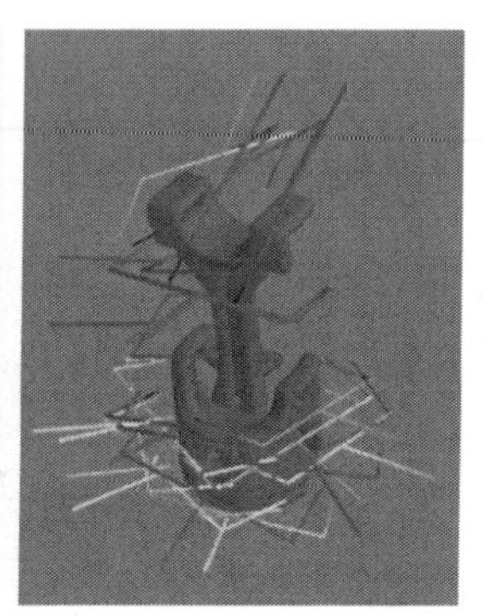

FIGURE 16.5

Simulation (left) of robot (Franka Emika Panda), camera (Kinect), and industrial part (piston). Detected grasp poses before (middle) and after (right) grasp candidate filtering are detected from point clouds by 6-DOF GraspNet [37].

properties, such as friction and inertia, are hard to model, leading to an unrealistic simulation environment and, therefore, unreliable results.

16.3.4 Pick-and-place results

While previous sections reported results on object pose estimation and grasp detection and execution, here results are reported on robotic pick-and-place actions. Again, grasping is simulated on a Franka Emika Panda robot in Gazebo. A Kinect v1 camera and PVN3D are chosen to estimate an object pose and execute a grasp action, followed by a placement action in a different location (see Fig. 16.6). A grasp pose is therefore predefined for each object and transformed from the estimated object pose. Table 16.3 reports results for the pick and place actions, for a selection of objects, based on the number of generated grasps, the percentage of successful grasps, and the percentage of grasps that could be accurately placed upright, with placement er-

Table 16.3 Results from pick-and-place simulation experiments from pose estimation with PVN3D [54]. The experiments are reported based on the number of generated grasps, grasp test: % of grasps that ended in successful holding of the object, placement test: % of grasps that were able to place the object upright with an (x, y) error < 5 cm and the average placement error in [cm], [deg].

Object	Generated grasps	Grasp test [%]	Placement test [%]	Placement error [cm], [deg]		
				x	y	yaw
Piston	88	83.0	60.2	0.9	2.5	25
Round peg	108	76.9	53.7	0.2	1.3	–
Square peg	108	72.2	42.6	0.9	1.4	46
Pendulum	54	87.0	61.1	0.4	2	2
Separator	58	50.0	36.2	1.1	3.8	2
Shaft	108	74.1	51.9	0.4	1.5	44

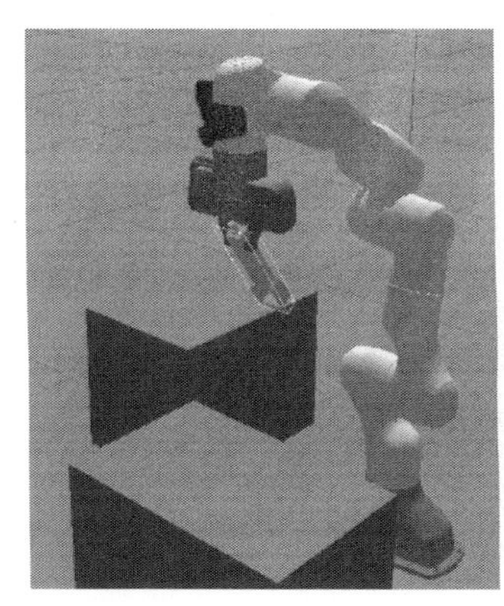

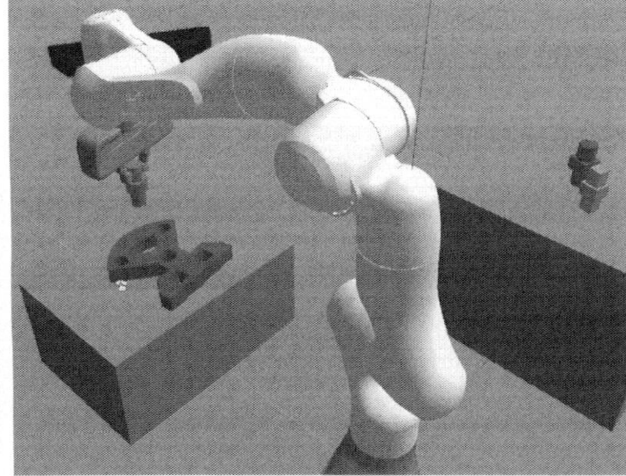

FIGURE 16.6

Simulation of robot grasping for a pick and place task (left) and a peg-in-hole manipulation task (right). Object pose is estimated from point cloud data by PVN3D [54].

ror < 5 cm. These results demonstrate that, while an object pose can be accurately estimated (see Section 16.3.2), this does not necessarily lead to a high grasp success (Grasp test in Table 16.3) or a high placement success (Placement test in Table 16.3). A likely cause of this is, again, that simulation is a poor representation of the real world, where physical properties (friction, inertia, etc.) are hard to model correctly.

Nevertheless, from all (simulated) evaluations of the different grasping approaches, valuable conclusions can be drawn. For example, the selection of a suitable grasp detector depends on several factors, such as the object models and their availability for training, suitability of the grasp detector for real-time control, and effort needed to select a single optimal grasp. This information can be obtained from simulated experiments, with realistic objects similar to their real world cases.

16.4 Manipulation benchmarking

Object manipulation represents the step after object grasping, where an object in the gripper or hand is manipulated toward a certain task or goal (see Fig. 16.6). Accurate grasps and grasp poses are crucial for this, which implies that neither the object falls out of the gripper nor slips too much inside the gripper. With an accurate estimate of initial object-pose, the in-hand pose of the object is assumed to be within certain constraints and the final placement of an object can be done with a reasonable certainty [78]. Robotic object manipulation tasks other than placement, such as peg-in-hole, hole-on-peg, and bolt screwing require additional sensors and advanced control methods to be successfully executed [6]. State of the art works that provide a general framework and benchmark for the aforementioned assembly tasks, utilizing motion-priors like spiraling around the hole for peg-insertion tasks and back-and-forth spinning for screwing tasks, can be found in [79] and [80]. These methods are thoroughly tested on various combinations of both compliant-arm and compliant-hand and fingers, both with and without contact sensors. The benefits of using various force-profiles as cues for driving the manipulation toward a more accurate and robust assembly are described in detail. Recent work also combines visual and force sensing with deep reinforcement learning to teach policies for contact-rich manipulation tasks [21]. By self-supervision, a compact and multimodal representation of the sensory inputs is learned, which can then be used to improve the sample efficiency of policy learning. Evaluation of the method is done on a peg insertion task, which shows that it generalizes over varying geometries, configurations, and clearances, while being robust to external perturbations. Bekiroglu et al. [81] developed a benchmark protocol specifically for the evaluation of robot grasping algorithms, describing in detail how to set up the workspace, object choices and placements, grasp execution, and scoring. Robustness of grasps is taken into account by describing a series of motions after an object is grasped. The ACRV picking benchmark [82] is another benchmark and consists of a set of 42 common objects, a widely available shelf, and exact guidelines for object arrangement using stencils. A well-defined evaluation protocol enables the comparison of complete robotic systems, including perception and manipulation, instead of sub-systems only. A benchmark for robot learning is RLBench [83], which features 100 completely unique, hand-designed tasks aimed to accelerate progress in a number of vision-guided manipulation research areas, including reinforcement learning, imitation learning, multitask learning, geometric computer vision, and few-shot learning.

Besides scientific works that push benchmarking forward in robotics, international organized challenges are another effort that drives scientific progress in robot object manipulation. Well-known initiatives are RoboCup@Work,[5] Amazon Picking Challenge (APC), European Robotic League (ERL) – Professional[6] and the AD-

[5] https://atwork.robocup.org.

[6] https://www.eu-robotics.net/robotics_league/erl-professional.

vanced Agile ProducTion (ADAPT) competition,[7] and competitions organized at international events, such as the Robotic Grasping and Manipulation Competition[8] at the IEEE/RSJ International Conference on Intelligent Robots and Systems and the Assembly Challenge[9] at the World Robot Summit.

16.5 Data sets

Robotic grasping is a popular topic in research and several surveys can be found that compare different approaches [16,32,33,38]. Most works present a large variation in objects such that the learned grasp model can handle most if not all objects in the data set. In fact, the importance of the training data and its generation should not be underestimated as it has a crucial role in the learning of a suitable grasp model. Generating grasps for objects that are unknown should thus be captured by the variance in the data set. Several works have tried to accommodate this by either a data set with large number of objects [74,84] or by creating a data set where all objects cover a statistical large variation in geometry [85]. This variation of objects in a data set can also be identified from their type, as in many cases household objects, kitchen utensils and tools are represented. Few works target specifically other part categories, for example, industrial parts as in [86]. The choice of object naturally needs to follow the capabilities of a robot gripper (e.g., object weight, size, and shape) and its configuration (e.g., parallel gripper or multifinger), which affects the model and representation of a grasp as well.

One important property to distinguish between different data sets is their method of generating training data, real or simulated. Early approaches used real objects and high definition sensor systems to record data sets leading to vast amounts of data (typically few gigabytes) that are freely available to utilize [20,35,40,73,75]. This implies that the data set is fixed and a pretrained grasp model only takes the objects in the data set into account, unless more images of real objects are added. From an industrial perspective, the question whether pretrained models are suitable for a company's particular industrial parts is difficult to answer. Moreover, training grasp models based on the industrial parts themselves might not be possible simply due to a lack of resources or due to the unavailability of (digital) part models. The resources required to commit to such effort are, for example, human effort and the required technical skills, and the computational infrastructure. Simulated data sets are a more recent approach, where digital tools (e.g., CAD or physics simulators) are utilized to generate object models. Training data can then be prepared by importing a CAD model into the simulation environment and generating different images or point clouds under varying conditions (e.g., viewpoint, lighting, color, noise, etc.).

[7] https://metricsproject.eu/agile-production.

[8] https://rpal.cse.usf.edu/competition_iros2020.

[9] https://worldrobotsummit.org/en/wrs2020/challenge/industrial/assembly.html.

Table 16.4 Overview of grasping data sets that utilize simulation for the generation of object training data. The variables in the **Grasp representation** column are: for **EGAD**; object position [x, y, z], rotation around the vertical axis θ, desired width of the gripper w and grasp quality q, for **Jacquard**; center of a rectangle [x, y], its height and width [h, w] and its orientation relative to the horizontal axis of the image θ, for **Zhao et al.**; location of the grasp relative to the object's geometric center [x, y, z] and planar rotation of the gripper about an axis orthogonal to the table surface θ.

Data set (year)	Parts	Categories (number)	Simulator	Modality	Grasp representation	Gripper
ACRONYM [84] (2020)	8872	household (2331)	FleX [89] PyRender [90]	Depth	6D	parallel
Dex-Net 4.0 [51] (2019)	1664	basic shapes, household	CAD	Depth	6D	suction, parallel
EGAD [84] (2020)	2000	geometrically diverse	–	Depth	[x, y, z, θ, w, q]	parallel
6-DOF GraspNet [37] (2019)	206	household (6)	FleX [89]	Depth	6D	parallel
Jacquard [74] (2018)	11000	household	PyBullet [91], Blender [92]	RGB-D	[x, y, h, w, θ]	parallel
Kappler et al. [87] (2015)	700	household, tools (87)	openRAVE [93]	RGB-D	6D	Barrett hand [28]
Veres et al. [88] (2017)	–	household (64)	V-REP [94]	RGB-D	6D	Barrett hand [28]
VR-grasping [95] (2018)	101	household (7)	PyBullet [91]	RGB-D	6D	2-finger
Zhao et al. [86] (2019)	1011	industrial parts	V-REP [94], PyBullet [91]	Depth	[x, y, z, θ]	parallel

The advantage is that only a CAD model is required and the type, properties and number of objects can be tuned by changing simulation parameters. Table 16.4 lists grasping data sets that utilize simulation to generate objects, instead of images or point clouds from real objects.

Noteworthy examples in Table 16.4 are the following. The approach of Zhao et al. [86] particularly focused on industrial parts and the industrial applications of assembly and palletizing. The data set includes more than 1000 CAD models of industrial parts, such as gears, brackets, and screws, collected from an online hardware shop. The Evolved Grasping Analysis Data set (EGAD) comprises over 2000 generated objects that are geometrically diverse and range from simple to complex shapes, with varying grasp complexity. A set of 49 diverse 3D-printable evaluation objects within the data set are selected to encourage reproducible testing of robotic grasping systems. Dexterity Network (Dex-Net[10]) has seen several iterations that have focused

[10] https://berkeleyautomation.github.io/dex-net.

on cloud-based data generation and training (Dex-Net 1.0 [48] with 10000 3D object models), gripper-specific models (Dex-Net 2.0 [49] and Dex-Net 3.0 [50]) and ambidextrous grasping (i.e., two or more heterogeneous grippers in Dex-net 4.0 [51]). As can be seen in Table 16.4, most works focus on parallel grippers and only few data sets consider variations, such as the BarettHand [87,88] or a suction gripper [51]. Other differences between the data sets are the sensor modality (RGB-D vs. depth perception), the representation of a grasp and the simulation environment and physics engine utilized for realistic dynamic grasp simulation.

16.6 Conclusion

In this chapter we have provided an overview of state of the art approaches in robotic grasping, with particular emphasis on deep learning. In context to the application area of agile production, the overview addresses several requirements and limitations to be taken into account (Section 16.1), such as data collection and computational resources. The principal part of the chapter (Section 16.2) presents different techniques for grasp detection, divided by grasp detection with RGB-D data and point cloud data. Evaluation of robotic grasping and the potential follow step of manipulation is covered in Section 16.3 and Section 16.4, respectively, where two methods are selected for evaluation on custom objects. The simulated grasp experiments demonstrate the capabilities and limitations of current state of the art grasping approaches. Regarding object pose estimation, while an accurate object pose can be estimated, high resolution object (CAD) models are required and objects should not be (partially) occluded or in a cluttered scene. Regarding grasp detection, as a large number of grasp poses will be generated for each object, their suitability still needs to be assessed in a filtering step, to remove unwanted grasps or select a most suitable grasp. In both cases, similar drawbacks can be identified regarding data collection and computation time. Large data sets of object models are required for training, leading to long computation times for model generation. In addition, the inference time of the trained models is high, implying real-time control based on them is not possible. Finally, the chapter describes different measures used to evaluate pose estimation and grasp detection actions, as well as several protocols in benchmarking grasps that are common in the field. Following, in Section 16.5, we provide a list of different data sets that generate object models by utilizing simulation.

The state of the art in robot grasping is improving rapidly, with most algorithms and data sets provided open source. The benefits of deep learning, compared to traditional approaches in grasping, are eminent, as the manual step of feature engineering is avoided and, instead, a learning based approach provides the generation of a grasping model. Despite this, different aspects require further research to tackle their limitations. First, methods need to be investigated that can better grasp unknown objects. Potential solutions to this are the automatic generation of object data sets, based on the types of objects to be expected or where object models include a wide variation of object geometries. Second, the computation time for both training a grasping model and its inference, should be suitable for real world applications. Current state of the art models is heavy to train and execute, and typically utilize computing clusters. Considering the application area of agile production, such computation power cannot be accessed. Potential solutions to this could, again, target the data generation step, where objects to be grasped are only included in the data set, or the object models contain wide variety in object geometry. To conclude this robotic grasping overview and analysis, deep learning as tool for robotic grasping offers great benefits, compared to traditional approaches in generating grasping models.

References

[1] V. Villani, F. Pini, F. Leali, C. Secchi, Survey on human–robot collaboration in industrial settings: safety, intuitive interfaces and applications, Mechatronics 55 (2018) 248–266.

[2] E. Matheson, R. Minto, E.G. Zampieri, M. Faccio, G. Rosati, Human–robot collaboration in manufacturing applications: a review, Robotics 8 (4) (2019) 100.

[3] J. Wang, Y. Ma, L. Zhang, R.X. Gao, D. Wu, Deep learning for smart manufacturing: methods and applications, Journal of Manufacturing Systems 48 (2018) 144–156.

[4] A. Kusiak, Smart manufacturing must embrace big data, Nature 544 (7648) (2017) 23–25.

[5] T. Brogårdh, Robot control overview: an industrial perspective, Modeling, Identification and Control 30 (3) (2009) 167.

[6] B. Siciliano, L. Sciavicco, L. Villani, G. Oriolo, Robotics: Modelling, Planning and Control, Springer Science & Business Media, 2010.

[7] IFR, How connected robots are transforming manufacturing, Tech. Rep., International Federation of Robotics, https://ifr.org/papers, 2020.

[8] C. Zhang, W. Xu, J. Liu, Z. Liu, Z. Zhou, D.T. Pham, Digital twin-enabled reconfigurable modeling for smart manufacturing systems, International Journal of Computer Integrated Manufacturing (2019) 1–25.

[9] Z. Zhu, H. Hu, Robot learning from demonstration in robotic assembly: a survey, Robotics 7 (2) (2018) 17.

[10] E. De Coninck, T. Verbelen, P. Van Molle, P. Simoens, B. Dhoedt, Learning robots to grasp by demonstration, Robotics and Autonomous Systems 127 (2020) 103474.

[11] A. Angleraud, R. Codd-Downey, M. Netzev, Q. Houbre, R. Pieters, Virtual teaching for assembly tasks planning, in: IEEE International Conference on Human–Machine Systems (ICHMS), 2020, pp. 1–6.

[12] A. Hietanen, R. Pieters, M. Lanz, J. Latokartano, J.-K. Kämäräinen, AR-based interaction for human–robot collaborative manufacturing, Robotics and Computer-Integrated Manufacturing 63 (2020) 101891.

[13] S. El Zaatari, M. Marei, W. Li, Z. Usman, Cobot programming for collaborative industrial tasks: an overview, Robotics and Autonomous Systems 116 (2019) 162–180.

[14] A. Chowdhury, A. Ahtinen, R. Pieters, K. Vaananen, User experience goals for designing industrial human–cobot collaboration: a case study of franka panda robot, in: Nordic Conference on Human–Computer Interaction (NordiCHI), 2020, pp. 1–13.

[15] H.A. Pierson, M.S. Gashler, Deep learning in robotics: a review of recent research, Advanced Robotics 31 (16) (2017) 821–835.

[16] G. Du, K. Wang, S. Lian, Vision-based robotic grasping from object localization, pose estimation, grasp detection to motion planning: a review, arXiv preprint, arXiv:1905.06658, 2019.

[17] J. Mahler, K. Goldberg, Learning deep policies for robot bin picking by simulating robust grasping sequences, in: Conference on Robot Learning, 2018, pp. 515–524.

[18] P. Wang, H. Liu, L. Wang, R.X. Gao, Deep learning-based human motion recognition for predictive context-aware human–robot collaboration, CIRP Annals 67 (1) (2018) 17–20.

[19] ISO-15066:2016, Robots and Robotic Devices — Collaborative Robots, Standard, International Organization for Standardization, 2016.

[20] S. Levine, P. Pastor, A. Krizhevsky, J. Ibarz, D. Quillen, Learning hand-eye coordination for robotic grasping with deep learning and large-scale data collection, The International Journal of Robotics Research 37 (4–5) (2018) 421–436.

[21] M.A. Lee, Y. Zhu, P. Zachares, M. Tan, K. Srinivasan, S. Savarese, L. Fei-Fei, A. Garg, J. Bohg, Making sense of vision and touch: learning multimodal representations for contact-rich tasks, IEEE Transactions on Robotics 36 (3) (2020) 582–596.

[22] J. Bütepage, A. Ghadirzadeh, Ö.Ö. Karadag, M. Björkman, D. Kragic, Imitating by generating: deep generative models for imitation of interactive tasks, Frontiers in Robotics and AI 7 (2020) 47.

[23] A. Ghadirzadeh, X. Chen, W. Yin, Z. Yi, M. Björkman, D. Kragic, Human-centered collaborative robots with deep reinforcement learning, arXiv preprint, arXiv:2007.01009, 2020.

[24] ISO-10218-1/2:2011, Robots and Robotic Devices – Safety Requirements for Industrial Robots – Part 1: Robots/Part 2: Robot Systems and Integration, Standard, International Organization for Standardization, 2011.

[25] R.-J. Halme, M. Lanz, J. Kämäräinen, R. Pieters, J. Latokartano, A. Hietanen, Review of vision-based safety systems for human-robot collaboration, Procedia CIRP 72 (2018) 111–116.

[26] L. Birglen, T. Schlicht, A statistical review of industrial robotic grippers, Robotics and Computer-Integrated Manufacturing 49 (2018) 88–97.

[27] M. Honarpardaz, M. Tarkian, J. Ölvander, X. Feng, Finger design automation for industrial robot grippers: a review, Robotics and Autonomous Systems 87 (2017) 104–119.

[28] Barrett Technology, BarrettHand – multi-fingered programmable grasper, https://advanced.barrett.com/barretthand. (Accessed 23 November 2020).

[29] M. Netzev, A. Angleraud, R. Pieters, Soft robotic gripper with compliant cell stacks for industrial part handling, IEEE Robotics and Automation Letters 5 (4) (2020) 6821–6828.

[30] V. Limère, H.V. Landeghem, M. Goetschalckx, E.-H. Aghezzaf, L.F. McGinnis, Optimising part feeding in the automotive assembly industry: deciding between kitting and line stocking, International Journal of Production 50 (15) (2012) 4046–4060.

[31] A. Hietanen, J. Latokartano, A. Foi, R. Pieters, V. Kyrki, M. Lanz, J.-K. Kämäräinen, Benchmarking 6D object pose estimation for robotics, arXiv preprint, arXiv:1906.02783, 2019.

[32] A. Sahbani, S. El-Khoury, P. Bidaud, An overview of 3D object grasp synthesis algorithms, Robotics and Autonomous Systems 60 (3) (2012) 326–336.

[33] S. Caldera, A. Rassau, D. Chai, Review of deep learning methods in robotic grasp detection, Multimodal Technologies and Interaction 2 (3) (2018) 57.

[34] Y. Jiang, S. Moseson, A. Saxena, Efficient grasping from RGBD images: learning using a new rectangle representation, in: IEEE International Conference on Robotics and Automation (ICRA), 2011, pp. 3304–3311.

[35] A. ten Pas, M. Gualtieri, K. Saenko, R. Platt, Grasp pose detection in point clouds, The International Journal of Robotics Research 36 (13–14) (2017) 1455–1473.
[36] H. Liang, X. Ma, S. Li, M. Görner, S. Tang, B. Fang, F. Sun, J. Zhang, PointNetGPD: detecting grasp configurations from point sets, in: IEEE International Conference on Robotics and Automation (ICRA), 2019, pp. 3629–3635.
[37] A. Mousavian, C. Eppner, D. Fox, 6-DOF GraspNet: variational grasp generation for object manipulation, in: IEEE International Conference on Computer Vision (ICCV), 2019, pp. 2901–2910.
[38] J. Bohg, A. Morales, T. Asfour, D. Kragic, Data-driven grasp synthesis—a survey, IEEE Transactions on Robotics 30 (2) (2013) 289–309.
[39] R. Szeliski, Computer Vision: Algorithms and Applications, Springer Science & Business Media, 2010.
[40] S. Hinterstoisser, V. Lepetit, S. Ilic, S. Holzer, G. Bradski, K. Konolige, N. Navab, Model based training, detection and pose estimation of texture-less 3D objects in heavily cluttered scenes, in: Asian Conference on Computer Vision (ACCV), 2012, pp. 548–562.
[41] Y. Xiang, T. Schmidt, V. Narayanan, D. Fox, PoseCNN: a convolutional neural network for 6D object pose estimation in cluttered scenes, arXiv preprint, arXiv:1711.00199, 2017.
[42] C. Capellen, M. Schwarz, S. Behnke, ConvPoseCNN: dense convolutional 6D object pose estimation, arXiv preprint, arXiv:1912.07333, 2019.
[43] C. Song, J. Song, Q. Huang, Hybridpose: 6D object pose estimation under hybrid representations, in: IEEE/CVF Conference on Computer Vision and Pattern Recognition (CVPR), 2020, pp. 431–440.
[44] S. Peng, Y. Liu, Q. Huang, X. Zhou, H. Bao, PVNet: pixel-wise voting network for 6DOF pose estimation, in: IEEE/CVF Conference on Computer Vision and Pattern Recognition (CVPR), 2019, pp. 4561–4570.
[45] H. Wang, S. Sridhar, J. Huang, J. Valentin, S. Song, L.J. Guibas, Normalized object coordinate space for category-level 6D object pose and size estimation, in: IEEE/CVF Conference on Computer Vision and Pattern Recognition (CVPR), 2019, pp. 2642–2651.
[46] K. He, G. Gkioxari, P. Dollár, R. Girshick, Mask R-CNN, in: IEEE International Conference on Computer Vision (ICCV), 2017, pp. 2961–2969.
[47] F.T. Pokorny, D. Kragic, Classical grasp quality evaluation: new algorithms and theory, in: IEEE/RSJ International Conference on Intelligent Robots and Systems (IROS), 2013, pp. 3493–3500.
[48] J. Mahler, F.T. Pokorny, B. Hou, M. Roderick, M. Laskey, M. Aubry, K. Kohlhoff, T. Kröger, J. Kuffner, K. Goldberg, Dex-Net 1.0: a cloud-based network of 3D objects for robust grasp planning using a multi-armed bandit model with correlated rewards, in: IEEE International Conference on Robotics and Automation (ICRA), 2016, pp. 1957–1964.
[49] J. Mahler, J. Liang, S. Niyaz, M. Laskey, R. Doan, X. Liu, J.A. Ojea, K. Goldberg, Dex-Net 2.0: deep learning to plan robust grasps with synthetic point clouds and analytic grasp metrics, arXiv preprint, arXiv:1703.09312, 2017.
[50] J. Mahler, M. Matl, X. Liu, A. Li, D. Gealy, K. Goldberg, Dex-Net 3.0: computing robust vacuum suction grasp targets in point clouds using a new analytic model and deep learning, in: IEEE International Conference on Robotics and Automation (ICRA), 2018, pp. 1–8.
[51] J. Mahler, M. Matl, V. Satish, M. Danielczuk, B. DeRose, S. McKinley, K. Goldberg, Learning ambidextrous robot grasping policies, Science Robotics 4 (26) (2019).

[52] J. Redmon, A. Angelova, Real-time grasp detection using convolutional neural networks, in: IEEE International Conference on Robotics and Automation (ICRA), 2015, pp. 1316–1322.
[53] D. Guo, T. Kong, F. Sun, H. Liu, Object discovery and grasp detection with a shared convolutional neural network, in: IEEE International Conference on Robotics and Automation (ICRA), 2016, pp. 2038–2043.
[54] Y. He, W. Sun, H. Huang, J. Liu, H. Fan, J. Sun, PVN3D: a deep point-wise 3D keypoints voting network for 6DOF pose estimation, in: IEEE/CVF Conference on Computer Vision and Pattern Recognition (CVPR), 2020, pp. 11632–11641.
[55] H. Zhao, J. Shi, X. Qi, X. Wang, J. Jia, Pyramid scene parsing network, in: IEEE/CVF Conference on Computer Vision and Pattern Recognition (CVPR), 2017, pp. 2881–2890.
[56] C.R. Qi, L. Yi, H. Su, L.J. Guibas, PointNet++: deep hierarchical feature learning on point sets in a metric space, in: Advances in Neural Information Processing Systems, 2017, pp. 5099–5108.
[57] C. Wang, D. Xu, Y. Zhu, R. Martín-Martín, C. Lu, L. Fei-Fei, S. Savarese, DenseFusion: 6D object pose estimation by iterative dense fusion, in: IEEE/CVF Conference on Computer Vision and Pattern Recognition (CVPR), 2019, pp. 3343–3352.
[58] D. Comaniciu, P. Meer, Mean shift: a robust approach toward feature space analysis, IEEE Transactions on Pattern Analysis and Machine Intelligence 24 (5) (2002) 603–619.
[59] Y. Guo, H. Wang, Q. Hu, H. Liu, L. Liu, M. Bennamoun, Deep learning for 3D point clouds: a survey, IEEE Transactions on Pattern Analysis and Machine Intelligence (2020).
[60] C.R. Qi, H. Su, K. Mo, L.J. Guibas, PointNet: deep learning on point sets for 3D classification and segmentation, in: IEEE Conference on Computer Vision and Pattern Recognition (CVPR), 2017, pp. 652–660.
[61] A. Zeng, S. Song, M. Nießner, M. Fisher, J. Xiao, T. Funkhouser, 3DMatch: learning local geometric descriptors from RGB-D reconstructions, in: IEEE Conference on Computer Vision and Pattern Recognition (CVPR), 2017, pp. 1802–1811.
[62] Q.-H. Pham, M.A. Uy, B.-S. Hua, D.T. Nguyen, G. Roig, S.-K. Yeung, LCD: learned cross-domain descriptors for 2D-3D matching, in: AAAI Conference on Artificial Intelligence, 2020, pp. 11856–11864.
[63] B. Zhao, H. Zhang, X. Lan, H. Wang, Z. Tian, N. Zheng, REGNet: region-based grasp network for single-shot grasp detection in point clouds, arXiv preprint, arXiv:2002.12647, 2020.
[64] A. Hertz, R. Hanocka, R. Giryes, D. Cohen-Or, PointGMM: a neural GMM network for point clouds, in: IEEE/CVF Conference on Computer Vision and Pattern Recognition (CVPR), 2020, pp. 12054–12063.
[65] B.S. Zapata-Impata, P. Gil, J. Pomares, F. Torres, Fast geometry-based computation of grasping points on three-dimensional point clouds, International Journal of Advanced Robotic Systems 16 (1) (2019).
[66] D. Morrison, P. Corke, J. Leitner, Closing the loop for robotic grasping: a real-time, generative grasp synthesis approach, in: Robotics: Science and Systems (RSS), 2018.
[67] D.P. Kingma, M. Welling, Auto-encoding variational Bayes, arXiv preprint, arXiv:1312.6114, 2013.
[68] T. Hodaň, J. Matas, Š. Obdržálek, On evaluation of 6D object pose estimation, in: European Conference on Computer Vision (ECCV), 2016, pp. 606–619.
[69] C. Eppner, A. Mousavian, D. Fox, A billion ways to grasp: an evaluation of grasp sampling schemes on a dense, physics-based grasp data set, arXiv preprint, arXiv:1912.05604, 2019.

[70] F. Bottarel, G. Vezzani, U. Pattacini, L. Natale, GRASPA 1.0: GRASPA is a robot arm grasping performance benchmark, IEEE Robotics and Automation Letters 5 (2) (2020) 836–843.
[71] A. Yershova, S. Jain, S.M. Lavalle, J.C. Mitchell, Generating uniform incremental grids on SO(3) using the Hopf fibration, The International Journal of Robotics Research 29 (7) (2010) 801–812.
[72] A. Bicchi, V. Kumar, Robotic grasping and contact: a review, in: IEEE International Conference on Robotics and Automation (ICRA), Vol. 1, 2000, pp. 348–353.
[73] B. Calli, A. Singh, A. Walsman, S. Srinivasa, P. Abbeel, A.M. Dollar, The YCB object and model set: towards common benchmarks for manipulation research, in: International Conference on Advanced Robotics (ICAR), 2015, pp. 510–517.
[74] A. Depierre, E. Dellandréa, L. Chen, Jacquard: a large scale dataset for robotic grasp detection, in: IEEE/RSJ International Conference on Intelligent Robots and Systems (IROS), 2018, pp. 3511–3516.
[75] H.-S. Fang, C. Wang, M. Gou, C. Lu, GraspNet-1Billion: a large-scale benchmark for general object grasping, in: IEEE/CVF Conference on Computer Vision and Pattern Recognition (CVPR), 2020, pp. 11444–11453.
[76] K. Collins, A. Palmer, K. Rathmill, The development of a European benchmark for the comparison of assembly robot programming systems, in: Robot Technology and Applications, Springer, 1985, pp. 187–199.
[77] A.X. Chang, T. Funkhouser, L. Guibas, P. Hanrahan, Q. Huang, Z. Li, S. Savarese, M. Savva, S. Song, H. Su, et al., ShapeNet: an information-rich 3D model repository, arXiv preprint, arXiv:1512.03012, 2015.
[78] C. Mitash, R. Shome, B. Wen, A. Boularias, K. Bekris, Task-driven perception and manipulation for constrained placement of unknown objects, IEEE Robotics and Automation Letters 5 (4) (2020) 5605–5612.
[79] K. Van Wyk, M. Culleton, J. Falco, K. Kelly, Comparative peg-in-hole testing of a force-based manipulation controlled robotic hand, IEEE Transactions on Robotics 34 (2) (2018) 542–549.
[80] J. Watson, A. Miller, N. Correll, Autonomous industrial assembly using force, torque, and RGB-D sensing, Advanced Robotics 34 (7–8) (2020) 546–559.
[81] Y. Bekiroglu, N. Marturi, M.A. Roa, K.J.M. Adjigble, T. Pardi, C. Grimm, R. Balasubramanian, K. Hang, R. Stolkin, Benchmarking protocol for grasp planning algorithms, IEEE Robotics and Automation Letters 5 (2) (2019) 315–322.
[82] J. Leitner, A.W. Tow, N. Sünderhauf, J.E. Dean, J.W. Durham, M. Cooper, M. Eich, C. Lehnert, R. Mangels, C. McCool, et al., The ACRV picking benchmark: a robotic shelf picking benchmark to foster reproducible research, in: IEEE International Conference on Robotics and Automation (ICRA), 2017, pp. 4705–4712.
[83] S. James, Z. Ma, D.R. Arrojo, A.J. Davison, RLBench: the robot learning benchmark & learning environment, IEEE Robotics and Automation Letters 5 (2) (2020) 3019–3026.
[84] C. Eppner, A. Mousavian, D. Fox, ACRONYM: a large-scale grasp dataset based on simulation, arXiv preprint, arXiv:2011.09584, 2020.
[85] D. Morrison, P. Corke, J. Leitner, EGAD! An evolved grasping analysis dataset for diversity and reproducibility in robotic manipulation, IEEE Robotics and Automation Letters (2020).
[86] J. Zhao, J. Liang, O. Kroemer, Towards precise robotic grasping by probabilistic post-grasp displacement estimation, arXiv preprint, arXiv:1909.02129, 2019.
[87] D. Kappler, J. Bohg, S. Schaal, Leveraging big data for grasp planning, in: IEEE International Conference on Robotics and Automation (ICRA), 2015, pp. 4304–4311.

[88] M. Veres, M. Moussa, G.W. Taylor, An integrated simulator and dataset that combines grasping and vision for deep learning, arXiv preprint, arXiv:1702.02103, 2017.
[89] M. Macklin, M. Müller, N. Chentanez, T.-Y. Kim, Unified particle physics for real-time applications, ACM Transactions on Graphics 33 (4) (2014) 1–12.
[90] M. Matl, Pyrender, https://github.com/mmatl/pyrender, 2019.
[91] E. Coumans, Y. Bai, PyBullet, a python module for physics simulation for games, robotics and machine learning, http://pybullet.org, 2016–2019.
[92] Blender Online Community, Blender – a 3D modelling and rendering package, http://www.blender.org, 2016.
[93] R. Diankov, J. Kuffner, OpenRAVE: a planning architecture for autonomous robotics, Tech. Rep. CMU-RI-TR-08-34 79, Robotics Institute, Pittsburgh, PA, 2008.
[94] E. Rohmer, S.P. Singh, M. Freese, V-REP: a versatile and scalable robot simulation framework, in: IEEE/RSJ International Conference on Intelligent Robots and Systems (IROS), 2013, pp. 1321–1326.
[95] X. Yan, J. Hsu, M. Khansari, Y. Bai, A. Pathak, A. Gupta, J. Davidson, H. Lee, Learning 6-DOF grasping interaction via deep geometry-aware 3D representations, in: IEEE International Conference on Robotics and Automation (ICRA), 2018, pp. 1–9.

CHAPTER

Deep learning in multiagent systems

17

Lukas Esterle

Department of Electrical and Computer Engineering, Aarhus University, Aarhus, Denmark

17.1 Introduction

Autonomous software agents are able to perform specific tasks, and make individual decisions on which they can base their autonomous actions to deal with various perturbations in their environment. These autonomous software agents are often embedded in physical devices allowing them to even interact with the physical world. With the rising ubiquity of computing systems, actions, reactions, and interactions of these systems are required to become autonomous and without or very limited human intervention. This autonomy can be brought about autonomous software agents, able to perform specific tasks, and make individual decisions on which they can base their autonomous actions to deal with various perturbations in their environment. These autonomous software agents are often embedded in physical devices allowing them to even interact with the physical world, so-called cyber-physical systems [1].

These Cyber-physical Systems (CPSs) bring about their very specific challenges on learning and decision making in autonomous agents [2,3]. Where pure software systems operate within a somewhat restricted space of what kind of situations they might experience, CPSs are exposed to almost limitless potential interactions and situations. This is not only due to their exposure to the continuously changing dynamic physical environment but also due new agents potentially joining and affecting the environment constantly. At the same time, a CPS is often not bound to single locations, allowing it to change its own environment and perception of the same. While decisions can be based on simple heuristics in specific situations, with advances in machine learning, different learning techniques have been utilized within autonomous agents. Learning during runtime, brings an obvious advantage as the system can learn about changing conditions and new, emerging situations that have not been considered during design time.

The autonomous actions of an agent may also lead to intended as well as unintended interactions with other autonomous agents. These actions can be directly or indirectly affect others. A direct interaction, for example, would be one physical system controlled by an autonomous agent pushing another systems. An indirect action would be an autonomous CPS manipulating the environment (e.g., pushing a box) forcing another agent to adapt its own behavior. Such unintentional interactions

Deep Learning for Robot Perception and Cognition. https://doi.org/10.1016/B978-0-32-385787-1.00022-1

pose different challenges to all participants, we will discuss such challenges in this chapter in more detail. Multiple software agents cooperating intentionally in a shared environment toward a common goal are considered a *multiagent system* (MAS) [4].

Apart from planned and restricted MAS, autonomous agents are required to constantly self-integrate with other agents if operating in an open environment. This means, they need to adjust and adapt in order to improve at runtime given their changing environment [5,6]. Again, deep learning can be one of the potential approaches utilized by individual agents to learn about the behavior and actions of others in order for them to make decisions on how to interact with one another. There are numerous examples where this needs to be taken into account and specifically for cyber-physical multiagent systems such as industry 4.0 and additive manufacturing, dynamic and agile production systems, smart transportation including autonomous driving cars, or multirobot operations (e.g., autonomous search-and-rescue operations with multiple stakeholders). While deep learning can be of use in autonomous agents as well as entire multiagent systems, learning occurs in the individual agent itself. However, we can differentiate between individual learning and cooperative and collaborative learning, that is, whether an agent can learn by itself or another agent is required to generate, calibrate, or refine knowledge.

In the following, we will first set the scene for autonomous agents and multiagent systems and discuss different properties of the environment and the agents themselves. We will use this opportunity to also explore some early challenges these properties can impose on the agents and their learning processes. Afterwards, we explore the challenges faced by agents able to interact with others and able to change their own position and point of view. Thereafter, we discuss the difference between individual learning and cooperative/collaborative learning in autonomous agents and multiagent systems. We will discuss different approaches and ongoing research directions and how they can be utilized in multiagent systems. We will conclude this chapter with an outlook and open challenges specifically for autonomous agents.

17.2 Setting the scene

Before we discuss the different approaches of deep learning in autonomous agents and multiagent systems, we should define the different aspects in more detail. In the Introduction, we briefly mentioned several of them. However, many of the properties affecting autonomous agents and entire multiagent systems are in tension or interact with each other.

Properties are present for the individual agents, the multiagent system, as well as the environment in which they operate. We will first discuss properties for the environment in which the agents operate. For the purpose of the discussion of the respective properties are independent of whether this environment is physical or virtual. Nevertheless, considering a physical environment often allows to easily visualize the arising constraints. We differentiate between the overall environment and the area of operation. The environment may be equal to the area of operation but also might extend far beyond it. The area of operation is usually limited by the agents capacity to move around or perceive and actuate on its environment. Some of the following properties closely relate but have subtle differences. We specifically try to highlight these differences.

- **Constraint versus unconstrained:** The environment in which our agents operate may be constrained or unconstrained. There are two sides to this limitation or openness. On one side, the area of operation of the may be constrained for the agents but the environment itself is still unconstrained, allowing external, uncontrolled effects influence the agents within their area of operation. External effects can be of natural cause, such as wind or temperature, but also other agents or physical elements entering the area of operation. In such a case, agents could be affected by *outside* effects but could not leave their area of operation. On the other side, if the environment is unconstrained and the area of operation is as well open, agents can leave and change their initial area of operation. This enables them to explore new locations, potentially discover new resources and information supporting them reaching their initial task. Furthermore, we may consider new agents joining the current area of operation and supporting existing agents in their task in an unconstrained environment. Conversely, agents operating in the area may also leave the area of operation and be unavailable in collaborative tasks.
- **Known versus unknown:** The environment, in which our agents operate, may contain all kinds of information for our agents to learn about. Nevertheless, we may also imagine situations where we do not consider agents to learn completely new information about their environment. Let us consider a simulation in which agents forage for food. The *known/unknown* property may refer to whether the agents know exactly "what" all the different food sources are, though they might not know "where" they are. Furthermore, the environment does not contain any adversaries or any competition in the foraging process. In such a case, we can argue that all aspects not primarily necessary for the foraging process are *known* to the agents. In an alternative scenario of foraging agents, the agents might be able to discover new, initially unknown food sources. For example, agents now can not only forage specific berries but also apples and pears. This not only requires the agents to be able to identify those fruits and determine them as food but the different fruit has to be available in the environment to begin with. One might now consider that *unknowns* can only appear in *unconstrained* environments (see above). However, even in a constrained environment, an agent might not be familiar with all aspects (e.g., all types of fruit in our example scenario above). Similarly, even in an *unconstrained* environment, an agent might be familiar with everything it is able to perceive. In addition, we might also want to consider other agents as such elements and, therefore, to be *known* or *unknown*.
- **Stationary versus nonstationary:** The environment, in which our agents operate, may be changing or remain fixed and unchanged. While this, on one hand, relates very closely to *constrained and unconstrained* environments, when it comes to agents or elements entering or leaving the area of operation. On the other hand, a nonstationary environment can also be within a constrained environment where

elements change their physical location, or even in-place, change their appearance or behavior. Consider a physical, but constrained environment, where cars can move around, buildings are being raised as well as razed, trees grow and change their physical size and appearance, or flowers wilt. All of these aspects change the environment passively. We could also consider blocks that the agents can move around within this constrained environment, actively changing the conditions. Importantly, we do not expect completely random or erratic changes in the environment. For example, buildings will not suddenly disappear altogether.

After the main properties for the environment, we also want to dive into different properties for agents and multiagent systems. In many cases, these properties apply to both cases. We will highlight those properties that only apply individual agents or only to multiagent systems, comprising multiple individual agents.

- **Static versus dynamic:** For individual agents as well as entire multiagent systems we can consider them to be either *static or dynamic*. We refer to static agents if they do not change their behavior or their actions performed in the environment. In contrast, dynamic agents (or multiagent systems for that matter) may change their behavior or actions on the same input. Importantly, we consider a dynamic agent to have a different behavior on the same inputs. This usually happens when the agent explores its own options by either random exploration or because learning mechanisms steer the exploration.
- **Benevolence versus indifferent versus oblivious versus malevolent:** The property of behavior of an agent or of a multiagent system is always in conjunction with another, external agent. We consider *malevolent* agents to work actively against the goals of another agent. In contrast, *benevolent* agents will try to support the actions of anther agent when possible. An *indifferent* agent will neither actively work against nor with another agent toward the other agents goal. One might say, the agent does not care what is going on around it. Importantly, a *dynamic* agent might change its behavior toward others between *indifferent to benevolent to malevolent*. Only an *oblivious* agent is simply unaware of other agents in its environment. An *oblivious* agent can usually not suddenly become aware of others and change its behavior, however, by loss of perception (e.g., sensor failure) an agent can easily become oblivious of its surroundings required to operate *obliviously*.
- **Homogeneity versus heterogeneity:** We can consider this property of *homogeneity and heterogeneity* at the level of multiagent systems and how individual agents within the system behave, appear, or operate. If all agents appear, behave, or operate in the same fashion, we consider this an *homogenous* multiagent system. If we now consider the previously discussed property of *dynamic* agents with respect to behavior and appearance. Under the assumption that underlying change mechanism is susceptible to external influences not perceived by all agents at all times, the entire multiagent system inherently is heterogeneous as the appearance and/or behavior of each agent within the system can change independently from each other, that is, not all agents will change their behavior/appearance in the

same way at the exact same time. However, we can also consider this property on a meta-level and ask if the underlying mechanism triggering the change within the agent is the same across all agents or not, that is, would two agents change their behavior/appearance the same way if they were subject to the same stimuli? Are they all using deep learning or even having different deep neural networks supporting their decision making?

- **Dependence versus independence:** Individual agents may operate completely *independent* without requiring other agents to achieve their individual goals. In contrast, *dependent* agents require other agents, either within or outside their own multiagent system, to achieve their individual goals. Consider an agent, with to goal to build a tower, to rely on another agent to provide bricks for such a structure. With respect to deep learning, can the agents learn independently or do they rely on the feedback or support of others?

17.3 Challenges

There are several challenges for autonomous agents and multiagent systems. Even more so, if the agent(s) are embodied in a physical device and deployed in a physical environment [2,3]. We can distinguish between spatiotemporal challenges for individual agents and challenges in collaboration and cooperation among multiple agents. Finally, there is also an obvious resource challenges when it comes to deep learning in autonomous agents.

A *temporal challenge* for autonomous agents can arise from a distortion between the real-world and the time it takes the agent to process the perceived information. For example, if a smart-camera [7] processes a frame to detect a specific object currently within its field of view, but upon completing this task, the object has moved outside the field of view of the camera. Similar in collaborating multiagent systems, time required to learn and/or make decisions can be a limiting factor as the time-window requiring a decision might have passed.

Spatial challenges arise usually due to the physical pose of the agent or through the limited range of perception. We can again consider a smart-camera deployed in a real world environment. If the camera is oriented in the wrong direction, it might not be able to perceive specific events at all. Even if the camera perceives the event, given its viewpoint, it might perceive an event differently than from another viewpoint. The spatial challenge is related to the situatedness concept in psychology, theorizing that the mind is intertwined with the current environment [8].

The spatiotemporal challenges above often lead to the samples not being generated from an independent and identical distribution (IID)—these samples are referred to as being non-IID [9]. Kairouz et al. [10] identify four challenges leading to deviating data from being identically distributed. We follow this discussion and relate it specifically to autonomous agents and multiagent systems. For different agents i, we can define different problems when drawing samples based on a conditional probability $P_i(L|\boldsymbol{x})P_i(\boldsymbol{x})$ and $P_i(\boldsymbol{x}|L)P_i(L)$ for feature vector $\boldsymbol{x}$ and a label L.

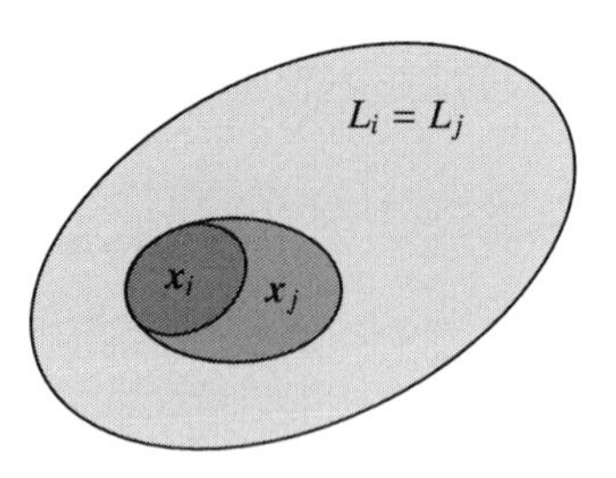

FIGURE 17.1

Euler diagram of conditional probabilities. L_i, L_j, $\boldsymbol{x}_i$, and $\boldsymbol{x}_j$ are shorthand for the probabilities of labels and features of agent i and j, $P_i(L)$, $P_j(L)$, $P_i(\boldsymbol{x})$, and $P_j(\boldsymbol{x})$, respectively. Agent i and j have the same set of labels, but agent j is able to precess more features; nevertheless, the conditional probability distribution.

1. **Feature distribution skew (covariate shift)**: *Characteristic of a feature for a specific label may vary.* Consider an agent interacting with different humans via gestures or voice recognition. Independent of the individual characteristics of the input signal, the label should be the same when the semantics of the signal is the same.
   However, we can also consider this for two agents i and j where the marginal distribution $P_i(\boldsymbol{x}) \neq P_j(\boldsymbol{x})$ varies between the agents, nevertheless $P_i(L|\boldsymbol{x}) = P_j(L|\boldsymbol{x})$. This is illustrated in Fig. 17.1 where both agents have the same set of labels but the set of features process by both differs.
2. **Label distribution skew (prior probability shift)**: *Quantity and quality of labels may vary among agents.* This is often owed to the situatedness of the agent, where due to the position of the agent, the available labels vary among multiple agents. Consider an agent observing a truck lane and one that is prohibiting trucks. The agent observing the truck lane might occasionally see normal cars, whereas the other agent should never encounter a truck. Even more so, animals native to specific islands will probably not be encountered anywhere else in the wild. Here, the marginal distribution of $P_i(L) \neq P_j(L)$ varies among two agents i and j though $P_i(\boldsymbol{x}|L) = P_j(\boldsymbol{x}|L)$ as illustrated in Fig. 17.2.
3. **Same label, different features (concept shift)**: *Features defining a label may vary across different agents.* Independent of the location of the agent and the individual perception, the resulting label should be consistent among the agents when the semantic of the perceived information is the same. Consider different types of cows around the world. Whether the cow is black and white, just brown or another color, the agent should recognize it as a cow rather than a sheep. This applies also to other aspects such as humans, cars, houses, or trees, to name a few. These factors can heavily vary based on the geographical location of the agent but can also be affected by environmental conditions such as snow, rain, or time of day. Here, we can argue that the conditional distributions $P_i(\boldsymbol{x}|L) \neq P_j(\boldsymbol{x}|L)$ even though the $P_i(L) = P_j(L)$. This is illustrated in Fig. 17.3.

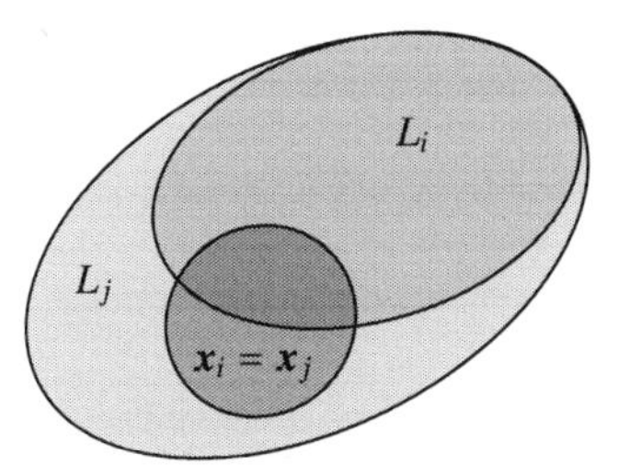

FIGURE 17.2

Euler diagram of conditional probabilities. L_i, L_j, x_i, and x_j are shorthand for the probabilities of labels and features of agent i and j, $P_i(L)$, $P_j(L)$, $P_i(x)$, and $P_j(x)$, respectively. The agents i and j cannot assign the same labels to the available features.

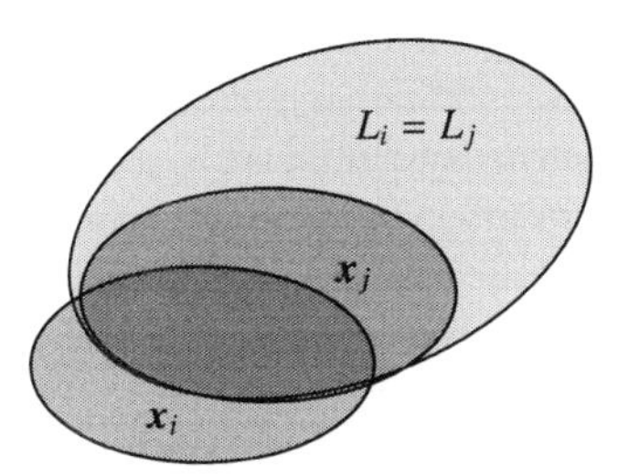

FIGURE 17.3

Euler diagram of conditional probabilities. L_i, L_j, x_i, and x_j are shorthand for the probabilities of labels and features of agent i and j, $P_i(L)$, $P_j(L)$, $P_i(x)$, and $P_j(x)$, respectively. Conditional distribution $P_i(L|x) \neq P_j(L|x)$ differs among the agent even though they have a common $P(L)$.

4. **Different label, same features (concept shift)**: *Same features may be defined by different labels across individual agents.* Labels may vary due to regional, cultural, application, or even personal preferences. As in our previous example, a cow should be classified by all agents, the individual breed, though, might only be identified by some of the agents. Importantly, an agent might only identify the label of the species being aware of the corresponding breed whereas another might have a problem connecting its own label with these sublabels appropriately. While this is usually not a problem with agents maintained by a common operator, it can lead to misinterpretations when agents, not handled by the same operator, try to collaborate. We can consider the conditional distributions $P_i(L|x) \neq P_j(L|x)$ even though the $P_i(x) = P_j(x)$ as illustrate in Fig. 17.4.
5. **Quantity skew or unbalancedness**: Some agents might be able to gather more information than others. Consider the cameras in a surveillance network at an airport. A camera observing the main entrance to the international terminal will gather considerably more data of human faces compare to a camera observing the entrance to the gate at the very end of the domestic terminal.

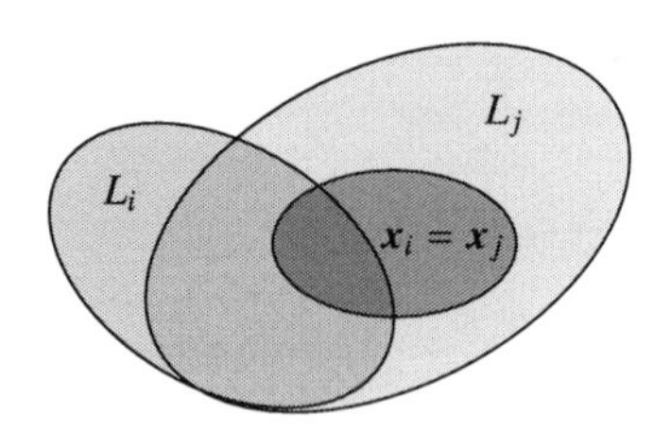

FIGURE 17.4

Euler diagram of conditional probabilities. L_i, L_j, $\boldsymbol{x}_i$, and $\boldsymbol{x}_j$ are shorthand for the probabilities of labels and features of agent i and j, $P_i(L)$, $P_j(L)$, $P_i(\boldsymbol{x})$, and $P_j(\boldsymbol{x})$, respectively. Different labels while the same features are being utilized. This can lead to a change in the conditional distributions among agents.

In addition to the non-IID challenge of deep learning that is also prevalent among multiagent systems, there are additional challenges when multiple autonomous agents interact in a common environment. In such cases, and specifically in the cyber-physical domain, each agent needs to be aware of the respective temporal and spatial actions and interactions of the other agents. Here, an awareness of *where*, *when,* and *how* another agent perceives and interacts with the environment become essential to enable an agent for successful autonomous self-integration and interaction. Lacking such capabilities in combination with a limitation of generating causal connections among cause and effect might lead to agents becoming superstitious [11].

Finally, each agent faces a challenge regarding resources when facing a trade-off between achieved performance, and the ability to learn and improve over time. Because autonomous agents are often resource constrained, an agent needs to decide whether it should spend its limited resources on performing actions it already knows work satisfactory in a given situation or it should explore new options and actions. However, *a priori* it is unclear to the agent, whether this exploration will lead to actions that are below the required satisfactory performance level or improve its performance. This exploration versus exploitation dilemma is well known and intensively researched in the machine learning community and specifically around multiarmed bandit problem solvers [12–14].

17.4 Deep learning in multiagent systems

There are numerous aspects where deep learning can be utilized in autonomous agents and multiagent systems. Specifically in cooperating benevolent agents, able to change their behavior dynamically and operate independently. Fig. 17.5 illustrates different agents embedded in a shared environment. The different arrows indicate different types of learning that can overall be classified in two groups: *individual* and *collaborative* learning. *Individual* learning contains *Self*, which represents learning of the agent about itself; *Direct* learning references the agent to learn about its environment on its own without support or information from others; *Transfer* learning aims

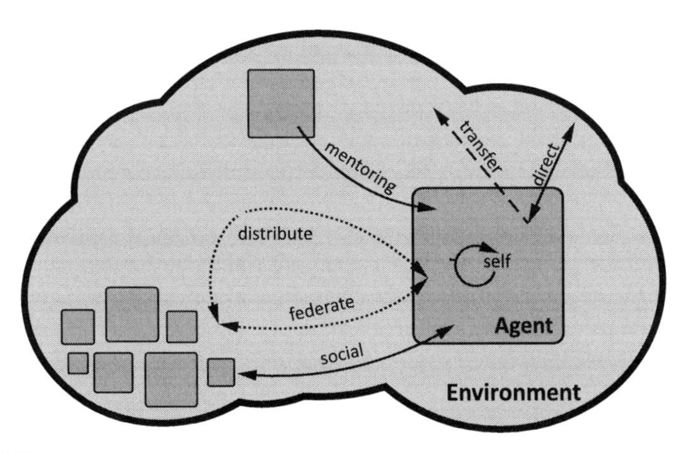

FIGURE 17.5

A selection of different types of learning in autonomous agents and multi-agent systems. Agents, illustrated as blue squares (dark gray in print version), are embedded and operate in an, potentially physical, environment. Arrows originating or ending in the environment, indicate learning or applying knowledge from the environment. Whereas arrows originating and ending within agents or next to a group of agents indicate learning from, with, or about these agents.

to apply previously learned information to a new problem or a new situation. These three types can be considered learning individually without other agent supporting or influencing the learning process. This leaves out the great potential of interactions and collaborations in multiagent systems.

The second type of *collaborative* learning involves *Mentored* learning, where one system, already having gained the required knowledge, directly supporting the learning process and "teaching" another agent; *Social* learning where both systems can observe each other and learn from each other equally; *Distributed* learning, where systems learn independently about something in their environment but utilize their inference mutually; and *federated* learning actively adjusts the knowledge bases in all participating agents based on everyone's input.

17.4.1 Individual learning

When we talk about individual learning in this chapter, we refer to learning techniques that do not rely on active interaction between two agents. The learning agent can receive feedback from the environment and utilize this feedback signal to guide its learning. This might also be through another agent. Let's assume the learner tries to improve a specific action. This action interferes with the actions of another agent prompting a different reaction of said agent. The received feedback in form of such a changed action of the other agent can be utilized as feedback signal from the environment. Key in individual learning is, there is no other agent actively trying to support, hinder, or affect the learning process of the learning agent.

17.4.1.1 Direct learning

When we discussing direct learning, we typically consider direct feedback from the environment on the agent. This represents the typical deep learning based on feedback signals. We will not go into the details of direct individual learning as this is handled at other places in the book, Chapters 2–7. While we do not dive into the details of direct learning in this chapter, we want to highlight two aspects relevant to deep learning in agents and multiagent systems. Typically, direct learning utilizes the stream of information and updates its learning continuously. This can result in selecting and executing suboptimal actions on the environment, leading to interference with the environment potentially affecting future learning. To overcome this problem, Ernst et al. [15] proposed *batch reinforcement learning*. Here, the agent is confronted with a batch of samples at once. The goal now is to find the (potentially) optimal solution in the form of a policy for the agent given the batch samples without interfering with the environment directly. Different and adaptive batch sizes enable agents to train their individual models more efficiently over time [16]. More recently, Liu et al. [17] proposed a batch learning algorithm which requires little exploration and provides strong guarantees for good results without strong prior assumptions on the distribution of the batch data.

The second aspect of direct reinforcement learning is the fact that agents are often confronted with multiple objectives, opposing each other. Agents face a trade-off between different objectives and have to find the optimal policies depending on the preferred objective. In many cases, the reward function for the different objectives is conflated and weights are being used to create a single, scalarized reward allowing the agent to select the relevant objective [18]. Respective reward functions often take the form of

$$rew = (1 - \omega) \times r_1 + \omega \times r_2 \tag{17.1}$$

where ω is the weighting factor of the reward function and r_1 and r_2 are the rewards for the first and the second objective, respectively. However, selecting the best weights to get Pareto-efficient trade-offs can be cumbersome and is often performed *a priori* by the operator [19]. Pareto active learning allows to the agent to explore the trade-off space to predict Pareto-optimal set of solutions in an iterative manner [20]. Van Moffaert et al. [21] use an iterative feedback to from multiple agents to identify local weights resulting in global, Pareto-optimal outcomes [21]. More so, their approach ensured that the outcomes are evenly spread across the Pareto-frontier allowing for a more granular and more informed selection of the weights.

While an autonomous agent wants to learn about its environment, there is also a benefit of learning about itself. There is a close relation to direct learning but with a reflectivity to it.

17.4.1.2 Learning about self

The first aspect an autonomous system might want to learn about is its direct environment and how it can be manipulated and affected by the actions of the agent. Later on or in parallel, an agent might also want to learn the limits and extent of its

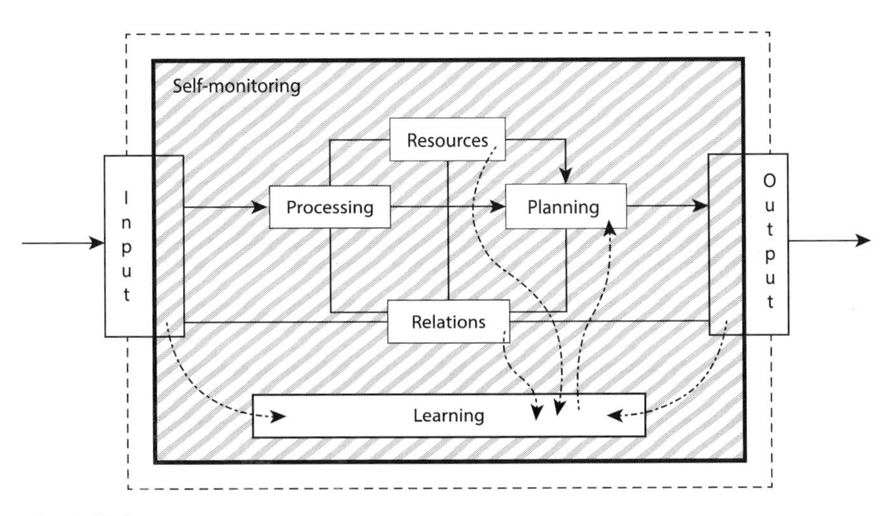

FIGURE 17.6

Learning about the self in an autonomous agent system. Core is self-monitoring, allowing the agent to monitor its own resources, processing, and planning. In a learning module, it can improve its behavior. If there are sensors and actuators (e.g., cyber-physical system), these can be incorporated in the learning. Additionally, a relation between the input and output, the resources, processing, and planning needs to be established. The results of the learning module can be incorporated in the planning of next and future actions.

own capabilities. In the literature, this is referred to as self-awareness [22,23], self-reflection [24], or situatedness [8]. A self-improving, self-aware, or self-reflective system has a common underlying goal: to learn about themselves and their actions in order to improve over time. This requires the autonomous agent to monitor its own perceived information as well as its *resources*, *planning* of its next steps, and the actions performed. Importantly, only by putting all these things in *relation*, a system can *learn* how to improve these aspects. Input and output is put into relation while considering the utilized resources and the developed plans. The information about resources and their relation to input and output is utilized in the planning stage. Learning will track the changes in input, output, their relations, and the available resources to generate better plans in the future. Fig. 17.6 shows these aspects: an agent, illustrated by the dashed rectangle, requires a self-monitoring module (hatched area) observing and monitoring the input, processing, available resources, planning and decision making as well as the generated output. Furthermore, putting all in relation and learning about these aspects might need to be monitored as well. Putting the perceived input in relation with the performed actions has been discussed as causality awareness and the ability of a system to build such causality relations based on inputs and outputs [25,26].

Different levels of consciousness and self-awareness in psychology [27] have been mapped to levels of computational self-awareness [28,29]. These levels are stimulus awareness as the lowest form, enabling systems to be aware of a stimulus;

time awareness, allowing the system to track stimuli over time; interaction awareness, allowing a system to assign the origin of the stimulus and reason about their different interactions; goal awareness, enabling systems to reason about their individual goals; and metaself-awareness, giving the system an awareness of its different levels of awareness. This last level is important when we want to enable the system to reason about the usefulness and sensitivity of a specific level. In Fig. 17.6, these different levels are embedded within the self-monitoring. What levels are utilized by the agent is either decided by the designer before deployment or by the agent using an implementation of the metaself-awareness level.

With all this, an important question arises whether a system should spend resources on learning to improve rather than performing actions and process information. This might lead to metadecision making processes that take decisions during runtime and switch resources from processing the main task to processes that enable the agent to learn about themselves. Brown proposed the B.R.A.I.N.S. model to tackle this problem [30]. Here, the system utilizes well-established approaches when resources are scarce, specifically if time or computational power can not be utilized for learning tasks. When resources become available, the system can utilize them to learn better policies and actions. One can imagine exploiting previously discussed batch reinforcement techniques for such tasks to train the models with the previously gathered data. However, the system will not replace the old models immediately but keep them in parallel and allow the system to reflect upon these new models and policies. Over time, when improved models have been established, old models and policies can be replaced for better ones and the overall performance can be improved given the goals of the agent. A remaining challenge is the autonomous validation and verification of new models and policies before replacing old ones.

17.4.1.3 Transfer learning

Transfer learning aims to solve a new task by applying knowledge developed and learned using a different, though potentially related, task [31]. This means that an agent needs to adapt its knowledge to fit the new purpose.

At this point, we do not go into the details of knowledge transfer and transfer learning as this is discussed in detail in Chapter 8 as well as in various surveys and books [31–35]. However, in this chapter, we want to highlight the important aspects of transferring knowledge in autonomous agents and multiagent systems and their respective challenges. Specifically, we consider two types of transferring knowledge. First, transferring knowledge within the same agent to achieve a new task, we call this *task transfer*. Second, transferring knowledge among two agents to accomplish the same task, we call this *cross-agent transfer*.

Transfer learning is usually considered as *inductive transfer learning*, when the domains of both tasks are the same but the task itself differs or *transductive transfer learning*, when the domains differ but the task remains the same [36]. For both cases, when an agent performs a *task transfer*, the agent needs to be able to understand the similarities of the domain and underlying task. The goal of the agent is to adapt its own knowledge using and adaptation function f_1 in order for it to fit the new task

β. Pan and Yang [36] and more recently Zhuang et al. [33] give a great overview on approaches to adapt the initial knowledge base and how to apply it to new domains.

In contrast, in *cross-agent transfer*, when the same task from the same domain is transferred among agents, there might not need to be any adaptation, as both agents are aiming to solve the same task. However, this is only correct, if the agents can have the exact same viewpoint on the task and experience the task equally. If this is not the case, there needs to be an adaptation between the different agents with respect to their individual perceptions of the task. While one might argue, that there is no difference between *cross-agent transfer* and more established *task transfer*. This is essentially correct from a standpoint of transfer learning as in both cases, knowledge needs to be adapted and adjusted for a different task. However, in *cross-agent transfer* an essential question arises about which agent should adapt the knowledge. Independent, the adapting agent is required to understand the differences of perception of the task and the domain between itself and the other agent. This requires a form of self-awareness as discussed earlier in this section as well as the understanding of the other agent receiving the knowledge [37,38]. While Lazaric et al. [39] proposed to exchange batch samples to identify the required change, this might not be feasible for multiple agents with different perceptions of the task. An alternative to this can be concurrent learning on both agents in order to identify the required change in the available knowledge.

Finally, while *task transfer* only requires an individual agent without any other agents involved, *cross-agent transfer* requires the individual agents to collaborate and interact. In the following, we will discuss collaborative approaches for deep learning in multiagent systems in more detail. An illustration of this is given in Fig. 17.7.

17.4.2 Collaborative and cooperative learning

In collaborative and cooperative machine learning in multiagent systems, an agent learns together with other rather than independently. Cooperative deep learning relies on additional input and feedback from other agents within the multiagent system. We can consider this as division of labor where we have one active learner combined with another agent. In collaborative learning on the other hand, agents collaborate on a single task and combine their skills, knowledge, and resources to learn something together [40].

17.4.2.1 Mentoring

Mentoring among autonomous agents considers at least two agents, a mentor and a mentee. The mentor represents a knowledgeable agent and an agent yet to become experienced in a task.

The active mentoring process can be considered in two ways. First, mentoring can be considered as transfer learning discussed earlier in this chapter as *cross-agent transfer*. In this case, the mentor directly and actively transfers knowledge to the mentee. However, even if the mentee is applying this expertise on the same problem, the individual situation of the mentee requires it to adjust the knowledge to the

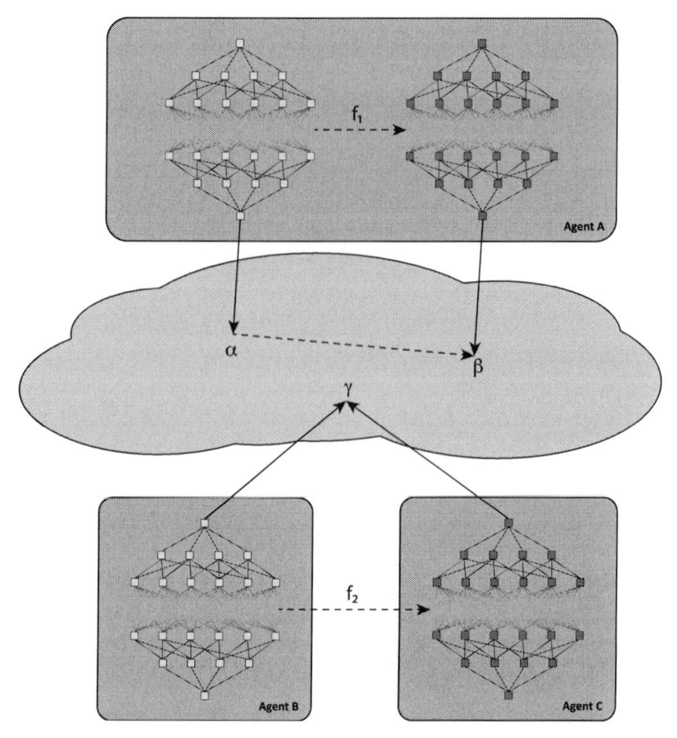

FIGURE 17.7

Two types of transfer learning to accomplish tasks within an environment. At the top, Agent A knows how to accomplishes task α and transfers this knowledge to tackle task β. In contrast, in the bottom, Agent B knows how to tackle task γ and transfers this knowledge to Agent C to enable it to tackle the same task.

problem. One may consider autonomous sensor nodes observing and experiencing the world from different viewpoints. The mentor might build up its knowledge based on this individual situation and the gained knowledge is not directly applicable from another position in the physical environment.

Second, the mentor is guiding the mentee in its learning process. Here, the mentee receives active feedback from the mentor on the learning outcomes. This process is also often seen in nature and among humans in teacher–student relationships. While biological systems cannot exchange knowledge and experiences directly, a drift in expectations, assumptions, and overall, experiences emerges. In technical systems, we are not limited by this and the mentee can potentially transfer the observed data together with the inference outcomes to receive feedback from the mentor. This requires a combination of the previously discussed *cross-agent transfer* and *task transfer* where the mentor uses task transfer to put itself into the shoes of the mentee and is able to provide feedback accordingly.

In contrast, the passive mentoring approach considers the mentor demonstrating tasks and the mentee observing these actions and replicating them in order to achieve the same outcome. However, this passive approach has several drawbacks as the mentor needs to perform an observable action and the mentee needs to be able to perceive this action. Furthermore, there are potential problems with the verification of the perceived actions and drifts of knowledge generation do to obscured and blocked information in the perception (i.e., the agent cannot "see" everything).

Using intelligent computational agents as mentors for humans, Baylor [41] argues there is a need of three main requirements for agents: (1) regulated intelligence, (2) the existence of a persona, and (3) pedagogical control. Regulated intelligence refers to the agent focusing on specific tasks or problems but not being *too* intelligent for the human to feel overwhelmed by it. Personas are required to allow humans to relate to the intelligent agent as a mentor, consider it a as an expert in the respective field and trustworthy enough to accept its teachings. Finally, pedagogical control limits the pedagogical interventions of the mentor to allow the human to learn by itself. With the mentee also being an agent, this changes slightly as the regulated intelligence is not needed to be imperfect to avoid overwhelming the mentee. Furthermore, two computational agents are not necessarily required to appeal to each other and, therefore, do not need personas. With respect to the pedagogical control, the mentor can use the actual reward information of the mentee to guide its own feedback. This can vary from simple feedback about the correctness of completing the task up to proposed changes in the adaption functions. Finally, among multiple agents operating as mentor and mentee, a question of trust arises. If mentor and mentee are not operated by the same stakeholders as is often the case in open multiagent systems, both agents have to establish a trust relationship in order to operate successfully and to avoid malevolent behavior of either the mentee or the mentor.

17.4.2.2 Social learning

In contrast to single agents interacting and forming a teacher–learner or mentoring relationship, social groups interact with each other and as a result, knowledge emerges from within the community. This is important as social learning is often considered as simply learning from another in a social context. However, in multiagent systems, any learning happens in a social context, even in individual direct learning the environment can be influenced unintendedly by other agents. Social learning, however, allows each participant of the learning process to continuously influence the generated communal knowledge. This communal outcome is often in the form of norms, rules, and values. Monitoring and enforcing these norms enables self-governance of a community [42] and even multiagent systems [43,44].

Thus, in this chapter we follow the argument of Reed et al. [45] who argues that social learning requires three fundamental changes have to occur within the participating learners:

1. The *knowledge of the individual needs to change*: for social learning to occur, the knowledge of the individual needs to change. Whether the learner itself was part

of shaping this knowledge is irrelevant. This means, a new agent can be influenced by social learning where the information has been shaped beforehand.

2. The *knowledge is situated in social community*: the knowledge does not reside in a single individual and can be shaped and influenced by every participant of the community. In the real world, there are several aspects that may affect the ability to influence the outcome of social learning such as social status or reputation of the individual. In artificial societies, such as a multiagent system, reputation and trust mechanisms can be implemented [46,47]. At the same time, agents may also contribute equally to shaping these norms and rules. If we consider social norms, they are not defined centrally and might even change with the social environment, even if the physical environment does not. For example, we treat our neighbors different than our family. Sometimes, we even treat different neighbors differently depending on previous social interactions
3. Social knowledge is *created through social interaction*: different types of social relations or interaction exist, ranging from simple movement to directed actions, intended or accidental, repeated or single actions [48]. Importantly, the social knowledge generated in the social learning process emerges from those interactions rather than a traditional feedback loop with an underlying ground truth.

In all social learning, there is an important aspect of *understanding other agents* in order to receive feedback from these agents. Both, explicit and implicit feedback can be utilized to improve the knowledge of the individual. Weber [49] defines social actions as a form of social interaction that is directed at another agent, Barnes et al. [50] map those social actions toward goal-directed actions of autonomous agents and for multiagent systems. However, in many cases, social learning in multiagent system is either restricted to single agent interactions such as described earlier in *mentoring*, allowing the learner to choose from different behavioral options [51,52], or single behavioral options but relying on multiple mentoring/teaching agents [53].

17.4.2.3 Federated learning

Federated deep learning is an approach enabling distributed and decentralized systems to share learned knowledge without sharing the underlying training data [54,55]. The approach allows to combine knowledge from several devices that started from the same knowledge base but learned and improved individually. Over time, the individual devices will share their knowledge at certain points in time to adjust and improve collaboratively. Fig. 17.8 illustrates this idea where 3 different agents are training individual networks. Through a federated learning technique and a federated aggregation function f, they can combine their experiences and generate a common network that can be used to update local networks on the individual device. Initial versions of federated learning used to share the actual gradients of the deep neural network. Using this information form a random subset among all devices, a single device (i.e., a server), would aggregate this information based on the amount of training data used to generate the respective gradients [56]. McMahan et al. [57] later showed that when starting from the same deep neural network, sharing the update weights is equivalent to sharing the gradients from a random subset of learning agents. The

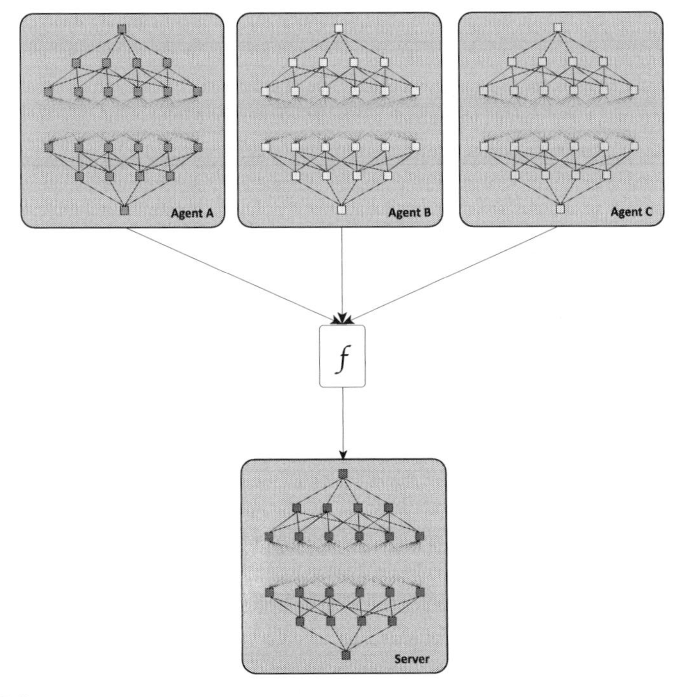

FIGURE 17.8

Federated learning with multiple networks. Three networks are combined in a single neural network using federated function f. Here, the weights of the three networks are averaged.

common averaging function looks as follows:

$$\boldsymbol{w}_{t+1} = \sum_{k=1}^{K} \frac{n_k}{n} \boldsymbol{w}_{t+1}^{k}, \tag{17.2}$$

where w_{t+1} is the common averaged weights at time $t+1$ in the common model, and n and n_k are the number of samples in total and for the individual agent k, respectively. In other words, a single device has to aggregate the weights across all agents. However, the update is proportional to the amount of samples utilized and only a subset of all agent is expected to apply the individual update steps w_{t+1}^{k}. This is a clear advantage as agents can learn locally and combine their knowledge without sharing the gradients or the underlying training data.

Since the data and the updates itself are not shared, federated learning does not assume IID data for the training among the different agents. This means each agent can train the network using its very local experience and contribute to a common network, allowing other agents to learn from experiences that they might not have encountered themselves actively. A challenging aspect, specifically for autonomous agents, is to ensure the networks are trained with the same underlying goal function and start from a common initialization.

17.4.2.4 Distributed learning and edge intelligence

Autonomous agents in a multiagent system are inherently operating in a distributed fashion and in most cases without a central controller coordinating information or the agents themselves. When considering deep learning, aspects of distributed deep learning [58,59] and edge intelligence [60–62] come to mind.

Distributed deep learning aims to reduce the amount of required processing on single devices. To achieve this, learning tasks are distributed across different devices. Deep neural networks can be split along layers or extract subnetworks entirely. While the computational burden is reduced for the individual devices, a communication overhead is introduced as the systems have to exchange the results of their calculations. Approaches such as PipeDream [63] help developers and practitioners to optimize and streamline the communication within their distributed learning system. Utilizing graph neural networks [64], discussed in Chapter 4, subnetworks or even individual nodes can be utilized to be distributed among multiple devices or agents [65]. While the distribution of work among multiple agents has substantial benefits, the coordination of the relevant work is needs to be defined *a priori*. This can become problematic when the available resources vary over time and across the agents and require dynamic assignment during runtime. Such a change can lead to overheads in communication and coordination, making the initially gained benefit in performance obsolete.

An alternative to simply distributing tasks among agents is to distribute networks with varying depth to different devices [66]. The different depths of networks introduce a trade-off between required resources and accuracy of the network on the inference task [67]. Devices with little resources available, utilize shallow networks to perform the inference tasks. If the inference task does not prove the expected or required confidence, deeper networks are consulted to achieve higher confidence. In order to minimize the training and communication effort, we can consider networks with different depths to utilize common core structures and train as one network with different exit branches initially. We then distribute the common core and the individual branches to different devices based on their available resources. This allows the device to only communicate the information resulting at the layer where the different networks diverge. An illustration for this is given in Fig. 17.9.

Following this idea, we can also partition larger networks and distributed it across the different devices and perform pre-processing locally before forwarding the information to devices with more processing power. After larger parts of the network have been processed, the last step could be done by the actual actuator, making very local decision and potentially overcome localization problems arising from non-IID data [62]. An example of splitting a deep neural network and distributing the workload to the edge and the cloud is given in Fig. 17.10. Utilizing pooling layers before transmission can help to reduce the amount of data transferred between different devices even further [68,61]. When agents are distributed across multiple physical

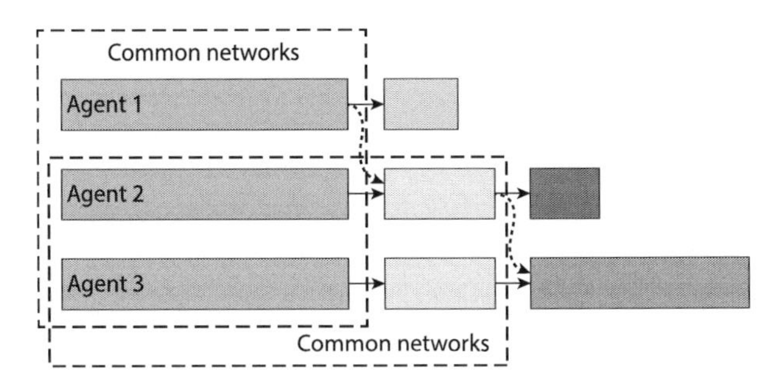

FIGURE 17.9

Multibranched network distributed over multiple agents and/or devices. Each line represents a single agent and its network and each block in the line illustrates a block of the network. The first block (blue) is common across all agents while the second and third agent share an additional part (light green). Each agent has its unique end of the network (i.e., lower layers in the network). If results of one network are insufficient, agents can forward their intermediate results from common network blocks to other agents (dashed arrow). (For interpretation of the colors in the figure, the reader is referred to the web version of this chapter.)

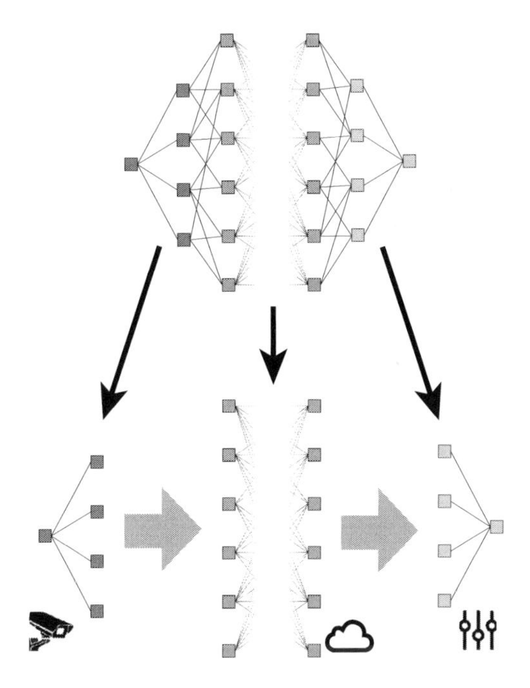

FIGURE 17.10

Edge intelligence and distributed learning using deep neural networks as an example. The network is split into different parts and outcomes are utilized as needed. Here, the sensor, a camera, performs some preprocessing before transferring the initial analysis to the cloud. In the cloud, the core work is performed utilizing the vast resources. Results are forwarded to the controller allowing for individual action selection based on the local device.

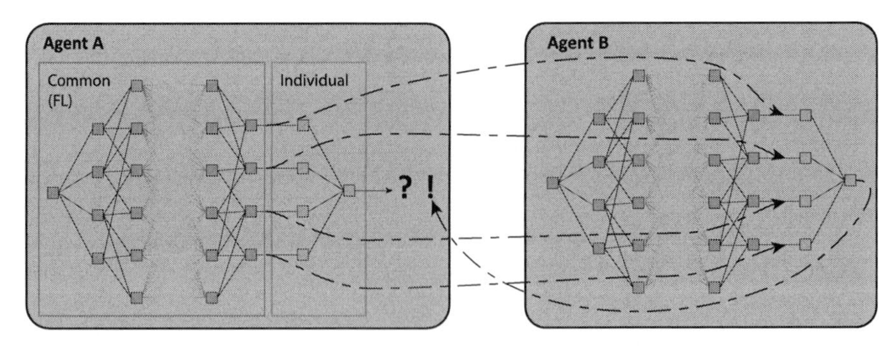

FIGURE 17.11

Collaborative deep learning among two agents. Both agents share a core part of the network and specialize individually at the end of the network. If the inference task does not achieve expected performance, the agent can request feedback from another agent to refine, confirm, or confute the result (dashed arrows). (For interpretation of the colors in the figure, the reader is referred to the web version of this chapter.)

devices, such an approach allows for localized preprocessing and post-processing. The outcome is highly dependent on the initial network and the final part of the network. The core of the network will in most cases be shared through a single server infrastructure (i.e., the cloud).

When taking this further and involve previously discussed approaches such as federated learning, social learning, and self-reflection, we can consider different future directions enabling autonomous agents to capitalize on distributed learning. One approach could be in the direction to utilize federated learning to learn common network parts and utilize local networks to perform individual inference steps [69]. This reduces individual learning efforts and even allows for common updates of the core part of the network. Exploiting the trade-off between deeper and shallower individual networks, individual devices can achieve more accurate, but more expensive, inference or faster, but potentially inaccurate, inference outcomes. By employing self-learning techniques, enabling the agent to reason about its resources, this decision could be made before the agent performs the inference task at hand. Additionally, in a collaborative multiagent setting, individual agents could collaborate in order to refine their individual inference outcome. This is illustrated in Fig. 17.11 where Agent A requests feedback from Agent B. Because both agents share a common network, Agent B only receives outcomes from the initial common network (dashed arrow) and only needs to execute the remaining individual part of the network (in orange).

17.5 Conclusion

In this chapter, we analyzed the different aspects of deep learning in autonomous agents and specifically multiagent systems. We identified the different properties of the environment and clarified that various settings may bring about different challenges to the agents. Specifically, open environments, where agents may join at any time can be come a pitfall as new agents may influence the environment unexpectedly. Furthermore, we also argued about the different properties of the agents themselves, their capabilities, and how they behave toward others.

When discussing the challenges, we focused on non-IID data and its consequences for multiagent systems. We recapped and summarized the different types of data also affecting agents operating in a common environment. We further briefly discussed the challenges coming about cyber-physical implementations of autonomous multiagent systems and their learning approaches.

Coming to deep learning in multiagent systems, we differentiated between agents learning *alone*, without the help or support of others, and the cooperative approaches, where agents work *together*. Throughout the discussion it became apparent that this distinction is not always easy or even entirely possible. For example, when discussing transfer learning, single agents may transfer knowledge across tasks but multiple agents might also transfer across the agents. Furthermore, there is a close relationship between different collaborative approaches such as social, federated, and distributed learning. In many cases, the distinction of the individual approaches is not always clear.

However, there are several hard challenges remaining when it comes to deep learning in multiagent systems:

- **Deliberate and random interaction:** Without explicit communication, an agent can only make assumptions whether another agent interacts deliberately or whether the other agents actions only contribute to the own goals by chance. To determine deliberate collaboration and interaction, individual agents will require the ability to dynamically evaluate these interactions. However, due to the infinite number of potential interactions, the agent needs to be able to develop such evaluation autonomously during runtime. Utilizing building blocks for such tests can be one step toward the right direction, but requires individual, per-agent setup. Understanding whether an interaction is accidental or deliberate can be important as accidental interaction can affect the other agent negatively. While a self-interested agent might ignore that fact and benefit briefly, it can have negative impacts in the long run.
- **Active cooperation**: The ability of an agent to decide to deliberately collaborate with another agent is directly related to the previous point. While such help and collaboration can be the by-product while pursuing own goals, active cooperation can have long-term benefits. However, an agent performs active cooperation by performing additional tasks for the other agents (i.e., tasks that do not progress its own goals), will potentially affect its own goals negatively. An agent might support another with the expectation of compensations but even if the agents can agree on this, it might be difficult to enforce such agreements in open systems with agents from different stakeholders.
- **Adversarial agents:** In open environments with self-interested agents, individuals need to be able to avoid exploitation through others. Being able to identify

interactions is already hard; however, enabling an agent to identify malevolent behavior becomes even more complex. While blunt malevolent or adversarial behavior might be easy to detect, an agent might also deliberately deceive other agents before performing malevolent actions.

- **Goal development and selection**: In open multiagent systems, agents might work indefinitely toward their individual goals. But what happens to the agents when they achieve their goals and would become idle. Can we enable agents to make decisions on when they do not need to work toward their goal any longer? And even when they are not finished with their tasks, can we allow agents to decide to pursue other goals in-between? Can we devise approaches that allow the agent to diverge from its primary task without being deliberately told to do so by a stakeholder in order to support other agents or a social community? An approach for this can be explainable AI with a human in the loop where the agent requests feedback from the human about potential changes in individual goals.

References

[1] Edward A. Lee, Cyber physical systems: design challenges, in: Proceedings of the International Symposium on Object and Component-Oriented Real-Time Distributed Computing, 2008, pp. 363–369.

[2] Lukas Esterle, Radu Grosu, Cyber-physical systems: challenge of the 21st century, e & i Elektrotechnik und Informationstechnik 133 (7) (2016) 299–303.

[3] Kirstie Bellman, Christopher Landauer, Nikil Dutt, Lukas Esterle, Andreas Herkersdorf, Axel Jantsch, Nima TaheriNejad, Peter R. Lewis, Marco Platzner, Kalle Tammemäe, Self-aware cyber-physical systems, ACM Transactions on Cyber-Physical Systems 4 (4) (2020).

[4] Michael Wooldridge, An Introduction to Multiagent Systems, John Wiley & Sons, 2009.

[5] Kirstie Bellman, Sven Tomforde, Rolf P. Würtz, Interwoven systems: self-improving systems integration, in: Proceedings of the International Conference on Self-Adaptive and Self-Organizing Systems Workshops, IEEE Computer Society, USA, 2014, pp. 123–127.

[6] Kirstie L. Bellman, Jean Botev, Ada Diaconescu, Lukas Esterle, Christian Gruhl, Christopher Landauer, Peter R. Lewis, Anthony Stein, Sven Tomforde, Rolf P. Würtz, Self-improving system integration – status and challenges after five years of SISSY, in: Proceedings of the International Conference on Self-Organizing and Self-Adaptive Systems Workshop, 2018, pp. 160–167.

[7] Bernhard Rinner, Wayne Wolf, An introduction to distributed smart cameras, Proceedings of the IEEE 96 (10) (2008) 1565–1575.

[8] Matthew Costello, Situatedness, Springer New York, New York, NY, 2014, pp. 1757–1762.

[9] Kevin Hsieh, Amar Phanishayee, Onur Mutlu, Phillip B. Gibbons, The non-IID data quagmire of decentralized machine learning, CoRR, arXiv:1910.00189 [abs], 2019.

[10] Peter Kairouz, H. Brendan McMahan, Brendan Avent, Aurélien Bellet Mehdi Bennis, Arjun Nitin Bhagoji, Keith Bonawitz, Zachary Charles, Graham Cormode, Rachel Cummings, Rafael G.L. D'Oliveira, Salim El Rouayheb, David Evans, Josh Gardner, Zachary Garrett, Adrià Gascón, Badih Ghazi, Phillip B. Gibbons, Marco Gruteser, Zaïd Harchaoui, Chaoyang He, Lie He, Zhouyuan Huo, Ben Hutchinson, Justin Hsu, Martin Jaggi,

Tara Javidi, Gauri Joshi, Mikhail Khodak, Jakub Konecný, Aleksandra Korolova, Farinaz Koushanfar, Sanmi Koyejo, Tancrède Lepoint, Yang Liu, Prateek Mittal, Mehryar Mohri, Richard Nock, Ayfer Özgür, Rasmus Pagh, Mariana Raykova, Hang Qi, Daniel Ramage, Ramesh Raskar, Dawn Song, Weikang Song, Sebastian U. Stich, Ziteng Sun, Ananda Theertha Suresh, Florian Tramèr, Praneeth Vepakomma, Jianyu Wang, Li Xiong, Zheng Xu, Qiang Yang, Felix X. Yu, Han Yu, Sen Zhao, Advances and open problems in federated learning, CoRR, arXiv:1912.04977 [abs], 2019.

[11] Chloe M. Barnes, Lukas Esterle, John N.A. Brown, "When you believe in things that you don't understand": the effect of cross-generational habits on self-improving system integration, in: 2019 IEEE 4th International Workshops on Foundations and Applications of Self* Systems (FAS* W), IEEE, 2019, pp. 28–31.

[12] Donald A. Berry, Bert Fristedt, Bandit Problems: Sequential Allocation of Experiments, Chapman and Hall, London, 1985.

[13] Peter Auer, Nicolo Cesa-Bianchi, Paul Fischer, Finite-time analysis of the multiarmed bandit problem, Machine Learning 47 (2–3) (2002) 235–256.

[14] Matteo Gagliolo, Jürgen Schmidhuber, Algorithm portfolio selection as a bandit problem with unbounded losses, Annals of Mathematics and Artificial Intelligence 61 (2) (2011) 49–86.

[15] Damien Ernst, Pierre Geurts, Louis Wehenkel, Tree-based batch mode reinforcement learning, Journal of Machine Learning Research 6 (18) (2005) 503–556.

[16] Sascha Lange, Thomas Gabel, Martin Riedmiller, Batch Reinforcement Learning, Springer, Berlin Heidelberg, 2012, pp. 45–73.

[17] Yao Liu, Adith Swaminathan, Alekh Agarwal, Emma Brunskill, Provably good batch reinforcement learning without great exploration, in: Advances in Neural Information Processing Systems, vol. 33, Curran Associates, Inc., 2020, pp. 1264–1274.

[18] Yaochu Jin, Multi-Objective Machine Learning, vol. 16, Springer Science & Business Media, 2006.

[19] Indraneel Das, John E. Dennis, A closer look at drawbacks of minimizing weighted sums of objectives for Pareto set generation in multicriteria optimization problems, Structural Optimization 14 (1) (1997) 63–69.

[20] Marcela Zuluaga, Guillaume Sergent, Andreas Krause, Markus Püschel, Active learning for multi-objective optimization, in: Proceedings of the International Conference on Machine Learning, 2013, pp. 462–470.

[21] Kristof Van Moffaert, Tim Brys, Arjun Chandra, Lukas Esterle, Peter R. Lewis, Ann Nowé, A novel adaptive weight selection algorithm for multi-objective multi-agent reinforcement learning, in: Proceedings of the International Joint Conference on Neural Networks, IEEE, 2014, pp. 2306–2314.

[22] Samuel Kounev, Peter Lewis, Kirstie L. Bellman, Nelly Bencomo, Javier Camara, Ada Diaconescu, Lukas Esterle, Kurt Geihs, Holger Giese, Sebastian Götz, et al., The notion of self-aware computing, in: Self-Aware Computing Systems, Springer, 2017, pp. 3–16.

[23] Peter R. Lewis, Arjun Chandra, Shaun Parsons, Edward Robinson, Kyrre Glette, Rami Bahsoon, Jim Torresen, Xin Yao, A survey of self-awareness and its application in computing systems, in: Proceedings of the International Conference on Self-Adaptive and Self-Organizing Systems Workshops, IEEE, 2011, pp. 102–107.

[24] Sven Tomforde, Jörg Hähner, Sebastian von Mammen, Christian Gruhl, Bernhard Sick, Kurt Geihs, "Know thyself"-computational self-reflection in intelligent technical systems, in: Proceedings of the International Conference on Self-Adaptive and Self-Organizing Systems Workshops, IEEE, 2014, pp. 150–159.

[25] Judea Pearl, Dana Mackenzie, The Book of Why, Basic Books, New York, NY, 2018.
[26] Judea Pearl, The seven tools of causal inference, with reflections on machine learning, Communications of the ACM 62 (3) (2019) 54–60.
[27] Alain Morin, Levels of consciousness and self-awareness: a comparison and integration of various neurocognitive views, Consciousness and Cognition 15 (2) (2006) 358–371.
[28] Peter Lewis, Kirstie L. Bellman, Christopher Landauer, Lukas Esterle, Kyrre Glette, Ada Diaconescu, Holger Giese, Towards a framework for the levels and aspects of self-aware computing systems, in: Self-Aware Computing Systems, Springer International Publishing, 2017, pp. 51–85.
[29] Tao Chen, Funmilade Faniyi, Rami Bahsoon, Peter R. Lewis, Xin Yao, Leandro L. Minku, Lukas Esterle, The handbook of engineering self-aware and self-expressive systems, arXiv preprint, arXiv:1409.1793, 2014.
[30] John N.A. Brown, Psychology and Neurology: The Surprisingly Simple Science of Using Your Brain, Springer International Publishing, Cham, 2016, pp. 103–118.
[31] Lisa Torrey, Jude Shavlik, Transfer learning, in: Handbook of Research on Machine Learning Applications and Trends: Algorithms, Methods, and Techniques, IGI Global, 2010, pp. 242–264.
[32] Chuanqi Tan, Fuchun Sun, Tao Kong, Wenchang Zhang, Chao Yang, Chunfang Liu, A survey on deep transfer learning, in: Proceedings of the International Conference on Artificial Neural Networks, Springer, 2018, pp. 270–279.
[33] Fuzhen Zhuang, Zhiyuan Qi, Keyu Duan, Dongbo Xi, Yongchun Zhu, Hengshu Zhu, Hui Xiong, Qing He, A comprehensive survey on transfer learning, arXiv preprint, arXiv: 1911.02685, 2019.
[34] Leandro L. Minku, Transfer learning in non-stationary environments, in: Learning from Data Streams in Evolving Environments, Springer, 2019, pp. 13–37.
[35] Zhuangdi Zhu, Kaixiang Lin, Jiayu Zhou, Transfer learning in deep reinforcement learning: a survey, arXiv preprint, arXiv:2009.07888, 2020.
[36] Sinno Jialin Pan, Qiang Yang, A survey on transfer learning, IEEE Transactions on Knowledge and Data Engineering 22 (10) (2009) 1345–1359.
[37] Lukas Esterle, John N.A. Brown, Levels of networked self-awareness, in: 2018 IEEE 3rd International Workshop on Foundations and Applications of Self* Systems (FAS* W), IEEE, 2018, pp. 237–238.
[38] Lukas Esterle, John N.A. Brown, I think therefore you are: models for interaction in collectives of self-aware cyber-physical systems, ACM Transactions on Cyber-Physical Systems 4 (4) (2020).
[39] Alessandro Lazaric, Marcello Restelli, Andrea Bonarini, Transfer of samples in batch reinforcement learning, in: Proceedings of the International Conference on Machine Learning, 2008, pp. 544–551.
[40] Pierre Dillenbourg (Ed.), Collaborative Learning: Cognitive and Computational Approaches, Advances in Learning and Instruction Series, second ed., Emerald Publishing, Limited, 1999.
[41] Amy Baylor, Beyond butlers: intelligent agents as mentors, Journal of Educational Computing Research 22 (4) (2000) 373–382.
[42] Elinor Ostrom, Governing the Commons: The Evolution of Institutions for Collective Action, Cambridge University Press, 1990.
[43] Alexander Artikis, Marek Sergot, Jeremy Pitt, Specifying norm-governed computational societies, ACM Transactions on Computational Logic 10 (1) (2009).

[44] Jeremy Pitt, Interactional justice and self-governance of open self-organising systems, in: Proceedings of the International Conference on Self-Adaptive and Self-Organizing Systems, IEEE, 2017, pp. 31–40.

[45] Mark S. Reed, Anna C. Evely, Georgina Cundill, Ioan Fazey, Jayne Glass, Adele Laing, Jens Newig, Brad Parrish, Christina Prell, Chris Raymond, et al., What is social learning?, Ecology and Society 15 (4) (2010).

[46] Trung Dong Huynh, Nicholas R. Jennings, Nigel R. Shadbolt, An integrated trust and reputation model for open multi-agent systems, Autonomous Agents and Multi-Agent Systems 13 (2) (2006) 119–154.

[47] Sarah Edenhofer, Sven Tomforde, Jan Kantert, Lukas Klejnowski, Yvonne Bernard, Jörg Hähner, Christian Müller-Schloer, Trust communities: an open, self-organised social infrastructure of autonomous agents, in: Trustworthy Open Self-Organising Systems, Springer, 2016, pp. 127–152.

[48] Piotr Sztompka, Socjologia – Analiza Społeczeństwa, Znak, 2002, 656 pp.

[49] Max Weber, Economy and Society: An Outline of Interpretive Sociology, vol. 1, Univ. of California Press, 1978.

[50] Chloe M. Barnes, Anikó Ekárt, Peter R. Lewis, Social action in socially situated agents, in: Proceedings of the International Conference on Self-Adaptive and Self-Organizing Systems, IEEE, 2019, pp. 97–106.

[51] Ben P. Jolley, James M. Borg, Alastair Channon, Analysis of social learning strategies when discovering and maintaining behaviours inaccessible to incremental genetic evolution, in: Proceedings of the International Conference on Simulation of Adaptive Behavior, Springer, 2016, pp. 293–304.

[52] Jason Noble, Daniel W. Franks, Social learning in a multi-agent system, Computing and Informatics 22 (6) (2012) 561–574.

[53] Jean Oh, Stephen F. Smith, A few good agents: multi-agent social learning, in: Proceedings of the International Joint Conference on Autonomous Agents and Multiagent Systems, 2008, pp. 339–346.

[54] Jakub Konečnỳ, Brendan McMahan, Daniel Ramage, Federated optimization: distributed optimization beyond the datacenter, arXiv preprint, arXiv:1511.03575, 2015.

[55] Jakub Konečný, H. Brendan McMahan, Felix X. Yu, Peter Richtarik, Ananda Theertha Suresh, Dave Bacon, Federated learning: strategies for improving communication efficiency, in: NIPS Workshop on Private Multi-Party Machine Learning, 2016, URL https://arxiv.org/abs/1610.05492.

[56] Reza Shokri, Vitaly Shmatikov, Privacy-preserving deep learning, in: Proceedings of the Conference on Computer and Communications Security, 2015, pp. 1310–1321.

[57] Brendan McMahan, Eider Moore, Daniel Ramage, Seth Hampson, Blaise Aguera y Arcas, Communication-efficient learning of deep networks from decentralized data, in: Artificial Intelligence and Statistics, PMLR, 2017, pp. 1273–1282.

[58] Tal Ben-Nun, Torsten Hoefler, Demystifying parallel and distributed deep learning: an in-depth concurrency analysis, ACM Computing Surveys (CSUR) 52 (4) (2019) 1–43.

[59] Matthias Langer, Zhen He, Wenny Rahayu, Yanbo Xue, Distributed training of deep learning models: a taxonomic perspective, IEEE Transactions on Parallel and Distributed Systems 31 (12) (2020) 2802–2818.

[60] Chaoyun Zhang, Paul Patras, Hamed Haddadi, Deep learning in mobile and wireless networking: a survey, IEEE Communications Surveys and Tutorials 21 (3) (2019) 2224–2287.

[61] Zhi Zhou, Xu Chen, En Li, Liekang Zeng, Ke Luo, Junshan Zhang, Edge intelligence: paving the last mile of artificial intelligence with edge computing, Proceedings of the IEEE 107 (8) (2019) 1738–1762.

[62] Jiasi Chen, Xukan Ran, Deep learning with edge computing: a review, Proceedings of the IEEE 107 (8) (2019) 1655–1674.

[63] Aaron Harlap, Deepak Narayanan, Amar Phanishayee, Vivek Seshadri, Nikhil Devanur, Greg Ganger, Phil Gibbons, Pipedream: fast and efficient pipeline parallel dnn training, arXiv preprint, arXiv:1806.03377, 2018.

[64] Michael M. Bronstein, Joan Bruna, Yann LeCun, Arthur Szlam, Pierre Vandergheynst, Geometric deep learning: going beyond Euclidean data, IEEE Signal Processing Magazine 34 (4) (2017) 18–42.

[65] Simone Scardapane, Indro Spinelli, Paolo Di Lorenzo, Distributed graph convolutional networks, arXiv preprint, arXiv:2007.06281, 2020.

[66] Surat Teerapittayanon, Bradley McDanel, Hsiang-Tsung Kung, Branchynet: fast inference via early exiting from deep neural networks, in: Proceedings of the International Conference on Pattern Recognition, IEEE, 2016, pp. 2464–2469.

[67] Chun-Fu Chen, Quanfu Fan, Neil Mallinar, Tom Sercu, Rogerio Feris, Big-little net: an efficient multi-scale feature representation for visual and speech recognition, arXiv preprint, arXiv:1807.03848, 2018.

[68] Surat Teerapittayanon, Bradley McDanel, Hsiang-Tsung Kung, Distributed deep neural networks over the cloud, the edge and end devices, in: Proceedings of the International Conference on Distributed Computing Systems, IEEE, 2017, pp. 328–339.

[69] Xinqian Zhang, Ming Hu, Jun Xia, Tongquan Wei, Mingsong Chen, Shiyan Hu, Efficient federated learning for cloud-based AIoT applications, IEEE Transactions on Computer-Aided Design of Integrated Circuits and Systems (2020).

CHAPTER

Simulation environments 18

Charalampos Symeonidis and Nikos Nikolaidis
Department of Informatics, Aristotle University of Thessaloniki, Thessaloniki, Greece

18.1 Introduction

Simulation is a highly interdisciplinary field, widely used in most academic research domains, including engineering, computer science, economics, as well as in industrial experimentation and testing. Particularly in robotics, simulation has been established as an important tool, allowing to study and experiment with the structure, characteristics, and function of a robotic system. Until the early 2000s, robotic simulators were restricted to 2D environments, without being able to fully simulate the complexity of the real world. A typical example is CARMEN [1], a collection of open-source software modules for mobile robot control, which provides navigation functions for base control, sensor reading and logging, localization, path planning, mapping, and obstacle avoidance. The simulator is capable of performing 2D simulations as it considers only 2D kinematic models. MissionLab [2] is another 2D simulator, which takes high-level military-style plans and executes them with teams of real or simulated robotic vehicles. MissionLab allows the control and interaction of multiple robots in both simulation and actual robotics platforms.

The need to simulate more complex robotic systems with multiple sensors (e.g., camera, GPS, Lidar) in 3D realistic environments along with the advances in computer hardware and computer graphics, pushed for the development of visually and physically realistic simulators. SimRobot [3] was one of the first simulators able to simulate arbitrary user-defined robots in three-dimensional space, while also including a physical model based on rigid body dynamics. Similarly, UberSim [4] is another early 3D simulation framework focused on more vision-centric robotic systems. One of the most popular and highly established robotic simulation frameworks is Gazebo [5]. Over the years, Gazebo has managed to form an active community that enabled it to evolve and mature. In addition to its physics and graphics fidelity, Gazebo provides the necessary tools for the easy integration of robotic systems, designed in the Robot Operating System (ROS) [6], to the simulation framework.

Recent advances in robotics and the major leap in the performance of perception, cognition and decision-making algorithms due to deep learning, improved the way robotic systems understand and interact with their environments, thus reaching high autonomy levels. Simulators that are specific for the development of autonomous systems such as autonomous cars [7] and UAVs [8] were also introduced, harnessing the

Deep Learning for Robot Perception and Cognition. https://doi.org/10.1016/B978-0-32-385787-1.00023-3

progress of the current generation gaming platforms [9], [10] regarding photorealistic graphics and high-fidelity physics. These simulators enabled the automatic generation of realistic and large scale synthetic data [11], suitable for use in training deep learning algorithms.

18.1.1 Robotic simulators architecture

Modern robotic simulators tend to follow a common architecture, which is depicted in Fig. 18.1, consisting of the following components:

- **Graphics/rendering and physics engines:** A graphics/rendering engine is the software that provides the user with high-level rendering functions, such as mesh drawing, scene graph, camera handling, shading, etc. A typical example of an open-source rendering engine is OGRE [12]. A physics engine simulates physical interactions of objects in the scene. This simulation is used to update the visual representation of an object in the rendering engine, frame-by-frame. ODE [13] and BULLET [14] are two open-source physics engines that provide high-fidelity physics. Some modern simulators rely on high-end game engines such as Unreal Engine 4 (UE4) [9] or Unity [10] that provide high-level physical fidelity, cutting-edge graphics, distributed architectures, scriptable environments, and continuous evolution and support.
- **Simulation framework:** A simulation framework is the software that binds the rendering engine and the physics engine together into a usable software and also provides facilities pertinent to robotics simulation such as robotic models, sensors, and interfaces. In most cases, Application Programming Interfaces (APIs) have been established as a way of interacting with the robot or with the simulation in

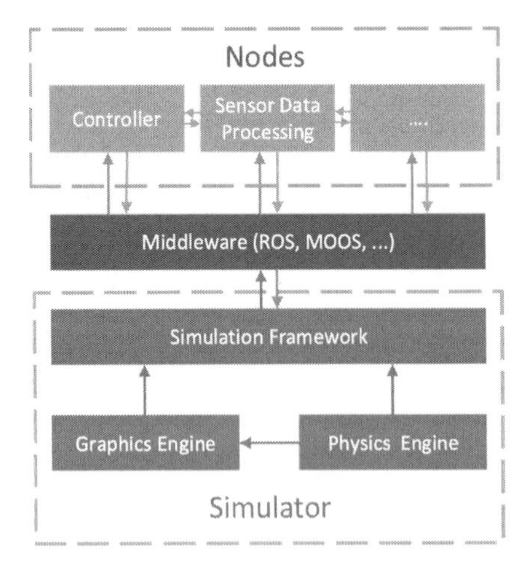

FIGURE 18.1

A typical structure of a simulator.

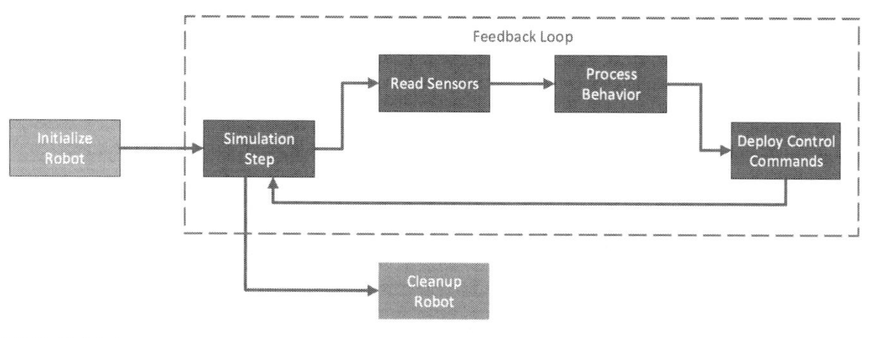

FIGURE 18.2

Diagram of a typical simulation feedback loop.

general. The process of controlling a robotic system can be summarized as a high-frequency repeating loop, which consists of retrieving information from sensors, processing the behavior of the robotic system based on the sensor input, and finally sending control commands back to the robotic system. A diagram of this process is depicted in Fig. 18.2. APIs provide the means to both retrieve information from sensors and to assign control commands.

The implementation of a robotic system in a simulator must be accomplished with as minimal effort as possible. Most of the times, robotic systems use a robotic middleware for their implementation in real-life scenarios. Thus, many simulators support the use of such robotic middleware by usually implementing it as a layer above their existing APIs, through wrapper libraries. A robotic system implemented through a middleware in the simulator, ideally makes it independent from it. Robot Operating System (ROS) [6], Mission Oriented Operating Suite (MOOS) [15], and Yet Another Robot Platform (YARP) [16] are typical examples of robotic middleware. ROS, the most widely used middleware, allows the creation of nodes, each of which may be a controller for a particular device. The nodes communicate through a common message-passing interface often without prior knowledge of each other. In the case of simulation, the simulator itself is often treated as a separate node, allowing thus the control algorithms to be independent of the selected simulation software. In addition, different control algorithms under consideration may be written into separate nodes, allowing them to be swapped and interchanged with ease.

18.1.2 Simulation types

The design, implementation, and testing of robotic models in real world are costly, time-consuming, complicated, and often not reproducible. Simulation is a valuable tool for easing those processes, allowing the integration of tests with the physical platforms to be carried out only in the last stages of their development. By applying virtual simulation technologies at different abstraction levels, several types of simulation can be defined. Two of the most common are:

- **Software-In-the-Loop (SIL)** [17] simulation, evaluates the implementation of a developed system on general-purpose hardware, instead of the specific hardware where the system is purposed to run (e.g., embedded systems, industrial computers, etc.). Usually SIL simulation allows to quickly test algorithms, as the simulation speed is much faster than the speed of the real world. Though the SIL simulation can't reach high levels of credibility, it is a valuable approach for testing a system in the earlier stages of its development.
- **Hardware-In-the-Loop (HIL)** [18] simulation aims to reproduce the real operating environment of the system as faithfully as possible. This is done by using real-time simulation computers and running the robotic system on the exact hardware it is purposed to be deployed. HIL simulation provides much more credibility that SIL simulation and is usually performed in the later development and testing stages.

18.1.3 Qualitative characteristics

Selecting a simulation framework for a specific task is not as straightforward as one might think. Below we highlight some important aspects for consideration:

- **Graphical fidelity:** The visual accuracy and realism of the simulator. This characteristic is of great importance for robotic systems that make use of computer vision methods, or when the simulator is to be used for the generation of data sets involving visual data (images, videos).
- **Physical fidelity:** The fidelity in simulating the physics of the simulated environment, including the robotic system. Usually robotic simulators offer solutions for computing rigid body dynamics and detecting and handing collisions and contacts.
- **Level of Graphical User Interface (GUI):** Whether the simulator provides user-friendly approaches for user-interaction (e.g., editors, menus, etc.) and not only command-line based interaction.
- **License:** Simulators are distributed under various software licenses. Free software licenses, such as Apache, MIT, and GNU GPL licenses, pose few to no restrictions to end-user regarding the use, modification, and redistribution of the software. On the other hand, simulators may be distributed under non-commercial or proprietary licenses. In some cases, a simulator's modules, libraries or assets may have separate licenses, of different types.
- **Operating system:** Some simulators are designed for specific systems, while other provide cross-platform support. A special category are simulators running on the cloud, attaining implicitly cross-platform support.
- **Available content:** Some simulators tend to provide rich content, including a variety of sensors, robotic models, 3D/2D assets such as object and terrain models, and even predefined scenarios, while others rely more on 3rd party content. Commonly, 3D assets in a scene are polygon meshes for representing objects and terrains, where textures (2D images) are used to define their appearance. 3D objects can be defined as Filmbox (.fbx), Collada (.dae), Wavefront OBJ (.obj), STL (.stl), and PLY Polygon format (.ply) files. A robotic model, compared to other

3D objects in a scene, follows a more complex structure as it is usually defined as an articulated object, consisting of many rigid body entities connected with joints. Sensors can also be attached to the model. Each rigid body entity and joint can have separate visual and physical properties. Their architecture and properties (e.g., physical, visual, and collision properties) are usually described in XML format files such as the SDF[1] and URDF[2] file formats.
- **Content generation tools:** A simulator may provide built-in tools for creating robotic models from scratch, or rely on generating content on 3rd party software (e.g., Blender[3]) and just provide tools for importing them.
- **Level of maturity:** Whether the simulator is already widely used and validated.
- **External Agent Support:** In order to control robots/objects using agent-based methodologies [19] (e.g., controlling pedestrians and traffic in autonomous car simulations), the simulator should feature a distributed architecture at the control level.
- **Scripting Language:** Simulators usually allow simulation interaction through a programming language (e.g., C/C++, Python, MATLAB®, etc.).
- **Robotic Middleware Support:** Some simulators provide the necessary tools and libraries to link themselves with common robotic frameworks (ROS, YARP, MOOS, etc.).
- **HIL Simulation Support:** Obviously, support of HIL simulation provides greater advantages over simulators that support only SIL simulation.

18.2 Robotic simulators

In this section a wide range of robotic simulators will be covered. In Table 18.1, technical details of the described simulation environments are shown. Pros and cons of five established simulators are described in Table 18.2.

18.2.1 Gazebo

Gazebo [20], [5] is one of the most popular open-source simulation platforms for academic and research purposes in the robotics domain. It was initially developed as part of the Player/Stage [21] project. During the project, a networked device server named Player and a simulator named Stage were developed. Player was designed to provide an interface to a variety of robot and sensor hardware, while Stage was developed for simulating large populations of mobile robots mostly in indoor environments. The need to perform simulations for robotic vehicles in outdoor environments, reproducing accurately all the dynamics a robot may encounter, led to the development of

[1] http://sdformat.org/.
[2] https://github.com/ros/urdf.
[3] https://www.blender.org/.

Table 18.1 Features summary of the robotic simulators covered in this chapter.

Simulation environment	Operating system	License	Programming language	API support	Physics engine	Graphics engine	ROS \ ROS 2 support
Gazebo	Linux	Apache 2.0	C++	C++	ODE, DART, Simbody, BULLET Physics	OGRE	Yes
AirSim	Windows, Linux, MacOS	MIT License	C++	C++, Python	Unreal Engine 4, Unity	Unreal Engine 4, Unity	Yes*
Webots	Windows, Linux, MacOS	Apache License 2.0	C/C++	C/C++, Python, Java, MATLAB	ODE	WREN	Yes
CARLA	Windows, Linux	Code: MIT License Assets: CC-BY License	C++	C++, Python	Unreal Engine 4	Unreal Engine 4	Yes*
CoppeliaSim	Windows, Linux, MacOS	Dual licensed (GPL or commercial)	Lua, C/C++	C/C++, Python, Java, MATLAB, Lua, Octave	BULLET Physics, ODE, Vortex Studio, Newton Dynamics	Custom OpenGL3 rendering engine	Yes
MORSE	Windows, Linux, MacOS	3-clause BSD license	Python	Python	BULLET	Blender Game engine	Yes*
ARGoS	Linux, MacOS	MIT license	C++	C++	Custom physics engine	Custom rendering engine	Yes*
USARSim	Windows, Linux, MacOS	GNU GPLv2	–	C/C++, Java	Karma physics Engine	Unreal Engine 2	Yes*
Isaac Sim	Linux	–	–	C, C♯, Python	Unity, Unreal Engine 4	Unity, Unreal Engine 4	Yes*
RoboDK	Windows, Linux, MacOS Android	Commercial license, free trialware license	Python	Python, C♯, C++, MATLAB	No	Custom OpenGL rendering engine	No

* *ROS only.*

Table 18.2 Pros and cons of the described simulators.

Simulators	Pros	Cons
Gazebo	• Widely used / Large Community • Seamless integration with ROS/ROS2 • Cloud simulation capabilities	• Mediocre visual realism • Computationally intensive multirobot simulation
Webots	• Fast GPU-bound rendering • Vast amount of robotic models • Detailed documentation with examples • Cloud simulation capabilities	• Few domain randomization tools • Insufficient tools for designing custom environments
AirSim	• High-level of visual realism • Access to a variety of custom 3D assets which can be easily imported • Support of external flight controllers	• Most provided environments are in binary format • Computationally intensive
CARLA	• High-level of visual realism • Large amount of provided 3D assets • Tools for scene reconfiguration (e.g., traffic managements tools)	• Computationally intensive
CoppeliaSim	• Multiple physics engines • 6 programming approaches • Fast and exact algorithms for collision detection and minimum distance calculation between various geometric representations (e.g., meshes, octrees, point clouds, etc.)	• Some plugins/modules are distributed under commercial licenses

Gazebo. For years after its release, the complexity of simulating rigid body dynamics in a 3D environment, restricted Gazebo to run mostly on high performance computers, allowing only a few robots to be simultaneously available in a scene. However, being an open source software, Gazebo managed to establish an active community, becoming more computationally efficient and accomplishing several improvements and updates over the years. Overall, Gazebo has received many major and minor release versions, including its latest major version (v.11) released in January 2020. Images from various environments and robotic models in Gazebo are depicted in Fig. 18.3.

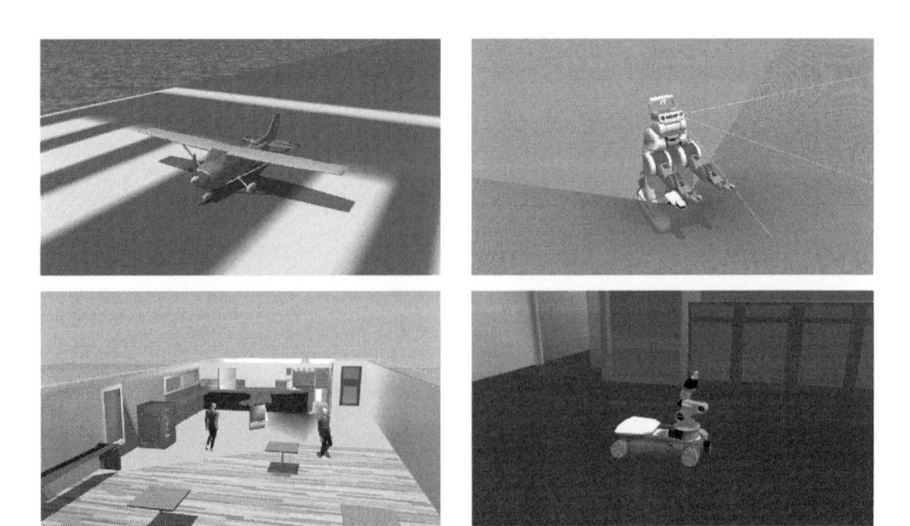

FIGURE 18.3

Various environments and robotic models in Gazebo.

18.2.1.1 Architecture

Gazebo uses a distributed architecture with separate libraries for physics simulation, rendering, user interface, communication, and sensor data generation. It provides two executable programs for running simulations, a server and a client named *gzserver* and *gzclient,* respectively. *Gzserver* is responsible for simulating the physics, rendering, and sensors, while *gzclient* provides a graphical interface to visualize and interact with the simulation. The client and server communicate using the gazebo communication library. A topic-based publish/subscribe model of interprocess communication is followed. The clients can access server's data through a shared memory. Each simulation object in Gazebo can be associated with one or more controllers that process commands for controlling the object and generate its corresponding state. Controllers can be defined as ROS nodes or as C++ plugins (see Section 18.2.1.2). The data generated by the controllers are published into the shared memory using Gazebo interfaces. The interfaces of other processes can read the data from the shared memory, thus allowing interprocess communication between the robot controlling software and Gazebo, independently of the programming language or the computer hardware platform.

An abstracted diagram of Gazebo's architecture is depicted in Fig. 18.4. Below, is a brief overview of Gazebo's main components:

- **The Gazebo master** is a topic name server. It provides name lookup, and topic management. A single master can handle multiple physics simulations, sensor generators, and GUIs.
- **The communication library**, which is used by almost all subsequent libraries, acts as the communication and transport mechanism for Gazebo. Gazebo uses

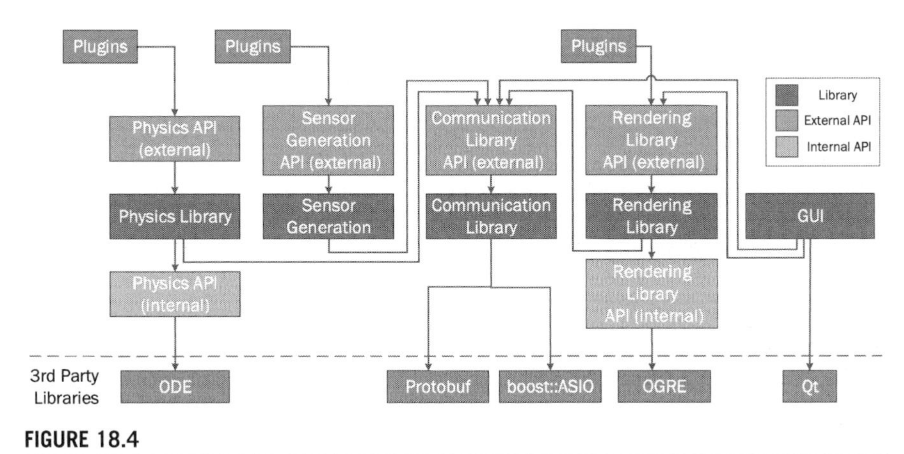

FIGURE 18.4

A visualization of the Gazebo structure.

the open source *Google Protobuf*[4] for message serialization and *boost::ASIO*[5] as its transport mechanism. It utilizes the publish/subscribe communication scheme. Messages are sent via publishers on named channels, called topics. On the other side, subscribers receive callbacks when messages arrive. In summary, to send messages one must publish messages using a publisher on a named topic, and to receive messages one must subscribe to a named topic using a subscriber. The easiest way to communicate with Gazebo over TCP/IP sockets is to link against the Gazebo libraries, and use the provided functions.

- **The physics library** provides a simple and generic interface to fundamental simulation components, including rigid bodies, collision shapes, and joints for representing articulation constraints. This interface has been integrated with four open-source physics engines, namely Open Dynamics Engine (ODE) [13], Dynamic Animation and Robotics Toolkit (DART) [22], Simbody [23], and BULLET [14].
- **The graphical user interface (GUI) library** uses Qt[6] to create graphical widgets for users to interact with the simulation. The user may control the flow of time by pausing or changing the time step size via GUI widgets. The user may also modify the scene, for example, add, move, and remove the scene's objects. Finally, tools for visualizing and logging simulated sensor data are provided.
- **The rendering library** uses Object-Oriented Graphics Rendering Engine (OGRE) [12] to provide a simple interface for rendering 3D scenes to both the GUI and sensor libraries. It achieves an adequate level of visual realism, including lighting, textures, and sky simulation.

[4] https://developers.google.com/protocol-buffers/.
[5] https://www.boost.org/.
[6] https://www.qt.io/.

- **The sensor generation library** implements various types of sensors, including camera (RGB, Depth, etc.), contact, IMU, LIDAR, GPS, Radio Frequency Identification (RFID), and force/torque sensors. Gazebo also provides tools both to modify the existing sensors (e.g., modify the noise levels) or create new ones. During a simulation the sensor generation library listens to world state updates and produces output specified by the instantiated sensors.

18.2.1.2 Plugins

A plugin in Gazebo can be defined as a block of code that is compiled as a shared library and inserted into the simulation. The plugin has direct access to all Gazebo's functionalities through its standard C++ classes. There are six types of plugins namely the World, Model, Sensor, System, Visual, and GUI plugins. Each plugin type is managed by a different component of Gazebo. For example, a Model plugin is attached to a specific robotic model and controls it. Similarly, a Sensor plugin is attached to a specific sensor. Plugins allow developers to control almost any aspect of Gazebo, they are self-contained routines that are easily shared and can be inserted and removed from a running system.

18.2.1.3 Robotic models

A robotic model in Gazebo, which may contain a single or multiple meshes, must always be accompanied by a Simulation Description Format (SDF) file. The SDF file, which is written in an XML format, is used by Gazebo to describe properties of both robotic models and environments. In the case of robotic models, an SDF file must contain the set of links and joints that the model consists of. The links, which represent the rigid parts of the robot, and the joints, which represent connections between two links, define the kinematic and dynamic properties of the robot. Each link is related to several properties, such as visual appearance, collision regions, inertia matrix, material type, etc. The structure of a model in SDF file format is depicted in Fig. 18.5. Due to the fact that Gazebo is very popular in the robotics domain, a large number of research and industrial robotic models have been designed, ready to be deployed in simulation. Some of the most well-known available robotic models are Willow Garage's PR2, ActivMedia Pioneer 2-DXE, iRobot Create, Parrot Bebop, and TurtleBot.

18.2.1.4 ROS/ ROS 2 support

Another big advantage of Gazebo compared to other simulators is its easy and seamless integration with ROS. A set of ROS packages, named *gazebo_ros_pkgs*,[7] provide wrappers around the stand-alone Gazebo. Those packages work as a bridge between Gazebo's C++ API and ROS, providing the necessary interfaces to simulate a robot in Gazebo using ROS messages, services and dynamic reconfigure. Though Gazebo is currently compatible with ROS 2, not all previous functionalities from ROS have been ported yet.

[7] https://github.com/ros-simulation/gazebo_ros_pkgs.

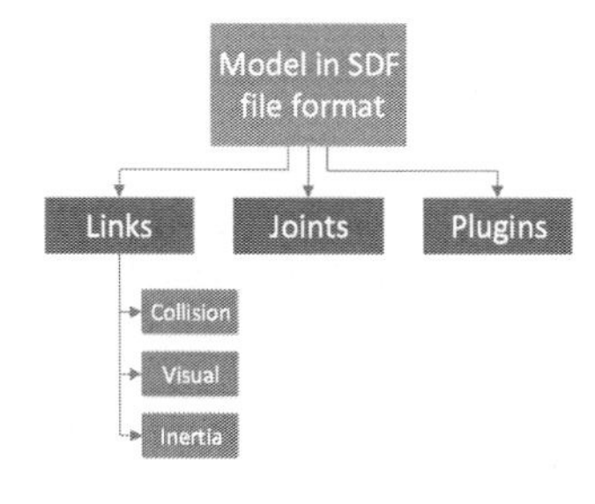

FIGURE 18.5

Structure of an SDF file.

18.2.1.5 Cloud simulation

Gazebo claims to support simulation deployment on cloud machines, for example, Amazon AWS[8], through CloudSim [24], simplifying the process of running simulations, while also providing a consistent and uniform environment. Gzweb [25] is a WebGL client, which allows the end-user to interact with the simulation environment through a web browser. Cloud simulation is a way for Gazebo to attain cross-platform support, even for mobile devices.

18.2.1.6 Research works

Gazebo has been extensively used in the development and testing of robotic systems in academic research. For example, in [26], a software and hardware system for autonomous mobile robot navigation in uneven and unstructured indoor environments was proposed. The system incorporated perception and navigation capabilities, using wheel odometry, 2D laser, and RGB-D data as input. Gazebo provided a valuable tool for the authors in order to conduct extensive experiments, demonstrating the efficacy and efficiency of the system. In [27], the authors used Gazebo to implement their proposed architecture for autonomous and cooperative cinematography planning of outdoor events using drones. Though Gazebo is a general-use simulator, able to be utilized for a variety of simulations, there might be specific tasks where it fails to provide a straightforward solution. In [28], the authors created an extension of Gazebo for underwater scenarios, named Unmanned Underwater Vehicle (UUV) simulator. This extension allows the simulation of multiple underwater robots and intervention tasks using robotic manipulators. The authors in [29] integrated an established reinforcement learning toolkit, named OpenAI Gym [30], to Gazebo, in order to address the lack of Deep Reinforcement Learning (DRL) simulation environments. Similarly, the authors in [31] designed a reinforcement learning framework with more realistic environments within Gazebo and validated it with a continuous UAV landing task.

[8] https://aws.amazon.com/.

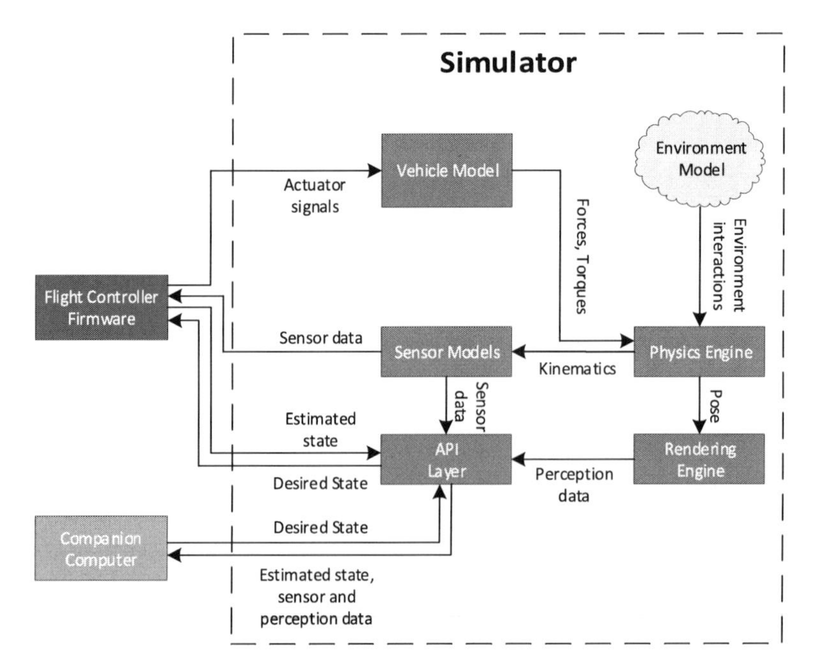

FIGURE 18.6

AirSim structure.

18.2.2 AirSim

AirSim [8] is a simulator for drones and cars, built on top of powerful game engines, namely Unreal Engine 4 [9] and Unity [10]. It is an open-source, cross-platform software developed by Microsoft and distributed under the MIT License. Due to the fact that AirSim uses a game engine as its rendering platform, it provides cutting edge graphics features such as physically based materials, photometric lights, planar reflections, ray traced distance field shadows, lit translucency, etc. Among its qualities, is that it supports SIL simulations with popular flight controllers, including PX4 [32] and ROSflight [33], as well as HIL simulations. Based on its development team, AirSims's goal is to provide a platform for AI research to experiment with deep learning, computer vision, and reinforcement learning algorithms for autonomous vehicles. AirSim uses C++ and Python APIs to achieve programmatic control and is also compatible with ROS.

18.2.2.1 Architecture

AirSim is implemented as a plugin in existing game engines, taking advantage of several qualities of those engines regarding rendering and physics simulation, user interface, and computational efficiency. The AirSim's structure is depicted in Fig. 18.6. The main components of AirSim are briefly reported below:

- **Vehicle model:** AirSim provides an interface to define a vehicle as a rigid body that may have an arbitrary number of actuators generating forces and torques. The vehicle model includes several parameters such as mass, inertia, coefficients for linear and angular drag, coefficients of friction and restitution, etc., which are used by the physics engine to compute rigid body dynamics.
- **Environmental model:** A vehicle in real world is exposed to various physical phenomena, most of which are computationally expensive to simulate. AirSim is focused on modeling accurately gravity, air-density, air pressure, and magnetic fields.
- **Sensors models:** AirSim offers a variety of sensors such as accelerometer, gyroscope, barometer, magnetometer, GPS, LIDAR, IMU, and cameras. The sensor models are implemented as C++ header-only libraries and can be independently used outside AirSim. Sensor models are expressed as abstract interfaces so they can easily be modified.
- **Rendering Engine:** Unreal Engine 4 (UE4) and Unity were selected as AirSim's rendering platforms. Both platforms can realistically depict all kinds of environments including cities, mountains, tropical and seaside environments (see Section 18.2.2.2). Since depicting such environments may occasionally be computationally intensive, Unreal Engine 4 and Unity allow users to downscale the visual quality (e.g., resolution, culling distance, etc.) in favor of attaining real-time simulations.
- **Physics Engine:** AirSim uses a body's position, orientation, linear velocity, linear acceleration, angular velocity and angular acceleration to express its kinematic state. The goal of the physics engine is to compute the next kinematic state for each body given the forces and torques acting on it. The physics engine can run its update loop at high frequency (up to 1000 Hz) which is desirable for enabling real-time simulation scenarios.
- **API layer:** AirSim exposes APIs to interact with the vehicle in the simulation using Python or C++. Through APIs one can retrieve information from sensors (e.g., images, kinematic state), be notified for detected collisions, control the vehicle, and interfere with the simulation (e.g., pause/continue the simulation, change the weather or time of day, etc.). The APIs are available as part of a separate, independent cross-platform library. Thus one can write and test code in the simulator, and later execute it on the real vehicle. The APIs use *msgpack-rpc*[9] protocol over TCP/IP. When AirSim starts, it opens a port and listens for incoming requests. The client connects to the port and sends RPC calls using the *msgpack* serialization format.

18.2.2.2 Environments and models

AirSim provides a plethora of environments, mostly in binary format, meaning that they cannot be modified neither in UE4 nor in Unity editor due to copyright reasons

[9] https://msgpack.org/.

FIGURE 18.7

AirSim's environments.

regarding their content. Images depicting some of those environments are shown in Fig. 18.7. Both of these engines have digital stores (e.g., UE Marketplace,[10] Unity AssetStore[11]), allowing the purchase of assets needed for the simulations. However, custom environments and 3D models in common formats, such as Filmbox (.fbx), Wavefront OBJ (.obj), and STL (.stl) file formats, can also be imported in the editors regardless of their source. Both editors provide editing tools for the imported assets.

18.2.2.3 Research works

AirSim has proven to be a valuable tool in academic research for developing autonomous cars and drones, especially for the latter. In [34], the authors proposed a method for following high-level navigation instructions by mapping directly from images, instructions, and pose estimates to continuous low-level velocity commands for real-time drone control. To this end, the Grounded semantic mapping network, a fully-differentiable neural network architecture was utilized. The network builds an explicit semantic map in the world reference frame by incorporating a pinhole camera. The authors selected AirSim as their simulation framework for validating their approach. In [35], a UAV safe landing pipeline was proposed, incorporating vision-based methods for human detection/avoidance and landing site detection. The pipeline was evaluated using AirSim in realistic environments designed in Unreal Engine 4.

AirSim has also been utilized for large-scale synthetic data set generation, due to its visual and physical fidelity. TartanAir [11] is a data set, generated using AirSim and Unreal Engine 4, for robot navigation tasks. The data set consists of 1037 annotated long motion sequences (500–4000 frames per sequence) collected from 30 diverse environments. Mid-Air [36] is also a synthetic data set collected using AirSim, for low altitude drone navigation in unstructured environments (e.g., forest, country). It contains more than 420k video frames and includes multimodal data regarding the flight (e.g., visual, GPS, IMU data).

[10] https://www.unrealengine.com/marketplace/.

[11] https://assetstore.unity.com/.

FIGURE 18.8

Webots GUI.

18.2.3 Webots

Webots [37] is an open-source and multiplatform simulator able to provide a complete development framework to model, program, and simulate robotic systems. It has been designed for professional use, and is widely used in industry, education and research. Until December 2018, the simulator used to be a commercial software, developed by Cyberbotics. At that time, Webots was released under the free and open-source Apache 2 license and is now publicly available on a GitHub repository.[12] Webots supports a wide variety of robotic simulations including two-wheeled table robots, industrial arms, bipeds, multilegged robots, modular robots, automobiles, flying drones, autonomous underwater vehicles, tracked robots, aerospace vehicles, etc. The simulator is available for Windows, Linux, and macOS and is compatible with both ROS and ROS2.

18.2.3.1 Architecture

Webots follows a similar architecture to that of most modern simulators. At its core, Webots incorporates ODE as its physics engine and a modern GUI based on QT. Though older versions of Webots used OGRE for 3D rendering, Cyberbotics implemented their own graphics engine, WREN, based on OpenGL 3.3, as a way to optimize rendering and improve its level of visual fidelity. WREN is a GPU-bound engine that implements a physically-based rendering pipeline, achieving high-quality graphics realism. Webots GUI is composed of four principal windows: the 3D window that displays and allows the user to interact with the 3D simulation, the scene tree which is a hierarchical representation of the current world, the text editor that allows the user to edit source code, and the console which displays both compilation and controller outputs. This GUI is depicted in Fig. 18.8.

[12] https://github.com/cyberbotics/webots/.

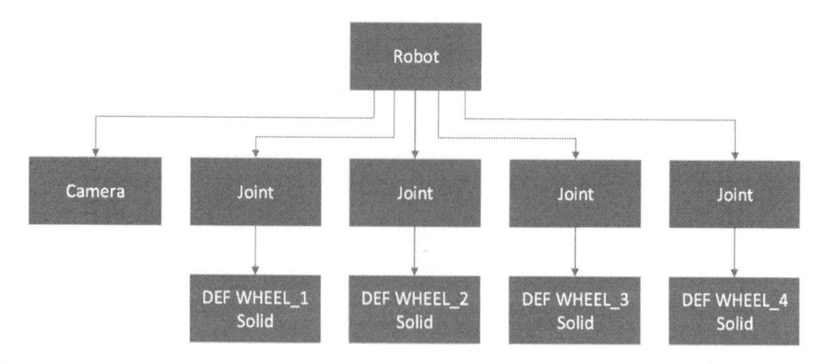

FIGURE 18.9

Hierarchical structure of a four-wheeled robot with a camera using nodes in Webots.

Webots allows programming interactions though APIs in various programming languages including C/C++, Java, Python, and MATLAB. The interfaces are implemented through a TCP/IP protocol. A major drawback of TCP/IP interfacing is that image transportation (e.g., retrieval of an image from a simulated camera device) becomes a very network intensive process. To overcome this, the development team suggest the use of a high speed network and small resolution, compressed, camera images. The APIs allow the interaction both with the robotic systems, as well as with the environment. More specifically, a Controller in Webots is defined as a program controlling the behavior of a robot and running on its own process. A Controller can be set as a Supervisor, allowing it to have access to some extra capabilities not available in real robots. Supervisors are very useful for scripting a simulation and performing various operations that a normal user could do manually from the Webots GUI (moving objects, pausing/restarting a simulation, etc.).

A simulation scene in Webots follows an hierarchical structure composed of nodes. The scene itself is defined as World through a .wbt file. Some basic nodes are the Solid, Device, Robot, Joint, and Motor nodes. Webots allows the creation of new nodes, or the extension of existing ones. Custom nodes can be flexibly defined though PROTO files. A physical object is defined as a Solid node. Solid nodes can be linked together by Joint nodes. Usually a robot in Webots is defined as a combination of multiple Solid, Sensor, and Joint nodes. Its root node should be a Robot node. An example of a high-level structure of a four-wheeled robot with a camera is shown in Fig. 18.9.

18.2.3.2 Environments and models

Webots contains a wide-range of robotic models, 3D assets, and complete environments. Each licensed robotic model integrated in the simulator is well documented and a demo environment is designed in order to demonstrate its capabilities and behavior. Some of the those models are depicted in Fig. 18.10. New robotic models can

FIGURE 18.10

Sample Webots robotic models. From left to right: DJI' Mavic 2 PRO, KUKA's youBott, Adept's Pioneer 3-DX, BlueBotics's Shrimp, Universal Robots UR5e, NASA Sojourner.

FIGURE 18.11

Webots environments.

be modeled as a PROTO file. In addition, Webots allows to import robotic models from Blender through plugins.[13]

A variety of environments are designed in Webots including industrial environments, halls, cities, etc. (Fig. 18.11). The simulator contains numerous 3D assets to customize existing environments or create new ones. 3D objects in common formats (AutoDesk 3DS (.3ds), Filmbox (.fbx), STL (.stl), Wavefront OBJ (.obj), Extensible 3D (.x3d), etc.) can be easily imported in the simulator. Finally, Webots allows the

[13] https://github.com/cyberbotics/blender-webots-exporter.

import of environments for automobile simulations generated from OpenStreetMap[14] maps. An open source traffic simulation package named SUMO (Simulation of Urban MObility) [38] is implemented within Webots, making it able to generate traffic using a large number of vehicles in real-time.

18.2.3.3 Research works

Webots has been a valuable tool in autonomous vehicles research. In [39] the authors presented a hierarchical Model Predictive Control (MPC) scheme for on-road convoy control of autonomous vehicles. The desired formation was modeled as a virtual structure evolving curvilinearly along a center line, and vehicle configurations were expressed as curvilinear relative longitudinal and lateral offsets from the virtual center. The authors selected Webots as their simulation framework, demonstrating the effectiveness of their approach in handling realistic on-road conditions such as curvy segments, other traffic participants and obstacles. In [40], the authors presented a simulation-based adversarial test generation framework for autonomous vehicles, which is built upon Webots. The proposed framework can be used to evaluate the closed-loop properties of an autonomous driving system model, which may include machine learning components, within a virtual environment.

Due to limitations in employing deep reinforcement learning methods in most simulators, including Weebots, the authors in [41] presented an open-source framework, named Deepbots, that combines the established OpenAI Gym interface, used by DRL researchers, with Webots in order to provide a standardized way to employ DRL in various robotics scenarios. According to the authors, Deepbots aims to enable researchers to easily develop DRL methods in Webots by handling all the low-level details and reducing the required development effort.

18.2.4 CARLA

CARLA (Car Learning to Act) [7] is a simulator developed as a tool for autonomous driving research, including the development, training, and validation of autonomous urban driving systems. It is an open platform implemented as an open-source layer over Unreal Engine 4, providing a high level of flexibility and realism in rendering and physics simulation. Another reason the simulator stands out for its visual realism, is because its content has been created from scratch by a dedicated team of digital artists employed for this purpose. It includes urban layouts, a multitude of vehicle models, buildings, pedestrians, street signs, etc. CARLA is designed to reflect the behavior of urban environments, providing tools to automate and modify the traffic density, the behavior of pedestrians, the weather, and the time of day. CARLA is ROS-compatible through packages developed for this purpose.

[14] https://www.openstreetmap.org.

18.2.4.1 Architecture

CARLA follows a server-client system architecture, where the server runs the simulation and renders the scene. The simulator provides a simple interface between the world and an agent that interacts with the world. A client API is implemented in Python and is responsible for the interaction between the autonomous agent and the server via sockets. The client sends to the server control commands to the vehicle, including steering, accelerating, and braking commands as well as meta-commands that control the behavior of the server and are used for resetting the simulation, changing the properties of the environment (e.g., weather conditions, illumination, density of cars, and pedestrians, etc.) and modifying the sensor suite. In return, the client receives sensor readings. The simulator incorporates cameras, GNSS, LIDAR, IMU, and RSS as sensors. The camera can retrieve RGB, depth, and segmentation images. In addition, a set of detectors are implemented including obstacle, collision, and lane invasion detectors.

18.2.4.2 Environments and models

As CARLA is focused on simulating urban environments, its content is composed of static 3D models such as buildings, vegetation, traffic signs, infrastructure as well as dynamic objects such as vehicles and pedestrians. CARLA supports map generation through RoadRunner,[15] a powerful software from Vector Zero to create 3D scenes. In addition, CARLA allows its maps to provide subleveling capabilities, meaning that their assets can be hierarchically structured, and thus handled (added/removed) through layers. CARLA assets and environments are shown in Fig. 18.12. Finally, the simulator provides extensive documentation regarding the import of custom 3D assets in the Filmbox (.fbx) file format.

18.2.4.3 Research works

CARLA has been used for training/evaluation of novel research methods related to autonomous cars. In [42], the authors proposed a conditional imitation learning formulation that is able to traverse intersections based on high-level navigational commands. CARLA proved to be a valuable tool for the authors in order to corroborate design decisions and evaluate the proposed approach in a dynamic urban environment with traffic. In [43] the authors used CARLA to collect a large amount of synthetic data for LiDAR point cloud object detection, in order to determinate how well networks trained on synthetic data perform on real data. Their work shows that models trained on large synthetic data sets and later fine-tuned using few real data can achieve similar results to models trained with a large amount of real world data from scratch. The authors concluded that synthetic data can lower the time and cost needed for the acquisition and labeling of large-scale real world data. In [44], the authors used CARLA to create data under various conditions in order to evaluate their proposed method for visual simultaneous localization and mapping (vSLAM).

[15] https://www.mathworks.com/products/roadrunner.html/.

FIGURE 18.12

Examples of assets and environments from CARLA.

In particular, the proposed method, enables vSLAM to stably estimate the camera motion by excluding points using a mask produced by semantic segmentation, thus achieving significantly higher accuracy compared to other state-of-the-art methods.

18.2.5 CoppeliaSim

CoppeliaSim [45], formerly known as V-REP, is a general-use, cross-platform, robotic simulator aimed to be used for fast algorithm development, factory automation simulations, fast prototyping and verification, remote monitoring, safety double-checking, etc. The core libraries of the simulator are distributed as open-source under a GNU GPL license. However several additional plugins and libraries show differences in the way they are distributed and licensed.

18.2.5.1 Overview and features

CoppeliaSim supports four different physics engines, namely BULLET, ODE, the Vortex Studio engine and the Newton Dynamics engine. In addition, the simulator supports six programming/coding approaches, each having particular advantages over the others, but all being mutually compatible. In particular, CoppeliaSim supports:

- Embedded Lua scripts, for easy and flexible customization of a simulation.
- Add-ons or sandbox scripts written in Lua programming language, allowing to quickly customize the simulator itself, offering a more generic, simulator-bound functionality.

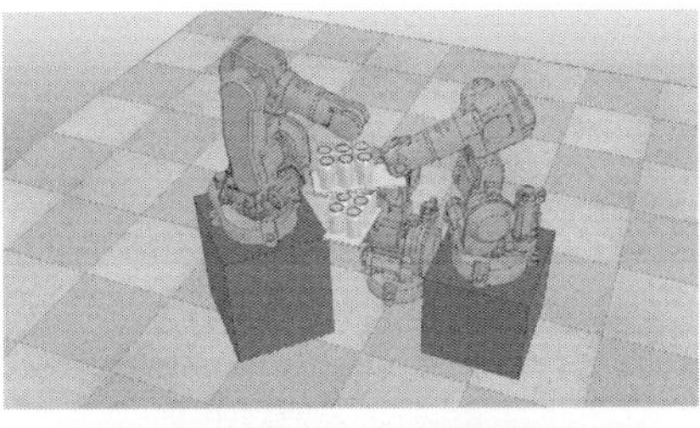
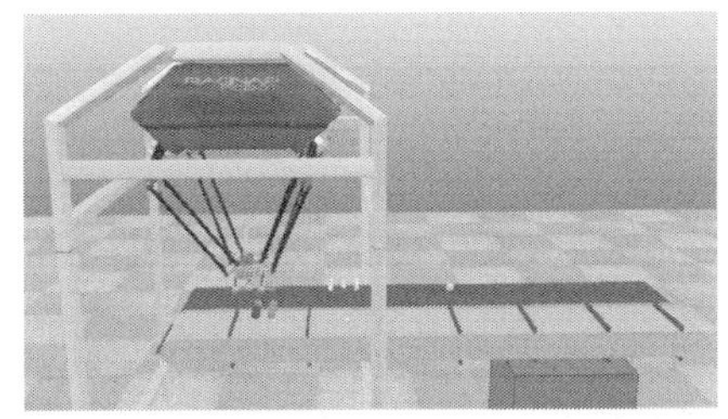

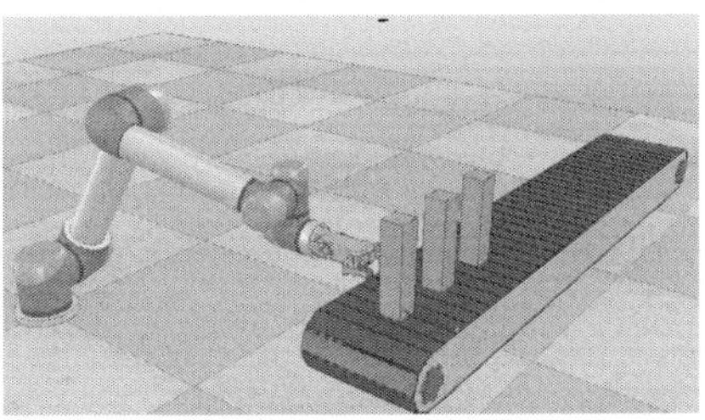

FIGURE 18.13

Various CoppeliaSim robotic models.

- C++ plugins which can be used to provide a simulation with customized Lua commands, or possibly to provide the simulator with a special functionality requiring either a fast calculation capability, a specific interface to a hardware device, or a special communication interface with the outside world.
- Remote APIs using a variety of programming languages for scripting including C++, Python, Java, MATLAB, Octave, and Lua.
- ROS nodes allowing an external application (e.g., located on a robot, another machine, etc.) to connect to CoppeliaSim via ROS.
- BlueZero nodes allowing an external application (e.g., located on a robot, another machine, etc.) to connect to CoppeliaSim via BlueZero.

CoppeliaSim features include: (1) the ability to simulate dynamic particles (e.g., realistic simulation of water and air through a system of particles), (2) a large variety of robots, including bipedal, hexapod, wheeled, flying and snake-like robots, (3) a large number of robot actuators and sensors, (4) the support of both inverse (IK) and forward (FK) kinematics calculation providing embeddable versions of the IK/FK algorithms, (5) fast and exact implementations for minimum distance calculation, collision detection, etc. Images of several robotics models in CoppeliaSim are shown in Fig. 18.13.

18.2.5.2 Research works

CoppeliaSim and V-REP have been vastly used as simulation tools in methods related to reinforcement learning. In [46] the authors successfully applied deep reinforcement learning for mapless navigation by taking the sparse 10-dimensional range findings and the target position with respect to the mobile robot coordinate frame as input and generate continuous steering commands as output. Experiments in real and virtual environments using V-REP showed that the proposed mapless motion planner can navigate the mobile robot to the desired targets without colliding with any obstacles. In [47] a model-free reinforcement learning (Q-learning) framework was proposed for combining grasping and pushing. The framework couples two networks, one for pushing and another for grasping, trained entirely in a self-supervised manner by trial and error, where rewards are provided from successful grasps. The corresponding policy encourages pushing motions that enable future grasps, while also encouraging grasps that can leverage past pushes. Experiments both in real world and in virtual environments using V-REP showed that the proposed method achieves better grasping success rates and picking efficiencies than baseline alternatives.

The simulator has also been used in research on human-robot collaboration in industrial setups. In [48] the authors proposed a Time-of-Flight sensor ring device as a form of intrinsic range sensing for collision avoidance. The simulation results obtained using V-REP showed that there is a significant benefit in terms of safety, performance, and productivity during the Human Robot Collaboration (HRB) task in comparison to conventional methods used in industry. In [49] a robot supervision framework applicable to human-robot collaborative tasks was proposed. The described system adaptively controls the movements of a robot by monitoring the human actions and its surrounding workspace. The framework was designed and simulated in V-REP using the collaborative robot UR 10[16] and the industrial robot Adept Viper 650.[17]

18.2.6 Other simulators

18.2.6.1 MORSE

MORSE [50] is a generic open-source simulator for academic robotics. The simulator is available on Github[18] under a BSD-3 clause license. Its development process has ended in 2020, and since then the simulator has not been officially updated or maintained. Regarding rendering and physics simulation, MORSE relies on the Blender Game Engine, a 3D game engine integrated in the open-source Blender modeling toolkit. The Blender Game Engine uses the BULLET physics engine for physics simulation. MORSE uses Blender's flexible graphical interface, named Logic Bricks, for scripting the behavior of the objects in a scene. Python can also be used for programmatic interaction, since this is the language that Blender natively uses. However, coding components that require a faster processing time, can be implemented in C/C++ and interfaced with Python using SWIG[19] wrappers. Due to the fact that Blender is one of the most common modeling tools, almost any 3D model can be imported and used into MORSE. Nonetheless, a variety or robotic platforms

[16] https://www.universal-robots.com/products/ur10-robot/.

[17] https://industrial.omron.eu/en/products/viper.

[18] https://github.com/morse-simulator/morse.

[19] http://www.swig.org/.

are already provided, including the Pioneer 3-DX[20] platform, the Yamaha RMAX[21] unmanned aerial vehicle, a NeoBotix[22] mobile platform, etc. Robotic arms are also available such as the Mitsubishi PA-10[23] and KUKA LWR,[24] both implemented with inverse kinematics. Some of the sensors implemented in MORSE are cameras, gyroscopes, GPS, accelerometers, thermometers, lasers, and proximity detectors. Finally, MORSE supports a wide-range of robotic middlewares including YARP, Pocolibs,[25] and ROS. The structure of MORSE, allows different middlewares to run in parallel in a simulation (e.g., one robot can be controlled with YARP, while another one is managed by a software running Pocolibs).

18.2.6.2 ARGoS

ARGoS [51] is an open-source physics-based simulator designed to simulate large-scale robot swarms. ARGoS is released under the MIT license and is publicly available on GitHub.[26] Compared to other robotic simulators, ARGoS is unique due to the fact that it incorporates multiple physics engines that can run in parallel during a simulation experiment. More precisely, ARGoS natively offers four kinds of physics engines: (i) a 3D dynamics engine based on ODE, (ii) a custom 3D particle engine, (iii) a 2D dynamics engine based on the open source physics engine library Chipmunk,[27] and (iv) a custom 2D kinematics engine. The physics engines can be manually assigned to robotic models in a simulation. It is also possible to divide the physical space in regions, and assign a physics engine per region. In addition, ARGoS offers three types of visualization: (i) an interactive graphical user interface based on Qt and OpenGL, (ii) a high-quality rendering engine based on the well-known ray-tracing software POV-Ray,[28] and (iii) a text-based visualization designed for interaction with plotting programs such as GNUPlot.[29] According to [51], the multithreaded architecture of ARGoS is very scalable, exhibiting low run-times and high speedup on multicore CPUs, allowing the authors to simulate 10,000 robots 40% faster than real-time, using multiple 2D dynamics physics engines. In [52] the authors presented a self-organized tasks allocation method, which is based on the delay experienced by the robots working on one subtask when waiting for input from another subtask. The method does not require the robots to communicate. The authors validated their proposed method in different two-dimensional environments simulated in ARGoS. In [53] the authors proposed a distributed SLAM system that incorporates a

[20] https://www.generationrobots.com/en/402395-robot-mobile-pioneer-3-dx.html.
[21] https://www.yamahamotorsports.com/motorsports/pages/precision-agriculture-rmax.
[22] https://www.neobotix-robots.com/homepage.
[23] https://robotik.dfki-bremen.de/en/research/robot-systems/mitsubishi-pa-10-7c.html.
[24] https://www.kuka.com/en-de/products/robot-systems/industrial-robots/lbr-iiwa.
[25] https://www.openrobots.org/wiki/pocolibs.
[26] https://github.com/ilpincy/argos3.
[27] https://chipmunk-physics.net/.
[28] http://www.povray.org/.
[29] http://www.gnuplot.info/.

novel outlier rejection mechanism for loop-closing. The proposed method was evaluated on simulations using ARGos as well as on benchmarking data sets and field experiments.

18.2.6.3 USARSim

USARSim (Urban Search and Rescue Simulation) [54] is an open source simulator built upon Unreal Engine 2 aimed for research and educational purposes. Although USARSim was originally developed for simulating urban search and rescue operations, it is actually a general purpose multirobot simulator that can be extended to model arbitrary application scenarios. USARSim is available for Windows, Linux, and MacOS. Though the simulator has not been updated or maintained since 2009, it is still used in the robotics domain, being one of the first simulation environments to rely on a game engine for simulation. Physical simulation in Unreal Engine 2 relies on the Karma physics engine. Karma processes rigid body movements and allows the simulation of motors, wheels, springs, hinges, and joints. From these base modules, complicated objects can be built through compounding. Robots and environment objects are created in 3rd party modeling applications. Worlds can be also created using an included utility. Programmatic control can be achieved using (i) the GameBot [55] interface, which allows the control of a virtual robot through a TCP/IP socket using a variety of programming languages (e.g., C++, Java, etc.), (ii) Pyro [56], a programming framework for robot control using Python, (iii) the Mobility Open Architecture Simulation and Tools system (MOAST) [57] or (iv) the Player interface. The controller (e.g., Player, MOAST) connects to the Unreal server and sends commands to USARSim to create and control a robot model or retrieve sensor data. The simulator contains several default maps, but more can be created by the users. A variety of sensors are incorporated such as touch sensors, sound sensors, cameras, and lasers. Several works have incorporated USARSim for performing simulations, in particular for algorithms related to search and rescue missions. In [58] the authors used Multicriteria Decision Making (MCDM) to define exploration strategies in the search and rescue domain. Those strategies were implemented and evaluated in USARSim environments. In [59], design strategies for Wireless Sensor Networks (WSNs) deployment in post-disaster monitoring tasks using Unmanned Aerial Vehicles (UAVs) are presented. USARSim was used to demonstrate the effectiveness of the corresponding strategies.

18.2.6.4 Nvidia's Isaac Sim

Isaac Sim [60] simulator is part of Nvidia's open source robotic AI development platform (Isaac SDK) for simulation, navigation, and manipulation tasks. Currently Isaac Sim is built on top of Unreal Engine 4 or Unity platforms. A new version of the simulator is under development using Nvidia's Omniverse platform, a powerful, multi-GPU, real-time simulation and collaboration platform for 3D production pipelines based on Pixar's Universal Scene Description (USD) and NVIDIA RTX. This version is currently distributed only as Early-Access. Robots in Isaac Sim are tightly related to the tools and frameworks in Isaac SDK, allowing to easily transfer algorithms and data between physical and virtual robots. Isaac SDK in particular is a software framework that lets the user to build modular robotics applications. It enables high-performance data processing and deep learning for intelligent robots. Robotics applications developed on Isaac SDK's robot engine can seamlessly run on edge devices like the NVIDIA Jetson AGX Xavier[30] and NVIDIA Jetson Nano,[31] as well as on workstations with modest NVIDIA GPUs. The robot engine allows developers to break down a complex robotics scenario into smaller building blocks. SDK features can be customized to fit a specific use case by configuring prepackaged components and even adding new features by developing custom components. Isaac SDK comes with a collection of high-performance algorithms, called GEMs, to accelerate the development of challenging robotics applications, including algorithms for planning and perception, navigation, and manipulation use cases. GEMs also provide support for key hardware components and robotic peripherals. Indeed, the simulator incorporates cameras, LIDAR and IMU as sensors. Isaac Sim provides an infinite stream of procedurally generated, fully annotated training data for machine learning. It allows the control of various randomization parameters in a scene including lighting, object materials, object poses, and camera properties, in order to increase the visual diversity of the data sets. The camera can retrieve RGB, stereo, depth, and segmentation images. As URDF is one of the most common formats for robotic models, Isaac Sim has an embedded URDF loader, which allows the import and simulation of their joints and movements. The URDF models for several mobile bases and manipulators have already been fully tested.

18.2.6.5 RoboDK

RoboDK is a cross-platform commercial industrial robot simulator, which enables robot programs to be created, simulated, and generated outside the production environment and without the presence of the robot. This process, named offline programming, has several advantages such as the removal of production downtime and allows the study and testing of multiple manufacturing tasks including milling, welding, pick and place, packaging and labeling, palletizing, painting, robot calibration, etc. Compared to other simulators, RoboDK can be installed on a wider-range of devices including even Android phones, Apple iPhones, and Raspberry Pis. Robots can be manipulated either visually, using the integrated 3D simulation environment, or by scripts in Python, MATLAB, C++, and C♯. Regardless of the chosen method, the software can generate programs in various robot-specific languages, for example, RAPID,[32] KRL [61], etc. In addition, RoboDK offers a built-in G-code interpreter for execution of tool-path related projects. Similar functionalities are also provided

[30] https://developer.nvidia.com/embedded/jetson-agx-xavier-developer-kit.

[31] https://developer.nvidia.com/embedded/jetson-nano-developer-kit.

[32] https://developercenter.robotstudio.com/.

by other industrial robotic simulators such as the FANUC's ROBOGUIDE[33] and ABB's RobotStudio.[34]

18.3 Conclusions

Simulation environments are nowadays widely used for the study, experimentation and testing of robotic systems, reducing the costs, complexity, and time needed for their real-world deployment. The introduction of deep learning approaches in robotic perception, cognition, and planning has increased the need for efficient and realistic simulation tools since they can provide diverse and large volume data for algorithm training and testing.

Luckily, numerous robotic and autonomous systems simulators, most of them open-source, are available to the research and education community. In this chapter, we presented the typical architecture of modern simulators, as well as their main functional components. Furthermore, we highlighted a number of qualitative characteristics, which one may consider when selecting a simulation framework. In addition, we presented various robotic simulators, focusing on five well-established and actively maintained ones. Details were provided regarding their architecture, interfaces, licensing, distribution, graphical and physical fidelity and functionalities. For each simulator, a small selection of scientific papers that utilize it for the corresponding research, is provided.

Each simulation framework has different characteristics and advantages over the rest, making the selection of an appropriate solution for a certain application, a nontrivial task. We hope that the chapter will facilitate this task by providing comprehensive descriptions of the currently available robotic simulation frameworks.

References

[1] S. Thrun, D. Fox, W. Burgard, F. Dellaert, Robust Monte Carlo localization for mobile robots, Artificial Intelligence 128 (1–2) (2001) 99–141.

[2] D.C. Mackenzie, R.C. Arkin, J.M. Cameron, Multiagent mission specification an execution, Autonomous Robots 4 (1) (1997) 29–52.

[3] T. Laue, K. Spiess, T. Röfer, SimRobot – A general physical robot simulator and its application in Robocup, in: RoboCup 2005: Robot Soccer World Cup IX, 2006, pp. 173–183.

[4] J. Go, B. Browning, M. Veloso, Accurate and flexible simulation for dynamic, vision-centric robots, in: Proceedings of International Joint Conference on Autonomous Agents and Multi-Agent Systems (AAMAS), vol. 3, 2004, pp. 1388–1389.

[33] https://www.fanuc.eu/dk/en/robots/accessories/roboguide.

[34] https://new.abb.com/products/robotics/robotstudio.

[5] N. Koenig, A. Howard, Design and use paradigms for Gazebo, an open-source multi-robot simulator, in: Proceedings of the IEEE/RSJ International Conference on Intelligent Robots and Systems (IROS), vol. 3, 2004, pp. 2149–2154.
[6] M. Quigley, Ros: an open-source Robot Operating System, in: ICRA Workshop Open Source Software, 2009.
[7] A. Dosovitskiy, G. Ros, F. Codevilla, A. Lopez, V. Koltun, CARLA: an open urban driving simulator, in: Proceedings of the 1st Annual Conference on Robot Learning, vol. 78, 2017, pp. 1–16.
[8] S. Shah, D. Dey, C. Lovett, A. Kapoor, Airsim: high-fidelity visual and physical simulation for autonomous vehicles, in: Field and Service Robotics, 2017, pp. 621–635.
[9] B. Karis, Epic Games, Real Shading in Unreal Engine 4, 2013.
[10] Unity Technologies, Unity, available: http://unity.com/, 2004.
[11] W. Wang, D. Zhu, X. Wang, Y. Hu, Y. Qiu, C. Wang, Y. Hu, A. Kapoor, S. Scherer, Tartanair: a dataset to push the limits of visual SLAM, in: Proceedings of the IEEE/RSJ International Conference on Intelligent Robots and Systems (IROS), 2020, pp. 4909–4916.
[12] K. Liu, M. Stilman, Object-oriented graphics rendering engine, in: Proceedings of the International Conference on Computer Science and Software Engineering, vol. 2, 2008, pp. 1027–1030.
[13] R. Smith, Open dynamics engine ODE, Online, available: http://www.ode.org/, 2015.
[14] E. Coumans, Bullet physics engine for rigid body dynamics, Online, available: http://bulletphysics.org/, 2015.
[15] P.M. Newman, MOOS — a Mission Oriented Operating Suite, Tech. Rep. OE2003-07, Department of Ocean Engineering, MIT, 2002.
[16] G. Metta, P. Fitzpatrick, L. Natale, YARP: yet another robot platform, International Journal of Advanced Robotic Systems 1238 (1) (2006) 43–48.
[17] S. Demers, P. Gopalakrishnan, A generic solution to software-in-the-loop, in: Proceedings of the IEEE Military Communications Conference (MILCOM), 2007, pp. 1–6.
[18] R. Isermann, J. Schaffnit, S. Sinsel, Hardware-in-the-loop simulation for the design and testing of engine-control systems, Control Engineering Practice 7 (5) (1999) 643–653.
[19] P.M. Boesch, F. Ciari, Agent-based simulation of autonomous cars, in: Proceedings of the American Control Conference (ACC), 2015, pp. 2588–2592.
[20] Robot simulation made easy, Online, available: http://gazebosim.org/.
[21] B.P. Gerkey, R.T. Vaughan, A. Howard, The Player/Stage project: tools for multi-robot and distributed sensor systems, in: Proceedings of the International Conference on Advanced Robotics (ICAR), 2003, pp. 317–323.
[22] J. Lee, X.M. Grey, S. Ha, T. Kunz, S. Jain, Y. Ye, S.S. Srinivasa, M. Stilman, C.K. Liu, DART: dynamic animation and robotics toolkit, The Journal of Open Source Software 3 (22) (2018) 500.
[23] M.A. Sherman, A. Seth, S.L. Delp, Simbody: multibody dynamics for biomedical research, Procedia IUTAM 2 (2011) 241–261.
[24] R.N. Calheiros, R. Ranjan, A. Beloglazov, C.A.F. De Rose, R. Buyya, Cloudsim: a toolkit for modeling and simulation of cloud computing environments and evaluation of resource provisioning algorithms, Software, Practice & Experience 41 (1) (2011) 23–50.
[25] Gzweb, Online, available: http://gazebosim.org/gzweb/.
[26] C. Wang, L. Meng, S. She, M.I. Mitchell, T. Li, F. Tung, W. Wan, M.Q.H. Meng, C. de Silva, Autonomous mobile robot navigation in uneven and unstructured indoor environment, in: Proceedings of the International Conference on Intelligent Robots and Systems (IROS), 2017, pp. 109–116.

[27] A. Torres-Gonzalez, J. Capitan, R. Cunha, A. Ollero, I. Mademlis, A multidrone approach for autonomous cinematography planning, in: Proceedings of the Iberian Robotics Conference (ROBOT 2017), 2018, pp. 337–349.
[28] M.M.M. Manhães, S.A. Scherer, M. Voss, L.R. Douat, T. Rauschenbach, U.U.V. Simulator, A Gazebo-based package for underwater intervention and multi-robot simulation, in: OCEANS 2016 MTS/IEEE Monterey, 2016, pp. 1–8.
[29] I. Zamora, N.G. Lopez, V.M. Vilches, A.H. Cordero, Extending the OpenAI Gym for robotics: a toolkit for reinforcement learning using ROS and Gazebo, ArXiv, 2016.
[30] G. Brockman, V. Cheung, L. Pettersson, J. Schneider, J. Schulman, J. Tang, W. Zaremba, OpenAI Gym, arXiv:1606.01540 [abs], 2016.
[31] A. Rodriguez-Ramos, C. Sampedro, H. Bavle, P.D.L. Puente, P. Campoy, A deep reinforcement learning strategy for UAV autonomous landing on a moving platform, Journal of Intelligent & Robotic Systems 93 (2019) 351–366.
[32] L. Meier, D. Honegger, M. Pollefeys, PX4: a node-based multithreaded open source robotics framework for deeply embedded platforms, in: Proceedings of the IEEE International Conference on Robotics and Automation (ICRA), 2015, pp. 6235–6240.
[33] J.S. Jackson, G.S. Ellingson, T.W. McLain, ROSflight: a lightweight, inexpensive MAV research and development tool, in: Proceedings of the International Conference on Unmanned Aircraft Systems, 2016, pp. 758–762.
[34] V. Blukis, N. Brukhim, A. Bennett, R.A. Knepper, Y. Artzi, Following high-level navigation instructions on a simulated quadcopter with imitation learning, ArXiv, 2018.
[35] C. Symeonidis, E. Kakaletsis, I. Mademlis, N. Nikolaidis, A. Tefas, I. Pitas, Vision-based UAV safe landing exploiting lightweight deep neural networks, in: Proceedings of the International Conference on Image and Graphics Processing (ICIGP), 2021.
[36] M. Fonder, M. Van Droogenbroeck, Mid-Air: a multi-modal dataset for extremely low altitude drone flights, in: Proceedings of the IEEE/CVF Conference on Computer Vision and Pattern Recognition (CVPR) Workshops, 2019.
[37] Webots: robot simulator, Online, available: https://cyberbotics.com/.
[38] P.A. Lopez, M. Behrisch, L. Bieker-Walz, J. Erdmann, Y.P. Flötteröd, R. Hilbrich, L. Lücken, J. Rummel, P. Wagner, E. Wiessner, Microscopic traffic simulation using SUMO, in: Proceedings of the International Conference on Intelligent Transportation Systems (ITSC), 2018, pp. 2575–2582.
[39] X. Qian, A. de La Fortelle, F. Moutarde, A hierarchical model predictive control framework for on-road formation control of autonomous vehicles, in: Proceedings of the IEEE Intelligent Vehicles Symposium (IV), 2016, pp. 376–381.
[40] C.E. Tuncali, G. Fainekos, H. Ito, J. Kapinski, Simulation-based adversarial test generation for autonomous vehicles with machine learning components, in: Proceedings of the IEEE Intelligent Vehicles Symposium (IV), 2018, pp. 1555–1562.
[41] M. Kirtas, K. Tsampazis, N. Passalis, A. Tefas, Deepbots: a Webots-based deep reinforcement learning framework for robotics, in: Artificial Intelligence Applications and Innovations, Springer International Publishing, 2020, pp. 64–75.
[42] F. Codevilla, M. Müller, A. López, V. Koltun, A. Dosovitskiy, End-to-end driving via conditional imitation learning, in: Proceedings of the IEEE International Conference on Robotics and Automation (ICRA), 2018, pp. 4693–4700.
[43] D. Dworak, F. Ciepiela, J. Derbisz, I. Izzat, M. Komorkiewicz, M. Wójcik, Performance of LiDAR object detection deep learning architectures based on artificially generated point cloud data from CARLA simulator, in: Proceedings of the International Conference on Methods and Models in Automation and Robotics (MMAR), 2019, pp. 600–605.

[44] M. Kaneko, K. Iwami, T. Ogawa, T. Yamasaki, K. Aizawa, Mask-SLAM: robust feature-based monocular SLAM by masking using semantic segmentation, in: IEEE/CVF Conference on Computer Vision and Pattern Recognition Workshops (CVPRW), 2018, pp. 371–3718.
[45] E. Rohmer, S.P.N. Singh, M. Freese, CoppeliaSim (formerly V-REP): a versatile and scalable robot simulation framework, in: Proceedings of the International Conference on Intelligent Robots and Systems (IROS), 2013, pp. 1321–1326.
[46] L. Tai, G. Paolo, M. Liu, Virtual-to-real deep reinforcement learning: continuous control of mobile robots for mapless navigation, in: Proceedings of the IEEE/RSJ International Conference on Intelligent Robots and Systems (IROS), 2017, pp. 31–36.
[47] A. Zeng, S. Song, S. Welker, J. Lee, A. Rodriguez, T. Funkhouser, Learning synergies between pushing and grasping with self-supervised deep reinforcement learning, in: Proceedings of the IEEE/RSJ International Conference on Intelligent Robots and Systems (IROS), 2018, pp. 4238–4245.
[48] S. Kumar, C. Savur, F. Sahin, Dynamic awareness of an industrial robotic arm using Time-of-Flight laser-ranging sensors, in: Proceedings of the International Conference on Systems, Man, and Cybernetics (SMC), 2018, pp. 2850–2857.
[49] S. Kumar, F. Sahin, A framework for an adaptive human-robot collaboration approach through perception-based real-time adjustments of robot behavior in industry, in: Proceedings of System of Systems Engineering Conference (SoSE), 2017, pp. 1–6.
[50] G. Echeverria, S. Lemaignan, A. Degroote, S. Lacroix, M. Karg, P. Koch, C. Lesire, S. Stinckwich, Simulating complex robotic scenarios with MORSE, in: Simulation, Modeling, and Programming for Autonomous Robots (SIMPAR), 2012, pp. 197–208.
[51] C. Pinciroli, V. Trianni, R. O'Grady, G. Pini, A. Brutschy, M. Brambilla, N. Mathews, E. Ferrante, G.A. Di Caro, F. Ducatelle, M. Birattari, L.M. Gambardella, M. Dorig, ARGoS: a modular, parallel, multi-engine simulator for multi-robot systems, Swarm Intelligence 6 (4) (2012) 271–295.
[52] A. Brutschy, G. Pini, C. Pinciroli, M. Birattari, M. Dorigo, Self-organized task allocation to sequentially interdependent tasks in swarm robotics, Autonomous Agents and Multi-Agent Systems 28 (2012) 101–125.
[53] P. Lajoie, B. Ramtoula, Y. Chang, L. Carlone, G. Beltrame, Door-slam: distributed, online, and outlier resilient SLAM for robotic teams, IEEE Robotics and Automation Letters 5 (2) (2020) 1656–1663.
[54] S. Carpin, M. Lewis, J. Wang, S. Balakirsky, C. Scrapper, USARSim: a robot simulator for research and education, in: Proceedings of the IEEE International Conference on Robotics and Automation, 2007, pp. 1400–1405.
[55] G. Kaminka, M. Veloso, S. Schaffer, C. Sollitto, R. Adobbati, A. Marshall, A. Scholer, S. Tejada, Gamebots: a flexible test bed for multiagent team research, Communications of the ACM 45 (2002) 43–45.
[56] D. Blank, D. Kumar, L. Meeden, H. Yanco Pyro, A Python-based versatile programming environment for teaching robotics, Journal on Educational Resources in Computing 4 (08 2004).
[57] S. Balakirsky, C. Scrapper, E. Messina, Mobility open architecture simulation and tools environment, in: Proceedings of the International Conference on Integration of Knowledge Intensive Multi-Agent Systems, 2005, pp. 175–180.
[58] N. Basilico, F. Amigoni, Exploration strategies based on multi-criteria decision making for searching environments in rescue operations, Autonomous Robots 31 (2011) 401–417.

[59] G. Tuna, T.V. Mumcu, K. Gulez, V.C. Gungor, H. Erturk, Unmanned aerial vehicle-aided wireless sensor network deployment system for post-disaster monitoring, vol. 304, 2012, pp. 298–305.

[60] NVIDIA, NVIDIA ISAAC platform for robotics, Online, available: https://www.nvidia.com/en-us/deep-learning-ai/industries/robotics/.

[61] D.G. Bobrow, T. Winograd, An overview of KRL, a Knowledge Representation Language, Cognitive Science 1 (1) (1977) 3–46.

CHAPTER

Biosignal time-series analysis

19

Serkan Kiranyaz[a], Turker Ince[c], Muhammad E.H. Chowdhury[a], Aysen Degerli[b], and Moncef Gabbouj[b]

[a]*Department of Electrical Engineering, Qatar University, Doha, Qatar*

[b]*Faculty of Information Technology and Communication Sciences, Tampere University, Tampere, Finland*

[c]*Department of Electrical and Electronics Engineering, Izmir University of Economics, Izmir, Turkey*

19.1 Introduction

There is an urgent need for robust machine learning (ML) paradigms that can achieve utmost accuracy, reliability, feasibility, and learning capability for addressing critical healthcare problems that threaten millions of human lives nowadays. Current deep learning technologies have yielded numerous breakthroughs such as AI victories over humans in games, revolutionized computer vision, translation, and pattern recognition. However, when the data is scarce, which is a commonality especially for most of the biomedical data sets their training becomes ambiguous and sometimes, infeasible. Biosignal time-series analysis further poses significant challenges such as the inevitable presence of artifacts and morphological degradations, high variations of bioevents among individuals, and scarcity of the labeled data. Especially for decision making and diagnosis purposes, numerous technical challenges should be addressed in advance. For instance, from deep networks achieving a quick, or even a real-time output can be infeasible for low-power devices even though biosignal data are usually acquired in real-time and sometimes from many sources. The latter brings the concept of multimodality, which naturally raises the issue of "decision fusion." Despite the presence of several AI victories all over, unfortunately, the truth is that there is not a well-established method or a *de facto* solution for the aforementioned challenges. However, certain smart approaches are properly adapted to the constraints encountered. In this chapter, we shall present the *state-of-the-art* solutions for three critical healthcare problems by ML-based time-series analysis. The first one is about cardiac arrhythmia, which is an irregularity or abnormality of heart rhythm or heartbeat that can lead to increased morbidity and Sudden Cardiac Death (SCD). Centers for Disease Control and Prevention have estimated SCD rates at 610,000 per year in the USA alone, that is, one in four deaths. About 80% of SCD, resulting from ventricular arrhythmias and other arrhythmias including atrial fibrillation/atrial flutter caused 112,000 deaths in 2013 alone. Electrocardiogram (ECG)

Deep Learning for Robot Perception and Cognition. https://doi.org/10.1016/B978-0-32-385787-1.00024-5

is the primary tool for monitoring heart rhythm. Recent advances in heart monitoring systems rely on past abnormal beats to train their classifier to recognize normal and abnormal beats for patients who have already been diagnosed with cardiac arrhythmia. Unfortunately, there are no robust solutions to detect abnormalities from healthy individuals, for whom symptoms have just emerged. Such conditions may persist from a few weeks to several months before the patient finally seeks medical help. Such a time-loss may be crucial as early diagnosis and treatment most often lead to improved chances of full and fast recovery. In the next section, we shall start with the problem of *patient-specific* ECG beat classification where the objective is to discriminate the arrhythmic beats from the normal (healthy) beats of an individual patient. So, this section will answer the ultimate question of how to design person-specific, real-time, and accurate monitoring of ECG signals. We shall then move on to the recent solution of a related problem, an early warning system that can alert an individual the instant his/her heart deviates from its normal rhythm. This is significantly more challenging than the first problem since, in the absence of arrhythmic beats for a specific individual, none of the person-specific methods including the most advanced ones using ML can be trained and used.

Finally, we turn our focus to a recent Coronavirus Disease 2019 (COVID-19) pandemic related problem, and in particular, present a novel early mortality prediction tool for COVID-19 patients using ML-based time series analysis. COVID-19 has spread fast all over the world from Wuhan, China [1–5] and it has infected more than 65 million people and cause the death of 1.51 million people as of 4 December 2020 [6]. Severe acute respiratory syndrome coronavirus 2 (SARS-CoV-2) is the COVID-19 disease, which was declared a pandemic by the World Health Organization (WHO) on 11 March 2020 [5]. The hospital capacity in most of the highly infected countries is overwhelmed and the hospitals are facing difficulties with staff, ventilators, personal protective equipment (PPE) along with other life-support equipment [7,8]. Although 81% of COVID-19 patients have a mild illness, the rest of the 19% of patients can be moderate and severe, and mortality among that patient's groups is very high. Recent studies demonstrated that 26.1%–32.0% of patients infected with COVID-19 are at high risk of critical illness and critical patients have a high fatality rate of 61.5%, which also intensifies with the age and other comorbidities of the patients [9]. It is very important to prioritize high-risk patients for hospital treatment while low-risk patients can be treated in home-setting or outpatients stations. Therefore, an effective tool is necessary to assess the course of the disease efficiently to allocate hospital resources by identifying key patient variables as prognostic variables that can help to reduce the mortality rate.

19.2 ECG classification and advance warning for arrhythmia

Each year more than 7 million people die from cardiac arrhythmias, which are certain abnormalities of heart rhythm that can lead to increase morbidity and sudden cardiac death. Centers for Disease Control and Prevention (CDCP) have estimated sudden cardiac death (SCD) rates at more than 600,000 per year [10]. About 80% of SCD, resulting from ventricular arrhythmias [11] and other arrhythmias including Atrial fibrillation/Atrial flutter resulted in 112,000 deaths in 2013 [12]. Electrocardiogram (ECG) of healthy individuals may have morphological variabilities/abnormalities at baseline [13] with variable prevalence. Similarly, patients following myocardial infarction may develop persistent abnormal ECG changes. Electrocardiogram (ECG) signal analysis can extract important information about the cardiac electrophysiology and the activity of the autonomic nervous system. ECG and clinical history are often used to detect ischemia (whose symptoms include chest pain in patients) and remain the first and most significant diagnostic test. The changes that reflect on ECG caused by severe ischemia and infarction are hyperacute changes within the T waves, elevation, and/or depression of the ST segment, variation in the QRS complex, and T waves inversions of the ECG signal.

Several techniques have been proposed for ECG beat analysis and classification, such as frequency analysis [14], wavelet transform [15], and filter banks [16], statistical [17] and heuristic approaches [18], hidden Markov models [19], Support Vector Machines (SVMs) [20], Artificial Neural Networks (ANNs) [21], and mixture-of-experts method [22]. However, none of these methods has achieved a certain performance level for clinical usage due to three major reasons:

i) high interpatient variations are present on ECG signals, which render the use of general-purpose modeling and analysis,
ii) for the ECG signal characteristics of a healthy person, the pattern of QRS complex, P waves, and R-R intervals may not be identical among the beats under different circumstances [10], and
iii) the discrimination capability of the extracted "hand-crafted" features, which are usually fixed varies significantly among numerous methods used.

Fig. 19.1 shows some typical ECG beats where different subjects in the benchmark MIT-BIH arrhythmia data set [34] have entirely different normal (N) beats (as annotated by cardiologists). However, the same normal beats show a high level of similarities to other subject's abnormal beats (i.e., compare the left (normal) and right (abnormal) beats in Fig. 19.1). Due to the aforementioned reasons, the past classification approaches usually exhibit a common drawback of having an inconsistent performance when, for instance, classifying a new patient's ECG signal. This makes them unreliable to be widely used clinically or in practice, and they tend to have high variations in their accuracy and efficiency for real databases [23,24].

For ECG classification, several approaches [35–40] based on deep learning have recently been proposed. The deep Convolutional Neural Networks (CNNs) with high complexity have been used which require a massive size labeled ECG data for training (e.g. > 50K beats). Moreover, such a deep network cannot be trained for a single patient due to the scarcity of the labeled data. They are naturally not patient-specific approaches, and thus suffer from the aforementioned drawbacks. Furthermore, they strictly depend on parallelized hardware for training and classification; hence, this makes them inapplicable on low-power or mobile devices such as Holter monitors.

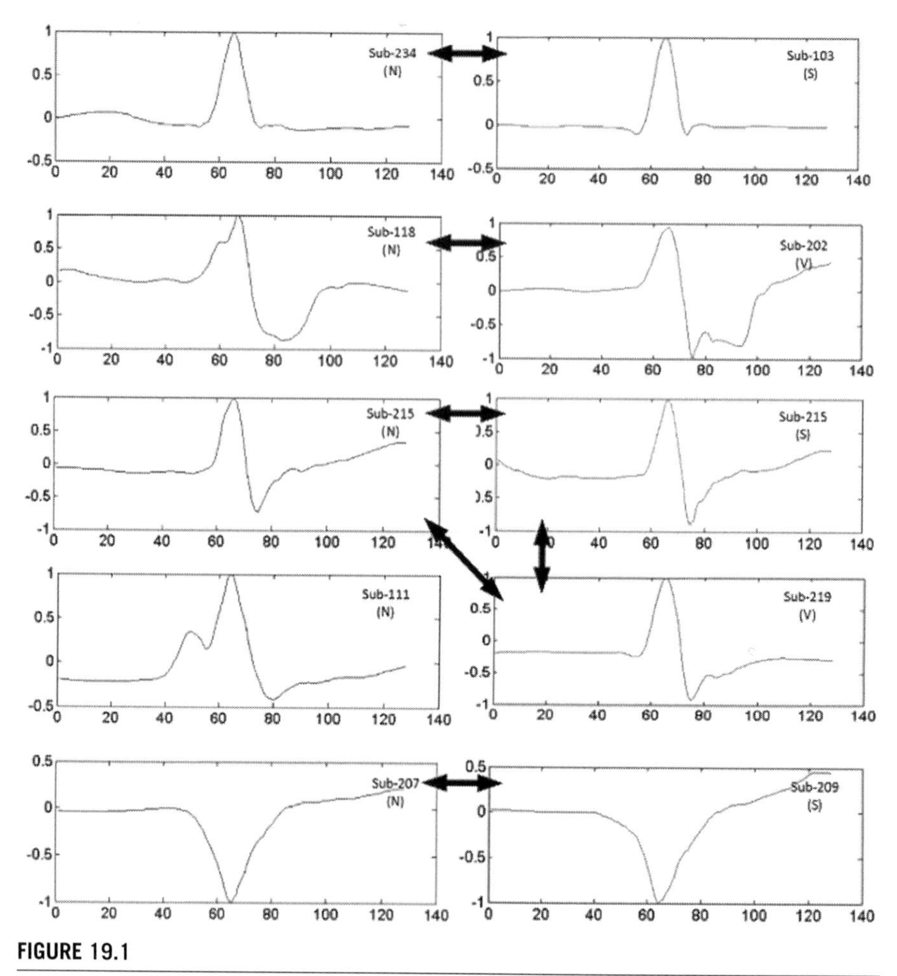

FIGURE 19.1

Normal (N) versus Abnormal (Supraventricular, S and Ventricular ectopic, V) beats from different subjects in MIT-BIH data set [34]. Typical ECG beats are shown where different subjects in the data base have entirely different normal (N) beats (as annotated by cardiologists). However, the same normal beats show a high level of similarities to other subject's abnormal beats (i.e., compare the left (normal) and right (abnormal) beats.

Another major problem is the lack of common practice when a particular method is evaluated over a benchmark data set since most of these methods vary the choice of the train and test data. To address this need, the *Association for the Advancement of Medical Instrumentation* (AAMI) recommends certain standards for evaluating the performance of the arrhythmia detection methods [25]. However, despite the numerous methods proposed in the literature, only a few [22,26–30,32], and [33] have followed the AAMI recommended standards in their approaches and less than half of them actually evaluated over all the patient data in the MIT-BIH arrhythmia database [34] although this is a crucial point for a fair comparison.

Such major deficiencies have recently brought the attention to several patient-specific designs and among them, in particular, the studies in [22,24,27–32], and [33] have demonstrated significant performance improvements over the global (one classifier for all) ECG classification methods thanks to their ability to adapt or optimize the classifier kernel according to each patient's ECG signal. However, they are typical ECG beat classifier systems that require a certain duration of training samples (e.g. 5 minutes) containing both normal and abnormal beats of a patient. In the absence of the latter, no classifier can be trained properly, and thus be used for early detection for advanced warning of abnormal beats, if and when they occur, for an otherwise healthy person with no history of cardiac problems. This is basically a "chicken and egg" problem where, in order to detect abnormal beats and accurately discriminate them from normal beats, one needs to already have a certain number of them available to learn their characteristics.

In the next section, the landmark study, [32] that achieved the *state-of-the-art* performance level in patient-specific ECG classification will be presented. Then we shall draw the focus on the significantly more challenging problem of advance warning for arrhythmia and present a real-time and personalized solution.

19.2.1 Patient-specific ECG classification by 1D convolutional neural networks

1D Convolutional Neural Networks (CNNs) can be utilized as a robust solution to the problem of accurate patient-specific electrocardiogram (ECG) classification and monitoring, especially on mobile or wearable health monitoring devices. By training with a relatively small part (<1%) of the whole available data, 1D CNNs can automatically learn to extract patient-specific features from the raw ECG signals and accurately detect cardiac arrhythmias. This architecture suits well to the philosophy of a "patient-specific" approach as opposed to the other global methods that employ the same handcrafted features and/or pre/post-processing blocks for all patients under all circumstances, and hence cannot optimally capture the characteristics of the underlying signal.

19.2.1.1 ECG data

The ECG records from the benchmark MIT/BIH arrhythmia database [34] are used to evaluate the performance of the 1D CNN-based patient-specific ECG classifier. From the database, a total of 44 out of 48 recordings (excluding 4 records with paced heartbeats) that contain two-channel 30-min duration ECG signals are selected for training and testing. The originally acquired ECG signals are first band-pass filtered at 0.1-100 Hz and then digitized at a sampling rate of 360 Hz. The signals in the database are annotated for both timing and beat class information, which is verified by independent experts. In this section, the same data partitioning used in [28] and [29] is applied, which also complies with the AAMI ECAR-1987 recommended practice

[25]. The first 20 records (numbered in the range of 100 to 124), which include representative samples of routine clinical recordings, are used to select representative beats to be included in the common training data. The remaining 24 records (numbered in the range of 200 to 234) contain uncommon but clinically significant arrhythmias such as ventricular, junctional, and supraventricular arrhythmias [42]. This second group presents more challenging arrhythmia cases for testing the performance of this method. For evaluating the performance of the 1D CNN classifier, a total of 83648 beats from all 44 records are used as test patterns. According to the AAMI recommendation, each ECG beat can be classified into the following five heartbeat types: N (beats originating in the sinus mode), S (supraventricular ectopic beats), V (ventricular ectopic beats), F (fusion beats), and Q (unclassifiable beats). In this study, we did not consider the beat detection process in ECG data and directly utilized the expert verified labels as the ground truth.

19.2.1.2 Methodology

The general view of the framework based on adaptive 1D CNNs is shown in Fig. 19.2 where the preprocessed time-domain ECG data frames of 64 or 128 samples in single and beat-trio channels are classified. As a supervised approach, there is a need for labeled ECG data by expert cardiologists. However, as shown in the experimental evaluation, only 1% of the total beats are sufficient for training the compact 1D CNNs.

After downsampling of the raw ECG data (originally at 360 Hz), both 64 and 128 sample-resolution representations are used for the evaluation. Sample raw ECG beats from the patient record 200 following downsampling and normalization are shown in Fig. 19.3. For the first channel to be fed into the CNN's input layer neuron, an equal number of samples with respect to the R peak of the beat are symmetrically taken to learn the morphological structure of the beat. To learn the temporal characteristics of each beat, a beat-trio including its neighbor beats is fed into the second channel of the CNN's input layer. Therefore the relative position of the R-peaks in the beat-trio formation can indicate some ECG anomalies related to the timing information such as the presence of an APC (S) beat. In addition to the time-domain representation of each ECG beat, its frequency domain representation by FFT magnitude and phase will also be considered as the third and fourth input channels of the 1D CNN. The purpose is to evaluate the performance gain—if any—through applying the extended raw-data representation.

The training data for the patient-specific classifier includes both global (common to each patient) and local (patient-specific) training patterns. While the first 5-minute segment from each patient's ECG record forms the local training data part that enables the classifier to adapt to the patient's ECG characteristics, the global data set contains a randomly selected small number of representative beats from each class, which allows the classifier to learn from other arrhythmia patterns that do not exist in the local training data set. This practice conforms to the AAMI recommended procedure allowing the usage of at most 5- minute section from the beginning of each patient's recording for training [25].

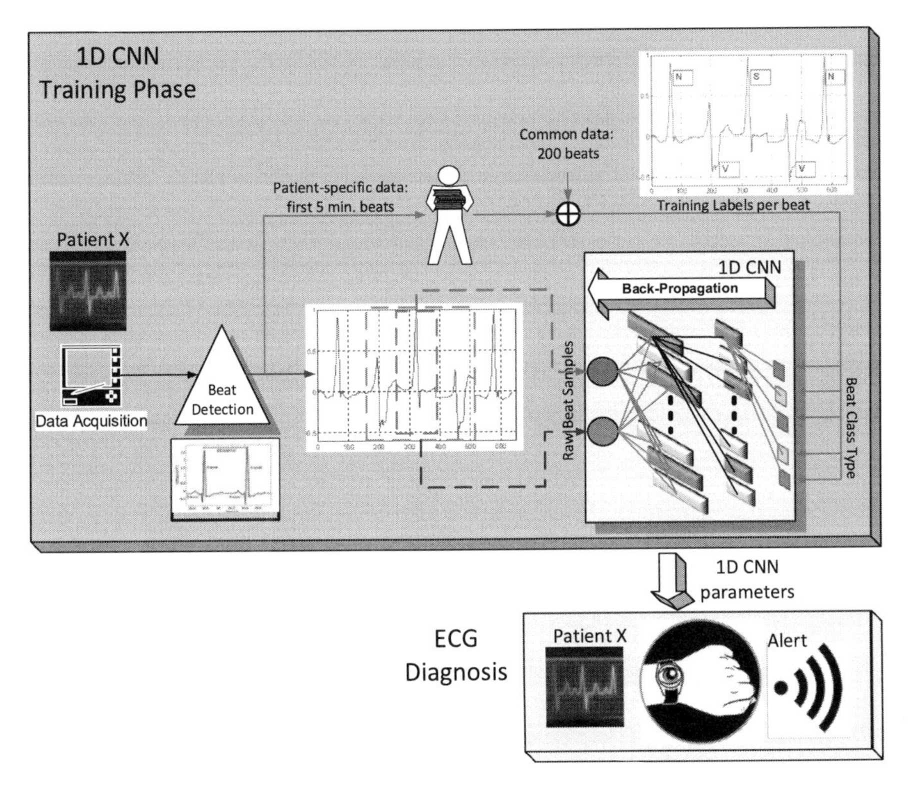

FIGURE 19.2

Overview of the approach in training (offline) and real-time classification phases [32].

This method applies adaptive 1D CNNs to each patient's raw ECG data to learn the best features for classifying cardiac arrhythmia. Similar to their conventional 2D counterparts, in 1D CNNs the input layer is a passive layer where the raw 1D signal is fed to, while the output layer is a dense layer with the number of neurons equal to the number of classes. The dense layers are identical to multilayer perceptrons, and thus sometimes called as, "MLP layers." A 1D-CNN configuration can be set *a priori* by the following hyperparameters:

1) the number of hidden CNN and MLP layers/neurons;
2) the kernel size in each CNN layer.;
3) pooling (down-sampling) factor in each CNN layer.;
4) the pooling and activation functions.

The fundamental difference between 1D and 2D CNNs is that the 2D kernels and feature maps in 2D conventional CNNs are replaced by 1D counterparts, the 1D arrays. This is a significant computational advantage for 1D CNNs because 1D convolutions are significantly faster than the 2D counterparts. In each variant, the

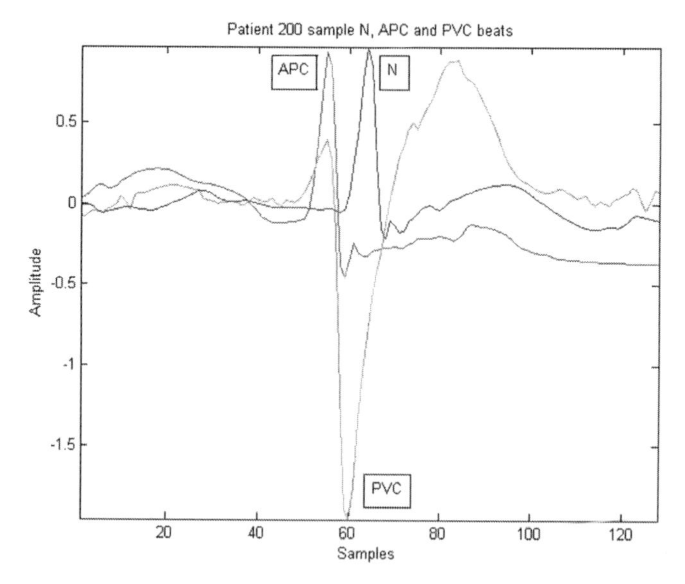

FIGURE 19.3

Sample beat representations of patient 200 following downsampling and normalization.

convolutional (CNN) layers process the raw data and "learn to extract" such features that are used in the classification task performed by dense layers. As a result, both feature extraction and classification operations are *fused* into one process that can be optimized to maximize the classification performance.

The adaptive 1D CNN implementation used in this study is illustrated in Fig. 19.4. In this illustration, the 1D filters have a size of 1×3 and the pooling (subsampling) factor is 2 where the kth neuron in the hidden CNN layer, l, first performs a sequence of convolutions, the sum of which is passed through the activation function, f, followed by the subsampling operation. In this implementation the two distinct layers (convolution and pooling) are fused into a single layer, called the "CNN layer." As a next step, the CNN layers process the raw 1D data and "learn to extract" such features, which are used in the classification task performed by the dense layers (or MLP layers). In this way, both feature extraction and classification operations are *fused* into one process that can be optimized to maximize the classification performance. This is also an adaptive implementation since the CNN topology will allow variations in the input layer dimension such that the subsampling factor of the output CNN layer is tuned adaptively. Forward and back-propagation in CNN layers will be detailed next.

In each CNN-layer, 1D forward propagation (1D-FP) is expressed as follows:

$$x_k^l = b_k^l + \sum_{i=1}^{N_{l-1}} conv1D\left(w_{ik}^{l-1}, s_i^{l-1}\right) \tag{19.1}$$

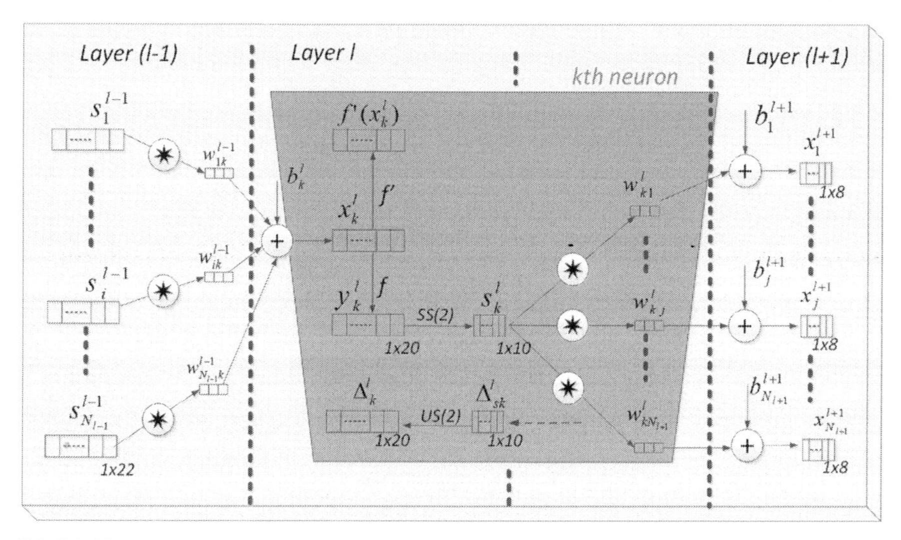

FIGURE 19.4

Three consecutive hidden CNN layers of a 1D CNN [32].

where x_k^l is defined as the input, b_k^l is defined as the bias of the kth neuron at layer l, s_i^{l-1} is the output of the ith neuron at layer $l-1$, w_{ik}^{l-1} is the kernel from the ith neuron at layer $l-1$ to the kth neuron at layer l. $conv1D(.,.)$ is used to perform "in-valid" 1D convolution without zero-padding. Therefore the dimension of the input array, x_k^l, is less than the dimension of the output arrays, s_i^{l-1}. The intermediate output, y_k^l, can be expressed by passing the input x_k^l through the activation function, $f(.)$, as

$$y_k^l = f\left(x_k^l\right) \quad \text{and} \quad s_k^l = y_k^l \downarrow ss \tag{19.2}$$

where s_k^l stands for the output of the kth neuron of the layer, l, and "$\downarrow ss$" represents the downsampling operation with a scalar factor, ss.

The back-propagation (BP) algorithm can be summarized as follows. Back propagating the error starts from the output MLP-layer. Assume $l = 1$ for the input layer and $l = L$ for the output layer. Let N_L be the number of classes in the database; then, for an input vector p, and its target and output vectors, $\boldsymbol{t}^p$ and $[y_1^L, \cdots, y_{N_L}^L]'$, respectively. With that, in the output layer for the input p; the mean-squared error (MSE), E_p, can be expressed as follows:

$$E_p = \text{MSE}\left(\boldsymbol{t}^p, \left[y_1^L, \cdots, y_{N_L}^L\right]'\right) = \sum_{i=1}^{N_L}\left(y_i^L - t_i^p\right)^2. \tag{19.3}$$

To find the derivative of E_p by each network parameter, the delta error, $\Delta_k^l = \frac{\partial E}{\partial x_k^l}$ should be computed. Specifically, for updating the bias of that neuron and all weights of the neurons in the preceding layer, one can use the chain rule of derivatives as

$$\frac{\partial E}{\partial w_{ik}^{l-1}} = \Delta_k^l y_i^{l-1} \quad \text{and} \quad \frac{\partial E}{\partial b_k^l} = \Delta_k^l \tag{19.4}$$

So, from the first MLP layer to the last CNN layer, the regular (scalar) BP is simply performed as

$$\frac{\partial E}{\partial s_k^l} = \Delta s_k^l = \sum_{i=1}^{N_{l+1}} \frac{\partial E}{\partial x_i^{l+1}} \frac{\partial x_i^{l+1}}{\partial s_k^l} = \sum_{i=1}^{N_{l+1}} \Delta_i^{l+1} w_{ki}^l. \tag{19.5}$$

Once the first BP is performed from the next layer, $l+1$, to the current layer, l, then one can carry on the BP to the input delta of the CNN layer l, Δ_k^l. Let zero order up-sampled map be: $\text{us}_k^l = \text{up}\left(s_k^l\right)$, then the delta error can be expressed as follows:

$$\Delta_k^l = \frac{\partial E}{\partial y_k^l} \frac{\partial y_k^l}{\partial x_k^l} = \frac{\partial E}{\partial \text{us}_k^l} \frac{\partial \text{us}_k^l}{\partial y_k^l} f'\left(x_k^l\right) = \text{up}\left(\Delta s_k^l\right) \beta f'\left(x_k^l\right) \tag{19.6}$$

where $\beta = (ss)^{-1}$. Then the BP of the delta error $\left(\Delta s_k^l \underbrace{\Sigma} \Delta_i^{l+1}\right)$ can be expressed as

$$\Delta s_k^l = \sum_{i=1}^{N_{l+1}} conv1Dz\left(\Delta_i^{l+1}, rev\left(w_{ki}^l\right)\right) \tag{19.7}$$

where $rev(.)$ is used to *reverse* the array and $conv1Dz(.,.)$ is used to perform full 1D convolution with zero-padding. The weight and bias sensitivities can be expressed as follows:

$$\frac{\partial E}{\partial w_{ik}^l} = conv1D\left(s_k^l, \Delta_i^{l+1}\right) \quad \text{and} \quad \frac{\partial E}{\partial b_k^l} = \sum_n \Delta_k^l(n). \tag{19.8}$$

When the weight and bias sensitivities are computed, they can be used to update biases and weights with the learning factor, ε as

$$w_{ik}^{l-1}(t+1) = w_{ik}^{l-1}(t) - \varepsilon \frac{\partial E}{\partial w_{ik}^{l-1}} \quad \text{and} \quad b_k^l(t+1) = b_k^l(t) - \varepsilon \frac{\partial E}{\partial b_k^l}. \tag{19.9}$$

19.2.1.3 Results

For experimental evaluation of the presented ECG arrhythmia classification framework, a shallow 1D CNN with only two convolutional (CNN) layers and two dense (MLP) layers is designed and tested over the commonly used MIT/BIH ECG arrhythmia data set. Through this system, we show that a computational efficient 1D CNN classifier can achieve a superior arrhythmia detection and classification performance without the need for deep learners. In all experiments, the number of neurons in the two hidden convolutional layers are 32 and 16, respectively, and 10 neurons are on the hidden dense layer. The output (MLP) layer size is set as 5 (i.e., the number of classes), and the input (CNN) layer size is either 2 (*base*) or 4 (*extended*) based on the input data representation. For 64 and 128 sample beat representations, the kernel sizes are set to 9 and 15, and the subsampling factors are set to 4 and 6, respectively.

A short training session is employed for the adaptive implementation of the 1D CNN for all experiments. The BP training algorithm is used and the stopping criterion of the BP is set to the earliest of the maximum number of iterations (50) or the minimum train classification error level of 3%. This early stopping is used to avoid overfitting of the learner to a training set. The adaptive learning rate method with global adaptation is used with an initial learning factor, ε, of 0.001. Based on the train MSE in the current iteration, ε is either increased slightly by 5% or reduced by 30% for the next iteration. Since the BP is an iterative gradient-descent-based technique, its performance is quite dependent on the initial (random) setting of the network parameters (kernels, weights, and biases). In this study, 10 individual BP runs for each patient in the database are performed and the average classification performance is reported.

The training data for the patient-specific classifier includes both global (common to each patient) and local (patient-specific) training patterns, which is parallel to the training and testing data used in [28] and [29]. While the first 5-minute segment from each patient's ECG record forms the local training data, the global data part contains a total of randomly selected 245 representative beats, including balanced samples of five classes from the first 20 patients' records in the database. The rest of each patient's record, in which 24 out of 44 records are completely new to the classifier, are used as testing examples to evaluate the trained classifier's generalization performance. Due to the significant differences in the number of beats for each class in the training and testing data sets, more suitable performance metrics such as *sensitivity*, *specificity*, and *positive predictivity* are used as performance metrics.

Table 19.1 presents the resulting confusion matrix of the adaptive 1D CNN for ECG arrhythmia detection over the complete benchmark MIT/BIH database. These matrices are reported for all 44 records (including the first 20 patient records from which the relatively small number of global training patterns are selected) and the more challenging 24 patient records of the test data set. For a fair and accurate performance comparison, the three other competing patient-specific methods, [22,27], and [28], all of which comply with the AAMI standards are selected. In the literature, few works comply with the AAMI standards, that is, [24] and [26]; however, due to significant differences in the training and testing data set partitioning, it is not possible to directly compare their results with the above selected methods. The VEB and SVEB classification performance results according to the AAMI recommendations are presented in Table 19.2. The following data partitioning of the MIT/BIH database is performed for comparing the classifiers' generalization ability. For the VEB detection problem, data set 1 is formed by including 11 test patient records (200, 202, 210, 213, 214, 219, 221, 228, 231, 233, and 234) which are common for all competing methods. Similarly for SVEB detection, a total of 14 common test patient records

Table 19.1 Summary of the ECG arrhythmia classification results for all 44 records (bottom) and the 24 test records (top) in the MIT/BIH arrhythmia database. The previous results from [28] are shown in parentheses.

		Classification result				
		N	S	V	F	Q
Ground truth	N	40963	807	350	67	4
		(40532)	(776)	(382)	(56)	(20)
	S	625	1440	149	14	1
		(672)	(1441)	(197)	(5)	(5)
	V	114	69	4247	39	2
		(392)	(299)	(4022)	(75)	(32)
	F	82	4	70	497	0
		(164)	(26)	(46)	(378)	(2)
	Q	6	2	5	0	0
		(6)	(0)	(1)	(1)	(0)

		Classification result				
		N	S	V	F	Q
Ground truth	N	73539	824	368	69	5
		(73019)	(991)	(513)	(98)	(29)
	S	837	1568	178	15	2
		(686)	(1568)	(205)	(5)	(6)
	V	230	72	5277	39	4
		(462)	(333)	(4993)	(79)	(32)
	F	92	4	73	503	0
		(168)	(28)	(48)	(379)	(2)
	Q	31	2	5	0	4
		(8)	(1)	(3)	(1)	(1)

(with the addition of records 212, 222, and 232) is used. The second and third data sets contain the whole (24) test patient records (labeled as 2xx) and the entire benchmark database with all 44 records, respectively. Note that the second data set consists of challenging and clinically significant arrhythmias such as ventricular, junctional, and supraventricular arrhythmias, and the classifier's performance is compared to [27] and [28] over this data set.

From the results in Table 19.2, it can be observed that the sensitivity and positive predictivity rates for the SVEB detection problem are comparably lower than VEB detection. On the other hand, similar high specificity performances are obtained for both VEB and SVEB detection. This significant performance difference between SVEB versus VEB detection is because SVEB class is severely under-represented in the training data (190 SVEB versus 1277 VEB cases) and as a result more SVEB beats are misclassified as normal beats. Moreover, for several patients particularly on

Table 19.2 VEB and SVEB detection performance comparison with the competing four patient-specific algorithms from the literature.

Methods	VEB				SVEB			
	Acc	*Sen*	*Spe*	*Ppr*	*Acc*	*Sen*	*Spe*	*Ppr*
Hu et al. [22][1]	94.8	78.9	96.8	75.8	N/A	N/A	N/A	N/A
Jiang and Kong [27][1]	98.8	94.3	99.4	95.8	**97.5**	74.9	98.8	78.8
Ince et al. [28][1]	97.9	90.3	98.8	92.2	96.1	**81.8**	98.5	63.4
1D CNN[1]	**98.9**	**95.9**	**99.4**	**96.2**	96.4	68.8	**99.5**	**79.2**
Jiang and Kong [27][2]	98.1	86.6	**99.3**	**93.3**	**96.6**	50.6	**98.8**	**67.9**
Ince et al. [28][2]	97.6	83.4	98.1	87.4	96.1	62.1	98.5	56.7
1D CNN[2]	**98.6**	**95**	98.1	89.5	96.4	**64.6**	98.6	62.1
Ince et al. [28][3]	98.3	84.6	98.7	87.4	97.4	**63.5**	99.0	53.7
1D CNN[3]	**99**	**93.9**	**98.9**	**90.6**	**97.6**	60.3	**99.2**	**63.5**

[1] *The comparison results are based on 11 common recordings for VEB detection and 14 common recordings for SVEB detection.*
[2] *The VEB and SVEB detection results are compared for 24 common testing records only.*
[3] *The VEB and SVEB detection results of the 1D CNN system for all training and testing records.*

Table 19.3 Confusion matrix of the ECG arrhythmia classification for patient 202.

Ground truth	Classification result					
		N	S	V	F	Q
	N	1435	258	10	0	0
	S	11	15	28	0	0
	V	4	1	10	0	0
	F	1	0	0	0	0
	Q	0	0	0	0	0

the more challenging test data set (data set 2) both the patient-specific and the global parts of the training data do not successfully characterize most of the SVEB beats and some of the VEB beats (e.g. patients 201, 202, 209, 222, and 232). Consequently, this increases the chances of the trained classifier model confuse such beats more often. Consider the example, the confusion matrix for patient 202 given in Table 19.3 where 258 normal (N) beats are misclassified as SVEB beats and 28 SVEB beats are confused with VEB beats. Fig. 19.5 illustrates a sample of four five-beat intervals selected from the test part of the patient 202's record. For this patient case, since the training data set does not include the anomalies as shown about the Normal and SVEB beats the classifier can easily misclassify them. Particularly, the plot in the bottom-left has a misclassified SVEB beat as VEB beat due to its morphological anomaly, and the SVEB beat in the bottom-right plot was misclassified as N beat due to its strong resemblance to the pattern of N beats. This proves the importance of the global training data selection for improved learning of the classifier. Rather than

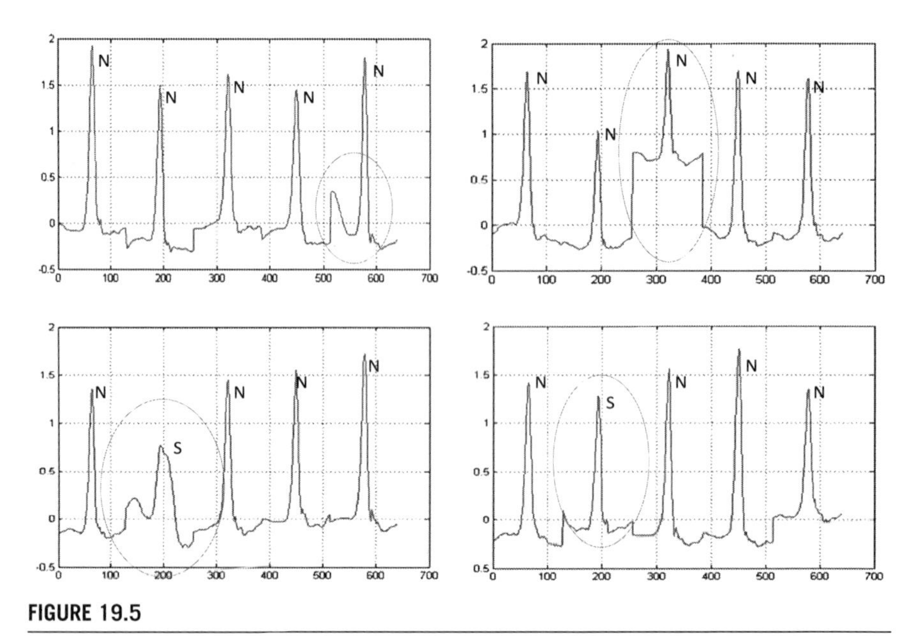

FIGURE 19.5

Four 5-beats intervals from the test section of patient 202's ECG record with the ground-truth labels.

the random selection of a balanced number of beats from each class as done in this study, an optimized selection of representative beats from each class can be done to minimize such an inevitable source of classification errors.

To test the robustness of the 1D CNN approach against the variations of the signal resolution along with the network configuration, we evaluate by varying the beat resolution (64 versus 128 samples) along with the CNN parameters and also the raw data representation (*base* versus *extended*). We observed that mostly comparable performances with the base performance level are achieved for all measures on both VEB and SVEB classification while the performance drop becomes somewhat significantly (~8%) only for the positive predictivity of the SVEB classification. Similar performances are obtained for other measures. Based on the test results, the effects of input data variations over the ECG arrhythmia detection performance are usually insignificant. Therefore, extended raw data representations, varying input data resolutions along with CNN parameters can be conveniently used within the 1D CNN classifier without any significant performance drop from the base level. However, for the case of 64 samples per beat resolution instead of 128 samples, there is a considerable gain in computational complexity due to smaller length 1D convolutions during both BP and FP phases. For the baseline implementation (using 128 samples beat resolution base raw data) of the presented ECG arrhythmia classification framework, the average time for 1 BP iteration per beat is about 4.24 msec. This time increases to 21.2 msec for a single CPU (core) implementation. Considering a complete BP run with a maximum of 50 iterations over the training data set for each patient, assuming 300 beats for the first 5-minute patient-specific data, the maximum time for training would be $600 \times 50 \times 21.2$ msec = 10.6 minutes. On the other hand, for comparison, the average training time for the previous patient-specific study [28] was about 24 minutes including the feature extraction and post-processing operations. In our implementation with OpenMP, a significantly smaller training time of fewer than 2 minutes per patient is required and it needs to be performed offline. For the 64 samples beat resolution, the average time for one BP iteration per beat reduces to 3.93 msec. The significantly low computational cost for the online ECG arrhythmia classification time is the important advantage of this system over the existing competing systems. Specifically, for the single-core implementation, the total time for an FP of a single ECG beat to obtain its classification output is about 0.58 msec and 0.74 msec for 64 and 128 sample beat resolutions, respectively. This is indeed an insignificant time complexity (more than 1000× faster than the real-time requirement) and naturally allows real-time implementation even on low-power ECG devices.

19.2.2 Personalized advance warning system for cardiac arrhythmias

As discussed earlier, despite several techniques that have appeared in the literature for ECG classification and analysis, reliability and robustness still stay a major issue. Especially for generic and global approaches where a single classifier is trained over a large ECG data set, this is the main cause of failures. Fig. 19.1 shows some typical ECG beats where different subjects in the benchmark MIT-BIH arrhythmia data set have entirely different normal (N) beats, which may, however, show a high level of morphological similarities to other subject's abnormal beats (e.g., arrows in the figure below) and vice versa. Obviously, a single classifier will inevitably confuse the patterns of normal/abnormal beats since the same pattern can exist in different beat types from different patients. As discussed earlier, the patient-specific (personalized) approach is the only logical way to address this crucial drawback. This is the main reason why the patient-specific ECG classification methods, [27–30,32], and [33], recently proposed have achieved *state-of-the-art* performances. However, such methods require *a priori* knowledge (labels) of both normal and abnormal beats of the patient. Therefore, the cardiologist's labeling is a strict requirement of all such methods and that is why they can only be applied to cardiac patients with known arrhythmia to classify their "future" ECG records automatically. In brief, they are mere diagnostic tools only for cardiac patients that aim to ease the burden of the MDs and cardiologist to classify long ECG records. For the case of "healthy" people with no history of arrhythmia, obviously, none can be used as a heart-health monitoring and advance warning system due to the simple fact that no "personal" abnormal ECG data yet exist. Even though a massive number of people die from cardiac arrhythmias each year, this is probably why only a few "personalized" methodology has ever been proposed up to date. In this section, we draw the focus on the recent method [41], which exhibited a real-time approach for *personalized* ECG monitoring. The main objective is the early (or first time) detection of arrhythmia and aim to:

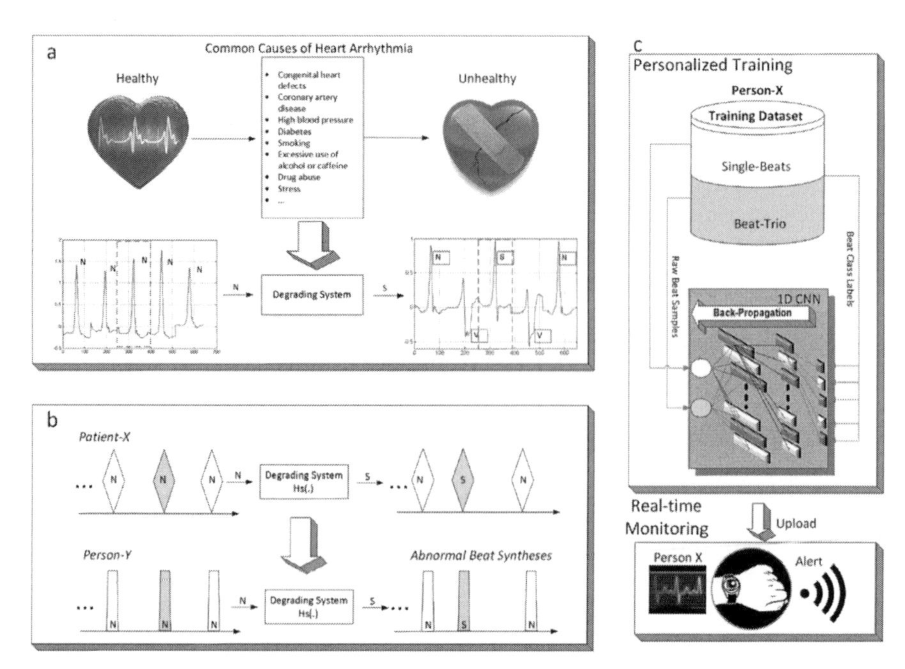

FIGURE 19.6

[41]. (a) Modeling common causes of cardiac arrhythmia in the signal domain by a degradation system. (b) A symbolic illustration of an abnormal *S* beat synthesis for *Person-Y* using the degradation system designed from *Patient-X's data*. (c) Illustration of the overall system, where a dedicated CNN is trained by back-propagation over the training data set created for *Person-X* (top) and then used for heart monitoring and advance warning system (bottom) for *Person-X*.

1) use abnormal beat syntheses (ABS) filters to model the common causes of cardiac arrhythmia,
2) synthesize potential arrhythmic beats for a "healthy person" using the ABS filters applied over the regular (or average) ECG beat,
3) create the "personalized" training data set using (1) and (2) for a healthy person,
4) train the personal classifier, the 1D CNN, over the personalized training data set,
5) use the personal classifier in real-time to detect any real arrhythmia –if and when occurs.

Fig. 19.6 illustrates the steps of the systematic approach in [41] where (a library of) ABS filters are generated in (a) to model the common causes of cardiac arrhythmias in the signal domain. In other words, each ABS filter captures the "degradation model" that occurred from a normal beat to an arrhythmic beat. This captured model can then be used to synthesize a potential arrhythmic beat from the normal beat of the healthy person as shown in (b). The idea behind this is the following: If one day the same (common) degradation occurs in that person's heart, the classifier that was already trained over the synthesized beat will be able to detect (classify) the real—degraded—beat accurately. Note that although the degradation model is common, the synthesized beat is personalized for the person, because it is generated from his/her normal beat. Fig. 19.6(b) presents an illustrative example for this fact, that is, the degrading model captured from a patient's (Patient-X) Supraventricular (S) beat is used to synthesize the (personal) S beat for a healthy person (Person-Y). This example also explains clearly why a generic method will usually fail—because a single classifier trained over the S beat of Patient-X will not accurately classify the S beat of Person-Y due to the morphological differences.

Ultimately, once a sufficient number of degradations are captured, then they are used to synthesize "potential" abnormal beats for a healthy person. Once again note that all the synthesized abnormal beats are "personalized," that is, dedicated for the healthy person since the input signal used now belongs to the healthy person's normal ECG beat. The ABS filter will degrade it just like the way the "common cause" has degraded the normal beat of the patient. But the output beat, the synthesized abnormal beat, is now likely to be one of the potential abnormal beats of the healthy person providing that (s)he has the same/similar cause for degradation.

19.2.2.1 ABS filter

The abnormal beat synthesis (ABS) is a linear and time-invariant (LTI) system model that characterizes the degradation of a normal ECG beat to an arrhythmic (abnormal) beat. Such a model basically features a single cause of arrhythmia that transforms the healthy heart with continuous normal beats to an unhealthy one that produces the arrhythmic beats. In a conventional input-output LTI system, the output is computed by linear convolution as

$$b[n] = h[n] * a[n] = \sum_{m=0}^{M-1} a[n-m]\,h[m]\,(M \le N) \tag{19.10}$$

where a and b are N-length input and output signals corresponding to (regular) normal and abnormal beats, respectively, and h is the M-length filter coefficients of the LTI system. Note that the convolution output will have the length, $N + M - 1$ where we only consider the first N samples since the output signal (abnormal beat) has the same length as the input. This can be written in the matrix equation as follows:

$$\begin{bmatrix} a[0] & 0 & 0 & 0 & \dots & 0 \\ a[1] & a[0] & 0 & 0 & \dots & 0 \\ a[2] & a[1] & a[0] & 0 & \dots & 0 \\ \dots & \dots & \dots & \dots & \dots & \dots \\ a[N-1] & a[N-2] & a[N-3] & \dots & \dots & a[N-M] \end{bmatrix} \begin{bmatrix} h[0] \\ h[1] \\ \dots \\ \dots \\ h[M-1] \end{bmatrix}$$

$$= \begin{bmatrix} b[0] \\ b[1] \\ \cdots \\ \cdots \\ b[N-1] \end{bmatrix} \tag{19.11}$$

or equivalently in a linear system equation,

$$Ax = b \tag{19.12}$$

where A is the $N \times M$ matrix with the shifted input signal samples, $\boldsymbol{x}$ is the column vector of filter coefficients, ($x(i) = h[i]$) and $\boldsymbol{b}$ is the column vector of the output signal, b. The Least Squares (LS) solution of this equation, x_{LS}, can be expressed as follows:

$$x_{LS} = \min_{x \in R^M} \|b - Ax\|^2 = \left(A^T A\right)^{-1} A^T b \tag{19.13}$$

However, matrix A may not be of full rank, that is, $rank(A) = r < M$, in which case, $A^T A$ will be singular and the inverse cannot be computed. To address this problem, we make use of the Singular Value Decomposition (SVD) of A as follows:

$$A = U \Sigma V^T = \sum_{i=1}^{r} \sigma_i u_i v_i^T \tag{19.14}$$

where $\boldsymbol{U}$ and $\boldsymbol{V}$ are $N \times N$ and $M \times M$ orthogonal matrices which hold the eigenvectors of the square matrices, AA^T and $A^T A$, respectively, as the column vectors. The $N \times M$ matrix, Σ, can be expressed as

$$\Sigma = \begin{bmatrix} \sigma_1 & 0 & 0 & \ldots & 0 & 0 & \ldots & 0 \\ 0 & \sigma_2 & 0 & \ldots & 0 & 0 & \ldots & 0 \\ 0 & 0 & \sigma_3 & \ldots & 0 & 0 & \ldots & 0 \\ \ldots & \ldots & \ldots & \ldots & \ldots & \ldots & \ldots & 0 \\ 0 & 0 & 0 & \ldots & \sigma_r & 0 & \ldots & 0 \\ 0 & 0 & 0 & \ldots & 0 & 0 & \ldots & 0 \\ \ldots & \ldots & \ldots & \ldots & \ldots & \ldots & \ldots & \ldots \\ 0 & 0 & 0 & 0 & 0 & 0 & \ldots & 0 \end{bmatrix} \tag{19.15}$$

where $\sigma_1 > \sigma_2 > \ldots > \sigma_r$ are the singular values or equivalently the eigenvalues of matrices, $A^T A$ and AA^T. This can yield the LS solution, x_{LS}, regardless of whether or not A is singular,

$$x_{LS} = V \Sigma^{-1} U^T b = \sum_{i=1}^{r} \frac{1}{\sigma_i} v_i u_i^T b. \tag{19.16}$$

However, the LS solution, x_{LS}, can still yield large values, the so-called "explosion" of the LS solution, due to noisy values in matrix A (the input) or vector b (the output), or both since in general, they are both susceptible to noise. A crucial disadvantage *therein* is that the smaller nonzero singular values will result in an even larger explosion of x_{LS}. To prevent this, we shall *regularize* the LS solution by optimizing the LS error together with the magnitude of the LS solution as

$$x_{RLS} = \min_{x \in R^M} \left(\|b - Ax\|^2 + \lambda^2 \|x\|^2 \right) \tag{19.17}$$

where λ is the regularization parameter. It is straightforward to show that this joint optimization can be expressed as

$$x_{RLS} = \min_{x \in R^M} \left\| \begin{pmatrix} b \\ 0 \end{pmatrix} - \begin{pmatrix} A \\ \lambda I \end{pmatrix} x \right\|^2 = \min_{x \in R^M} \|b_\lambda - A_\lambda x\|^2 \tag{19.18}$$

where A_λ is now an $M \times (M+N)$ full-rank matrix ($r = M$) and therefore, the LS solution over A_λ can be obtained by using Eq. (19.13) as

$$x_{LS}(\lambda) = \min_{x \in R^M} \|b_\lambda - A_\lambda x\|^2 = \left(A_\lambda^T A_\lambda\right)^{-1} A_\lambda^T b_\lambda = \left(A^T A + \lambda^2 I\right)^{-1} A^T b \tag{19.19}$$

The ith eigenvector of $\left(A_\lambda^T A_\lambda\right)$ can be obtained by solving

$$A_\lambda^T A_\lambda v_i = \left(A^T A + \lambda^2 I\right) v_i = \left(\sigma_i^2 + \lambda^2\right) v_i. \tag{19.20}$$

So it is clear that matrix $A_\lambda^T A_\lambda$ has the same eigenvector, v_i, as matrix $A^T A$ but a larger eigenvalue, $\left(\sigma_i^2 + \lambda^2\right)$. Therefore using the orthogonality of the eigenvectors, one can write the eigenvector decomposition of $A_\lambda^T A_\lambda$, and its inverse as follows:

$$\begin{aligned} A_\lambda^T A_\lambda &= \sum_{j=1}^{M} \left(\sigma_j^2 + \lambda^2\right) v_j v_j^T = V \Lambda V^T, \\ \left(A_\lambda^T A_\lambda\right)^{-1} &= \sum_{j=1}^{M} \frac{1}{\left(\sigma_j^2 + \lambda^2\right)} v_j v_j^T = V \Lambda^{-1} V^T. \end{aligned} \tag{19.21}$$

Using Eqs. (19.14) and (19.16) above yields the regularized LS solution, x_{RLS}, expressed as

$$\begin{aligned} x_{RLS} = x_{LS}(\lambda) &= \left(A_\lambda^T A_\lambda\right)^{1} A^T b \\ &= \left(\sum_{j=1}^{M} \frac{1}{\sigma_j^2 + \lambda^2} v_j v_j^T\right) \left(\sum_{i=1}^{r} \sigma_i v_i u_i^T\right) b \\ &= \left(\sum_{j=1}^{r} \frac{\sigma_j}{\sigma_j^2 + \lambda^2} v_j u_j^T b\right). \end{aligned} \tag{19.22}$$

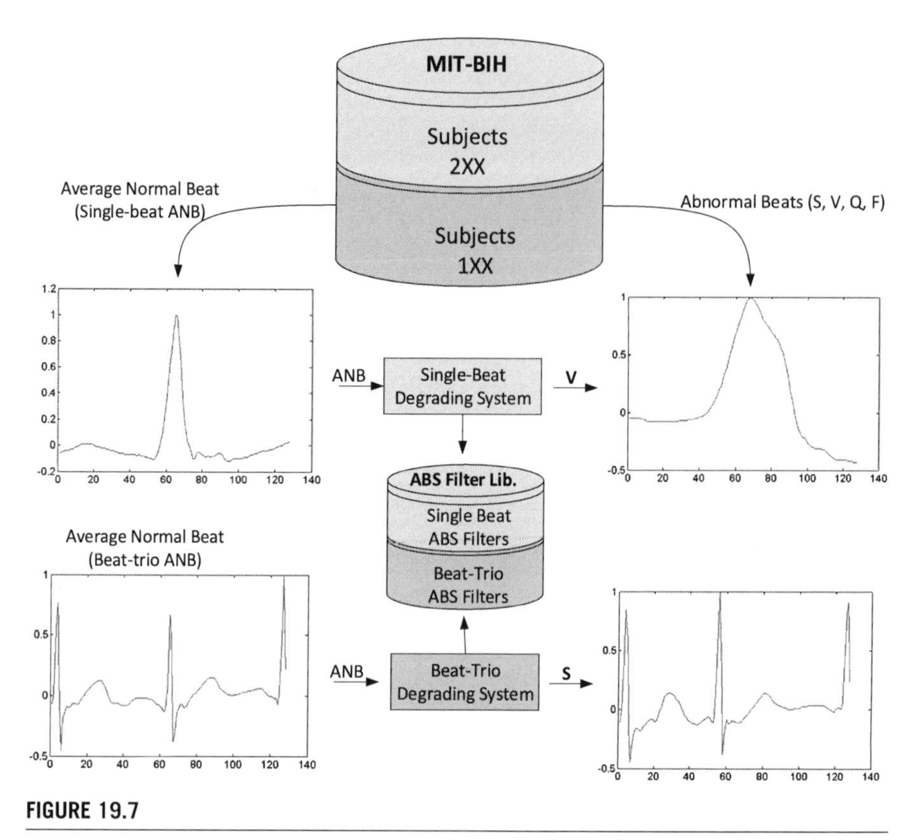

FIGURE 19.7

Illustration of two ABS filter designs: the top one is for a V-beat in single-beat representation and the bottom one is for an S-beat in a beat-trio representation.

Comparing regularized LS solution in Eq. (19.22) with the LS solution in Eq. (19.16), the explosive effects of those noise-like singular values over the solution can now be significantly suppressed with a practical choice of λ, for example, $0.1 \leq \lambda \leq 0.5$. As a result, one can design an ABS filter using Eq. (19.22) for each pair of normal-abnormal ECG beats and as illustrated in Fig. 19.7, a library of ABS filters can be designed for each representation (single-beat and beat-trio).

19.2.2.2 ABS filter selection

In [41], more than 6000 ABS filters from the train partition of the MIT-BIH database are extracted since each abnormal beat in a patient's record was used to create one ABS filter; however, as shown in Fig. 19.8, many of these filters have very similar coefficients indicating a similar type of degradations (or the common cause) encountered. Therefore, those filters with different coefficients have been selected, and thus, those redundant filters with similar coefficients were removed. The normalized variance was used as the criterion to perform this selection and as a result, 345 filters were selected.

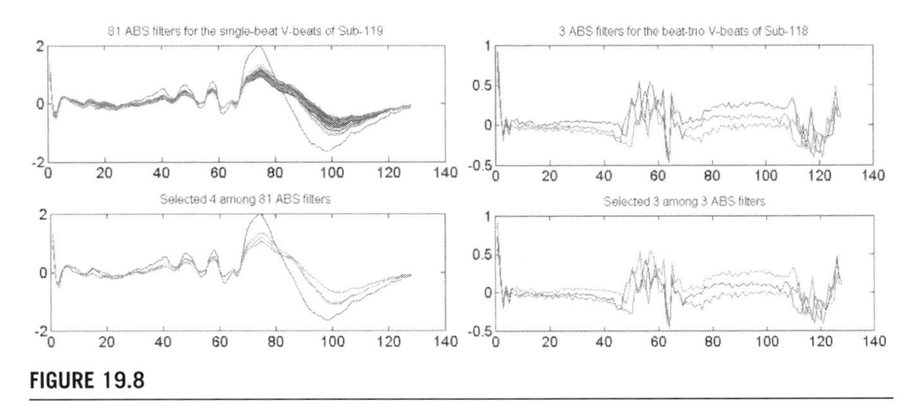

FIGURE 19.8

ABS filters (total and selected) for V beats for subjects 119 (left) and 118 (right).

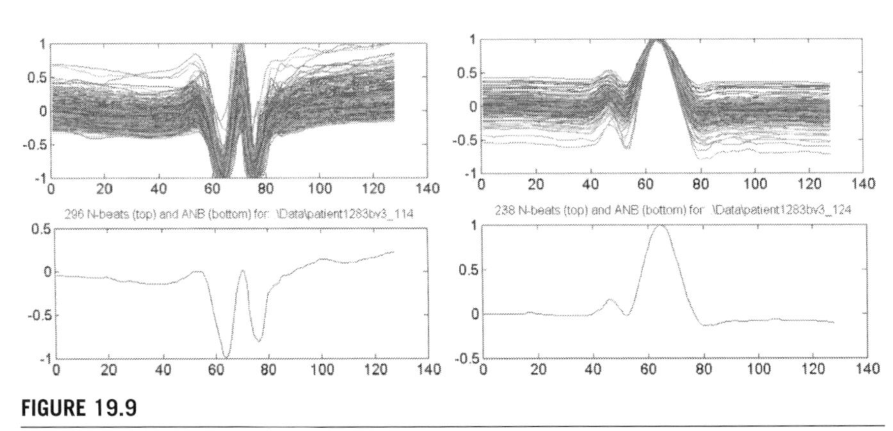

FIGURE 19.9

Beat-trio (top) and single-beat (bottom) ANB selections for subjects 114 (top) and 124 (bottom).

19.2.2.3 Evaluation of ABS filters

The evaluation of the ABS was performed over the benchmark MIT/BIH data set [25]. For each beat, there are two beat representations: single beat and beat-trio with three consecutive beats. In order to extract the normal beat of a patient, the so-called *average normal beat* (ANB) is formed from the first 5 minutes of the ECG record. The ANB beat is the one selected among the N-beats during this period, which resembles the most to the average of the N-beats. Fig. 19.9 shows two typical ANB selections for the two subjects with IDs 114 and 124.

Over the training partition of the database 345 filters were created in the ABS filter library and they were basically to synthesize abnormal beats for the subjects in the test partition. For the parameter selection, the authors set $\lambda = 0.2$ and the minimum variances for filter selection are, 0.1 and 0.25 for the S and V type abnormalities, respectively. The reason for different variance settings is the scarcity of S beats compared to V beats, hence the final number of filters for each of them is balanced.

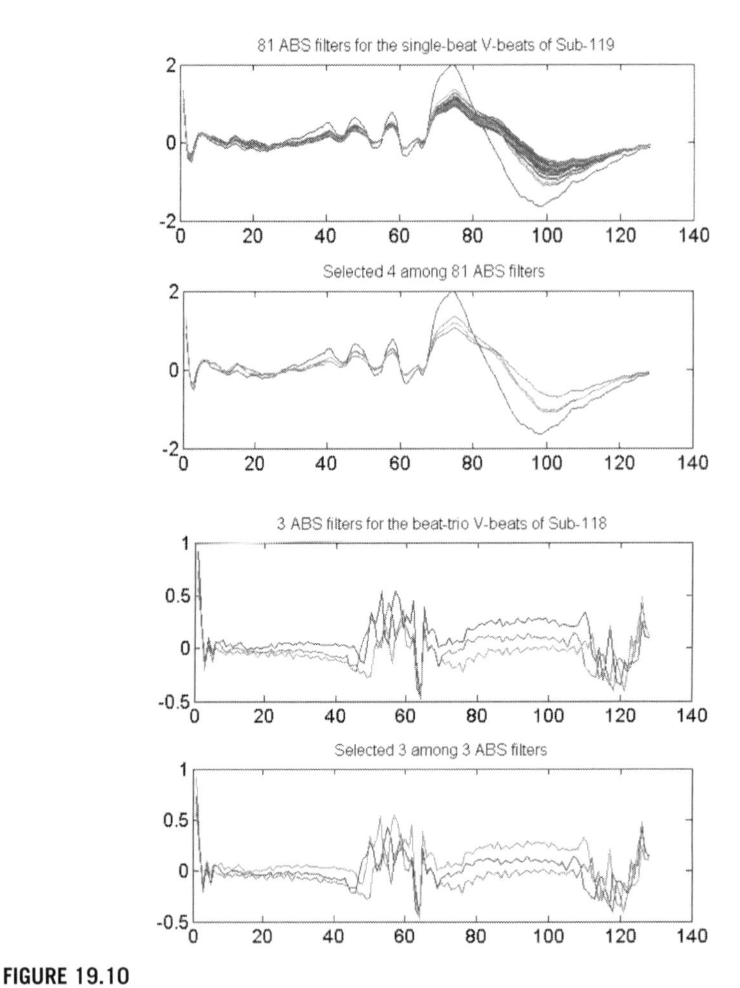

FIGURE 19.10

ABS filters (total and selected) for V beats for subjects 119 (left) and 118 (right).

This variance setting yields 239 filters from the common part and 106 filters from the patient-specific (i.e., first 5 minutes) part. There is no selection for Q and F type anomalies since only a few ($13+7=20$) of them exist in the training partition. So, all ABS filters for Q and F type abnormal beats are kept in the library. Fig. 19.8 presents the ABS filters formed for the V beats from the patient-specific training data of the subjects with IDs 118 and 119. Note that for subject 119 only 4 out of 81 ABS filters are selected due to the high similarity among the filter coefficients while all filters are selected for subject 118 (Fig. 19.10).

Once the ABS filter library is formed, then for each subject in the test partition the personalized training data set will be generated as explained earlier and illustrated in Fig. 19.7. Fig. 19.11 shows real abnormal beats in the patient-specific part and the

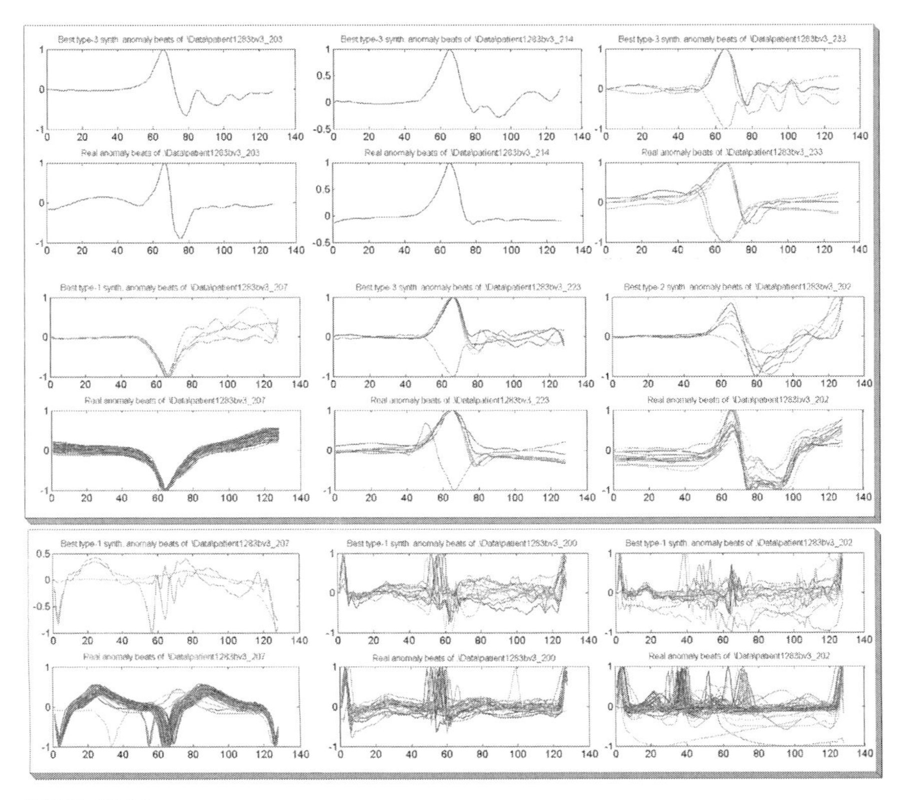

FIGURE 19.11

Real beats and their most similar synthesized beats of some subjects in the test partition. In both single-beat (upper block) and beat-trio (lower block) representations, the top subplots show the synthesized beats.

most similar abnormal beats synthesized by the ABS filters from the corresponding ANB. These are the typical results obtained showing that the abnormal beat pattern (S, V, or Q) can indeed be synthesized with a proper ABS filter as long as that filter has already been formed in advance. Some other typical examples with a higher number of real abnormal beats (e.g., >60) can be seen in Fig. 19.12 where MSE plots are also drawn for quantitative evaluation. The plots basically show the (dis) similarity scores in terms of minimum MSE between the real abnormal versus ANB (blue plot) and synthesized beats (red). It is obvious that most of the synthesized beats by ABS filters are highly similar to real abnormal beats than the ANB. The opposite is also true. The synthesized abnormal beats are usually more different than the N beats.

Fig. 19.13 presents two plots for subjects 220 and 223 showing the synthesized (top) versus N beats (middle) and the MSE plots (bottom) in beat-trio representation. It is clear that the majority of N beats are more similar to the ANB than the most similar synthesized beat. However, for some N beats this is not the case especially

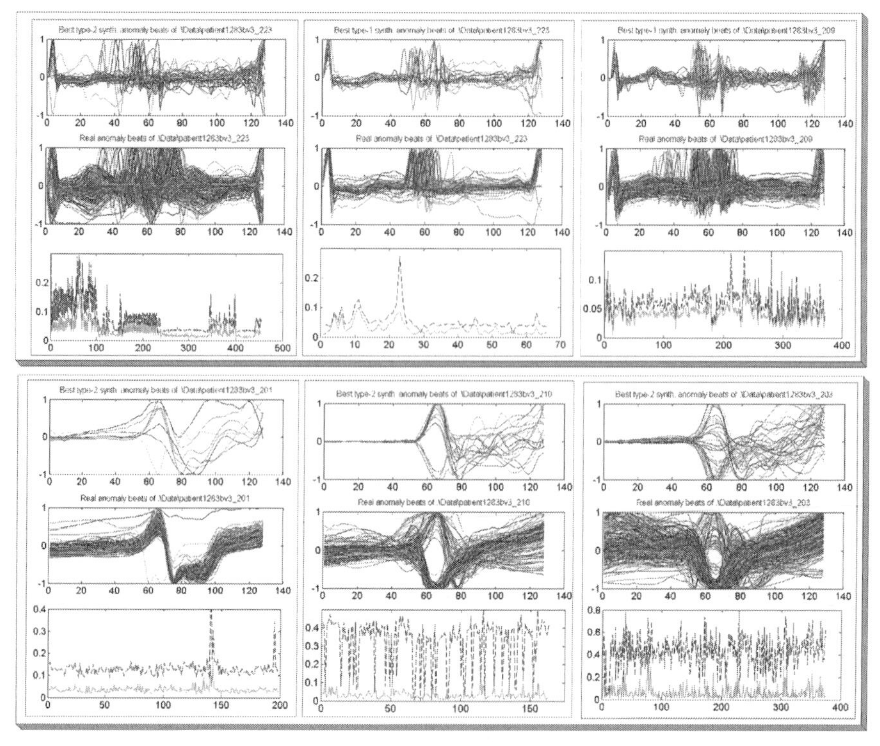

FIGURE 19.12

Real beats and their most-similar synthesized beats of some subjects in the test partition. In both single-beat (lower block) and beat-trio (upper block) representations, the top subplots show the synthesized beats and the bottom plots show minimum MSE of real vs. ANB (blue plot) and real versus synthesized beats (red) for each real abnormal beat. (For interpretation of the colors in the figure, the reader is referred to the web version of this chapter.)

when the N beat has a similar pattern to S or V beats. This is also apparent for some subjects. For example, in Fig. 19.14, the ground-truth labels of subjects 200 and 201 are shown for some beats in the patient-specific part. As the red arrows indicate those beats (V beat for subject 200 and 6 N beats for subject 201) show typical characteristics of S beats. These and some other similar cases where the S beat characteristics are quite visible on the beats with different ground-truth labels are shown in beat-trio representations in Fig. 19.15. It is evident that such beats will naturally have the highest similarity to a (synthesized or real) S beat and this is one of the reasons for having relatively low sensitivity scores for S beats in recent patient-specific ECG classification works such as [27–30]. In some patients, this is also true for the V beats which have high similarities to N beats and vice versa. Such ambiguous cases eventually yield false-positives for abnormal beat detection or misclassifications among abnormal beats, for example, 103 N beats of Subject-203 (out of 422) and 35 V beats of Subject 207 (out of 101) are most similar to synthesized S beats.

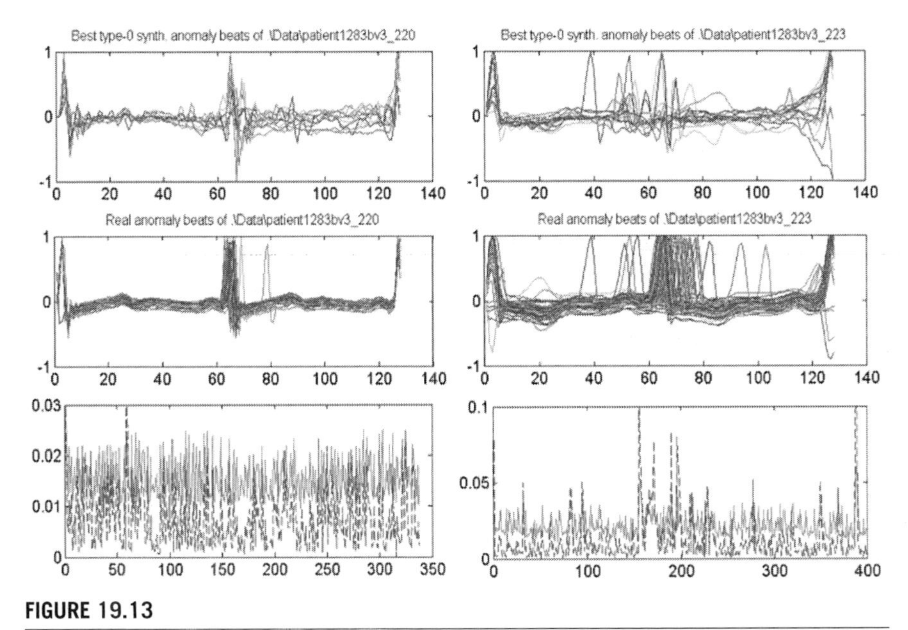

FIGURE 19.13

N beats and their most-similar synthesized beats of subjects 220 (left) and 223 (right) in beat-trio representation.

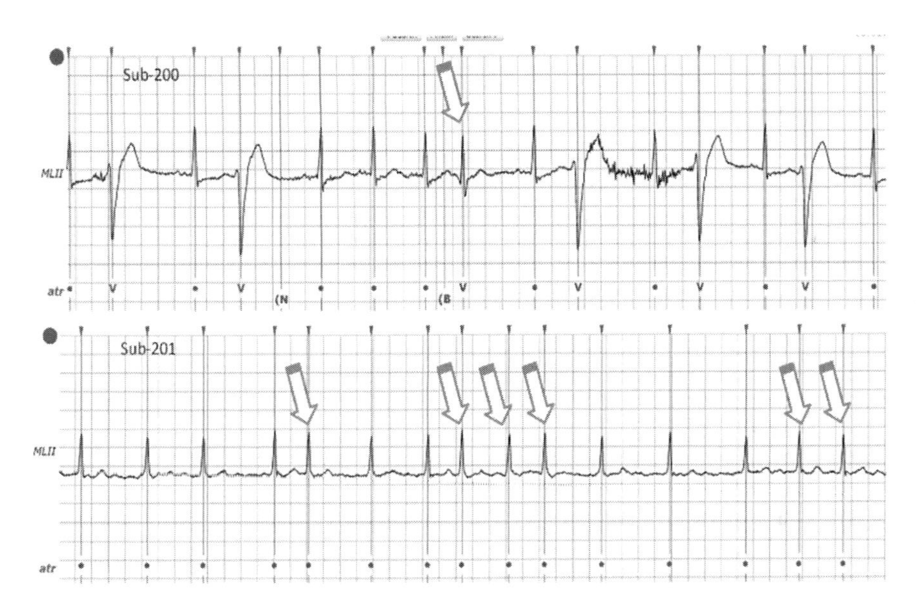

FIGURE 19.14

Red arrows show some typical S beats with different ground-truth labels: V beat for subject 200 (top) and N beats for subject 201 (bottom). (For interpretation of the colors in the figure, the reader is referred to the web version of this chapter.)

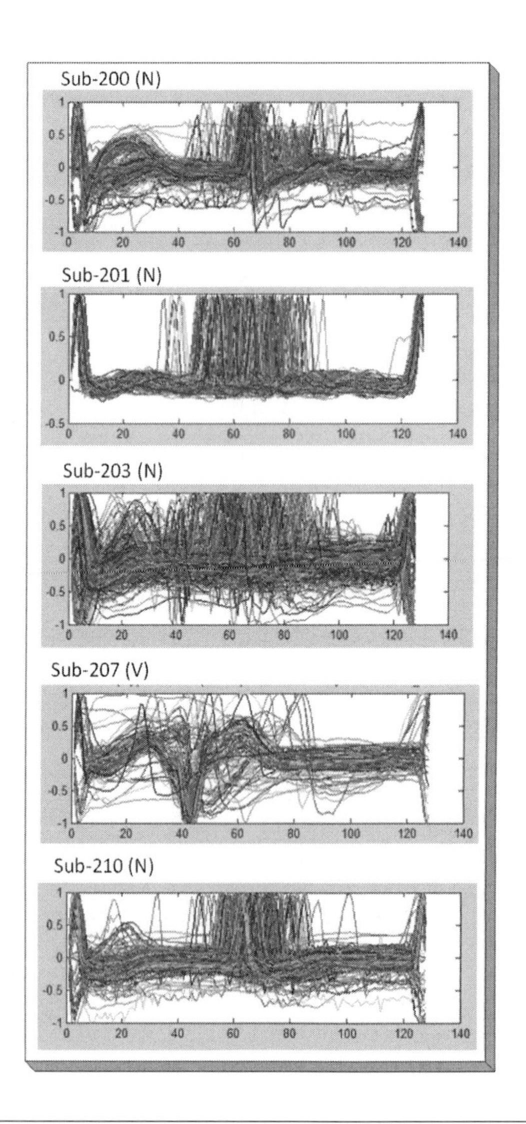

FIGURE 19.15

Cumulative plots of N (subjects 200, 201, 203, 210) and V (subject 207) beats in beat-trio representation.

Table 19.4 presents the Confusion Matrix (CM) cumulated by the classification results of 10 CNNs each of which is trained in a distinct backpropagation (BP) run. Table 19.5 shows the 2×2 deducted CM using the standard performance metrics, that is, accuracy (Acc) and false-alarm rate (FAR), and it was reported that $Acc = 80.1\%$ and $FAR = 0.43\%$. The average probability of missing the first abnormal beat is therefore 0.199 and the average probability of missing three consecutive abnormal beats is around 0.0079. Therefore this method yields a detection of at least one arrhythmic

Table 19.4 Cumulated CM for the test data set over 10 runs.

		Ground truth				
		N	S	V	F	Q
Real	N	272202	6025	2643	253	20
	S	866	9058	11437	92	0
	V	277	1923	12734	60	0
	F	23	17	204	2	0
	Q	2	7	452	3	0

Table 19.5 2×2 confusion matrix deducted from the CM in Table 19.4.

		Ground truth	
		N	A
Real	N	272202	8941
	A	1168	35989

beat during the first three occurrences of arrhythmic beats with a high probability (>99.2%).

This approach also has highlighted some potential mislabeling in the MIT-BIH data set. For example, Fig. 19.16 presents typical false-alarm cases of the approach for 6 patients (the bottom-6 subplots) in beat-trio representation. When compared with the average N beat (ANB) on the top, those typical false alarms that are classified as arrhythmic beats by the approach indeed exhibit strong characteristics of V and S beats. Especially some of these N beats have similar characteristics of typical premature supraventricular contractions or premature ventricular contractions. Even if they are not mislabeled, obviously the detection of such highly variant or degraded N beats will still be desirable since they may indicate a forthcoming arrhythmia or a temporary degradation— a severe variation of the heart condition.

19.3 Early prediction of mortality risk for COVID-19 patients

19.3.1 Introduction and motivation

Since the outbreak of the disease, COVID-19 has spread rapidly throughout the world and caused the death of 5.24 million people. It has been reported from a study of 2449 patients that the COVID-19 pandemic has overwhelmed the hospital capacity with 20% to 31% utilization and the intensive care unit (ICU) with 4.9% to 11.5% admission [43]. Considering the significant impact of the early prediction tool for detecting the mortality risk of COVID-19 patients, it is vital to stratify the prime biomarkers, which can help to identify the progression of the disease at admission. An early detec-

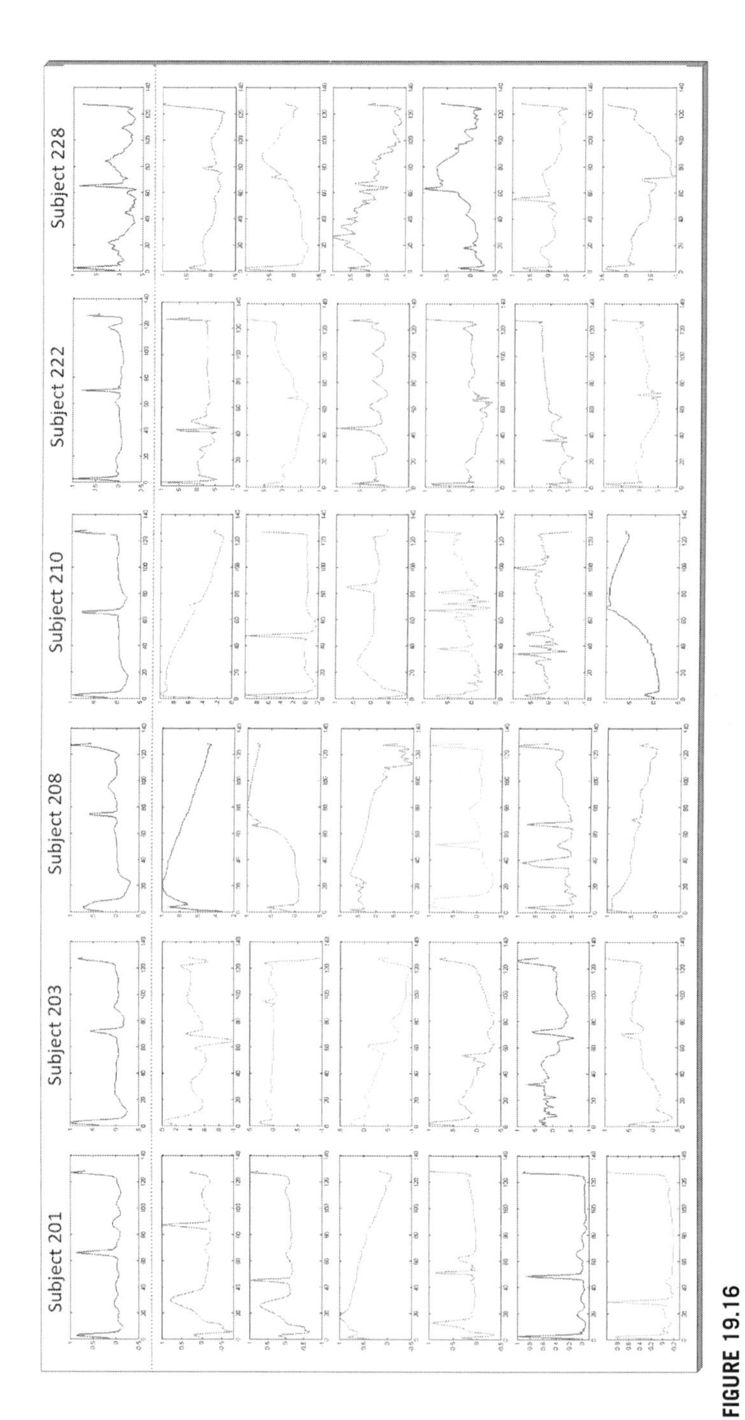

FIGURE 19.16

ANBs (top-most) and typical false alarm beats (bottom 6 plots) for subjects: 201, 203, 208, 210, 222, and 228 (from right to left).

tion technique can significantly reduce the burden on the healthcare system by proper utilization of healthcare resources through the prioritization of the patients at high risk to reduce the mortality rate. Several machine learning-based models were proposed for the early detection of the mortality risk of high-risk COVID-19 patients. However, these models were not developed either using state-of-the-art machine learning models or most of them are not using scoring technique to quantitatively assess the importance of the biomarkers and death probability with the most important predictors. It is important for the healthcare professional and policymakers to identify the patients at high risk at admission to ensure proper utilization of healthcare resources and treatment plans. This becomes predominantly important during the pandemic due to the sudden surge of patients in the hospital. Moreover, the horizontal monitoring of the critical patients during the hospital stay using a simple, easy-to-use, and reliable scoring system will be extremely valuable for healthcare professionals. On the other hand, very low or low-risk patients can be easily monitored from the ambulatory facilities or at home by self-quarantine to reduce the patient admissions at the hospital to minimize the load on healthcare facilities.

A simple, easy-to-use but reliable machine learning model-based scoring system for the risk severity stratification technique is presented in this section, which can calculate the risk of death of the COVID-19 patients using important biomarkers collected at the hospital admission [44]. This will help to stratify them into appropriate risk groups so that the healthcare professionals can provide necessary health support according to their health condition. For this purpose, an early warning system is developed using a machine learning model-based nomogram score and used for identifying the most contributory biomarkers for the classification of death and survival of COVID-19 patients among 76 biomarkers available in the cohort of 375 COVID-19 patients. Machine learning models were used to identify the top-ranked biomarkers and best performing classification model to stratify the patient outcome and a multivariable logistic regression-based nomogram-based scoring system was then developed and validated to predict the mortality risk of COVID-19 patients. A mobile application was developed to help the medical professional to use the scoring system to easily monitor the death risk of the patients longitudinally.

Several researchers reported that blood biomarkers can provide crucial health status information of the COVID-19 patients to assess the risk of the disease and mortality. Kuwait Progression Indicator (KPI) scoring system is a prognostic model for predicting the progression of COVID-19 severity [45,46], which was created using quantifiable laboratory values. The model was not developed using the patient's reported sign-symptoms and biased-scoring system. The KPI method stratifies COVID-19 infected patients into a low and high-risk group if the KPI is <-7 and >16, respectively. However, the disease advancement risk becomes uncertain with scores between -6 to 15 even though there could be a large number of patients in this intermediate group, which limits the usability of this scoring system for different risk group patients. ANDC (age, neutrophil-to-lymphocyte ratio (NLR), D-dimer, and C-reactive protein (CRP)) scoring system was developed by Weng et al. [47] using a cohort of 301 patients as an early prediction score, which can predict the mortality risk for COVID-19 patients. Age, neutrophil-to-lymphocyte ratio (NLR), D-dimer, and C-reactive protein (CRP) were identified as key-biomarkers for the mortality prediction for COVID-19 patients using Least Absolute Shrinkage and Selection Operator (LASSO) regression (Weng et al., 2020). A nomogram with an integrated ANDC scoring system with two cut-off values to categorize the COVID-19 patients into low, moderate, and high-risk groups and their corresponding probability of death. The model showed reasonable performance for the development data set however performance degrades for the validation set. The ANDC scoring system has two cut-off values of 59 and 101 and the corresponding death probability of <5% for the low-risk group and >50% for high-risk groups, respectively, and an intermediate group called moderate risk group of 5% to 50% death probability with an ANDC score between 59 and 101. Xie et al. [48] proposed a prognostic model using a cohort of 444 COVID-19 patients to identify the key predictors of COVID-19 death risk using age, SpO2, lymphocyte count, and lactate dehydrogenase. The model showed an AUC validation of 0.89 and 0.98 for internal and external respectively. However, it was reported that the model was over and underfitting for high and low-risk patients of the external validation set, respectively.

A novel machine learning model was proposed by Yan et al. [49] using lactic dehydrogenase (LDH), high-sensitivity C-reactive protein (hs-CRP), and lymphocyte, three biomarkers to envisage the mortality risk of COVID-19 patents on an average 10 days in advance with an accuracy of 90%. It was reported that LDH alone can play a major role in the prediction of COVID-19 severity in most cases and the patients can be categorized as high-risk patients who require the special attention of medical doctors. However, this work did not propose any quantitative scoring system to help the medical doctor to predict the patients' severity and categorize them accordingly. Therefore, this work is a scholarly important contribution, however, not useful to the medical doctor in the stratification decision. In addition to that the claim of early prediction on an average 10 days earlier is not a very strong matrix as a good number of patients admitted on the date of death while other patients admitted even 32 days earlier than the date of the patient outcome. Thus, this average information will not reflect the performance of the model at the individual level. A clinical investigation carried out on 82 COVID-19 patients has demonstrated that 100%, 89%, 80.5%, 78%, and 31.7% of COVID-19 patients died with respiratory, cardiovascular, sepsis, hepatic, and renal failure, respectively. The patients with increased CRP and D-dimers were predominantly observed among 100% and 97.1% patients, respectively, who have the outcome of death. D-dimer was found to be greater than 1 μg/L at admission for the patients who died later [50–52]. However, this study was carried out on a small group of patients and there was no scoring system developed in this study. A prospective study was carried out on 1096 patients admitted to Jaber Hospital, Kuwait between 24 February to 20 April 2020, and the KPI model was developed and validated with 375 patients' data from [49] as external validation. The KPI calculator is made available publicly at covidkscore.com and the model works with individual cut-off values for six KPI biomarkers. It was therefore difficult to compare the patient's condition longitudinally [45].

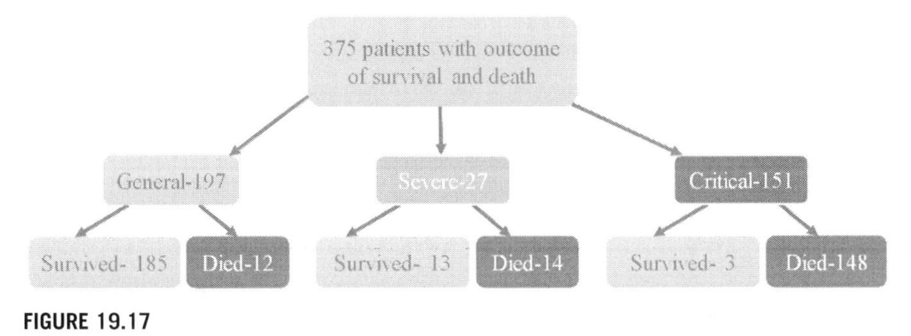

FIGURE 19.17

Outcome tree of the COVID-19 positive patients with the initial condition at admission.

19.3.2 Methodology

There are multiple blood samples collected during the hospital stay periods of the patients, however, the blood sample collected at admission were used for the model development and validation, and later on, the longitudinal study was carried out to see whether the model is able to predict the patient's outcome correctly at different time points or not.

19.3.2.1 Study participants

The blood samples were collected from 375 patients from the Tongji Hospital, Wuhan, China between 10 January and 18 February 2020. Blood samples were retrospectively examined to detect consistent and pertinent markers to predict death risk. Demographic, epidemiological, laboratory, clinical, and mortality outcomes were recorded. This data set was originally collected by Yan et al. [49]. The authors have shared the data set publicly. The retrospective clinical study was originally permitted by the Ethics Committee of the Tongji Hospital, Wuhan. The patients with some conditions (age<18 years, breastfeeding/pregnant women, and subjects with 20% or more data missing) were discarded from the study. Among the 375 patients, 49.9%, 13.9%, 3.7%, 2.1%, 1.9%, and 0.5% patients had cough, fever, dyspnea, fatigue, muscular soreness, and chest distress, respectively.

19.3.2.2 Statistical analysis

174 patients were died out of 375 patients, while 201 patients were recovered. The outcome of the patients based on their condition at admission was summarized in Fig. 19.17. Among 375 patients, 197 patients were general, 27 were severe, and 151 were critical. The minimum, median, and maximum hospital stays were 0, 12, and 35 days, respectively, from the hospital admission to the outcome event (death/discharge).

Stata/MP v13.0 software was used to conduct Wilcoxon and chi-square tests for continuous variables and age variable, respectively. Gender was shown in percentage while continuous variables were reported with sample size, missing data length, average, and standard deviation of the variables for the death and survived patients. A P-value <0.05 was used for statistically significant variables. Gender, age, and 14 biomarkers out of 76 biomarkers along with outcomes were summarized in Table 19.6. The selection criteria of these 14 biomarkers were illustrated in the next section. Table 19.6 shows a summary of the demographic, outcomes, and clinical characteristics of COVID-19 patients. Patients' characteristics: Wuhan residents (37.9%), contact-traced (0.5%,) familial cluster (6.4%), health workers (1.9%), Huanan Seafood Market (0.5%), and no contact history (52.5%). The male and female patients were 59.7% and 40.3% respectively. The average age and standard deviation of the patients are 58.8±16.5 years. Two different feature selection techniques were used to identify the top-ranked 10 features, which resulted in 15 different features that were found to be most contributing to the early prediction of death (Fig. 19.18).

Table 19.6 reported the top-ranked 16 features and it was observed that LDH, lymphocyte (%), hs-CRP, neutrophils (%), gender, monocyte (%), age, eosinophil (%), high sensitivity cardiac troponin I, INR, albumin, and NT-proBNP were significantly different ($P < 0.05$) among the groups while serum chloride, serum sodium, MCHC, and APTT were not significantly different. Twelve statistically significant variables were investigated further to identify the most useful set of predictors.

19.3.2.3 Imputation and feature selection

It is evident from Table 19.6 that some data are missing for some patients. One of the approaches to deal with this issue could be discarding those patients however this will reduce the data set significantly. It could be done if thousands of patients' data were available in the data set. Moreover, deleting rows with missing data can lead to loss of valuable information which could be helpful for the analysis and can also result in prejudiced estimates [53]. As there are only 375 patients available in the data set, discarding missing data rows can affect the model performance significantly.

Several popular data imputation techniques are reported in the literature and imputation by two popular techniques such as by −1 and data imputation using a technique called "multiple imputations using chained equations" (MICE) for imputing clinical data [54]. In the "−1 imputation technique," the missing data for different variables are replaced by −1 while the MICE technique predicts the missing data using a regression model developed from the available variables in the data set. In the MICE technique, continuous variables were predicted using predictive mean matching while the binary variables are predicted using logistic regression [53]. In this work, clinical data of 76 biomarkers from 375 COVID-19 patients were imputed using two techniques to compare their performance: imputation by "−1" and MICE technique. 'Z-score' normalization was applied to data imputed using both techniques [55,56].

The clinical data set of 76 parameters imputed using two techniques were independently normalized and Multi-Tree Extreme Gradient Boost (XGBoost) technique [57] was applied to identify the top-10 parameters for each of the imputation techniques. In the XGBoost, each feature importance is calculated from its accumulated use in the decision tree, which showed promising performance in clinical applications [49,58]. In this study, default XGBoost parameters, (i.e., tree estimators = 150,

Table 19.6 Characteristic of the COVID-19 patients for the top-ranked features.

	Characteristics	Survived	Death	Total	P-value
1	Gender				<0.00001
	• Male (%)	98(49%)	126(72%)	224(60%)	
	• Female (%)	103(51%)	48(28%)	151(40%)	
2	Age				<0.0001
	• N(missing)	201(0)	174(0)	375(0)	
	• Mean ± SD	50.2±15	68.8±11.8	58.8±16.5	
3	Lactate dehydrogenase				<0.0001
	• N(missing)	193(8)	163(11)	356(19)	
	• Mean ± SD	271±102	642±341	441±305	
4	Neutrophils (%)				<0.0001
	• N(missing)	194(7)	162(12)	356(19)	
	• Mean ± SD	65.7±13.8	87±9.86	75.4±16.1	
5	Lymphocyte (%)				<0.0001
	• N(missing)	194(7)	162(12)	356(19)	
	• Mean ± SD	24.8±11.4	7.6±6.22	17±12.7	
6	High sensitivity C-reactive protein				<0.0001
	• N(missing)	194(7)	159(15)	353(22)	
	• Mean ± SD	36±44	127±75.5	77±75.4	
7	Serum sodium				0.12
	• N(missing)	193(8)	161(13)	354(21)	
	• Mean ± SD	138.9±3.38	139.9±8.37	139.3±6.18	
8	Eosinophil (%)				<0.0001
	• N(missing)	194(7)	162(12)	356(19)	
	• Mean ± SD	0.7±.94	0.11±0.38	0.44±.79	
9	Serum chloride				0.52
	• N(missing)	193(8)	161(13)	354(21)	
	• Mean ± SD	100.8±3.8	101.5±8.56	101.1±6.42	
10	Monocyte (%)				<0.0001
	• N(missing)	194(7)	152(12)	356(19)	
	• Mean ± SD	8.4±3.15	5.1±4.31	6.9±4.08	
11	International standard ratio				<0.0001
	• N(missing)	189(12)	163(11)	352(23)	
	• Mean ± SD	1.055±.086	1.37(1.01)	1.2±.709	
12	Activation of partial thromboplastin time				0.23
	• N(missing)	165(36)	133(41)	298(77)	
	• Mean ± SD	40.1±5.7	41.9±11.4	41±8.7	

continued on next page

Table 19.6 *(continued)*

	Characteristics	Survived	Death	Total	P-value
13	Hypersensitive cardiac troponin I				<0.0001
	• N(missing)	141(60)	146(28)	287(88)	
	• Mean ± SD	12±53.3	1391±5748	714±414	
14	Brain natriuretic peptide precursor (NT-proBNP)				<0.0001
	• N(missing)	128(73)	139(35)	267(108)	
	• Mean ± SD	1039±6620	2806±5906	1959±6308	
15	Albumin				<0.0001
	• N(missing)	193(8)	163(11)	356(19)	
	• Mean ± SD	37.1±4.53	30.3±4.22	34±5.57	
16	Mean corpuscular hemoglobin concentration				0.023
	• N(missing)	194(7)	162(12)	356(19)	
	• Mean ± SD	343±13.9	346±18.7	345±16.3	
17	Outcome (%)	201(54%)	174(46%)	375	

maximum depth = 4, learning rate = 0.2, subsample ratio of the training instances = 0.9 and regularization parameter, α was set to 1) were used to avoid overfitting limited data set along with large feature set [49,59].

19.3.2.4 Development and validation of classification model

Several different classification models were investigated in this study: multilayer perceptron (MLP), random forest, support vector machine (SVM), and logistic regression. Top-ranked 10 features using two imputation techniques were investigated to identify the best performing feature combinations. Therefore each feature was individually and in combination (top 2 and top 3, top 4, etc.) were investigated using different machine learning models. Since this work is not only focused to classify the death and survival group using an important feature matrix but also intended to develop a prediction scoring system using the nomogram technique, it was important to identify a common classification and nomogram development technique. Logistic regression is a supervised machine learning (ML) method dedicated to classification tasks [60], which is a very useful tool for binary classification [61]. Moreover, the logistic function uses sigmoid function, which converts linear inputs into a probability of 0 to 1, makes it suitable for clinical application. Another important reason for choosing a supervised logistic regression classifier [62] as it can also be used for nomogram development. This will ensure the coherent performance of classification and regression models.

The different combinations of features as discussed earlier were validated using 5-fold cross-validation and receiver operating characteristics (ROC) curves were used to calculate the area under the curve (AUC) for the predictor variables separately and also in combination. The AUC values for different combinations of top-ranked features were investigated for the binary classification. Several performance metrics including AUC values, specificity, sensitivity, negative likelihood ratio (NLR), and positive likelihood ratio (PLR) were calculated to evaluate the models. For the five unseen test folds, the overall confusion matrix was calculated and the per-class metrics were computed.

$$Sensitivity = \frac{TP}{TP + FN}, \tag{19.23}$$

$$Specificity = \frac{TN}{TN + FP}, \tag{19.24}$$

$$PLR = \frac{Sensitivity}{1 - Specificity}, \tag{19.25}$$

$$LR = \frac{1 - Sensitivity}{Specificity}. \tag{19.26}$$

In Eqs. (19.23)–(19.26), true negative (TN), false negative (FN), true positive (TP), and false-positive (FP) were used to refer to the survivors identified as survivors, dead patients identified as dead, deceased patients incorrectly recognized as survivors, and the survivors erroneously recognized as deceased patients, respectively.

19.3.2.5 Development and validation of nomogram based scoring system

A nomogram-based scoring system was developed using Nomolog [63], which is an add-on feature of Stata/MP v13.0 software. Nomolog utilizes multivariate logistic regression in the form of a binary regression where the outcome variable can have two possible values. The outcome variable is typically labeled as "1" and "0" where "1" represents death and "0" represents survival. The definition of odds, and linear prediction (LP), and the probability of death were reported in Eqs. (19.27)–(19.30), where the probability and odds can range within 0–1 and 0–∞, respectively. The predictors (binary and continuous variables) are linearly related to the logarithm of odds (LP) for logistic regression. The death probability and LP can be related by Eq. (19.30):

$$odds = \frac{P}{1 - P}, \tag{19.27}$$

$$LP = \ln(odds) = b_0 + b_1x_1 + b_2x_2 + \cdots + b_nx_n, \tag{19.28}$$

$$e^{b_0 + b_1x_1 + b_2x_2 + \cdots + b_nx_n} = e^{LP}, \tag{19.29}$$

$$P = \frac{e^{LP}}{1 + e^{LP}} = \frac{1}{1 + e^{-LP}}. \tag{19.30}$$

The best performing feature sets with the best AUC were used for the development of multivariate logistic regression model-based nomogram. The entire data set was divided into 70% training set and 30% validation set to evaluate the nomogram-based model. Development calibration curve and internal validation curve was plotted to compare the actual and predicted death probability of COVID-19 patients. Stata/MP v13.0 was used to develop the nomogram, where each parameter was represented by a horizontal line and a range of values depending on the range of the parameter. Another two horizontal lines are representing the total score and the death probability for an individual patient. A vertical line drawn from the value of each parameter horizontal line to a corresponding score axis allows obtaining the score for the parameters, which can be added to produce a total point. A perpendicular line can be drawn to the score axis to obtain the corresponding death risk percentage.

19.3.3 Results and discussion

This section elaborates the performance comparison of two different imputation schemes along with the classification performance for the best set of features using a logistics regression classifier. The best classification model was used with the top contributing features to develop a multivariate regression model and nomogram based scoring system and the performance of the scoring system was validated with internal and external sets.

19.3.3.1 Performance evaluation of the classification model

The best performing classification model for predicting the risk of death for COVID-19 patients from the blood sample biomarkers at hospital admission was compared for the data set imputed with −1 and MICE imputation techniques. Two different imputed data sets were used to identify the top-10 parameters to predict the risk of death. Top-ranked features (individually and in combination) were investigated to get the best-performing combination (Fig. 19.18). It was evident from Fig. 19.19 that five top-ranked features showed an AUC value of 0.97 for the MICE imputed data while −1 imputed data with 3 top-ranked feature produces an AUC of 0.95. Table 19.7 demonstrations the performance matrix and confusion matrix for the different groupings of features for −1 and MICE imputed data.

It is clear from Table 19.7 that the MICE imputation technique outperforms the −1 imputation technique. The top-ranked 5-features using the MICE imputation technique are: **L**actate dehydrogenase (LDH), **N**eutrophils (%), **L**ymphocyte (%), high sensitivity **C**-reactive protein (hsCRP), and **A**ge (LNLCA). Logistic regression-based nomogram was developed using these 5-features and a scoring system was developed, which was validated using longitudinal data to predict the death probability of the COVID-19 patients during the hospital stay.

As reported in the earlier studies on the COVID family diseases (SARS [64], MERS [65], and COVID-19 [66]), a key indicator of death was age. This can be understood easily as the elderly population are typically immunocompromised and have multiple medical conditions and are more prone to severe illness due to COVID-19 [47]. Yan et al. [9] reported that LDH significantly increases for patient with severe lung diseases [47] and the result of this study is consistent with these findings, which can suggest increased activity due to lung injury. Liu et al. [67] showed that elevated NLR (Neutrophil-to-Lymphocyte Ratio) indicates an increased risk of death

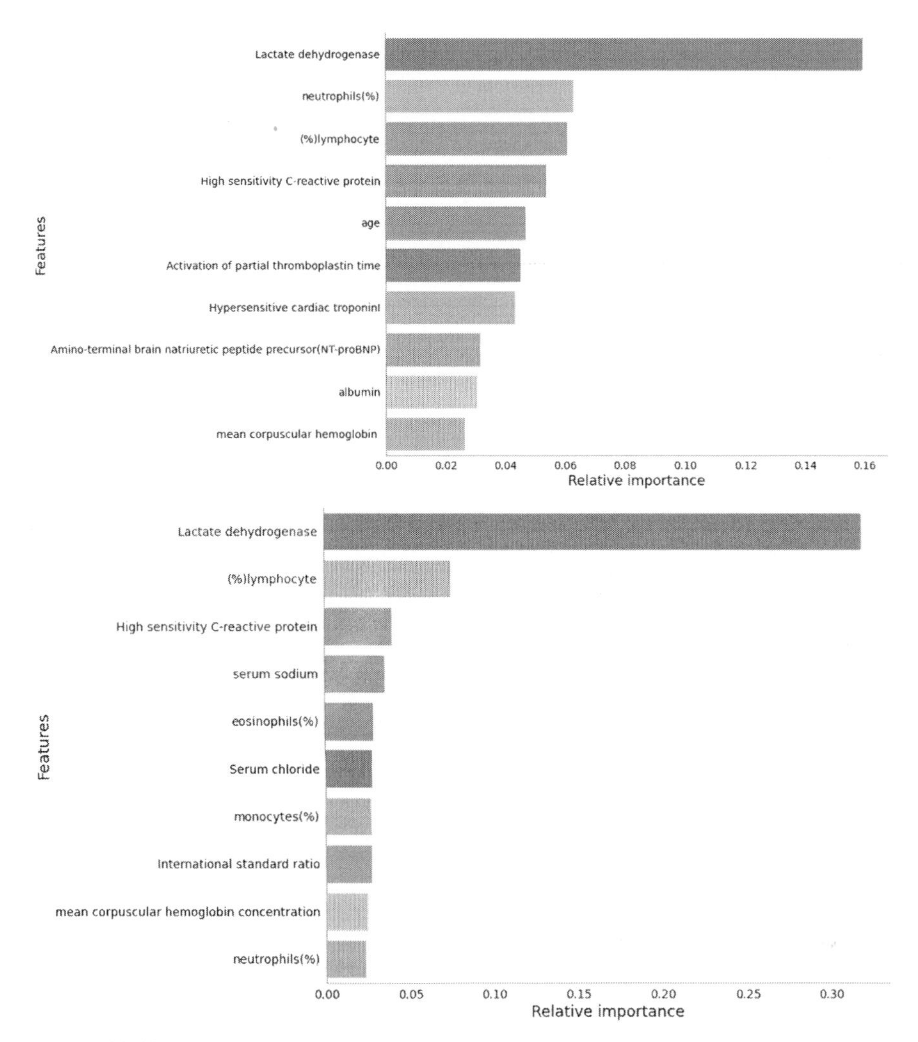

FIGURE 19.18

Comparison of the top-ranked 10 features identified from imputed and normalized data using MICE (top) and (−1) (bottom) techniques and Multi-Tree XGBoost algorithm.

for COVID-19 patients. Moreover, these parameters are critical immunity system parameters and play very vital role in infection reduction and body defense mechanism [68] and, therefore, reduction in their percentage are important indications of associated severity. Lu et al. [69] and other researchers showed that CRP can assist in predicting the mortality risk of COVID-19 by developing serious lung infection and secondary bacterial infection, who should be treated with antibiotics [49].

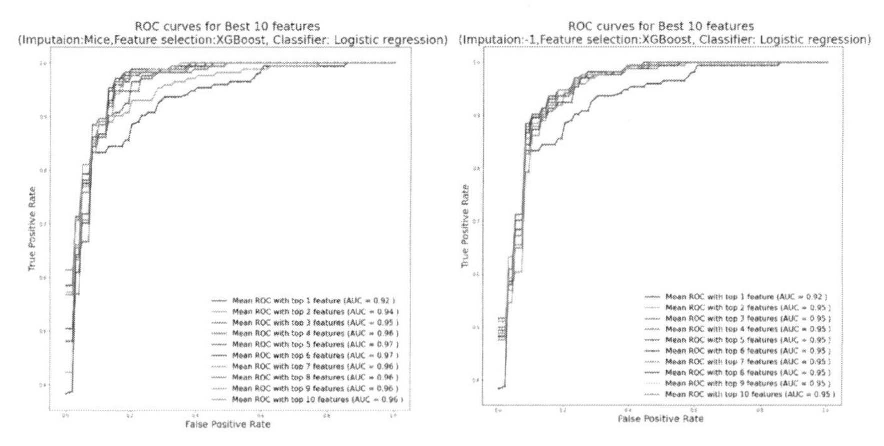

FIGURE 19.19

Comparison of ROC curves for top-1 to top-10 features sequentially adding together for the data imputed using (right) (−1) and (left) MICE. Note: Feature selection was done by XGBoost technique and the Classifier was Logistic regression in both the case. (For interpretation of the colors in the figure(s), the reader is referred to the web version of this chapter.)

19.3.3.2 Performance evaluation of the developed nomogram model

A novel machine learning pipeline is developed using statistical technique and multi-parameter logistic regression model based nomogram and scoring technique to quantify the death risk of COVID-19 patients. The regression model for the top-5 features was shown in Table 19.8. Table 19.8 shows the regression coefficient for each feature variable, z-value (ratio of regression coefficient and its standard error), standard error, and statistical significance with $p < 0.05$. A strong contributor is represented by a higher z-value (+ve/-ve) while a weak contributor is represented by a value of z close to zero. It can be noticed from Table 19.8 that age and Lactate dehydrogenase are strong contributors while neutrophils (%) is a comparatively weak predictor. Statistical significance of a regression coefficient can be obtained from p-value and a p-value less than 0.05 represents a significant relationship of X-variables to the Y-variable. Table 19.8 shows that Neutrophils (%) and Lymphocyte (%) are not statistically significant. However, Table 19.7 shows that the 5 variables outperform the 4 variables and so, these 5-variables were used for nomogram development.

The calibration graphs shown in Fig. 19.20 depict the reliability of the model for internal and external validation. It is evident from Fig. 19.20 that the external validation set model is matching well with an ACU = 0.99 on a complete unseen data set. The nomogram has eight rows and one to five rows are showing predictors while the last two rows are total score and death probability (Fig. 19.21). A score for each independent variable can be found by joining the variable's score to the total score axis and the total score is calculated by adding individual score from each variable. Finally, the death probability of COVID-19 is obtained by linking the total score to the probability line. The nomogram developed in this work is useful

Table 19.7 Comparison of the evaluation metrics and confusion matrix for top-performing features in combination using two different data imputation: (A) (−1), and (B) MICE imputation techniques.

	Weighted average (95°/o confidence interval)				Confusion matrix			
					Death		Survived	
	Sensitivity	Specificity	PLR	NLR	TP	FN	FP	TN
A (Imputation using −1)								
Top 1 feature	87±3.92	87.4±3.01	7.4±4.1	0.15±0.1	142	32	14	187
Top 2 features	88.04±3.13	88±3.5	8.1±5.1	0.14±0.08	148	26	17	184
Top 3 features	90±3.8	88.9±3.78	9.3±6.9	0.12±0.09	155	19	19	182
Top 4 features	**90.5±3.92**	**90.7±3.72**	**11.8±10.1**	**0.10±0.09**	**157**	**17**	**18**	**183**
Top 5 features	**90.1±3.6**	**90.03±3.5**	**10.5±7.9**	**0.11±0.086**	**155**	**19**	**18**	**183**
Top 6 features	90.08±2.7	90±2.4	9.63±5.1	0.11±0.06	154	20	19	182
Top 7 features	89.8±2.3	90.16±3.4	10.5±7.5	0.12±0.05	156	18	21	180
Top 8 features	89.3±3.6	89.1±3	8.96±5.5	0.12±0.08	155	19	21	180
Top 9 features	89.6±3.2	88.9±3.5	9.06±6.2	0.11±0.07	153	21	20	181
Top 10 features	89.01±3.3	89.01±4	9.46±7.3	0.13±0.083	154	20	21	180
B (Imputation using MICE)								
Top 1 feature	88.2±7.4	87.6±3.5	7.91±5.6	0.13±0.17	143	31	13	188
Top 2 features	87.7±4.4	87.01±3.5	7.37±4.6	0.14±0.11	145	29	17	184
Top 3 features	87.1±3.5	87±4.1	7.53±5.2	0.15±0.09	148	26	22	179
Top 4 features	89.2±2.8	89±3.2	8.93±5.6	0.12±0.07	155	19	22	179
Top 5 features	**92±2.6**	**92±3**	**13.52±10.6**	**0.09±0.06**	**160**	**14**	**16**	**185**
Top 6 features	**92.3±2.45**	**92±4.1**	**15.86±16.5**	**0.085±0.06**	**162**	**12**	**17**	**184**
Top 7 features	90.2±5	90.6±3.5	11.37±9.3	0.11±0.12	158	16	22	179
Top 8 features	89.9±4.8	90.2 ±3.8	11.02±9.3	0.11±0.11	158	16	23	178
Top 9 features	89.2±2.8	89.03±3.2	8.97±5.6	0.12±0.07	155	19	22	179
Top 10 features	88±3.4	89.6±3.7	9.82±7.5	0.14±0.08	156	18	23	178

Table 19.8 Multivariate logistic regression model to predict COVID-19 patients' death probability.

Outcome	Coef.	Std. Err.	z	$P > \|z\|$	[95% conf. interval]	
Lactate dehydrogenase	.0070514	.0017099	4.12	0.000	.0037001	.0104027
Neutrophils	−.0327053	.0568836	−0.57	0.565	−.1441951	.0787845
Lymphocyte	−.1624422	.0806231	−2.01	0.044	−.3204607	−.0044238
High sensitivity CRP	.0110451	.0043462	2.54	.011	.0025267	.0195635
Age	.0735038	.0185211	3.97	0.000	.0372032	.1098045
_cons	−3.662636	5.65169	−0.65	0.517	−14.73975	7.414473

for visual indication but equations are required to develop mathematical model for mobile or web-implementation and the death probability using Eqs. (19.27)–(19.30)

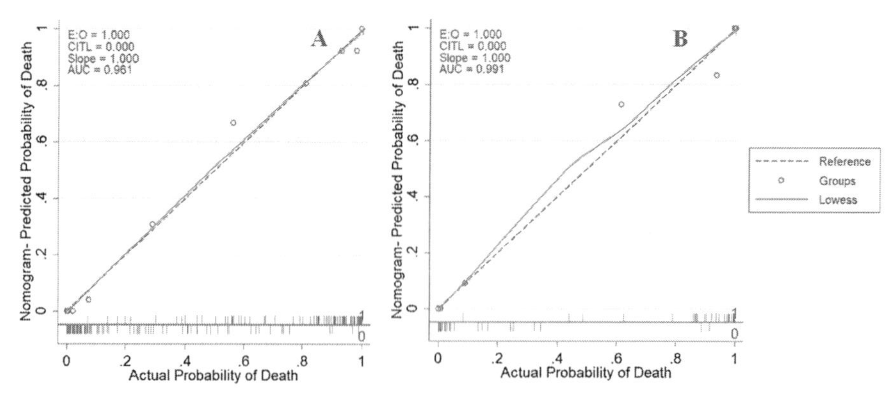

FIGURE 19.20

Developmental model (A) and internal validation (B) calibration curves to demonstrate the model performance on development data set and unseen test set in prediction death risk.

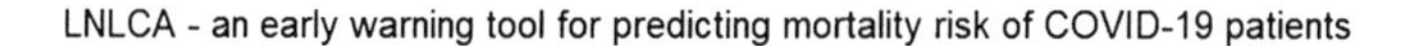

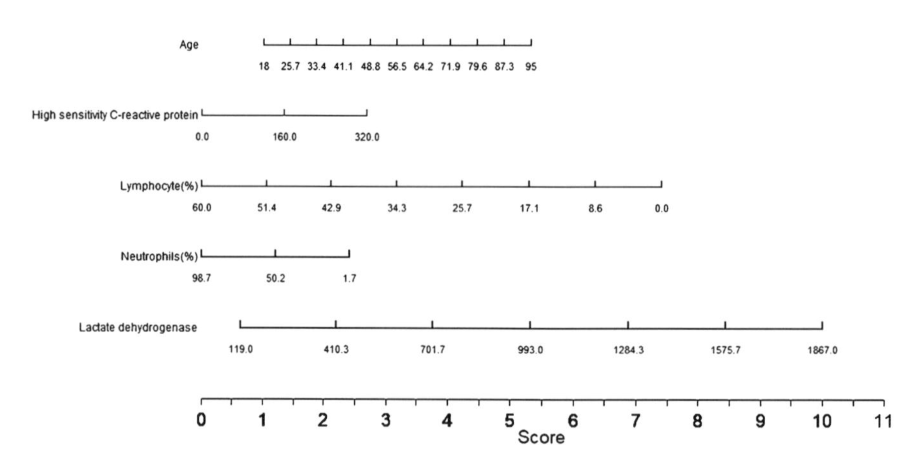

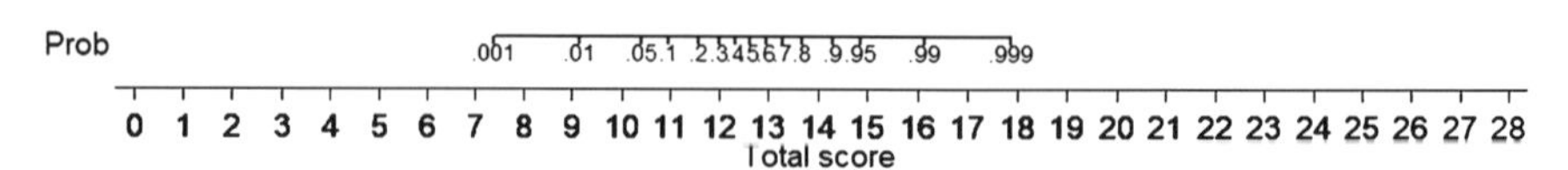

FIGURE 19.21

LNLCA Nomogram using logistic regression for predicting the death probability of the COVID-19 patients using LDH, Neutrophils (%), Lymphocytes (%), hsCRP, and age.

Table 19.9 LNLCA score and the corresponding death probability.

Patient group	Low			Moderate					High						
LNLCA score	7.45	9.2	**10.4**	10.95	11.6	11.99	12.4	**12.65**	12.95	13.3	13.7	14.3	14.8	16.2	17.85
Death probability	0.001	0.01	**0.05**	0.1	0.2	0.3	0.4	**0.5**	0.6	0.7	0.8	0.9	0.95	0.99	0.999

so that these can be used to develop mobile and/or web applications. The LNLCA score and death probability are expressed below:

$$\begin{aligned}
&\text{LNLCA Score (LP)}\\
&\quad = -3.662636 + 0.0735038 \times \text{age years} + 0.0110451 \times \text{hsCRP}\ \frac{\text{mg}}{\text{L}}\\
&\qquad - 0.1624422 \times \text{lymphocyte \%} - 0.0327053 \times \text{neutrophils \%}\\
&\qquad + 0.0070514 \times \text{lactate dehydrogenase}\left(\frac{\text{u}}{\text{L}}\right) \qquad (19.31)
\end{aligned}$$

$$\text{Death probability} = 1/(1 + \exp(-\text{Linear prediction})). \qquad (19.32)$$

Table 19.9 illustrates the LNLCA scores and corresponding death probability. The cut-off values of LNLCA score for low, moderate, and high-risk patients are 10.4 and 12.65, which correspond to 5%–50% death probability (based on [47]). The death probability of <5%, between 5% and 50%, and >50 % correspond to low, moderate, and high-risk groups, respectively.

Fig. 19.22 shows an example patient with age: 70 years, neutrophils (%): 89.2, lymphocyte (%): 6.7, LDH: 495, and hsCRP: 26.3 were used to calculate the death probability of the patient using nomogram based scoring system. The variables were recorded at admission which is 9 days before the actual outcome and the death probability was calculated to 80%. This reflects that the patient is at high risk of death which was calculated 9 days before the actual outcome and the patient died.

The training and testing subsets were subdivided into three risk groups using the LNLCA score and related them to the actual outcome to see what proportion of patients in different risk groups survived and died. It can be noticed from Table 19.10 that no patients in the low-risk group from training and testing cohorts have died whereas 88.1% (training set) and 94% (test set) died among the high-risk patients. More than 77% of the training and testing cohort were survived from the moderate group while more than 22% died. It can be seen that such a nomogram-based technique will be useful for medical doctors to categorize the patients in risk groups at admission to prioritize the resource utilization for the moderate and high-risk group patients.

19.3.3.3 Longitudinal validation of prognostic model

Since multiple blood samples were available for each subject in the data set, fifty-two COVID-19 patients' data, who were died at differing hospital stay were investigated to see how early the model can predict the death of these patients. Some patients died on the same day of admission while some were admitted 35 days earlier before the actual outcome.

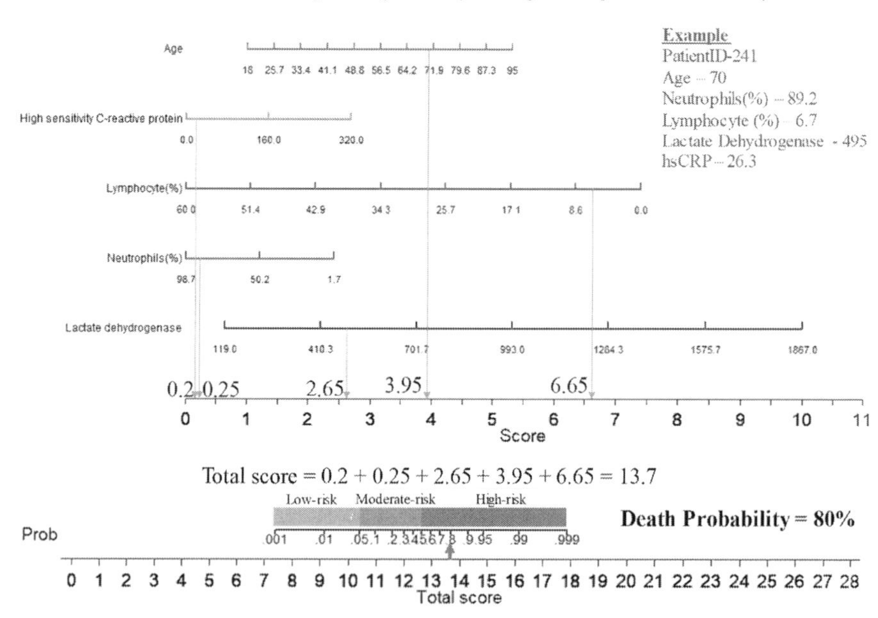

FIGURE 19.22

A example patient's data from test set was used to predict the risk of death at admission was evaluated and it was observed that it is identified as death outcome, 9-days earlier than the real outcome.

Table 19.10 Fisher exact probability test was used to associate the three risk-groups and real outcome for the training and testing sets.

	Outcome		Overall
	Alive	Death	
Risk category (Training Cohort)			
Low-risk	83 (100.0%)	0 (0%)	83 (100.0%)
Moderate-risk	41 (77.36%)	12 (22.64%)	53 (100.0%)
High-risk	15 (11.9%)	111 (88.1%)	126 (100.0%)
Overall	139 (53%)	123 (47%)	262 (100.0%)
Risk category (Testing Cohort)			
Low-risk	41 (100%)	0 (0%)	41 (100.0%)
Moderate-risk	17 (77.27%)	5 (22.73%)	22 (100.0%)
High-risk	3 (6%)	47 (94%)	50 (100.0%)
Overall	61 (54%)	52 (46%)	113 (100.0%)

P-value among the three groups is less than 0.001.
P-value of low-risk group vs moderate-risk group is 0.001 (training) and 0.0037 (testing).
The P-value of the low-risk group vs the high-risk group is less than 0.001.
The P-value of the moderate-risk group vs the high-risk group is less than 0.001.

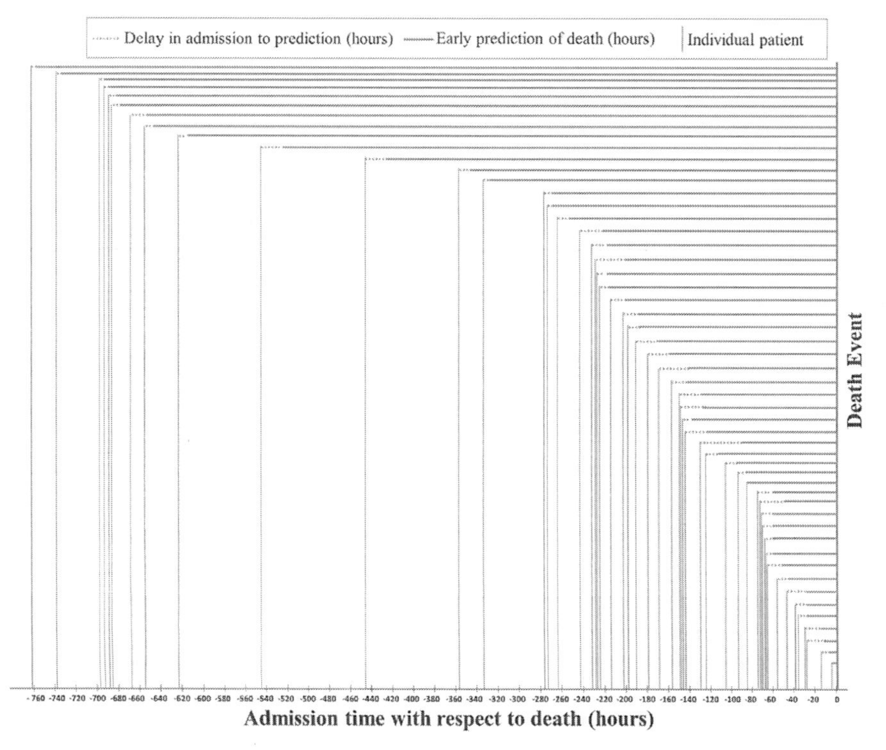

FIGURE 19.23

A graphical illustration of the event of actual outcome of 52 test-subjects along with their hospital admission and correct estimation of death outcome, where ``0" index shows the actual outcome and vertical lines show the hospital admission point and solid line starting point represents the correct estimation of death by the model while dotted line segment shows the delay between admission and the correct model prediction.

The maximum, minimum, median, and mean (±standard deviation) of hospital admission to the outcome for the test set were found to be 760.92, 3.68, 172.79, and 249.2 ± 227.55 hours, respectively. Fig. 19.23 reports the timing of death outcome for 52 test-patients along with their hospital admission and model's correct prediction point. Fig. 19.23 confirms that the LNLCA model manages to predict the onset of death for 52 test patients immediately after several hours of hospital admission. LNLCA model can predict the death of a patient even 31.5 days in advance of the actual outcome. The facility of such early prediction can help medical professionals to keep track of the patients' death risk and facilitate them with the appropriate healthcare facilities.

19.4 Conclusion

This chapter starts with the demonstration of a patient-specific ECG heartbeat classifier based on the adaptive 1D Convolutional Neural Networks (CNNs) that has significant advantages over the traditional ECG classification systems employing separate manually designed feature extraction and classification blocks together with some necessary preprocessing and postprocessing techniques. This compact patient-specific implementation has the advantages of not requiring any hand-crafted feature extraction, feature selection, or any type of preprocessing and postprocessing, which can be conveniently used in a real-time mobile heart monitoring and anomaly detection system. Besides the speed and computational efficiency achieved, the hardware implementation of this method can be easily achieved using the FPGAs as it only requires 1D convolutions (multiplications and additions). Additionally, the properly trained 1D CNN for an individual patient can solely be used to monitor and detect arrhythmias in a patient's long ECG record (i.e., Holter register) or it can be specifically trained for detection of atrial fibrillation with high accuracy. Based on the experimental evaluation using the benchmark MIT/BIH arrhythmia database, the base performance of this approach is comparable or better than the competing patient-specific methods for the VEB and SVEB detection. The robustness of the method is also checked for the variations of the input data resolution, CNN parameter settings, and raw data representations, and it is proven that no performance degradation results and even lower computational complexity can be achieved. Overall, the important contributions of this approach can be summarized as:

- By using the adaptive implementation of 1D CNNs, the feature extraction, feature selection, and classification modules can be efficiently combined and learned simultaneously.
- By directly learning the best possible features from the patient's ECG data, compact 1D CNNs can adapt to the significant variability in ECG measurements at different times and between different patients.
- The 1D CNN-based implementation has the advantage of removing the need for the hand-crafted feature extraction, feature selection, and any type of preprocessing and post-processing blocks. It can directly learn to detect the arrhythmias from the raw ECG data of a patient.
- The 1D CNNs have significantly lower computational time complexity and allow simpler hardware implementation than most of the competing methods, which suits well for real-time cardiac arrhythmia detection applications.

The arrhythmic ECG beats may have sometimes significant and sometimes insignificant differences to normal beats. This is why only the expertise and the trained eye of the cardiologists can detect the arrhythmic beats accurately. Therefore, the real challenge is to mimic an expert cardiologist in a classifier with proper training that requires both normal and abnormal beats of the patient. This challenge has already been addressed successfully in recent studies. However, for a "healthy person" with no history of cardiac arrhythmia, accurate detection of the first-time occurrence of the arrhythmic beat(s) poses the ultimate challenge that has never been addressed up to date. The method presented in the second section of this chapter is a systematic approach to synthesize potential abnormal beats that can be used to train the "personalized" classifier of the healthy person. Only then this "personalized" classifier can be used to detect any "real" arrhythmic beat(s) at the moment they occur. The method was tested over 63,341 ECG beats of the MIT-BIH benchmark data set and demonstrated that:

- Realistic arrhythmic beats that are morphologically and temporally similar to the real counterparts can be synthesized,
- A detection probability higher than 99.2% with a very low false-alarm rate can be achieved,
- The personalized cardiac health monitoring system can work on mobile devices in real-time.

Finally, in the last section, five key predictors (LDH, Neutrophils (%), Lymphocytes (%), hsCRP, and age) collected at admission are selected by the XGBoost feature ranking technique. A novel nomogram-based prognostic model to predict the death probability of COVID-19 patients with very reliable network performance is reported in this work. The reliability of the model was validated on developmental and validation sets with an AUC of 0.961 and 0.991, respectively. This easy-to-use, explainable and high-performing nomogram based scoring system can be very vital for stratifying COVID-19 patients into risk groups which can significantly help the healthcare professional to reduce the burden on the healthcare system. COVID-19 patients can be categorized into 3 risk groups using the LNLCA model based on their 5 parameters at admission so that low-risk group can be treated at home or outpatients while moderate patients can be treated at isolation center and high-risk group should be provided with highest medical facilities to reduce mortality. To ease the use of the model for clinicians, a web-based application has been developed using the LNLCA model [70].

References

[1] C. Huang, Y. Wang, X. Li, L. Ren, J. Zhao, Y. Hu, L. Zhang, G. Fan, J. Xu, X. Gu, et al., Clinical features of patients infected with 2019 novel coronavirus in Wuhan, China, The Lancet 395 (2020) 497–506.

[2] W.-j. Guan, Z.-y. Ni, Y. Hu, W.-h. Liang, C.-q. Ou, J.-x. He, L. Liu, H. Shan, C.-l. Lei, D.S.C. Hui, et al., Clinical characteristics of coronavirus disease 2019 in China, The New England Journal of Medicine 382 (2020) 1708–1720.

[3] D. Wang, B. Hu, C. Hu, F. Zhu, X. Liu, J. Zhang, B. Wang, H. Xiang, Z. Cheng, Y. Xiong, et al., Clinical characteristics of 138 hospitalized patients with 2019 novel coronavirus-infected pneumonia in Wuhan, China, JAMA 323 (2020) 1061–1069.

[4] Z. Wu, J.M. McGoogan, Characteristics of and important lessons from the coronavirus disease 2019 (COVID-19) outbreak in China: summary of a report of 72 314 cases from the Chinese Center for Disease Control and Prevention, JAMA 323 (2020) 1239–1242.

[5] W. H. Organization and others, Coronavirus disease 2019 (COVID-19): situation report, 70, 2020.

[6] M. Roser, H. Ritchie, E. Ortiz-Ospina, J. Hasell, Coronavirus pandemic (COVID-19), Our World in Data, 2020.

[7] G. Grasselli, A. Pesenti, M. Cecconi, Critical care utilization for the COVID-19 outbreak in Lombardy, Italy: early experience and forecast during an emergency response, JAMA 323 (2020) 1545–1546.

[8] S.M. Moghadas, A. Shoukat, M.C. Fitzpatrick, C.R. Wells, P. Sah, A. Pandey, J.D. Sachs, Z. Wang, L.A. Meyers, B.H. Singer, et al., Projecting hospital utilization during the COVID-19 outbreaks in the United States, Proceedings of the National Academy of Sciences 117 (2020) 9122–9126.

[9] L. Yan, H.-T. Zhang, J. Goncalves, Y. Xiao, M. Wang, Y. Guo, C. Sun, X. Tang, L. Jin, M. Zhang, et al., A machine learning-based model for survival prediction in patients with severe COVID-19 infection, MedRxiv (2020).

[10] Z.J. Zheng, J.B. Croft, W.H. Giles, State specific mortality from sudden cardiac death—United States, MMWR Morb Mortal Wkly Rep. 51, 2002, pp. 123–126.

[11] R. Mehra, Global public health problem of sudden cardiac death, Journal of Electrocardiology 40 (6 Suppl.) (2007), S118–22.

[12] GBD 2013 Mortality and Causes of Death Collaborators, Global, regional, and national age-sex specific all-cause and cause-specific mortality for 240 causes of death, 1990–2013: a systematic analysis for the Global Burden of Disease Study 2013, Lancet 385 (9963) (2013) 117–171.

[13] R.G. Hiss, L.E. Lamb, Electrocardiographic findings in 122,043 individuals, Circulation 25 (1962) 947–961.

[14] K. Minami, H. Nakajima, T. Toyoshima, Real-time discrimination of ventricular tachyarrhythmia with Fourier-transform neural network, IEEE Transactions on Biomedical Engineering 46 (2) (Feb. 1999) 179–185.

[15] O.T. Inan, et al., Robust neural-network based classification of PVCs using wavelet transform and timing interval features, IEEE Transactions on Biomedical Engineering 53 (12) (Dec. 2006) 2507–2515.

[16] X. Alfonso, T.Q. Nguyen, ECG beat detection using filter banks, IEEE Transactions on Biomedical Engineering 46 (2) (Feb. 1999) 192–202.

[17] J.L. Willems, E. Lesaffre, Comparison of multigroup logistic and linear discriminant ECG and VCG classification, Journal of Electrocardiology 20 (1987) 83–92.

[18] J.L. Talmon, Pattern Recognition of the ECG, Akademisch Proefscrift, Berlin, Germany, 1983.

[19] D.A. Coast, R.M. Stern, G.G. Cano, S.A. Briller, An approach to cardiac arrhythmia analysis using hidden Markov models, IEEE Transactions on Biomedical Engineering 37 (9) (Sep. 1990) 826–836.

[20] S. Osowski, L.T. Hoai, T. Markiewicz, Support vector machine based expert system for reliable heartbeat recognition, IEEE Transactions on Biomedical Engineering 51 (4) (Apr. 2004) 582–589.

[21] Y.H. Hu, W.J. Tompkins, J.L. Urrusti, V.X. Afonso, Applications of artificial neural networks for ECG signal detection and classification, Journal of Electrocardiology (1994) 66–73.

[22] Y. Hu, S. Palreddy, W.J. Tompkins, A patient-adaptable ECG beat classifier using a mixture of experts approach, IEEE Transactions on Biomedical Engineering 44 (9) (Sep. 1997) 891–900.

[23] S.C. Lee, Using a translation-invariant neural network to diagnose heart arrhythmia, in: IEEE Proc. Conf. Neural Information Processing Systems, Nov. 1989.
[24] P. de Chazal, R.B. Reilly, A patient-adapting heartbeat classifier using ECG morphology and heartbeat interval features, IEEE Transactions on Biomedical Engineering 53 (12) (Dec. 2006) 2535–2543.
[25] Recommended Practice for Testing and Reporting Performance Results of Ventricular Arrhythmia Detection Algorithms, Association for the Advancement of Medical Instrumentation, Arlington, VA, 1987.
[26] P. de Chazal, M. O'Dwyer, R.B. Reilly, Automatic classification of heartbeats using ECG morphology and heartbeat interval features, IEEE Transactions on Biomedical Engineering 51 (7) (Jul. 2004) 1196–1206.
[27] W. Jiang, S.G. Kong, Block-based neural networks for personalized ECG signal classification, IEEE Transactions on Neural Networks 18 (6) (Nov. 2007) 1750–1761.
[28] T. Ince, S. Kiranyaz, M. Gabbouj, A generic and robust system for automated patient-specific classification of electrocardiogram signals, IEEE Transactions on Biomedical Engineering 56 (5) (May 2009) 1415–1426.
[29] S. Kiranyaz, T. Ince, M. Gabbouj, Multi-Dimensional Particle Swarm Optimization for Machine Learning and Pattern Recognition, Springer, Aug. 2013, 383 pages.
[30] M. Llamedo, J.P. Martinez, An automatic patient-adapted ECG heartbeat classifier allowing expert assistance, IEEE Transactions on Biomedical Engineering 59 (8) (Aug 2012) 2312–2320.
[31] S. Kiranyaz, T. Ince, J. Pulkkinen, M. Gabbouj, Personalized long-term ECG classification: a systematic approach, Expert Systems with Applications (2011) 3220–3226.
[32] S. Kiranyaz, T. Ince, M. Gabbouj, Real-time patient-specific ECG classification by 1D convolutional neural networks, IEEE Transactions on Biomedical Engineering 63 (3) (Mar 2016) 664–675.
[33] S. Kiranyaz, T. Ince, M. Gabbouj, R. Hamila, Convolutional neural networks for patient-specific ECG classification, in: The 37th IEEE Engineering in Medicine and Biology Society Conference (EMBC'15), Milano, Italy, Aug. 2015.
[34] R. Mark, G. Moody, MIT-BIH arrhythmia database directory, Online, available: http://ecg.mit.edu/dbinfo.html.
[35] U.R. Acharya, H. Fujitad, O.S. Liha, Y. Hagiwaraa, J.H. Tana, M. Adam, Automated detection of arrhythmias using different intervals of tachycardia ECG segments with convolutional neural network, Information Sciences 405 (2017) 81–90.
[36] U.R. Acharya, et al., A deep convolutional neural network model to classify heartbeats, Computers in Biology and Medicine 89 (2017) 389–396.
[37] M. Zubair, J. Kim, C. Yoon, An automated ECG beat classification system using convolutional neural networks, in: 2016 6th Int. Conf. on IT Convergence and Security (ICITCS), 2016.
[38] S.S. Xu, M. Mak, C. Cheung, Towards end-to-end ECG classification with raw signal extraction and deep neural networks, IEEE Journal of Biomedical and Health Informatics 23 (4) (2019) 1574–1584.
[39] Y. Xia, Y. Xie, A novel wearable electrocardiogram classification system using convolutional neural networks and active learning, IEEE Access 7 (2019) 7989–8001.
[40] X. Xu, H. Liu, Ecg heartbeat classification using convolutional neural networks, IEEE Access 8 (2020) 8614–8619.
[41] S. Kiranyaz, T. Ince, M. Gabbouj, Personalized monitoring and advance warning system for cardiac arrhythmias, Scientific Reports – Nature 7 (Aug. 2017), https://doi.org/10.1038/s41598-017-09544-z (SREP-16-52549-T), http://rdcu.be/vfYE.

[42] G.B. Moody, R.G. Mark, The impact of the mit/bih arrhythmia database, IEEE Engineering in Medicine and Biology Magazine 20 (3) (2001) 45–50.
[43] T. C. D. C. COVID and R. Team, Severe Outcomes Among Patients with Coronavirus Disease 2019 (COVID-19) – United States, February 12 – March 16, 2020, MMWR Morb Mortal Wkly Rep., vol. 69, 2020, pp. 343–346.
[44] M.E.H. Chowdhury, T. Rahman, A. Khandakar, S. Al-Madeed, S.M. Zughaier, H. Hassen, M.T. Islam, et al., An early warning tool for predicting mortality risk of COVID-19 patients using machine learning, arXiv preprint, arXiv:2007.15559, 2020.
[45] M.H. Jamal, S.A. Doi, S. AlYouha, S. Almazeedi, M. Al-Haddad, A. Al-Muhaini, F. Al-Ghimlas, M.E. Chowdhury, d.S. Al-Sabah, A biomarker based severity progression indicator for COVID-19: the Kuwait prognosis indicator score, Biomarkers (2020) 1–21.
[46] S. Al Youha, S.A. Doi, M.H. Jamal, S. Almazeedi, M. Al Haddad, M. AlSeaidan, A.Y. Al-Muhaini, F. Al-Ghimlas, S.K. Al-Sabah, Validation of the Kuwait Progression Indicator Score for predicting progression of severity in COVID19, medRxiv (2020).
[47] Z. Weng, Q. Chen, S. Li, H. Li, Q. Zhang, S. Lu, L. Wu, L. Xiong, B. Mi, D. Liu, et al., ANDC: an early warning score to predict mortality risk for patients with Coronavirus Disease 2019, 2020.
[48] X. Jianfeng, H. Daniel, C. Hui, T.A. Simon, L. Shusheng, W. Guozheng, W. Yishan, K. Hanyujie, B. Laura, Z. Ruiqiang, et al., Development and external validation of a prognostic multivariable model on admission for hospitalized patients with Covid-19, https://www.medrxiv.org/content/medrxiv/early/2020/03/30/2020.03.28.20045997.full.pdf, 2020.
[49] L. Yan, H.-T. Zhang, J. Goncalves, Y. Xiao, M. Wang, Y. Guo, C. Sun, X. Tang, L. Jing, M. Zhang, et al., An interpretable mortality prediction model for COVID-19 patients, Nature Machine Intelligence (2020) 1–6.
[50] B. Zhang, X. Zhou, Y. Qiu, F. Feng, J. Feng, Y. Jia, H. Zhu, K. Hu, J. Liu, Z. Liu, et al., Clinical characteristics of 82 death cases with COVID-19, MedRxiv (2020).
[51] M.P. McRae, G.W. Simmons, N.J. Christodoulides, Z. Lu, S.K. Kang, D. Fenyo, T. Alcorn, I.P. Dapkins, I. Sharif, D. Vurmaz, et al., Clinical decision support tool and rapid point-of-care platform for determining disease severity in patients with COVID-19, Lab on a Chip, 2020.
[52] L. Zhang, X. Yan, Q. Fan, H. Liu, X. Liu, Z. Liu, Z. Zhang, D-dimer levels on admission to predict in-hospital mortality in patients with Covid-19, Journal of Thrombosis and Haemostasis 18 (2020) 1324–1329.
[53] H. Hegde, N. Shimpi, A. Panny, I. Glurich, P. Christie, A. Acharya, MICE vs PPCA: missing data imputation in healthcare, Informatics in Medicine Unlocked 17 (2019) 100275.
[54] S.v. Buuren, K. Groothuis-Oudshoorn, mice: multivariate imputation by chained equations in R, Journal of Statistical Software (2010) 1–68.
[55] A. Jain, K. Nandakumar, A. Ross, Score normalization in multimodal biometric systems, Pattern Recognition 38 (2005) 2270–2285.
[56] S. Patro, K.K. Sahu, Normalization: a preprocessing stage, arXiv preprint, arXiv:1503.06462, 2015.
[57] T. Chen, T. He, M. Benesty, V. Khotilovich, Y. Tang, Xgboost: extreme gradient boosting, R package version 0.4-2, p. 1–4, 2015.
[58] L. Torlay, M. Perrone-Bertolotti, E. Thomas, M. Baciu, Machine learning-XGBoost analysis of language networks to classify patients with epilepsy, Brain Informatics 4 (2017) 159–169.

[59] W. Li, Y. Yin, X. Quan, H. Zhang, Gene expression value prediction based on XGBoost algorithm, Frontiers in Genetics 10 (2019) 1077.

[60] R.P. Anderson, R. Jin, G.L. Grunkemeier, Understanding logistic regression analysis in clinical reports: an introduction, The Annals of Thoracic Surgery 75 (2003) 753–757.

[61] A. Ng, M. Jordan, On discriminative vs. generative classifiers: a comparison of logistic regression and naive Bayes, Advances in Neural Information Processing Systems 14 (2001) 841–848.

[62] S. Le Cessie, J.C. Van Houwelingen, Ridge estimators in logistic regression, Journal of the Royal Statistical Society Series C Applied Statistics 41 (1992) 191–201.

[63] A. Zlotnik, V. Abraira, A general-purpose nomogram generator for predictive logistic regression models, Stata Journal 15 (2015) 537–546.

[64] J.C.K. Chan, E.L.H. Tsui, V.C.W. Wong, H.A.S.C. Group, Prognostication in severe acute respiratory syndrome: a retrospective time-course analysis of 1312 laboratory-confirmed patients in Hong Kong, Respirology 12 (2007) 531–542.

[65] A. Assiri, J.A. Al-Tawfiq, A.A. Al-Rabeeah, F.A. Al-Rabiah, S. Al-Hajjar, A. Al-Barrak, H. Flemban, W.N. Al-Nassir, H.H. Balkhy, R.F. Al-Hakeem, et al., Epidemiological, demographic, and clinical characteristics of 47 cases of Middle East respiratory syndrome coronavirus disease from Saudi Arabia: a descriptive study, Lancet Infectious Diseases 13 (2013) 752–761.

[66] C. for Disease Control, Prevention and others, Interim clinical guidance for management of patients with confirmed 2019 novel coronavirus (2019-nCoV) Infection, Updated February, vol. 12, 2020.

[67] J. Liu, Y. Liu, P. Xiang, L. Pu, H. Xiong, C. Li, M. Zhang, J. Tan, Y. Xu, R. Song, et al., Neutrophil-to-lymphocyte ratio predicts severe illness patients with 2019 novel coronavirus in the early stage, MedRxiv (2020).

[68] I. Huang, R. Pranata, Lymphopenia in severe coronavirus disease-2019 (COVID-19): systematic review and meta-analysis, Journal of Intensive Care 8 (2020) 1–10.

[69] J. Lu, S. Hu, R. Fan, Z. Liu, X. Yin, Q. Wang, Q. Lv, Z. Cai, H. Li, Y. Hu, et al., ACP risk grade: a simple mortality index for patients with confirmed or suspected severe acute respiratory syndrome coronavirus 2 disease (COVID-19) during the early stage of outbreak in Wuhan, China, 2020.

[70] COVID 19 Mortality Prediction, https://www.openasapp.net/portal#!/client/app/2309220f-105e-4e7d-9fe0-51cbef5bb18c.

CHAPTER

Medical image analysis 20

Aysen Degerli[a], Mehmet Yamac[a], Mete Ahishali[a], Serkan Kiranyaz[b], and Moncef Gabbouj[a]

[a]*Faculty of Information Technology and Communication Sciences, Tampere University, Tampere, Finland*

[b]*Department of Electrical Engineering, Qatar University, Doha, Qatar*

20.1 Introduction

Artificial intelligence applications, particularly deep learning algorithms are the most promising area in medical image analysis that can provide a robust, accurate, automatic, and fast assessment of the health status of patients. Deep learning models can provide quantitative assessments of medical images, and achieve superior diagnosis over the medical doctors as helping them treating a patient most appropriately. Their wide range of applications enables researchers and medical doctors to work together to provide simple yet robust diagnostic tools to improve the quality of patients' lives. Furthermore, deep learning is increasingly utilized in many medical imaging techniques including X-ray and ultrasound imaging that also reduced the workload of medical doctors. In this chapter, solutions to medical imaging applications are described that play a vital role in the diagnosis of myocardial infarction and coronavirus disease 2019 (COVID-19), which are both life-threatening health problems. In fact, it is observed that patients with cardiovascular metabolic disease records are in the high-risk group for COVID-19 since the development and prognosis of COVID-19 pneumonia are largely affected by the cardiac history [1]. At the same time, several studies investigate cardiac injuries because of viral infection in COVID-19 patients. For example, the study in [2] explores the relation between cardiac injury and mortality in the hospitalized patients of Wuhan city of China where COVID-19 was first detected. Accordingly, there occurred cardiac injury in 82 (19.7%) of 416 patients during their hospitalization. For those patients, the mortality rate was higher than without cardiac injury patients, 42 (51.2%) versus 15 (4.5%) with $p < 0.001$, respectively.

The first study [3] presented on medical image analysis is early detection of myocardial infarction (MI), or commonly known as "heart attack." MI happens due to the blockage or plaque in one or more arteries that feed the heart muscle. It is the most life-threatening health problem worldwide from which around 35 million people suffer each year. This also makes it the leading cause of death worldwide and time passing after its onset is critical. Therefore, early detection is of utmost importance,

Deep Learning for Robot Perception and Cognition. https://doi.org/10.1016/B978-0-32-385787-1.00025-7

which can prevent fatal damages of the cardiac tissue. However, this is a challenging problem since electrocardiography (ECG) findings or biomedical marker values found in the blood may not yield a distinguishable symptom. Moreover, the pathological results are simply too late for early MI detection because they only indicate the presence of dead cells on the heart muscle [4]. Furthermore, ECG indicates only the late stages and cannot differentiate between MI and myocardial ischemia findings; therefore, it is not useful to detect the onset of MI from ECG recordings [5,6]. On the other hand, the biomedical marker values (cardiac enzymes) might be useful for early detection [4]; however, they yield a high false alarm rate (or low specificity) in general [7]. Accordingly, the focus of this study is particularly drawn on echocardiography for early MI detection. Early and fundamental signs of MI can be visible as the abnormality in one or several segments of the LV wall, where a segment may move "abnormally" or "nonuniformly." A regional wall motion abnormality (RWMA) can be categorized as a "weak motion," known as hypokinesia, "no motion," known as akinesia, or "out of sync," known as dyskinesia. The RWMA detection and categorization can be performed by echocardiography, which is an advantageous imaging technique with fast acquisition, easy accessibility and lowest risk options [8–10]. In this chapter, a three-stage approach for early MI detection is presented, where the left ventricle (LV) wall is segmented at each frame of an echocardiography recording by a deep learning model, and the characteristics of the segmented LV wall are obtained via feature engineering. At the last step, early MI detection is performed on the basis of the extracted LV wall features. The ground-truth of the unannotated LV wall is generated by pseudo-labeling method, which is a collaborative method between human and machine. The echocardiographic data set of the study is publicly available, which includes both MI detection and LV wall segmentation ground-truth labels.

The second study [11] presented in this chapter is the recognition of COVID-19 caused by severe acute respiratory syndrome coronavirus-2 (SARS-CoV-2). The highly infectious nature of the disease has caused many causalities and paralyzed mobility around the world. Therefore, World Health Organization declared a pandemic in March 2020 due to COVID-19. The disease is a danger for elderly people and people in high-risk groups since it may lead to hospitalization or even death [12]. In order to maintain public health, the spreading of the disease should be minimized by social distancing and isolation of the infected patients. This brings an urgent need for computer-aided diagnosis for automatic, fast, accurate, and robust COVID-19 recognition algorithms that can distinguish COVID-19 from other thoracic diseases. The computer-aided diagnosis can be provided by medical imaging techniques. In fact, chest X-ray imaging is the most useful technique for medical doctors to both diagnose and monitor COVID-19. Chest X-ray imaging has advantages as easy accessibility, fast acquisition, and less radiation exposure compared to other chest imaging techniques such as computed tomography [13]. Deep learning techniques have achieved state-of-the-art performance in many medical imaging tasks as an accurate and real-time diagnosis of diseases. In fact, deep learning algorithms have been already utilized for the detection and recognition of COVID-19. However, due to data scarcity, the generalization capabilities of these models have degraded that results in unreliable detection of COVID-19. On the other hand, alternative classifiers such as representation-based classification that can generalize from scarce data are in fact successful for COVID-19 recognition. However, their performance and speed cannot cope with the deep learning methods. Therefore, in order to overcome the issues of both deep learning models and representation-based classification methods, Convolutional Support Estimation Network (CSEN) is presented for the recognition of COVID-19 from bacterial pneumonia, viral pneumonia, and control group (healthy subjects) using chest X-ray images. The bridge between deep learning models and model-based classification is ensured by CSEN with a noniterative and real-time mapping of the input to sparse representation coefficient support. The mapping assures that the information for the classification decision in the representation-based methods is maintained. In order to utilize CSEN on a scarce and imbalanced data set, the features are extracted from a state-of-the-art pre-trained deep learning model, CheXNet [14] that is trained over chest X-ray images. Consequently, the performance of representation-based classification models and the convolutional support estimator models are investigated for the recognition of COVID-19 from other thoracic diseases and healthy subjects.

20.2 Early detection of myocardial infarction using echocardiography

In this section, early detection of myocardial infarction using echocardiography is discussed. Myocardial infarction, or commonly known as heart attack, is the major cause of death in the world [4,15,16]. It occurs because of myocardial ischemia, which is a blockage or plaque in one or more arteries feeding the myocardium layer of the heart. If ischemia is not detected in the early stage, then it damages the myocardium tissue; eventually, causing MI. Therefore, the early detection of MI plays a vital role in the prevention of further myocardial tissue damages. The symptoms of MI can be pain around the chest, arm, and shoulder, shortness of breath, dizziness, and discomfort from anxiety [15]. Unfortunately, these symptoms are not decisive enough for the diagnosis. The World Health Organization has defined several diagnostic indicators to detect MI: pathological observations, biochemical marker values, electrocardiography, and imaging techniques [17]. The conclusions of pathological observations can reveal the signs of myocardial cell death. However, they are not convenient in the early MI diagnosis since the findings can only be determined after the myocardium tissue is severely infarcted [4]. On the other hand, biomedical marker values, that is, cardiac enzymes found within the blood sample can detect MI in its early stages, but with a rather low specificity [4]. Measuring the electrical activity of the heart by ECG is not practical for MI detection, since there exist correlations in the ECG findings of MI and myocardial ischemia [16]. The earliest and most reliable sign of ischemia is the regional wall motion abnormality of the infarcted ventricular muscle. Consequently, the most useful tool to detect early MI is an imaging technique: echocardiography that can exhibit the RWMA noninvasively, fast, and with the lowest risk among the imaging options [10]. It is suitable for both clinical and research purposes as the developments in echocardiography support the diagnosis of coronary artery disease (CAD), and its treatment as it is performed on patients with known or suspected CAD since it can provide an extensive assessment of the cardiovascular structure and its function [9,10].

Echocardiography is a cardiac ultrasound that captures the sound waves reflected from the heart tissue to monitor in real-time from different views by changing the angle of the probe pointed to it. The four heart chambers: right ventricle, left ventricle, right atrium, and left atrium can be examined in more detail with ultrasound devices. In this chapter, the focus is early MI diagnosis on the LV wall using 2D echocardiography with the apical 4-chamber (A4C) view, where each chamber of the heart is visible. Early signs of MI are reflected as abnormalities in the LV wall characteristics. Echocardiography is widely used for the clinical evaluation of the LV wall characteristics, such as its dimension, volume, and motion [18]. However, the human-centric visual evaluation is very subjective and, therefore, highly operator-dependent [18]. Thus, to overcome this issue, highly robust, accurate, and automated diagnostic techniques should be developed.

During the last 20 years, the computer-aided diagnosis has become the supporting unit to develop more accurate and objective diagnosis as helping the cardiologists [19,20]. The emerging computer-aided solutions for the detection of MI in echocardiography are active contour-based models, deformation (strain) imaging, and machine learning algorithms. The active contour (or snake), proposed by Kass et al. [21], is an elastic curve, which is evolved by the external constraint forces and the internal image forces to find edges, contours, or to detect and localize specific objects in an image as it is placed nearby it. Snakes are used to track the extracted contours during motion. Therefore, several studies [22,23] have applied snake models to extract the endocardial boundary of the LV wall in echocardiography. However, the performance of the snake considerably fails as the quality of the echocardiography degrades. Therefore they are not suitable to extract the true endocardial boundary of the LV wall.

In echocardiography, strain imaging has become the focus of the main studies [24,25]. The strain is calculated by the LV wall displacement over time, and an infarcted muscle can be determined if all the segments on the LV wall move with the same velocity [24]. Therefore, many studies have adapted strain imaging in echocardiography to distinguish the active and passive myocardial movements. To perform strain analysis, several points on the LV wall are tracked during one cardiac cycle by the speckle tracking method. In order to track the speckles, block-based motion estimation algorithms are used. However, the motion estimation approaches are unreliable or even infeasible in the echocardiograhy, where the LV wall is missing due to low contrast or echocardiograhy is subjected to a high level of noise [24,26,27]. Therefore, the performance of the strain imaging highly depends on the motion estimation algorithm, which brings limitations to this method [24]. Consequently, strain imaging by speckle tracking is not a viable technique for MI detection [28].

Machine Learning (ML) algorithms have effectively and accurately solved many challenging problems in recent years. Especially, recent applications of ML on medical tasks have eased the diagnosis of many diseases and helped medical experts to save numerous hours in time-critical situations [29]. Many studies have proposed to use conventional and Deep Learning (DL) models in cardiology. The accurate segmentation of the LV wall with DL techniques can reveal the true characteristics of the wall muscle, such as its dimension, volume, and motion via echocardiography [18]. Primarily, outstanding performance in biomedical image segmentation has been achieved by U-Net on the available annotated data sets. Following its steps, the authors in [30] have modified the U-Net architecture as MFP-Unet to improve the segmentation performance for the LV wall extraction in echocardiography, which considers the semantic strengths intercalarily. In another study [31], authors have labeled an unannotated data set via a generative model that learns the inverse mapping of the segmentation mask of the echocardiography recording to use it for improving the segmentation masks generated by a segmentation network similar to U-Net. It is also shown on a large-scale data set in [32] that deep Convolutional Neural Networks (CNN) outperform the conventional segmentation methods. However, to the best of our knowledge, there has not been any published study that segments the entire LV wall enclosing endocardium, myocardium, and epicardium all at once. Nevertheless, due to the noisy nature and low spatial/temporal resolution of echocardiography recordings, segmentation of the LV wall is known to be a challenging problem. Several studies attempted to diagnose MI without any segmentation approach using conventional and DL algorithms. The authors of [33] extracted three types of features from 800 ultrasound images and ranked the features to feed a support vector machine classifier in order to detect MI with the best performance using a minimum number of features. In [34], the RWMA classification performances of two approaches: (1) a random forest classifier fed with hand-crafted features, and (2) a deep CNN model trained with hierarchical aggregation information of 2D transesophageal echocardiography are compared. Even though these approaches are simple and straightforward, the major limitation is the data scarcity in this domain. As it is already an obvious issue with DL methods, the need for a large data set for training brings limitations since the number of publicly available data sets is low. In fact, there is a growth in the publicly published echocardiographic data sets, such as CAMUS [32] and EchoNet-Dynamic [35]. However, these data sets are not for MI diagnosis, but for LV volume estimation, that is, ejection fraction measurement. Thus, there is no publicly available data set for MI detection in echocardiography.

In this chapter, a three-phase approach is presented, which first segments the entire LV wall in each frame of the echocardiography by a deep learning model, and then the characteristics of the segmented wall are utilized to perform early detection of MI robustly and accurately. The study introduces HMC-QU data set, which consists of A4C view echocardiography recordings created by a group of cardiologists in Hamad Medical Corporation (HMC) Hospital and researchers in Qatar University. The ground-truths of MI detection are given by cardiologists, whereas the LV segmentation ground-truths are created via collaborative approach between the deep

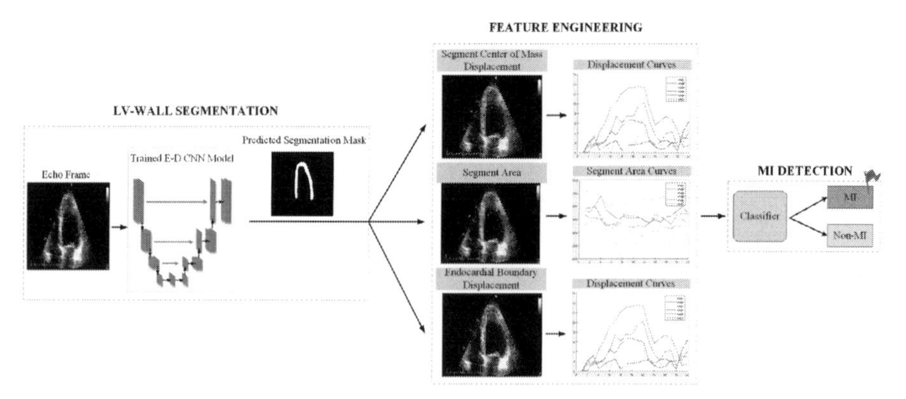

FIGURE 20.1

The three-staged system for early detection of MI. The system starts by segmenting the LV wall on each frame of the echocardiography using the trained deep network. Second, feature engineering is performed by analyzing the predicted LV wall segmentation masks. Lastly, MI is detected by feeding the extracted features to the conventional classifiers.

learning model and cardiologists. The generated segmentation masks are confirmed under expert supervision that saved the researchers numerous hours compared to the manual labeling process. An encoder-decoder CNN (E-D CNN), that is, U-Net segmentation model is trained with the generated ground-truth masks, and predicted the LV wall on each frame of each echocardiography recordings in the data set. Subsequently, the predicted LV wall is divided into standardized six-segments from which the characteristic features are extracted. Finally, the classifiers that are trained by the extracted features are utilized to detect early MI. Moreover, the outputs of the scheme provide advanced visualizations to cardiologists via the color-coded segment plot on the predicted LV wall, the displacement curves of segment center and LV endocardial points, and the segment area curve. This study further shares the 2D echocardiographic data set, HMC-QU[1] with the research community, that serves both MI detection and LV wall segmentation.

20.2.1 Methodology

In Fig. 20.1, the three stages of the developed scheme are illustrated: (1) segmentation of the LV wall, (2) feature engineering, and (3) early detection of MI. In order to perform LV wall segmentation, a novel pseudo-labeling technique is introduced to generate ground-truth segmentation masks for the LV wall. Then, the produced ground-truth masks are used to train the E-D CNN model, which segments the entire LV wall on each echocardiography frame. The characteristics of the predicted LV

[1] The HMC-QU data set is available at the repository https://www.kaggle.com/aysendegerli/hmcqu-dataset.

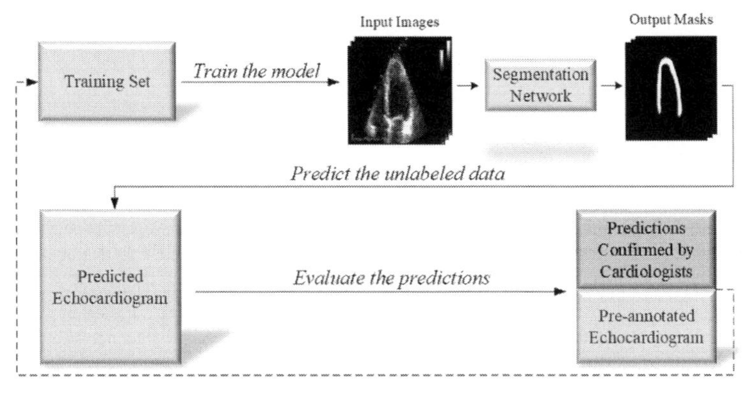

FIGURE 20.2

The ground-truth generation process of the benchmark dataset by the pseudo-labeling approach utilizing the deep E-D CNN segmentation network.

wall are extracted by feature engineering, which is then used to feed conventional classifiers for MI detection.

20.2.1.1 Pseudo-labeling technique for ground-truth generation

Supervised learning methods require ground-truth labels for the training process. In order to perform LV wall segmentation using the deep models, there is a need for ground-truth segmentation masks. However, the expert annotation for each echocardiography frame is not a practical labeling approach since there are numerous frames in the data set. To tackle this problem, a pseudo-labeling technique is introduced as depicted in Fig. 20.2, to perform ground-truth annotation of the benchmark data set automatically, and more accurately. The pseudo-labeling process is initialized with a few echocardiography recordings annotated by the cardiologists. Then, the annotated data is used to train a segmentation network: U-Net, proposed by [30]. Then the trained network is used to predict the next echocardiography frames that are not annotated. The predicted echocardiography recordings are evaluated visually by the cardiologists, and the confirmed ground-truths are added to the annotated data, whereas the incorrect ones are excluded. Therefore, the training set is expanded via additional ground-truths, then the model has trained again over the data. This process is executed until the whole data set is labeled.

The data set is annotated after eight iterations; thus, high-quality LV wall labels are assured. Post-processing is performed within the iterations in order to remove noise and false positives while preserving the shape and the size of the predicted LV wall. For the post-processing, morphological operations are performed, that is, erosion followed by dilation, using 1 as the kernel value, and 3×3 is the kernel size. As more iterations are performed, only the challenging echocardiography recordings remain. On the other hand, the performance of the segmentation network increases gradually as more training data is used to train it (see Fig. 20.3). Therefore, this technique can also be used for low-quality echocardiography detection, where some

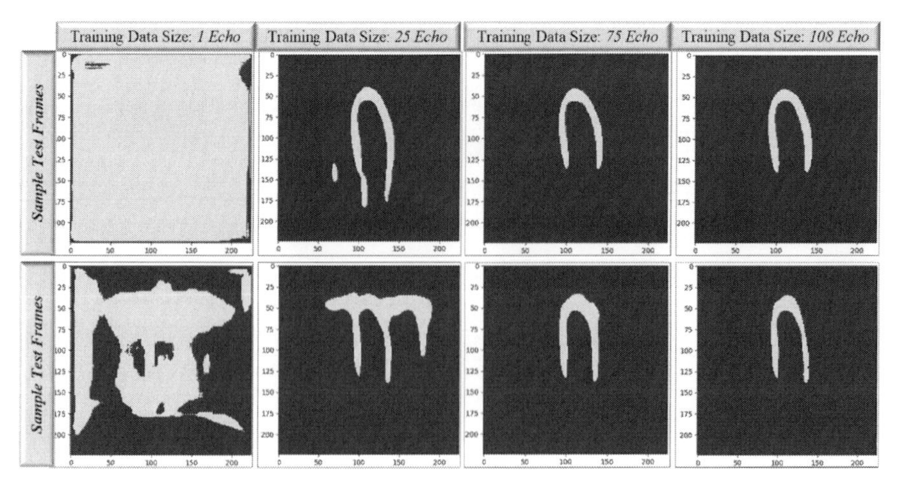

FIGURE 20.3

Sample test frames from two different patient's echocardiography recordings are taken to perform the ground-truth generation by the pseudo-labeling technique. The segmentation network gradually learns to segment the LV wall successfully as the training set expands.

parts of the LV wall are invisible, or the echocardiography is noisy. In fact, the study observes that this technique cannot segment 10 echocardiography recordings even after all the iterations. Further investigations confirm that these recordings are either too noisy or some parts of the LV wall are missing.

20.2.1.2 Segmentation of the LV wall

The segmentation of the LV wall is the primary stage of the system as shown in Fig. 20.1. After generating the ground-truth segmentation masks for the benchmark data set by the pseudo-labeling technique, the same network structure is used to predict the LV wall. The network architecture is inspired by U-Net [30], and the detail of its structure is given in Table 20.1. The network is trained from scratch with the annotated data, then the LV wall is predicted in all the frames by feeding each echocardiography recording into the trained network.

In order to perform further analysis on the predicted wall, it is divided into standardized segments as illustrated in Fig. 20.4. This study adapts the standardized model, which was recommended by the American Heart Association Writing Group on Myocardial Segmentation and Registration for Cardiac Imaging [36] in 2002. According to this model, the LV wall is divided into 7-segments for the A4C view. The division is performed by separating the endocardial boundary into two splines as left (from start to apical cap) and right (from apical cap to end). The length is defined as L and R for the left and right splines, respectively, as presented in Fig. 20.4. The visual representation of the LV wall segments is enhanced for cardiologists by the color-coded segmentation outputs as depicted in Fig. 20.4. Additionally, the endo-

Table 20.1 The structure details of the E-D CNN model, from the encoder block through the decoder block.

	Encoder block		Decoder block	
Filter size	Number of filters	Max pooling	Number of filters	Up sampling
3×3	32	2×2	512	2×2
3×3	64	2×2	256	2×2
3×3	128	2×2	128	2×2
3×3	256	2×2	64	2×2
3×3	512	2×2	32	2×2
3×3	1024	–	–	–

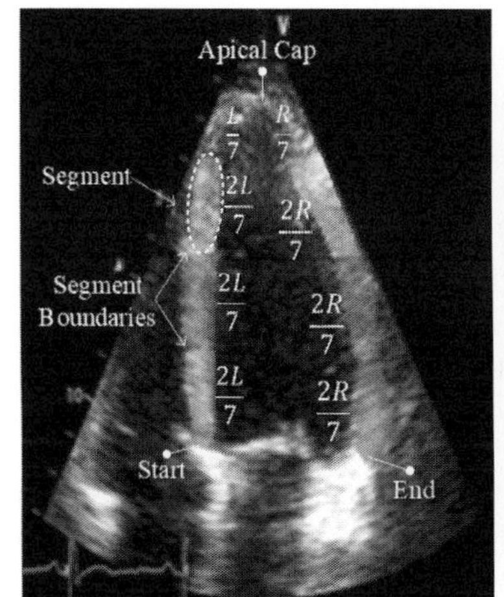

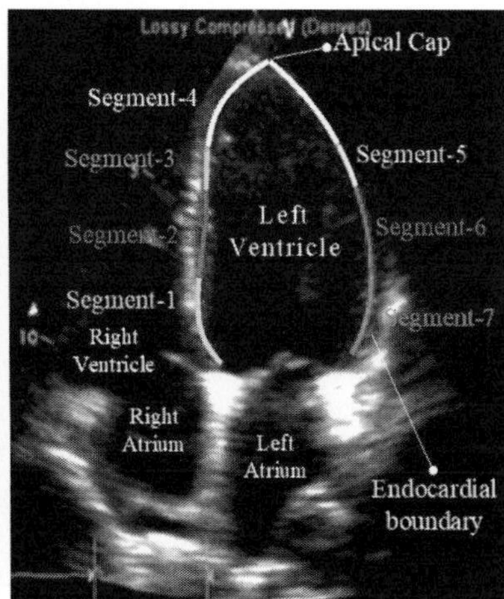

FIGURE 20.4

Right-side of the figure represents the four chambers of the heart, the myocardial segments on the LV, and the endocardial boundary of the LV wall. Left-side of the figure indicates the division ratios of the myocardial segments as the length of right part and the left part of the endocardial boundary are defined as R and L, respectively.

cardial boundary is extracted from the segmentation masks for endocardial boundary displacement in the next section for feature engineering.

20.2.1.3 Feature engineering

In this section, feature engineering is performed on the segments of the LV wall. The extracted features are fed to several conventional classifiers to detect MI. For feature engineering, only the six-segments are considered by excluding the 4th segment on the apical cap since it does not exhibit any inward motion activity in the A4C view [37]. In feature engineering, motion and area information are extracted from the predicted LV wall of each echocardiography recording (see Fig. 20.1). The aim is to mimic a typical diagnosis of a cardiologist who evaluates segments that show a lack of motion. Thus the defined features can evaluate the rate of displacement from the captured global motion of the LV wall.

After segmenting the LV wall, its endocardial boundary is further extracted, where there is a high contrast between the muscle and the chamber. The endocardial boundary segment displacements are calculated by Manhattan distance through the recording as follows in Eq. (20.1):

$$d_{L1} = |x^t - x^{t_r}| + |y^t - y^{t_r}| \tag{20.1}$$

where x and y are the pixel coordinates of current frame t, and reference frame t_r, which is the first frame of the cardiac cycle. Equally distanced N points $P \in \{(x_1, y_1), (x_2, y_2), \ldots, (x_n, y_n)\}$ are taken on the endocardial boundary in order to capture it more precisely. Then, the pairwise distances d_{s_t} of the endocardial points on each frame t for each segment s between the reference t_r and current t frames are calculated. Next, segment displacement is calculated as follows in Eq. (20.2):

$$d_{s_t} = \frac{1}{N} \sum_{n=1}^{N} |x_n^{t_r} - x_n^t| + |y_n^{t_r} - y_n^t| \tag{20.2}$$

Accordingly, the displacement curve d_s is calculated for each segment on the LV wall, from which certain motion features are extracted. In addition to the endocardial boundary, the displacement of the center points of each segment is also calculated. Therefore, Eq. (20.1) is still valid, but in this case, there is only one point to track on each segment. Then, the displacement curve of the segment center points is plotted for six segments. Lastly, as the advantage of the segmentation approach, where the LV wall is entirely extracted, the segment area information can be obtained. The segment area is defined as the total number of pixels inside one segment. In this way, the segment area curve is plotted. The overlapped regions in the consecutive frames yield information regarding segment motion, and thus, deformation since the region would be smaller for the infarcted segments compared to the normal ones.

In the feature engineering stage, three sets of features are extracted: (1) endocardial boundary motion, (2) segment center of mass motion, and (3) segment area. The cardiologists seek for infarcted segments in A4C view by visually assessing the LV wall, where there occurs an attenuated segment motion. Therefore, the motion features are defined for endocardial boundary and segment center of mass points as the maximum displacements of the segments in one cardiac cycle of echocardiography as shown in Fig. 20.5. Therefore, the motion feature (MF) is defined as in Eq. (20.3):

$$MF = \max(d_{s_t}) \tag{20.3}$$

In order to compute the area feature, the total number of pixels, P inside the overlapped segment regions of the consecutive frames are calculated as given in Eq. (20.4):

$$P_{s_t} = P_t \cap P_{t_r} \tag{20.4}$$

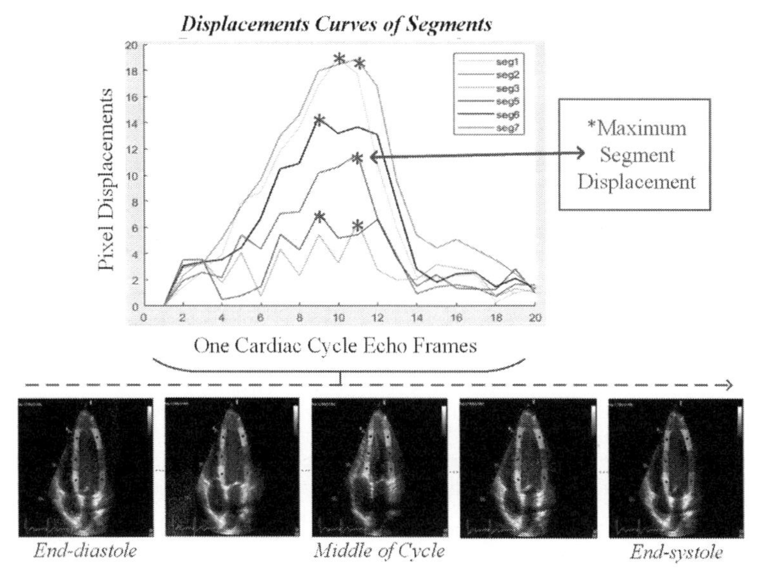

FIGURE 20.5

Displacement curves of each segment are plotted through one cardiac cycle of a patient's echocardiography, where the segments are tracked from end-diastole to end-systole frames. Then the motion features, that is, maximum displacements of the segments are extracted. (For interpretation of the colors in the figure(s), the reader is referred to the web version of this chapter.)

Table 20.2 The feature descriptions, which are extracted from the predicted LV wall segments for one cardiac cycle echocardiography.

Extracted features	Description
Motion Feature (#1)	The maximum displacement of endocardial boundary
Motion Feature (#2)	The maximum displacement of segment center
Area Feature	Minimum area intersection of segments

where P_{s_t} is the number of intersected pixels for segment s, at frame t, computed between the reference frame t_r. The area feature (AF) is defined in Eq. (20.5):

$$AF = \frac{P_{s_t}}{P_{t_r}} \tag{20.5}$$

where the minimum area of the intersected segment is defined as P_{s_t} and, P_{t_r} is the number of pixels in the reference frame segment region. The intersection between the segment regions reveals information regarding segment displacement since the larger intersected area shows a smaller segment movement in any direction. As a summary, Table 20.2 gives a brief description of the extracted features. Three features from six segments, a total of 18 features are extracted from one echocardiography recording.

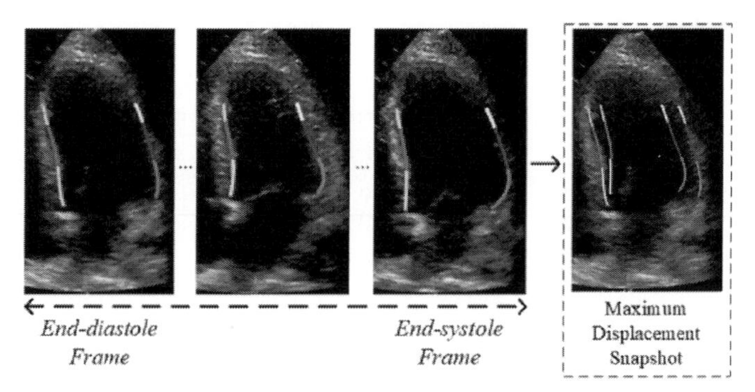

FIGURE 20.6

The color-coded segment visualization of the predicted endocardial boundary with end-diastole, middle, and end-systole frames of one cardiac cycle are consecutively illustrated. The maximum displacement snapshot includes the color-coded segments of the end-diastole and the frames where the maximum segment displacement occurs.

In addition, for a better and more objective assessment of MI, several bi-products are generated for cardiologists to help them in the diagnosis. These products are the color-coded LV wall and endocardial boundary segments, segment center, and endocardial boundary displacement curves, and segment area curves. Lastly, the maximum snapshots of the endocardial boundary are provided for each segment as illustrated in Fig. 20.6.

20.2.1.4 Myocardial infarction detection

The last stage of the developed scheme is MI detection using several supervised ML techniques for the binary classification that are Linear Discriminant Analysis (LDA), Decision Tree (DT), Random Forest (RF), and Support Vector Machine (SVM). In this stage, the selected algorithms seek for any pattern or inference from the extracted features. LDA performs classification by expanding the gap between intraclass and interclass variances. It can efficiently work on imbalanced data sets, where the samples from the different classes are not equal [38]. DT has a tree-like structure, where the data is fed through the branches (conjunctions) by passing the nodes (attributes), and the decision is made as it reaches the most suitable leaf (class label) [39]. The hierarchical structure of the DT is beneficial for the benchmark data set since it is suitable for data sets, which consists of a small number of samples. One of the limitations of DT is its tendency for overfitting. However, another classifier from the tree classifier family: RF [40] overcomes this limitation. RF is structured by individual trees that perform the classification task by minimizing the correlation within them. The final model is formed by majority voting within these individual trees, which is then used for the rest of the task. Lastly, SVM is utilized that separates classes by fitting a hyperplane [41]. The kernel trick can be used in SVM, which maps the data to higher dimensions; therefore the performance of the classification tasks can be improved.

This study does not utilize any DL models for MI detection since the benchmark data set is small and imbalanced for such deep models. The extracted features are the simple solution for the diagnosis as the approach mimics the cardiologists; therefore they are not suitable for complex structured deep models. The training-testing of the classifiers is performed in a 5-fold cross-validation scheme for fair performance evaluation.

20.2.2 Experimental evaluation

The performances are evaluated for both LV wall segmentation and MI detection stages. The segmentation of the LV wall is considered as a binary problem at a pixel-level, where the LV muscle pixels are positive-class, and the background pixels are negative-class. On the other hand, the evaluation of MI detection is done based on the non-MI subjects as negative-class, and MI subjects as positive-class. Then, the confusion matrices (CM) are formed by evaluating the predicted samples with respect to the ground-truth labels. Accordingly, the standard performance metrics are computed from the CM elements, which are true positive (TP), false positive (FP), true negative (TN), and false-negative (FN). *Sensitivity* is the rate of correctly predicted positive-class samples among all the positive-class members; *specificity* is the sensitivity of the negative-class; *precision* is the ratio of correctly predicted positive-class samples to all samples that are predicted as positive-class; *F1-score* is the harmonic average of precision and sensitivity; lastly, *accuracy* is the rate of correctly predicted samples to the total number of samples in the data set. The main objective of the analysis is to have high sensitivity and specificity in order not to miss any LV wall pixels or MI subjects.

20.2.2.1 HMC-QU data set

The HMC-QU benchmark data set is created by a collaboration between the cardiologists at the Hamad Medical Corporation Hospital, and the researchers at the Qatar University. The data set consists of 160 echocardiography recordings from A4C view obtained at HMC Hospital during 2018–2019. However, a subset of 109 echocardiography recordings is used in this study from 72 MI patients and 37 non-MI subjects including 2349 images. The remaining recordings are excluded since they do not have the entire LV wall to be evaluated by the cardiologists as shown in Fig. 20.7. The MI recordings are taken from the patients, who were admitted to the hospital with acute MI evidence, before or within 24 hours of coronary angioplasty for treatment. The normal (non-MI) recordings are taken from patients not admitted for MI, but they underwent check-ups in the hospital for other reasons. The local ethics board of HMC Hospital has approved the usage of the data set in February 2019. The cardiologists at the HMC Hospital have provided the ground-truth labels for the echocardiography unanimously. The recordings are labeled segmentwise as normal (1), hypokinetic (2), and akinetic (3). However, for a straightforward evaluation, the labels are downsized to two classes as 1-normal (non-MI), and 2-infarcted (MI). Table 20.3 shows the number of subjects per-segment and per-echocardiography, which reveals the challenge of the imbalance data set for this problem.

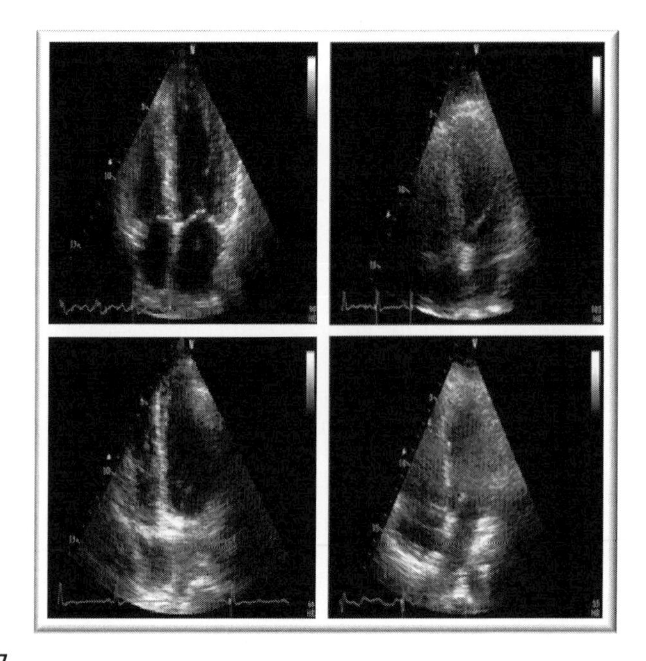

FIGURE 20.7

The sample frames from four different patient's echocardiography recordings, where the quality is low due to the presence of high noise level and low contrast acquisition. The LV wall is generally absent or some parts of it cannot be recognized.

Table 20.3 The number of samples in MI and non-MI class samples in the data set with respect to the LV wall segments.

LV wall segments	Number of MI patients	Number of non-MI subjects
Segment-1	24	85
Segment-2	43	66
Segment-3	59	50
Segment-5	44	65
Segment-6	25	84
Segment-7	15	94
Subject-based	72	37

Echocardiography recordings are acquired from the Phillips and GE Vivid (GE-Healthcare-USA) ultrasound machines. The temporal resolution of the recordings is 25 fps, and the spatial resolution varies from 422×636 to 768×1024 pixels. However, the recordings are resized to 224×224, which is a suitable input size for many pre-trained deep network structures. Lastly, each recording is analyzed in one cardiac cycle, where the end-diastole and end-systole frames are determined by the ECG

Table 20.4 The segmentation 5-fold CV performance results (%) on the test fold of the E-D CNN segmentation model.

CV folds	Sensitivity	Specificity	Precision	F1-score	Accuracy
Fold-1	94.90	99.70	93.66	94.26	99.49
Fold-2	96.53	99.67	92.31	94.37	99.54
Fold-3	96.35	99.63	91.63	93.93	99.50
Fold-4	97.61	99.17	83.75	90.15	99.10
Fold-5	93.23	99.75	94.24	93.73	99.48
Mean	**95.72**	**99.58**	**91.11**	**93.29**	**99.42**
Std	±1.69	±0.23	±4.25	±1.77	±0.18

recordings of the subjects. For the subjects without the ECG data, the cardiac cycle is determined by the frames, where the LV area is the largest and smallest.

20.2.2.2 LV wall segmentation experiments

The performance of the segmentation model is evaluated in a 5-fold cross-validation (CV) scheme, where the model is trained with 80% of the data and tested on the remaining 20% unseen echocardiography recordings. The E-D CNN model is trained using Adam optimizer [42] and cross-entropy loss function with a batch size of 32, a learning rate of 10-3 and 25 epochs at each fold. The model implementation is performed by Keras with Tensorflow backend on NVIDIA GeForce GTX 1080 Ti GPU with 128 GB memory.

The results of the LV wall segmentation are given in Table 20.4 for the CV both individual folds and their averages (mean). The standard deviation of each fold can also be depicted from the table, where it indicated the robustness of the model with a low variance value. The trained model performs a successful segmentation at a pixel-level with 99.42% accuracy by 93.29% F1-Score on average. It can be observed that the model can capture the LV muscle pixels with a sensitivity of 95.72% while keeping the false alarm rate (1-specificity) with an elegant performance by 0.42%. The segmentation is a challenging task for low-quality echocardiography, where the LV wall is partially invisible and there exists a high level of noise as shown in Fig. 20.8.

20.2.2.3 Myocardial infarction detection experiments

MI detection is evaluated on the same 5-fold CV as in the previous section. The performance of detection by feature engineering is evaluated on the train and test sets of each fold. The aforementioned features are extracted and fed to the classifiers: LDA, DT, RF, and SVM. The classifiers are trained by the features of the training set and evaluated at the test phase. The feature engineering and MI detection stages of the introduced scheme are implemented on MATLAB® with version 2020. A computer with 1.90 GHz CPU and 32 GB memory is used for the experiments. The SVM classifier is implemented by the Library for Support Vector Machines – LIBSVM [43]. The hyperparameters of the classifiers are set by the optimal parameters that yield the maximum F1-Score as follows: the discriminant type of LDA is pseudo-

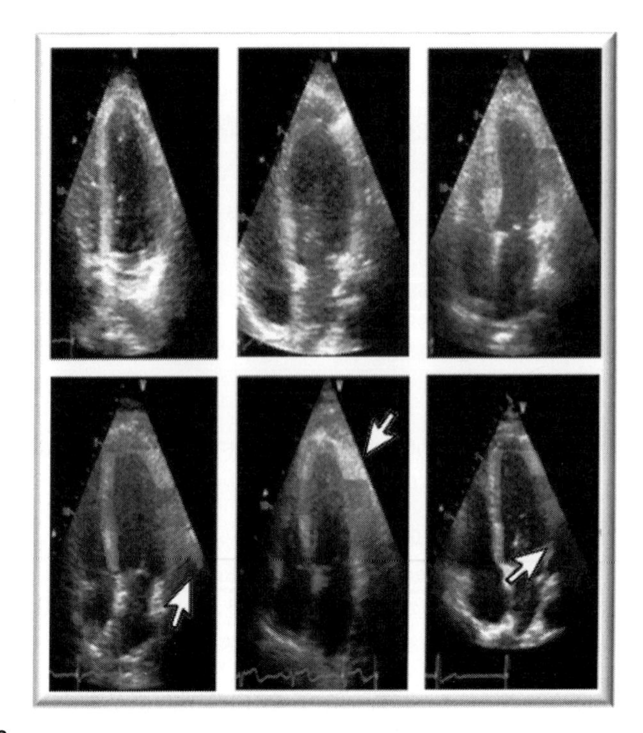

FIGURE 20.8

The sample low-quality echocardiography frames with successful (top-row) and failure (bottom-row) cases of LV wall segmentation predictions. The problematic regions in the failure cases are denoted by the arrow.

Table 20.5 MI detection performance results (%) of the classifiers, where 5-folds are averaged.

Classifier	Sensitivity	Specificity	Precision	F1-score	Accuracy
6-segment features					
LDA	78.51	70.10	83.89	80.67	75.65
DT	79.09	58.60	80.41	79.48	72.62
RF	80.26	**71.81**	**85.99**	82.57	77.47
SVM	**85.97**	70.10	85.52	**85.29**	**80.24**
5-segment features					
LDA	81.38	72.32	86.61	83.21	78.52
DT	79.09	62.60	81.72	80.00	73.53
RF	82.29	67.37	84.94	82.65	76.61
SVM	**83.09**	**74.03**	**86.85**	**84.83**	**80.24**

linear and its prior probability is uniform, DT pruning criterion is an impurity, the number of trees for RF is 10, and the SVM type is C-SVC with the kernel type of radial basis function (RBF), and cost value of 10.

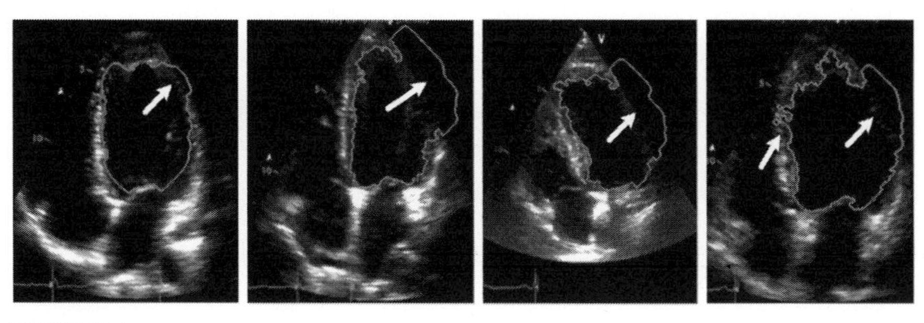

FIGURE 20.9

Executing the snake algorithm [44] over four sample echocardiography frames from different patients, where the extraction of LV wall endocardial boundary is a failure at the regions denoted by the white arrow.

Table 20.5 shows the averages of the classifiers that are evaluated in a 5-fold CV. The aim is to have a high sensitivity value so as to not miss any patient with MI. The SVM classifier reports the highest sensitivity of 85.97% with an elegant specificity of 74.03% and precision by 86.85%. For deeper analysis, the number of features are downsized by excluding the apex segment (eliminating segment 5) to investigate the effect of noise on the echocardiography, where it usually is on the apex part of the LV wall. The results with the 5-segment features show that the performance of the detection is boosted by the highest specificity of 74.03% and precision of 86.85%. Consequently, the noisy nature of the apex segments deteriorates the performance of the detection.

The low-quality nature of the benchmark data set brings limitations to other existing algorithms, which obstructs the comparison of the developed scheme against the existing methods. In fact, the improved version of the snake proposed by Chan–Vese [44] fails to detect the LV wall in the benchmark data set as depicted in Fig. 20.9. In addition to snake methods, speckle tracking-based methods are also not compatible with the temporal resolution, that is, 25 fps of the benchmark data set, where it is required to perform speckle tracking on echocardiography with a temporal resolution of 60 fps [24]. Even with a proper temporal resolution, the high noise level and lack of contrast present in the echocardiography would still limit the speckle tracking approach (see Fig. 20.7). Therefore, the introduced scheme is the only feasible technique in this domain, where the quality of echocardiography is low.

20.2.2.4 Computational complexity analysis

This section presents the execution times of the algorithm stages that are introduced in Fig. 20.1. Table 20.6 shows the elapsed time in milliseconds (ms) during the inference step of each algorithm stage in the experiments. The results show the execution time per echocardiography in one cardiac cycle. It can be observed from the table that the segmentation stage has the largest complexity, where the E-D CNN model execution is completed in 2.58 seconds to process one cardiac cycle echocardiography, which consists of 20–30 frames.

Table 20.6 The elapsed times (ms) to execute the stages of the algorithm in one cardiac cycle.

Algorithm stage	Implemented methods	Elapsed time (ms)
LV wall segmentation	E-D CNN	2579
Feature engineering	Area	169.6
	Segment center motion	176.9
	Endocardial motion	44.9
MI detection	LDA	1.3
	DT	0.5
	RF	7.0
	SVM	0.2

20.3 COVID-19 recognition from X-ray images via convolutional sparse support estimator based classifier

As of the day this book is written, the world has still been struggling with the COVID-19 pandemic. Currently, the epidemic has infected more than 100 million people, and the number of COVID-19 cases and its death toll is increasing day by day. The rate of spread of the pandemic is so high that it is expected that the number of cases in Europe could double every 3 days if social distance and other related measures are not taken [45]. Therefore, both rapid and accurate diagnosis is crucial to prevent this rapid spread. Today, the most used diagnosis method is known as reverse transcription polymerase chain reaction (RT-PCR). However, at the earliest, a case can be diagnosed within 24 hours by this method [46]. On the other hand, as reported by the World Health Organization, RT-PCR may give false results in COVID-19 cases due to reasons such as an insufficient quality sample taken from the patient or imprecise processing of the sample [47]. Computed tomography technology can be an alternative method that provides faster detection than RT-PCR, but the acquisition equipment is expensive and not easily available [48]. Another tool that medical professionals can use in both diagnosis and follow-up of the course of the disease is X-ray technology.

In a sudden outbreak, such as in the COVID-19 pandemic, it is crucial to quickly develop technological tools that can alleviate healthcare practitioners' burden. For example, a computer program that can distinguish COVID-19 cases from other types of diseases and healthy cases by leveraging patients' samples would be an essential tool for medical experts. Today, deep learning methods are reported as the best-performing methods in many classification and detection problems. Despite all these advantages, these methods require large amounts of data to achieve sufficient generalization capability; therefore satisfactory classification/detection performance. In the case of a sudden pandemic such as COVID-19 and similar ones, it may not be possible to clean and label such an extensive data set in a short time, that is, the time that could be very critical to prevent further spreading of the pandemic.

As an alternative supervised machine learning technique, representation-based classification [49–51], also known as dictionary-based classification, requires a much smaller training data than deep learning counterparts do in order to achieve a satisfactory classification performance. In order to create a representation-based classifier, a dictionary of the whole training set is created in advance. In this dictionary, samples are stacked in columns in such a way that the locations of the samples from the same class come together. If the predefined dictionary, $\mathbf{D}$, is representative enough, when a new test example, $\mathbf{y}$, is given, its representation in the dictionary, $\mathbf{y} = \mathbf{Dx}$ will be satisfied with a relatively small representation error. Thus, in such decomposition, it is expected that the representation coefficient vector $\mathbf{x}$ carries enough information about the test sample's class.

Sparse representation-based classification (SRC) [50] and collaborative representation-based classification (CRC) [49] are the most well-known dictionary-based classification methods. SRC method can achieve higher classification performance by generating a sparse solution for $\mathbf{y} = \mathbf{Dx}$. In fact, for an ideal dictionary and test sample, in the solution, $\hat{\mathbf{x}}$, found from $\mathbf{y} = \mathbf{Dx}$, only the elements corresponding to the indices of the columns where the samples with the same class of $\mathbf{y}$ are placed will be nonzero. Therefore, the location of nonzero elements, also known as the support set, is sufficient information to determine the sample's class. Current SRC-based techniques first obtain the elements of $\mathbf{x}$ by applying an iterative and relatively costly sparse recovery algorithm, then obtain the support of it (or class of the sample). However, obtaining the coefficients itself is redundant to determine the class. On the other hand, for the moderate-sized training data set, when the dictionary size increases, solving such a sparse recovery is cumbersome, and the performance deteriorates drastically.

The recent convolutional support estimation [52] technique introduces reconstruction-free, noniterative support estimation as bypassing the redundant signal recovery. Performing such a direct mapping allows a CNN-based solution that can use only relatively limited size training data. In this manner, CSEN presents a good alternative to traditional model-based representation-based classification and DL alternatives when there is a small and moderate size training data as it is in the case of the COVID-19 outbreak. The idea is instead of stacking all the training samples to a dictionary; some of it is spared for the dictionary and the rest to train CSEN-based classifier. This can be seen as a third alternative approach (an interpretable lightweight classifier network) to model-based solutions and black-box models.

20.3.1 Preliminaries

20.3.1.1 Sparse signal representation

Sparse Representation (SR) of a signal $\mathbf{s} \in \mathbb{R}^d$ refers to approximate it as a linear combination of a relatively small subset of the atoms in a predefined dictionary $\mathbf{\Phi} \in \mathbb{R}^{d \times n}$, that is, $\mathbf{s} = \mathbf{\Phi x}$. Defining a dictionary where the signal of interest's representation can be satisfied with the compact as a possible subset of waveforms dated back to Fourier's pioneering work [53]. Until Mallat work [54] on overcomplete dictionaries

($n >> d$), traditional compete dictionaries (or basis i.e., $d = n$) such as Discrete Cosine Transform (DCT), Wavelets, etc., had been excessively studied. Recently, sparse representation-based techniques have been a very attractive research topic in many application areas, including denoising [55], classification [56], compressive sensing [57,58], and signal encryption [59,60].

Let assume that a linear compression is also applied via a compression matrix, $\mathbf{A} \in \mathbb{R}^{m \times d}$ ($m << d$), that is,

$$\mathbf{y} = \mathbf{A}\mathbf{s} = \mathbf{A}\mathbf{\Phi}\mathbf{x} = \mathbf{D}\mathbf{x}, \tag{20.6}$$

where $\mathbf{D} \in \mathbb{R}^{m \times n} = \mathbf{A}\mathbf{\Phi}$. Elementary linear algebra tells us that Eq. (20.6) is an underdetermined linear system of equations, and it has infinitely many solutions. With an assumption that the desired signal is sparse in the predefined dictionary, the sparsest solution among all is expected,

$$\min_{\mathbf{x}} \ \|\mathbf{x}\|_0 \text{ subject to } \mathbf{D}\mathbf{x} = \mathbf{y}. \tag{20.7}$$

Eq. (20.7) is also known as *sparse representation* of $\mathbf{s}$ in subspace $\mathbf{D}$. Let $\mathbf{x}$ be an exactly k-sparse signal, that is, $\|\mathbf{x}\|_0 = k$. Then the representation defined in Eq. (20.7) is uniquely provided that $\mathbf{D}$ satisfies some properties [61]. However, the problem given in Eq. (20.7) is a nonconvex one and its solution is NP-hard. Fortunately, if the matrix $\mathbf{D}$ satisfies certain conditions [62,63], the equivalent solution to the solution of ℓ_0-minimization in Eq. (20.7) can be achieved by solving the following optimization problem:

$$\min_{\mathbf{x}} \ \|\mathbf{x}\|_1 \text{ subject to } \mathbf{D}\mathbf{x} = \mathbf{y}. \tag{20.8}$$

On the other hand, when the representation coefficient, $\mathbf{x}$, is not exactly sparse but compressible or/and the query, $\mathbf{y}$ is corrupted, the stable solution is still possible via the relaxed optimization problem: $\min_{\mathbf{x}} \{\|\mathbf{x}\|_1 \text{ s.t. } \|\mathbf{y} - \mathbf{D}\mathbf{x}\| \leq \epsilon\}$, where ϵ is a small constant [64,65]. The stable solution, $\hat{\mathbf{x}}$ refers to one with a relatively small reconstruction error, that is, $\left\|\mathbf{x} - \hat{\mathbf{x}}\right\| \leq \kappa$, where κ is a small constant.

Let us define sparse support set $\Lambda \subset \{1, 2, 3, ..., n\}$ as the set of indices that the nonzero coefficients are located, that is, $\Lambda := \{i : x_i \neq 0\}$. The process of locating the indices of nonzero elements of a sparse signal, $\mathbf{x}$, is known as sparse support prediction [66–68]. Recalling that sparse signal recovery refers to finding both the location (support set) and the value of these nonzero elements, there is an essential relationship between sparse signal recovery and sparse support prediction. A sparse signal with support can be easily resolved using least squares optimization [69,70].

20.3.1.2 Representation based classification

In recent years, representation-based classification, also known as dictionary-based classification methods, has attracted considerable attention. Particularly, the first of its kind, sparse representation-based method, has been applied successfully in a wide range of application areas including face recognition [50,51], human action

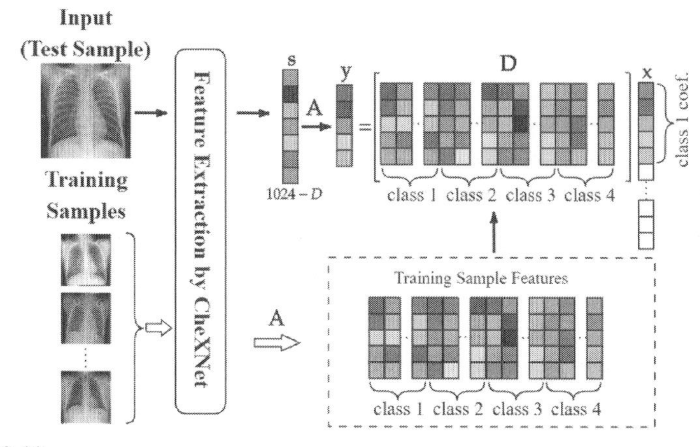

FIGURE 20.10

Dictionary design for representation-based classification.

recognition [71], and hyperspectral image segmentation [72]. In a nutshell, consider stacking n_i number of samples from the training data set that belongs to class i, that is, $\mathbf{\Phi^i} = \left[\phi_1{}^i, \phi_2{}^i, ..., \phi_{n_i}{}^i\right]$, where $\phi^i_j \in \mathbb{R}^d$ is the jth sample of class i. In a representation-based classification scheme, the sub-dictionaries C number of classes are concatenated to obtain one representative dictionary, $\mathbf{\Phi} = \left[\mathbf{\Phi^1}, \mathbf{\Phi^2}, ..., \mathbf{\Phi^C}\right] \in \mathbb{R}^{d \times n}$, where $\mathbf{\Phi^c}$ is subdictionary belonging class c. In most cases, the samples in the training dictionary are transferred to another subspace $\mathbf{A}$ such as PCA, that is, $\mathbf{D} = \mathbf{A\Phi} \in \mathbb{R}^{m \times n}$. Given a sufficiently representative dictionary, $\mathbf{D}$, any new test query $\mathbf{y}$, can be approximated on it with a relatively small reconstruction error, $\mathbf{e}$, that is, $\mathbf{y} = \mathbf{Dx} + \mathbf{e}$. Therefore one can expect from the representation coefficient vector, $\mathbf{x}$, to carry enough information about the class of $\mathbf{y}$.

In Fig. 20.10, the ideal setup for a dictionary, a query sample from class i, and its corresponding sparse code vector is picturized: The nonzero elements are obtained in places with the indices where the atoms from the same class of $\mathbf{y}$ are located in the dictionary. In this respect, support set, Λ, is sufficient information to determine the class of $\mathbf{y}$. However, in practice, a solution $\hat{\mathbf{x}}$ of $\mathbf{y} = \mathbf{Dx}$ is not exactly sparse, and the representation of the test sample in the dictionary is not necessarily perfect. It is due to the high inter and cross-class correlations, especially when the dictionary size grows. The author of [50] proposed a systematic four steps solution, which is known as sparse representation based classification, to obtain the class of $\mathbf{y}$: (i) normalization: normalize all the atoms in $\mathbf{D}$ and $\mathbf{y}$ to have unit ℓ_2-norm, (ii) sparse recovery: $\hat{\mathbf{x}} = \arg\min_{\mathbf{x}} \|\mathbf{x}\|_1$ s.t $\|\mathbf{y} - \mathbf{Dx}\|_2$, (iii) finding the residuals: $\mathbf{e_i} = \left\|\mathbf{y} - \mathbf{D_i}\hat{\mathbf{x}}_\mathbf{i}\right\|_2$, where $\hat{\mathbf{x}}_\mathbf{i}$ is the recovered elements that corresponds the class i, (iv) determination of class: $\text{class}\,(\mathbf{y}) = \arg\min\,(\mathbf{e_i})$.

Although the remarkable performances of SRC are reported for many applications [50,51,71,72], its recognition performance and computational complexity highly depend on sparse recovery algorithm (e.g., ℓ_1-minimization strategy). In the cases where dictionary size is very large, the computational complexity becomes cumbersome and recognition performance may start decreasing. On the other hand, if the number of classes is very few (the representation coefficient vector is not sparse enough), its performance drops drastically [52]. Alternatively, the authors of [49] proposed to replace computationally costly ℓ_1-minimization based recovery (or any other sparse recovery) with a conventional ℓ_2-minimization step; $\hat{\mathbf{x}} = \arg\min_{\mathbf{x}} \left\{ \|\mathbf{y} - \mathbf{D}\mathbf{x}\|_2^2 + \lambda \|\mathbf{x}\|_2^2 \right\}$. By doing this, the representation coefficient can be obtained in low-cost linear operation, that is, $\hat{\mathbf{x}} = \left(\mathbf{D}^T \mathbf{D} + \lambda \mathbf{I}_{n \times n}\right)^{-1} \mathbf{D}^T \mathbf{y}$. The technique is known as collaborative representation-based classification.

20.3.2 CSEN-based COVID-19 recognition system

20.3.2.1 Data set

Differentiating COVID-19 from other respiratory diseases is an overwhelming task not only for the machines but also for the medical experts. Therefore, this study collects chest X-ray images from healthy subjects (the samples that are not diagnosed with pneumonia), and patients with COVID-19, bacterial, and other viral pneumonia. The chest X-ray images belonging to the COVID-19 class was collected from a wide range of different sources. The major ones include Radiopaedia [73], Italian Society of Medical and Interventional Radiology (SIRM) COVID-19 Database [74], Chest Imaging (Spain) at thread reader [75], and some news portals and online articles [76]. The authors also collected and indexed a different number of X-ray images reported online from China, South Korea, the USA, Taiwan, Spain, Italy, and online newsportals (up to 20 April 2020). The data set includes 462 COVID-19 cases in a variety of age groups, gender, ethnicity, and country. The negative class samples include 1583 images that belong to healthy people (normal images) and 4280 images from bacterial and viral pneumonia classes. They are obtained from the Kaggle chest X-ray database [77]. Some of the images are from the data set are shown in Fig. 20.11.

20.3.2.2 Feature extraction via CheXNet

DL-based classifiers have become dominant in recent years. However, they generally require a large number of training samples to achieve sufficient generalization capabilities. Although in a sudden outbreak as COVID-19, it may not be possible to extract such a large enough data set in a short time. Nevertheless, DL-based classifiers can still be a valuable tool; especially, if they are trained in the same domain for a similar classification task. This study utilizes a pretrained deep network, CheXNet [14], to make discriminative feature extraction. CheXNet was trained for pneumonia detection problem using 112, 120 frontal-view chest X-ray images in the ChestX-ray14 data set [78]. ChexNet has the same structure as DenseNet [79] that consists of 121 layers.

The output of the average pooling layer of ChexNet is used, which is a 1024-size feature vector. Data normalization is applied in a way that feature vectors $\mathbf{s} \in \mathbb{R}^{d=1024}$ have zero mean and unit variance. Then, the dimensional reduction is applied using the PCA method, that is, $\mathbf{y} = \mathbf{A}\mathbf{y} \in \mathbb{R}^m$, where $\mathbf{A} \in \mathbb{R}^{m \times d}$ is the PCA matrix $(m < d)$.

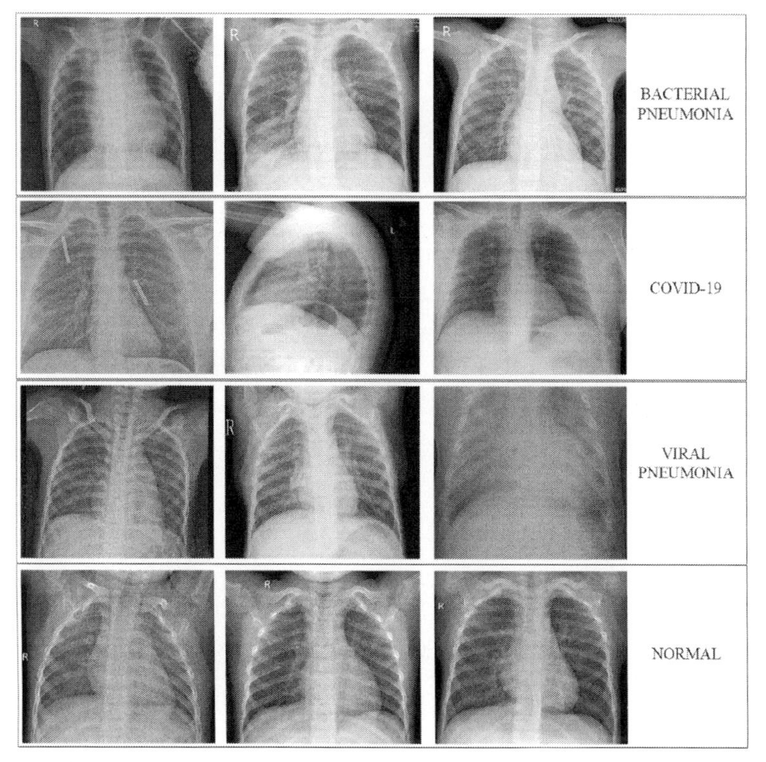

FIGURE 20.11

Chest X-ray samples from the data set.

Finally, query sample, **s** was one more time normalized to have zero mean, unit-variance for Multilayer Perceptron (MLP), Support Vector Machine, and k-Nearest Neighbor (k-NN) classifiers, and zero mean and unit-norm for SRC, CRC, and CSEN methods.

20.3.2.3 CSEN-based classifier

As mentioned above, among the dictionary-based methods, the sparse representation-based one is superior to the collaborative one in many applications by recovering the sparse representation vector via a computationally costly sparse recovery algorithm. However, such a costly signal recovery is redundant (even deteriorating when ground-truth representation coefficient is not sparse enough, for example, in the case of binary classification [52]) because the support set is sufficient information to determine the class of query sample. Recently proposed CNN-based support estimator, CSEN [52], is a remedy to the problem by estimating the support directly sparse representation coefficient, **x**.

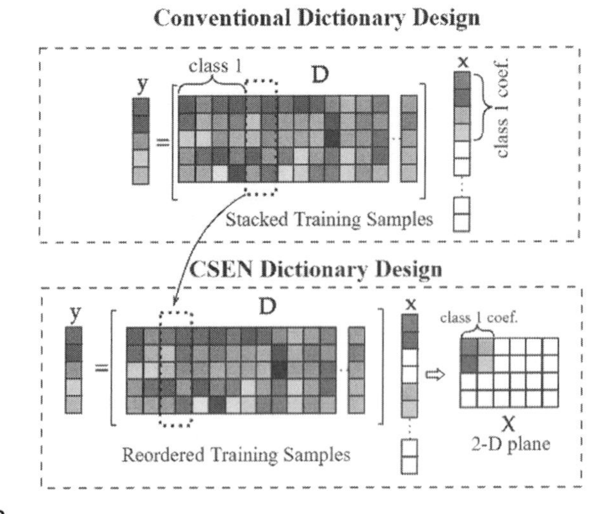

FIGURE 20.12

The traditional dictionary design versus the reordered dictionary for CSEN.

Given query and dictionary pair, $(\mathbf{y}, \mathbf{D})$, an ideal sparse support estimator should produce a binary mask $\mathbf{b} \in \{0, 1\}^n$:

$$b_i = 1 \quad \text{if } i \in \Lambda \tag{20.9}$$

which marks the ground-truth signal support, that is, $\Lambda = \{i \in \{1, 2, .., n\} : b_i = 1\}$. A CSEN network, $\mathcal{P}(\mathbf{y}, \mathbf{D})$ approximates this ideal situation by producing a probability such as a vector, $\mathbf{p}$, whose ith coefficient, $p_i \in [0, 1]$, yields a measure about how probable for this index is being in the support set. Therefore, one may easily estimate the support set by simple thresholding, that is, $\hat{\Lambda} = \{ i \in \{ 1, 2, .., n\} : p_i > \tau \}$, where τ is a fixed threshold.

In [52], CNN-based support estimator networks takes coarse estimation of sparse representation vector, $\tilde{\mathbf{x}}$, which is also known as a proxy sparse signal [80,81], and can be produced by $\tilde{\mathbf{x}} = \mathbf{B}\mathbf{y}$, where $\mathbf{B}$ is the denoiser matrix $((\mathbf{D}^T\mathbf{D} + \lambda\mathbf{I})^{-1}\mathbf{D}^T\mathbf{y}$ or simply $\mathbf{D}^T\mathbf{y}$). A CSEN is a fully convolutional neural network that produces the aforementioned probability as vector $\mathbf{p}$ from $\tilde{\mathbf{x}}$.

In order to use the original CSEN configurations in [52] the input $\tilde{\mathbf{x}}$ is reshaped to have a 2-D plane representation. Such a transformation can be easily done by relocating the atoms of the dictionary so that corresponding nonzero elements for a specific class come together in the 2-D plane. Fig. 20.12 illustrates such a reordering.

Having the $\tilde{\mathbf{x}}$, it is then convolved with N_1 number of kernels in order to produce first layer feature maps with biases $\alpha_{\mathbf{1}}$, that is,

$$\mathbf{f_1} = \{S_1(ReLu(\alpha_{\mathbf{1}}^{\mathbf{i}} + \mathbf{w}_{\mathbf{1}}^{\mathbf{i}} * \tilde{\mathbf{x}}))\}_{i=1}^{N_1}, \tag{20.10}$$

where $\mathbf{w_1^i}$ is the first layer's ith weight, α_1^i is the corresponding bias term, $S_1(.)$ is either identity or subsampling, and $ReLu(x) = \max(0, x)$. Hereafter, the following feature maps are obtained via

$$\mathbf{f_l^k} = S_l\left(ReLu\left(\alpha_l^k + \sum_i^{N_{l-1}} \mathbf{w_l^{ik}} * \mathbf{f_{l-1}^i}\right)\right), \tag{20.11}$$

where $\mathbf{f_l^k}$ is the kth feature map of the lth layer, $S_l(.)$ is either subsampling or identity operator according to pre-defined CSEN structure and N_l is the number of feature maps in this layer. Thus the learnable parameters of a CSEN can be listed as $\mathbf{\Theta_{CSEN}} = \left\{\{\mathbf{w}_1^i, \alpha_1^i\}_{i=1}^{N_1}, \{\mathbf{w}_2^i, \alpha_2^i\}_{i=1}^{N_2}, ...\{\mathbf{w}_L^i, \alpha_L^i\}_{i=1}^{N_L}\right\}$ for an L layer CSEN design.

End-to-end learning of CSEN-based classifier. As previously mentioned, for a representation-based classifier, collaborative responses of group members belonging to the same class are more important than the independent response of each coefficient in the solution of representation coefficient vector, $\hat{\mathbf{x}}$. Therefore instead of traditional ℓ_1-minimization, one can also use the following group ℓ_1-minimization:

$$\min_{\mathbf{x}} \left\{\|\mathbf{Dx} - \mathbf{y}\|_2^2 + \lambda \sum_{i=1}^{c} \|\mathbf{x_{Gi}}\|_2\right\} \tag{20.12}$$

where $\mathbf{x_{Gi}}$ is the group of coefficients that belongs to the ith class. In this respect, for a CSEN network, $\mathcal{P}_\Theta(.)$, the cost function can be cast as

$$E(\mathbf{x}) = \sum_p \left(\mathcal{P}_\Theta(\tilde{\mathbf{x}})_p - b_p\right)^2 + \lambda \sum_{i=1}^{c} \|\mathcal{P}_\Theta(\tilde{\mathbf{x}})_{Gi}\|_2, \tag{20.13}$$

where $\mathcal{P}_\Theta(\tilde{\mathbf{x}})_p$ is the pth coefficient in the network's output and b_p is the ground truth binary mask of the representation coefficient vector $\mathbf{x}$.

One may impose the cost in Eq. (20.13) to better estimate the support set. Then, using this support estimation, the class of the query can be easily detected. However, instead of having an intermediate step as a support estimation, the cost in Eq. (20.13) can be easily approximated by using a simple average pooling layer right after the convolutional one. In that way, direct estimation of the class is possible. In other words, this design enables us to train CSEN-based classifier end-to-end. In training, as a cost function, cross-entropy is used, and SoftMax Operation is applied after average pooling to determine the estimated class. The whole pipeline of the developed CSEN-based COVID-19 recognition is illustrated in Fig. 20.13.

20.3.2.4 Evaluation of the classifiers

In order to have a fair investigation, the same feature, $\mathbf{s}$ or its dimensionally reduced version $\mathbf{y}$ are used both in training and test phases of the following classifiers.

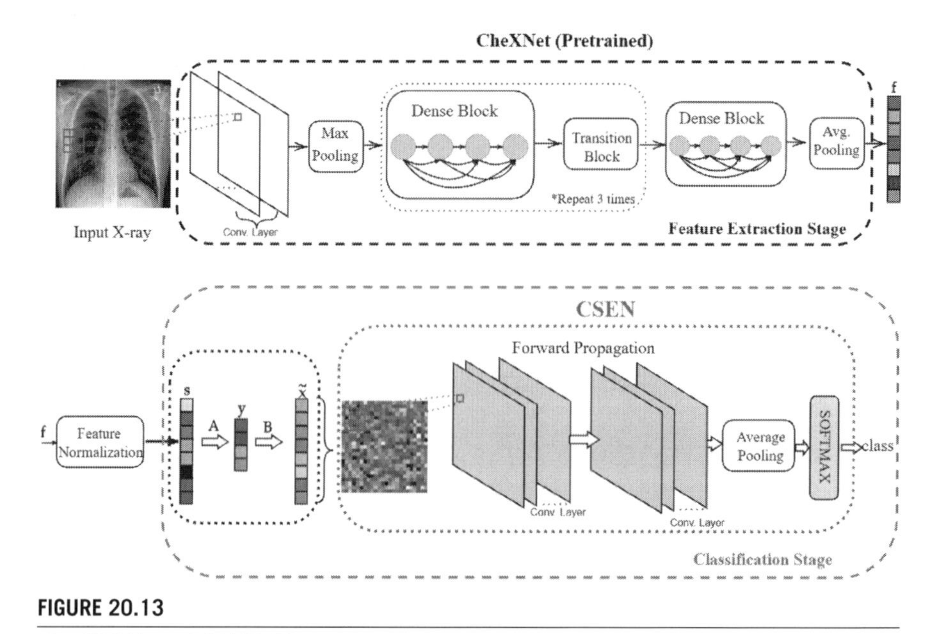

FIGURE 20.13

The CSEN based COVID-19 recognition pipeline consists of a pretrained network for the feature extraction, then it employs dimensional reduction with PCA matrix, $\mathbf{A}$, and the input of CSEN, $\tilde{\mathbf{x}}$, is obtained by the denoiser matrix: $\tilde{\mathbf{x}} = \mathbf{B}\mathbf{y}$ where $\mathbf{B} = \left(\mathbf{D}^T\mathbf{D} + \lambda\mathbf{I}\right)^{-1}\mathbf{D}^T$.

Multilayer perceptron. As an alternative to the dictionary-based classifier, one may directly train an MLP, which is a well-studied classifier, using the feature vector, $\mathbf{s}$ as the input. A four-layer MLP network is designed as shown in Fig. 20.14. This network configuration is the one giving the best performance among all other trials when combined with the following hyperparameters: learning rate, $\mathrm{lr} = 10^{-4}$, and moment updates $\beta_1 = 0.9$, $\beta_2 = 0.999$, and the network is trained in 50 epochs. The deeper ones result in overfitting, while the lighter ones have decreased recognition performance.

Collaborative representation-based classification. The traditional dictionary-based approaches are also investigated besides the CSEN-based classification approach. They are applied to PCA coefficients as shown in Fig. 20.15. Among them, the best performing one for this problem is determined as CRC [49]. It is a noniterative technique, therefore computationally more efficient compared to SRC one. Moreover, it can perform even better than the SRC technique in the case of the representation coefficient vector is not sparse enough, for example, the class number is relatively low like this task. In order to find the optimum trade-off parameter of the regularized least square solution, λ, a grid search is performed on the range of $[10^{-15}, 10^{-1}]$ with log scale using the validation set to be explained later. This trade-off parameter was found as $\lambda = 2 \times 10^{-12}$.

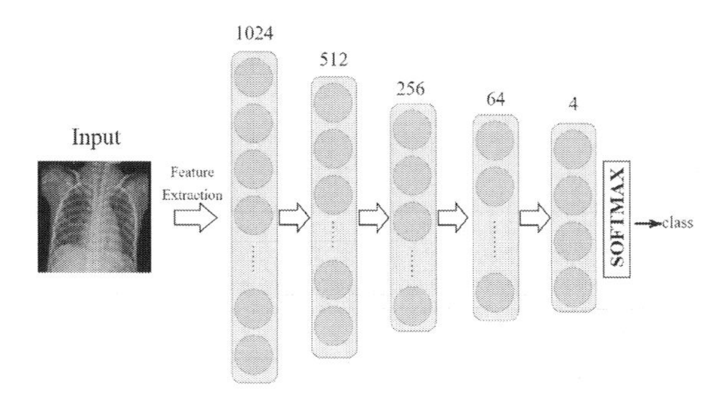

FIGURE 20.14

The 4-layers MLP configuration that takes the extracted features by CheXNet as input.

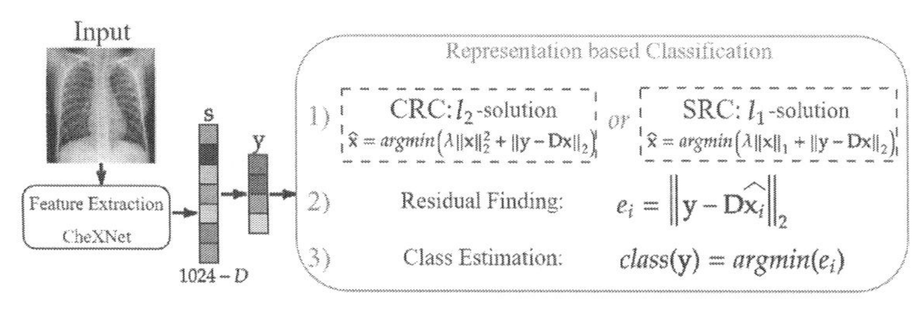

FIGURE 20.15

COVID-19 recognition frameworks of representation based classification approaches (CRC and SRC) using features extracted by the pretrained CheXNet model.

Support vector machines. SVM is a well-known classifier reported to be state-of-the-art for many classification tasks; especially, when the training data is not abundant. This is why, alternative to the dictionary-based classifiers and neural networks, SVM is also applied to query **y**. SVM topology can be both one-vs-one or one-vs-all. A grid search is performed to find the optimal topology and hyperparameters. The following kernel functions are evaluated: linear, radial basis function (rbf). Then other hyper-parameters are optimized: box constraint (C parameter) in the range of $[1, 10^3]$ with log a scale, and kernel scale (γ for the RBF kernel) in the range $[10^{-4}, 10^{-2}]$ with a log scale.

k-Nearest neighbor. k-NN is a traditional approach that is applied to **y**. Previously, a grid search was performed to validate the data for hyperparameter optimization. The search is performed on the following distance metrics: City-block, Hamming, Euclidean, Cosine, Correlation, Jaccard, Mahalanobis, Spearman metrics, standardized Euclidean, Chebyshev, and Minkowski. The k-value is searched in the range of $[1, 4416]$ with a log scale.

Table 20.7 Data set preparation: For a specific CV fold, the number of samples in the training set and test set after and before the data balancing.

Class	Data set size	# of training samples	Augmented training data set size	# of test samples
Bacterial pneumonia	2760	2208	2208	552
Viral pneumonia	1485	1188	2208	297
Normal	1579	1263	2208	316
COVID-19	462	370	2208	92
Total	6286	5029	8832	1257

20.3.3 Experimental evaluations

20.3.3.1 Experimental setup

As previously explained, the data set contains normal and three different pneumonia classes: bacterial, viral, and COVID-19. 5-fold cross-validation (CV) is applied for a robust evaluation, where in each fold, a randomly selected 80% portion of the data set is used as the training set, whereas the rest as the test set. This way, the performance over the whole data set can be calculated. During the training, the training set is augmented to have a balanced training data while the test set remains untouched. The data augmentation during the training set (data balancing) is performed in the following way: Image Data Generator by Keras is used to apply random rotation within the range of 10 degrees and random shifting within the interval $[-0.1, +1]$ both horizontally and vertically. In that way, for each CV fold, a total of 8832 images are obtained as the training set. The remaining 1257 images that is 20% of the data are used as test (unseen) sets for each fold. The data preparation is summarized in Table 20.7.

20.3.3.2 Experimental results

The compact support estimator networks, CSEN1 and CSEN2, proposed in [52] are investigated in the study. CSEN1 consists of two hidden layers. The first hidden layer includes 48 feature maps, while the second one has 24. CSEN2 includes one down-sampling and one upsampling layer between the 48 neuron CNN layer and 24 neuron CNN layer. Downsampling is performed via max-pooling, and upsampling is applied with a 24 neuron transpose convolution layer. The kernel size for each layer is set as 3×3 in both CSEN configurations.

In the dictionary (for CSEN), 625 samples from the augmented training set are used from each class. In this respect, the samples are reorder in a way that the nonzero coefficients that represent the samples from the same class come together in an ideal representation coefficient 50×50 matrix, $\mathbf{X}$ (the reshaped version of $\mathbf{x}$ on the 2-D plane). Then the remaining samples in the augmented training set were used to

Table 20.8 Computation times of different methods over 1257 test samples.

	CSEN1	CSEN2	ReconNet	MLP	CRC-light	CRC
Computational time (in sec.)	0.2196	0.2272	0.2993	0.2935	13.4176	40.7878

Table 20.9 The number of trainable parameters of each network.

	CSEN1	CSEN2	ReconNet	MLP
# of learnable parameters	11.089	16.297	22.914	672.836

train CSEN networks. PCA dimension reduction with $m = 512$ is applied for all methods except the MLP network. Therefore the dictionary, $\mathbf{D}$, has the dimensions of 2500×512, and the denoiser matrix $\mathbf{B} = \left(\mathbf{D}^T\mathbf{D} + \lambda\mathbf{I}\right)^{-1}\mathbf{D}^T$ has the dimensions of 512×2500.

One may question if the given CSEN architectures are the best possible ones for support estimation or support estimation-based classification problems. Since there is no competing deep network in the literature designed for sparse support estimator, the ReconNet [82] structure is chosen with 6 convolutional layers. It is chosen because it is one of the state-of-the-art network architecture for an akin problem, compressive sensing [57,58]. The dictionary that was used for CSEN-based classifier was also used for ReconNet for a fair investigation.

All of the training samples can be used in the dictionary for the CRC algorithm because it does not require further training. This is why the dictionary, $\mathbf{D}$ becomes a 512×8832 matrix and denoiser $\mathbf{B}$ becomes a 512×8832 matrix. Comparing to the lighter size dictionary used in CSEN and ReconNet designs, the extended size dictionary significantly increases the computational complexity and number of parameters to be recorded. Computational times and trainable parameters for different networks are presented in Table 20.8, and Table 20.9, respectively. In Table 20.9, the case where the same dictionary is used as the in CSENs and ReconNet (the so-called CRC-light) is also considered.

The performance metrics in this study is defined as follows: true positive: the number of correctly detected positive cases, true negative: the number of correctly detected negative class members, false positive: the number of misclassified negative cases as positive, and false negative: the number of missed positive cases,

$$\text{Sensitivity} = \frac{\text{TP}}{\text{TP} + \text{FN}}, \tag{20.14}$$

$$\text{Specificity} = \frac{\text{TN}}{\text{TN} + \text{FP}}, \tag{20.15}$$

$$\text{Precision} = \frac{\text{TP}}{\text{TP} + \text{FP}}, \tag{20.16}$$

Table 20.10 Performance evaluations of the classifiers. The best performances on COVID-19 recognition are highlighted.

	Class	k-NN	SVM	MLP	CRC	ReconNet	CSEN1	CSEN2
Sensitivity	Bacterial	0.623	0.611	0.609	0.758	0.590	0.656	0.659
	Viral	0.612	0.660	0.595	0.550	0.625	0.642	0.646
	Normal	0.899	0.904	0.900	0.922	0.891	0.906	0.904
	COVID-19	0.965	0.978	0.961	0.968	0.970	**0.985**	**0.985**
Specificity	Bacterial	0.898	0.913	0.883	0.869	0.902	0.901	0.900
	Viral	0.859	0.826	0.817	0.913	0.834	0.856	0.852
	Normal	0.904	0.944	0.944	0.931	0.927	0.932	0.934
	COVID-19	0.949	0.943	0.948	0.954	0.933	0.953	**0.957**
Accuracy	Bacterial	0.777	0.780	0.763	0.820	0.765	0.793	0.794
	Viral	0.801	0.787	0.765	0.827	0.785	0.805	0.803
	Normal	0.903	0.934	0.933	0.928	0.918	0.926	0.927
	COVID-19	0.950	0.945	0.949	0.955	0.936	0.955	**0.959**
F1-Score	Bacterial	0.711	0.710	0.693	0.787	0.688	0.736	0.737
	Viral	0.592	0.593	0.545	0.601	0.578	0.609	0.608
	Normal	0.823	0.873	0.870	0.866	0.845	0.860	0.861
	COVID-19	0.740	0.724	0.735	0.758	0.690	0.763	**0.778**
TN	Bacterial	3166	3219	3114	3063	3180	3177	3173
	Viral	4123	3965	3923	4385	4005	4109	4091
	Normal	4253	4444	4442	4380	4364	4388	4396
	COVID-19	5525	5489	5522	5554	5435	5548	**5572**
TP	Bacterial	1720	1687	1680	2091	1629	1810	1818
	Viral	909	979	884	816	928	954	959
	Normal	1420	1427	1421	1456	1407	1431	1428
	COVID-19	446	452	444	447	448	**455**	**455**
FP	Bacterial	360	307	412	463	346	349	353
	Viral	678	836	878	416	796	692	710
	Normal	454	263	265	327	343	319	311
	COVID-19	299	335	302	270	389	276	**252**
FN	Bacterial	1040	1073	1080	669	1131	950	942
	Viral	576	506	601	669	557	531	526
	Normal	159	152	158	123	172	148	151
	COVID-19	16	10	18	15	14	**7**	**7**

$$\text{Accuracy} = \frac{\text{TP} + \text{TN}}{\text{TN} + \text{TP} + \text{FP} + \text{FN}}, \tag{20.17}$$

$$\text{F1-score} = 2\frac{(\text{Precision} * \text{Sensitivity})}{(\text{Precision} + \text{Sensitivity})}. \tag{20.18}$$

The performance metrics of the implemented algorithms are listed in Table 20.10 for a deeper investigation. For the COVID-19 recognition task, among all algorithms,

Table 20.11 Cumulative confusion matrix of the developed COVID-19 detection method.

CSEN2		Predicted			
		Bacterial	Viral	Normal	COVID-19
Ground-truth	Bacterial	1818	636	180	126
	Viral	338	959	127	61
	Normal	15	71	1428	65
	COVID-19	0	3	4	455

Table 20.12 Performance of CRC algorithm when the same dictionary that is used for CSEN and ReconNet based methods.

	CRC-light		
	Accuracy	**Sensitivity**	**Specificity**
Bacterial	0.8129	0.7464	0.8650
Viral	0.8163	0.5461	0.8998
Normal	0.9267	0.9170	0.9299
COVID-19	0.9564	0.9394	0.9578

CSEN-based classifiers are the best-performing ones with 98.5% sensitivity and over 95% specificity. One should note that for the COVID-19 detection problem, false negative results are the most crucial since each one of them basically means that a COVID-19 positive case is missed. This is why the study seeks the best possible algorithm with the highest sensitivity. The overall confusion matrix of the developed COVID-19 scheme is also given in Table 20.11 for a more detailed analysis.

Instead of using the baseline CRC technique directly, using CSEN for classification has two clear advantages: (i) computational efficiency since only a subset of the training samples are used in the dictionary, and the rest is used to train learned SE, (ii) improved recognition accuracy, and (iii) memory efficiency. The computational efficiency comes from the fact that a larger size dictionary matrix (size of 512×8832) used in CRC, instead of a light dictionary matrix (size of 512×2500) requires more competition in matrix-vector multiplications. This phenomenon can be observed from Table 20.8 as CSEN designs are 200–300 times faster than CRC. Furthermore, saving the trainable parameters ($\sim 16k$) and a light dictionary matrix coefficients ($\sim 1280k$) in the test device (e.g., mobile phone) are more memory efficient compared to saving coefficients ($\sim 4521k$) of larger size dictionary used in CRC. Instead of using a larger size dictionary, when the same dictionary is used both in CRC and in CSEN approaches, the recognition performance gap between CSEN and CRC grows even larger. The performance evaluation of this light version of CRC is presented in Table 20.12.

20.4 Conclusion

In this chapter, deep learning methodologies are introduced for two different medical imaging tasks: early MI diagnosis and COVID-19 recognition. A three-stage scheme is introduced for early MI diagnosis by a primary tool, that is, echocardiography, which is suitable for the diagnosis of MI in its early stages. The major challenge of the study is the low-quality echocardiography, which is subjected to a high level of noise, lack of contrast, and LV wall is partially missing. As a summary of the study, the ground-truth segmentation masks of the entire LV wall at a pixel-level are generated using the pseudo-labeling approach. Then, a deep E-D CNN model is trained for the segmentation of the LV wall on each frame of each echocardiography recording in the HMC-QU data set. Lastly, the characteristics of the predicted LV wall are extracted in the feature engineering stage, and the features are utilized in several classifiers to diagnose MI. The introduced approach achieved a very small false alarm rate for the segmentation of the LV wall in the benchmark data set. In fact, experimental results on MI detection are quite promising with the low-quality data. Nevertheless, there is a still gap for further improvements. When compared to the diagnosis of a cardiologist, which has prior knowledge from the patients and additional echocardiographic views, the developed approach can be improved with additional input from the patients. The introduced scheme performs an objective evaluation with the quantitative measurements for the LV wall characteristics. It does not depend on the local wall tracking or motion estimation since the LV wall is segmented entirely. Therefore the method is robust to noise and artifacts. Additionally, the visualization outputs of the method: color-coded segments on the LV wall illustration and segment displacement, area curves offer an assistive tool for the cardiologists. Finally, the benchmark HMC-QU data set, which is the first 2D-echocardiographic data set that serves both LV wall segmentation and MI detection, is publicly available for the research community along with the ground-truth segmentation masks and MI labels.

As the second medical imaging task, an efficient and compact COVID-19 recognition system that can achieve top recognition sensitivity among other recent techniques is introduced. Compared to current COVID-19 technologies such as RT-PCR, the introduced approach is faster, and cheaper and has higher accuracy. Although DL-based solutions dominated the machine learning area in the last decade, the technology requires a large training set to achieve sufficient generalization capabilities in many tasks. However, in a sudden outbreak as COVID-19, it may not be possible to collect and process such a large training data in a very short time. As the chapter investigates, a dictionary-based classifier does not require an abundant data size to operate compared to the other (deep) machine learning solutions. However, the performance of dictionary-based classifiers comes short compared to neural network-based classifiers. This chapter explained a third alternative approach that fuses both paradigm and hence called convolutional support estimation network. CSEN can outperform existing solutions for both moderate and small-size data sets. It also reduces the computational complexity of dictionary-based solutions.

References

[1] B. Li, J. Yang, F. Zhao, L. Zhi, X. Wang, L. Liu, Z. Bi, Y. Zhao, Prevalence and impact of cardiovascular metabolic diseases on Covid-19 in China, Clinical Research in Cardiology 109 (5) (2020) 531–538.

[2] S. Shi, M. Qin, B. Shen, Y. Cai, T. Liu, F. Yang, W. Gong, X. Liu, J. Liang, Q. Zhao, et al., Association of cardiac injury with mortality in hospitalized patients with Covid-19 in Wuhan, China, JAMA Cardiology 5 (7) (2020) 802–810.

[3] A. Degerli, M. Zabihi, S. Kiranyaz, T. Hamid, R. Mazhar, R. Hamila, M. Gabbouj, Early detection of myocardial infarction in low-quality echocardiography, IEEE Access 9 (2021) 34442–34453, https://doi.org/10.1109/ACCESS.2021.3059595.

[4] K. Thygesen, J. Alpert, A. Jaffe, M. Simoons, B. Chaitman, H. White, Third universal definition of myocardial infarction, Circulation 126 (16) (2012) 2020–2035.

[5] K. Thygesen, et al., Fourth universal definition of myocardial infarction, Journal of the American College of Cardiology 72 (18) (2018) 2231–2264, https://doi.org/10.1016/j.jacc.2018.08.1038.

[6] M.T. Sadeghi, et al., Value of early rest myocardial perfusion imaging with spect in patients with chest pain and non-diagnostic ecg in emergency department, The International Journal of Cardiovascular Imaging 35 (5) (2019) 965–971, https://doi.org/10.1007/s10554-018-01518-0.

[7] A.E. Stillman, et al., Assessment of acute myocardial infarction: current status and recommendations from the North American society for cardiovascular imaging and the European society of cardiac radiology, The International Journal of Cardiovascular Imaging 27 (1) (2011) 7–24, https://doi.org/10.1007/s10554-010-9714-0.

[8] J.S. Gottdiener, et al., American society of echocardiography recommendations for use of echocardiography in clinical trials: a report from the American society of echocardiography's guidelines and standards committee and the task force on echocardiography in clinical trials, Journal of The American Society of Echocardiography 17 (10) (2004) 1086–1119, https://doi.org/10.1016/j.echo.2004.07.013, PMID:15452478.

[9] T.R. Porter, et al., Clinical applications of ultrasonic enhancing agents in echocardiography: 2018 American society of echocardiography guidelines update, Journal of The American Society of Echocardiography 31 (3) (2018) 241–274, https://doi.org/10.1016/j.echo.2017.11.013.

[10] Y.S. Chatzizisis, V.L. Murthy, S.D. Solomon, Echocardiographic evaluation of coronary artery disease, Coronary Artery Disease 24 (7) (2013) 613–623, https://doi.org/10.1097/MCA.0000000000000028.

[11] M. Yamac, M. Ahishali, A. Degerli, S. Kiranyaz, M.E.H. Chowdhury, M. Gabbouj, Convolutional sparse support estimator based Covid-19 recognition from x-ray images, IEEE Transactions on Neural Networks and Learning Systems 32 (5) (2021) 1810–1820.

[12] World Health Organization, Coronavirus disease 2019 (Covid-19): situation report, vol. 88, 2020.

[13] D.J. Brenner, E.J. Hall, Computed tomography—an increasing source of radiation exposure, New England Journal of Medicine 357 (22) (2007) 2277–2284.

[14] P. Rajpurkar, J. Irvin, K. Zhu, B. Yang, H. Mehta, T. Duan, D. Ding, A. Bagul, C. Langlotz, K. Shpanskaya, et al., Chexnet: radiologist-level pneumonia detection on chest x-rays with deep learning, arXiv preprint, arXiv:1711.05225, 2017.

[15] H.A. DeVon, J.J. Zerwic, Symptoms of acute coronary syndromes: are there gender differences? A review of the literature, Heart & Lung 31 (4) (2002) 235–245.

[16] K. Thygesen, J.S. Alpert, H.D. White, Universal definition of myocardial infarction, Circulation 116 (22) (2007) 2634–2653.
[17] G.B. Moody, R.G. Mark, The impact of the mit-bih arrhythmia database, IEEE Engineering in Medicine and Biology Magazine 20 (3) (2001) 45–50.
[18] V. Sudarshan, U.R. Acharya, E.Y.-K. Ng, C.S. Meng, R. San Tan, D.N. Ghista, Automated identification of infarcted myocardium tissue characterization using ultrasound images: a review, IEEE Reviews in Biomedical Engineering 8 (2014) 86–97.
[19] K. Doi, Diagnostic imaging over the last 50 years: research and development in medical imaging science and technology, Physics in Medicine and Biology 51 (13) (2006) R5.
[20] M.L. Giger, N. Karssemeijer, S.G. Armato, Computer-aided diagnosis in medical imaging, 2001.
[21] M. Kass, A. Witkin, D. Terzopoulos, Snakes: active contour models, International Journal of Computer Vision 1 (4) (1988) 321–331.
[22] S. Dong, G. Luo, G. Sun, K. Wang, H. Zhang, A left ventricular segmentation method on 3d echocardiography using deep learning and snake, in: 2016 Computing in Cardiology Conference (CinC), IEEE, 2016, pp. 473–476.
[23] A. Mishra, P. Dutta, M. Ghosh, A ga based approach for boundary detection of left ventricle with echocardiographic image sequences, Image and Vision Computing 21 (11) (2003) 967–976.
[24] M. Dandel, H. Lehmkuhl, C. Knosalla, N. Suramelashvili, R. Hetzer, Strain and strain rate imaging by echocardiography-basic concepts and clinical applicability, Current Cardiology Reviews 5 (2) (2009) 133–148.
[25] M. Bansal, L. Jeffriess, R. Leano, J. Mundy, T.H. Marwick, Assessment of myocardial viability at dobutamine echocardiography by deformation analysis using tissue velocity and speckle-tracking, JACC: Cardiovascular Imaging 3 (2) (2010) 121–131.
[26] J. Konrad, E. Dubois, Bayesian estimation of motion vector fields, IEEE Transactions on Pattern Analysis and Machine Intelligence 9 (1992) 910–927.
[27] R.C. Kordasiewicz, M.D. Gallant, S. Shirani, Affine motion prediction based on translational motion vectors, IEEE Transactions on Circuits and Systems for Video Technology 17 (10) (2007) 1388–1394.
[28] O.A. Smiseth, H. Torp, A. Opdahl, K.H. Haugaa, S. Urheim, Myocardial strain imaging: how useful is it in clinical decision making?, European Heart Journal 37 (15) (2016) 1196–1207.
[29] P. Bizopoulos, D. Koutsouris, Deep learning in cardiology, IEEE Reviews in Biomedical Engineering 12 (2018) 168–193.
[30] O. Ronneberger, P. Fischer, T. Brox, U-net: convolutional networks for biomedical image segmentation, in: International Conference on Medical Image Computing and Computer-Assisted Intervention, Springer, 2015, pp. 234–241.
[31] M.H. Jafari, H. Girgis, A.H. Abdi, Z. Liao, M. Pesteie, R. Rohling, K. Gin, T. Tsang, P. Abolmaesumi, Semi-supervised learning for cardiac left ventricle segmentation using conditional deep generative models as prior, in: 2019 IEEE 16th International Symposium on Biomedical Imaging (ISBI 2019), IEEE, 2019, pp. 649–652.
[32] S. Leclerc, E. Smistad, J. Pedrosa, A. Østvik, F. Cervenansky, F. Espinosa, T. Espeland, E.A.R. Berg, P.-M. Jodoin, T. Grenier, et al., Deep learning for segmentation using an open large-scale dataset in 2d echocardiography, IEEE Transactions on Medical Imaging 38 (9) (2019) 2198–2210.
[33] K.S. Vidya, E. Ng, U.R. Acharya, S.M. Chou, R. San Tan, D.N. Ghista, Computer-aided diagnosis of myocardial infarction using ultrasound images with dwt, glcm and hos methods: a comparative study, Computers in Biology and Medicine 62 (2015) 86–93.

[34] H.A. Omar, A. Patra, J.S. Domingos, P. Leeson, A.J. Noblel, Automated myocardial wall motion classification using handcrafted features vs a deep cnn-based mapping, in: Annual International Conference of the IEEE Engineering in Medicine and Biology Society, 2018, pp. 3140–3143.
[35] D. Ouyang, B. He, A. Ghorbani, C. Langlotz, P.A. Heidenreich, R.A. Harrington, D.H. Liang, E.A. Ashley, J.Y. Zou, Interpretable ai for beat-to-beat cardiac function assessment, medRxiv (2019) 19012419.
[36] R.M. Lang, M. Bierig, R.B. Devereux, F.A. Flachskampf, E. Foster, P.A. Pellikka, M.H. Picard, M.J. Roman, J. Seward, et al., Recommendations for chamber quantification: a report from the American society of echocardiography's guidelines and standards committee and the chamber quantification writing group, developed in conjunction with the European association of echocardiography, a branch of the European society of cardiology, Journal of the American Society of Echocardiography 18 (12) (2005) 1440–1463.
[37] R.M. Lang, L.P. Badano, V. Mor-Avi, J. Afilalo, A. Armstrong, L. Ernande, F.A. Flachskampf, E. Foster, S.A. Goldstein, T. Kuznetsova, et al., Recommendations for cardiac chamber quantification by echocardiography in adults: an update from the American society of echocardiography and the European association of cardiovascular imaging, European Heart Journal – Cardiovascular Imaging 16 (3) (2015) 233–271.
[38] S. Balakrishnama, A. Ganapathiraju, Linear discriminant analysis – a brief tutorial, vol. 18, Institute for Signal and Information Processing, 1998, pp. 1–8.
[39] P.A. Chou, Optimal partitioning for classification and regression trees, IEEE Transactions on Pattern Analysis and Machine Intelligence 4 (1991) 340–354.
[40] L. Breiman, Random forests, Machine Learning 45 (1) (2001) 5–32.
[41] N. Cristianini, J. Shawe-Taylor, et al., An Introduction to Support Vector Machines and Other Kernel-Based Learning Methods, Cambridge University Press, 2000.
[42] D.P. Kingma, J. Ba Adam, A method for stochastic optimization, arXiv preprint, arXiv: 1412.6980, 2014.
[43] C.-C. Chang, C.-J. Lin Libsvm, A library for support vector machines, ACM Transactions on Intelligent Systems and Technology 2 (3) (2011) 1–27.
[44] T.F. Chan, L.A. Vese, Active contours without edges, IEEE Transactions on Image Processing 10 (2) (2001) 266–277.
[45] L. Pellis, F. Scarabel, H.B. Stage, C.E. Overton, L.H. Chappell, K.A. Lythgoe, E. Fearon, E. Bennett, J. Curran-Sebastian, R. Das, et al., Challenges in control of Covid-19: short doubling time and long delay to effect of interventions, arXiv preprint, arXiv: 2004.00117, 2020.
[46] Y. Fang, H. Zhang, J. Xie, M. Lin, L. Ying, P. Pang, W. Ji, Sensitivity of chest ct for Covid-19: comparison to rt-pcr, Radiology (2020) 200432.
[47] W.H. Organization, et al., Laboratory testing for coronavirus disease 2019 (Covid-19) in suspected human cases: interim guidance, 2 March 2020, Tech. Rep., World Health Organization, march 2020.
[48] K.A. Erickson, K. Mackenzie, A. Marshall, Advanced but expensive technology. Balancing affordability with access in rural areas, Canadian Family Physician (Medecin de famille canadien) 39 (1993) 28–30.
[49] L. Zhang, M. Yang, X. Feng, Sparse representation or collaborative representation: which helps face recognition?, in: 2011 International Conference on Computer Vision, IEEE, 2011, pp. 471–478.
[50] J. Wright, A.Y. Yang, A. Ganesh, S.S. Sastry, Y. Ma, Robust face recognition via sparse representation, IEEE Transactions on Pattern Analysis and Machine Intelligence 31 (2) (2008) 210–227.

[51] J. Wright, Y. Ma, J. Mairal, G. Sapiro, T.S. Huang, S. Yan, Sparse representation for computer vision and pattern recognition, Proceedings of the IEEE 98 (6) (2010) 1031–1044.
[52] M. Yamac, M. Ahishali, S. Kiranyaz, M. Gabbouj, Convolutional sparse support estimator network (CSEN) from energy efficient support estimation to learning-aided compressive sensing, IEEE Transactions on Neural Networks and Learning Systems (2021) 1–15, Early Access.
[53] O.D. Escoda, L. Granai, P. Vandergheynst, Mémoire sur la propagation de la chaleur dans les corps solides, Nouveau Bulletin des Sciences de la Société Philomathique de Paris 6 (1808) 112–116.
[54] S.G. Mallat, Z. Zhang, Matching pursuits with time-frequency dictionaries, IEEE Transactions on Signal Processing 41 (12) (1993) 3397–3415.
[55] J.-L. Starck, E.J. Candès, D.L. Donoho, The curvelet transform for image denoising, IEEE Transactions on Image Processing 11 (6) (2002) 670–684.
[56] J. Yang, K. Yu, Y. Gong, T.S. Huang, et al., Linear spatial pyramid matching using sparse coding for image classification, in: CVPR, vol. 1, 2009, p. 6.
[57] D.L. Donoho, et al., Compressed sensing, IEEE Transactions on Information Theory 52 (4) (2006) 1289–1306.
[58] E.J. Candès, et al., Compressive sampling, in: Proceedings of the International Congress of Mathematicians, vol. 3, 2006, pp. 1433–1452.
[59] A. Orsdemir, H.O. Altun, G. Sharma, M.F. Bocko, On the security and robustness of encryption via compressed sensing, in: Proceedings of the IEEE Military Communications Conference, 2008, pp. 1–7.
[60] M. Yamaç, M. Ahishali, N. Passalis, J. Raitoharju, B. Sankur, M. Gabbouj, Multi-level reversible data anonymization via compressive sensing and data hiding, IEEE Transactions on Information Forensics and Security 16 (2021) 1014–1028, https://doi.org/10.1109/TIFS.2020.3026467.
[61] D.L. Donoho, M. Elad, Optimally sparse representation in general (nonorthogonal) dictionaries via l_1 minimization, Proceedings of the National Academy of Sciences 100 (5) (2003) 2197–2202.
[62] A. Cohen, W. Dahmen, R. DeVore, Compressed sensing and best k-term approximation, Journal of the American Mathematical Society 22 (1) (2009) 211–231.
[63] H. Rauhut, Compressive sensing and structured random matrices, Theoretical Foundations and Numerical Methods for Sparse Recovery 9 (2010) 1–92.
[64] S.S. Chen, D.L. Donoho, M.A. Saunders, Atomic decomposition by basis pursuit, SIAM Review 43 (1) (2001) 129–159.
[65] E.J. Candes, J.K. Romberg, T. Tao, Stable signal recovery from incomplete and inaccurate measurements, Communications on Pure and Applied Mathematics: A Journal Issued by the Courant Institute of Mathematical Sciences 59 (8) (2006) 1207–1223.
[66] W. Wang, M.J. Wainwright, K. Ramchandran, Information-theoretic limits on sparse support recovery: dense versus sparse measurements, in: 2008 IEEE International Symposium on Information Theory, IEEE, 2008, pp. 2197–2201.
[67] J. Haupt, R. Baraniuk, Robust support recovery using sparse compressive sensing matrices, in: 2011 45th Annual Conference on Information Sciences and Systems, IEEE, 2011, pp. 1–6.
[68] G. Reeves, M. Gastpar, Sampling bounds for sparse support recovery in the presence of noise, in: 2008 IEEE International Symposium on Information Theory, IEEE, 2008, pp. 2187–2191.
[69] J.A. Tropp, A.C. Gilbert, Signal recovery from random measurements via orthogonal matching pursuit, IEEE Transactions on Information Theory 53 (12) (2007) 4655–4666.

[70] D. Needell, J.A. Tropp, Cosamp: iterative signal recovery from incomplete and inaccurate samples, Applied and Computational Harmonic Analysis 26 (3) (2009) 301–321.

[71] T. Guha, R.K. Ward, Learning sparse representations for human action recognition, IEEE Transactions on Pattern Analysis and Machine Intelligence 34 (8) (2011) 1576–1588.

[72] W. Li, Q. Du, A survey on representation-based classification and detection in hyperspectral remote sensing imagery, Pattern Recognition Letters 83 (2016) 115–123.

[73] Covid-19, https://radiopaedia.org/playlists/25975?lang=us. (Accessed 27 February 2021), 2020.

[74] Covid-19 database, https://www.sirm.org/category/senza-categoria/covid-19/. (Accessed 27 February 2021), 2020.

[75] Covid-19 cxr, https://threadreaderapp.com/thread/1243928581983670272.html. (Accessed 27 February 2021), 2020.

[76] Covid-chestxray dataset, https://github.com/ieee8023/covid-chestxray-dataset. (Accessed 27 February 2021), 2020.

[77] P. Mooney, Chest x-ray images (pneumonia), https://www.kaggle.com/paultimothymooney/chest-xray-pneumonia. (Accessed 27 February 2021), 2018.

[78] X. Wang, Y. Peng, L. Lu, Z. Lu, M. Bagheri, R.M. Summers, Chestx-ray8: hospital-scale chest x-ray database and benchmarks on weakly-supervised classification and localization of common thorax diseases, in: Proceedings of the IEEE Conference on Computer Vision and Pattern Recognition, 2017, pp. 2097–2106.

[79] G. Huang, Z. Liu, L. Van Der Maaten, K.Q. Weinberger, Densely connected convolutional networks, in: Proceedings of the IEEE Conference on Computer Vision and Pattern Recognition, 2017, pp. 4700–4708.

[80] A. Değerli, S. Aslan, M. Yamac, B. Sankur, M. Gabbouj, Compressively sensed image recognition, in: 2018 7th European Workshop on Visual Information Processing (EUVIP), IEEE, 2018, pp. 1–6.

[81] S. Lohit, K. Kulkarni, P. Turaga, Direct inference on compressive measurements using convolutional neural networks, in: 2016 IEEE International Conference on Image Processing (ICIP), 2016, pp. 1913–1917.

[82] K. Kulkarni, S. Lohit, P. Turaga, R. Kerviche, A. Ashok, Reconnet: non-iterative reconstruction of images from compressively sensed measurements, in: Proceedings of the IEEE Conference on Computer Vision and Pattern Recognition, 2016, pp. 449–458.

CHAPTER

Deep learning for robotics examples using OpenDR 21

Manos Kirtas, Konstantinos Tsampazis, Pavlos Tosidis, Nikolaos Passalis, and Anastasios Tefas
Aristotle University of Thessaloniki, Thessaloniki, Greece

21.1 Introduction

Deep Learning (DL) provides powerful tools for a wide range of robotics-related tasks, ranging from solving complex perception problems to performing end-to-end control and planning [1,2]. However, DL tools can be notoriously difficult to be used by researchers in other fields, such as robotics. At the same time, a number of domain-specific additional limitations, for example, computational constraints of embedded devices that require specially-designed DL models [3], is further slowing down the adoption of DL in these fields. To address these limitations, the Open Deep Learning toolkit for Robotics (OpenDR)[1] has been developed, aiming at making DL tools for robotics easily accessible and lowering the access barrier, both for researchers and the industry, by providing fast, easy-to-use and open source implementations for DL methods that can be used in various robotics settings. At the same time, training DL tools for robotics often requires a vastly different paradigm compared to other tasks, dictating a cointegrated training and simulation pipeline that differs from traditional DL training approaches that are using static data sets.

This chapter provides a brief overview of the structure of the first version of the OpenDR toolkit (Section 21.2), along with some examples demonstrating how easily the OpenDR toolkit can be used for solving two common visual perception tasks. Then a framework that provides a solid approach for co-integrating simulation and training, the deepbots framework [4], is presented (Section 21.3). This framework can be used both in Reinforcement Learning (RL) setups, as well for other robotics-related tasks, for example, active perception [5,6]. Finally, Section 21.4 concludes this chapter.

[1] opendr.eu.

Deep Learning for Robot Perception and Cognition. https://doi.org/10.1016/B978-0-32-385787-1.00026-9

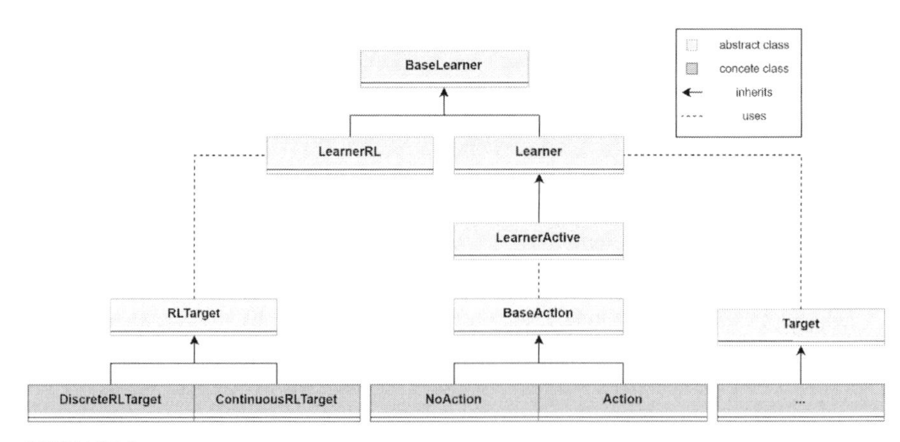

FIGURE 21.1

Class diagram of the Learner classes, along with their interactions with RLTarget, BaseAction, and Target classes, which are used as their output. The concrete classes inheriting from the Target class are omitted.

21.2 Structure of OpenDR toolkit and application examples

In this section, we briefly review the basic structure of the OpenDR toolkit v1.0 and then we proceed by presenting how it can be used to solve two common visual perception problems, that is, face recognition and pose estimation. OpenDR toolkit is organized with a hierarchy of classes to ensure a common interface between all algorithms/models provided by the toolkit. The abstract classes used for the implemented algorithms provided by OpenDR toolkit are based on the *BaseLearner* class, as shown in Fig. 21.1. Then all algorithms included in the toolkit inherit from one of the three more specialized *Learner* classes, namely the *Learner* class, the *LearnerRL* class, and the *LearnerActive* class. Each of these classes provides a different type of output. The generic *Learner* class uses the various concrete classes inherited the *Target* class, such as

- *Category* for classification problems,
- *MultiCategory* for multilabel classification problems,
- *VectorTarget* for multivalue regression or embedding learning,
- *BoundingBox* and *BoundingBox3D* for individual bounding boxes,
- *KeyPointList* for problems related to key point detection (e.g., pose estimation, body part detection, etc.),
- *Heatmap* for multiclass segmentation problems or multiclass problems that require heatmap annotations/outputs,
- *Map and SensorPose* for representing the output of SLAM algorithms.

The *LearnerRL* class is used for RL algorithms and provides its output using either the concrete class *DiscreteRLTarget* or the *ContinuousRLTarget*, inheriting from *RLTarget*. Finally, the *LearnerActive* class, which is used for active perception algo-

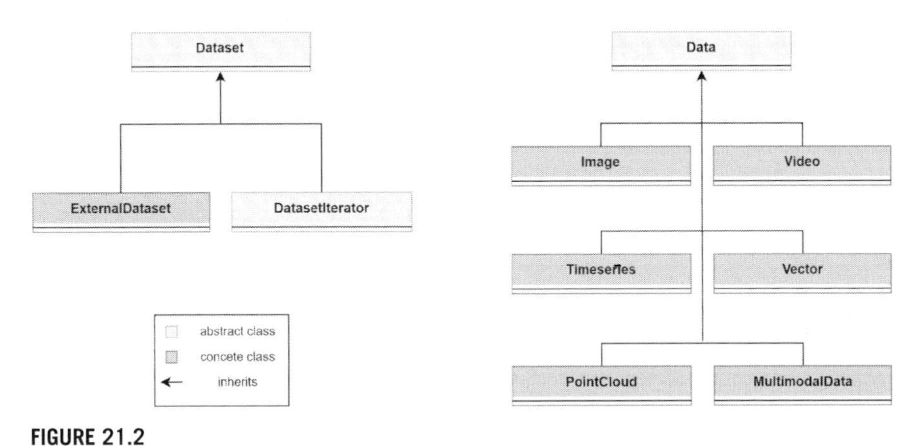

FIGURE 21.2

Class diagram of the Data set and Data classes.

rithms building upon the more generic *Learner*, uses the concrete classes *NoAction* and *Action* inherited from *BaseAction*. The aforementioned data classes are employed by OpenDR toolkit to return the results of the inference process. However, at the same time, they can be also used to represent training annotations. Most algorithms provided by OpenDR use the aforementioned *LearnerBased*-based classes, while providing concrete implementations according to the needs of each algorithm. End-users of the OpenDR toolkit just need to initialize an appropriate *Learner* for their use-case and just call the methods needed, as we will demonstrate later. The toolkit ensures that the functionality is easily accessed and used, with well-defined data structures and documentation to guide the users.

Furthermore, each algorithm provided by OpenDR receives data input using either the abstract *Data* class (mainly during inference) or the *Dataset* class (mainly for training and evaluation). These classes serve as a base for the more specific data and dataset classes respectively, as shown in Fig. 21.2. The models provided by OpenDR toolkit are able to handle data sets using the concrete class *ExternalDataset*, which is used for data set formats supported by OpenDR toolkit, for example, the Pascal Visual Object Classes (VOC) format [7]. Alternatively, models can be also trained/evaluated using any class that implements the *DatasetIterator* abstract class, which allows for supporting any other data set format, given the appropriate implementation of a *DatasetIterator*.

In the following sections, we provide examples of how OpenDR toolkit can be used for two common perception tasks, that is, face recognition and pose estimation. Face recognition is the task of matching a human face taken from an image or a video frame with a human face existing in a known database [8]. This is done by matching face features, typically provided by a DL model. Even through face recognition can be tackled as a classification problem, typically it is modeled as a retrieval problem. Therefore, the typical pipeline of a face recognition system is the following:

1. detect the faces that appear in an image or video frame,
2. extract a feature vector using a DL model from each face,
3. compare the extracted feature vectors against a database with known identities and find the closest match(es) for each detected face.

Backbone architectures with different complexity are provided in OpenDR toolkit, to allow for selecting the one that is more appropriate for the hardware platform where it will be deployed. Some of them aim for better accuracy, while others aim on improving their efficiency. Furthermore, a large variety of different approaches have been proposed in the literature to extract discriminative feature vectors from face images, such as ArcFace [9], CosFace [10], and SphereFace [11]. Note that optionally, a face alignment approach can be applied before the feature extraction step to reduce distribution shift phenomena and improve the accuracy of face recognition. Furthermore, OpenDR toolkit provides algorithms for performing face alignment, such as the Joint Face Detection and Alignment using Multitask Cascaded Convolutional Networks (MTCNN) [12]. Finally, face recognition is achieved by comparing the face features extracted from a face in an image or a video frame, to face features lying in a database of known faces. OpenDR toolkit also supports a classification-based approach, where a head network is trained on top of a pretrained backbone model for a small number of person identities. This method can be useful for applications that require recognizing only a small number of identities, avoiding the need to maintain a database and perform queries for each detected face.

The Face Recognition *Learner* class implemented in the OpenDR toolkit provides the necessary methods for training and evaluating a model, running inference on new images, saving and loading a model, as well as optimizing a given model. Numerous internal methods are implemented that are needed for the *Learner* class to work smoothly under the OpenDR toolkit's requirements. It also contains multiple backbone architectures, with some of them providing real-time performance on embedded devices, such as NVIDIA's Jetson TX2, with minor accuracy drop. The whole align-extract features-classify pipeline is included, while the functionality of the first step of the pipeline, that is, detecting a face in an image, is also included in the OpenDR toolkit.

Alongside the OpenDR toolkit's implementation several demos are provided, in order to showcase the usage of the developed algorithm in various use cases. Those include a webcam demo that detects and recognizes faces in a webcam feed, a training demo both for retrieval and classification, an inference demo using images and an evaluation demo. Furthermore, a demo utilizing the Webots robot simulator is included, in which a user can move a TIAGo robot, equipped with a camera, around a room using a keyboard. Inside the room exist three different 3D human models, two of which exist in the database of known IDs. While moving, face detection and recognition is applied to the camera feed and the results are shown in a window, as shown in Fig. 21.3.

FIGURE 21.3

Demonstration of a face recognition implementation using the Webots robot simulator. Green bounding boxes (white in print version) are used to depict the detected faces in the acquired frame (right part of the image, the identity of each person is also provided above each bounding box). The robot used for acquiring the input frames is shown in the center of the road (along with its orientation, provided by the corresponding arrows).

Using OpenDR toolkit for training and performing inference is straightforward as shown in Code Examples 21.1 and 21.2, respectively.[2] Note that using a face recognition algorithm within OpenDR toolkit is fairly simple; we just need to create a *FaceRecognition* object, load a pretrained data set, fit the model using the reference database of known identities (only for the first time the model is used, since the fitted model can be then reused) and then perform inference for a given image. Similarly, we can train a backbone model by creating two data sets, one for training and one for evaluation, fitting the model and then saving the resulting model for later use.

```
from opendr... import FaceRecognition
from opendr... import ExternalDataset

face_recognizer = FaceRecognition(backbone='mobilefacenet',
                   mode='backbone_only', device='cuda')
face_recognizer.load('./temp/mobilefacenet')
face_recognizer.fit_reference(path='./database/imgs',
                save_path='temp/')

img = ...
ID = face_recognizer.infer(img)
```

Code Example 21.1: Inference using an OpenDR-based face recognition model.

[2] Please refer to OpenDR's documentation for the latest interfaces, since they might be updated between releases.

```
from opendr... import FaceRecognition
from opendr... import ExternalDataset

face_recognizer = FacerRecognition(backbone='mobilefacenet',
                    network_head='arcface', iters=120, lr=0.1)


train_set = ExternalDataset(path='./imgs',
                  dataset_type='imagefolder')

eval_set = ExternalDataset(path='./data', dataset_type='lfw')
face_recognizer.fit(dataset=train_set, val_dataset=eval_set)
face_recognizer.save('./temp/mobilefacenet')
```

Code Example 21.2: Training and evaluating an OpenDR-based face recognition model.

OpenDR provides several implementations of well-known and established algorithms to perform 2D human pose estimation. We will first briefly introduce the task of pose estimation, before diving in the algorithms and implementations. Pose estimation is a general problem where the position and orientation of objects are detected through images or video. The provided algorithms in the OpenDR toolkit focus on human pose estimation. Human pose estimation is performed by localizing human joints in images, called keypoints. 2D pose estimation algorithms aim at identifying the locations of keypoints by estimating their x and y coordinates on the image. Note that there are several challenges during this process, such as the need for detecting any possible pose, providing invariance to occlusion of various parts of the body, clothing, and various lighting conditions, among others.

Pose estimation can be approached in two distinct ways called top-down and bottom-up. The first category of methods employs a person detector model to detect humans in the given input image. Then, for each person detected, the model returns a bounding box enclosing them. Each bounding box is cropped from the original image and the resulting images are fed into a pose estimator which then detects all keypoints constructing the pose. The second category of methods follows a bottom up approach by feeding images directly into the pose estimator and detecting all keypoints. The keypoints are then linked into separate poses with various methods. Top-down methods usually have better accuracy as a consequence of the higher resolution images, but suffer by an increase in inference time when multiple people are detected, due to the fact that for each detection a full forward pass of the pose estimator is required. Moreover, top-down methods rely on the person detector to provide good detections; persons that are not detected are not fed into the pose estimator at all. On the other hand, bottom-up methods have near constant inference time regardless of the number of people in the image, but have lower accuracy due to comparatively lower resolution of the images.

FIGURE 21.4

Demonstration of TIAGo robot mimicking arm and head pose of provided example image taken from the COCO dataset in the Webots robot simulator. Regular pose mimicking demo operates on a webcam.

OpenDR provides an implementation of the lightweight version of the OpenPose algorithm [13,14], which employs a more lightweight backbone architecture than the original implementation, providing near real-time performance with negligible accuracy drop. In general, the OpenPose algorithm is a bottom-up approach, which uses a nonparametric representation referred to as Part Affinity Field, that associates body parts with individuals in the image. This implementation provides near real-time performance on mobile platforms, such as the Jetson TX2 embedded system, while having a near constant inference time regardless of the number of persons that appear in an image, making it ideal for robotics implementations.

The implementations of the Lightweight OpenPose algorithm provided in OpenDR toolkit follow the abstract class structure introduced before, that is, they provide the appropriate implementations for basic methods used for training new models, evaluating them and running inference on new samples, etc. Additional methods implemented include methods for saving, loading, and optimizing a model. The most important method would be *infer* that implements the forward pass of the model, which provided with an image input, returns the poses detected in the form of keypoint coordinates on the image. This method handles all the required pre/post-processing required by the algorithms to enable users to easily get results and use them.

Accompanying the implementation, an array of demos is provided that showcase the usage of the developed algorithms in various use cases. Those include a webcam benchmarking demo to test the inference speed on a system with a webcam, a webcam demo that shows the video feed with the pose estimation results in a window for live testing, and a training and evaluation demo. Three demos using the Webots robot simulator and the TIAGo robot equipped with a camera are also provided to showcase more realistic robotics applications of the algorithm. The first Webots demo includes the TIAGo robot and a 3D human model in an enclosed arena, with the ability to drive TIAGo using the keyboard to move around the arena while performing pose estimation through the robot's camera and showing the results in a window. The second demo draws images from a webcam and performs pose estimation, whose results are used to move TIAGo's arms and head around to mimic the pose as seen in Fig. 21.4. The final demo, again performs pose estimation on webcam frames and uses the angle of the elbows detected to drive TIAGo's motors and move it around the arena.

Two examples of inference and training using the toolkit are provided in Code Examples 21.3 and 21.4, respectively. Note that using the pose estimator for inference requires only three lines of code, that is, creating a pose estimation leaner, loading a pretrained model, and then simply calling it on a given image:

```
from opendr... import LightweightOpenPoseLearner


pose_estimator = LightweightOpenPoseLearner(device="cuda")
pose_estimator.load("./path_to_trained_model")

img = ...
poses = pose_estimator.infer(img)
```

Code Example 21.3: Inference using an OpenDR-based pose estimation model.

Similarly, we can perform training by creating two datasets, one for training and one for valuation, fitting the model and then saving the resulting model for later use:

```
from opendr... import LightweightOpenPoseLearner
from opendr... import ExternalDataset


pose_estimator = LightweightOpenPoseLearner(batch_size=32,
                 device="cuda")

training_dataset = ExternalDataset(path="./path_to_data",
                dataset_type="COCO")
validation_dataset = ExternalDataset(path="./path_to_data",
                dataset_type="COCO")

pose_estimator.fit(training_dataset,
                val_dataset=validation_dataset,
                logging_path="./tensorboard_log_path")

pose_estimator.save("./path_to_save_model")
```

Code Example 21.4: Training using an OpenDR-based pose estimation model.

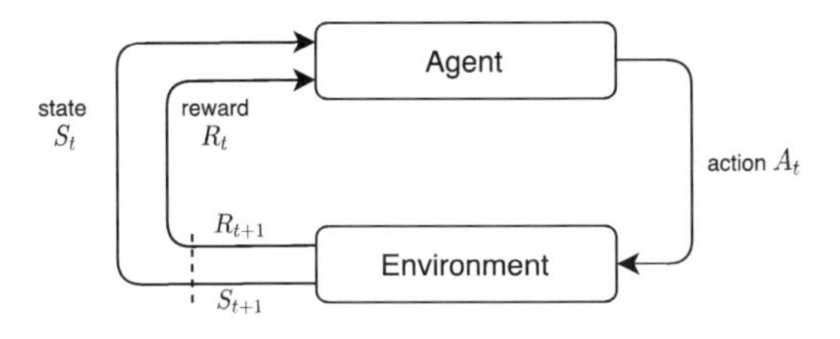

FIGURE 21.5

Agent-environment loop scheme.

21.3 Cointegration of simulation and training

Before Reinforcement Learning (RL) agents are applicable on real-world scenarios, they should first be trained in a simulated environment. During training, RL agents require a number of supporting procedures that are cost and time inefficient to be implemented in the real world. The training process requires an enormous amount of time to sufficiently explore the environment and manage to solve the task, often suffering from low sample efficiency [15]. Furthermore, during the initial stages of training, the agents take actions at random, potentially endangering the robot's hardware. To circumvent these restrictions, users usually first run training sessions on realistic simulators, such as Gazebo [16] and OpenRAVE [17] where the simulation can run at accelerated speeds and with no danger, only later to transfer the trained agents on physical robots. However, this poses additional challenges [18], due to the fact that simulated environments provide a varying degree of realism, so it is not always possible for the agent to observe and act in the real world exactly as it did during training. This led to the development of more realistic simulators, which further reduce the gap between the simulation and the real world, such as Webots [19]. Other than that, simulators provide easily parameterizable environments, which can be adjusted according to the needs of every different real life scenario.

To deeply understand the way that RL agents are deployed in the real world and integrated into robots, we have to understand the involved components ranging from simulation training to real life deployment. During the training, RL agents try to accomplish a given task with their actions in an unsupervised manner, taking into account the observation received from the environment and the given reward. Rewards guide the training procedure as the RL agent tries to maximize the future reward. Additionally, in most of the proposed algorithms, there is some randomization involved when the agent selects actions in order to explore the environment. If the observations, available actions and reward function are chosen carefully, the agent manages to accomplish the given task. The above procedure can be conceptually modeled by the *agent-environment loop* [20]. Accordingly, in each timestep, the agent chooses an *action*, and in turn, the environment returns the *observations* and the *reward* value (Fig.21.5).

Programmatically, the involved components (environment and agent) can be seen as ontologies, such us classes that interact, with both ontologies having attributes according to the given experiment. For example, the environment ontology represents this subset of the simulated world that is observable from the agent, while the attributes are referred to the state of both agent and environment. Finally, the interactions between the agent and the environment can be easily modeled as functions. The minimum required functions for the environment are:

- **observation:** Returns the observable state of the environment at the timestep t. For example, for a face recognition task, the observations would be the frame at timestep t from the robot's camera.
- **reward:** Returns the value of reward regarding the action taken by the agent at timestep $t-1$. For example, in a "find target" task the reward could be a function of the Euclidean distance from the target.
- **done:** Returns a Boolean-type value that encodes whether the agent accomplished the task. If the value is True, the environment should be reset in a initial or random state S_0.
- **reset:** Resets the environment to the S_0 state. The reset function is called either if the agent accomplished the task or if any other constraint, like time or episode failure constraints, are applied.

The first three functions can be implemented under a single function called *step* since it includes the appropriate values for the transition from S_t to S_{t+1} state. On the other hand, the agent is the ontology that includes the RL algorithm and it represents the control mechanism that enforces the actions on the robot/actor. Thus it is required to have a function that receives the current state S_t and the given reward in the previous timestep R_{t-1} in order to be processed by the learning algorithm, as well as to provide the action A_t for timestep t. The *agent-environment loop* works on a layer of abstraction that enables researchers to use it out-of-the-box in any different setup either physical or simulated, ignoring the technical details for both environment and agent.

Every simulator has its own unique way to represent a scenario. For example, in Webots, the associated components of a scene are represented in a hierarchical manner and more precisely as a tree structure. The root of the tree is called "world" and typically contains the appropriate nodes that simulate a scenario. For example, every robot, obstacle and human-like model are represented with nodes. This hierarchical structure enables us to add new functionalities by adding nodes on a specific parent node. Additionally, every node has properties defining its current state (e.g., position, rotation, speed, etc.). A useful property on a robot node is the *controller*, which defines the script that runs on a node. The controller enables users to manipulate the node's properties programmatically and give to users the ability to observe and interact with the simulated world. Conceptually, we can think that at each timestep the simulator executes controller scripts one by one, and then it calculates the next state of each node's properties according to the rules of the physics engine. Finally, there is a special kind of property named "supervisor" that when it is enabled, allows the "controller" script to have access on operational functionalities of the simulator which, for example, enables them to manipulate objects through the script, or access to nonchildren nodes and theirs attributes. For example, through the controller of a supervisor node, objects can be moved in the world or the distance between a robot and a target can be measured.

Cointegrating RL training in a simulator environment requires that the representation of the RL ontologies will be in compliance with simulation logic. In the remainder of this section, we present two different architectures to cointegrate the RL *agent-environment loop* scheme with Webots simulator making use of the open-source framework, called Deepbots [4], which is provided under the OpenDR toolkit. The employed architecture should be chosen according to the employed scenario since each of them targets different use cases. The first architecture involves only one node, which is a robot-supervisor node, and is preferable in scenarios with large observation space such as images captured by robot's high-fidelity camera, making it ideal for perception tasks that, for example, utilize the algorithms included in the OpenDR toolkit. On the other hand, the second architecture separates the robot from the supervisor in different nodes and it targets scenarios, which could include more than one robot, with smaller observations.

Deepbots aims to enable researchers to use RL in Webots and it has been created mostly for educational and research purposes. In essence, it acts as a middle-ware between Webots and the DRL algorithms, exposing an *agent-environment loop* interface with multiple levels of abstraction. The framework uses design patterns to achieve high code readability and reusability, allowing to easily incorporate it in most research pipelines. The aforementioned features come as an easy-to-install Python package that allows developers to efficiently implement environments that can be utilized by researchers or students in order to use their algorithms in realistic benchmarking. At the same time, deepbots provides ready-to-use standardized environments for well-known problems. Finally, the developed framework provides some extra tools for monitoring, for example, tensorboard logging and plotting, allowing to directly observe the training progress. Deepbots is available at https://github.com/aidudezzz/deepbots.

21.3.1 One-node architecture

In the simpler case of the one-node architecture, we can consider that the *agent-environment loop* is implemented in a single node. This node should be able not only to interact with the robot/actor, but also to operate as a supervisor in order to have access to higher level information and functionalities. Higher level functionalities are required, for example, to reset the simulator when needed. The one-node architecture comes with significant efficiency advantages, since the observations are collected by the node and are processed by its processing units. Additionally, one-node architecture can be easily used to implement high-end scenarios that demonstrate the capabilities of RL without extra effort on robot's communication methods. Although this architecture can be easily implemented avoiding any inter-node communication overhead, it comes with the restriction that every robot goes hand-in-hand with an agent. This limits its usefulness when it comes to multirobot scenarios that are

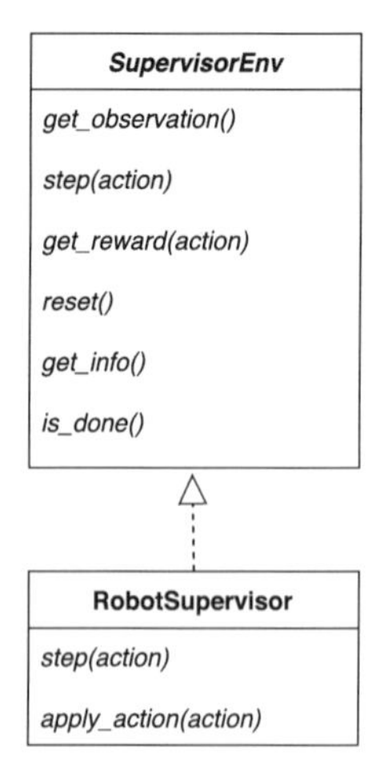

FIGURE 21.6

One-node architecture class diagram.

controlled by a central DL agent. Finally, the one-node architecture does not cover scenarios in which the processing units of an RL agent are separated (or distributed) from the actual robot unit, and actions are transmitted via a communication channel.

The one-node architecture can be depicted with a class diagram as presented in Fig. 21.6. The class diagram includes an abstract super-class (*SupervisorEnv*) that guides the development of the *agent-environment loop*, while the child class (*Robot-Supervisor*) hides the reset functionality, which is linked to the simulator and enforces users to use *step* function, which includes all the appropriate functionalities required by the simulator, as mentioned above. We can easily develop a class that inherits from one of those classes, and then add it to a robot node as a controller. In every timestep, we feed the RL agent with the observations, which come from the robot's sensors, and in turn the RL agent returns an *action* accordingly. First, the chosen *action* is passed to *apply_action* and then to the *step* function, in order to be processed by the RL agent. If the taken action results in solving the task, the simulator is reset. The Code Example 21.5 can give us some further insights.

21.3.2 Emitter-receiver architecture

In many practical robotics scenarios, robots are not equipped with any processing unit while they are responsible only for collecting the observations and sending them to a processing unit via a communication channel. In turn, the system, which is equipped with the processing unit, inference the DL model, which calculates the values of actions in order to be sent to the robot. This architecture enables users to lower the cost of the setup, since it does not require powerful hardware for each robot, but only to implement a communication protocol between the robots and the involved nodes of processing units. The *emitter-receiver* architecture follows the same *agent-*

```
class OneNodeBot(RobotSupervisor):
    def step(self, action):...
    def apply_action(action)...

env = OneNodeBot()
env = TensorboardLogger(env)

agent = DDPG(...)

for i in range(N_EPOCHS):
    is_done = False
    score = 0
    observ = env.reset()
    while not is_done:
        action = agent.choose_action(observ)
        env.aply_action(action)
        observ, reward, is_done, info = env.step(action)
        agent.remember(observ, action, reward, is_done)
        agent.learn()
        score += reward
```

Code Example 21.5: One-node controller code example.

environment loop, with the only difference being that the agent, which is responsible for choosing an action, runs on the supervisor node and the observations are acquired by the robot/actor node. This is possible in Webots using its built-in functionality for broadcasting messages with the *emitter* component and receiving messages from the scene with the *receiver* component. However, the application of this implementation scheme is hindered by its reliance on the available channel's bandwidth, which is sometimes not enough to efficiently transmit high volume data, such as high-resolution images. This limitation should be taken into account before choosing the *emitter-receiver* implementation scheme.

The *emitter-receiver* architecture can be schematically described by the workflow diagram in Fig. 21.7. Similarly as before, the simulator should be reset to the initial state S_0 with the difference being that in the *emitter-receiver* architecture the resetting functionality is included in the supervisor node. The robot, on the one hand, collects the first set of observations and by using the broadcasting functionality, that is, the emitter component, it broadcasts the message to the scene. On the other hand, the supervisor node receives the observations using the receiver component, which in turn are fed to the RL agent. If needed, the observations are augmented with extra information from the supervisor node, since it has access on additional information from the scene (e.g., distances between nodes and the position of the robot node). Additionally to the observations, the agent receives the action in time step $t-1$, the reward value and/or any other information of the state. Deepbots provides the users with the ability to use any of the existing python DL frameworks, for example, PyTorch, TensorFlow,

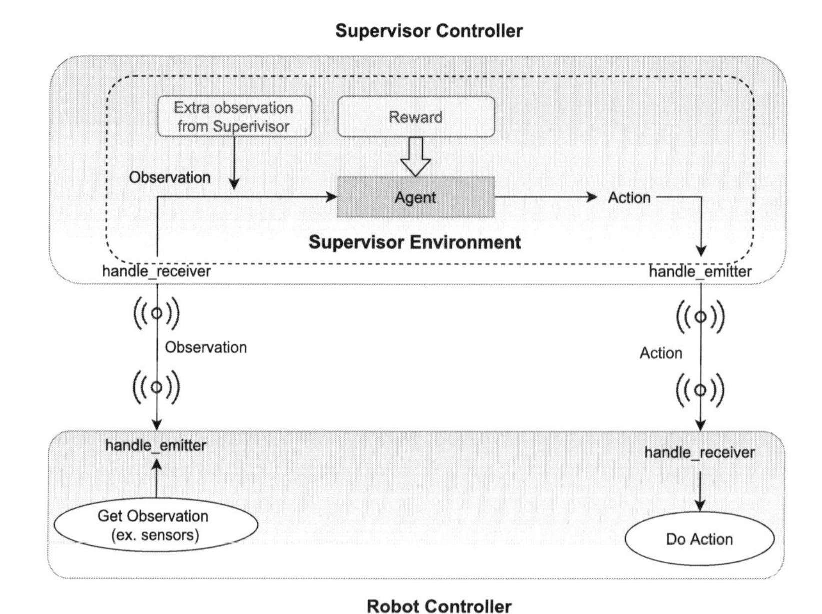

FIGURE 21.7

Emitter-receiver workflow diagram.

etc. Then the agent calculates the actions for S_{t+1}, which are transmitted to the robot node using the supervisor node's emitter. Finally, the robot/actor node has to perform the actions with its actuators. The aforementioned procedure is performed iteratively until the robot achieves its objective for a certain number of epochs/episodes, or any other condition is needed by the use-case.

Deepbots implements the *emitter-receiver* architecture using the partially abstract classes presented in Fig. 21.8. The implementation involves two different ontologies/classes in both sides of the communication channel. On the supervisor side, the user has to implement an *agent-environment loop* using the associated abstract methods and training/inference procedures. An example script for the supervisor's node controller is presented in Code Example 21.6. A typical implementation for the robot's side could be a deserialization process of the received data followed by an appropriate preprocessing to feed the action in the actuators. A demonstration script for this task is shown in Code Example 21.7. The Deepbots framework handles both the simulation's specific procedures, such as the stepping function for the simulation, as well as the communication between the supervisor and robot controllers hiding the low-level details from the users. It should be noted that the *emitter-receiver* scheme is beneficial in terms of code reusability and agility since it promotes the separation of DL agent procedures from the required robot's low-level operations to actually perform an action with its actuators. Furthermore, this architecture is similar to Robotic Operation System (ROS) [21] and can be integrated with it with minimal effort.

```
class FindTargetSupervisor(SupervisorEmitterReceiver):
    def get_observation(self):...
    def get_reward(self, action):...
    def is_done(self):...
    def reset(self):...
    def step(self, action):...

env = FindTargetSupervisor()
env = TensorboardLogger(env)

agent = DDPG(...)
for i in range(N_EPOCHS):
    is_done = False
    score = 0
    observ = env.reset()

    while not is_done:
        action = agent.choose_action(observ)
        observ, reward, is_done, info = env.step(action)
        agent.remember(observ, action, reward, is_done)
        agent.learn()
        score += reward
```

Code Example 21.6: Supervisor's controller example.

```
class FindTargetBot(RobotEmitterReceiver):
    def create_message(self):
        message = []
        for rfinder in self.rangefinders:
            message.append(rfinder.value())
        return message

    def use_message_data(self, message):
        gas = float(message[1])
        wheel = float(message[0])
        ...

robot_controller = FindTargetRobot()
robot_controller.run()
```

Code Example 21.7: Robot's controller example.

Deepbots uses a hierarchical approach with partially abstract classes to facilitate and guide the development process. In the *emitter-receiver* approach, the highest level class (*SuperivisorEnv*) on the supervisor's end represents the *agent-environment loop*, while lower-level classes implement the different communication protocols. On the

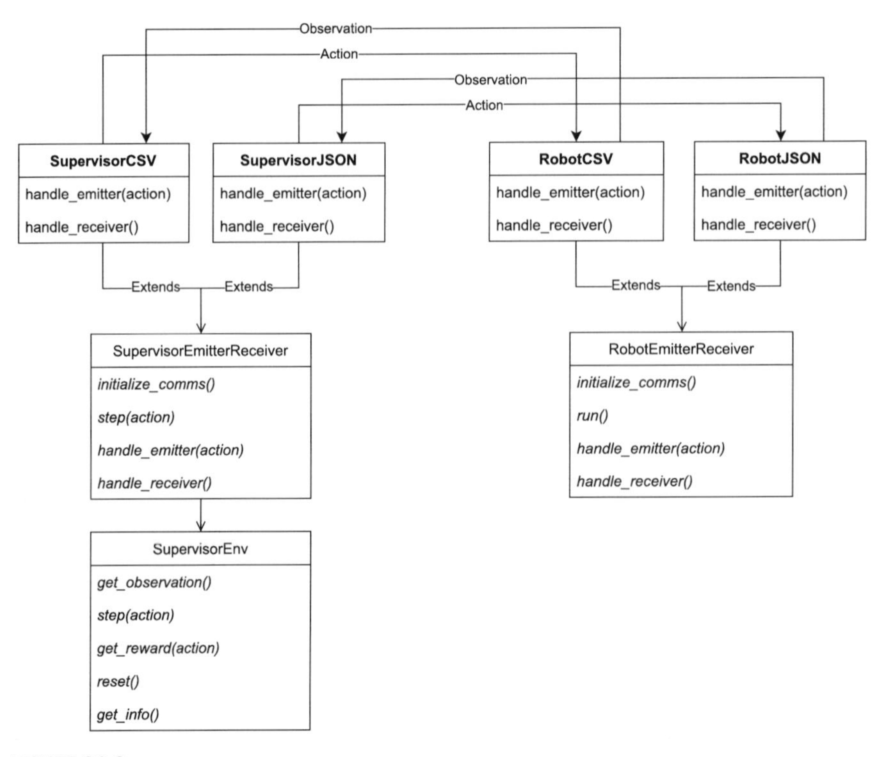

FIGURE 21.8

The emitter-receiver architecture class diagram.

robot's end the highest-level class (*RobotEmitterReceiver*) represents the minimum required functionalities, such as the procedures for emitting and receiving messages, while on the lowest-level classes of the hierarchy, the different communication protocols are implemented.

21.3.3 Design decisions

Cointegrating training with simulator environments requires a development effort to build an interface between the *agent-environment loop* and the scripting mechanism of the simulator. The implemented interface acts as a middle-ware so it should retain the abstractness of the *agent-environment loop*, while providing functionalities as well as the ability to be extended according to specific needs. More precisely, the Deepbots framework introduces different levels of abstractions to facilitate the development for both experienced and inexperienced users. Experienced users are able to use classes higher in class hierarchy without limiting themselves on the predefined procedures, such as the communication scheme. Inexperienced users can develop classes that they inherit from classes lower in hierarchy without worrying about implementation details, in order to remain focused on solving the task at hand. Additionally, retaining the *agent-environment loop* is beneficial since a number of utility functionalities can be incorporated in the environments. Deepbots uses *decorator* design pattern to enclose the *agent-environment loop* base operation providing in this way extra functionality, such as monitoring. This design pattern can be used to stack different functionalities on top of the *agent-environment loop*.

21.4 Concluding remarks

In this chapter we presented the OpenDR toolkit, along with two application examples on face recognition and pose estimation, demonstrating its ability to greatly simplify the task of applying DL models in robotics settings. Then we continued by delving into cointegrated simulation and training approaches that can be used when RL agents are applied in the context of robotics. It is worth mentioning that cointegrating simulation and training is not only limited to these applications, but holds all the credentials for enabling DL models to learn how to actively precept the environment and process not only static stimuli, but also more complex dynamic ones [5]. This is expected to further increase the accuracy of DL models in real robotics applications, moving away from the sterilized computer vision approaches that ignoring the temporal dynamics of the environment.

References

[1] H.A. Pierson, M.S. Gashler, Deep learning in robotics: a review of recent research, Advanced Robotics 31 (16) (2017) 821–835.

[2] N. Sünderhauf, O. Brock, W. Scheirer, R. Hadsell, D. Fox, J. Leitner, B. Upcroft, P. Abbeel, W. Burgard, M. Milford, et al., The limits and potentials of deep learning for robotics, The International Journal of Robotics Research 37 (4–5) (2018) 405–420.

[3] A.G. Howard, M. Zhu, B. Chen, D. Kalenichenko, W. Wang, T. Weyand, M. Andreetto, H. Adam, Mobilenets: efficient convolutional neural networks for mobile vision applications, arXiv preprint, arXiv:1704.04861, 2017.

[4] M. Kirtas, K. Tsampazis, N. Passalis, A. Tefas, Deepbots: a webots-based deep reinforcement learning framework for robotics, https://doi.org/10.1007/978-3-030-49186-4_6, 2020.

[5] Y. Aloimonos, Active Perception, Psychology Press, 2013.

[6] N. Passalis, A. Tefas, Leveraging active perception for improving embedding-based deep face recognition, in: Proceedings of the International Workshop on Multimedia Signal Processing, 2020, pp. 1–6.

[7] M. Everingham, L. Van Gool, C.K. Williams, J. Winn, A. Zisserman, The Pascal visual object classes (voc) challenge, International Journal of Computer Vision 88 (2) (2010) 303–338.

[8] O.M. Parkhi, A. Vedaldi, A. Zisserman, Deep Face Recognition (2015).

[9] J. Deng, J. Guo, N. Xue, S. Zafeiriou, Arcface: additive angular margin loss for deep face recognition, arXiv:1801.07698, 2019.

[10] H. Wang, Y. Wang, Z. Zhou, X. Ji, D. Gong, J. Zhou, Z. Li, W. Liu, Cosface: large margin cosine loss for deep face recognition, arXiv:1801.09414, 2018.
[11] W. Liu, Y. Wen, Z. Yu, M. Li, B. Raj, L. Song, Sphereface: deep hypersphere embedding for face recognition, arXiv:1704.08063, 2018.
[12] K. Zhang, Z. Zhang, Z. Li, Y. Qiao, Joint face detection and alignment using multitask cascaded convolutional networks, IEEE Signal Processing Letters 23 (10) (2016) 1499–1503, https://doi.org/10.1109/lsp.2016.2603342.
[13] Z. Cao, G. Hidalgo Martinez, T. Simon, S. Wei, Y.A. Sheikh, Openpose: realtime multi-person 2d pose estimation using part affinity fields, IEEE Transactions on Pattern Analysis and Machine Intelligence (2019).
[14] D. Osokin, Real-time 2d multi-person pose estimation on cpu: lightweight openpose, arXiv preprint, arXiv:1811.12004, 2018.
[15] D. Erhan, Y. Bengio, A. Courville, P. Vincent, Visualizing higher-layer features of a deep network, University of Montreal 1341 (3) (2009) 1.
[16] N. Koenig, A. Howard, Design and use paradigms for gazebo, an open-source multi-robot simulator, in: 2004 IEEE/RSJ International Conference on Intelligent Robots and Systems (IROS) (IEEE Cat. No. 04CH37566), vol. 3, 2004, pp. 2149–2154.
[17] R. Diankov, Automated construction of robotic manipulation programs, 2010.
[18] G. Dulac-Arnold, D. Mankowitz, T. Hester, Challenges of real-world reinforcement learning, arXiv:1904.12901, 2019.
[19] O. Michel, Cyberbotics ltd. webots™: professional mobile robot simulation, International Journal of Advanced Robotic Systems 1 (1) (2004) 5.
[20] G. Brockman, V. Cheung, L. Pettersson, J. Schneider, J. Schulman, J. Tang, W. Zaremba, OpenAI gym, CoRR, arXiv:1606.01540, 2016.
[21] Stanford Artificial Intelligence Laboratory et al., Robotic operating system, URL https://www.ros.org.

Index

A

B

C

D

E

F

G

L

N

O

P

Q

R

S